INTRODUCTION TO
ELECTRICAL ENGINEERING

THE OXFORD SERIES IN ELECTRICAL AND COMPUTER ENGINEERING

ADEL S. SEDRA, Series Editor

INTRODUCTION TO
ELECTRICAL ENGINEERING

Mulukutla S. Sarma
Northeastern University

New York   Oxford
OXFORD UNIVERSITY PRESS
2001

Oxford University Press

Oxford   New York
Athens   Auckland   Bangkok   Bogotá   Buenos Aires   Calcutta
Cape Town   Chennai   Dar es Salaam   Delhi   Florence   Hong Kong   Istanbul
Karachi   Kuala Lumpur   Madrid   Melbourne   Mexico City   Mumbai
Nairobi   Paris   São Paulo   Shanghai   Singapore   Taipei   Tokyo   Toronto   Warsaw

and associated companies in
Berlin   Ibadan

Published by Oxford University Press, Inc.,
198 Madison Avenue, New York, New York, 10016
http://www.oup-usa.org

Oxford is a registered trademark of Oxford University Press

Library of Congress Cataloging-in-Publication Data

Sarma, Mulukutla S., 1938–
       Introduction to electrical engineering / Mulukutla S. Sarma
           p.   cm. — (The Oxford series in electrical and computer engineering)
       ISBN 0-19-513604-7 (cloth)
       1. Electrical engineering.   I. Title.   II. Series.
   TK146.S18   2001
   621.3—dc21   00-020033

Acknowledgments—Table 1.2.2 is adapted from *Principles of Electrical Engineering (McGraw-Hill Series in Electrical Engineering)*, by Peyton Z. Peebles Jr. and Tayeb A. Giuma, reprinted with the permission of McGraw-Hill, 1991; figures 2.6.1, 2.6.2 are adapted from *Getting Started with MATLAB 5: Quick Introduction*, by Rudra Pratap, reprinted with the permission of Oxford University Press, 1998; figures 4.1.2–4.1.5, 4.2.1–4.2.3, 4.3.1–4.3.2, are adapted from *Electric Machines: Steady-State Theory and Dynamic Performance, Second Edition*, by Mulukutla S. Sarma, reprinted with the permission of Brooks/Cole Publishing, 1994; figure 4.6.1 is adapted from *Medical Instrumentation Application and Design*, by John G. Webster, reprinted with the permission of John Wiley & Sons, Inc., 1978; table 4.6.1 is adapted from "Electrical Safety in Industrial Plants," *IEEE Spectrum*, by Ralph Lee, reprinted with the permission of IEEE, 1971; figure P5.3.1 is reprinted with the permission of Fairchild Semiconductor Corporation; figures 5.6.1, 6.6.1, 9.5.1 are adapted from *Electrical Engineering: Principles and Applications*, by Allen R. Hambley, reprinted with the permission of Prentice Hall, 1997; figure 10.5.1 is adapted from *Power System Analysis and Design, Second Edition*, by Duncan J. Glover and Mulukutla S. Sarma, reprinted with the permission of Brooks/Cole Publishing, 1994; figures 11.1.2, 13.2.10 are adapted from *Introduction to Electrical Engineering, Second Edition*, by Clayton Paul, Syed A. Nasar, and Louis Unnewehr, reprinted with the permission of McGraw-Hill, 1992; figures E12.2.1(a,b), 12.2.2–12.2.5, 12.2.9– 12.2.10, 12.3.1–12.3.3, 12.4.1, E12.4.1, P12.1.2, P12.4.3, P12.4.8, P12.4.12, 13.1.1–13.1.8, 13.2.1–13.2.9, 13.2.11–13.2.16, 13.3.1–13.3.3, E13.3.2, 13.3.4, E13.3.3, 13.3.5–13.3.6 are adapted from *Electric Machines: Steady-State Theory and Dynamic Performance, Second Edition*, by Mulukutla S. Sarma, reprinted with the permission of Brooks/Cole Publishing, 1994; figure 13.3.12 is adapted from *Communication Systems Engineering*, by John G. Proakis and Masoud Salehi, reprinted with the permission of Prentice Hall, 1994; figures 13.4.1–13.4.7, E13.4.1(b), 13.4.8–13.4.12, E13.4.3, 13.4.13, 13.6.1 are adapted from *Electric Machines: Steady-State Theory and Dynamic Performance, Second Edition*, by Mulukutla S. Sarma Brooks/Cole Publishing, 1994; figures 14.2.8, 14.2.9 are adapted from *Electrical Engineering: Concepts and Applications, Second Edition*, by A. Bruce Carlson and David Gisser, reprinted with the permission of Prentice Hall, 1990; figure 15.0.1 is adapted from *Communication Systems, Third Edition*, by A. Bruce Carlson, reprinted with the permission of McGraw-Hill, 1986; figures 15.2.15, 15.2.31, 15.3.11 are adapted from *Communication Systems Engineering*, by John G. Proakis and Masoud Salehi, reprinted with the permission of Prentice Hall, 1994; figures 15.2.19, 15.2.27, 15.2.28, 15.2.30, 15.3.3, 15.3.4, 15.3.9, 15.3.10, 15.3.20 are adapted from *Principles of Electrical Engineering (McGraw-Hill Series in Electrical Engineering)*, by Peyton Z. Peebles Jr. and Tayeb A. Giuma, reprinted with the permission of McGraw-Hill, 1991; figures 16.1.1–16.1.3 are adapted from *Electric Machines: Steady-State Theory and Dynamic Performance, Second Edition*, by Mulukutla S. Sarma, reprinted with the permission of Brooks/Cole Publishing, 1994; table 16.1.3 is adapted from *Electric Machines: Steady-State Theory and Dynamic Performance, Second Edition*, by Mulukutla S. Sarma, reprinted with the permission of Brooks/Cole Publishing, 1994; table 16.1.4 is adapted from *Handbook of Electric Machines*, by S. A. Nasar, reprinted with the permission of McGraw-Hill, 1987; and figures 16.1.4–13.1.9, E16.1.1, 16.1.10–16.1.25 are adapted from *Electric Machines: Steady-State Theory and Dynamic Performance, Second Edition*, by Mulukutla S. Sarma, reprinted with the permission of Brooks/Cole Publishing, 1994.

Printing (last digit): 10 9 8 7 6 5 4 3 2 1

Printed in the United States of America
on acid-free paper

To my grandchildren

Puja Sree
Sruthi Lekha
Pallavi Devi

* * *

and those to come

CONTENTS

PART 5   CONTROL SYSTEMS

LIST OF CASE STUDIES AND COMPUTER-AIDED ANALYSIS

Case Studies

Computer-Aided Analysis

PREFACE

I. OBJECTIVES

The purpose of this text is to present a problem-oriented introductory survey text for the extraordinarily interesting electrical engineering discipline by arousing student enthusiasm while addressing the underlying concepts and methods behind various applications ranging from consumer gadgets and biomedical electronics to sophisticated instrumentation systems, computers, and multifarious electric machinery. The focus is on acquainting students majoring in all branches of engineering and science, especially in courses for *nonelectrical engineering majors*, with the nature of the subject and the potentialities of its techniques, while emphasizing the principles. Since principles and concepts are most effectively taught by means of a problem-oriented course, judicially selected topics are treated in sufficient depth so as to permit the assignment of adequately challenging problems, which tend to implant the relevant principles in students' minds.

In addition to an academic-year (two semesters or three quarters) introductory course traditionally offered to non-EE majors, the text is also suitable for a sophomore survey course given nowadays to electrical engineering majors in a number of universities. At a more rapid pace or through selectivity of topics, the introductory course could be offered in one semester to either electrical and computer engineering (ECE) or non-EE undergraduate majors. Although this book is written primarily for non-EE students, it is hoped that it will be of value to undergraduate ECE students (particularly for those who wish to take the Fundamentals of Engineering examination, which is a prerequisite for becoming licensed as a Professional Engineer), to graduate ECE students for their review in preparing for qualifying examinations, to meet the continuing-education needs of various professionals, and to serve as a reference text even after graduation.

II. MOTIVATION

This text is but a modest attempt to provide an exciting survey of topics inherent to the electrical and computer engineering discipline. Modern technology demands a team approach in which electrical engineers and nonelectrical engineers have to work together sharing a common technical vocabulary. Nonelectrical engineers must be introduced to the language of electrical engineers, just as the electrical engineers have to be sensitized to the relevance of nonelectrical topics.

The dilemma of whether electrical engineering and computer engineering should be separate courses of study, leading to distinctive degrees, seems to be happily resolving itself in the direction of togetherness. After all, computers are not only pervasive tools for engineers but also their product; hence there is a pressing need to weave together the fundamentals of both the electrical and the computer engineering areas into the new curricula.

An almost total lack of contact between freshmen and sophomore students and the Department of Electrical and Computer Engineering, as well as little or no exposure to electrical and computer engineering, seems to drive even the academically gifted students away from the program. An initial spark that may have motivated them to pursue electrical and computer engineering has to be nurtured in the early stages of their university education, thereby providing an inspiration to continue.

This text is based on almost 40 years of experience teaching a wide variety of courses to electrical as well as non-EE majors and, more particularly, on the need to answer many of the questions raised by so many of my students. I have always enjoyed engineering (teaching, research, and consultation); I earnestly hope that the readers will have as much fun and excitement in using this book as I have had in developing it.

III. PREREQUISITES AND BACKGROUND

The student will be assumed to have completed the basic college-level courses in algebra, trigonometry, basic physics, and elementary calculus. A knowledge of differential equations is helpful, but not mandatory. For a quick reference, some useful topics are included in the appendixes.

IV. ORGANIZATION AND FLEXIBILITY

The text is developed to be student-oriented, comprehensive, and up to date on the subject with necessary and sufficient detailed explanation at the level for which it is intended. The key word in the organization of the text is flexibility.

The book is divided into five parts in order to provide flexibility in meeting different circumstances, needs, and desires. A glance at the Table of Contents will show that Part 1 concerns itself with basic electric circuits, in which circuit concepts, analysis techniques, time-dependent analysis including transients, as well as three-phase circuits are covered. Part 2 deals with electronic analog and digital systems, in which analog and digital building blocks are considered along with operational amplifiers, semiconductor devices, integrated circuits, and digital circuits.

Part 3 is devoted to energy systems, in which ac power systems, magnetic circuits and transformers, principles of electromechanics, and rotating machines causing electromechanical energy conversion are presented. Part 4 deals with information systems, including the underlying principles of signal processing and communication systems. Finally, Part 5 presents control systems, which include the concepts of feedback control, digital control, and power semiconductor-controlled drives.

The text material is organized for optimum flexibility, so that certain topics may be omitted without loss of continuity when lack of time or interest dictates.

V. FEATURES

1. The readability of the text and the level of presentation, from the student's viewpoint, are given utmost priority. The quantity of subject matter, range of difficulty, coverage of topics, numerous illustrations, a large number of comprehensive worked-out examples, and a variety of end-of-chapter problems are given due consideration, to ensure that engineering is not a "plug-in" or "cookbook" profession, but one in which reasoning and creativity are of the highest importance.

2. Fundamental physical concepts, which underlie creative engineering and become the most valuable and permanent part of a student's background, have been highlighted while giving due attention to mathematical techniques. So as to accomplish this in a relatively short time, much thought has gone into rationalizing the theory and conveying in a concise manner the essential details concerning the nature of electrical and computer engineering. With a good grounding

in basic concepts, a very wide range of engineering systems can be understood, analyzed, and devised.

3. The theory has been developed from simple beginnings in such a manner that it can readily be extended to new and more complicated situations. The art of reducing a practical device to an appropriate mathematical model and recognizing its limitations has been adequately presented. Sufficient motivation is provided for the student to develop interest in the analytical procedures to be applied and to realize that all models, being approximate representations of reality, should be no more complicated than necessary for the application at hand.

4. Since the essence of engineering is the design of products useful to society, the end objective of each phase of preparatory study should be to increase the student's capability to design practical devices and systems to meet the needs of society. Toward that end, the student will be motivated to go through the sequence of understanding physical principles, processes, modeling, using analytical techniques, and, finally, designing.

5. Engineers habitually break systems up into their component blocks for ease of understanding. The building-block approach has been emphasized, particularly in Part II concerning analog and digital systems. For a designer using IC blocks in assembling the desired systems, the primary concern lies with their terminal characteristics while the internal construction of the blocks is of only secondary importance.

6. Considering the world of electronics today, both analog and digital technologies are given appropriate coverage. Since students are naturally interested in such things as op amps, integrated circuits, and microprocessors, modern topics that can be of great use in their career are emphasized in this text, thereby motivating the students further.

7. The electrical engineering profession focuses on information and energy, which are the two critical commodities of any modern society. In order to bring the message to the forefront for the students' attention, Parts III, IV, and V are dedicated to energy systems, information systems, and control systems, respectively. However, some of the material in Parts I and II is critical to the understanding of the latter.

An understanding of the principles of energy conversion, electric machines, and energy systems is important for all in order to solve the problems of energy, pollution, and poverty that face humanity today. It can be well argued that today's non-EEs are more likely to encounter electromechanical machines than some of the ECEs. Thus, it becomes essential to have sufficient breadth and depth in the study of electric machines by the non-ECEs.

Information systems have been responsible for the spectacular achievements in communication in recent decades. Concepts of control systems, which are not limited to any particular branch of engineering, are very useful to every engineer involved in the understanding of the dynamics of various types of systems.

8. Consistent with modern practice, the international (SI) system of units has been used throughout the text. In addition, a review of units, constants and conversion factors for the SI system can be found in Appendix C.

9. While solid-state electronics, automatic control, IC technology, and digital systems have become commonplace in the modern EE profession, some of the older, more traditional topics, such as electric machinery, power, and instrumentation, continue to form an integral part of the curriculum, as well as of the profession in real life. Due attention is accorded in this text to such topics as three-phase circuits and energy systems.

10. Appendixes provide useful information for quick reference on selected bibliography for supplementary reading, the SI system, mathematical relations, as well as a brief review of the Fundamentals of Engineering (FE) examination.

11. Engineers who acquire a basic knowledge of electric circuits, electronic analog and digital circuits, energy systems, information systems, and control systems will have a well-rounded background and be better prepared to join a team effort in analyzing and designing systems. Therein lies the justification for the Table of Contents and the organization of this text.

12. At the end of each chapter, the *learning objectives* of that chapter are listed so that the student can check whether he or she has accomplished each of the goals.

13. At the very end of each chapter, *Practical Application: A Case Study* has been included so that the reader can get motivated and excited about the subject matter and its relevance to practice.

14. Basic material introduced in this book is totally independent of any software that may accompany the usage of this book, and/or the laboratory associated with the course. The common software in usage, as of writing this book, consists of *Windows, Word Perfect, PSPICE, Math CAD,* and *MATLAB*. There are also other popular specialized simulation programs such as *Signal Processing Workstation (SPW)* in the area of analog and digital communications, *Very High Level Description Language (VHDL)* in the area of digital systems, *Electromagnetic Transients Program (EMTP)* in the field of power, and *SIMULINK* in the field of control. In practice, however, any combination of software that satisfies the need for word processing, graphics, editing, mathematical analysis, and analog as well as digital circuit analysis should be satisfactory.

In order to integrate computer-aided circuit analysis, two types of programs have been introduced in this text: A circuit simulator PSpice and a math solver MATLAB. Our purpose here is not to teach students how to use specific software packages, but to help them develop an analysis style that includes the intelligent use of computer tools. After all, these tools are an intrinsic part of the engineering environment, which can significantly enhance the student's understanding of circuit phenomena.

15. The basics, to which the reader is exposed in this text, will help him or her to select consultants—experts in specific areas—either in or out of house, who will provide the knowledge to solve a confronted problem. After all, no one can be expected to be an expert in all areas discussed in this text!

VI. PEDAGOGY

A. Outline

Beyond the overview meant as an orientation, the text is basically divided into five parts.

Part 1: Electric Circuits This part provides the basic circuit-analysis concepts and techniques that will be used throughout the subsequent parts of the text. Three-phase circuits have been introduced to develop the background needed for analyzing ac power systems. Basic notions of residential circuit wiring, including grounding and safety considerations, are presented.

Part 2: Electronic Analog and Digital Systems With the background of Part I, the student is then directed to analog and digital building blocks. Operational amplifiers are discussed as an especially important special case. After introducing digital system components, computer systems, and networks to the students, semiconductor devices, integrated circuits, transistor amplifiers, as well as digital circuits are presented. The discussion of device physics is kept to the necessary minimum, while emphasis is placed on obtaining powerful results from simple tools placed in students' hands and minds.

Part 3: Energy Systems With the background built on three-phase circuits in Part I, ac power systems are considered. Magnetic circuits and transformers are then presented, before the student is introduced to the principles of electromechanics and practical rotating machines that achieve electromechanical energy conversion.

Part 4: Information Systems Signal processing and communication systems (both analog and digital) are discussed using the block diagrams of systems engineering.

Part 5: Control Systems By focusing on control aspects, this part brings together the techniques and concepts of the previous parts in the design of systems to accomplish specific tasks. A section on power semiconductor-controlled drives is included in view of their recent importance. The basic concepts of feedback control systems are introduced, and finally the flavor of digital control systems is added.

Appendices The appendices provide ready-to-use information:

Appendix A: Selected bibliography for supplementary reading

Appendix B: Brief review of fundamentals of engineering (FE) examination

Appendix C: Technical terms, units, constants, and conversion factors for the SI system

Appendix D: Mathematical relations (used in the text)

Appendix E: Solution of simultaneous equations

Appendix F: Complex numbers

Appendix G: Fourier series

Appendix H: Laplace transforms

B. Chapter Introductions

Each chapter is introduced to the student stating the objective clearly, giving a sense of what to expect, and motivating the student with enough information to look forward to reading the chapter.

C. Chapter Endings

At the end of each chapter, the *learning objectives* of that chapter are listed so that the student can check whether he or she has accomplished each of the goals.

In order to motivate and excite the student, practical applications using electrical engineering principles are included. At the very end of each chapter, a relevant *Practical Application: A Case Study* is presented.

D. Illustrations

A large number of illustrations support the subject matter with the intent to motivate the student to pursue the topics further.

E. Examples

Numerous comprehensive examples are worked out in detail in the text, covering most of the theoretical points raised. An appropriate difficulty is chosen and sufficient stimulation is built in to go on to more challenging situations.

F. End-of-Chapter Problems

A good number of problems (identified with each section of every chapter), with properly graded levels of difficulty, are included at the end of each chapter, thereby allowing the instructor considerable flexibility. There are nearly a thousand problems in the book.

G. Preparation for the FE Exam

A brief review of the Fundamentals of Engineering (FE) examination is presented in Appendix B in order to aid the student who is preparing to take the FE examination in view of becoming a registered Professional Engineer (PE).

VII.   SUPPLEMENTS

A **Solutions Manual to Accompany Introduction to Electrical Engineering**, by M.S. Sarma (ISBN 019-514260-8), with complete detailed solutions (provided by the author) for all problems in the book is available to adopters.

MicroSoft PowerPoint Overheads to Accompany *Introduction to Electrical Engineering* (ISBN 019-514472-1) are free to adopters. Over 300 text figures and captions are available for classroom projection use.

A **web-site**, **MSSARMA.org**, will include interesting web links and enhancement materials, errata, a forum to communicate with the author, and more.

A **CD-ROM Disk is packaged with each new book**. The CD contains:

- **Complete Solutions for Students to 20% of the problems.** These solutions have been prepared by the author and are resident on the disk in Adobe Acrobat (.pdf) format. The problems with solutions on disk are marked with an asterisk next to the problem in the text.

- **The demonstration version of Electronics Workbench Multisim Version 6**, an innovative teaching and learning software product that is used to build circuits and to simulate and analyze their electrical behavior. This demonstration version includes **20 demo circuit files** built from circuit examples from this textbook. The CD also includes another **80 circuits** from the text that can be opened with the full student or educational versions of Multisim. These full versions can be obtained from Electronics Workbench at www.electronicsworkbench.com.

To extend the introduction to selected topics and provide additional practice, we recommend the following additional items:

- Circuits: *Allan's Circuits Problems* by Allan Kraus (ISBN 019-514248-9), which includes over 400 circuit analysis problems with complete solutions, many in MATLAB and SPICE form.

- Electronics: *KC's Problems and Solutions to Accompany Microelectronic Circuits* by K.C. Smith (019–511771-9), which includes over 400 electronics problems and their complete solutions.

- SPICE: *SPICE* by Gordon Roberts and Adel Sedra (ISBN 019-510842-6) features over 100 examples and numerous exercises for computer-aided analysis of microelectronic circuits.

- MATLAB: *Getting Started with MATLAB* by Rudra Pratap (ISBN 019-512947-4) provides a quick introduction to using this powerful software tool.

For more information or to order an examination copy of the above mentioned supplements contact Oxford University Press at *college@oup-usa.org*.

VIII.   ACKNOWLEDGMENTS

The author would like to thank the many people who helped bring this project to fruition. A number of reviewers greatly improved this text through their thoughtful comments and useful suggestions.

I am indebted to my editor, Peter C. Gordon, of Oxford University Press, who initiated this project and continued his support with skilled guidance, helpful suggestions, and great encouragement. The people at Oxford University Press, in particular, Senior Project Editor Karen Shapiro, have been most helpful in this undertaking. My sincere thanks are also due to Mrs. Sally Gupta, who did a superb job typing most of the manuscript.

I would also like to thank my wife, Savitri, for her continued encouragement and support, without which this project could not have been completed. It is with great pleasure and joy that I dedicate this work to my grandchildren.

Mulukutla S. Sarma
Northeastern University

OVERVIEW

What is electrical engineering? What is the scope of electrical engineering?

To answer the first question in a simple way, electrical engineering deals mainly with information systems and with power and energy systems. In the former, electrical means are used to transmit, store, and process information; while in the latter, bulk energy is transmitted from one place to another and power is converted from one form to another.

The second question is best answered by taking a look at the variety of periodicals published by the *Institute of Electrical and Electronics Engineers (IEEE)*, which is the largest technical society in the world with over 320,000 members in more than 140 countries worldwide. Table I lists 75 IEEE Society/Council periodicals along with three broad-scope publications.

The transactions and journals of the IEEE may be classified into broad categories of devices, circuits, electronics, computers, systems, and interdisciplinary areas. All areas of electrical engineering require a working knowledge of physics and mathematics, as well as engineering methodologies and supporting skills in communications and human relations. A closely related field is that of computer science.

Obviously, one cannot deal with all aspects of all of these areas. Instead, the general concepts and techniques will be emphasized in order to provide the reader with the necessary background needed to pursue specific topics in more detail. The purpose of this text is to present the basic theory and practice of electrical engineering to students with varied backgrounds and interests. After all, electrical engineering rests upon a few major principles and subprinciples.

Some of the areas of major concern and activity in the present society, as of writing this book, are:

- Protecting the environment
- Energy conservation
- Alternative energy sources
- Development of new materials
- Biotechnology
- Improved communications
- Computer codes and networking
- Expert systems

This text is but a modest introduction to the exciting field of electrical engineering. However, it is the ardent hope and fervent desire of the author that the book will help inspire the reader to apply the basic principles presented here to many of the interdisciplinary challenges, some of which are mentioned above.

TABLE I IEEE Publications

Publication	Pub ID
IEEE Society/Council Periodicals	
Aerospace & Electronic Systems Magazine	3161
Aerospace & Electronic Systems, Transactions on	1111
Annals of the History of Computing	3211
Antennas & Propagation, Transactions on	1041
Applied Superconductivity, Transactions on	1521
Automatic Control, Transactions on	1231
Biomedical Engineering, Transactions on	1191
Broadcasting, Transactions on	1011
Circuits and Devices Magazine	3131
Circuits & Systems, Part I, Transactions on	1561
Circuits & Systems, Part II, Transactions on	1571
Circuits & Systems for Video Technology, Transactions on	1531
Communications, Transactions on	1201
Communications Magazine	3021
Components, Hybrids, & Manufacturing Technology, Transactions on	1221
Computer Graphics & Applications Magazine	3061
Computer Magazine	3001
Computers, Transactions on	1161
Computer-Aided Design of Integrated Circuits and Systems, Transactions on	1391
Consumer Electronics, Transactions on	1021
Design & Test of Computers Magazine	3111
Education, Transactions on	1241
Electrical Insulation, Transactions on	1301
Electrical Insulation Magazine	3141
Electromagnetic Compatibility, Transactions on	1261
Electron Device Letters	3041
Electron Devices, Transactions on	1151
Electronic Materials, Journal of	4601
Energy Conversion, Transactions on	1421
Engineering in Medicine & Biology Magazine	3091
Engineering Management, Transactions on	1141
Engineering Management Review	3011
Expert Magazine	3151
Geoscience & Remote Sensing, Transactions on	1281
Image Processing, Transactions on	1551
Industrial Electronics, Transactions on	1131
Industry Applications, Transactions on	1321
Information Theory, Transactions on	1121
Instrumentation & Measurement, Transactions on	1101
Knowledge & Data Engineering, Transactions on	1471
Lightwave Technology, Journal of	4301
LTS (The Magazine of Lightwave Telecommunication Systems)	3191
Magnetics, Transactions on	1311
Medical Imaging, Transactions on	1381
Micro Magazine	3071
Microelectromechanical Systems, Journal of	4701
Microwave and Guided Wave Letters	1511
Microwave Theory & Techniques, Transactions on	1181
Network Magazine	3171
Neural Networks, Transactions on	1491
Nuclear Science, Transactions on	1061
Oceanic Engineering, Journal of	4201
Parallel & Distributed Systems, Transactions on	1501
Pattern Analysis & Machine Intelligence, Transactions on	1351
Photonics Technology Letters	1481
Plasma Science, Transactions on	1071
Power Delivery, Transactions on	1431

Continued

TABLE I Continued

Publication	Pub ID
Power Electronics, Transactions on	4501
Power Engineering Review	3081
Power Systems, Transactions on	1441
Professional Communication, Transactions on	1251
Quantum Electronics, Journal of	1341
Reliability, Transactions on	1091
Robotics & Automation, Transactions on	1461
Selected Areas in Communication, Journal of	1411
Semiconductor Manufacturing, Transactions on	1451
Signal Processing, Transactions on	1001
Signal Processing Magazine	3101
Software Engineering, Transactions on	1171
Software Magazine	3121
Solid-State Circuits, Journal of	4101
Systems, Man, & Cybernetics, Transactions on	1271
Technology & Society Magazine	1401
Ultrasonics, Ferroelectrics & Frequency Control, Transactions on	1211
Vehicular Technology, Transactions on	1081
Broad Scope Publications	
IEEE Spectrum	5001
Proceedings of the IEEE	5011
IEEE Potentials	5061

A historical perspective of electrical engineering, in chronological order, is furnished in Table II. A mere glance will thrill anyone, and give an idea of the ever-changing, fast-growing field of electrical engineering.

TABLE II Chronological Historical Perspective of Electrical Engineering

1750–1850	Coulomb's law (1785)
	Battery discovery by Volta
	Mathematical theories by Fourier and Laplace
	Ampere's law (1825)
	Ohm's law (1827)
	Faraday's law of induction (1831)
1850–1900	Kirchhoff's circuit laws (1857)
	Telegraphy: first transatlantic cables laid
	Maxwell's equations (1864)
	Cathode rays: Hittorf and Crookes (1869)
	Telephony: first telephone exchange in New Haven, Connecticut
	Edison opens first electric utility in New York City (1882): dc power systems
	Waterwheel-driven dc generator installed in Appleton, Wisconsin (1882)
	First transmission lines installed in Germany (1882), 2400 V dc, 59km
	Dc motor by Sprague (1884)
	Commercially practical transformer by Stanley (1885)
	Steinmetz's ac circuit analysis
	Tesla's papers on ac motors (1888)
	Radio waves: Hertz (1888)
	First single-phase ac transmission line in United States (1889): Ac power systems, Oregon City to Portland, 4 kV, 21 km
	First three-phase ac transmission line in Germany (1891), 12 kV, 179 km
	First three-phase ac transmission line in California (1893), 2.3 kV, 12 km
	Generators installed at Niagara Falls, New York
	Heaviside's operational calculus methods

1900–1920	Marconi's wireless telegraph system: transatlantic communication (1901)
	Photoelectric effect: Einstein (1904)
	Vacuum-tube electronics: Fleming (1904), DeForest (1906)
	First AM broadcasting station in Pittsburgh, Pennsylvania
	Regenerative amplifier: Armstrong (1912)
1920–1940	Television: Farnsworth, Zworykin (1924)
	Cathode-ray tubes by DuMont; experimental broadcasting
	Negative-feedback amplifier by Black (1927)
	Boolean-algebra application to switching circuits by Shannon (1937)
1940–1950	Major advances in electronics (World War II)
	Radar and microwave systems: Watson-Watts (1940)
	Operational amplifiers in analog computers
	FM communication systems for military applications
	System theory papers by Bode, Shannon, and Wiener
	ENIAC vacuum-tube digital computer at the University of Pennsylvania (1946)
	Transistor electronics: Shockley, Bardeen, and Brattain of Bell Labs (1947)
	Long-playing microgroove records (1948)
1950–1960	Transistor radios in mass production
	Solar cell: Pearson (1954)
	Digital computers (UNIVAC I, IBM, Philco); Fortran programming language
	First commercial nuclear power plant at Shippingport, Pennsylvania (1957)
	Integrated circuits by Kilby of Texas Instruments (1958)
1960–1970	Microelectronics: Hoerni's planar transistor from Fairchild Semiconductors
	Laser demonstrations by Maiman (1960)
	First communications satellite *Telstar I* launched (1962)
	MOS transistor: Hofstein and Heiman (1963)
	Digital communications
	765 kV AC power lines constructed (1969)
	Microprocessor: Hoff (1969)
1970–1980	Microcomputers; MOS technology; Hewlett-Packard calculator
	INTEL's 8080 microprocessor chip; semiconductor devices for memory
	Computer-aided design and manufacturing (CAD/CAM)
	Interactive computer graphics; software engineering
	Personal computers; IBM PC
	Artificial intelligence; robotics
	Fiber optics; biomedical electronic instruments; power electronics
1980–Present	Digital electronics; superconductors
	Neural networks; expert systems
	High-density memory chips; digital networks

INTRODUCTION TO
ELECTRICAL ENGINEERING

PART ONE

ELECTRICAL CIRCUITS

1 Circuit Concepts

Electric circuits, which are collections of *circuit elements* connected together, are the most fundamental structures of electrical engineering. A circuit is an interconnection of simple electrical devices that have at least one closed path in which current may flow. However, we may have to clarify to some of our readers what is meant by "current" and "electrical device," a task that we shall undertake shortly. Circuits are important in electrical engineering because they process electrical signals, which carry energy and information; a signal can be any time-varying electrical quantity. Engineering circuit analysis is a mathematical study of some useful interconnection of simple electrical devices. An electric circuit, as discussed in this book, is an idealized mathematical *model* of some physical circuit or phenomenon. The ideal circuit elements are the resistor, the inductor, the capacitor, and the voltage and current sources. The ideal circuit model helps us to *predict*, mathematically, the approximate behavior of the actual event. The models also provide insights into how to *design* a physical electric circuit to perform a desired task. Electrical engineering is concerned with the *analysis* and *design* of electric circuits, systems, and devices. In Chapter 1 we shall deal with the fundamental concepts that underlie all circuits.

Electrical quantities will be introduced first. Then the reader is directed to the lumped-circuit elements. Then Ohm's law and Kirchhoff's laws are presented. These laws are sufficient for analyzing and designing simple but illustrative practical circuits. Later, a brief introduction is given to meters and measurements. Finally, the analogy between electrical and other nonelectric physical systems is pointed out. The chapter ends with a case study of practical application.

1.1 ELECTRICAL QUANTITIES

In describing the operation of electric circuits, one should be familiar with such electrical quantities as charge, current, and voltage. The material of this section will serve as a review, since it will not be entirely new to most readers.

Charge and Electric Force

The proton has a charge of $+1.602 \; 10^{-19}$ coulombs (C), while the electron has a charge of -1.602×10^{-19} C. The neutron has zero charge. Electric charge and, more so, its movement are the most basic items of interest in electrical engineering. When many charged particles are collected together, larger charges and charge distributions occur. There may be point charges (C), line charges (C/m), surface charge distributions (C/m^2), and volume charge distributions (C/m^3).

A charge is responsible for an *electric field* and charges exert *forces* on each other. Like charges repel, whereas unlike charges attract. Such an electric force can be controlled and utilized for some useful purpose. *Coulomb's law* gives an expression to evaluate the electric force in newtons (N) exerted on one point charge by the other:

$$\text{Force on } Q_1 \text{ due to } Q_2 \; = \bar{F}_{21} = \frac{Q_1 Q_2}{4\pi \varepsilon_0 R^2} \bar{a}_{21} \tag{1.1.1a}$$

$$\text{Force on } Q_2 \text{ due to } Q_1 \; = \bar{F}_{12} = \frac{Q_2 Q_1}{4\pi \varepsilon_0 R^2} \bar{a}_{12} \tag{1.1.1b}$$

where Q_1 and Q_2 are the point charges (C); R is the separation in meters (m) between them; ε_0 is the permittivity of the free-space medium with units of C^2/N · m or, more commonly, farads per meter (F/m); and $\bar{a}_{21}$ and $\bar{a}_{12}$ are unit vectors along the line joining Q_1 and Q_2, as shown in Figure 1.1.1.

Equation (1.1.1) shows the following:

1. Forces $\bar{F}_{21}$ and $\bar{F}_{12}$ are experienced by Q_1 and Q_2, due to the presence of Q_2 and Q_1, respectively. They are equal in magnitude and opposite of each other in direction.
2. The magnitude of the force is proportional to the product of the charge magnitudes.
3. The magnitude of the force is inversely proportional to the square of the distance between the charges.
4. The magnitude of the force depends on the medium.
5. The direction of the force is along the line joining the charges.

Note that the SI system of units will be used throughout this text, and the student should be conversant with the conversion factors for the SI system.

The force per unit charge experienced by a small test charge placed in an electric field is known as the electric field intensity $\bar{E}$, whose units are given by N/C or, more commonly, volts per meter (V/m),

$$\bar{E} = \lim_{Q \to 0} \frac{\bar{F}}{Q} \tag{1.1.2}$$

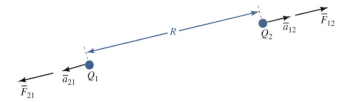

Figure 1.1.1 Illustration of Coulomb's law.

Equation (1.1.2) is the defining equation for the electric field intensity (with units of N/C or V/m), irrespective of the source of the electric field. One may then conclude:

$$\bar{F}_{21} = Q_1 \bar{E}_2 \tag{1.1.3a}$$

$$\bar{F}_{12} = Q_2 \bar{E}_1 \tag{1.1.3b}$$

where $\bar{E}_2$ is the electric field due to Q_2 at the location of Q_1, and $\bar{E}_1$ is the electric field due to Q_1 at the location of Q_2, given by

$$\bar{E}_2 = \frac{Q_2}{4\pi \varepsilon_0 R^2} \bar{a}_{21} \tag{1.1.4a}$$

$$\bar{E}_1 = \frac{Q_1}{4\pi \varepsilon_0 R^2} \bar{a}_{12} \tag{1.1.4b}$$

Note that the electric field intensity due to a positive point charge is directed everywhere radially away from the point charge, and its constant-magnitude surfaces are spherical surfaces centered at the point charge.

EXAMPLE 1.1.1

(a) A small region of an impure silicon crystal with dimensions 1.25×10^{-6} m $\times 10^{-3}$ m $\times 10^{-3}$ m has only the ions (with charge $+1.6 \ 10^{-19}$ C) present with a volume density of $10^{25}/m^3$. The rest of the crystal volume contains equal densities of electrons (with charge -1.6×10^{-19} C) and positive ions. Find the net total charge of the crystal.

(b) Consider the charge of part (a) as a point charge Q_1. Determine the force exerted by this on a charge $Q_2 = 3\mu C$ when the charges are separated by a distance of 2 m in free space, as shown in Figure E1.1.1.

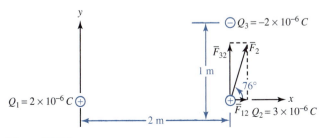

Figure E1.1.1

(c) If another charge $Q_3 = -2\mu C$ is added to the system 1 m above Q_2, as shown in Figure E1.1.1, calculate the force exerted on Q_2.

Solution

(a) In the region where both ions and free electrons exist, their opposite charges cancel, and the net charge density is zero. From the region containing ions only, the volume-charge density is given by

$$\rho = (10^{25})(1.6 \times 10^{-19}) = 1.6 \times 10^6 \ C/m^3$$

The net total charge is then calculated as:

$$Q = \rho v = (1.6 \times 10^6)(1.25 \times 10^{-6} \times 10^{-3} \times 10^{-3}) = 2 \times 10^{-6} \ C$$

(b) The rectangular coordinate system shown defines the locations of the charges: $Q_1 = 2 \times 10^{-6}$ C; $Q_2 = 3 \times 10^{-6}$ C. The force that Q_1 exerts on Q_2 is in the positive direction of x, given by Equation (1.1.1),

$$\bar{F}_{12} = \frac{(3 \times 10^{-6})(2 \times 10^{-6})}{4\pi(10^{-9}/36\pi)2^2} \bar{a}_x = \bar{a}_x \ 13.5 \times 10^{-3} \ N$$

This is the force experienced by Q_2 due to the effect of the electric field of Q_1. Note the value used for free-space permittivity, ε_0, as (8.854×10^{-12}), or approximately $10^{-9}/36\pi$ F/m. $\bar{a}_x$ is the unit vector in the positive x-direction.

(c) When Q_3 is added to the system, as shown in Figure E1.1.1, an additional force on Q_2 directed in the positive y-direction occurs (since Q_3 and Q_2 are of opposite sign),

$$\bar{F}_{32} = \frac{(3 \times 10^{-6})(-2 \times 10^{-6})}{4\pi(10^{-9}36\pi)1^2}(-\bar{a}_y) = \bar{a}_y \ 54 \times 10^{-3} \ N$$

The resultant force $\bar{F}_2$ acting on Q_2 is the superposition of $\bar{F}_{12}$ and $\bar{F}_{32}$ due to Q_1 and Q_3, respectively.

The vector combination of $\bar{F}_{12}$ and $\bar{F}_{32}$ is given by:

$$\bar{F}_2 = \sqrt{F_{12}^2 + F_{32}^2} \angle \tan^{-1} \frac{\bar{F}_{32}}{\bar{F}_{12}}$$

$$= \sqrt{13.5^2 + 54^2} \times 10^{-3} \angle \tan^{-1} \frac{54}{13.5}$$

$$= 55.7 \times 10^{-3} \ \angle 76° \ N$$

Conductors and Insulators

In order to put charge in motion so that it becomes an electric current, one must provide a path through which it can flow easily by the movement of electrons. Materials through which charge flows readily are called *conductors*. Examples include most metals, such as silver, gold, copper, and aluminum. Copper is used extensively for the conductive paths on electric circuit boards and for the fabrication of electrical wires.

Insulators are materials that do not allow charge to move easily. Examples include glass, plastic, ceramics, and rubber. Electric current cannot be made to flow through an insulator, since a charge has great difficulty moving through it. One sees insulating (or *dielectric*) materials often wrapped around the center conducting core of a wire.

Although the term resistance will be formally defined later, one can say qualitatively that a conductor has a very low resistance to the flow of charge, whereas an insulator has a very high resistance to the flow of charge. Charge-conducting abilities of various materials vary in a wide range. *Semiconductors* fall in the middle between conductors and insulators, and have a moderate resistance to the flow of charge. Examples include silicon, germanium, and gallium arsenide.

Current and Magnetic Force

The rate of movement of net positive charge per unit of time through a cross section of a conductor is known as *current*,

$$i(t) = \frac{dq}{dt} \tag{1.1.5}$$

The SI unit of current is the ampere (A), which represents 1 coulomb per second. In most metallic conductors, such as copper wires, current is exclusively the movement of free electrons in the wire. Since electrons are negative, and since the direction designated for the current is that of the net positive charge movement, the charges in the wire are thus moving in the direction opposite to the direction of the current designation. The net charge transferred at a particular time is the net area under the current–time curve from the beginning of time to the present,

$$q(t) = \int_{-\infty}^{t} i(\tau)\, d\tau \tag{1.1.6}$$

While Coulomb's law has to do with the electric force associated with two charged bodies, *Ampere's law of force* is concerned with magnetic forces associated with two loops of wire carrying currents by virtue of the motion of charges in the loops. Note that isolated current elements do not exist without sources and sinks of charges at their ends; magnetic monopoles do not exist. Figure 1.1.2 shows two loops of wire in freespace carrying currents I_1 and I_2.

Considering a differential element $d\bar{l}_1$ of loop 1 and a differential element $d\bar{l}_2$ of loop 2, the differential magnetic forces $d\bar{F}_{21}$ and $d\bar{F}_{12}$ experienced by the differential current elements $I_1\, d\bar{l}_1$, and $I_2\, d\bar{l}_2$, due to I_2 and I_1, respectively, are given by

$$d\bar{F}_{21} = I_1\, d\bar{l}_1 \times \left(\frac{\mu_0}{4\pi} \frac{I_2 d\bar{l}_2 \times \bar{a}_{21}}{R^2} \right) \tag{1.1.7a}$$

$$d\bar{F}_{12} = I_2\, d\bar{l}_2 \times \left(\frac{\mu_0}{4\pi} \frac{I_1 d\bar{l}_1 \times \bar{a}_{12}}{R^2} \right) \tag{1.1.7b}$$

where $\bar{a}_{21}$ and $\bar{a}_{12}$ are unit vectors along the line joining the two current elements, R is the distance between the centers of the elements, μ_0 is the permeability of free space with units of N/A^2 or commonly known as henrys per meter (H/m). Equation (1.1.7) reveals the following:

1. The magnitude of the force is proportional to the product of the two currents and the product of the lengths of the two current elements.

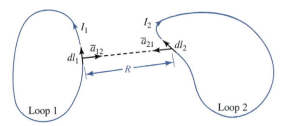

Figure 1.1.2 Illustration of Ampere's law (of force).

2. The magnitude of the force is inversely proportional to the square of the distance between the current elements.

3. To determine the direction of, say, the force acting on the current element $I_1 \, d\bar{l}_1$, the *cross product* $d\bar{l}_2 \times \bar{a}_{21}$ must be found. Then crossing $d\bar{l}_1$ with the resulting vector will yield the direction of $d\bar{F}_{21}$.

4. Each current element is acted upon by a *magnetic field* due to the other current element,

$$d\bar{F}_{21} = I_1 \, d\bar{l}_1 \times \bar{B}_2 \tag{1.1.8a}$$

$$d\bar{F}_{12} = I_2 d\bar{l}_2 \times \bar{B}_1 \tag{1.1.8b}$$

where $\bar{B}$ is known as the *magnetic flux density vector* with units of N/A · m, commonly known as webers per square meter (Wb/m^2) or tesla (T).

Current distribution is the source of magnetic field, just as charge distribution is the source of electric field. As a consequence of Equations (1.1.7) and (1.1.8), it can be seen that

$$\bar{B}_2 = \frac{\mu_0}{4\pi} \, I_2 \, d\bar{l}_2 \times \bar{a}_{21} \tag{1.1.9a}$$

$$\bar{B}_1 = \frac{\mu_0}{4\pi} \frac{I_1 \, d\bar{l}_1 \times \bar{a}_{12}}{R^2} \tag{1.1.9b}$$

which depend on the medium parameter. Equation (1.1.9) is known as the *Biot–Savart law*. Equation (1.1.8) can be expressed in terms of moving charge, since current is due to the flow of charges. With $I = dq/dt$ and $d\bar{l} = \bar{v} \, dt$, where $\bar{v}$ is the velocity, Equation (1.1.8) can be rewritten as

$$d\bar{F} = \left(\frac{dq}{dt}\right)(\bar{v} \, dt) \times \bar{B} = dq \, (\bar{v} \times \bar{B}) \tag{1.1.10}$$

Thus it follows that the force $\bar{F}$ experienced by a test charge q moving with a velocity $\bar{v}$ in a magnetic field of flux density $\bar{B}$ is given by

$$\bar{F} = q \, (\bar{v} \times \bar{B}) \tag{1.1.11}$$

The expression for the total force acting on a test charge q moving with velocity $\bar{v}$ in a region characterized by electric field intensity $\bar{E}$ and a magnetic field of flux density $\bar{B}$ is

$$\bar{F} = \bar{F}_E + \bar{F}_M = q \, (\bar{E} + \bar{v} \times \bar{B}) \tag{1.1.12}$$

which is known as the *Lorentz force equation*.

EXAMPLE 1.1.2

Figure E1.1.2 (a) gives a plot of $q(t)$ as a function of time t.

   (a) Obtain the plot of $i(t)$.

   (b) Find the average value of the current over the time interval of 1 to 7 seconds.

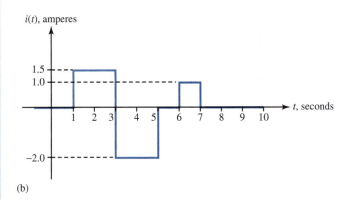

Figure E1.1.2 **(a)** Plot of $q(t)$. **(b)** Plot of $i(t)$.

Solution

   (a) Applying Equation (1.1.5) and interpreting the first derivative as the slope, one obtains the plot shown in Figure E1.1.2(b).

   (b) $I_{av} = (1/T) \int_0^T i\, dt$. Interpreting the integral as the area enclosed under the curve, one gets:

$$I_{av} = \frac{1}{(7-1)}[(1.5 \times 2) - (2.0 \times 2) + (0 \times 1) + (1 \times 1)] = 0$$

Note that the net charge transferred during the interval of 1 to 7 seconds is zero in this case.

EXAMPLE 1.1.3

Consider an infinitesimal length of 10^{-6} m of wire whose center is located at the point $(1, 0, 0)$, carrying a current of 2 A in the positive direction of x.

(a) Find the magnetic flux density due to the current element at the point $(0, 2, 2)$.

(b) Let another current element (of length 10^{-3} m) be located at the point $(0, 2, 2)$, carrying a current of 1 A in the direction of $(-\bar{a}_y + \bar{a}_z)$. Evaluate the force on this current element due to the other element located at $(1, 0, 0)$.

Solution

(a) $I_1 d\bar{l}_1 = 2 \times 10^{-6} \bar{a}_x$. The unit vector $\bar{a}_{12}$ is given by

$$\bar{a}_{12} = \frac{(0 - 1)\bar{a}_x + (2 - 0)\bar{a}_y + (2 - 0)\bar{a}_z}{\sqrt{1^2 + 2^2 + 2^2}}$$

$$= \frac{(-\bar{a}_x + 2\bar{a}_y + 2\bar{a}_z)}{3}$$

Using the Biot–Savart law, Equation (1.1.9), one gets

$$[\bar{B}_1]_{(0,2,2)} = \frac{\mu_0}{4\pi} \frac{I_1 \, d\bar{l}_1 \times \bar{a}_{12}}{R^2}$$

where μ_0 is the free-space permeability constant given in SI units as $4\pi \times 10^{-7}$ H/m, and R^2 in this case is $\{(0 - 1)^2 + (2 - 0)^2 + (2 - 0)^2\}$, or 9. Hence,

$$[\bar{B}_1]_{(0,2,2)} = \frac{4\pi \times 10^{-7}}{4\pi} \left[\frac{(2 \times 10^{-6} \bar{a}_x) \times (-\bar{a}_x + 2\bar{a}_y + 2\bar{a}_z)}{9 \times 3} \right]$$

$$= \frac{10^{-7}}{27} \times 4 \times 10^{-6} (\bar{a}_z - \bar{a}_y) \text{ Wb/m}^2$$

$$= 0.15 \times 10^{-13} (\bar{a}_z - \bar{a}_y) \text{ T}$$

(b) $I_2 \, d\bar{l}_2 = 10^{-3} (-\bar{a}_y + \bar{a}_z)$

$$d\bar{F}_{12} = I_2 d\bar{l}_2 \times \bar{B}_1$$

$$= \left[10^{-3} (-\bar{a}_y + \bar{a}_z) \right] \times \left[0.15 \times 10^{-13} (\bar{a}_z - \bar{a}_y) \right] = 0$$

Note that the force is zero since the current element $I_2 \, d\bar{l}_2$ and the field $\bar{B}_1$ due to $I_1 \, d\bar{l}_1$ at $(0, 2, 2)$ are in the same direction.

The Biot–Savart law can be extended to find the magnetic flux density due to a current-carrying filamentary wire of any length and shape by dividing the wire into a number of infinitesimal elements and using superposition. The net force experienced by a current loop can be similarly evaluated by superposition.

Electric Potential and Voltage

When electrical forces act on a particle, it will possess potential *energy*. In order to describe the potential energy that a particle will have at a point x, the *electric potential* at point x is defined as

$$v(x) = \frac{dw(x)}{dq} \tag{1.1.13}$$

where $w(x)$ is the potential energy that a particle with charge q has when it is located at the position x. The zero point of potential energy can be chosen arbitrarily since only differences in energy have practical meaning. The point where electric potential is zero is known as the *reference point* or *ground point*, with respect to which potentials at other points are then described. The *potential difference* is known as the *voltage* expressed in volts (V) or joules per coulomb (J/C). If the potential at B is higher than that at A,

$$v_{BA} = v_B - v_A \tag{1.1.14}$$

which is positive. Obviously voltages can be either positive or negative numbers, and it follows that

$$v_{BA} = -v_{AB} \tag{1.1.15}$$

The voltage at point A, designated as v_A, is then the potential at point A with respect to the ground.

Energy and Power

If a charge dq gives up energy dw when going from point a to point b, then the voltage across those points is defined as

$$v = \frac{dw}{dq} \tag{1.1.16}$$

If dw/dq is positive, point a is at the higher potential. The voltage between two points is the work per unit positive charge required to move that charge between the two points. If dw and dq have the same sign, then energy is *delivered* by a positive charge going from a to b (or a negative charge going the other way). Conversely, charged particles *gain* energy inside a *source* where dw and dq have opposite polarities.

The *load* and *source* conventions are shown in Figure 1.1.3, in which point a is at a higher potential than point b. The load *receives* or *absorbs* energy because a positive charge goes in the direction of the current arrow from higher to lower potential. The source has a capacity to *supply* energy. The *voltage source* is sometimes known as an *electromotive force*, or *emf*, to convey the notation that it is a force that drives the current through the circuit.

The *instantaneous power* p is defined as the rate of doing work or the rate of change of energy dw/dt,

$$p = \frac{dw}{dt} = \left(\frac{dw}{dq}\right)\left(\frac{dq}{dt}\right) = vi \tag{1.1.17}$$

The electric power consumed or produced by a circuit element is given by its voltage–current product, expressed in volt-amperes (VA) or watts (W). The energy over a time interval is found by integrating power,

$$w = \int_0^T p \, dt \tag{1.1.18}$$

which is expressed in watt-seconds or joules (J), or commonly in electric utility bills in kilowatt-hours (kWh). Note that 1 kWh equals 3.6×10^6 J.

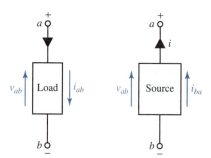

Figure 1.1.3 Load and source conventions.

A typical 12-V automobile battery, storing about 5 megajoules (MJ) of energy, is connected to a 4-A headlight system.

    (a) Find the power delivered to the headlight system.

    (b) Calculate the energy consumed in 1 hour of operation.

    (c) Express the auto-battery capacity in ampere-hours (Ah) and compute how long the headlight system can be operated before the battery is completely discharged.

Solution

    (a) Power delivered: $P = VI = 12 \cdot 4 = 48 \text{W}$.

    (b) Assuming V and I remain constant, the energy consumed in 1 hour will equal

$$W = 48(60 \times 60) = 172.8 \times 10^3 \text{J} = 172.8 \text{kJ}$$

    (c) 1 Ah $= (1\,\text{C/s})(3600\,\text{s}) = 3600\text{C}$. For the battery in question, $5 \times 10^6 \text{J}/12\,\text{V} = 0.417 \times 10^6 \text{C}$. Thus the auto-battery capacity is $0.417 \times 10^6/3600 \cong 116\,\text{Ah}$. Without completely discharging the battery, the headlight system can be operated for $116/4 = 29$ hours.

Sources and Loads

A source–load combination is represented in Figure 1.1.4. A *node* is a point at which two or more components or devices are connected together. A part of a circuit containing only one component, source, or device between two nodes is known as a *branch*. A voltage *rise* indicates an electric source, with the charge being raised to a higher potential, whereas a voltage *drop* indicates a load, with a charge going to a lower potential. The voltage *across* the source is the same as the voltage across the load in Figure 1.1.4. The current delivered by the source goes *through* the load. Ideally, with no losses, the power ($p = vi$) delivered by the source is consumed by the load.

    When current flows out of the positive terminal of an electric source, it implies that non-electric energy has been transformed into electric energy. Examples include mechanical energy transformed into electric energy as in the case of a generator source, chemical energy changed into electric energy as in the case of a battery source, and solar energy converted into electric energy as in the case of a solar-cell source. On the other hand, when current flows in the direction of voltage drop, it implies that electric energy is transformed into nonelectric energy. Examples include electric energy converted into thermal energy as in the case of an electric heater, electric energy transformed into mechanical energy as in the case of motor load, and electric energy changed into chemical energy as in the case of a charging battery.

Batteries and ac outlets are the familiar electric sources. These are *voltage sources*. An *ideal voltage source* is one whose terminal voltage v is a specified function of time, regardless of the current i through the source. An ideal battery has a constant voltage V with respect to time, as shown in Figure 1.1.5(a). It is known as a dc source, because $i = I$ is a direct current. Figure 1.1.5(b) shows the symbol and time variation for a *sinusoidal voltage source* with $v = V_m \cos \omega t$. The positive sign on the source symbol indicates instantaneous polarity of the terminal at the higher potential whenever $\cos \omega t$ is positive. A sinusoidal source is generally termed an ac source because such a voltage source tends to produce an alternating current.

The concept of an *ideal current source*, although less familiar but useful as we shall see later, is defined as one whose current i is a specified function of time, regardless of the voltage across its terminals. The circuit symbols and the corresponding i–v curves for the ideal voltage and current sources are shown in Figure 1.1.6.

Even though ideal sources could theoretically produce infinite energy, one should recognize that infinite values are physically impossible. Various circuit laws and device representations or *models* are approximations of physical reality, and significant limitations of the idealized concepts or models need to be recognized. Simplified representations or models for physical devices are the most powerful tools in electrical engineering. As for ideal sources, the concept of constant V or constant I for dc sources and the general idea of v or i being a specified function of time should be understood.

When the source voltage or current is independent of all other voltages and currents, such sources are known as *independent sources*. There are *dependent* or *controlled sources*, whose

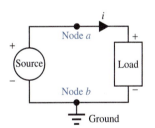

Figure 1.1.4 Source–load combination.

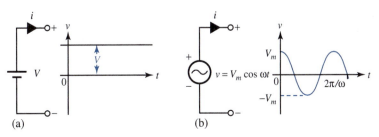

Figure 1.1.5 Voltage sources. **(a)** Ideal dc source (battery). **(b)** Ideal sinusoidal ac source.

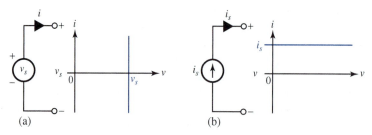

Figure 1.1.6 Circuit symbols and i–v curves. **(a)** Ideal voltage source. **(b)** Ideal current source.

voltage or current does depend on the value of some other voltage or current. As an example, a *voltage amplifier* producing an output voltage $v_{out} = Av_{in}$, where v_{in} is the input voltage and A is the constant-voltage amplification factor, is shown in Figure 1.1.7, along with its controlled-source model using the diamond-shaped symbol. Current sources controlled by a current or voltage will also be considered eventually.

Waveforms

We are often interested in *waveforms*, which may not be constant in time. Of particular interest is a *periodic waveform*, which is a *time-varying waveform* repeating itself over intervals of time $T > 0$.

$$f(t) = f(t \pm nT) \qquad n = 1, 2, 3, \cdots \qquad (1.1.19)$$

The repetition time T of the waveform is called the *period* of the waveform. For a waveform to be periodic, it must continue indefinitely in time. The dc waveform of Figure 1.1.5(a) can be considered to be periodic with an infinite period. The *frequency* of a periodic waveform is the reciprocal of its period,

$$f = \frac{1}{T} \text{ Hertz (Hz)} \qquad (1.1.20)$$

A sinusoidal or cosinusoidal waveform is typically described by

$$f(t) = A \sin(\omega t + \phi) \qquad (1.1.21)$$

where A is the amplitude, ϕ is the phase offset, and $\omega = 2\pi f = 2\pi/T$ is the radian frequency of the wave. When $\phi = 0$, a sinusoidal wave results, and when $\phi = 90°$, a cosinusoidal wave results. The *average value* of a periodic waveform is the net positive area under the curve for one period, divided by the period,

$$F_{av} = \frac{1}{T} \int_0^T f(t) \, dt \qquad (1.1.22)$$

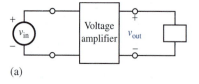

(a)

(b)

Figure 1.1.7 Voltage amplifier and its controlled-source model.

The *effective*, or *root-mean square* (rms), value is the square root of the average of $f^2(t)$,

$$F_{rms} = \sqrt{\frac{1}{T} \int_0^T f^2(t)\, dt} \qquad (1.1.23)$$

Determining the square of the function $f(t)$, then finding the mean (average) value, and finally taking the square root yields the rms value, known as effective value. This concept will be seen to be useful in comparing the effectiveness of different sources in delivering power to a resistor. The effective value of a periodic current, for example, is a constant, or dc value, which delivers the same average power to a resistor, as will be seen later.

For the special case of a dc waveform, the following holds:

$$f(t) = F; \qquad F_{av} = F_{rms} = F \qquad (1.1.24)$$

For the sinusoid or cosinusoid, it can be seen that

$$f(t) = A\,\sin(\omega t + \phi); \qquad F_{av} = 0; \qquad F_{rms} = A/\sqrt{2} \cong 0.707\,A \qquad (1.1.25)$$

The student is encouraged to show the preceding results using graphical and analytical means. Other common types of waveforms are *exponential* in nature,

$$f(t) = Ae^{-t/\tau} \qquad (1.1.26a)$$

$$f(t) = A(1 - e^{-t/\tau}) \qquad (1.1.26b)$$

where τ is known as the *time constant*. After a time of one time constant has elapsed, looking at Equation (1.1.26a), the value of the waveform will be reduced to 37% of its initial value; Equation (1.1.26b) shows that the value will rise to 63% of its final value. The student is encouraged to study the functions graphically and deduce the results.

EXAMPLE 1.1.5

A periodic current waveform in a rectifier is shown in Figure E1.1.5. The wave is sinusoidal for $\pi/3 \le \omega t \le \pi$, and is zero for the rest of the cycle. Calculate the rms and average values of the current.

Figure E1.1.5

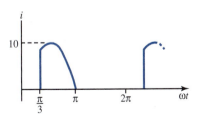

Solution

$$I_{rms} = \sqrt{\frac{1}{2\pi}\left[\int_0^{\pi/3} i^2\, d(\omega t) + \int_{\pi/3}^{\pi} i^2\, d(\omega t) + \int_{\pi}^{2\pi} i^2\, d(\omega t) \right]}$$

Notice that ωt rather than t is chosen as the variable for convenience; $\omega = 2\pi f = 2\pi/T$; and integration is performed over three discrete intervals because of the discontinuous current function. Since $i = 0$ for $0 \le \omega t < \pi/3$ and $\pi \le \omega t \le 2\pi$,

$$I_{rms} = \sqrt{\frac{1}{2\pi} \int_{\pi/3}^{\pi} 10^2 \sin^2 \omega t \, d(\omega t)} = 4.49 \text{ A}$$

$$I_{av} = \frac{1}{2\pi} \int_{\pi/3}^{\pi} 10 \sin \omega t \, d(\omega t) = 2.39 \text{ A}$$

Note that the base is the entire period 2π, even though the current is zero for a substantial part of the period.

1.2 LUMPED-CIRCUIT ELEMENTS

Electric *circuits* or *networks* are formed by interconnecting various devices, sources, and components. Although the effects of each element (such as heating effects, electric-field effects, or magnetic-field effects) are distributed throughout space, one often lumps them together as *lumped elements*. The *passive* components are the *resistance R* representing the heating effect, the *capacitance C* representing the electric-field effect, and the *inductance L* representing the magnetic-field effect. Their characteristics will be presented in this section. The capacitor models the relation between voltage and current due to changes in the accumulation of electric charge, and the inductor models the relation due to changes in magnetic flux linkages, as will be seen later. While these phenomena are generally distributed throughout an electric circuit, under certain conditions they can be considered to be concentrated at certain points and can therefore be represented by lumped parameters.

Resistance

An *ideal resistor* is a circuit element with the property that the current through it is linearly proportional to the potential difference across its terminals,

$$i = v/R = Gv, \text{ or } v = iR \tag{1.2.1}$$

which is known as *Ohm's law*, published in 1827. R is known as the resistance of the resistor with the SI unit of *ohms* (Ω), and G is the reciprocal of resistance called *conductance*, with the SI unit of *siemens* (S). The circuit symbols of fixed and variable resistors are shown in Figure 1.2.1, along with an illustration of Ohm's law. Most resistors used in practice are good approximations to *linear* resistors for large ranges of current, and their i–v characteristic (current versus voltage plot) is a straight line.

The value of resistance is determined mainly by the physical dimensions and the *resistivity* ρ of the material of which the resistor is composed. For a bar of resistive material of length l and cross-sectional area A the resistance is given by

$$R = \frac{\rho l}{A} = \frac{l}{\sigma A} \tag{1.2.2}$$

where ρ is the resistivity of the material in ohm-meters ($\Omega \cdot$ m), and σ is the *conductivity* of the material in S/m, which is the reciprocal of the resistivity. Metal wires are often considered as ideal

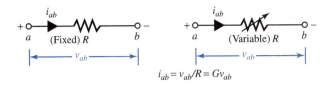

Figure 1.2.1 Circuit symbols of fixed and variable resistors and illustration of Ohm's law.

$$i_{ab} = v_{ab}/R = Gv_{ab}$$

TABLE 1.2.1 Resistivity of Some Materials

Type	Material	$\rho(\Omega \cdot m)$
Conductors (at 20°C)	Silver	16×10^{-9}
	Copper	17×10^{-9}
	Gold	24×10^{-9}
	Aluminum	28×10^{-9}
	Tungsten	55×10^{-9}
	Brass	67×10^{-9}
	Sodium	0.04×10^{-6}
	Stainless steel	0.91×10^{-6}
	Iron	0.1×10^{-6}
	Nichrome	1×10^{-6}
	Carbon	35×10^{-6}
	Seawater	0.25
Semiconductors (at 27°C or 300 K)	Germanium	0.46
	Silicon	2.3×10^3
Insulators	Rubber	1×10^{12}
	Polystyrene	1×10^{15}

conductors with zero resistance as a good approximation. Table 1.2.1 lists values of ρ for some materials.

The resistivity of conductor metals varies linearly over normal operating temperatures according to

$$\rho_{T2} = \rho_{T1} \left(\frac{T_2 + T}{T_1 + T} \right) \tag{1.2.3}$$

where ρ_{T2} and ρ_{T1} are resistivities at temperatures T_2 and T_1, respectively, and T is a temperature constant that depends on the conductor material. All temperatures are in degrees Celsius. The conductor resistance also depends on other factors, such as spiraling, frequency (the skin effect which causes the ac resistance to be slightly higher than the dc resistance), and current magnitude in the case of magnetic conductors (e.g., steel conductors used for shield wires).

Practical resistors are manufactured in standard values, various resistance tolerances, several power ratings (as will be explained shortly), and in a number of different forms of construction. The three basic construction techniques are *composition* type, which uses carbon or graphite and is molded into a cylindrical shape, *wire-wound* type, in which a length of enamel-coated wire is wrapped around an insulating cylinder, and *metal-film* type, in which a thin layer of metal is vacuum deposited. Table 1.2.2 illustrates the standard color-coded bands used for evaluating resistance and their interpretation for the common carbon composition type. Sometimes a fifth band is also present to indicate reliability. Black is the least reliable color and orange is 1000 times more reliable than black.

For resistors ranging from 1 to 9.1 Ω, the standard resistance values are listed in Table 1.2.3. Other available values can be obtained by multiplying the values shown in Table 1.2.3 by factors

TABLE 1.2.2 Standard Color-Coded Bands for Evaluating Resistance and Their Interpretation

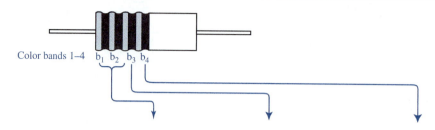

Color bands 1–4 b_1 b_2 b_3 b_4

Color of Band	Digit of Band	Multiplier	% Tolerance in Actual Value
Black	0	10^0	—
Brown	1	10^1	—
Red	2	10^2	—
Orange	3	10^3	—
Yellow	4	10^4	—
Green	5	10^5	—
Blue	6	10^6	—
Violet	7	10^7	—
Grey	8	10^8	—
White	9	—	—
Gold	—	10^{-1}	± 5%
Silver	—	10^{-2}	± 10%
Black or no color	—	—	± 20%

Resistance value $= (10b_1 + b_2) \times 10^{b_3}$ Ω.

of 10 ranging from 10 Ω to about 22×10^6 Ω. For example, 8.2 Ω, 82 Ω, 820 Ω, . . . , 820 kΩ are standard available values.

The maximum allowable power dissipation or *power rating* is typically specified for commercial resistors. A common power rating for resistors used in electronic circuits is ¼ W; other ratings such as ⅛, ½, 1, and 2 W are available with composition-type resistors, whereas larger ratings are also available with other types. Variable resistors, known as *potentiometers*, with a movable contact are commonly found in rotary or linear form. Wire-wound potentiometers may have higher power ratings up to 1000 W.

The advent of integrated circuits has given rise to *packaged resistance arrays* fabricated by using film technology. These packages are better suited for automated manufacturing and are usually less costly than discrete resistors in large production runs.

An important property of the resistor is its ability to convert energy from electrical form into heat. The manufacturer generally states the maximum power dissipation of the resistor in watts. If more power than this is converted to heat by the resistor, the resistor will be damaged due to overheating. The instantaneous power absorbed by the resistor is given by

$$p(t) = v(t)i(t) = i^2 R = v^2/R = v^2 G \tag{1.2.4}$$

where v is the voltage drop across the resistance and i is the current through the resistance. It can be shown (see Problem 1.2.13) that the average value of Equation (1.2.4) is given by

$$P_{av} = V_{rms}I_{rms} = I_{rms}^2 R = V_{rms}^2/R = V_{rms}^2 G \tag{1.2.5}$$

for periodically varying current and voltage as a function of time. Equation (1.2.5) gives the expression for the power converted to heat by the resistor.

TABLE 1.2.3 Standard Available Values of Resistors

1.0	1.5	2.2	3.3	4.7	6.8
1.1	1.6	2.4	3.6	5.1	7.5
1.2	1.8	2.7	3.9	5.6	8.2
1.3	2.0	3.0	4.3	6.2	9.1

Series and *parallel* combinations of resistors occur very often. Figure 1.2.2 illustrates these combinations.

Figure 1.2.2(a) shows two resistors R_1 and R_2 in series sharing the voltage v in direct proportion to their values, while the same current i flows through both of them,

$$v = v_{AC} = v_{AB} + v_{BC} = iR_1 + iR_2 = i(R_1 + R_2) = iR_{eq}$$

or, when R_1 and R_2 are in series,

$$R_{eq} = R_1 + R_2 \tag{1.2.6}$$

Figure 1.2.2(b) shows two resistors in parallel sharing the current i in inverse proportion to their values, while the same voltage v is applied across each of them. At node B,

$$i = i_1 + i_2 = \frac{v}{R_1} + \frac{v}{R_2} = v\left(\frac{1}{R_1} + \frac{1}{R_2}\right) = v\left/\left(\frac{R_1R_2}{R_1 + R_2}\right)\right. = \frac{v}{R_{eq}}$$

or, when R_1 and R_2 are in parallel,

$$R_{eq} = \frac{R_1R_2}{R_1 + R_2} \quad \text{or} \quad \frac{1}{R_{eq}} = \frac{1}{R_1} + \frac{1}{R_2} \quad \text{or} \quad G_{eq} = G_1 + G_2 \tag{1.2.7}$$

Notice the *voltage division* shown in Figure 1.2.2(a), and the *current division* in Figure 1.2.2(b).

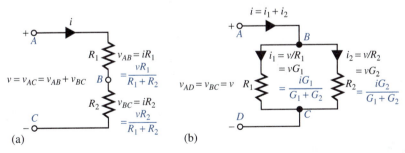

Figure 1.2.2 Resistances in series and in parallel. **(a)** R_1 and R_2 in series. **(b)** R_1 and R_2 in parallel.

EXAMPLE 1.2.1

A no. 14 gauge copper wire, commonly used in extension cords, has a circular wire diameter of 64.1 mils, where 1 mil = 0.001 inch.

(a) Determine the resistance of a 100-ft-long wire at 20°C.

(b) If such a 2-wire system is connected to a 110-V (rms) residential source outlet in order to power a household appliance drawing a current of 1 A (rms), find the rms voltage at the load terminals.

(c) Compute the power dissipated due to the extension cord.

(d) Repeat part (a) at 50°C, given that the temperature constant for copper is 241.5°C.

Solution

(a) $d = 64.1$ mils $= 64.1 \times 10^{-3}$ in $= 64.1 \times 10^{-3} \times 2.54$ cm/1 in $\times$ 1 m/100 cm $= 1.628 \times 10^{-3}$ m. From Table 1.2.1, ρ of copper at 20°C is 17×10^{-9} m,

$$l = 100 \text{ ft} = 100 \text{ ft} \times \frac{12 \text{ in}}{1 \text{ ft}} \times \frac{2.54 \text{ cm}}{1 \text{ in}} \times \frac{1 \text{ m}}{100 \text{ cm}} = 30.48 \text{ m}$$

$$A = \frac{\pi d^2}{4} = \frac{\pi (1.628 \times 10^{-3})^2}{4} = 2.08 \times 10^{-6} \text{ m}^2$$

Per Equation (1.2.2),

$$R_{20°C} = \frac{17 \times 10^{-9} \times 30.48}{2.08 \times 10^{-6}} \cong 0.25 \ \Omega$$

(b) Rms voltage at load terminals, $V = 110 - (0.25)2 = 109.5$ V (rms). Note that two 100-ft-long wires are needed for the power to be supplied.

(c) Power dissipated, per Equation (1.2.5), $P = (1)^2(0.25)(2) = 0.5$ W.

(d) Per Equation (1.2.3),

$$\rho_{50°C} = \rho_{20°C} \left(\frac{50 + 241.5}{20 + 241.5} \right) = \frac{17 \times 10^{-9} \times 291.5}{261.5} = 18.95 \times 10^{-9} \ \Omega \cdot \text{m}$$

Hence,

$$R_{50°C} = \frac{18.95 \times 10^{-9} \times 30.48}{2.08 \times 10^{-6}} \cong 0.28 \ \Omega$$

EXAMPLE 1.2.2

(a) Consider a series–parallel combination of resistors as shown in Figure E1.2.2(a). Find the equivalent resistance as seen from terminals A–B.

(b) Determine the current I and power P delivered by a 10-V dc voltage source applied at terminals A–B, with A being at higher potential than B.

(c) Replace the voltage source by an equivalent current source at terminals A–B.

(d) Show the current and voltage distribution clearly in all branches of the original circuit configuration.

Solution

(a) The circuit is reduced as illustrated in Figure E1.2.2(b).

(b) $I = 5$ A; $P = VI = I^2R = V^2/R = 50$ W [see Figure E1.2.2(c)].

(c) See Figure E1.2.2(d).

(d) See Figure E1.2.2(e).

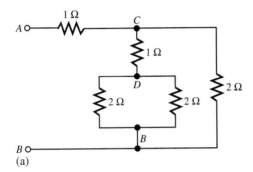

(a)

Figure E1.2.2

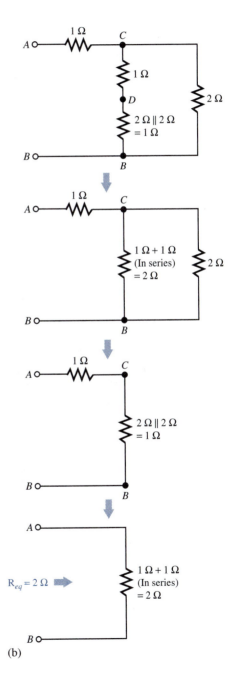

(b)

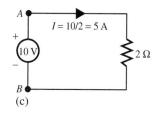

(c)

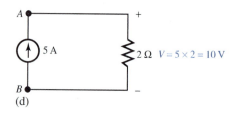

(d)

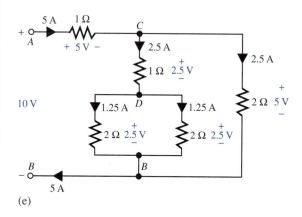

(e)

Maximum Power Transfer

In order to investigate the power transfer between a practical source and a load connected to it, let us consider Figure 1.2.3, in which a constant voltage source v with a known internal resistance R_S is connected to a variable load resistance R_L. Note that when R_L is equal to zero, it is called a *short circuit*, in which case v_L becomes zero and i_L is equal to v/R_S. When R_L approaches infinity, it is called an *open circuit*, in which case i_L becomes zero and v_L is equal to v. One is generally interested to find the value of the load resistance that will absorb maximum power from the source.

The power P_L absorbed by the load is given by

$$P_L = i_L^2 R_L \tag{1.2.8}$$

where the load current i_L is given by

$$i_L = \frac{v^2}{R_S + R_L} \tag{1.2.9}$$

Substituting Equation (1.2.9) in Equation (1.2.8), one gets

$$P_L = \frac{v^2}{(R_S + R_L)^2} R_L \tag{1.2.10}$$

For given fixed values of v and R_S, in order to find the value of R_L that maximizes the power absorbed by the load, one sets the first derivative dP_L/dR_L equal to zero,

$$\frac{dP_L}{dR_L} = \frac{v^2(R_L + R_S)^2 - 2v^2 R_L(R_L + R_S)}{(R_L + R_S)^4} = 0 \tag{1.2.11}$$

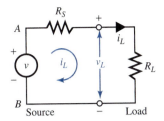

Source   Load

Figure 1.2.3 Power transfer between source and load. *Note:* $R_L = 0$ implies short circuit; $v_L = 0$ and $i_L = \frac{v}{R_S}$ and $R_L \to \bullet$ implies open circuit; $i_L = 0$ and $v_L = v$.

which leads to the following equation:

$$(R_L + R_S)^2 - 2R_L(R_L + R_S) = 0 \tag{1.2.12}$$

The solution of Equation (1.2.12) is given by

$$R_L = R_S \tag{1.2.13}$$

That is to say, in order to transfer maximum power to a load, the load resistance must be *matched* to the source resistance or, in other words, they should be equal to each other.

A problem related to power transfer is that of *source loading*. Figure 1.2.4(a) illustrates a practical voltage source (i.e., an ideal voltage source along with a series internal source resistance) connected to a load resistance; Figure 1.2.4(b) shows a practical current source (i.e., an ideal current source along with a parallel or shunt internal source resistance) connected to a load resistance. It follows from Figure 1.2.4(a) that

$$v_L = v - v_{\text{int}} = v - i_L R_S \tag{1.2.14}$$

where v_{int} is the internal voltage drop within the source, which depends on the amount of current drawn by the load. As seen from Equation (1.2.14), the voltage actually seen by the load v_L is somewhat lower than the *open-circuit voltage* of the source. When the load resistance R_L is infinitely large, the load current i_L goes to zero, and the load voltage v_L is then equal to the open-circuit voltage of the source v. Hence, it is desirable to have as small an internal resistance as possible in a practical voltage source.

From Figure 1.2.4(b) it follows that

$$i_L = i - i_{\text{int}} = i - \frac{v_L}{R_S} \tag{1.2.15}$$

where i_{int} is the internal current drawn away from the load because of the presence of the internal source resistance. Thus the load will receive only part of the *short-circuit current* available from the source. When the load resistance R_L is zero, the load voltage v_L goes to zero, and the load

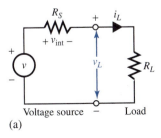

Voltage source   Load
(a)

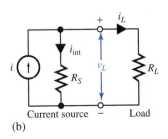

Current source   Load
(b)

Figure 1.2.4 Source-loading effects.

current i_L is then equal to the short-circuit current of the source i. Hence, it is desirable to have as large an internal resistance as possible in a practical current source.

Capacitance

An *ideal capacitor* is an energy-storage circuit element (with no loss associated with it) representing the electric-field effect. The capacitance in farads (F) is defined by

$$C = q/v \qquad (1.2.16)$$

where q is the charge on each conductor, and v is the potential difference between the two perfect conductors. With v being proportional to q, C is a constant determined by the geometric configuration of the two conductors. Figure 1.2.5(a) illustrates a two-conductor system carrying $+q$ and $-q$ charges, respectively, forming a capacitor.

The general circuit symbol for a capacitor is shown in Figure 1.2.5(b), where the current entering one terminal of the capacitor is equal to the rate of buildup of charge on the plate attached to that terminal,

$$i(t) = \frac{dq}{dt} = C\frac{dv}{dt} \qquad (1.2.17)$$

in which C is assumed to be a constant and not a function of time (which it could be, if the separation distance between the plates changed with time).

The terminal v–i relationship of a capacitor can be obtained by integrating both sides of Equation (1.2.17),

$$v(t) = \frac{1}{C}\int_{-\infty}^{t} i(\tau)\, d\tau \qquad (1.2.18)$$

which may be rewritten as

$$v(t) = \frac{1}{C}\int_{0}^{t} i(\tau)\, d\tau + \frac{1}{C}\int_{-\infty}^{0} i(\tau)\, d\tau = \frac{1}{C}\int_{0}^{t} i(\tau)\, d\tau + v(0) \qquad (1.2.19)$$

where $v(0)$ is the *initial* capacitor voltage at $t = 0$.

The instantaneous power delivered to the capacitor is given by

$$p(t) = v(t)i(t) = C\,v(t)\,\frac{dv(t)}{dt} \qquad (1.2.20)$$

whose average value can be shown (see Problem 1.2.13) to be zero for sinusoidally varying current and voltage as a function of time. The energy stored in a capacitor at a particular time is found by integrating,

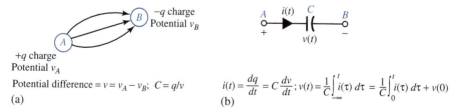

$-q$ charge
Potential v_B

$+q$ charge
Potential v_A

Potential difference $= v = v_A - v_B$; $C = q/v$
(a)

$i(t) = \dfrac{dq}{dt} = C\dfrac{dv}{dt}$; $v(t) = \dfrac{1}{C}\displaystyle\int_{-\infty}^{t} i(\tau)\, d\tau = \dfrac{1}{C}\displaystyle\int_{0}^{t} i(\tau)\, d\tau + v(0)$
(b)

Figure 1.2.5 Capacitor. **(a)** Two perfect conductors carrying $+q$ and $-q$ charges. **(b)** Circuit symbol.

$$w(t) = \int_{-\infty}^{t} p(\tau)\, d\tau = C \int_{-\infty}^{t} v(\tau)\frac{dv(\tau)}{d\tau} = \frac{1}{2}Cv^2(t) - \frac{1}{2}Cv^2(-\infty) \qquad (1.2.21)$$

Assuming the capacitor voltage to be zero at $t = -\infty$, the stored energy in the capacitor at some time t is given by

$$w(t) = \frac{1}{2}Cv^2(t) \qquad (1.2.22)$$

which depends only on the voltage of the capacitor at that time, and represents the stored energy in the electric field between the plates due to the separation of charges.

If the voltage across the capacitor does not change with time, no current flows, as seen from Equation (1.2.17). Thus the capacitor acts like an open circuit, and the following relations hold:

$$C = \frac{Q}{V}; \quad I = 0, \quad W = \frac{1}{2}CV^2 \qquad (1.2.23)$$

An ideal capacitor, once charged and disconnected, the current being zero, will retain a potential difference for an indefinite length of time. Also, the voltage across a capacitor cannot change value instantaneously, while an instantaneous change in the capacitor current is quite possible. The student is encouraged to reason through and justify the statement made here by recalling Equation (1.2.17).

Series and parallel combinations of capacitors are often encountered. Figure 1.2.6 illustrates these.

It follows from Figure 1.2.6(a),

$$v = v_{AC} = v_{AB} + v_{BC}$$

$$\frac{dv}{dt} = \frac{dv_{AB}}{dt} + \frac{dv_{BC}}{dt} = \frac{i}{C_1} + \frac{i}{C_2} = i\left(\frac{C_1 + C_2}{C_1 C_2}\right) = \frac{i}{C_{eq}}$$

or, when C_1 and C_2 are in series,

$$C_{eq} = \frac{C_1 C_2}{C_1 + C_2} \quad \text{or} \quad \frac{1}{C_{eq}} = \frac{1}{C_1} + \frac{1}{C_2} \qquad (1.2.24)$$

Referring to Figure 1.2.6(b), one gets

$$i = i_1 + i_2 = C_1\frac{dv}{dt} + C_2\frac{dv}{dt} = (C_1 + C_2)\frac{dv}{dt} = C_{eq}\frac{dv}{dt}$$

or, when C_1 and C_2 are in parallel,

$$C_{eq} = C_1 + C_2 \qquad (1.2.25)$$

Note that capacitors in parallel combine as resistors in series, and capacitors in series combine as resistors in parallel.

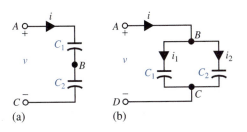

(a) (b)

Figure 1.2.6 Capacitors in series and in parallel. (a) C_1 and C_2 in series. (b) C_1 and C_2 in parallel.

The *working voltage* for a capacitor is generally specified by the manufacturer, thereby giving the maximum voltage that can safely be applied between the capacitor terminals. Exceeding this limit may result in the breakdown of the insulation and then the formation of an electric arc between the capacitor plates. Unintentional or *parasitic* capacitances that occur due to the proximity of circuit elements may have serious effects on the circuit behavior.

Physical capacitors are often made of tightly rolled sheets of metal film, with a dielectric (paper or nylon) sandwiched in between, in order to increase their capacitance values (or ability to store energy) for a given size. Table 1.2.4 lists the range of general-purpose capacitances together with the maximum voltages and frequencies for different types of dielectric materials. Practical capacitors come in a wide range of values, shapes, sizes, voltage ratings, and constructions. Both fixed and adjustable devices are available. Larger capacitors are of the electrolytic type, using aluminum oxide as the dielectric.

TABLE 1.2.4 Characteristics of General-Purpose Capacitors

Material	Capacitance Range	Maximum Voltage Range (V)	Frequency Range (Hz)
Mica	1 pF to 0.1 μF	50–600	10^3–10^{10}
Ceramic	10 pF to 1 μF	50–1600	10^3–10^{10}
Mylar	0.001 F to 10 μF	50–600	10^2–10^8
Paper	10 pF to 50 μF	50–400	10^2–10^8
Electrolytic	0.1 μF to 0.2 F	3–600	10–10^4

Note: 1 pF $= 10^{-12}$ F; 1 μF $= 10^{-6}$ F.

EXAMPLE 1.2.3

(a) Consider a 5-μF capacitor to which a voltage $v(t)$ is applied, shown in Figure E1.2.3(a), top. Sketch the capacitor current and stored energy as a function of time.

(b) Let a current source $i(t)$ be attached to the 5-μF capacitor instead of the voltage source of part (a), shown in Figure E1.2.3(b), top. Sketch the capacitor voltage and energy stored as a function of time.

(c) If three identical 5-μF capacitors with an initial voltage of 1 mV are connected (i) in series and (ii) in parallel, find the equivalent capacitances for both cases.

Solution

(a) From Figure E1.2.3(a) it follows that

$$v(t) = 0 , \qquad t \leq -1 \, \mu s$$
$$= 5(t + 1) \text{ mV}, \qquad -1 \leq t \leq 1 \, \mu s$$
$$= 10 \text{ mV}, \qquad 1 \leq t \leq 3 \, \mu s$$
$$= -10(t - 4) \text{ mV}, \qquad 3 \leq t \leq 4 \, \mu s$$
$$= 0 , \qquad 4 \leq t \, \mu s$$

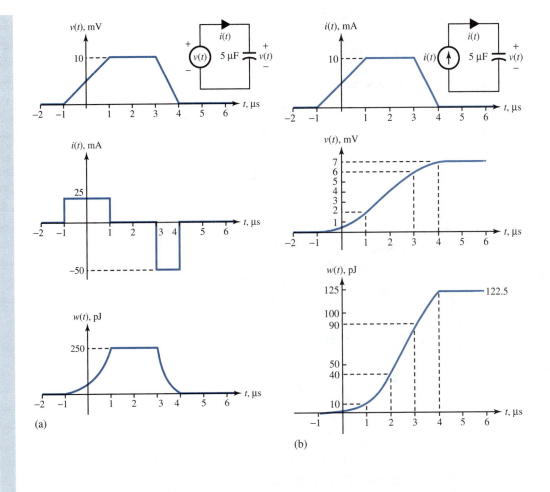

(a)

(b)

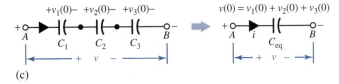

(c)

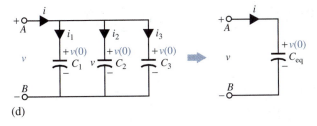

(d)

Figure E1.2.3

Since

$$i(t) = C\frac{dv}{dt} = (5 \times 10^{-6})\frac{dv}{dt}$$

it follows that

$$
\begin{aligned}
i(t) &= 0, & t \leq -1 \ \mu s \\
&= 25 \text{ mA}, & -1 \leq t \leq 1 \ \mu s \\
&= 0, & 1 \leq t \leq 3 \ \mu s \\
&= -50 \text{ mA}, & 3 \leq t \leq 4 \ \mu s \\
&= 0, & 4 \leq t \ \mu s
\end{aligned}
$$

which is sketched in the center of Figure E1.2.3(a).

Since the energy stored at any instant is

$$w(t) = \frac{1}{2}Cv^2(t) = \frac{1}{2}(5 \times 10^{-6})v^2(t)$$

it follows that:

$$
\begin{aligned}
w(t) &= 0, & t \leq -1 \ \mu s \\
&= 62.5 \ (t^2 + 2t + 1) \text{ pJ}, & -1 \leq t \leq 1 \ \mu s \\
&= 250 \text{ pJ}, & 1 \leq t \leq 3 \ \mu s \\
&= 250 \ (t^2 - 8t + 16) \text{ pJ}, & 3 \leq t \leq 4 \ \mu s \\
&= 0, & 4 \leq t \ \mu s
\end{aligned}
$$

which is sketched at the bottom of Figure E1.2.3(a).

(b) From Figure E1.2.3(b) it follows that

$$
\begin{aligned}
i(t) &= 0, & t \leq -1 \ \mu s \\
&= 5 \ (t + 1) \text{ mA}, & -1 \leq t \leq 1 \ \mu s \\
&= 10 \text{ mA}, & 1 \leq t \leq 3 \ \mu s \\
&= -10 \ (t - 4) \text{ mA}, & 3 \leq t \leq 4 \ \mu s \\
&= 0, & 4 \leq t \ \mu s
\end{aligned}
$$

Since

$$v(t) = \frac{1}{C} \int_{-\infty}^{t} i(\tau) \, d\tau = \frac{1}{5 \times 10^{-6}} \int_{-\infty}^{t} i(\tau) \, d\tau$$

it follows that

$$
\begin{aligned}
v(t) &= 0, & t \leq -1 \ \mu s \\
&= \left(\frac{t^2}{2} + t + \frac{1}{2}\right) \text{ mV}, & -1 \leq t \leq 1 \ \mu s
\end{aligned}
$$

$$= 2t \text{ mV}, \qquad\qquad 1 \leq t \leq 3 \ \mu s$$

$$= -t^2 + 8t - 9 \text{ mV}, \qquad 3 \leq t \leq 4 \ \mu s$$

$$= 7 \text{ mV}, \qquad\qquad 4 \leq t \ \mu s$$

which is sketched in the center of Figure E1.2.3(b).

Since the energy stored at any instant is

$$w(t) = \frac{1}{2} C v^2(t) = \frac{1}{2} (5 \times 10^{-6}) v^2(t)$$

it follows that

$$w(t) = 0, \qquad\qquad\qquad t \leq -1 \ \mu s$$

$$= 2.5 \left(\frac{t^2}{2} + t + \frac{1}{2} \right)^2 \text{ pJ}, \qquad -1 \leq t \leq 1 \ \mu s$$

$$= 10 t^2 \text{ pJ}, \qquad\qquad\qquad 1 \leq t \leq 3 \ \mu s$$

$$= 2.5 (-t^2 + 8t - 9)^2 \text{ pJ}, \qquad 3 \leq t \leq 4 \ \mu s$$

$$= 122.5 \text{pJ}, \qquad\qquad\qquad 4 \leq t \ \mu s$$

which is sketched at the bottom of Figure E1.2.3(b).

(c) (i)

$$\frac{1}{C_{eq}} = \frac{1}{C_1} + \frac{1}{C_2} + \frac{1}{C_3} = \frac{3}{5 \times 10^{-6}}, \qquad \text{or} \qquad C_{eq} = \frac{5}{3} \times 10^{-6} \text{F} = \frac{5}{3} \ \mu \text{F},$$

with an initial voltage $v(0) = 3$ mV [Figure E1.2.3(c)].

(ii)

$$C_{eq} = C_1 + C_2 + C_3 = 3 \times 5 \times 10^{-6} \text{ F} = 15 \ \mu \text{F}$$

with an initial voltage $v(0) = 1$ mV [Figure E1.2.3(d)].

Inductance

An *ideal inductor* is also an energy-storage circuit element (with no loss associated with it) like a capacitor, but representing the magnetic-field effect. The inductance in henrys (H) is defined by

$$L = \frac{\lambda}{i} = \frac{N\psi}{i} \tag{1.2.26}$$

where λ is the magnetic-flux linkage in weber-turns (Wb·t), N is the number of turns of the coil, and $N\psi$ is the magnetic flux in webers (Wb) produced by the current i in amperes (A). Figure 1.2.7(a) illustrates a single inductive coil or an inductor of N turns carrying a current i that is linked by its own flux.

The general circuit symbol for an inductor is shown in Figure 1.2.7(b). According to Faraday's law of induction, one can write

$$v(t) = \frac{d\lambda}{dt} = \frac{d(N\psi)}{dt} = N \frac{d\psi}{dt} = \frac{d(Li)}{dt} = L \frac{di}{dt} \tag{1.2.27}$$

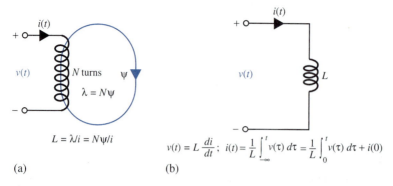

$$L = \lambda/i = N\psi/i$$

$$v(t) = L\frac{di}{dt}\,; \quad i(t) = \frac{1}{L}\int_{-\infty}^{t} v(\tau)\,d\tau = \frac{1}{L}\int_{0}^{t} v(\tau)\,d\tau + i(0)$$

(a)   (b)

Figure 1.2.7 An inductor. (a) A single inductive coil of N turns. (b) Circuit symbol.

where L is assumed to be a constant and not a function of time (which it could be if the physical shape of the coil changed with time). Mathematically, by looking at Equations (1.2.17) and (1.2.27), the inductor is the *dual* of the capacitor. That is to say, the terminal relationship for one circuit element can be obtained from that of the other by interchanging v and i, and also by interchanging L and C.

The terminal i–v relationship of an inductor can be obtained by integrating both sides of Equation (1.2.27),

$$i(t) = \frac{1}{L}\int_{-\infty}^{t} v(\tau)\,d\tau \qquad (1.2.28)$$

which may be rewritten as

$$i(t) = \frac{1}{L}\int_{0}^{t} v(\tau)\,d\tau + \frac{1}{L}\int_{-\infty}^{0} v(\tau)\,d\tau = \frac{1}{L}\int_{0}^{t} v(\tau)\,d\tau + i(0) \qquad (1.2.29)$$

where $i(0)$ is the initial inductor current at $t = 0$.

The instantaneous power delivered to the inductor is given by

$$p(t) = v(t)i(t) = Li(t)\frac{di(t)}{dt} \qquad (1.2.30)$$

whose average value can be shown (see Problem 1.2.13) to be zero for sinusoidally varying current and voltage as a function of time. The energy stored in an inductor at a particular time is found by integrating,

$$w(t) = \int_{-\infty}^{t} p(\tau)\,d\tau = \frac{1}{2}Li^2(t) - \frac{1}{2}Li^2(-\infty) \qquad (1.2.31)$$

Assuming the inductor current to be zero at $t = -\infty$, the stored energy in the inductor at some time t is given by

$$w(t) = \frac{1}{2}Li^2(t) \qquad (1.2.32)$$

which depends only on the inductor current at that time, and represents the stored energy in the magnetic field produced by the current carried by the coil.

If the current flowing through the coil does not change with time, no voltage across the coil exists, as seen from Equation (1.2.27). The following relations hold:

$$L = \frac{\lambda}{I}; \qquad V = 0; \qquad W = \frac{1}{2}LI^2 \qquad (1.2.33)$$

Under dc conditions, an ideal inductor acts like an ideal wire, or short circuit. Note that the current through an inductor cannot change value instantaneously. However, there is no reason to rule out an instantaneous change in the value of the inductor voltage. The student should justify the statements made here by recalling Equation (1.2.27).

If the medium in the flux path has a linear magnetic characteristic (i.e., constant permeability), then the relationship between the flux linkages λ and the current i is *linear*, and the slope of the linear λ–i characteristic gives the *self-inductance*, defined as flux linkage per ampere by Equation (1.2.26). While the inductance in general is a function of the geometry and permeability of the material medium, in a linear system it is independent of voltage, current, and frequency. If the inductor coil is wound around a ferrous core such as iron, the λ–i relationship will be *nonlinear* and even multivalued because of hysteresis. In such a case the inductance becomes a function of the current, and the inductor is said to be nonlinear. However, we shall consider only linear inductors here.

Series and parallel combinations of inductors are often encountered. Figure 1.2.8 illustrates these.

By invoking the principle of *duality*, it can be seen that the inductors in series combine like resistors in series and capacitors in parallel; the inductors in parallel combine like resistors in parallel and capacitors in series. Thus, when L_1 and L_2 are in series,

$$L_{eq} = L_1 + L_2 \tag{1.2.34}$$

and when L_1 and L_2 are in parallel,

$$L_{eq} = \frac{L_1 L_2}{L_1 + L_2} \qquad \text{or} \qquad \frac{1}{L_{eq}} = \frac{1}{L_1} + \frac{1}{L_2} \tag{1.2.35}$$

A practical inductor may have considerable resistance in the wire of a coil, and sizable capacitances may exist between various turns. A possible *model* for a practical inductor could be a combination of ideal elements: a combination of resistance and inductance in series, with a capacitance in parallel. Techniques for modeling real circuit elements will be used extensively in later chapters.

Practical inductors range from about 0.1 μH to hundreds of millihenrys. Some, meant for special applications in power supplies, can have values as large as several henrys. In general, the larger the inductance, the lower its frequency is in its usage. The smallest inductance values are generally used at radio frequencies. Although inductors have many applications, the total demand does not even remotely approach the consumption of resistors and capacitors. Inductors generally tend to be rather bulky and expensive, especially in low-frequency applications. Industry-wide standardization for inductors is not done to the same degree as for more frequently used devices such as resistors and capacitors.

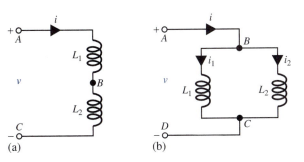

(a) (b)

Figure 1.2.8 Inductors in series and parallel. **(a)** L_1 and L_2 in series. **(b)** L_1 and L_2 in parallel.

EXAMPLE 1.2.4

(a) Consider a 5-μH inductor to which a current source $i(t)$ is attached, as shown in Figure E1.2.3(b). Sketch the inductor voltage and stored energy as a function of time.

(b) Let a voltage source $v(t)$ shown in Figure E1.2.3(a) be applied to the 5-μH inductor instead of the current source in part (a). Sketch the inductor current and energy stored as a function of time.

(c) If three identical 5-μH inductors with initial current of 1 mA are connected (i) in series and (ii) in parallel, find the equivalent inductance for each case.

Solution

(a) From the principle of duality and for the given values, it follows that the inductor-voltage waveform is the same as the capacitor-current waveform of Example 1.2.3(a), in which $i(t)$ is to be replaced by $v(t)$, and 25 and -50 mA are to be replaced by 25 and -50 mV. The stored energy $w(t)$ is the same as in the solution of Example 1.2.3(a).

(b) The solution is the same as that of Example 1.2.3(b), except that $v(t)$ in mV is to be replaced by $i(t)$ in mA.

(c) (i) Looking at the solution of Example 1.2.3(c), part (ii),

$$L_{eq} = L_1 + L_2 + L_3 = 3 \times 5 \times 10^{-6} \text{ H} = 15 \ \mu\text{H}$$

with an initial current $i(0) = 1$ mA.

(ii) Following the solution of Example 1.2.3(c), part (i),

$$\frac{1}{L_{eq}} = \frac{1}{L_1} + \frac{1}{L_2} + \frac{1}{L_3} = \frac{3}{5 \times 10^{-6}} \quad or \quad L_{eq} = \frac{5}{3} \times 10^{-6} \text{ H} = \frac{5}{3} \ \mu\text{H}$$

with an initial current $i(0) = 3$ mA.

When more than one loop or circuit is present, the flux produced by the current in one loop may link another loop, thereby inducing a current in that loop. Such loops are said to be mutually coupled, and there exists a *mutual inductance* between such loops. The mutual inductance between two circuits is defined as the flux linkage produced in one circuit by a current of 1 ampere in the other circuit. Let us now consider a pair of mutually coupled inductors, as shown in Figure 1.2.9. The self-inductances L_{11} and L_{22} of inductors 1 and 2, respectively, are given by

$$L_{11} = \frac{\lambda_{11}}{i_1} \tag{1.2.36}$$

and

$$L_{22} = \frac{\lambda_{22}}{i_2} \tag{1.2.37}$$

where λ_{11} is the flux linkage of inductor 1 produced by its own current i_1, and λ_{22} is the flux linkage of inductor 2 produced by its own current i_2. The mutual inductances L_{12} and L_{21} are given by

$$L_{12} = \frac{\lambda_{12}}{i_2} \tag{1.2.38}$$

and

$$L_{21} = \frac{\lambda_{21}}{i_1} \tag{1.2.39}$$

where λ_{12} is the flux linkage of inductor 1 produced by the current i_2 in inductor 2, and λ_{21} is the flux linkage of inductor 2 produced by the current i_1 in inductor 1.

If a current of i_1 flows in inductor 1 while the current in inductor 2 is zero, the equivalent fluxes are given by

$$\psi_{11} = \frac{\lambda_{11}}{N_1} \tag{1.2.40}$$

and

$$\psi_{21} = \frac{\lambda_{21}}{N_2} \tag{1.2.41}$$

where N_1 and N_2 are the number of turns of inductors 1 and 2, respectively. That part of the flux of inductor 1 that does not link any turn of inductor 2 is known as the equivalent *leakage flux* of inductor 1,

$$\psi_{l1} = \psi_{11} - \psi_{21} \tag{1.2.42}$$

Similarly,

$$\psi_{l2} = \psi_{22} - \psi_{12} \tag{1.2.43}$$

The *coefficient of coupling* is given by

$$k = \sqrt{k_1 k_2} \tag{1.2.44}$$

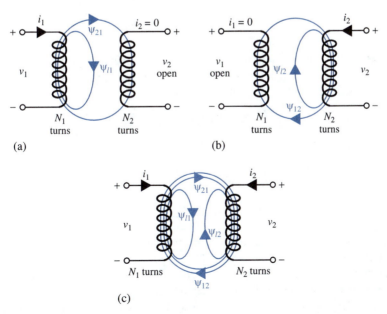

Figure 1.2.9 Mutually coupled inductors.

where $k_1 = \psi_{21}/\psi_{11}$ and $k_2 = \psi_{12}/\psi_{22}$. When k approaches unity, the two inductors are said to be tightly coupled; and when k is much less than unity, they are said to be loosely coupled. While the coefficient of coupling can never exceed unity, it may be as high as 0.998 for iron-core transformers; it may be smaller than 0.5 for air-core transformers.

When there are only two inductively coupled circuits, the symbol M is frequently used to represent the mutual inductance. It can be shown that the mutual inductance between two electric circuits coupled by a homogeneous medium of constant permeability is reciprocal,

$$M = L_{12} = L_{21} = k\sqrt{L_{11}L_{22}} \tag{1.2.45}$$

The energy considerations that lead to such a conclusion are taken up in Problem 1.2.30 as an exercise for the student.

Let us next consider the energy stored in a pair of mutually coupled inductors,

$$W_m = \frac{i_1\lambda_1}{2} + \frac{i_2\lambda_2}{2} \tag{1.2.46}$$

where λ_1 and λ_2 are the total flux linkages of inductors 1 and 2, respectively, and subscript m denotes association with the magnetic field. Equation (1.2.46) may be rewritten as

$$W_m = \frac{i_1}{2}(\lambda_{11} + \lambda_{12}) + \frac{i_2}{2}(\lambda_{22} + \lambda_{21})$$

$$= \frac{1}{2}L_{11}i_1^2 + \frac{1}{2}L_{12}i_1i_2 + \frac{1}{2}L_{22}i_2^2 + \frac{1}{2}L_{21}i_1i_2$$

or

$$W_m = \frac{1}{2}L_{11}i_1^2 + Mi_1i_2 + \frac{1}{2}L_{22}i_2^2 \tag{1.2.47}$$

Equation (1.2.47) is valid whether the inductances are constant or variable, so long as the magnetic field is confined to a uniform medium of constant permeability.

Where there are n coupled circuits, the energy stored in the magnetic field can be expressed as

$$W_m = \sum_{j=1}^{n} \sum_{k=1}^{n} \frac{1}{2}L_{jk}i_ji_k \tag{1.2.48}$$

Going back to the pair of mutually coupled inductors shown in Figure 1.2.9, the flux-linkage relations and the voltage equations for circuits 1 and 2 are given by the following equations, while the resistances associated with the coils are neglected:

$$\lambda_1 = \lambda_{11} + \lambda_{12} = L_{11}i_1 + L_{12}i_2 = L_{11}i_1 + Mi_2 \tag{1.2.49}$$

$$\lambda_2 = \lambda_{21} + \lambda_{22} = L_{21}i_1 + L_{22}i_2 = Mi_1 + L_{22}i_2 \tag{1.2.50}$$

$$v_1 = \frac{d\lambda_1}{dt} = L_{11}\frac{di_1}{dt} + M\frac{di_2}{dt} \tag{1.2.51}$$

$$v_2 = \frac{d\lambda_2}{dt} = M\frac{di_1}{dt} + L_{22}\frac{di_2}{dt} \tag{1.2.52}$$

For the terminal voltage and current assignments shown in Figure 1.2.9, the coil windings are such that the fluxes produced by currents i_1 and i_2 are additive in nature, and in such a case the algebraic sign of the mutual voltage term is positive, as in Equations (1.2.51) and (1.2.52).

In order to avoid drawing detailed sketches of windings showing the sense in which each coil is wound, a *dot convention* is developed, according to which the pair of mutually coupled inductors of Figure 1.2.9 are represented by the system shown in Figure 1.2.10. The notation is such that a current i entering a dotted (undotted) terminal in one coil induces a voltage $M[di/dt]$ with a positive polarity at the dotted (undotted) terminal of the other coil. If the two currents i_1 and i_2 were to be entering (or leaving) the dotted terminals, the adopted convention is such that the fluxes produced by i_1 and i_2 will be aiding each other, and the mutual and self-inductance terms for each terminal pair will have the same sign; otherwise they will have opposite signs.

Although just a pair of mutually coupled inductors are considered here for the sake of simplicity, complicated magnetic coupling situations do occur in practice. For example, Figure 1.2.11 shows the coupling between coils 1 and 2, 1 and 3, and 2 and 3.

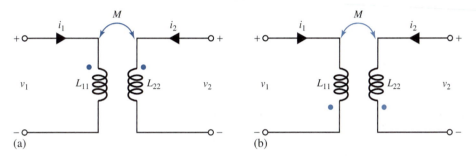

Figure 1.2.10 Dot notation for a pair of mutually coupled inductors. (a) Dots on upper terminals. (b) Dots on lower terminals.

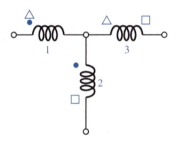

Figure 1.2.11 Polarity markings for complicated magnetic coupling situations.

EXAMPLE 1.2.5

Referring to the circuit of Figure 1.2.8, let

$$L_{11} = L_{22} = 0.1 \text{ H}$$

and

$$M = 10 \text{ mH}$$

Determine v_1 and v_2 if:

(a) $i_1 = 10$ mA and $i_2 = 0$.

(b) $i_1 = 0$ and $i_2 = 10 \sin 100t$ mA.

(c) $i_1 = 0.1 \cos t$ A and $i_2 = 0.3 \sin(t + 30°)$ A.

Also find the energy stored in each of the above cases at $t = 0$.

Solution

$$L_{11} = L_{22} = 0.1\text{H}; \qquad M = 10 \text{ mH} = 10 \times 10^{-3} \text{ H}$$

$$v_1 = L_{11}\frac{di_1}{dt} + M\frac{di_2}{dt}; \qquad v_2 = M\frac{di_1}{dt} + L_{22}\frac{di_2}{dt}$$

$$W_m = \text{energy stored} = \frac{1}{2}L_{11}i_1^2 + L_{12}i_1i_2 + \frac{1}{2}L_{22}i_2^2$$

(a) Since both i_1 and i_2 are constant and not a function of time,

$$v_1 = 0; \qquad v_2 = 0$$

$$W_m = \tfrac{1}{2}(0.1)(10 \times 10^{-3})^2 + 0 + 0 = 5 \times 10^{-6} \text{ J} = 5 \text{ } \mu\text{J}$$

(b) $v_1 = 0 + 10 \times 10^{-3}(10 \times 100 \cos 100t)10^{-3} = 10 \cos 100t \text{ mV}$

$v_2 = 0 + 0.1(10 \times 100 \cos 100t)10^{-3} = 100 \cos 100t \text{ mV} = 0.1 \cos 100t \text{ V}$

$W_m = 0 + 0 + \tfrac{1}{2}(0.1)(100 \sin^2 100t)10^{-6}; \qquad W_m = 0 \text{ at } t = 0$

(c) $v_1 = 0.1(-0.1 \sin t) + 10 \times 10^{-3}[0.3 \cos(t + 30°)] = -10 \sin t + 3 \cos(t + 30°) \text{ mV}$

$v_2 = 10 \times 10^{-3} (-0.01 \sin t) + 0.1[0.3 \cos(t + 30°)] = -\sin t + 30 \cos(t + 30°) \text{ mV}$

$W_m = \tfrac{1}{2}(0.1)(0.01 \cos^2 t) + (10 \times 10^{-3})(0.1 \cos t)[0.3 \sin(t + 30°)] + \tfrac{1}{2}(0.1)$
$[0.09 \sin^2(t + 30°)]; \qquad \text{at } t = 0$

$W_m = \tfrac{1}{2}(0.1)(0.01) + 10 \times 10^{-3}(0.1)(0.15) + \tfrac{1}{2}(0.1)\left(\frac{0.09}{4}\right) = 1.775 \text{ } \mu\text{J}$

Transformer

A transformer is basically a static device in which two or more stationary electric circuits are coupled magnetically, the windings being linked by a common time-varying magnetic flux. All that is really necessary for transformer action to take place is for the two coils to be so positioned that some of the flux produced by a current in one coil links some of the turns of the other coil. Some air-core transformers employed in communications equipment are no more elaborate than this. However, the construction of transformers utilized in power-system networks is much more elaborate to minimize energy loss, to produce a large flux in the ferromagnetic core by a current in any one coil, and to see that as much of that flux as possible links as many of the turns as possible of the other coils on the core.

An elementary model of a two-winding core-type transformer is shown in Figure 1.2.12. Essentially it consists of two windings interlinked by a mutual magnetic field. The winding that is excited or energized by connecting it to an input source is usually referred to as the *primary* winding, whereas the other, to which the electric load is connected and from which the output energy is taken, is known as the *secondary* winding. Depending on the voltage level at which the winding is operated, the windings are classified as HV (*high voltage*) and LV (*low voltage*) windings. The terminology of *step-up* or *step-down transformer* is also common if the main purpose of the transformer is to raise or lower the voltage level. In a step-up transformer, the primary is a low-voltage winding whereas the secondary is a high-voltage winding. The opposite is true for a step-down transformer.

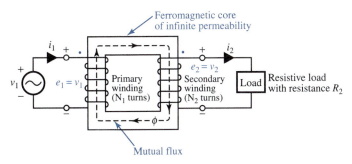

Figure 1.2.12 Elementary model of a two-winding core-type transformer (ideal transformer).

An *ideal transformer* is one that has no losses (associated with iron or copper) and no leakage fluxes (i.e., all the flux in the core links both the primary and the secondary windings). The winding resistances are negligible. While these properties are never actually achieved in practical transformers, they are, however, approached closely. When a time-varying voltage v_1 is applied to the N_1-turn primary winding (assumed to have zero resistance), a core flux ϕ is established and a counter emf e_1 with the polarity shown in Figure 1.2.12 is developed such that e_1 is equal to v_1. Because there is no leakage flux with an ideal transformer, the flux ϕ also links all N_2 turns of the secondary winding and produces an induced emf e_2, according to Faraday's law of induction. Since $v_1 = e_1 = d\lambda_1/dt = N_1\,d\phi/dt$ and $v_2 = e_2 = d\lambda_2/dt = N_2\,d\phi/dt$, it follows from Figure 1.2.12 that

$$\frac{v_1}{v_2} = \frac{e_1}{e_2} = \frac{N_1}{N_2} = a \tag{1.2.53}$$

where a is the *turns ratio*. Thus, in an ideal transformer, voltages are transformed in the direct ratio of the turns. For the case of an ideal transformer, since the instantaneous power input equals the instantaneous power output, it follows that

$$v_1 i_1 = v_2 i_2 \qquad \text{or} \qquad \frac{i_1}{i_2} = \frac{v_2}{v_1} = \frac{N_2}{N_1} = \frac{1}{a} \tag{1.2.54}$$

which implies that currents are transformed in the inverse ratio of the turns.

Equivalent circuits viewed from the source terminals, when the transformer is ideal, are shown in Figure 1.2.13. As seen from Figure 1.2.13(a), since $v_1 = (N_1/N_2)v_2$, $i_1 = (N_2/N_1)i_2$, and $v_2 = i_2 R_L$, it follows that

$$\frac{v_1}{i_1} = \left(\frac{N_1}{N_2}\right)^2 R_L = a^2 R_L = R'_L \tag{1.2.55}$$

where R'_L is the secondary-load resistance *referred to* the primary side. The consequence of Equation (1.2.55) is that a resistance R_L in the secondary circuit can be replaced by an *equivalent* resistance R'_L in the primary circuit in so far as the effect at the source terminals is concerned. The reflected resistance through a transformer can be very useful in *resistance matching* for *maximum power transfer*, as we shall see in the following example. Note that the circuits shown in Figure 1.2.13 are indistinguishable viewed from the source terminals.

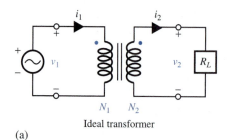

Figure 1.2.13 Equivalent circuits viewed from source terminals when the transformer is ideal.

N_1 N_2

Ideal transformer

(a)

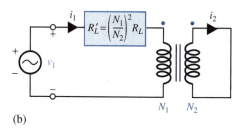

N_1 N_2

(b)

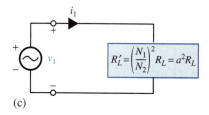

(c)

EXAMPLE 1.2.6

Consider a source of voltage $v(t) = 10\sqrt{2} \sin 2t$ V, with an internal resistance of 1800 Ω. A transformer that can be considered ideal is used to couple a 50-Ω resistive load to the source.

(a) Determine the primary to secondary turns ratio of the transformer required to ensure maximum power transfer by matching the load and source resistances.

(b) Find the average power delivered to the load.

Solution

By considering a constant voltage source (with a given internal resistance R_S) connected to a variable-load resistance R_L, as shown in Figure E1.2.6(a), for a value of R_L equal to R_S given by Equation (1.2.13), the maximum power transfer to the load resistance would occur when the load resistance is matched with the source resistance.

(a) For maximum power transfer to the load, R'_L (i.e., R_L referred to the primary side of the transformer) should be equal to R_S, which is given to be 1800 Ω. Hence,

$$R'_L = a^2 R_L = 50a^2 = 1800 \quad \text{or} \quad a^2 = 36, \text{ or } a = 6.$$

Thus $N_1/N_2 = 6$ [see Figure E1.2.6(b)].

(b) By voltage division [see Figure E1.2.6(c)] one gets

$$v'_L = \frac{1800}{1800 + 1800}(10\sqrt{2}\ \sin\ 2t) = (5\sqrt{2}\ \sin\ 2t)\ \text{V}$$

which has an rms value of 5 V. Hence, the average power delivered to the load resistance (R'_L or R_L) is

$$P_{av} = \frac{(V'_{L\ RMS})^2}{1800} = \frac{25}{1800}\ \text{W} \cong 13.9\ \text{mW}$$

Note that v_L across R_L is $(5\sqrt{2}/6)\ \sin\ 2t$ V, and i_L through R_L is $(5\sqrt{2}/6 \times 50)\ \sin\ 2t$ A. The rms value of i_L is then 5/300 A, and the rms value of v_L is $5/6$ V. Thus,

$$P_{av} = \frac{5}{300} \times \frac{5}{6}\ \text{W} \cong 13.9\ \text{mW}$$

which is also the same as

$$P_{av} = I^2_{L\ RMS}\ R_L = V^2_{L\ RMS}/R_L$$

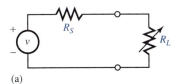

(a)

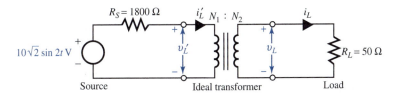

(b)

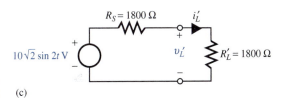

(c)

Figure E1.2.6

1.3 KIRCHHOFF'S LAWS

The basic laws that must be satisfied among circuit currents and circuit voltages are known as *Kirchhoff's current law* (KCL) and *Kirchhoff's voltage law* (KVL). These are fundamental for the systematic analysis of electric circuits.

KCL states that, at any node of any circuit and at any instant of time, the sum of all currents entering the node is equal to the sum of all currents leaving the node. That is, the *algebraic* sum of all currents (entering or leaving) at any node is zero, or no node can accumulate or store charge. Figure 1.3.1 illustrates Kirchhoff's current law, in which at node a,

$$i_1 - i_2 + i_3 + i_4 - i_5 = 0 \qquad \text{or} \qquad -i_1 + i_2 - i_3 - i_4 + i_5 = 0$$

$$\text{or} \qquad i_1 + i_3 + i_4 = i_2 + i_5 \qquad\qquad\qquad (1.3.1)$$

Note that so long as one is consistent, it does not matter whether the currents directed toward the node are considered positive or negative.

KVL states that the *algebraic* sum of the voltages (drops or rises) encountered in traversing any *loop* (which is a *closed path* through a circuit in which no electric element or node is encountered more than once) of a circuit in a specified direction must be zero. In other words, the sum of the voltage rises is equal to the sum of the voltage drops in a loop. A loop that contains no other loops is known as a *mesh*. KVL implies that moving charge around a path and returning to the starting point should require no net expenditure of energy. Figure 1.3.2 illustrates the Kirchhoff's voltage law.

For the mesh shown in Figure 1.3.2, which depicts a portion of a network, starting at node a and returning back to it while traversing the closed path $abcdea$ in either clockwise or anticlockwise direction, Kirchhoff's voltage law yields

$$-v_1 + v_2 - v_3 - v_4 + v_5 = 0 \qquad \text{or} \qquad v_1 - v_2 + v_3 + v_4 - v_5 = 0$$

$$\text{or} \qquad v_1 + v_3 + v_4 = v_2 + v_5 \qquad\qquad\qquad (1.3.2)$$

Note that so long as one is consistent, it does not matter whether the voltage drops are considered positive or negative. Also notice that the currents labeled in Figure 1.3.2 satisfy KCL at each of the nodes.

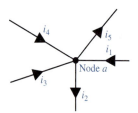

Figure 1.3.1 Illustration of Kirchhoff's current law.

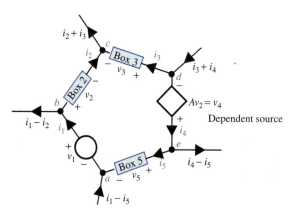

Figure 1.3.2 Illustration of Kirchhoff's voltage law.

In a network consisting of one or more energy sources and one or more circuit elements, the *cause-and-effect* relationship in *circuit theory* can be studied utilizing Kirchhoff's laws and volt-ampere relationships of the circuit elements. While the cause is usually the voltage or current source exciting the circuit, the effect is the voltages and currents existing in various parts of the network.

EXAMPLE 1.3.1

Consider the circuit shown in Figure E1.3.1 and determine the unknown currents using KCL.

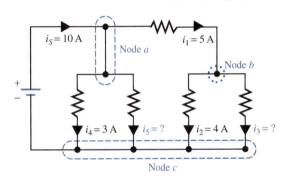

Figure E1.3.1

Solution

Let us assign a + sign for currents entering the node and a − sign for currents leaving the node.
   Applying KCL at node a, we get

$$+ i_S - i_1 - i_4 - i_5 = 0$$

or

$$10 - 5 - 3 - i_5 = 0 \quad \text{or} \quad i_5 = 2 \text{ A}$$

Applying KCL at node b, we get

$$+ i_1 - i_2 - i_3 = 0$$

or

$$5 - 4 - i_3 = 0 \quad \text{or} \quad i_3 = 1 \text{ A}$$

The student is encouraged to rework this problem by:

(a) Assigning a − sign for currents entering the node and a + sign for currents leaving the node; and

(b) Applying the statement that the sum of the currents entering a node is equal to the sum of the currents leaving that node.

EXAMPLE 1.3.2

For the circuit shown in Figure E1.3.2, use KCL and KVL to determine i_1, i_2, v_{bd} and v_x. Also, find v_{eb}.

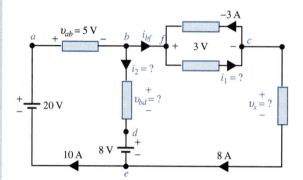

Figure E1.3.2

Solution

Using KCL at node c, we get

$$i_1 = 8 + (-3) = 5 \text{ A}$$

Using KCL at node f, we have

$$i_{bf} = i_1 - (-3) = 5 + 3 = 8 \text{ A}$$

Applying KCL at node b, we get

$$10 = i_2 + i_{bf} = i_2 + 8 \quad \text{or} \quad i_2 = 2 \text{ A}$$

Using KVL around the loop $abdea$ in the clockwise direction, we have

$$v_{ab} + v_{bd} + v_{de} + v_{ea} = 0$$

or

$$5 + v_{bd} + 8 - 20 = 0 \quad \text{or} \quad v_{bd} = 20 - 8 - 5 = 7 \text{ V}$$

Note that in writing KVL equations with + and − polarity symbols, we write the voltage with a positive sign if the + is encountered before the − and with a negative sign if the − is encountered first as we move around the loop.

Applying KVL around the loop $abfcea$ in the clockwise direction, we get

$$v_{ab} + v_{bf} + v_{fc} + v_{ce} + v_{ea} = 0$$

or

$$5 + 0 + 3 + v_x + (-20) = 0 \quad \text{or} \quad v_x = 20 - 3 - 5 = 12 \text{ V}$$

Note that a direct connection between b and f implies ideal connection, and hence no voltage between these points.

The student is encouraged to rewrite the loop equations by traversing the closed path in the anticlockwise direction.

Noting that $v_{eb} = v_{ed} + v_{db}$, we have

$$v_{eb} = -8 - 7 = -15 \text{ V}$$

Alternatively,

$$v_{eb} = v_{ea} + v_{ab} = -20 + 5 = -15 \text{ V}$$

or

$$v_{eb} = v_{ec} + v_{cf} + v_{fb} = -v_x + v_{cf} + 0 = -12 - 3 = -15 \text{ V}$$

The student should observe that $v_{be} = -v_{eb} = 15$ V and node b is at a higher potential than node e.

EXAMPLE 1.3.3

Referring to Figure 1.3.2, let boxes 2, 3, and 5 consist of a 0.2-H inductor, a 5-Ω resistor, and a 0.1-F capacitor, respectively. Given $A = 5$, and $v_1 = 10 \sin 10t$, $i_2 = 5 \sin 10t$, and $i_3 = 2 \sin 10t - 4 \cos 10t$, find i_5.

Solution

From Equation (1.2.19),

$$v_2 = L\frac{di_2}{dt} = 0.2\frac{d}{dt}(5 \sin 10t) = 10 \cos 10t \text{ V}$$

From Equation (1.2.1),

$$v_3 = Ri_3 = 5(2 \sin 10t - 4 \cos 10t) = 10 \sin 10t - 20 \cos 10t \text{ V}$$

From Equation (1.3.2),

$$v_5 = v_1 + v_3 + v_4 - v_2 = v_1 + v_3 + 5v_2 - v_2$$
$$= 10 \sin 10t + (10 \sin 10t - 20 \cos 10t) + 4(10 \cos 10t)$$
$$= 20 \sin 10t + 20 \cos 10t \text{ V}$$

From Equation (1.2.9),

$$i_5 = C\frac{dv_5}{dt} = 0.1\frac{d}{dt}(20 \sin 10t + 20 \cos 10t)$$
$$= 20 \cos 10t - 20 \sin 10t \text{ A}$$

Note the consistency of voltage polarities and current directions in Figure 1.3.2.

EXAMPLE 1.3.4

Consider the network shown in Figure E1.3.4(a).

(a) Find the voltage drops across the resistors and mark them with their polarities on the circuit diagram.

(b) Check whether the KVL is satisfied, and determine V_{bf} and V_{ec}.

(c) Show that the conservation of power is satisfied by the circuit.

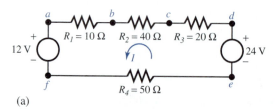

(a)

Figure E1.3.4

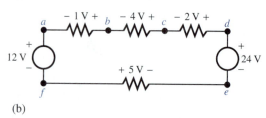

(b)

Solution

(a) From Ohm's law, the current I is given by

$$I = \frac{24 - 12}{10 + 40 + 20 + 50} = \frac{12}{120} = 0.1 \text{ A}$$

Therefore, voltage drops across the resistors are calculated as follows:

$$V_{ba} = I R_1 = 0.1 \times 10 = 1 \text{ V}$$
$$V_{cb} = I R_2 = 0.1 \times 40 = 4 \text{ V}$$
$$V_{dc} = I R_3 = 0.1 \times 20 = 2 \text{ V}$$
$$V_{fe} = I R_4 = 0.1 \times 50 = 5 \text{ V}$$

These are shown in Figure E1.3.4(b) with their polarities. Note that capital letters are used here for dc voltages and currents.

(b) Applying the KVL for the closed path $edcbafe$, we get

$$- 24 + 2 + 4 + 1 + 12 + 5 = 0$$

which confirms that the KVL is satisfied,

$$V_{bf} = V_{ba} + V_{af} = 1 + 12 = 13 \text{ V}$$
$$V_{ec} = V_{ed} + V_{dc} = -24 + 2 = -22 \text{ V}$$

(c) Power delivered by 24-V source $= 24 \times 0.1 = 2.4$ W

Power delivered by 12-V source $= -12 \times 0.1 = -1.2$ W

Power absorbed by resistor $R_1 = (0.1)^2 \times 10 = 0.1$ W

Power absorbed by resistor $R_2 = (0.1)^2 \times 40 = 0.4$ W

Power absorbed by resistor $R_3 = (0.1)^2 \times 20 = 0.2$ W

Power absorbed by resistor $R_4 = (0.1)^2 \times 50 = 0.5$ W

Power delivered by sources $= 2.4 - 1.2 = 1.2$ W

Power absorbed by resistors R_1, R_2, R_3, and $R_4 = 0.1 + 0.4 + 0.2 + 0.5 = 1.2$ W

The conservation of power is satisfied by the circuit.

EXAMPLE 1.3.5

Given the network in Figure E1.3.5,

   (a) Find the currents through resistors R_1, R_2, and R_3.

   (b) Compute the voltage V_1.

   (c) Show that the conservation of power is satisfied by the circuit.

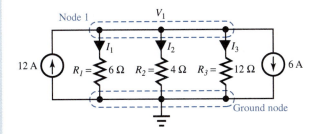

Figure E1.3.5

Solution

   (a) Applying the KCL at node 1, we have

$$12 = \frac{V_1}{R_1} + \frac{V_1}{R_2} + \frac{V_1}{R_3} + 6 = V_1 \left(\frac{1}{6} + \frac{1}{4} + \frac{1}{12} \right) + 6 = \frac{V_1}{2} + 6$$

Therefore, $V_1 = 12$ V. Then,

$$I_1 = \frac{12}{6} = 2 \text{ A}$$

$$I_2 = \frac{12}{4} = 3 \text{ A}$$

$$I_3 = \frac{12}{12} = 1 \text{ A}$$

   (b) $V_1 = 12$ V

   (c) Power delivered by 12-A source $= 12 \times 12 = 144$ W

      Power delivered by 6-A source $= -6 \times 12 = -72$ W

      Power absorbed by resistor $R_1 = I_1^2 R_1 = (2)^2 6 = 24$ W

Power absorbed by resistor $R_2 = I_2^2 R_2 = (3)^2 4 = 36$ W

Power absorbed by resistor $R_3 = I_3^2 R_3 = (1)^2 12 = 12$ W

Power delivered by sources $= 144 - 72 = 72$ W

Power absorbed by resistors R_1, R_2, and $R_3 = 24 + 36 + 12 = 72$ W

The conservation of power is satisfied by the circuit.

EXAMPLE 1.3.6

Consider the network shown in Figure E1.3.6 containing a voltage-controlled source producing the controlled current $i_c = gv$, where g is a constant with units of conductance, and the control voltage happens to be the terminal voltage in this case.

(a) Obtain an expression for $R_{eq} = v/i$.

(b) For (i) $gR = \frac{1}{2}$, (ii) $gR = 1$, and (iii) $gR = 2$, find R_{eq} and interpret what it means in each case.

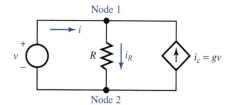

Node 1

Node 2

Figure E1.3.6

Solution

(a) Applying the KCL at node 1, we get

$$i + i_c = i_R = \frac{v}{R}$$

Therefore,

$$i = \frac{v}{R} - i_c = \frac{v}{R} - gv = \frac{1 - gR}{R}v$$

Then,

$$R_{eq} = \frac{v}{i} = \frac{R}{1 - gR}$$

(b) (i) For $gR = \frac{1}{2}$, $R_{eq} = 2R$. The equivalent resistance is greater than R; the internal controlled source provides part of the current through R, thereby reducing the input current i for a given value of v. When $i < v/R$, the equivalent resistance is greater than R.

(ii) For $gR = 1$, $R_{eq} \to \infty$. The internal controlled source provides all of the current through R, thereby reducing the input current i to be zero for a given value of v.

(iii) For $gR = 2$, $R_{eq} = -R$, which is a negative equivalent resistance. This means that the controlled source provides more current than that going through R; the current direction of i is reversed when $v > 0$. However, the relation $v = R_{eq}i$ is satisfied at the input terminals.

1.4   METERS AND MEASUREMENTS

The subject of electrical measurements is such a large one that entire books have been written on the topic. Only a few basic principles will be introduced here. Practical measurements are made with real instruments, which in general disturb the operation of a circuit to some extent when they are connected. Measurements may be affected by *noise*, which is undesirable randomly varying signals.

Voltmeter

In order to measure the potential difference between two terminals or nodes of a circuit, a voltmeter is connected *across* these two points. A practical voltmeter can usually be modeled as a parallel combination of an ideal voltmeter (through which no current flows) and a shunt resistance R_V, as shown in Figure 1.4.1. The internal resistance R_V of an ideal voltmeter is infinite, while its value in practice is of the order of several million ohms. There are what are known as dc voltmeters and ac voltmeters. An ac voltmeter usually measures the rms value of the time-varying voltage.

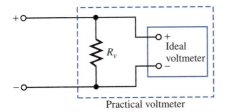

Figure 1.4.1

Practical voltmeter

EXAMPLE 1.4.1

An electromechanical voltmeter with internal resistance of 1 kΩ and an electronic voltmeter with internal resistance of 10 MΩ are used separately to measure the potential difference between A and the ground of the circuit shown in Figure E1.4.1. Calculate the voltages that will be indicated by each of the two instruments and the percentage error in each case.

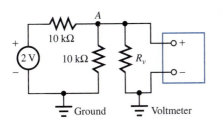

Figure E1.4.1

Solution

When no instrument is connected, by voltage division, $V + AG = 1$ V. With internal resistance R_V of the instrument, the voltmeter reading will be

$$V_A = 2.0 \left(\frac{R_V \| 10^4}{R_V \| 10^4 + 10^4} \right) = 2 \left(\frac{R_V \cdot 10^4}{R_V + 10^4} \right) \Big/ \left(\frac{R_V \cdot 10^4}{R_V + 10^4} + 10^4 \right)$$

$$= 2 \frac{R_V \cdot 10^4}{2 \times 10^4 \, R_V + 10^8}$$

With $R_V = 1000$,

$$V_A = \frac{2 \times 10^7}{2 \times 10^7 + 10^8} = \frac{2}{2 + 10} = \frac{2}{12} = 0.1667 \text{ V}$$

for which the percent error is

$$\frac{1 - 0.1667}{1} \times 100 = 83.33\%$$

With $R_V = 10^7 \, \Omega$,

$$V_A = \frac{2 \times 10^{11}}{2 \times 10^{11} + 10^8} = \frac{2000}{2000 + 1} = 0.9995 \text{ V}$$

for which the percent error is

$$\frac{1 - 0.9995}{1} \times 100 = 0.05\%$$

One can see why electronic voltmeters which have relatively very large internal resistance are often used.

Ammeter

In order to measure the current through a wire or line of a circuit, an ammeter is connected *in series* with the line. A practical ammeter can usually be modeled as a series combination of an ideal ammeter and an internal resistance R_I. The potential difference between the two terminals of an ideal ammeter is zero, which corresponds to zero internal resistance. There are what are known as dc ammeters and ac ammeters. An ac ammeter usually measures the rms value of the time-varying current. Note that for the ammeter to be inserted for measuring current, the circuit has to be broken, whereas for the voltmeter to be connected for measuring voltage, the circuit need not be disassembled.

Multimeters that measure multiple ranges of voltage and current are available in practice. *Ohmmeters* measure the dc resistance by the use of Ohm's law. A multimeter with scales for volts, ohms, and milliamperes is known as VOM. An ohmmeter should not be used to measure the resistance of an electronic component that might be damaged by the sensing current.

Instrument Transformers

These are generally of two types, *potential transformers* (PTs) and *current transformers* (CTs). They are designed in such a way that the former may be regarded as having an ideal potential ratio, whereas the latter has an ideal current ratio. The accuracy of measurement is quite important for ITs that are commonly used in ac circuits to supply instruments, protective relays, and control devices.

PTs are employed to step down the voltage to a suitable level, whereas CTs (connected in series with the line) are used to step down the current for metering purposes. Often the primary of a CT is not an integral part of the transformer itself, but is a part of the line whose current is being measured. In addition to providing a desirable low current in the metering circuit, the CT isolates the meter from the line, which may be at a high potential. Note that the secondary terminals of a CT should *never* be open-circuited under load. The student is encouraged to reason and justify this precaution. One of the most useful instruments for measuring currents in the ampere range is the *clip-on ammeter* combining the CT with one-turn primary and the measurement functions.

Oscilloscope

To measure time-varying signals (voltages and currents), an instrument known as an *oscilloscope* is employed. It can be used as a practical electronic voltmeter which displays a graph of voltage as a function of time. Such a display allows one not only to read off the voltage at any instant of time, but also to observe the general behavior of the voltage as a function of time. The horizontal and vertical scales of the display are set by the oscilloscope's controls, such as 5 ms per each horizontal division and 50 V per each vertical division. For periodic waveforms, the moving light spot repeatedly graphs the same repetitive shape, and the stationary waveform is seen. For nonperiodic cases, a common way of handling is to cause the oscilloscope to make only one single graph, representing the voltage over a single short time period. This is known as *single-sweep operation*. Since the display lasts for only a very short time, it may be photographed for later inspection.

   Digital meters are generally more accurate and can be equipped with more scales and broader ranges than *analog meters*. On the other hand, analog meters are generally less expensive and give an entire range or scale of reading, which often could be very informative. A digital oscilloscope represents the combination of analog and digital technologies. By digital sampling techniques, the oscilloscope trace is digitized and stored in the digital memory included with the digital oscilloscope. Digital oscilloscopes are generally more costly than analog ones, but their capability in the analysis and processing of signals is vastly superior.

Wheatstone Bridge

Null measurements are made with bridge circuits and related configurations. They differ from direct measurements in that the quantity being measured is compared with a known reference quantity. The balancing strategy avoids undesirable interaction effects and generally results in more accurate measurement than the direct one.

   By far the most common is the Wheatstone bridge designed for precise measurement of resistance. Figure 1.4.2 shows the basic circuit in which the measurement of an unknown resistance R_x is performed by balancing the variable resistances R_a and R_b until no current flows through meter A. Under this null condition,

$$R_x = \frac{R_a}{R_b} \cdot R_s \tag{1.4.1}$$

where R_s is the known standard resistance. There are other bridge-circuit configurations to measure inductance and capacitance. Typical instruments utilizing bridge circuits are found in strain gauges measuring stress and in temperature measuring systems with thermocouples and thermistors.

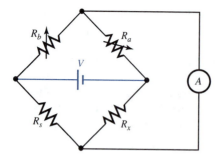

Figure 1.4.2 Basic Wheatstone bridge circuit for resistance measurement.

Redraw the Wheatstone bridge circuit of Figure 1.4.2 and show that Equation (1.4.1) holds good for the null condition when the meter A reads zero current.

Solution

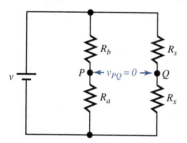

Figure E1.4.2 Figure 1.4.2 redrawn.

Under null condition, $V_{PQ} = 0$, or P and Q are at the same potential. Using the voltage division principle,

$$\left(\frac{V}{R_b + R_a}\right) R_b = \left(\frac{V}{R_s + R_x}\right) R_s$$

yielding $R_x R_b = R_a R_s$, which is the same as Equation (1.4.1).

1.5 ANALOGY BETWEEN ELECTRICAL AND OTHER NONELECTRIC PHYSICAL SYSTEMS

Systems such as those encountered in mechanics, thermodynamics, and hydraulics can be represented by analogous electric networks, from the response of which the system characteristics can be investigated. The *analogy*, of course, is a mathematical one: that is, two systems are analogous to each other if they are described by similar equations. The analogous electric quantities for a mechanical system are listed in Table 1.5.1.

Consider a tank filled with water, as shown in Figure 1.5.1, with input flow rate F_i and output flow rate $F_o = h/R$, where h is the fluid level or head and R is related to the diameter of the pipe, denoting the fluid resistance. Let A be the cross-sectional area of the tank. We may think of the fluid as being analogous to charge, and the fluid flow as being analogous to current. Then, in effect, the water tank acts as a capacitor storing charge, which is fluid in this case. This analogy is illustrated in Table 1.5.2.

TABLE 1.5.1 Mechanical-Electrical Analogs

Quantity	Unit	Symbol	Mathematical Relation	Force–Current Analog	Force–Voltage Analog
Force	N	$f(t)$	$\cdots$	$i(t)$	$v(t)$
Velocity	m/s	$u(t)$	$\cdots$	$v(t)$	$i(t)$
Mass	kg	$\xrightarrow{f}\ \boxed{M}$ $u \longrightarrow$	$f = M\frac{du}{dt}$ (Newton's law)	C	L
Compliance (= 1/stiffness)	m/N	$C_m = \frac{1}{K}$ $\text{\textsf{m}} \longrightarrow f$ $u \longrightarrow$	$f = \frac{1}{C_m}\int u\, dt$ (Hooke's law)	L	C
Viscous friction or damping	$N \cdot s/m$	$\xrightarrow{f}\ \overset{B}{\dashv|}$ $u \longrightarrow$	$f = Bu$	$G = 1/R$	R

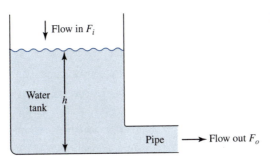

Figure 1.5.1 Simple hydraulic system.

Flow in F_i

Water tank h

Pipe ⟶ Flow out F_o

TABLE 1.5.2 Analogy Between Electrical and Hydraulic Systems

Quantity	Hydraulic System	Electrical System
Flow	Output flow rate F_o	Current i
Potential	Fluid level h	Voltage v
Resistance	Fluid resistance R	Electrical resistance R
Energy storage element	Fluid storage parameter A	Capacitance C
Volume of fluid (or charge)	$V = Ah$	$q = Cv$

Next, let us consider heat flow from an enclosure with a heating system to the outside of the enclosure, depending upon the temperature difference ΔT between the inside and outside of the enclosure. The heat capacity of the enclosure is analogous to capacitance, in the sense that the enclosure retains part of the heat produced by the heating system. One can then infer the analogy between electrical and thermal systems given in Table 1.5.3.

The heat flow, per Newton's law of cooling in a very simplified form, can be considered to be proportional to the rate of change of temperature with respect to distance. An approximate linear relationship between heat flow and change in temperature can be expressed as

TABLE 1.5.3 Analogy Between Thermal and Electrical Systems

Thermal System	Electrical System
Heat	Charge
Heat flow	Current
Temperature difference	Voltage
Ambient temperature	Ground reference
Heat capacity	Capacitance
Thermal resistance	Electrical resistance

$$\text{Heat flow} \cong \frac{k}{\Delta x}\Delta T \qquad (1.5.1)$$

where k is a constant, and $k/\Delta x$ in thermal systems is analogous to conductance in electrical systems. Then Newton's law of cooling, in a very simplified form, can be seen to be a thermal version of Ohm's law.

1.6 LEARNING OBJECTIVES

The *learning objectives* of this chapter are summarized here, so that the student can check whether he or she has accomplished each of the following.

- Review of basic electrical quantities.
- Application of Coulomb's law, Ampere's law, and the Biot–Savart law.
- Energy and power computations in a circuit consisting of a source and a load.
- Calculation of average and RMS values for periodic waveforms, and time constant for exponential waveforms.
- i–v relationships for ideal resistors, capacitors, and inductors; duality principle.
- Reduction of series and parallel combinations of resistors, capacitors, and inductors.
- Solution of simple voltage and current divider circuits.
- Computation of power absorbed by a resistor, and energy stored in a capacitor or inductor.
- Maximum power transfer and matched load.
- Volt-ampere equations and energy stored in coupled inductors.
- Ideal transformer and its properties.
- Application of Kirchhoff's current and voltage laws to circuits.
- Measurement of basic electrical parameters.
- Analogy between electrical and other nonelectric physical systems.

1.7 PRACTICAL APPLICATION: A CASE STUDY

Resistance Strain Gauge

Mechanical and civil engineers routinely employ the dependence of resistance on the physical dimensions of a conductor to measure strain. A strain gauge is a device that is bonded to the surface of an object, and whose resistance varies as a function of the surface strain experienced by the object. Strain gauges can be used to measure strain, stress, force, torque, and pressure.

The resistance of a conductor with a circular cross-sectional area A, length l, and conductivity σ is given by Equation (1.2.2),

$$R = \frac{l}{\sigma A}$$

Depending on the compression or elongation as a consequence of an external force, the length changes, and hence the resistance changes. The relationship between those changes is given by the gauge factor G,

$$G = \frac{\Delta R/R}{\Delta l/l}$$

in which the factor $\Delta l/l$, the fractional change in length of an object, is known as the strain. Alternatively, the change in resistance due to an applied strain $\varepsilon (= \Delta l/l)$ is given by

$$\Delta R = R_0 G \varepsilon$$

where R_0 is the zero-strain resistance, that is, the resistance of the strain gauge under no strain. A typical gauge has $R_0 = 350 \ \Omega$ and $G = 2$. Then for a strain of 1%, the change in resistance is $\Delta R = 7 \ \Omega$. A Wheatstone bridge as presented in Section 1.4 is usually employed to measure the small resistance changes associated with precise strain determination.

A typical strain gauge, shown in Figure 1.7.1, consists of a metal foil (such as nickel–copper alloy) which is formed by a photoetching process in multiple conductors aligned with the direction of the strain to be measured. The conductors are usually bonded to a thin backing made out of a tough flexible plastic. The backing film, in turn, is attached to the test structure by a suitable adhesive.

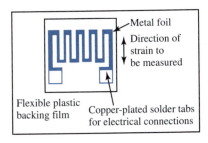

Figure 1.7.1 Resistance strain gauge and circuit symbol.

PROBLEMS

1.1.1 Consider two 1-C charges separated by 1 m in free space. Show that the force exerted on each is about one million tons.

*1.1.2 Point charges, each of $\sqrt{4\pi\varepsilon_0}$ C, are located at the vertices of an equilateral triangle of side a. Determine the electric force on each charge.

1.1.3 Two charges of equal magnitude 5 μC but opposite sign are separated by a distance of 10 m. Find the net force experienced by a positive charge

$q = 2 \ \mu C$ that is placed midway between the two charges.

1.1.4 The electric field intensity due to a point charge in free space is given to be

$$(-\bar{a}_x - \bar{a}_y + \bar{a}_z)/\sqrt{12}\text{V/m at } (0,0,1)$$

$$\text{and} \qquad 6 \bar{a}_z \text{ at } (2,2,0)$$

Determine the location and the value of the point charge.

*Complete solutions for problems marked with an asterisk can be found on the CD-ROM packaged with this book.

1.1.5 A wire with $n = 10^{30}$ electrons/m^3 has an area of cross section $A = 1$ mm^2 and carries a current $i = 50$ mA. Compute the number of electrons that pass a given point in 1 s, and find their average velocity.

1.1.6 A beam containing two types of charged particles is moving from A to B. Particles of type I with charge $+3q$, and those of type II with charge $-2q$ (where $-q$ is the charge of an electron given by -1.6×10^{-19} C) flow at rates of 5×10^{15}/s and 10×10^{15}/s, respectively. Evaluate the current flowing in the direction from B to A.

1.1.7 A charge $q(t) = 50 + 1.0t$ C flows into an electric component. Find the current flow.

*1.1.8 A charge variation with time is given in Figure P1.1.8. Draw the corresponding current variation with time.

1.1.9 A current $i(t) = 20 \cos(2\pi \times 60)t$ A flows through a wire. Find the charge flowing, and the number of electrons per second that are passing some point in the wire.

1.1.10 Consider a current element $I_1 \, d\bar{l}_1 = 10 \, dz \bar{a}_z$ kA located at $(0,0,1)$ and another $I_2 \, d\bar{l}_2 = 5dx \bar{a}_x$ kA located at $(0,1,0)$. Compute $d\bar{F}_{21}$ and $d\bar{F}_{12}$ experienced by elements 1 and 2, respectively.

1.1.11 Given $\bar{B} = (y\bar{a}_x - x\bar{a}_y)/(x^2 + y^2)$ T, determine the magnetic force on the current element $I \, d\bar{l} = 5 \times 0.001\bar{a}_z$ A located at $(3,4,2)$.

*1.1.12 In a magnetic field $\bar{B} = B_0(\bar{a}_x - 2\bar{a}_y + 2\bar{a}_z)$ T at a point, let a test charge have a velocity of $v_0(\bar{a}_x + \bar{a}_y - \bar{a}_z)$. Find the electric field $\bar{E}$ at that point if the acceleration experienced by the test charge is zero.

1.1.13 Consider an infinitely long, straight wire (in free space) situated along the z-axis and carrying current of I A in the positive z-direction. Obtain an expression for $\bar{B}$ everywhere. (*Hint:* Consider a circular coordinate system and apply the Biot–Savart law.)

1.1.14 A magnetic force exists between two adjacent, parallel current-carrying wires. Let I_1 and I_2 be the currents carried by the wires, and r the separation between them. Making use of the result of Problem 1.1.13, find the force between the wires.

1.1.15 A point charge $Q_1 = -5$ nC is located at $(6, 0, 0)$. Compute the voltage v_{ba} between two points $a(1, 0, 0)$ and $b(5, 0, 0)$. Comment on whether point a is at a higher potential with respect to point b.

1.1.16 A charge of 0.1 C passes through an electric source of 6 V from its negative to its positive terminals. Find the change in energy received by the charge. Comment on whether the charge has gained or lost energy, and also on the sign to be assigned to the change of energy.

1.1.17 The voltage at terminal a relative to terminal b of an electric component is $v(t) = 20 \cos 120\pi t$ V. A current $i(t) = -4 \sin 120\pi t$ A flows into terminal a. From time t_1 to t_2, determine the total energy flowing into the component. In particular, find the energy absorbed when $t_2 = t_1 + 1/15$.

*1.1.18 Obtain the instantaneous power flow into the component of Problem 1.1.17, and comment on the sign associated with the power.

1.1.19 A residence is supplied with a voltage $v(t) = 110\sqrt{2} \cos 120\pi t$ V and a current $i(t) = 10\sqrt{2}\cos 120\pi t$ A. If an electric meter is used to measure the average power, find the meter reading, assuming that the averaging is done over some multiple of $1/60$ s.

1.1.20 A 12-V, 115-Ah automobile storage battery is used to light a 6-W bulb. Assuming the battery to be a constant-voltage source, find how long the bulb can be lighted before the battery is completely discharged. Also, find the total energy stored in the battery before it is connected to the bulb.

1.2.1 In English units the conductor cross-sectional area is expressed in circular mils (cmil). A circle with diameter d mil has an area of $(\pi/4)d^2$ sq. mil,

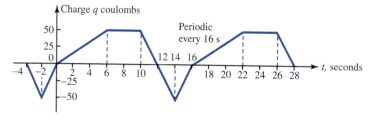

Charge q coulombs

Periodic every 16 s

Figure P1.1.8

or d^2 cmil. The handbook for aluminum electrical conductors lists a dc resistance of 0.01558 Ω per 1000 ft at 20°C for Marigold conductor, whose size is 1113 kcmil.

(a) Verify the dc resistance assuming an increase in resistance of 3% for spiraling of the strands.

(b) Calculate the dc resistance at 50°C, given that the temperature constant for aluminum is 228.1°C.

(c) If the 60-Hz resistance of 0.0956 Ω/mile at 50°C is listed in the handbook, determine the percentage increase due to skin effect or frequency.

1.2.2 MCM is the abbreviation for 1 kcmil. (See Problem 1.2.1 for a definition of cmil.) Data for commercial-base aluminum electrical conductors list a 60-Hz resistance of 0.0880 Ω/km at 75°C for a 795-MCM conductor.

(a) Determine the cross-sectional conducting area of this conductor in m².

(b) Calculate the 60-Hz resistance of this conductor in Ω/km at 50°C, given a temperature constant of 228.1°C for aluminum.

*1.2.3 A copper conductor has 12 strands with each strand diameter of 0.1328 in. For this conductor, find the total copper cross-sectional area in cmil (see Problem 1.2.1 for definition of cmil), and calculate the dc resistance at 20°C in (ohms/km), assuming a 2% increase in resistance due to spiraling.

1.2.4 A handbook lists the 60-Hz resistance at 50°C of a 900-kcmil aluminum conductor as 0.1185 Ω/mile. If four such conductors are used in parallel to form a line, determine the 60-Hz resistance of this line in Ω/km per phase at 50°C.

1.2.5 Determine R_{eq} for the circuit shown in Figure P1.2.5 as seen from terminals A–B.

1.2.6 Viewed from terminals A–B, calculate R_{eq} for the circuit given in Figure P1.2.6.

1.2.7 Find R_{eq} for the circuit of Figure P1.2.7.

*1.2.8 Determine R_{eq} for the circuit of Figure P1.2.8 as seen from terminals A–B.

1.2.9 A greatly simplified model of an audio system is shown in Figure P1.2.9. In order to transfer maximum power to the speaker, one should select equal values of R_L and R_S. Not knowing that

Figure P1.2.5

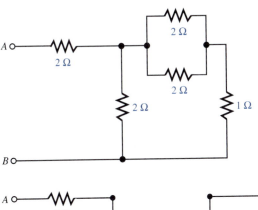

Figure P1.2.6

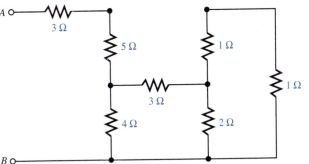

the internal resistance of the amplifier is $R_S = 8\,\Omega$, one has connected a mismatched speaker with $R_L = 16\,\Omega$. Determine how much more power could be delivered to the speaker if its resistance were matched to that of the amplifier.

1.2.10 For the circuit of Figure P1.2.10:

(a) Find an expression for the power absorbed by the load as a function of R_L.

(b) Plot the power dissipated by the load as a function of the load resistance, and determine the value of R_L corresponding to the maximum power absorbed by the load.

1.2.11 A practical voltage source is represented by an ideal voltage source of 30 V along with a series internal source resistance of $1.2\,\Omega$. Compute the smallest load resistance that can be connected to the practical source such that the load voltage

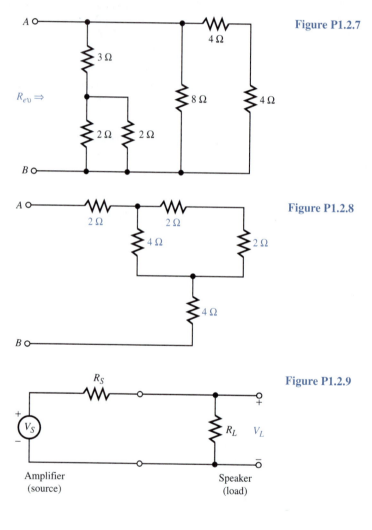

Figure P1.2.7

Figure P1.2.8

Figure P1.2.9

Amplifier (source)

Speaker (load)

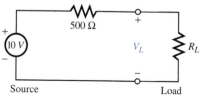

Figure P1.2.10

Source   Load

would not drop by more than 2% with respect to the source open-circuit voltage.

*1.2.12 A practical current source is represented by an ideal current source of 200 mA along with a shunt internal source resistance of 12 kΩ. Determine the percentage drop in load current with respect to the source short-circuit current when a 200-Ω load is connected to the practical source.

1.2.13 Let $v(t) = V_{max} \cos \omega t$ be applied to (a) a pure resistor, (b) a pure capacitor (with zero initial capacitor voltage, and (c) a pure inductor (with zero initial inductor current). Find the average power absorbed by each element.

1.2.14 If $v(t) = 120\sqrt{2} \sin 2\pi \times 60t$ V is applied to terminals A–B of problems 1.2.5, 1.2.6, 1.2.7, and 1.2.8, determine the power in kW converted to heat in each case.

1.2.15 With a direct current of I A, the power expended as heat in a resistor of $R\Omega$ is constant, independent of time, and equal to I^2R. Consider Problem 1.2.14 and find in each case the effective value of the current to give rise to the same heating effect as in the ac case, thereby justifying that the rms value is also known as the effective value for periodic waveforms.

1.2.16 Consider Problem 1.2.14 and obtain in each case a replacement of the voltage source by an equivalent current source at terminals A–B.

1.2.17 Consider Problem 1.2.5. Let $V_{AB} = 120$ V (rms).

Show the current and voltage distribution clearly in all branches of the original circuit configuration.

*1.2.18 Determine the voltages V_x using voltage division and equivalent resistor reductions for the circuits shown in Figure P1.2.18.

1.2.19 Find the currents I_x using current division and equivalent resistor reductions for the networks given in Figure P1.2.19.

1.2.20 Considering the circuit shown in Figure P1.2.20, sketch $v(t)$ and the energy stored in the capacitor as a function of time.

1.2.21 For the capacitor shown in Figure P1.2.21 connected to a voltage source, sketch $i(t)$ and $w(t)$.

*1.2.22 The energy stored in a 2-μF capacitor is given by $w_c(t) = 9e^{-2t} \mu$J for $t \geq 0$. Find the capacitor voltage and current at $t = 1$ s.

1.2.23 For a parallel-plate capacitor with plates of area A m^2 and separation d m in air, the capacitance in farads may be computed from the approximate relation

$$C \approx \varepsilon_0 \frac{A}{d} = \frac{8.854 \times 10^{-12} A}{d}$$

Compute the area of each plate needed to develop $C = 1$ pF for $d = 1$ m. (You can appreciate why large values of capacitance are constructed as electrolytic capacitors, and modern integrated-circuit technology is utilized to obtain a wide variety of capacitance values in an extremely small space.)

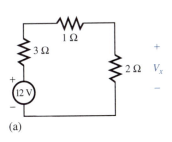

(a)

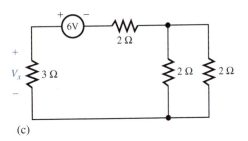

(c)

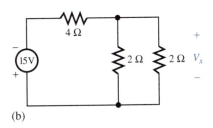

(b)

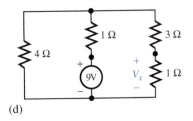

(d)

Figure P1.2.18

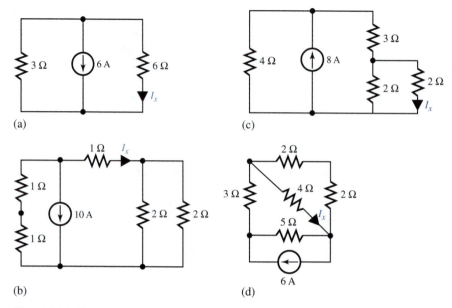

(a)

(c)

(b)

(d)

Figure P1.2.19

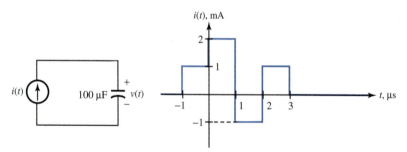

Figure P1.2.20

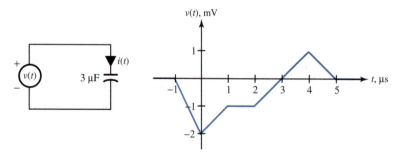

Figure P1.2.21

1.2.24 Determine the equivalent capacitance at terminals A–B for the circuit configurations shown in Figure P1.2.24.

1.2.25 For the circuit shown in Figure P1.2.25, sketch $i(t)$ and $w(t)$. See Problem 1.2.21 and check whether the duality principle is satisfied.

1.2.26 For the circuit given in Figure P1.2.26, sketch $v(t)$ and $w(t)$. See Problem 1.2.20 and check whether the duality principle is satisfied.

*1.2.27 The energy stored in a 2-μH inductor is given by $w_L(t) = 9e^{-2t}$ μJ for $t \geq 0$. Find the inductor current and voltage at $t = 1$ s. Compare the results of this problem with those of Problem 1.2.22 and comment.

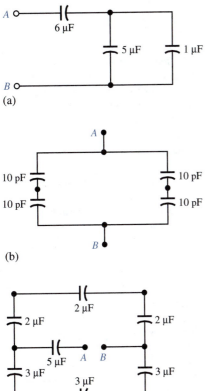

Figure P1.2.24

(a)

(b)

(c)

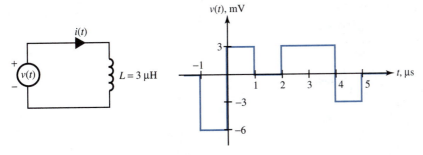

Figure P1.2.25

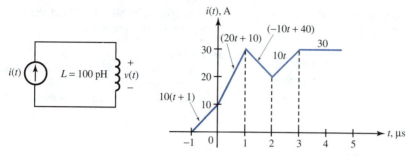

Figure P1.2.26

1.2.28 The inductance per unit length in H/m for parallel-plate infinitely long conductors in air is given by $L = \mu_0 d/w = 4\pi \times 10^{-7} d/w$, where d and w are in meters. Compute L (per unit length) for $d = 1$ m and $w = 0.113$ m. See Problem 1.2.23 and show that the product of inductance (per unit length) and capacitance (per unit length) is $\mu_0 \varepsilon_0$.

1.2.29 Determine the duals for the circuit configurations of Problem 1.2.24 and determine the equivalent inductance at terminals $A–B$ for each case.

1.2.30 Consider a pair of coupled coils as shown in Figure 1.2.10 of the text, with currents, voltages, and polarity dots as indicated. Show that the mutual inductance is $L_{12} = L_{21} = M$ by following these steps:

(a) Starting at time t_0 with $i_1(t_0) = i_2(t_0) = 0$, maintain $i_2 = 0$ and increase i_1 until, at time $t_1, i_1(t_1) = I_1$ and $i_2(t_1) = 0$. Determine the energy accumulated during this time. Now maintaining $i_1 = I_1$, increase i_2 until at time $t_2, i_2(t_2) = I_2$. Find the corresponding energy accumulated and the total energy stored at time t_2.

(b) Repeat the process in the reverse order, allowing the currents to reach their final values. Compare the expressions obtained for the total energy stored and obtain the desired result.

1.2.31 For the coupled coils shown in Figure P1.2.31, a dot has been arbitrarily assigned to a terminal as

indicated. Following the dot convention presented in the text, place the other dot in the remaining coil and justify your answer with an explanation. Comment on whether the polarities are consistent with Lenz's law.

*1.2.32 For the configurations of the coupled coils shown in Figure P1.2.32, obtain the voltage equations for v_1 and v_2.

1.2.33 The self-inductances of two coupled coils are L_{11} and L_{22}, and the mutual inductance between them is M. Show that the effective inductance of the two coils in series is given by

$$L_{\text{series}} = L_{11} + L_{22} \pm 2M$$

and the effective inductance of the two coils in parallel is given by

$$L_{\text{parallel}} = \frac{L_{11}L_{22} - M^2}{L_{11} \mp 2M + L_{22}}$$

Specify the conditions corresponding to different signs of the term $2M$.

1.2.34 For the coupled inductors shown in Figure P1.2.34, neglecting the coil resistances, write the volt-ampere relations.

1.2.35 Consider an amplifier as a voltage source with an internal resistance of 72 Ω. Find the turns ratio of the ideal transformer such that maximum power is delivered when the amplifier is connected to an 8-Ω speaker through an $N_1 : N_2$ transformer.

1.2.36 For the circuit shown in Figure P1.2.36, determine $v_{\text{out}}(t)$.

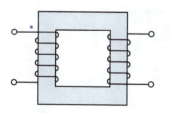

Figure P1.2.31

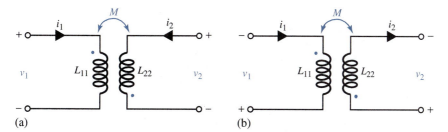

Figure P1.2.32

Figure P1.2.34

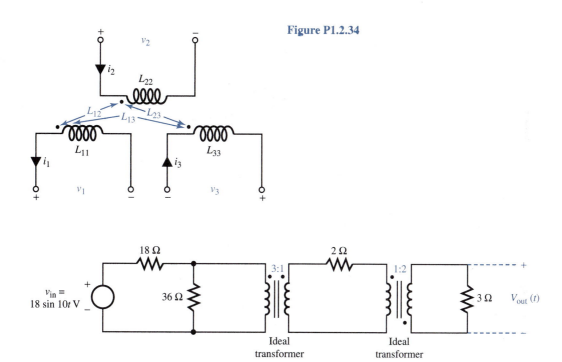

Figure P1.2.36

*1.2.37 A 60-Hz, 100-kVA, 2400/240-V (rms) transformer is used as a step-down transformer from a transmission line to a distribution system. Consider the transformer to be ideal.

(a) Find the turns ratio.

(b) What secondary load resistance will cause the transformer to be fully loaded at rated voltage (i.e., delivering the rated kVA), and what is the corresponding primary current?

(c) Determine the value of the load resistance referred to the primary side of the transformer.

1.2.38 A transformer is rated 10 kVA, 220:110 V (rms). Consider it an ideal transformer.

(a) Compute the turns ratio and the winding current ratings.

(b) If a 2-Ω load resistance is connected across the 110-V winding, what are the currents in the high-voltage and low-voltage windings when rated voltage is applied to the 220-V primary?

(c) Find the equivalent load resistance referred to the 220-V side.

1.3.1 Some element voltages and currents are given in the network configuration of Figure P1.3.1. Determine the remaining voltages and currents. Also calculate the power delivered to each element as well as the algebraic sum of powers of all elements, and comment on your result while identifying sources and sinks.

*1.3.2 Calculate the voltage v in the circuit given in Figure P1.3.2.

1.3.3 Determine v, i, and the power delivered to elements in the network given in Figure P1.3.3. Check whether conservation of power is satisfied by the circuit.

1.3.4 For a part of the network shown in Figure P1.3.4, given that $i_1 = 4$ A; $i_3(t) = 5e^{-t}$, and $i_4(t) = 10 \cos 2t$, find v_1, v_2, v_3, v_4, i_2, and i_5.

1.3.5 For the circuit given in Figure P1.3.5, given that $V_{AC} = 10$ V and $V_{BD} = 20$ V, determine V_1

and V_2. Show that the conservation of power is satisfied by the circuit.

*1.3.6 The current sources in Figure P1.3.6 are given to be $I_A = 30$ A and $I_B = 50$ A. For the values of $R_1 = 20\ \Omega$, $R_2 = 40\ \Omega$, and $R_3 = 80\ \Omega$, find:

(a) The voltage V.

(b) The currents I_1, I_2, and I_3.

(c) The power supplied by the current sources and check whether conservation of power is satisfied.

1.3.7 Show how the conservation of power is satisfied by the circuit of Figure P1.3.7.

1.3.8 Given that $V_0 = 10$ V, determine I_S in the circuit drawn in Figure P1.3.8.

Figure P1.3.1

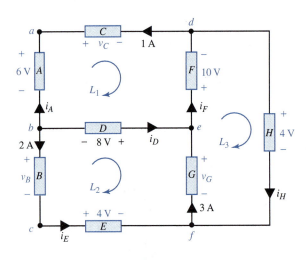

Figure P1.3.2

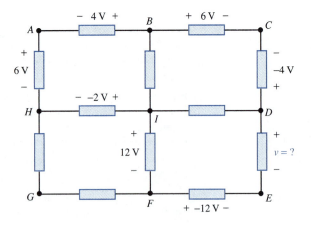

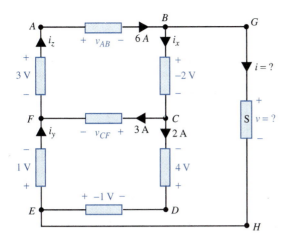

Figure P1.3.3

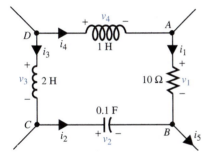

Figure P1.3.4

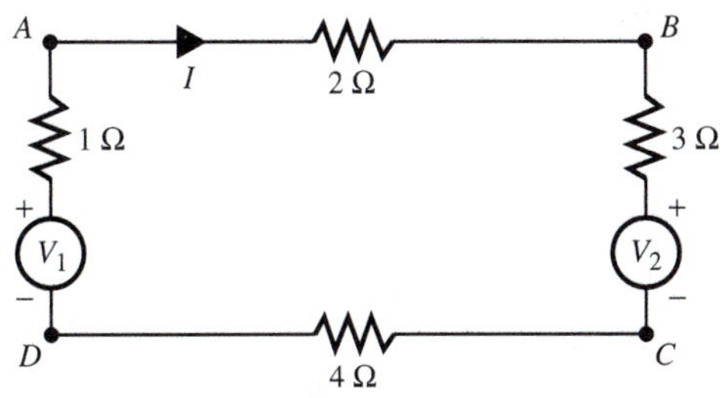

Figure P1.3.5

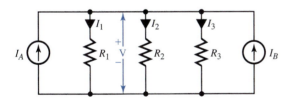

Figure P1.3.6

1.3.9 Consider the circuit shown in Figure P1.3.9.

(a) Given $v(t) = 10e^{-t}$ V, find the current source $i_s(t)$ needed.

(b) Given $i(t) = 10e^{-t}$ A, find the voltage source $v_s(t)$ needed.

1.3.10 An operational amplifier stage is typically represented by the circuit of Figure P1.3.10. For the values given, determine V_{out} and the power supplied by the 2.5-V source.

1.4.1 A voltmeter with a full scale of 100 V has a probable error of 0.1% of full scale. When this

meter is employed to measure 100 V, find the percent of probable error that can exist.

1.4.2 A current of 65 A is measured with an analog ammeter having a probable error of ±0.5% of full scale of 100 A. Find the maximum probable percentage error in the measurement.

1.4.3 Error specifications on a 10-A digital ammeter are given as 0.07% of the reading, 0.05% of full scale, 0.005% of the reading per degree Celsius, and 0.002% of full scale per degree Celsius. The 10-A analog meter error specification is given as 0.5% of full scale and 0.001% of the reading per

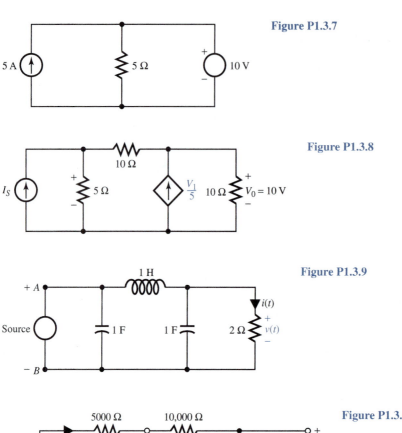

Figure P1.3.7

Figure P1.3.8

Figure P1.3.9

Figure P1.3.10

degree Celsius. If the temperature at the time of measurement is 20°C above ambient, compare the percent error of both meters when measuring a current of 5 A.

1.4.4 A DMM (digital multimeter) reads true rms values of current. If the peak value of each of the following periodic current waves is 5 A, find the meter reading for: (a) a sine wave, (b) a square wave, (c) a triangular wave, and (d) a sawtooth wave.

1.4.5 Consider the bridge circuit given in Figure P1.4.5 with $R_1 = 24$ kΩ, $R_2 = 48$ kΩ, and $R_3 = 10$ kΩ. Find R_4 when the bridge is balanced with $V_1 = 0$.

*1.4.6 In the Wheatstone bridge circuit shown in Figure P1.4.6, $R_1 = 16$ Ω, $R_2 = 8$ Ω, and $R_3 = 40$ Ω; R_4

is the unknown resistance. R_M is the galvanometer resistance of 6 Ω. If no current is detected by the galvanometer, when a 24-V source with a 12-Ω internal resistance is applied across terminals a–b, find: (a) R_4, and (b) the current through R_4.

1.4.7 Three waveforms seen on an oscilloscope are shown in Figure P1.4.7. If the horizontal scale is set to 50 ms per division (500 ms for the entire screen width), and the vertical scale is set to 5 mV per division (±25 mV for the entire screen height with zero voltage at the center), determine: (i) the maximum value of the voltage, and (ii) the frequency for each of the waveforms.

Figure P1.4.5

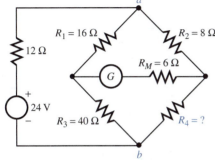

Figure P1.4.6

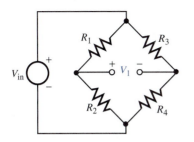

Figure P1.4.7 (a) Sinusoidal wave. (b) Rectangular wave. (c) Sawtooth wave.

2 Circuit Analysis Techniques

In Chapter 1 the basic electric circuit concepts were presented. In this chapter we consider some circuit analysis techniques, since one needs not only basic knowledge but also practical and efficient techniques for solving problems associated with circuit operations.

One simplifying technique often used in complex circuit problems is that of breaking the circuit into pieces of manageable size and analyzing individually the pieces that may be already familiar. *Equivalent* circuits are introduced which utilize Thévenin's and Norton's theorems to replace a voltage source by a current source or vice versa. *Nodal* and *loop* analysis methods are then presented. Later the principles of *superposition* and *linearity* are discussed. Also, *wye–delta* transformation is put forth as a tool for network reduction. Finally, computer-aided circuit analyses with SPICE and MATLAB are introduced. The chapter ends with a case study of practical application.

2.1 THÉVENIN AND NORTON EQUIVALENT CIRCUITS

For a *linear* portion of a circuit consisting of ideal sources and linear resistors, the volt–ampere (*v–i*) relationship at any two accessible terminals can be expressed by the linear equation

$$v = Ai + B \tag{2.1.1}$$

where A and B are two constants. The Thévenin equivalent circuit at any two terminals a and b (to replace the linear portion of the circuit) is given by

$$v = R_{Th}i + v_{oc} \tag{2.1.2}$$

where it can be seen that

$$R_{Th} = v/i|_{v_{oc}=0} \tag{2.1.3}$$

and

$$v_{oc} = v|_{i=0} \tag{2.1.4}$$

Thus, v_{oc} is known as the *open-circuit voltage* (or Thévenin voltage) with $i = 0$, and R_{Th} is the *Thévenin equivalent resistance* (as seen from the terminals a–b) with $v_{oc} = 0$. Equation (2.1.4) accounts for the ideal sources present in that linear portion of the circuit, as shown in Figure 2.1.1(a), whereas Equation (2.1.3) implies deactivating or zeroing all ideal sources (i.e., replacing voltage sources by short circuits and current sources by open circuits). The model with the voltage source v_{oc} in series with R_{Th} is known as *Thévenin equivalent circuit*, as shown in Figure 2.1.1(b).

Equation (2.1.1) may be rewritten as

$$i = \frac{v}{A} - \frac{B}{A} = \frac{v}{R_{Th}} - \frac{v_{oc}}{R_{Th}} = \frac{v}{R_{Th}} - i_{sc} \tag{2.1.5}$$

which is represented by the *Norton equivalent circuit* with a current source i_{sc} in parallel with R_{Th}, as shown in Figure 2.1.1(c). Notice that with $v = 0$, $i = -i_{sc}$. Also, $i_{sc} = v_{oc}/R_{Th}$, or $v_{oc} = i_{sc}R_{Th}$.

Besides representing complete one-ports (or two-terminal networks), Thévenin and Norton equivalents can be applied to portions of a network (with respect to any two terminals) to simplify intermediate calculations. Moreover, successive conversions back and forth between the two equivalents often save considerable labor in circuit analysis with multiple sources. *Source transformations* can be used effectively by replacing the voltage source V with a series resistance R by an equivalent current source $I (= V/R)$ in parallel with the same resistance R, or vice versa.

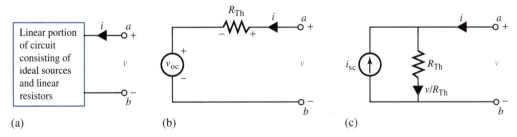

Figure 2.1.1 Equivalent circuits. **(a)** Two-terminal or one-port network. **(b)** Thévenin equivalent circuit. **(c)** Norton equivalent circuit.

EXAMPLE 2.1.1

Consider the circuit shown in Figure E2.1.1(a). Reduce the portion of the circuit to the left of terminals a–b to (a) a Thévenin equivalent and (b) a Norton equivalent. Find the current through $R = 16\ \Omega$, and comment on whether resistance matching is accomplished for maximum power transfer.

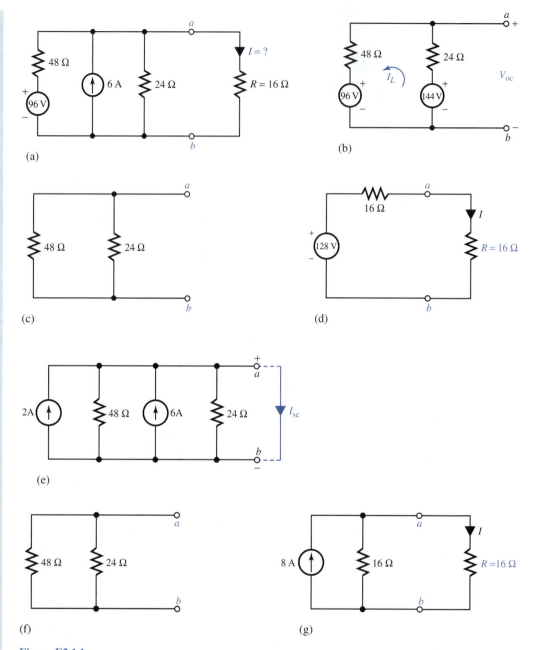

Figure E2.1.1

Solution

The 6-A source with 24 Ω in parallel can be replaced by a voltage source of 6 × 24 = 144 V with 24 Ω in series. Thus, by using source transformation, in terms of voltage sources, the equivalent circuit to the left of terminals *a–b* is shown in Figure E2.1.1(b).

(a) KVL: $144 - 24I_L - 48I_L - 96 = 0$, or $72I_L = 48$, or $I_L = \frac{2}{3}$ A

$$V_{oc} = 144 - 24(\frac{2}{3}) = 128 \text{ V}$$

Deactivating or zeroing all ideal sources, i.e., replacing voltage sources by short circuits in the present case, the circuit of Figure E2.1.1(b) reduces to that shown in Figure E2.1.1(c).

Viewed from terminals a–b, the 48-Ω resistor and the 24-Ω resistor are in parallel,

$$R_{Th} = 48\|24 = \frac{48 \times 24}{48 + 24} = 16 \ \Omega$$

Thus, the Thévenin equivalent to the left of terminals a–b, attached with the 16-Ω resistor, is shown in Figure E2.1.1(d). Note that the Thévenin equivalent of any linear circuit consists of a single Thévenin voltage source in series with a single equivalent Thévenin resistance.

The current in the 16-Ω resistor to the right of terminals a–b can now be found,

$$I = 128/32 = 4 \text{ A}$$

(b) The 96-V source with 48 Ω in series can be replaced by a current source of $96/48 = 2$ A with a parallel resistance of 48 Ω. Thus, by using source transformation, in terms of current sources, the equivalent circuit to the left of terminals a–b is given in Figure E2.1.1(e).

Shorting terminals a–b, one can find I_{sc}, $I_{sc} = 8$ A. Replacing current sources by open circuits, viewed from terminals a–b, $R_{Th} = 48\|24 = 16 \ \Omega$, which is the same as in part (a). The circuit of Figure E2.1.1(e) to the left of terminals a–b reduces to that shown in Figure E2.1.1(f).

Thus, the Norton equivalent to the left of terminals a–b, attached with the 16-Ω resistor, is given in Figure E2.1.1(g). Note that the Norton equivalent of any linear circuit consists of a single current source in parallel with a single equivalent Thévenin resistance.

The current in the 16-Ω resistor to the right of terminals a–b can now be found. $I = 4$ A, which is the same as in part (a).

The equivalent source resistance, also known as the output resistance, is the same as the load resistance of 16 Ω in the present case. Hence, resistance matching is accomplished for maximum power transfer.

EXAMPLE 2.1.2

Consider the circuit of Figure E2.1.2(a), including a dependent source. Obtain the Thévenin equivalent at terminals a–b.

Figure E2.1.2

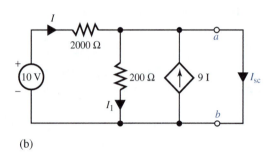

(a)

(b)

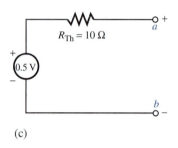

(c)

First, the open-circuit voltage at terminals a–b is to be found.

KCL at node a: $\quad I + 9I = I_1$, or $I_1 = 10I$

KVL for the left-hand mesh: $\quad 2000I + 200I_1 = 10$, or $4000I = 10$, or $I = 1/400$ A

$$V_{oc} = 200I_1 = 200(1/400) = 0.5 \text{ V}$$

Because of the presence of a dependent source, in order to find R_{Th}, one needs to determine I_{sc} after shorting terminals a–b, as shown in Figure E2.1.2(b).

Note that $I_1 = 0$, since $V_{ab} = 0$.

KCL at node a: $\quad I_{sc} = 9I + I = 10I$

KVL for the outer loop: $\quad 2000I = 10$, or $I = 1/200$ A

$$I_{sc} = 10(1/200) = 1/20 \text{ A}$$

Hence the equivalent Thévenin resistance R_{Th} viewed from terminals a–b is

$$R_{Th} = \frac{V_{oc}}{I_{sc}} = \frac{0.5}{1/20} = 10\Omega$$

Thus, the Thévenin equivalent is given in Figure E2.1.2(c).

The preceding examples illustrate how a complex network could be reduced to a simple representation at an output port. The effect of load on the terminal behavior or the effect of an output load on the network can easily be evaluated. Thévenin and Norton equivalent circuits help us in matching, for example, the speakers to the amplifier output in a stereo system. Such equivalent circuit concepts permit us to represent the entire system (generation and distribution) connected to a receptacle (plug or outlet) in a much simpler model with the open-circuit voltage as the measured voltage at the receptacle itself.

When a system of sources is so large that its voltage and frequency remain constant regardless of the power delivered or absorbed, it is known as an *infinite bus*. Such a bus (node) has a voltage and a frequency that are unaffected by external disturbances. The infinite bus is treated as an ideal voltage source. Even though, for simplicity, only resistive networks are considered in this section, the concept of equivalent circuits is also employed in ac sinusoidal steady-state circuit analysis of networks consisting of inductors and capacitors, as we shall see in Chapter 3.

2.2   NODE-VOLTAGE AND MESH-CURRENT ANALYSES

The node-voltage and mesh-current methods, which complement each other, are well-ordered systematic methods of analysis for solving complicated network problems. The former is based on the KCL equations, whereas the KVL equations form the basis for the latter. In both methods an appropriate number of simultaneous algebraic equations are developed. The unknown nodal voltages are found in the nodal method, whereas the unknown mesh currents are calculated in the loop (or mesh) method. A decision to use one or the other method of analysis is usually based on the number of equations needed for each method.

Even though, for simplicity, only resistive networks with dc voltages are considered in this section, the methods themselves are applicable to more general cases with time-varying sources, inductors, capacitors, and other circuit elements.

Nodal-Voltage Method

A set of node-voltage variables that implicitly satisfy the KVL equations is selected in order to formulate circuit equations in this nodal method of analysis. A *reference* (datum) node is chosen arbitrarily based on convenience, and from each of the remaining nodes to the reference node, the voltage drops are defined as *node-voltage* variables. The circuit is then described completely by the necessary number of KCL equations whose solution yields the unknown nodal voltages from which the voltage and the current in every circuit element can be determined. Thus, the number of simultaneous equations to be solved will be equal to one less than the number of network nodes. All voltage sources in series with resistances are replaced by equivalent current sources with conductances in parallel. In general, resistances may be replaced by their corresponding conductances for convenience. Note that the nodal-voltage method is a general method of network analysis that can be applied to any network.

Let us illustrate the method by considering the simple, but typical, example shown in Figure 2.2.1. By replacing the voltage sources with series resistances by their equivalent current sources with shunt conductances, Figure 2.2.1 is redrawn as Figure 2.2.2, in which one can identify three nodes, *A, B,* and *O.*

Notice that the voltages V_{AO}, V_{BO}, and V_{AB} satisfy the KVL relation:

$$V_{AB} + V_{BO} - V_{AO} = 0, \qquad \text{or} \qquad V_{AB} = V_{AO} - V_{BO} = V_A - V_B \qquad (2.2.1)$$

where the node voltages V_A and V_B are the voltage drops from A to O and B to O, respectively. With node O as reference, and with V_A and V_B as the node-voltage unknown variables, one can write the two independent KCL equations:

Node *A:* $V_A G_1 + (V_A - V_B)G_3 = I_1,$ or $(G_1 + G_3)V_A - G_3 V_B = I_1$ (2.2.2)

Node *B:* $V_B G_2 - (V_A - V_B)G_3 = I_2,$ or $-G_3 V_A + (G_2 + G_3)V_B = I_2$ (2.2.3)

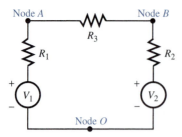

Figure 2.2.1 Circuit for illustration of nodal-voltage method.

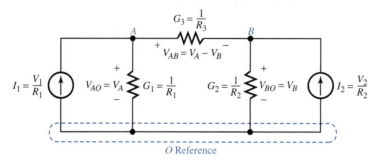

Figure 2.2.2 Redrawn Figure 2.2.1 for node-voltage method of analysis.

An examination of these equations reveals a pattern that will allow nodal equations to be written directly by inspection by following the rules given here for a network containing no dependent sources.

1. For the equation of node A, the coefficient of V_A is the positive sum of the conductances connected to node A; the coefficient of V_B is the negative sum of the conductances connected between nodes A and B. The right-hand side of the equation is the sum of the current sources feeding into node A.

2. For the equation of node B, a similar situation exists. Notice the coefficient of V_B to be the positive sum of the conductances connected to node B; the coefficient of V_A is the negative sum of the conductances connected between B and A. The right-hand side of the equation is the sum of the current sources feeding into node B.

Such a formal systematic procedure will result in a set of N independent equations of the following form for a network with $(N + 1)$ nodes containing no dependent sources:

$$
\begin{aligned}
G_{11}V_1 &- G_{12}V_2 - \cdots - G_{1N}V_N = I_1 \\
-G_{21}V_1 &+ G_{22}V_2 - \cdots - G_{2N}V_N = I_2 \\
&\quad\vdots \qquad\qquad\qquad\qquad\quad \vdots \\
-G_{N1}V_1 &- G_{N2}V_2 - \cdots + G_{NN}V_N = I_N
\end{aligned}
\tag{2.2.4}
$$

where G_{NN} is the sum of all conductances connected to node N, $G_{JK} = G_{KJ}$ is the sum of all conductances connected between nodes J and K, and I_N is the sum of all current sources entering node N. By solving the equations for the unknown node voltages, other voltages and currents in the circuit can easily be determined.

EXAMPLE 2.2.1

By means of nodal analysis, find the current delivered by the 10-V source and the voltage across the 10-Ω resistance in the circuit shown in Figure E2.2.1(a).

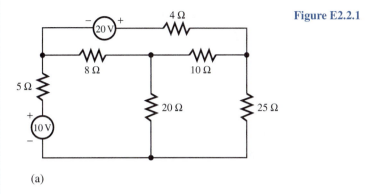

(a)

Figure E2.2.1

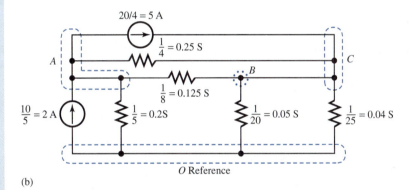

(b)

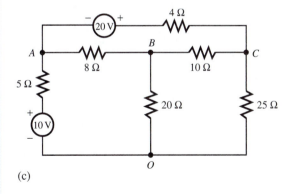

(c)

STEP 1: Replace all voltage sources with series resistances by their corresponding Norton equivalents consisting of current sources with shunt conductances. The given circuit is redrawn in Figure E2.2.1(b) by replacing all resistors by their equivalent conductances.

STEP 2: Identify the nodes and choose a convenient reference node O. This is also shown in Figure E2.2.1(b).

STEP 3: In terms of unknown node-voltage variables, write the KCL equations at all nodes (except, of course, the reference node) by following rules 1 and 2 for nodal equations given in this section.

$$\text{Node } A: \quad (0.2 + 0.125 + 0.25)V_A - 0.125V_B - 0.25V_C \quad = 2 - 5 = -3$$
$$\text{Node } B: \quad -0.125V_A + (0.125 + 0.05 + 0.1)V_B - 0.1V_C \quad = 0$$
$$\text{Node } C: \quad -0.25V_A - 0.1V_B + (0.25 + 0.1 + 0.04)V_C \quad = 5$$

Rearranging, one gets

$$
\begin{aligned}
0.575\,V_A \quad - \quad 0.125\,V_B \quad - \quad 0.25\,V_C \;&=\; -3 \\
-0.125\,V_A \quad + \quad 0.275\,V_B \quad - \quad 0.1\,V_C \;&=\; 0 \\
-0.25\,V_A \quad - \quad 0.1\,V_B \quad + \quad 0.39\,V_C \;&=\; 5
\end{aligned}
$$

STEP 4: Simultaneously solve the independent equations for the unknown nodal voltages by Gauss elimination or Cramer's rule. In our example, the solution yields

$$V_A = 4.34 \text{ V}; \qquad V_B = 8.43V; \qquad V_C = 17.77 \text{ V}$$

STEP 5: Obtain the desired voltages and currents by the application of KVL and Ohm's law. To find the current I in the 10-V source, since it does not appear in Figure E2.2.1(b) redrawn for nodal analysis, one has to go back to the original circuit and identify the equivalence between nodes A and O, as shown in Figure E2.2.1(c).

Now one can solve for I, delivered by the 10-V source,

$$V_A = 4.34 = -5I + 10 \qquad \text{or} \qquad I = \frac{5.66}{5} = 1.132 \text{ A}$$

The voltage across the 10-Ω resistance is $V_B - V_C = 8.43 - 17.77 = -9.34$ V. The negative sign indicates that node C is at a higher potential than node B with respect to the reference node O.

Nodal analysis deals routinely with current sources. When we have voltage sources along with series resistances, the source-transformation technique may be used effectively to convert the voltage source to a current source, as seen in Example 2.2.1. However, in cases where we have *constrained nodes*, that is, the difference in potential between the two node voltages is constrained by a voltage source, the concept of a *supernode* becomes useful for the circuit analysis, as shown in the following illustrative example.

EXAMPLE 2.2.2

For the network shown in Figure E2.2.2, find the current in each resistor by means of nodal analysis.

Solution

Note that the reference node is chosen at one end of an independent voltage source, so that the node voltage V_A is known at the start,

$$V_A = 12 \text{ V}$$

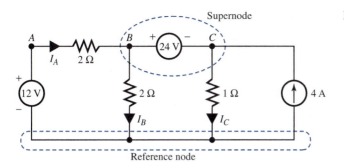

Figure E2.2.2

Supernode

Reference node

Note that we cannot express the branch current in the voltage source as a function of V_B and V_C. Here we have constrained nodes B and C. Nodal voltages V_B and V_C are not independent. They are related by the constrained equation

$$V_B - V_C = 24 \text{ V}$$

Let us now form a *supernode*, which includes the voltage source and the two nodes B and C, as shown in Figure E2.2.2. KCL must hold for this supernode, that is, the algebraic sum of the currents entering or leaving the supernode must be zero. Thus one valid equation for the network is given by

$$I_A - I_B - I_C + 4 = 0 \qquad \text{or} \qquad \frac{12 - V_B}{2} - \frac{V_B}{2} - \frac{V_C}{1} + 4 = 0$$

which reduces to

$$V_B + V_C = 10$$

This equation together with the supernode constraint equation yields

$$V_B = 17 \text{ V} \qquad \text{and} \qquad V_C = -7 \text{ V}$$

The currents in the resistors are thus given by

$$I_A = \frac{12 - V_B}{2} = \frac{12 - 17}{2} = -2.5 \text{ A}$$

$$I_B = \frac{V_B}{2} = \frac{17}{2} = 8.5 \text{ A}$$

$$I_C = \frac{V_C}{1} = \frac{-7}{1} = -7 \text{ A}$$

Mesh-Current Method

This complements the nodal-voltage method of circuit analysis. A set of independent *mesh-current* variables that implicitly satisfy the KCL equations is selected in order to formulate circuit equations in this mesh analysis. An *elementary loop*, or a *mesh*, is easily identified as one of the "window panes" of the whole circuit. However, it must be noted that not all circuits can be laid out to contain only meshes as in the case of planar networks. Those which cannot are called nonplanar circuits, for which the mesh analysis cannot be applied, but the nodal analysis can be employed.

A mesh current is a fictitious current, which is defined as the one circulating around a mesh of the circuit in a certain direction. While the direction is quite arbitrary, a clockwise direction is traditionally chosen. Branch currents can be found in terms of mesh currents, whose solution is obtained from the independent simultaneous equations. The number of necessary equations in the mesh-analysis method is equal to the number of independent loops or meshes.

All current sources with shunt conductances will be replaced by their corresponding Thévenin equivalents consisting of voltage sources with series resistances. Let us illustrate the method by considering a simple, but typical, example, as shown in Figure 2.2.3.

Replacing the current source with shunt resistance by the Thévenin equivalent, Figure 2.2.3 is redrawn as Figure 2.2.4, in which one can identify two elementary loops, or independent meshes.

By assigning loop or mesh-current variables I_1 and I_2, as shown in Figure 2.2.4, both in the clockwise direction, one can write the KVL equations for the two closed paths (loops) $ABDA$ and $BCDB$,

Loop $ABDA$: $I_1 R_1 + (I_1 - I_2)R_2 = V_1 - V_2$ or $(R_1 + R_2)I_1 - R_2 I_2 = V_1 - V_2$ (2.2.5)

Loop $BCDB$: $I_2 R_3 + (I_2 - I_1)R_2 = V_2 - V_3$ or $-R_2 I_1 + (R_2 + R_3)I_2 = V_2 - V_3$ (2.2.6)

Notice that current I_1 exists in R_1 and R_2 in the direction indicated; I_2 exists in R_2 and R_3 in the direction indicated; hence, the net current in R_2 is $I_1 - I_2$ directed from B to D. An examination of Equations (2.2.5) and (2.2.6) reveals a pattern that will allow loop equations to be written directly by inspection by following these rules:

1. In the first loop equation with mesh current I_1, the coefficient of I_1 is the sum of the resistances in that mesh; the coefficient of I_2 is the negative sum of the resistances common to both meshes. The right-hand side of the equation is the algebraic sum of the source voltage rises taken in the direction of I_1.

2. Similar statements can be made for the second loop with mesh current I_2. (See also the similarity in setting up the equations for the mesh-current and nodal-voltage methods of analysis.)

Such a formal systematic procedure will yield a set of N independent equations of the following form for a network with N independent meshes containing no dependent sources:

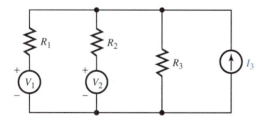

Figure 2.2.3 Circuit for illustration of mesh-current method.

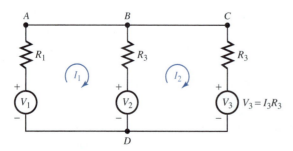

Figure 2.2.4 Redrawn Figure 2.2.3 for mesh-current method of analysis.

$$R_{11}I_1 \quad - \quad R_{12}I_2 \quad - \quad \cdots \quad - \quad R_{1N}I_N \quad = \quad V_1$$
$$-R_{21}I_1 \quad + \quad R_{22}V_2 \quad - \quad \cdots \quad - \quad R_{2N}I_N \quad = \quad V_2$$
$$\vdots \qquad\qquad\qquad\qquad\qquad\qquad \vdots$$
$$-R_{N1}I_1 \quad - \quad R_{N2}V_2 \quad - \quad \cdots \quad + \quad R_{NN}I_N \quad = \quad V_N$$

$$(2.2.7)$$

where R_{NN} is the sum of all resistances contained in mesh N, $R_{JK} = R_{KJ}$ is the sum of all resistances common to both meshes J and K, and V_N is the algebraic sum of the source-voltage rises in mesh N, taken in the direction of I_N.

By solving the equations for the unknown mesh currents, other currents and voltages in the circuit elements can be determined easily.

EXAMPLE 2.2.3

By means of mesh-current analysis, obtain the current in the 10-V source and the voltage across the 10-Ω resistor in the circuit of Example 2.2.1.

Solution

STEP 1: Replace all current sources with shunt resistances by their corresponding Thévenin equivalents consisting of voltage sources with series resistances. Conductances included in the circuit are replaced by their equivalent resistances.

In this example, since there are no current sources and conductances, the circuit of Figure E2.2.1(a) is redrawn as Figure E2.2.3 for convenience.

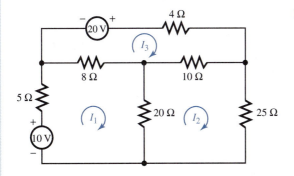

Figure E2.2.3

STEP 2: Identify elementary loops (meshes) and choose a mesh-current variable for each elementary loop, with all loop currents in the same clockwise direction. Mesh currents I_1, I_2, and I_3 are shown in Figure E2.2.3.

STEP 3: In terms of unknown mesh-current variables, write the KVL equations for all meshes by following the rules for mesh analysis.

Loop 1 with mesh current I_1: $\qquad (5+8+20)I_1 - 20I_2 - 8I_3 = 10$

Loop 2 with mesh current I_2: $\quad -20I_1 + (20+10+25)I_2 - 10I_3 = 0$

Loop 3 with mesh current I_3: $\qquad -8I_1 - 10I_2 + (4+10+8)I_3 = 20$

Rearranging, one gets

$$33I_1 - 20I_2 - 8I_3 = 10$$
$$-20I_1 + 55I_2 - 10I_3 = 0$$
$$-8I_1 - 10I_2 + 22I_3 = 20$$

STEP 4: Simultaneously solve the independent equations for the unknown mesh currents by Gauss elimination or Cramer's rule.

In this example the solution yields

$$I_1 = 1.132 \text{ A}; \qquad I_2 = 0.711 \text{ A}; \qquad I_3 = 1.645 \text{ A}$$

The current through the 10-V source is $I_1 = 1.132$ A, which is the same as in Example 2.2.1. The voltage across the 10-Ω resistor is $V_{BC} = 10(I_2 - I_3) = 10(0.711 - 1.645) = -9.34$ V, which is the same as in Example 2.2.1.

Looking at Examples 2.2.1 and 2.2.3, it can be seen that there is no specific advantage for either method since the number of equations needed for the solution is three in either case. Such may not be the case in a number of other problems, in which case one should choose judiciously the more convenient method, usually with the lower number of equations to be solved.

The mesh-current method deals routinely with voltage sources. When we have current sources with shunt conductances, the source-transformation technique may be used effectively to convert the current source to a voltage source. However, in cases where we have *constrained meshes*, that is, the two mesh currents are constrained by a current source, the concept of a *supermesh* becomes useful for the circuit analysis, as shown in the following illustrative example.

EXAMPLE 2.2.4

For the network shown in Figure E2.2.4, find the current delivered by the 10-V source and the voltage across the 3-Ω resistor by means of mesh-current analysis.

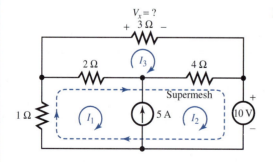

Figure E2.2.4

Solution

Note that we cannot express the voltage across the current source in terms of the mesh currents I_1 and I_2. The current source does, however, *constrain* the mesh currents by the following equation:

$$I_2 - I_1 = 5$$

Let us now form a *supermesh*, which includes meshes 1 and 2, as shown in Figure E2.2.4. We now write a KVL equation around the periphery of meshes 1 and 2 combined. This yields

$$1I_1 + 2(I_1 - I_3) + 4(I_2 - I_3) + 4(I_2 - I_3) + 10 = 0$$

Next we write a KVL equation for mesh 3,

$$3I_3 + 4(I_3 - I_2) + 2(I_3 - I_1) = 0$$

Now we have the three linearly independent equations needed to find the three mesh currents I_1, I_2, and I_3. The solution of the three simultaneous equations yields

$$I_1 = \frac{-25A}{9} \text{ A}; \qquad I_2 = \frac{20}{9} \text{ A}; \qquad I_3 = \frac{70}{27} \text{ A}$$

The current delivered by the 10-V source is $-I_2$, or $-20/9$ A. That is to say, the 10-V source is absorbing the current 20/9 A.

The voltage across the 3-Ω resistor is $V_x = 3I_3 = 3(70/27) = 70/9 = 7.78$ V.

Node-Voltage and Mesh-Current Equations with Controlled Sources

Since a controlled source acts at its terminals in the same manner as does an independent source, source conversion and application of KCL and KVL relations are treated identically for both types of sources. Because the strength of a controlled source depends on the value of a voltage or current elsewhere in the network, a *constraint* equation is written for each controlled source. After combining the constraint equations with the loop or nodal equations based on treating all sources as independent sources, the resultant set of equations are solved for the unknown current or voltage variables.

EXAMPLE 2.2.5

Consider the circuit in Figure E2.2.5(a), which include a controlled source, and find the current in the 5-V source and the voltage across the 5-Ω resistor by using (a) the loop-current method and (b) the node-voltage method.

Solution

(a) Loop-Current Method: The voltage-controlled current source and its parallel resistance are converted into a voltage-controlled voltage source and series resistance. When you are source transforming dependent sources, note that the identity of the control variable (i.e., the location in the circuit) must be retained. The converted circuit is shown in Figure E2.2.5(b) with the chosen loop currents I_1 and I_2.

The KVL equations are

For loop carrying I_1: $\qquad (10 + 4 + 2)I_1 - 2I_2 = 5$

For loop carrying I_2: $\qquad -2I_1 + (2 + 10 + 5)I_2 = -5V_1$

The constraint equation is

$$V_1 = (I_1 - I_2)2$$

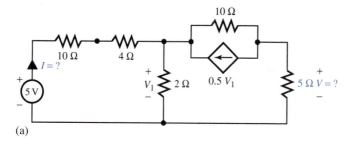

Figure E2.2.5

(a)

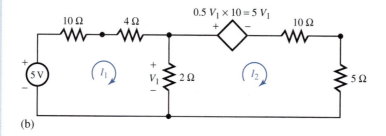

(b)

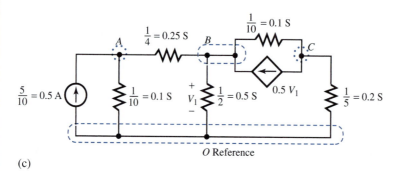

O Reference

(c)

Combining the constraint equation with the loop equations, one gets

$$16I_1 - 2I_2 = 5; \qquad -2I_1 + 17I_2 = -10(I_1 - I_2), \qquad \text{or} \qquad 8I_1 + 7I_2 = 0$$

from which

$$I_1 = 35/128 \ A; \qquad I_2 = -5/16 \ A$$

Thus, the current through the 5-V source is $I = I_1 = 35/128 = 0.273$ A, and the voltage across the 5-Ω resistor is $V = 5I_2 = 5(-5/16) = -1.563$ V.

(b) Node-Voltage Method: The 5-V voltage source with its 10-Ω series resistor is replaced by its Norton equivalent. Resistances are converted into conductances and the circuit is redrawn in Figure E2.2.5(c) with the nodes shown.

The nodal equations are

A : $\qquad (0.1 + 0.25)V_A - 0.25V_B = 0.5$

B : $\qquad -0.25V_A + (0.25 + 0.5 + 0.1)V_B - 0.1V_C = 0.5V_1$

C : $\qquad -0.1V_B + (0.1 + 0.2)V_C = -0.5V_1$

The constraint equation is

$$V_1 = V_B$$

Combining these with the nodal equations already written, one has

$$
\begin{aligned}
0.35\,V_A &- 0.25\,V_B & &= 0.5 \\
-0.25\,V_A &+ 0.35\,V_B &- 0.1\,V_C &= 0 \\
&0.4\,V_B &+ 0.3\,V_C &= 0
\end{aligned}
$$

Solving, one gets

$$V_A = 2.266V; \qquad V_B = 1.173; \qquad V_C = -1.564 \text{ V}$$

Notice that $V_C = -1.564$ V is the voltage V across the 5-Ω resistor, which is almost the same as that found in part (a).

In order to find the current I through the 5-V source, one needs to go back to the original circuit and recognize that

$$5 - 10I = V_A = 2.266 \qquad \text{or} \qquad I = 0.273 \text{ A}$$

which is the same as that found in part (a).

2.3 SUPERPOSITION AND LINEARITY

Mathematically a function is said to be linear if it satisfies two properties: *homogeneity (proportionality or scaling)* and *additivity (superposition)*,

$$f(Kx) = Kf(x) \qquad \text{(homogeneity)} \tag{2.3.1}$$

where K is a scalar constant, and

$$f(x_1 + x_2) = f(x_1) + f(x_2) \qquad \text{(additivity)} \tag{2.3.2}$$

Linearity requires both additivity and homogeneity. For a linear circuit or system in which excitations x_1 and x_2 produce responses y_1 and y_2, respectively, the application of K_1x_1 and K_2x_2 together (i.e., $K_1x_1 + K_2x_2$) results in a response of $(K_1y_1 + K_2y_2)$, where K_1 and K_2 are constants. With the cause-and-effect relation between the excitation and the response, all *linear* systems satisfy the principle of *superposition*. A circuit consisting of independent sources, linear dependent sources, and linear elements is said to be a linear circuit. Note that a resistive element is linear. Capacitors and inductors are also circuit elements that have a linear input–output relationship provided that their initial stored energy is zero. Nonzero initial conditions are to be treated as independent sources.

In electric circuits, the excitations are provided by the voltage and current sources, whereas the responses are in terms of element voltages and currents. All circuits containing only ideal resistances, capacitances, inductances, and sources are linear circuits (described by linear differential equations). For a linear network consisting of several *independent* sources, according to the principle of superposition, the net response in any element is the algebraic sum of the individual responses produced by each of the independent sources acting only by itself. While each independent source acting on the network is considered separately by itself, the other independent sources are suppressed; that is to say, voltage sources are replaced by short circuits and current sources are replaced by open circuits, thereby reducing the source strength to zero. The effect of any dependent sources, however, must be included in evaluating the response due to each of the independent sources, as illustrated in the following example.

EXAMPLE 2.3.1

Determine the voltage across the 20-Ω resistor in the following circuit of Figure E2.3.1 (a) with the application of superposition.

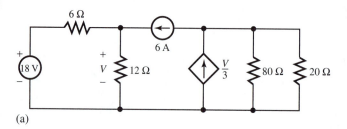

Figure E2.3.1

(a)

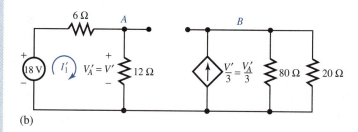

(b)

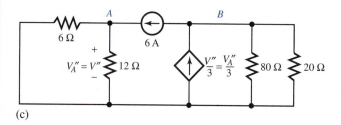

(c)

Solution

Let us suppress the independent sources in turn, recognizing that there are two independent sources. First, by replacing the independent current source with an open circuit, the circuit is drawn in Figure E2.3.1(b). Notice the designation of V' across the 12-Ω resistor and $V'/3$ as the dependent current source for this case. At node B,

$$\left(\frac{1}{80} + \frac{1}{20}\right) V'_B = \frac{V'_A}{3} \quad \text{or} \quad V'_B = \frac{V'_A}{48}$$

For the mesh on the left-hand side, $(6 + 12)I'_1 = 18$, or $I'_1 = 1$ A. But, $I'_1 = V'_A/12$, or $V'_A = 12$ V. The voltage across the 20-Ω resistor from this part of the solution is

$$V'_B = \frac{12}{48} = \frac{1}{4} \text{ V}$$

Next, by replacing the independent voltage source with a short circuit, the circuit is shown in Figure E2.3.1(c). Notice the designation of V'' across the 12-Ω resistor and $V''/3$ as the dependent current source for this case. At node A,

$$\left(\frac{1}{6} + \frac{1}{12}\right) V_A'' = 6 \qquad \text{or} \qquad V_A'' = 24 \text{ V}$$

and at node B,

$$\left(\frac{1}{80} + \frac{1}{20}\right) V_B'' = \frac{V_A''}{3} - 6 = \frac{24}{3} - 6 = 2 \qquad \text{or} \qquad V_B'' = 32 \text{ V}$$

Thus, the voltage across the 20-Ω resistor for this part of the solution is

$$V_B'' = 32 \text{ V}$$

Then the total net response, by superposition, is

$$V_B = V_B' + V_B'' = \frac{1}{4} + 32 = 32.25 \text{ V}$$

The *principle of superposition* is indeed a powerful tool for analyzing a wide range of linear systems in electrical, mechanical, civil, or industrial engineering.

2.4 WYE-DELTA TRANSFORMATION

Certain network configurations cannot be reduced or simplified by series–parallel combinations alone. In some such cases wye–delta (Y–Δ) transformation can be used to replace three resistors in wye configuration by three resistors in delta configuration, or vice versa, so that the networks are equivalent in so far as the terminals (A, B, C) are concerned, as shown in Figure 2.4.1. For equivalence, it can be shown that (see Problem 2.4.1)

$$R_A = \frac{R_{AB} R_{CA}}{R_{AB} + R_{BC} + R_{CA}}; \qquad R_B = \frac{R_{AB} R_{BC}}{R_{AB} + R_{BC} + R_{CA}};$$

$$R_C = \frac{R_{CA} R_{BC}}{R_{AB} + R_{BC} + R_{CA}} \tag{2.4.1}$$

$$R_{AB} = \frac{R_A R_B + R_B R_C + R_C R_A}{R_C}; \qquad R_{BC} = \frac{R_A R_B + R_B R_C + R_C R_A}{R_A};$$

$$R_{CA} = \frac{R_A R_B + R_B R_C + R_C R_A}{R_B} \tag{2.4.2}$$

For the simple case when $R_A = R_B = R_C = R_Y$, and $R_{AB} = R_{BC} = R_{CA} = R_\Delta$, Equations (2.4.1) and (2.4.2) become

$$R_Y = \frac{R_\Delta}{3} \tag{2.4.3}$$

$$R_\Delta = 3R_Y \tag{2.4.4}$$

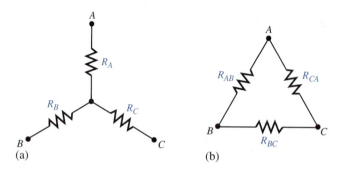

Figure 2.4.1 Wye–delta transformation. **(a)** Wye configuration. **(b)** Delta configuration.

EXAMPLE 2.4.1

Use delta–wye transformation for network reduction and determine the current through the 12-Ω resistor in the circuit of Figure E2.4.1(a).

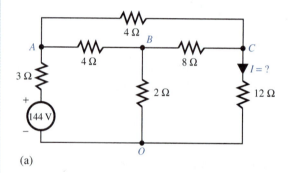

Figure E2.4.1

(a)

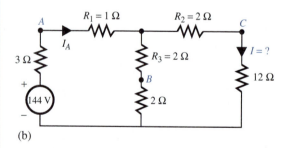

(b)

Solution

The delta-connected portion between terminals A–B–C is replaced by an equivalent wye connection [see Equation (2.4.1)] with

$$R_1 = \frac{4 \times 4}{4 + 8 + 4} = 1 \ \Omega$$

$$R_2 = \frac{4 \times 8}{4 + 8 + 4} = 2 \ \Omega$$

$$R_3 = \frac{4 \times 8}{4 + 8 + 4} = 2 \ \Omega$$

The circuit is redrawn in Figure E2.4.1(b).

Using the KVL equation,

$$I_A = \frac{144}{(3+1) + (4 \| 14)} = \frac{81}{4} \ A$$

By current division,

$$I = \frac{81}{4} \times \frac{4}{18} = \frac{9}{2} = 4.5 \ A$$

2.5 COMPUTER-AIDED CIRCUIT ANALYSIS: SPICE

A word of caution is appropriate if this is the student's first experience with simulation. Just as the proliferation of calculators did not eliminate the need to understand the theory of mathematics, circuit simulation programs do not eliminate the need to understand circuit theory. However, computer-aided tools can free the engineer from tedious calculations, thereby freeing more time for doing the kind of creative work a computer cannot do.

A circuit-analysis program known as SPICE, an acronym for simulation program with integrated circuit emphasis, is introduced in this section. The original SPICE program was developed in the early 1970s at the University of California at Berkeley. Since that time, various SPICE-based commercial products have been developed for personal computer and workstation platforms.[1]

A block diagram summarizing the major features of a SPICE-based circuit simulation program is shown in Figure 2.5.1. Micro Sim Corporation has developed a design center in

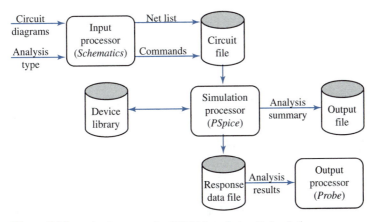

Figure 2.5.1 Major features of a SPICE-based circuit simulation program.

---

[1]For supplementary reading on SPICE, the student is encouraged to refer to G. Roberts and A. Sedra, *SPICE*, 2nd ed., published by Oxford University Press (1997), and to P. Tuinenga, *SPICE*, 3rd ed., published by Prentice Hall (1995).

which the input processor is called *Schematics*, the simulation processor is a version of SPICE called *PSpice*, and the output processor is called *PROBE*. These three programs, working together, create a graphical environment in which the circuit diagram and the analysis objectives are entered using *Schematics*, the circuit is analyzed using *PSpice*, and the resulting circuit responses are viewed using *PROBE*. A student's version of these programs is widely available and is used in this book.

The first step for describing the circuit is to number the circuit nodes. The reference node (or ground node) is labeled as zero (0), and in PSpice syntax the other node names can be numbers or letters. In order to describe the circuit, statements are written with a separate statement for each circuit element. The name of an element must begin with a particular letter identifying the kind of circuit element. Some of these are listed here:

R   Resistor

V   Independent voltage source

I   Independent current source

G   Voltage-controlled current source

E   Voltage-controlled voltage source

F   Current-controlled current source

H   Current-controlled voltage source

While the original SPICE recognized only uppercase letters, PSpice is actually case insensitive. Because PSpice does not recognize subscripts, R_1, for example, will be represented by R1, and so on. The name of each circuit element must be unique. Numerical values can be specified in the following forms:

$$4567 \quad \text{or} \quad 4567.0 \quad \text{or} \quad 4.567 \text{ E3}$$

SPICE uses the following scale factor designations:

$$
\begin{array}{lll}
T = 1E12 & G = 1E9 & MEG = 1E6 \\
K = 1E3 & M = 1E-3 & U = IE-6 \\
N = 1E-9 & P = 1E-12 & F = 1E-15
\end{array}
$$

Sometimes, for clarity, additional letters following a numerical value may be used; but these are ignored by SPICE. For example, 4.4 KOHMS is recognized as the value 4400, and "ohms" is ignored by the program. Comment statements are identified by an asterisk (*) in the first column, and these are helpful for making the program meaningful to users. PSpice also allows inserting comments on any line by starting the comment with a semicolon. Figure 2.5.2 shows the four types of controlled sources and their corresponding PSpice statements.

While SPICE is capable of several types of analysis, here we illustrate how to solve resistive circuits containing dc sources using the DC command. PSpice can *sweep* the value of the source, when the starting value, the end value, and the increment between values are given. If the starting and end values are the same, the solution is carried out for only a single value of the source.

Next we give an example of PSpice analysis. Note that SPICE has capabilities far beyond what we use in this section, and clearly, one can easily solve complex networks by using programs like PSpice.

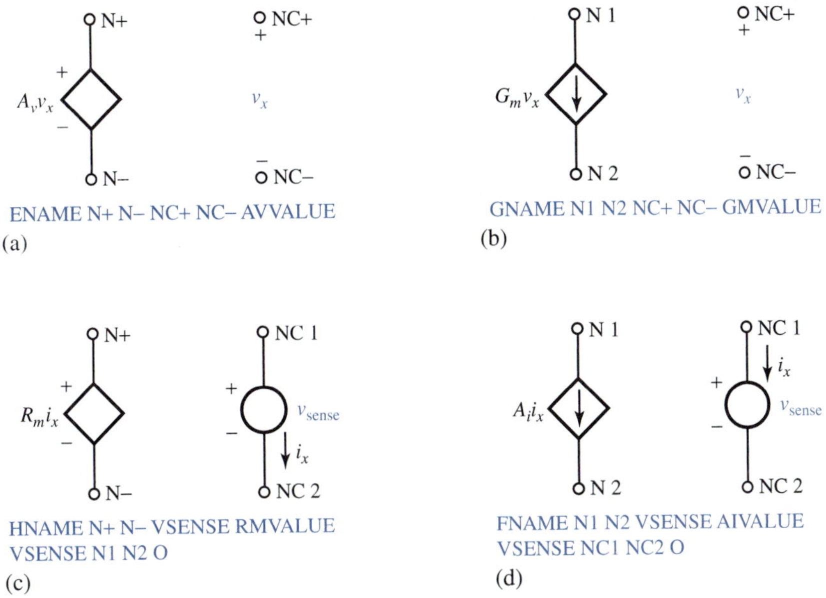

Figure 2.5.2 Four types of controlled sources and their corresponding PSpice statements. **(a)** Voltage-controlled voltage source. **(b)** Voltage-controlled current source. **(c)** Current-controlled voltage source. **(d)** Current-controlled current source.

EXAMPLE 2.5.1

Develop and execute a PSpice program to solve for the current I_2 in Figure E2.5.1(a).

Solution

Figure E2.5.1(b) is drawn showing the node numbers, and adding a voltage source of zero value in series with R_1, because there is a current-controlled source. The program is as follows:

```
EXAMPLE E2.5.1(a) A Title Identifying the Program.
* THE CIRCUIT DIAGRAM IS GIVEN IN FIGURE E2.5.1(b); a comment
statement
* CIRCUIT DESCRIPTION WITH COMPONENT STATEMENTS
IS 0 1 3
R1 1 4 5
R2 1 2 10
R3 2 0 2
R4 3 0 5
HCCVS 2 3 VSENSE 2
VSENSE 4 3 0
* ANALYSIS REQUEST
• DC IS 3 3 1
* OUTPUT REQUEST
• PRINT DC I(R2) V(1) V(2) V(3)
• END ; an end statement
```

After executing the program, from the output file, $I_2 = I(R2) = 0.692$ A.

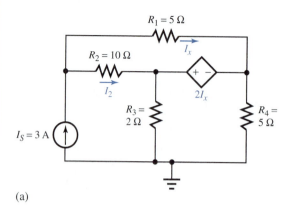

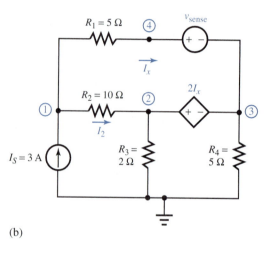

(a)

(b)

Figure E2.5.1 (a) Circuit. (b) Redrawn circuit for developing a PSpice program.

2.6 COMPUTER-AIDED CIRCUIT ANALYSIS: MATLAB

This text does not teach MATLAB; it assumes that the student is familiar with it through previous work. Also, the book does not depend on a student having MATLAB. MATLAB, however, provides an enhancement to the learning experience if it is available. If it is not, the problems involving MATLAB can simply be skipped, and the remainder of the text still makes sense. If one wants to get a quick introduction, the book entitled *Getting Started with MATLAB 5* by R. Pratap, listed under Selected Bibliography for Supplemental Reading for Computer-Aided Circuit Analysis, may be a good source.

MATLAB (MATrix LABoratory), a product of The Math Works, Inc., is a software package for high-performance numerical computation and visualization. It is simple, powerful, and for most purposes quite fast with its easy-to-learn and easy-to-use language. It provides an interactive environment with hundreds of built-in functions for technical computation, graphics, and animation. MATLAB also provides easy extensibility with its own high-level programming language.

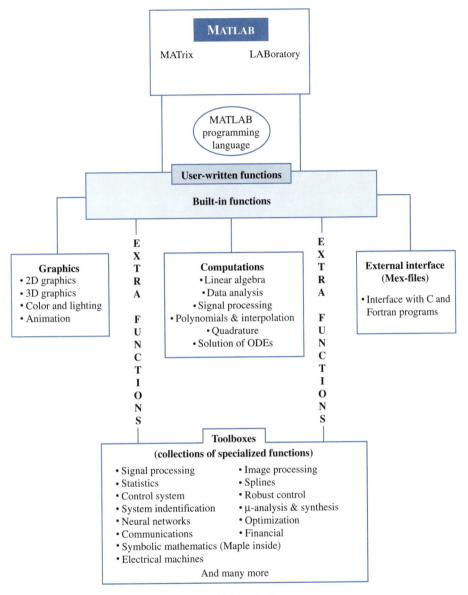

Figure 2.6.1 Schematic diagram of MATLAB's main features.

Figure 2.6.1 illustrates MATLAB's main features. The built-in functions with their state-of-the-art algorithms provide excellent tools for linear algebra computations, data analysis, signal processing, optimization, numerical solution of ordinary differential equations (ODE), numerical integration (Quadrature), and many other types of scientific and engineering computations. Numerous functions are also available for 2D and 3D graphics as well as for animation. Users can also write their own functions, which then behave just like the built-in functions. MATLAB even provides an external interface to run Fortran and C programs. Optional "toolboxes," which are collections of specialized functions for particular applications, are also available. For example, the author of this text has developed "Electrical Machines Toolbox" for the analysis and design of electrical machines.

The MATLAB environment consists of a command window, a figure window, and a platform-dependent edit window, as illustrated in Figure 2.6.2. The command window, which is the main window, is characterized by the MATLAB command prompt >>. All commands, including those for running user-written programs, are typed in this window at the MATLAB prompt. The graphics window or the figure window receives the output of all graphics commands typed in the command window. The user can create as many figure windows as the system memory would allow. The edit window is where one writes, edits, creates, and saves one's own programs in files called M-files. Most programs that are written in MATLAB are saved as M-files, and all built-in functions in MATLAB are M-files.

Let us now take an illustrative example in circuits to solve a set of simultaneous equations with the use of MATLAB.

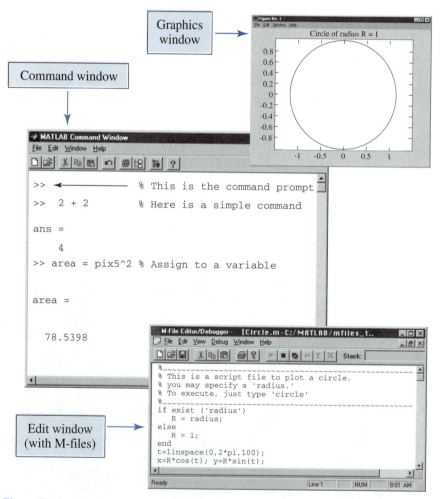

Figure 2.6.2 Illustration of a command window, a figure window, and a platform-dependent edit window in the MATLAB environment.

EXAMPLE 2.6.1

Consider the circuit shown in Figure E2.6.1 and identify the connection equations to be the following:

$$\begin{array}{llll} \text{Node } A: & I_S + I_1 + I_4 = 0; & \text{Loop 1:} & -V_S + V_1 + V_2 = 0 \\ \text{Node } B: & -I_1 + I_2 + I_3 = 0; & \text{Loop 2:} & -V_1 + V_4 - V_3 = 0 \\ \text{Node } C: & -I_3 - I_4 + I_5 = 0; & \text{Loop 3:} & -V_2 + V_3 + V_5 = 0 \end{array}$$

The element equations are given by

$$\begin{array}{lll} V_S = 15; & V_1 = 60I_1; & V_2 = 90I_2 \\ V_3 = 50I_3; & V_4 = 90I_4; & V_5 = 60I_5 \end{array}$$

Solve these 12 simultaneous equations by using MATLAB and find the voltage across the 50-Ω resistor in the circuit. Also evaluate the total power dissipated in the circuit.

Figure E2.6.1 Circuit for Example 2.6.1.

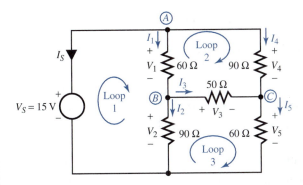

Solution

The M-file and answers are as follows.

```
function example261
clc

% Given Connection Equations
eqn01 = 'Is + I1 + I4 = 0';
eqn02 = '-I1 + I2 + I3 = 0';
eqn03 = '-I3 - I4 + I5 = 0';

eqn04 = '-Vs + V1 + V2 = 0';
eqn05 = '-V1 - V3 + V4 = 0';
eqn06 = '-V2 + V3 + V5 = 0';

% Element Equations
eqn07 = 'Vs = 15';
eqn08 = 'V1 = 60*I1';
eqn09 = 'V2 = 00*I2';
eqn10 = 'V3 = 50*I3';
eqn11 = 'V4 = 90*I4';
eqn12 = 'V5 = 60*I5';
```

```
% Solve Equations
sol = solve (eqn01, eqn02, eqn03, eqn04, eqn05, eqn06, . . .
              eqn07, eqn08, eqn09, eqn10, eqn11, eqn12, . . .
              'I1, I2, I3, I4, I5, Is, V1, V2, V3, V4, V5, Vs');

% Answers
V3 = eval (sol. V3)
Is = eval (sol. Is)
eval (sol.I1*sol.V1 +sol.I2*sol.V2+sol.I3*sol.V3+sol.I4*sol.V4+
sol.I5*sol.V5)
V3 = 1.2295
IS = -0.2049
ans = 3.0738
```

2.7 LEARNING OBJECTIVES

The learning objectives of this chapter are summarized here, so that the student can check whether he or she has accomplished each of the following.

- Obtaining Thévenin equivalent circuit for a two-terminal (or one-port) network with or without dependent sources.
- Obtaining Norton equivalent circuit for a two-terminal (or one-port) network with or without dependent sources.
- Nodal-voltage method of network analysis, including the concept of a supernode.
- Mesh-current method of network analysis, including the concept of a supermesh.
- Node-voltage and mesh-current equations with controlled sources and their constraint equations.
- Analysis of linear circuits, containing more than one source, by using the principle of superposition.
- Wye–delta transformation for resistive network reduction.
- Computer-aided circuit analysis using SPICE and MATLAB.

2.8 PRACTICAL APPLICATION: A CASE STUDY

Jump Starting a Car

Voltage and current in an electric network are easily measured. They obey Kirchoff's laws, KCL and KVL, and facilitate the monitoring of energy flow. For these reasons, voltage and current are used by engineers in order to describe the state of an electric network.

When a car battery is weak, say 11 V in a 12-V system, in order to jump-start that car, we bring in another car with its engine running and its alternator charging its battery. Let the healthy and strong battery have a voltage of 13 V. According to the recommended practice, one should first connect the positive terminals with the red jumper cable, as shown in Figure 2.8.1, and then complete the circuit between the negative terminals with the aid of the black jumper cable. Note that the negative terminal of any car battery is always connected to its auto chasis.

Applying KVL in Figure 2.8.1, we have

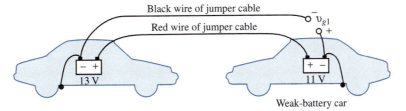

Figure 2.8.1 Jumper cable connections for jump starting a car with a weak battery.

$$v_{g1} - 13 + 11 = 0 \qquad \text{or} \qquad v_{g1} = 2 \text{ V}$$

where v_{g1} is the voltage across the airgap, or the voltage existing between the black jumper cable and the negative terminal of the weak battery.

Now suppose one makes, by mistake, incorrect connections, as shown in Figure 2.8.2. Note that the red jumper cable is connected between the positive terminal of the strong battery and the negative terminal of the weak battery. Application of the KVL now fields

$$v_{g2} - 13 - 11 = 0 \qquad \text{or} \qquad v_{g2} = 24 \text{ V}$$

where v_{g2} is the gap voltage with incorrect connections. With such a large voltage difference, when one tries to complete the black jumper cable connection, it presents a danger to both batteries and to the person making the connections.

Energy to Start an Engine

A simplified circuit model for an automotive starter circuit is shown in Figure 2.8.3. Let the car battery voltage be 12.5 V and let the automobile starter motor draw 60 A when turning over the engine. If the engine starts after 10 seconds, we can easily calculate the power to the starter motor, which is the same as the power out of the battery,

$$P = VI = 12.5 \times 60 = 750 \text{ W}$$

The energy required to start the engine can be computed as

$$W = 750 \times 10 = 7500 \text{ J}$$

Thus, simple circuit models can be used to simulate various physical phenomena of practical interest. They can then be analyzed by circuit-analysis techniques to yield meaningful solutions rather easily.

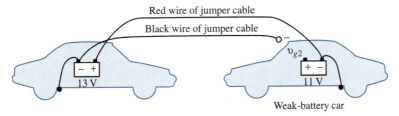

Figure 2.8.2 Incorrect connections for jump starting a car with a weak battery.

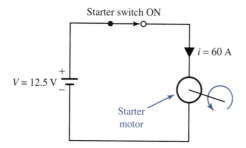

Starter switch ON

$i = 60$ A

$V = 12.5$ V

Starter motor

Figure 2.8.3 Simplified circuit model for the automotive starter circuit.

PROBLEMS

2.1.1 (a) Determine the Thévenin and Norton equivalent circuits as viewed by the load resistance R in the network of Figure P2.1.1.

(b) Find the value of R if the power dissipated by R is to be a maximum.

(c) Obtain the value of the power in part (b).

2.1.2 Reduce the circuit of Figure P2.1.2 to a Thévenin and a Norton equivalent circuit.

*2.1.3 Find the Thévenin and Norton equivalent circuits for the configuration of Figure P2.1.3 as viewed from terminals a–b.

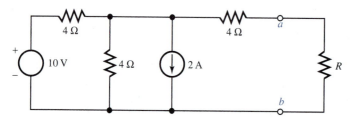

$4\,\Omega$

10 V

$4\,\Omega$

2 A

$4\,\Omega$

a

R

b

Figure P2.1.1

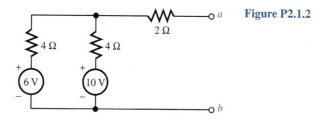

$4\,\Omega$

$4\,\Omega$

$2\,\Omega$

a

6 V

10 V

b

Figure P2.1.2

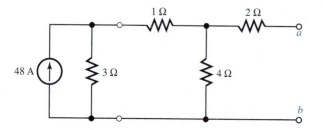

$1\,\Omega$

$2\,\Omega$

a

48 A

$3\,\Omega$

$4\,\Omega$

b

Figure P2.1.3

2.1.4 Obtain the Thévenin and Norton equivalent circuits for the portion of the circuit to the left of terminals a–b in Figure P2.1.4, and find the current in the 200-Ω resistance.

2.1.5 Determine the voltage across the 20-Ω load resistance in the circuit of Figure P2.1.5 by the use of the Thévenin equivalent circuit.

2.1.6 Find the current in the 5-Ω resistance of the circuit of Figure P2.1.6 by employing the Norton equivalent circuit.

*2.1.7 Obtain the voltage across the 3-kΩ resistor of the circuit (transistor amplifier stage) given in Figure

P2.1.7 by the use of the Thévenin equivalent circuit.

2.1.8 Reduce the circuit of Figure P2.1.8 to a Thévenin and a Norton equivalent circuit with respect to terminals a–b.

2.1.9 (a) Redraw the circuit in Figure P2.1.9 by replacing the portion to the left of terminals a–b with its Thévenin equivalent.

(b) Redraw the circuit of Figure P2.1.9 by replacing the portion to the right of terminals a'–b' with its Thévenin equivalent.

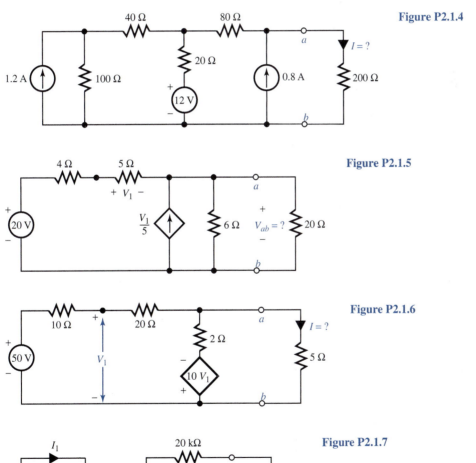

Figure P2.1.4

Figure P2.1.5

Figure P2.1.6

Figure P2.1.7

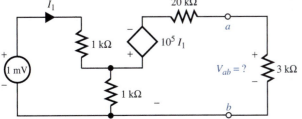

2.1.10 (a) Consider the Wheatstone bridge circuit given in Figure P2.1.10(a) and find the Thévenin equivalent with respect to terminals *a–b*.

(b) Suppose a source with resistance is connected across *a–b*, as shown in Figure P2.1.10(b). Then find the current I_{ab}.

2.2.1 In the circuit given in Figure P2.2.1, determine the current *I* through the 2-Ω resistor by (a) the nodal-voltage method, and (b) mesh-current analysis.

2.2.2 Consider the circuit of Figure P2.2.2 and rearrange it such that only one loop equation is required to solve for the current *I*.

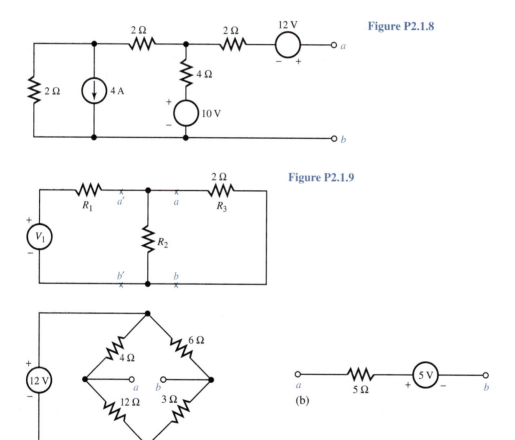

Figure P2.1.8

Figure P2.1.9

(a)

Figure P2.1.10

(b)

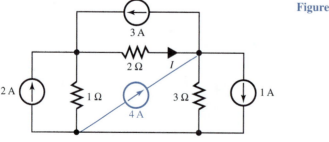

Figure P2.2.1

2.2.3 Use the node-voltage method to find the current I through the 5-Ω resistor of the circuit of Figure P2.2.3.

2.2.4 Use the node-voltage method to determine the voltage across the 12-Ω resistor of the circuit given in Figure P2.2.4. Verify by mesh analysis.

2.2.5 Determine the current I through the 10-Ω resistor of the circuit of Figure P2.2.5 by employing the node-voltage method. Check by mesh analysis.

*2.2.6 (a) Find the voltage across the 8-A current source in the circuit of Figure P2.2.6 with the use of nodal analysis.

(b) Determine the current in the 0.5-Ω resistor of the circuit by mesh analysis.

2.2.7 By using the mesh-current method, determine the voltage across the 1-A current source of the circuit of Figure P2.2.7, and verify by nodal analysis.

2.2.8 Find the current I_1 through the 20-Ω resistor of the circuit of Figure P2.2.8 by both mesh and nodal analyses.

2.2.9 Determine the voltage V in the circuit of Figure P2.2.9 by nodal analysis and verify by mesh analysis.

2.2.10 Find the current I in the circuit of Figure P2.2.10 by mesh analysis and verify by nodal analysis.

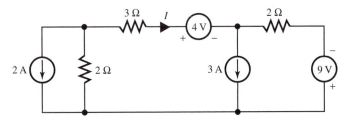

Figure P2.2.2

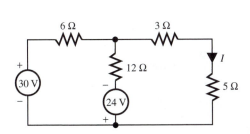

Figure P2.2.3

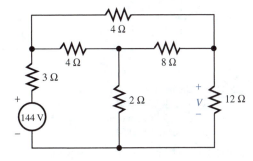

Figure P2.2.4

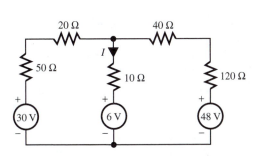

Figure P2.2.5

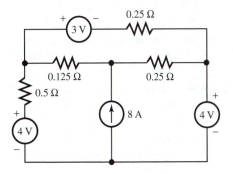

Figure P2.2.6

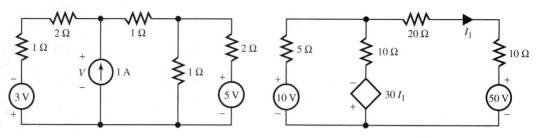

Figure P2.2.7

Figure P2.2.8

Figure P2.2.9

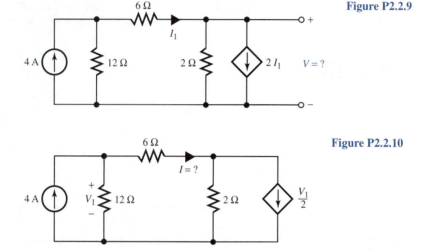

Figure P2.2.10

2.2.11 For the network of Figure P2.2.11, find the nodal voltages V_1, V_2, and V_3 by means of nodal analysis, using the concept of a supernode. Verify by mesh-current analysis.

*__2.2.12__ Use nodal analysis and the supernode concept to find V_2 in the circuit shown in Figure P2.2.12. Verify by mesh-current analysis, by using source transformation and by using the concept of a supermesh.

2.2.13 Use mesh-current analysis and the supermesh concept to find V_0 in the circuit of Figure P2.2.13. Verify by nodal analysis.

2.2.14 For the network shown in Figure P2.2.14, find V_x across the 3-Ω resistor by using mesh current analysis. Verify by means of nodal analysis.

2.3.1 Consider the circuit of Problem 2.2.1 and find the current I through the 2-Ω resistor by the principle of superposition.

2.3.2 Solve Problem 2.2.3 by the application of superposition.

2.3.3 Solve Problem 2.2.5 by the application of superposition.

2.3.4 Solve Problem 2.2.6 by the application of superposition.

2.3.5 Solve Problem 2.2.7 by the application of superposition.

*__2.3.6__ Solve Problem 2.2.8 by the application of superposition.

2.4.1 Show that Equations (2.4.1) and (2.4.2) are true.

*__2.4.2__ Determine R_S in the circuit of Figure P2.4.2 such that it is matched at terminals a–b, and find the power delivered by the voltage source.

2.4.3 Find the power delivered by the source in the circuit given in Figure P2.4.3. Use network reduction by wye–delta transformation.

2.5.1 Develop and execute a PSpice program to analyze the circuit shown in Figure P2.5.1 to evaluate the node voltages and the current through each element.

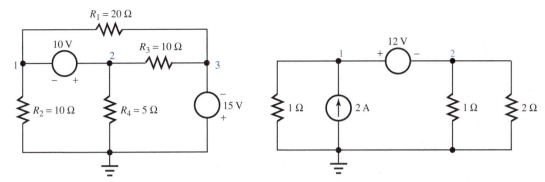

Figure P2.2.11

Figure P2.2.12

Figure P2.2.13

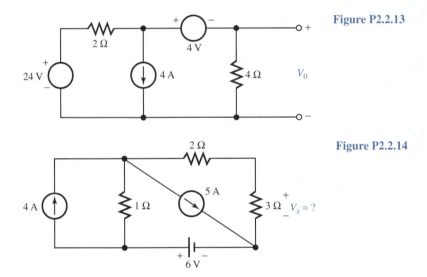

Figure P2.2.14

2.5.2 Develop and execute a PSpice program to find the node voltages and the current through each element of the circuit given in Figure P2.5.2.

*2.5.3 For the circuit shown in Figure P2.5.3, develop and execute a PSpice program to obtain the node voltages and the current through each element.

2.5.4 For the circuit given in Figure P2.5.4, develop and execute a PSpice program to solve for the node voltages.

2.5.5 Write and execute a PSpice program to analyze the resistor bridge circuit shown in Figure P2.5.5 to solve for the node voltages and the voltage-source current. Then find the voltage across the 50-Ω resistor and the total power supplied by the source.

2.6.1 The current through a 2.5-mH indicator is a damped sine given by $i(t) = 10e^{-500t} \sin 2000t$.

With the aid of MATLAB, plot the waveforms of the inductor current $i(t)$, with voltage $v(t) = L \, di/dt$, power $p(t) = vi$, and energy $w(t) = \int_0^t p(\tau) \, d\tau$. Starting at $t = 0$, the plots should include at least one cycle and at least 20 points per cycle.

*2.6.2 An interface circuit consisting of R_1 and R_2 is designed between the source and the load, as illustrated in Figure P2.6.2 such that the load sees a Thévenin resistance of 50 Ω between terminals C and D, while simultaneously the source sees a load resistance of 300 Ω between A and B. Using MATLAB, find R_1, and R_2.

Hint: solve the two nonlinear equations given by

$$\frac{(R_1 + 300)R_2}{R_1 + 300 + R_2} = 50; \quad R_1 + \frac{50R_2}{R_2 + 50} = 300$$

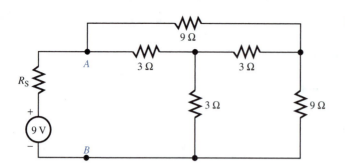

Figure P2.4.2

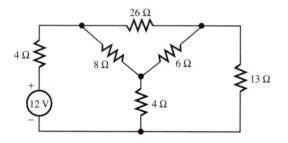

Figure P2.4.3

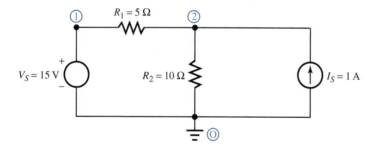

Figure P2.5.1

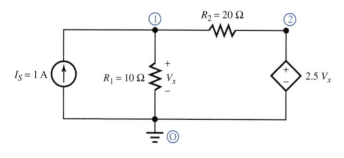

Figure P2.5.2

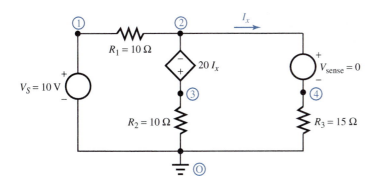

Figure P2.5.3

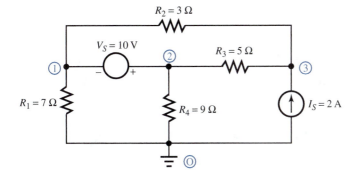

Figure P2.5.4

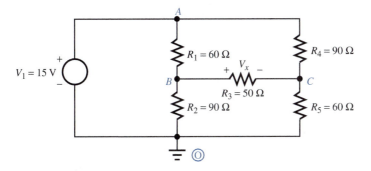

Figure P2.5.5

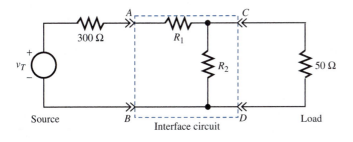

Figure P2.6.2

3 Time-Dependent Circuit Analysis

The response of networks to time-varying sources is considered in this chapter. The special case of sinusoidal signals is of particular importance, because the low-frequency signals (i.e., currents and voltages) that appear in electric power systems as well as the high-frequency signals in communications are usually sinusoidal. The powerful technique known as *phasor analysis,* which involves the use of complex numbers, is one of the electrical engineer's most important tools developed to solve steady-state ac circuit problems. Since a periodic signal can be expressed as a sum of sinusoids through a *Fourier series*, and superposition applies to linear systems, phasor analysis will be used to determine the steady-state response of any linear system excited by a periodic signal. Thus the superposition principle allows the phasor technique to be extended to determine the system response of a linear system. The *sinusoidal steady-state response of linear circuits* is presented in Section 3.1. The response when the excitation is suddenly applied or suddenly changed is examined next in Section 3.2. The total response of a system containing energy-storage elements (capacitors and inductors) is analyzed in terms of natural and forced responses (or transient and steady-state responses). The *Laplace transformation,* which provides a systematic algebraic approach for determining both the forced and the natural components of a network response, is then taken up in Section 3.3. The concept of a *transfer function* is also introduced along with its application to solve circuit problems. Then, in Section 3.4 the network response to sinusoidal signals of variable frequency is investigated. Also, *two-port networks* and block diagrams, in terms of their input–output characteristics, are dealt with in this chapter. Finally, computer-aided circuit simulations using PSpice and PROBE, as well as MATLAB are illustrated in Sections 3.5 and 3.6. The chapter ends with a case study of practical application.

3.1 SINUSOIDAL STEADY-STATE PHASOR ANALYSIS

The kind of response of a physical system in an applied excitation depends in general on the type of excitation, the elements in the system and their interconnection, and also on the past history of the system. The total response generally consists of a *forced response* determined by the particular excitation and its effects on the system elements, and a *natural response* dictated by the system elements and their interaction. The natural response caused by the energy storage elements in circuits with nonzero resistance is always *transient*; but the forced response caused by the sources can have a *transient* and a *steady-state* component. The *boundary conditions* (usually *initial conditions*), representing the effect of past history in the total response, decide the amplitude of the natural response and reflect the degree of mismatch between the original state and the steady-state response. However, when excitations are periodic or when they are applied for lengthy durations, as in the case of many applications, the solution for the forced response is all that is needed, whereas that for the natural response becomes unnecessary. When a linear circuit is driven by a sinusoidal voltage or current source, all steady-state voltages and currents in the circuit are sinusoids with the same frequency as that of the source. This condition is known as the *sinusoidal steady state*. Sinusoidal excitation refers to excitation whose waveform is sinusoidal (or cosinusoidal). Circuits excited by constant currents or voltages are called dc circuits, whereas those excited by sinusoidal currents or voltages are known as ac circuits.

Sinusoids can be expressed in terms of exponential functions with the use of Euler's identity,

$$e^{j\theta} = \cos\theta + j\sin\theta \tag{3.1.1}$$

$$e^{-j\theta} = \cos\theta - j\sin\theta \tag{3.1.2}$$

$$\cos\theta = \frac{e^{j\theta} + e^{-j\theta}}{2} \tag{3.1.3}$$

$$\sin\theta = \frac{e^{j\theta} - e^{-j\theta}}{2j} \tag{3.1.4}$$

where j represents the imaginary number $\sqrt{-1}$. The reader is expected to be conversant with complex numbers.

If we are able to find the response to exponential excitations, $e^{j\theta}$ or $e^{-j\theta}$, we can use the principle of superposition in order to evaluate the sinusoidal steady-state response. With this in mind let us now study the response to exponential excitations.

Responses to Exponential Excitations

Let us consider Ae^{st} as a typical exponential excitation in which A is a constant and s is a *complex-frequency variable* with a dimension of 1/second such that the exponent st becomes dimensionless.

The variable s can assume real, imaginary, or complex values. The time-invariant dc source is represented by setting $s = 0$. The use of $s = j\omega$ would imply sinusoidal excitation.

Note that Ae^{st} is the only function for which a linear combination of

$$K_1 Ae^{st} + K_2 \frac{d}{dt}\left(Ae^{st}\right) + K_3 \int Ae^{st}$$

in which K_1, K_2, and K_3 are constants has the same shape or waveform as the original signal. Therefore, if the excitation to a linear system is Ae^{st}, then the response will have the same waveform.

Recall the volt–ampere relationships (for ideal elements) with time-varying excitation.
For the resistor R:

$$v_R = Ri_R \tag{3.1.5}$$

$$i_R = Gv_R \tag{3.1.6}$$

For the inductor L:

$$v_L = L\frac{di_L}{dt} \tag{3.1.7}$$

$$i_L = \frac{1}{L}\int v_L\,dt \tag{3.1.8}$$

For the capacitor C:

$$v_C = \frac{1}{C}\int i_C\,dt \tag{3.1.9}$$

$$i_C = C\frac{dv_C}{dt} \tag{3.1.10}$$

With exponential excitation in which $\mathbf{v}(t) = Ve^{st}$ and $\mathbf{i}(t) = Ie^{st}$, it can be seen that the following holds good because exponential excitations produce exponential responses with the same exponents. (Notationwise, note that $v(t)$ and $i(t)$ represent the *real-valued* signals, whereas $\mathbf{v}(t)$ and $\mathbf{i}(t)$ represent *complex-valued* signals.)
For R:

$$V_R = RI_R \tag{3.1.11}$$

$$I_R = GV_R \tag{3.1.12}$$

For L:

$$V_L = (sL)I_L \tag{3.1.13}$$

$$I_L = (1/sL)V_L \tag{3.1.14}$$

For C:

$$V_C = (1/sC)I_C \tag{3.1.15}$$

$$I_C = (sC)V_C \tag{3.1.16}$$

The preceding equations resemble the Ohm's law relation. The quantities R, sL, and $1/sC$ have the dimension of ohms, whereas G, $1/sL$, and sC have the dimension of siemens, or 1/ohm. The ratio of voltage to current in the frequency domain at a pair of terminals is known as the *impedance,* designated by $Z(s)$, whereas that of current to voltage is called the *admittance,* designated by

$Y(s)$. Note that both the impedance and the admittance are in general functions of the variable s, and they are reciprocal of each other. Such expressions as Equations (3.1.11) through (3.1.16) relate the amplitudes of the exponential voltages and currents, and are the *frequency-domain representations* of the elements. Networks drawn using impedance or admittance symbols are known as *transformed networks,* which play a significant role in finding the network response, as shown in the following examples.

EXAMPLE 3.1.1

Consider an *RLC* series circuit excited by $\mathbf{v}(t) = Ve^{st}$ in the time domain. Assume no initial capacitor voltage or inductive current at $t = 0$. Draw the transformed network in the s-domain and solve for the frequency-domain forced response of the resultant current. Then find the t-domain forced response $\mathbf{i}(t)$.

Solution

The forced response is produced by the particular excitation applied. The KVL equation for the circuit of Figure E3.1.1 (a) is

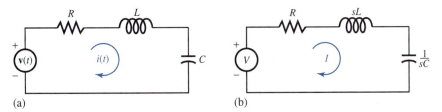

(a)   (b)

Figure E3.1.1 *RLC* series circuit with $\mathbf{v}(t) = Ve^{st}$. **(a)** Time domain. **(b)** Transformed network in s-domain.

$$v(t) = Ri(t) + L\frac{di(t)}{dt} + \frac{1}{C}\int_{-\infty}^{t} i(\tau)\, d\tau$$

The corresponding transformed network in the s-domain is shown in Figure E3.3.1(b), for which the following KVL relation holds. (Note that the initial capacitor voltage at $t = 0$ is assumed to be zero.)

$$V = RI + LsI + \frac{1}{Cs}I$$

Solving for I, one gets

$$I = \frac{V}{R + Ls + (1/Cs)} = \frac{V}{Z(s)}$$

where $Z(s)$ can be seen to be the addition of each series impedance of the elements. The time function corresponding to the frequency-domain response is given by

$$\mathbf{i}(t) = Ie^{st} = \frac{V}{R + Ls + (1/Cs)}e^{st}$$

which is also an exponential with the same exponent contained in $\mathbf{v}(t)$.

EXAMPLE 3.1.2

Consider a *GLC* parallel circuit excited by $\mathbf{i}(t) = Ie^{st}$ in the time domain. Assume no initial inductive current or capacitive voltage at $t = 0$. Draw the transformed network in the frequency s-domain and solve for the frequency-domain forced response of the resultant voltage. Then find the t-domain forced response $\mathbf{v}(t)$.

Solution

The forced response is produced by the particular excitation applied. The KCL equation for the circuit of Figure E3.1.2(a) is

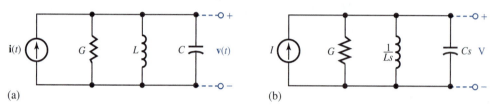

(a) (b)

Figure E3.1.2 *GLC* parallel circuit with $\mathbf{i}(t) = Ie^{st}$. **(a)** Time domain. **(b)** Transformed network in s-domain.

$$i(t) = Gv(t) + \frac{1}{L}\int_{-\infty}^{t} v(\tau)\,d\tau + C\frac{dv(t)}{dt}$$

The corresponding transformed network in the s-domain is shown in Figure E3.1.2(b), for which the following KCL relation holds. (Note that the initial inductive current at $t = 0$ is assumed to be zero.)

$$I = GV + \frac{1}{Ls}V + CsV$$

Solving for V, one gets

$$V = \frac{I}{G + (1/Ls) + Cs} = \frac{I}{Y(s)}$$

where $Y(s)$ can be seen to be the addition of each parallel admittance of the elements. The time function corresponding to the frequency-domain response is given by

$$\mathbf{v}(t) = Ve^{st} = \frac{I}{G + (1/Ls) + Cs}e^{st}$$

which is also an exponential with the same exponent contained in $\mathbf{i}(t)$.

Note that impedances in series are combined like resistances in series, whereas admittances in parallel are combined like conductances in parallel. Series–parallel impedance/admittance combinations can be handled in the same way as series–parallel resistor/conductance combinations. Notice that in the dc case when $s = 0$, the impedance of the capacitor $1/Cs$ tends to be infinite, signifying an open circuit, whereas the impedance of the inductor Ls becomes zero, signifying a short circuit.

Forced Response to Sinusoidal Excitation

Consider an excitation of the form

$$v(t) = V_m\cos(\omega t + \phi) \tag{3.1.17}$$

where V_m is the peak amplitude and ϕ is the phase angle. This may be expressed in terms of exponential functions as

$$v(t) = \frac{V_m}{2} \left[e^{j(\omega t + \phi)} + e^{-j(\omega t + \phi)} \right] = \bar{V}_a e^{j\omega t} + \bar{V}_b e^{-j\omega t} \qquad (3.1.18)$$

where $\bar{V}_a = (V_m/2) e^{j\phi}$ and $\bar{V}_b = (V_m/2) e^{-j\phi}$. Note that $\bar{V}_a$ and $\bar{V}_b$ are complex numbers.

Even though Equation (3.1.17) is a cosine function that is considered here, recall that any sine, cosine, or combination of sine and cosine waves of the same frequency can be written as a cosine wave with a phase angle. Some useful trigonometric identities are as follows:

$$\sin(\omega t + \phi) = \cos\left(\omega t + \phi - \frac{\pi}{2}\right) \qquad (3.1.19)$$

$$A\cos\omega t + B\sin\omega t = \sqrt{A^2 + B^2} \cos(\omega t - \phi) \qquad (3.1.20)$$

where $\phi = \tan^{-1}(B/A)$.

By expressing the sinusoidal excitation as the sum of two exponentials, as in Equation (3.1.18), the method developed to find the response to exponential excitations can easily be extended with the principle of superposition to obtain the forced response to sinusoidal excitation, as illustrated in the following example.

EXAMPLE 3.1.3

Consider an *RLC* series circuit excited by $v(t) = V_m \cos \omega t$ in the time domain. By using superposition, solve for the time-domain forced response of the resultant current through the frequency-domain approach.

Solution

Figures E3.1.3(a) and E3.1.3(b) are equivalent in the time domain. Figure E3.1.3(c) shows the transformed networks with the use of superposition. It follows then

$$\bar{I}_1 = \frac{V_m/2}{R + j\omega L + (1/j\omega C)} = \frac{V_m/2}{R + j[\omega L - (1/\omega C)]} = I_1 e^{j\theta_1}$$

$$\bar{I}_2 = \frac{V_m/2}{R - j\omega L + (1/j\omega C)} = \frac{V_m/2}{R - j[\omega L - (1/\omega C)]} = I_2 e^{j\theta_2}$$

where

$$I_1 = I_2 = \frac{V_m/2}{R^2 + [\omega L - (1/\omega C)]^2}; \qquad \theta_1 = -\theta_2 = -\tan^{-1}\frac{\omega L - (1/\omega C)}{R}$$

The student is encouraged to use a knowledge of complex numbers and check these results. The corresponding time functions are given by

$$i_1(t) = I_1 e^{j\theta_1} e^{j\omega t} \qquad \text{and} \qquad i_2(t) = I_2 e^{-j\theta_1} e^{-j\omega t}$$

By superposition, the total response is

$$i(t) = \frac{V_m}{\left[R^2 + \{\omega L - (1/\omega C)\}^2\right]^{1/2}} \cos(\omega t + \theta_1) = I_m \cos(\omega t + \theta_1)$$

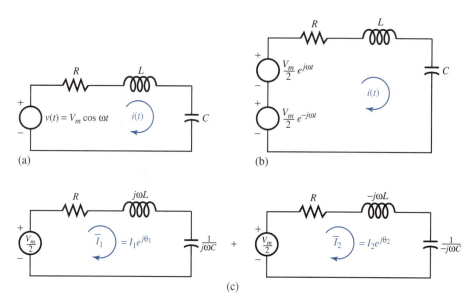

Figure E3.1.3 *RLC* series circuit with sinusoidal excitation. **(a)** Time-domain circuit with a sinusoidal excitation. **(b)** Time-domain circuit with exponential excitations. **(c)** Transformed networks (one with $s = j\omega$ and the other with $s = -j\omega$).

In view of the redundancy that is found in the information contained in $\bar{I}_1$ and $\bar{I}_2$ as seen from Example 3.1.3, only one component needs to be considered for the purpose of finding the sinusoidal steady-state response. Notice that an exponential excitation of the form $\mathbf{v}(t) = (V_m e^{j\phi})e^{j\omega t} = \bar{V}_m e^{j\omega t}$ produces an exponential response $\mathbf{i}(t) = (I_m e^{j\theta}) e^{j\omega t} = \bar{I}_m e^{j\omega t}$, whereas a sinusoidal excitation of the form $v(t) = V_m \cos(\omega t + \phi)$ produces a sinusoidal response $i(t) = I_m \cos(\omega t + \theta)$. The complex terms $\bar{V}_m = V_m e^{j\phi}$ and $\bar{I}_m = I_m e^{j\theta}$ are generally known as *phasors*, with the additional understanding that a function such as $\mathbf{v}(t)$ or $\mathbf{i}(t)$ can be interpreted graphically in terms of a rotating phasor in the counterclockwise direction (considered positive for positive ω and positive t). When the frequency of rotation becomes a constant equal to ω rad/s, the projection of a rotating phasor on the real (horizontal) axis varies as $\cos \omega t$, whereas its projection on the imaginary (vertical) axis varies as $\sin \omega t$.

The use of a single exponential function with $s = j\omega$ to imply sinusoidal excitation (and response) leads to the following volt–ampere relations.

For *R*:

$$\bar{V}_R = R\bar{I}_R \tag{3.1.21}$$

$$\bar{I}_R = G\bar{V}_R \tag{3.1.22}$$

For *L*:

$$\bar{V}_L = j\omega L\bar{I}_L = jX_L\bar{I}_L \tag{3.1.23}$$

$$\bar{I}_L = (1/j\omega L)\,\bar{V}_L = jB_L\bar{V}_L \tag{3.1.24}$$

For *C*:

$$\bar{V}_C = (1/j\omega C)\,\bar{I}_C = jX_C\bar{I}_C \tag{3.1.25}$$

$$\bar{I}_C = j\omega C \bar{V}_C = jB_C \bar{V}_C \qquad (3.1.26)$$

where $X_L = \omega L$ is the *inductive reactance*, $X_C = -1/\omega C$ is the *capacitive reactance*, $B_L = -1/\omega L$ is the *inductive susceptance*, and $B_C = \omega C$ is the *capacitive susceptance*. (Notice that the fact $1/j = -j$ has been used.)

The general impedance and admittance functions with $s = j\omega$ for sinusoidal excitation are given by

$$\bar{Z}(j\omega) = R + jX = \frac{1}{\bar{Y}(j\omega)} \qquad (3.1.27)$$

$$\bar{Y}(j\omega) = G + jB = \frac{1}{\bar{Z}(j\omega)} \qquad (3.1.28)$$

where the real part is either the resistance R or the conductance G, and the imaginary part is either the *reactance X* or the *susceptance B*.

A positive value of X or a negative value of B indicates inductive behavior, whereas capacitive behavior is indicated by a negative value of X or a positive value of B. Further, the following KVL and KCL equations hold:

$$\bar{V} = \bar{Z}\bar{I} = (R + jX)\bar{I} \qquad (3.1.29)$$
$$\bar{I} = \bar{Y}\bar{V} = (G + jB)\bar{V} \qquad (3.1.30)$$

Phasor Method

For sinusoidal excitations of the *same frequency*, the forced or steady-state responses are better found by the technique known as the *phasor method*. Time functions are transformed to the phasor representations of the sinusoids. For example, current and voltage in the time domain are given by the forms

$$i = \sqrt{2}\, I_{rms} \cos(\omega t + \alpha) = \text{Re}\left[\sqrt{2}\, I_{rms} e^{j\alpha} e^{j\omega t}\right] \qquad (3.1.31)$$

$$v = \sqrt{2}\, V_{rms} \cos(\omega t + \beta) = \text{Re}\left[\sqrt{2}\, V_{rms} e^{j\beta} e^{j\omega t}\right] \qquad (3.1.32)$$

and where Re stands for the "real part of"; their corresponding phasors in the frequency domain are defined by

$$\bar{I} = I_{rms} e^{j\alpha} = I_{rms} \angle \alpha \qquad (3.1.33)$$
$$\bar{V} = V_{rms} e^{j\beta} = V_{rms} \angle \beta \qquad (3.1.34)$$

Notice that the magnitudes of the phasors are chosen for convenience to be the rms values of the original functions (rather than the peak amplitudes), and angles are given by the argument of the cosine function at $t = 0$. The student should observe that phasors are referenced here to *cosine* functions. Therefore, the conversion of sine functions into equivalent cosine functions makes it more convenient for expressing phasor representations of sine functions. The phasor volt-ampere equations for R, L, and C are given by Equations (3.1.21) through (3.1.26), whereas the *phasor operators* $\bar{Z}$ and $\bar{Y}$ are given by Equations (3.1.27) and (3.1.28). The KVL and KCL equations (3.1.29) and (3.1.30) hold in phasor form.

EXAMPLE 3.1.4

Let $\omega = 2\pi \times 60$ rad/s corresponding to a frequency of 60 Hz.

(a) Consider $v(t) = 100\sqrt{2}\cos(\omega t + 30°)$ V and $i(t) = 10\sqrt{2}\sin(\omega t + 30°)$ A. Find the corresponding phasors $\bar{V}$ and $\bar{I}$ by choosing the rms values for the phasor magnitudes.

(b) Consider the phasors $\bar{V}$ and $\bar{I}$ as obtained in part (a), and show how to obtain the time-domain functions.

Solution

(a)
$$v(t) = 100\sqrt{2}\ \cos(\omega t + 30°) \text{ V} = \text{Re}\left[100\sqrt{2}\ e^{j\ 30°}e^{j\omega t}\right]$$

Suppressing the explicit time variation and choosing the rms value for the phasor magnitude, the phasor representation in the frequency domain then becomes

$$\bar{V} = 100\angle 30° = 100e^{j\ 30°} = 100e^{j\pi/6} = 100\,(\cos 30° + j\sin 30°) \text{ V}$$

Similarly, for the current:

$$\bar{I} = 10\angle -60° = 10e^{-j\ 60°} = 10e^{-j\pi/3} = 10\,(\cos 60° - j\sin 60°) \text{ A}$$

since $\sin(\omega t + 30°) = \cos(\omega t + 30° - 90°) = \cos(\omega t - 60°)$.

(b) Now that $\bar{V} = 100\angle 30°$, the corresponding time-domain function $v(t)$ is obtained as $v(t) = \text{Re}\ [\sqrt{2}\ \bar{V}e^{j\omega t}] = \text{Re}\ [\sqrt{2}\ 100e^{j\ 30°}e^{j\omega t}]$, or $v(t) = \text{Re}\ [100\sqrt{2}\ e^{j\ (\omega t + 30°)}] = 100\sqrt{2}\cos(\omega t + 30°)$ V. Note that the multiplicative factor of $\sqrt{2}$ is used to obtain the peak amplitude, since for sinusoids the peak value is $\sqrt{2}$ times the rms value.

Similarly, $i(t) = \text{Re}\ [10\sqrt{2}\ e^{j\ (\omega t - 60°)}] = 10\sqrt{2}\cos(\omega t - 60°) = 10\sqrt{2}\sin (\omega t + 30°)$ A, where $\omega = 2\pi f = 2\pi \times 60$ rad/s in this case.

For the three linear time-invariant passive elements R (pure resistance), L (pure inductance), and C (pure capacitance), the relationships between voltage and current in the time domain as well as in the frequency domain are shown in Figure 3.1.1.

Phasors, being complex numbers, can be represented in the complex plane in the conventional polar form as an arrow, having a length corresponding to the magnitude of the phasor, and an angle (with respect to the positive real axis) that is the phase of the phasor. In a *phasor diagram*, the various phasor quantities corresponding to a given network may be combined in such a way that one or both of Kirchhoff's laws are satisfied. In general the phasor method of analyzing circuits is credited to Charles Proteus Steinmetz (1865–1923), a well-known electrical engineer with the General Electric Company in the early part of the 20th century.

The phasor diagram may be drawn in a number of ways, such as in a polar (or ray) form, with all phasors originating at the origin, or in a polygonal form, with one phasor located at the end of another, or in a combination of these, depending on the convenience and the point to be made. Such diagrams provide geometrical insight into the voltage and current relationships in a network. They are particularly helpful in visualizing steady-state phenomena in the analysis of networks with sinusoidal signals.

Time Domain	Phasors in Frequency Domain

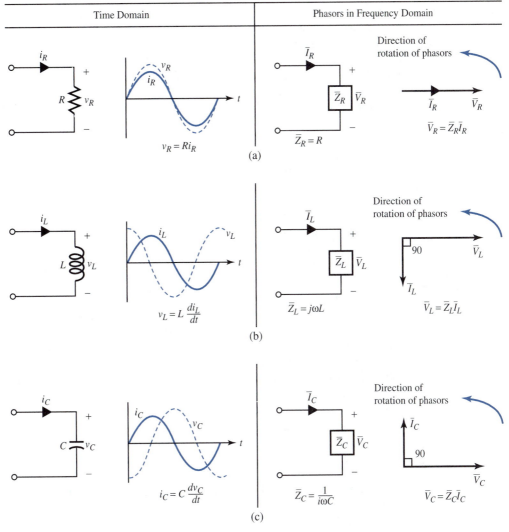

Figure 3.1.1 Voltage and current relationships in time domain and frequency domain for elements R, L, and C. (a) Current is in phase with voltage in a purely resistive circuit (unity power factor). (b) Current lags voltage by 90° with a pure inductor (zero power factor lagging). (c) Current leads voltage by 90° with a pure capacitor (zero power factor leading).

In constructing a phasor diagram, each sinusoidal voltage and current is represented by a phasor of length equal to the rms value of the sinusoid, and with an angular displacement from the positive real axis, which is the angle of the equivalent cosine function at $t = 0$. This use of the cosine is arbitrary, and so also the use of the rms value. Formulation in terms of the sine function could just as well have been chosen, and the amplitude rather than the rms value could have been chosen for the magnitude. While phasor diagrams can be drawn to scale, they are usually sketched as a visual check of the algebraic solution of a problem, especially since the KVL and KCL equations can be shown as graphical addition.

The *reference* of a phasor diagram is the line $\theta = 0$. It is not really necessary that any voltage or current phasor coincide with the reference, even though it is often more convenient for the analysis when a given voltage or current phasor is taken as the reference. Since a phasor diagram is a frozen picture at one instant of time, giving the relative locations of various phasors involved, and the whole diagram of phasors is assumed to be rotating counterclockwise at a constant frequency, different phasors may be made to be the reference simply by rotating the entire phasor diagram in either the clockwise or the counterclockwise direction.

From the viewpoints of ease and convenience, it would be a matter of common sense to choose the current phasor as the reference in the case of series circuits, in which all the elements are connected in series, and the common quantity for all the elements involved is the current. Similarly for the case of parallel circuits, in which all the elements are connected in parallel and the common quantity for all the elements involved is the voltage, a good choice for the reference is the voltage phasor. As for the series–parallel circuits, no firm rule applies to all situations.

Referring to Figure 3.1.1(a), in a purely resistive circuit, notice that the current is in phase with the voltage. Observe the waveforms of i_R and v_R in the time domain and their relative locations in the phasor domain, notice the phasors $\bar{V}_R$ chosen here as a reference, and $\bar{I}_R$, which is in phase with $\bar{V}_R$. The phase angle between voltage and current is zero degrees, and the cosine of that angle, namely, unity or 1 in this case, is known as the *power factor*. Thus a purely resistive circuit is said to have unity power factor.

In the case of a pure inductor, as shown in Figure 3.1.1(b), the current *lags* (behind) the voltage by 90°, or one can also say that the voltage *leads* the current by 90°. Observe the current and voltage waveforms in the time domain along with their relative positions, as well as the relative phasor locations of $\bar{V}_L$ and $\bar{I}_L$ in the phasor domain, with $\bar{V}_L$ chosen here as a reference. The phase angle between voltage and current is 90°, and the cosine of that angle being zero, the power factor for the case of a pure inductor is said to be zero power factor lagging, since the current lags the voltage by 90°.

For the case of a pure capacitor, as illustrated in Figure 3.1.1(c), the current leads the voltage by 90°, or one can also say that the voltage lags the curent by 90°. Notice the current and voltage waveforms in the time domain along with their relative positions, as well as the relative phasor locations of $\bar{V}_C$ and $\bar{I}_C$ in the phasor domain, with $\bar{V}_C$ chosen here as a reference. The phase angle between voltage and current is 90°, and the cosine of that angle being zero, the power factor for the case of a pure capacitor is said to be zero power factor leading, since the current leads the voltage by 90°.

The circuit analysis techniques presented in Chapter 2 (where only resistive networks are considered for the sake of simplicity) apply to ac circuits using the phasor method. However, the constant voltages and currents in dc circuits are replaced by phasor voltages and currents in ac circuits. Similarly, resistances and conductances are replaced by the complex quantities for impedance and admittance. Nodal and mesh analyses, being well-organized and systematic methods, are applied to ac circuits along with the concepts of equivalent circuits, superposition, and wye–delta transformation.

Power and Power Factor in ac Circuits

Power is the rate of change of energy with respect to time. The unit of power is a watt (W), which is a joule per second (J/s). The use of rms or effective values of voltage and current allows the average power to be found from phasor quantities. Let us consider a circuit consisting of an impedance $Z\angle\phi = R + jX$ excited by an applied voltage of $v(t) = \sqrt{2}\,V_{rms}\cos(\omega t + \phi)$, producing a current of $i(t) = \sqrt{2}\,I_{rms}\cos\omega t$. The corresponding voltage and current phasors are then given by $V\angle\phi$ and $I\angle0°$, which satisfy the Ohm's-law relation $\bar{V}/\bar{I} = V\angle\phi/I\angle0° = Z\angle\phi$.

The instantaneous power $p(t)$ supplied to the network by the source is

$$p(t) = v(t) \cdot i(t) = \sqrt{2}\, V_{rms}\cos(\omega t + \phi) \cdot \sqrt{2}\, I_{rms} \cos \omega t \qquad (3.1.35)$$

which can be rearranged as follows by using trigonometric relations:

$$p(t) = V_{rms}I_{rms} \cos \phi\, (1 + \cos 2\omega t) - V_{rms}I_{rms} \sin \phi \sin 2\omega t$$

$$= V_{rms}I_{rms} \cos \phi + V_{rms}I_{rms} \cos(2\omega t + \phi) \qquad (3.1.36)$$

A typical plot of $p(t)$ is shown in Figure 3.1.2, revealing that it is the sum of an average component, $V_{rms}I_{rms} \cos \phi$, which is a constant that is time independent, and a sinusoidal component, $V_{rms}I_{rms} \cos(2\omega t + \phi)$, which oscillates at a frequency double that of the original source frequency and has zero average value. The average component represents the electric power delivered to the circuit, whereas the sinusoidal component reveals that the energy is stored over one part of the period and released over another, thereby denoting no net delivery of electric energy. It can be seen that the power $p(t)$ is pulsating in time and its *time-average* value P is given by

$$P_{av} = V_{rms}\, I_{rms} \cos \phi \qquad (3.1.37)$$

since the time-average values of the terms $\cos 2\omega t$ and $\sin 2\omega t$ are zero. Note that ϕ is the angle associated with the impedance, and is also the phase angle between the voltage and the current. The term $\cos \phi$ is called the *power factor*. An inductive circuit, in which the current lags the voltage, is said to have a *lagging power factor*, whereas a capacitive circuit, in which the current leads the voltage, is said to have a *leading power factor*. Notice that the power factor associated with a purely resistive load is unity, whereas that of a purely inductive load is zero (lagging) and that of a purely capacitive load is zero (leading).

Equation (3.1.37), representing the average power absorbed by the entire circuit, known as the *real power* or active power, may be rewritten as

$$P = V_{rms}\, I_{rms} \cos \phi = I_{rms}^2 R \qquad (3.1.38)$$

where ϕ is the phase angle between voltage and current. Equation (3.1.38) can be identified as the average power taken by the resistance alone, since

$$p_R(t) = v_R(t)\, i(t) = i^2 R = \left(\sqrt{2}\, I_{rms} \cos \omega t\right)^2 R = I_{rms}^2 R(1 + \cos 2\omega t) \quad (3.1.39)$$

One should recognize that the other two circuit elements, pure inductance and pure capacitance, do not contribute to the average power, but affect the instantaneous power.

For a pure inductor L,

$$p_L(t) = v_L(t)\, i(t) = iL\frac{di}{dt}$$

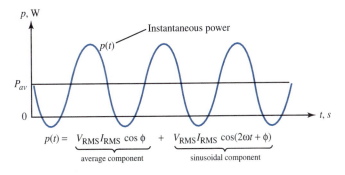

$$p(t) = \underbrace{V_{RMS}I_{RMS} \cos \phi}_{\text{average component}} + \underbrace{V_{RMS}I_{RMS} \cos(2\omega t + \phi)}_{\text{sinusoidal component}}$$

Figure 3.1.2 Typical plot of instantaneous power $p(t)$ and average power P_{av}.

$$= \left(\sqrt{2}\, I_{\text{rms}} \cos \omega t\right) L \left(-\sqrt{2}\, \omega I_{\text{rms}} \sin \omega t\right) = -I_{\text{rms}}^2 X_L \sin 2\omega t \qquad (3.1.40)$$

where $X_L = \omega L$ is the inductive reactance.

For a pure capacitor C,

$$p_C(t) = v_C(t)\, i(t) = i\frac{1}{C}\int i\, dt$$

$$= \left(\sqrt{2}\, I_{\text{rms}} \cos \omega t\right) \frac{1}{C} \left(\frac{\sqrt{2}}{\omega}\, I_{\text{rms}} \sin \omega t\right) = -I_{\text{rms}}^2 X_C \sin 2\omega t \qquad (3.1.41)$$

where $X_C = -1/\omega C$ is the capacitive reactance.

The amplitude of power oscillation associated with either inductive or capacitive reactance is thus seen to be

$$Q = I_{\text{rms}}^2 X = V_{\text{rms}} I_{\text{rms}} \sin \phi \qquad (3.1.42)$$

which is known as the *reactive power* or imaginary power. Note that ϕ is the phase angle between voltage and current. The unit of real power is watts (W), whereas that of reactive power is reactive volt amperes (VAR). The *complex power* is given by

$$\bar{S} = S\angle\phi = P + jQ = V_{\text{RMS}} I_{\text{RMS}}(\cos\phi + j\sin\phi) = \bar{V}_{\text{rms}} \bar{I}_{\text{rms}}^* \qquad (3.1.43)$$

where $\bar{I}_{\text{rms}}^*$ is the complex conjugate of $\bar{I}$. The magnitude of $\bar{S}$, given by $\sqrt{P^2 + Q^2}$, is known as the *apparent power*, with units of volt-amperes (VA). The concept of a *power triangle* is illustrated in Figure 3.1.3. The power factor is given by P/S.

The condition for *maximum power transfer* to a load impedance $\bar{Z}_L \ (= R_L + jX_L)$ connected to a voltage source with an impedance of $\bar{Z}_S \ (= R_S + jX_S)$ (as illustrated in Figure 3.1.4) can be shown to be

$$\bar{Z}_L = \bar{Z}_S^* \qquad \text{or} \qquad R_L = R_S \qquad \text{and} \qquad X_L = -X_S \qquad (3.1.44)$$

When Equation (3.1.44) is satisfied, the load and the source are said to be *matched*. If source and load are purely resistive, Equation (3.1.44) reduces to $R_L = R_S$.

In the phasor method of analysis, the student should recall and appropriately apply the circuit-analysis techniques learned in Chapter 2, which include nodal and mesh analyses, Thévenin and Norton equivalents, source transformations, superposition, and wye–delta transformation.

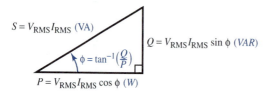

Figure 3.1.3 Power triangle.

$S = V_{\text{RMS}} I_{\text{RMS}}$ (VA)

$Q = V_{\text{RMS}} I_{\text{RMS}} \sin \phi$ (VAR)

$\phi = \tan^{-1}\left(\frac{Q}{P}\right)$

$P = V_{\text{RMS}} I_{\text{RMS}} \cos \phi$ (W)

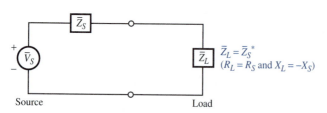

Figure 3.1.4 Illustration of maximum power transfer.

$\bar{Z}_L = \bar{Z}_S^*$
$(R_L = R_S \text{ and } X_L = -X_S)$

Source   Load

EXAMPLE 3.1.5

Consider an RLC series circuit excited by $v(t) = \left(100\sqrt{2}\cos 10t\right)$ V, with $R = 20\ \Omega$, $L = 1$ H, and $C = 0.1$ F. Use the phasor method to find the steady-state response current in the circuit. Sketch the corresponding phasor diagram showing the circuit-element voltages and current. Also draw the power triangle of the load.

Solution

The time-domain circuit and the corresponding frequency-domain circuit are shown in Figure E3.1.5. The KVL equation is

$$\bar{V} = \bar{V}_R + \bar{V}_L + \bar{V}_C = \bar{I}\left(R + j\omega L + \frac{1}{j\omega C}\right)$$

Note that $\omega = 10$ rad/s in our example, and rms values are chosen for the phasor magnitudes. Thus,

$$100\angle 0° = \bar{I}\ (20 + j10 - j1) \qquad \text{or} \qquad \bar{I} = \frac{100\angle 0°}{20 + j9} = \frac{100\angle 0°}{21.93\angle 24.23°} = 4.56\angle -24.23°\ \text{A}$$

Then $i(t) = \text{Re}\left[\sqrt{2}\ \bar{I}\ e^{j\omega t}\right]$, which is

$$i(t) = \sqrt{2}\ (4.56)\cos(10t - 24.23°)\ \text{A}.$$

The phasor diagram is shown in Figure E3.1.5(c).

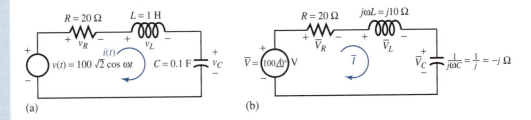

(a)   (b)

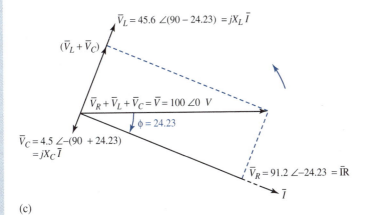

(c)

Figure E3.1.5 (a) Time-domain circuit. (b) Frequency-domain circuit. (c) Phasor diagram. (d) Power triangle.

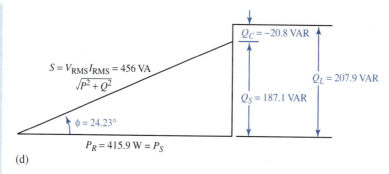

(d)

Figure E3.1.5 Continued

Note that the power factor of the circuit is $\cos(24.23°) = 0.912$ lagging, since the current is lagging the voltage in this inductive load.

$$P_R = V_{R\,\text{rms}}^2/R = I_{\text{rms}}^2 R = (4.56)^2\,20 = 415.9\ \text{W}$$

A resistor absorbs real power.

$$Q_L = V_{L\,\text{rms}}^2/X_L = I_{\text{rms}}^2 X_L = (4.56)^2\,10 = 207.9\ \text{VAR}$$

An inductor absorbs positive reactive power.

$$Q_C = V_{C\,\text{rms}}^2/X_C = I_{\text{rms}}^2 X_C = (4.56)^2\,(-1) = -20.8\ \text{VAR}$$

A capacitor absorbs negative reactive power (or delivers positive reactive power).

The power triangle is shown in Figure E3.1.5(d). Note that P_S ($=P_R$) is delivered by the source; and Q_S ($= Q_L + Q_C$) is delivered by the source. Note that Q_C is negative.

EXAMPLE 3.1.6

Draw the phasor diagrams for an RLC series circuit supplied by a sinusoidal voltage source with a lagging power factor and a GLC parallel circuit supplied by a sinusoidal current source with a leading power factor.

Solution

For the RLC series circuit in the frequency domain in Figure E3.1.6(a), the phasor diagram for the case of a lagging power factor is shown in Figure E3.1.6(b).

For the GLC parallel circuit in the frequency domain in Figure E3.1.6(c), the phasor diagram for the case of a leading power factor is shown in Figure E3.1.6(d).

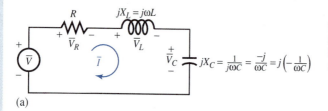

Figure E3.1.6 (a) RLC series circuit in frequency domain. **(b)** Phasor diagram for lagging power factor. **(c)** GLC parallel circuit in frequency domain. **(d)** Phasor diagram for leading power factor.

(a)

Figure E3.1.6 Continued

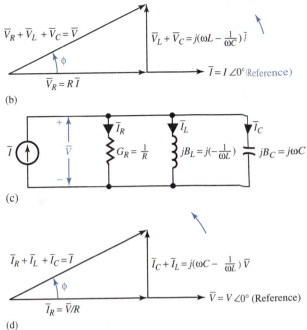

(b)

(c)

(d)

EXAMPLE 3.1.7

Two single-phase 60-Hz sinusoidal-source generators (with negligible internal impedances) are supplying to a common load of 10 kW at 0.8 power factor lagging. The impedance of the feeder connecting the generator G_1 to the load is $1.4 + j1.6$ Ω, whereas that of the feeder connecting the generator G_2 to the load is $0.8 + j1.0$ Ω. If the generator G_1, operating at a terminal voltage of 462 V (rms), supplies 5 kW at 0.8 power factor lagging, determine:

(a) The voltage at the load terminals;

(b) The terminal voltage of generator G_2; and

(c) The real power and the reactive power output of the generator G_2.

Solution

(a) From Equation (3.1.37) applied to G_1, $462 I_1(0.8) = 5 \times 10^3$, or $I_1 = 13.53$ A. With $\bar{V}_1$ taken as reference, the phasor expression for $\bar{I}_1$ is given by $\bar{I}_1 = 13.53\angle - cos^{-1} 0.8 = 13.53\angle - 36.9°$ A, since G_1 supplies at 0.8 lagging power factor.

The KVL equation yields

$$\bar{V}_L = 462\angle 0° - (13.53\angle - 36.9°)(1.4 + j\ 1.6) = 433.8\angle - 0.78° \text{ V}$$

The magnitude of the voltage at the load terminals is 433.8 V (rms).

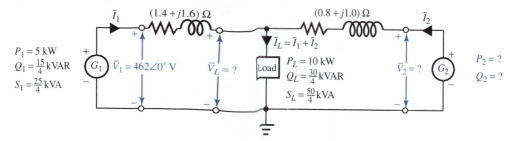

Figure E3.1.7

(b) From Equation (3.1.37) applied to the load, $433.8 I_L(0.8) = 10 \times 10^3$, or $I_L = 28.8$ A. As a phasor, with $\bar{V}_1$ as reference, $\bar{I}_1 = 28.8\angle - (0.78° + 36.9°) = 28.8\angle - 37.68°$ A.

The KCL equation yields:

$$\bar{I}_2 = \bar{I}_L - \bar{I}_1 = 15.22\angle - 38.33°$$

The KVL equation yields

$$\bar{V}_2 = \bar{V}_L + (15.22\angle - 38.38°)(0.8 + j\ 1.0) = 4.52.75\angle - 0.22° \text{ V}$$

The terminal voltage of G_2 is thus 452.75 V (rms).

(c) The power triangle values for G_1 and the load are shown in Figure E3.1.7 so that the student can check it out.

$$P_2 = V_2 I_2 \cos \phi_2 = 452.75 \times 15.22 \times \cos(38.38° - 0.22°)$$
$$= 5418 \text{ W, or } 5.418 \text{ kW}$$

(Note that ϕ_2 is the angle between phasors $\bar{V}_2$ and $\bar{I}_2$.)

$$Q_2 = V_2 I_2 \sin \phi_2 = 452.75 \times 15.22 \ \sin(38.38° - 0.22°)$$
$$= 4258 \text{ VAR, or } 4.258 \text{ kVAR}$$

The conservation of power (real and reactive) is satisfied as follows:

$$P_L + \text{loss in feeder resistances} = 10 + \frac{(13.53)^2\ 1.4}{1000} + \frac{(15.22)^2\ 0.8}{1000} \cong 10.4 \text{ kW}$$

which is the real power delivered by G_1 and G_2, which in turn is the same as $P_1 + P_2 = 5 + 5.4 = 10.4$ kW.

$$Q_L + (I^2 X_L) \text{ of feeders} = \frac{30}{4} + \frac{(13.53)^2\ 1.6}{1000} + \frac{(15.22)^2\ 1.0}{1000} \cong 8 \text{ kVAR}$$

which is the same as the reactive power delivered by G_1 and G_2, which in turn is the same as

$$Q_1 + Q_2 = \frac{15}{4} + 4.26 \cong 8 \text{ kVAR}$$

EXAMPLE 3.1.8

(a) Find the Thévenin equivalent of the circuit shown in Figure E3.1.8(a) at the terminals *A–B*.

(b) Determine the impedance that must be connected to the terminals *A–B* so that it is matched.

(c) Evaluate the maximum power that can be transferred to the matched impedance at the terminals *A–B*.

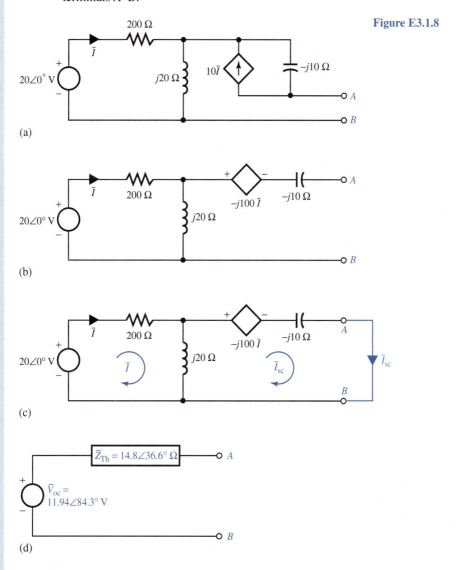

Figure E3.1.8

(a)

(b)

(c)

(d)

Solution

The circuit is redrawn with the controlled current source replaced by its equivalent controlled voltage source, as shown in Figure E3.1.8(b).

(a) Since the Thévenin impedance is the ratio of the open-circuit voltage to the short-circuit current, calculation with respect to terminals A–B is shown next. Let us first calculate the open-circuit voltage $\bar{V}_{oc}$.

The KVL equation for the left-hand loop is

$$(200 + j\,20)\bar{I} = 20\angle0°, \qquad \text{or} \qquad \bar{I} = \frac{20\angle0°}{200 + j\,20}$$

Then,

$$\bar{V}_{oc} = j\,20\,I - (-j\,100\bar{I}) = j\,120\bar{I} = \frac{(j\,120)\,(20)}{200 + j\,20} = 1.19 + j\,11.88$$

$$= 11.94\angle84.3° \text{ V}$$

The short-circuit current $\bar{I}_{sc}$ is found from the circuit of Figure E3.1.8(c).

The KVL equations are given by

$$(200 + j\,20)\,\bar{I} - j\,20\bar{I}_{sc} = 20\angle0°$$

$$- j\,20\bar{I} + \bar{I}_{sc}\,(j\,20 - j\,10) = -(-j\,100\bar{I})$$

Solving for $\bar{I}_{sc}$, one gets

$$\bar{I}_{sc} = 12\bar{I} = \frac{240}{200 - j\,220} = 0.807\angle47.7° \text{ A}$$

The Thévenin impedance at terminals A–B is then

$$\bar{Z}_{Th} = \frac{\bar{V}_{oc}}{\bar{I}_{sc}} = \frac{11.94\angle84.3°}{0.807\angle47.7°} = 14.8\angle36.6° \ \Omega = 11.88 + j\,8.82\ \Omega$$

The Thévenin equivalent circuit at the A–B terminals is shown in Figure E3.1.8(d).

(b) The impedance to be connected at terminals A–B for matching is given by

$$\bar{Z} = \bar{Z}_{Th}^{*} = 14.8\angle-36.6°\ \Omega = R_{Th} - jX_{Th} = 11.88 - j\,8.82\ \Omega$$

(c) The maximum power transfer to the matched impedance $\bar{Z}$ is given by

$$P_{max} = \left(\frac{V_{oc}}{2R_{Th}}\right)^2 R_{Th} = \frac{V_{oc}^2}{4R_{Th}} = \frac{11.94^2}{4 \times 11.88} = 3.0 \text{ W}$$

EXAMPLE 3.1.9

A single-phase source delivers 100 kW to a load operating at 0.8 lagging power factor. In order to improve the system's efficiency, power factor improvement (correction) is achieved through connecting a capacitor in parallel with the load. Calculate the reactive power to be delivered by the capacitor (considered ideal) in order to raise the source power factor to 0.95 lagging, and draw the power triangles. Also find the value of the capacitance, if the source voltage and frequency are 100 V (rms) and 60 Hz, respectively. Assume the source voltage to be a constant and neglect the source impedance as well as the line impedance between source and load.

Solution

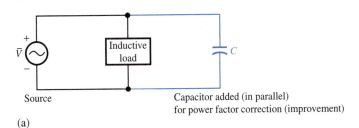

Source

Capacitor added (in parallel)
for power factor correction (improvement)

(a)

Figure E3.1.9 (a) Circuit.
(b) Power triangles.

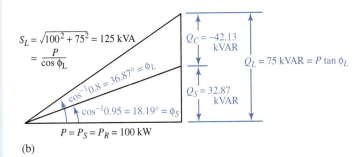

(b)

The circuit and the power triangles are shown in Figure E3.1.9. The real power $P = P_S = P_R$ delivered by the source and absorbed by the load is not changed when the capacitor (considered ideal) is connected in parallel with the load. After the capacitor is connected, noting that Q_C is negative, $Q_S = Q_L + Q_C = 100 \tan(18.19) = 32.87$ kVAR,

$$S_S = \frac{P}{\cos \phi_S} = \frac{100}{0.95} = 105.3 \text{ kVa}$$

(Note that the power factor correction reduces the current supplied by the generator significantly.) So

$$Q_C = Q_S - Q_L = 32.87 - 75 = -43.13 \text{ kVAR}$$

Thus the capacitor is delivering 43.13 kVAR to the system (or absorbing negative kVAR). Then

$$\frac{V_{\text{rms}}^2}{X_C} = \frac{100^2}{X_C} = -43.13 \times 10^3 \quad \text{or} \quad X_C = -0.232 \ \Omega = -\frac{1}{\omega C}$$

or

$$C = \frac{1}{2\pi \times 60 \times 0.232} = 0.0114 \text{ F} \quad \text{or} \quad 11.4 \text{ mF}$$

Fourier Series

The phasor method of circuit analysis can be extended (by using the principle of superposition) to find the response in linear systems due to nonsinusoidal, periodic source functions. A periodic function (t) with period T can be expressed in *Fourier series*,

$$f(t) = \sum_{n=0}^{\infty} a_n \cos n\omega t + \sum_{n=1}^{\infty} b_n \sin n\omega t \qquad (3.1.45)$$

where $\omega = 2\pi f = 2\pi/T$ is the *fundamental* angular frequency, a_0 is the *average* ordinate or the *dc* component of the wave, $(a_1 \cos \omega t + b_1 \sin \omega t)$ is the *fundamental component*, and $(a_n \cos n\omega t + b_n \sin n\omega t)$, for $n \geq 2$, is the *nth harmonic* component of the function.

The *Fourier coefficients* are evaluated as follows:

$$a_0 = \frac{1}{T} \int_0^T f(t) \, dt \tag{3.1.46}$$

$$a_n = \frac{2}{T} \int_0^T f(t) \cos n\omega t \, dt, \qquad n \geq 1 \tag{3.1.47}$$

$$b_n = \frac{2}{T} \int_0^T f(t) \sin n\omega t \, dt, \qquad n \geq 1 \tag{3.1.48}$$

The *exponential form* of the Fourier series can be shown to be given by

$$f(t) = \sum_{n=-\infty}^{\infty} \bar{C}_n e^{jn\omega t} \tag{3.1.49}$$

where

$$\bar{C}_n = \frac{1}{T} \int_0^T f(t) e^{-jn\omega t} \, dt \tag{3.1.50}$$

Even though the exact representation of the nonsinusoidal, periodic wave requires an infinite number of terms in the Fourier series, a good approximation for engineering purposes can be obtained with comparatively few first terms, since the amplitude of the harmonics decreases progressively as the order of the harmonic increases.

The system response is determined by the principle of superposition, and the phasor technique yields the steady-state response. Each frequency component of the response is produced by the corresponding harmonic of the excitation. The sum of these responses becomes the Fourier series of the system response. Note that while the phasor method is employed to determine each frequency component, the individual time functions must be used in forming the series for the system response. Such a method of analysis is applicable to all linear systems.

EXAMPLE 3.1.10

(a) Find the Fourier series for the square wave shown in Figure E3.1.10(a).

(b) Let a voltage source having the waveform of part (a) with a peak value of 100 V and a frequency of 10 Hz be applied to an *RC* series network with $R = 20 \, \Omega$ and $C = 0.1$ F. Determine the first five nonzero terms of the Fourier series of $v_C(t)$.

Solution

(a) From Equation (3.1.46),

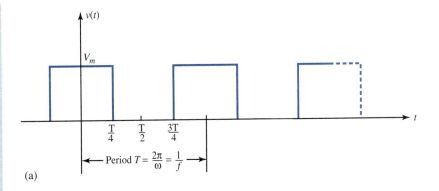

(a)

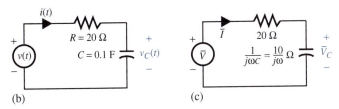

(b)   (c)

Figure E3.1.10 (a) Square wave. (b) Time-domain circuit. (c) Frequency-domain circuit.

$$a_0 = \frac{1}{T} \int_0^T f(t) \ dt = \frac{1}{T} \left(\int_0^{T/4} V_m \ dt + \int_{3T/4}^T V_m \ dt \right) = \frac{V_m}{2}$$

From Equation (3.1.47),

$$a_n = \frac{2}{T} \left(\int_0^{T/4} V_m \ \cos \ n\omega t \ dt + \int_{\pi/4}^T V_m \ \cos \ n\omega t \ dt \right)$$

and

$$a_n = \begin{cases} 0, & \text{for even } n \\ \pm \dfrac{2V_m}{n\pi}, & \text{for odd } n, \text{ where the algebraic sign is } + \text{ for } n = 1 \\ & \text{and changes alternately for each successive term.} \end{cases}$$

From Equation (3.1.48),

$$b_n = \frac{2}{T} \left(\int_0^{T/4} V_m \ \sin \ n\omega t \ dt + \int_{3T/4}^T V_m \ \sin \ n\omega t \ dt \right) = 0$$

which can also be seen from *symmetry* of the square wave with respect to the chosen origin.

The Fourier series is then

$$v(t) = \frac{V_m}{2} + \frac{2V_m}{\pi} \cos \omega t - \frac{2V_m}{3\pi} \cos 3\omega t + \frac{2V_m}{5\pi} \cos 5\omega t - \frac{2V_m}{7\pi} \cos 7\omega t + \cdots$$

(b) The time-domain and frequency-domain circuits are shown in Figures E3.1.10(b) and (c). Note that the capacitive impedance is expressed in terms of ω, since the frequency of each term of the Fourier series is different. The general phasor expressions for $\bar{I}$ and $\bar{V}_C$ are given by

$$\bar{I} = \frac{\bar{V}}{20 + (10/j\omega)}$$

$$\bar{V}_C = \left(\frac{10}{j\omega}\right)\bar{I} = \frac{\bar{V}}{1 + 2j\omega}$$

Treating each Fourier-series term separately, we have the following.

1. For the dc case, $\omega = 0$; $V_{C\text{ dc}} = 50$ V.

2. For the fundamental component,

$$v_1(t) = 63.7 \cos(20\pi t), \qquad \bar{V}_1 = \frac{63.7}{\sqrt{2}}\angle 0°$$

$$\bar{V}_{C1} = \frac{(63.7/\sqrt{2})\angle 0°}{1 + j(40\pi)} = \frac{0.51}{\sqrt{2}}\angle -89.5° \text{ V}$$

and

$$v_{C1} = 0.51 \cos(20\pi t - 89.5°) \text{ V}$$

3. For the third harmonic ($\omega = 3 \times 20\pi$) component,

$$v_3(t) = -21.2 \cos(60\,\pi t), \qquad \bar{V}_3 = -\frac{21.2}{\sqrt{2}}\angle 0°$$

$$\bar{V}_{C3} = \frac{-(21.2/\sqrt{2})\angle 0°}{1 + j\,2(60\pi)} = -\frac{0.06}{\sqrt{2}}\angle -89.8° \text{ V}$$

and

$$v_{C3}(t) = -0.06 \cos(60\pi t - 89.8°) \text{ V}$$

4. For the fifth harmonic ($\omega = 5 \times 20\pi$) component,

$$v_5(t) = 12.73 \cos(100\pi t), \qquad \bar{V}_5 = \frac{12.73}{\sqrt{2}}\angle 0°$$

$$\bar{V}_{C5} = \frac{\left(12.73/\sqrt{2}\right)\angle 0°}{1 + j\,2(100\pi)} = \frac{0.02}{\sqrt{2}}\angle -89.9° \text{ V}$$

and

$$v_{C5}(t) = 0.02 \cos(100\pi t - 89.9°) \text{ V}$$

5. For the seventh harmonic ($\omega = 7 \times 20\pi$) component,

$$v_7(t) = -9.1 \cos(140\pi t), \qquad \bar{V}_7 = -\frac{9.1}{\sqrt{2}}\angle 0°$$

$$\bar{V}_{C7} = \frac{-(9.1/\sqrt{2})\angle 0°}{1 + j\,2(140\pi)} = -\frac{0.01}{\sqrt{2}}\angle -89.9° \text{ V}$$

and

$$v_{C7}(t) = -0.01 \cos(140\pi t - 89.9°) \text{ V}$$

Thus,

$$v_C(t) = 50 + 0.51 \cos(20\pi t - 89.5°) - 0.06 \cos(60\pi t - 89.8°)$$
$$+ 0.02 \cos(100\pi t - 89.9°) - 0.01 \cos(140\pi t - 89.9°) \text{ V}$$

in which the first five nonzero terms of the Fourier series are shown.

3.2 TRANSIENTS IN CIRCUITS

The *total response* of a system to an excitation that is suddenly applied or changed consists of the sum of the *steady-state* and *transient* responses, or of *natural* and *forced* responses. The forced response is the component of the system response that is due to the applied excitation. The transition from the response prior to the change in excitation to the response produced by the excitation is indicated by the natural (transient) response, which is characteristic of all systems containing energy-storage components.

Let us consider the *RL* circuit excited by a voltage source $v(t)$, as shown in Figure 3.2.1. The KVL equation around the loop, for $t > 0$, is given by

$$\frac{L\,di\,(t)}{dt} + Ri\,(t) = v\,(t) \tag{3.2.1}$$

Dividing both sides by L, we have

$$\frac{di\,(t)}{dt} + \frac{R}{L}i\,(t) = \frac{v\,(t)}{L} \tag{3.2.2}$$

which is a first-order ordinary differential equation with constant coefficients. Note that the highest derivative in Equation (3.2.2) is of first order, and the coefficients 1 and R/L are independent of time t. Equation (3.2.2) is also linear since the unknown $i(t)$ and any of its derivatives are not raised to a power other than unity or do not appear as products of each other. Let us then consider a general form of a linear, first-order ordinary differential equation with constant coefficients,

$$\frac{dx\,(t)}{dt} + ax\,(t) = f\,(t) \tag{3.2.3}$$

in which a is a constant, $x(t)$ is the unknown, and $f(t)$ is the known forcing function. By rewriting Equation (3.2.3) as

$$\frac{dx\,(t)}{dt} + ax\,(t) = 0 + f\,(t) \tag{3.2.4}$$

and using the superposition property of linearity, one can think of

$$x(t) = x_{tr}(t) + x_{ss}(t) \tag{3.2.5}$$

where $x_{tr}(t)$ is the transient solution which satisfies the homogeneous differential equation with $f(t) = 0$,

$$\frac{dx_{tr}\,(t)}{dt} + ax_{tr}\,(t) = 0 \tag{3.2.6}$$

and $x_{ss}(t)$ is the steady-state solution (or a particular solution) which satisfies the inhomogeneous differential equation for a particular $f(t)$,

$$\frac{dx_{ss}\,(t)}{dt} + ax_{ss}\,(t) = f\,(t) \tag{3.2.7}$$

A possible form for $x_{tr}(t)$ in order to satisfy Equation (3.2.6) is the exponential function e^{st}. Note,

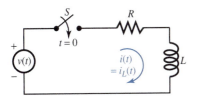

Figure 3.2.1 *RL* circuit excited by $v(t)$.

however, that the function e^{st} used here in finding transient solutions is not the generalized phasor function used earlier for exponential excitations. Thus let us try a solution of the form

$$x_{tr}(t) = Ae^{st} \tag{3.2.8}$$

in which A and s are constants yet to be determined. Substituting Equation (3.2.8) into Equation (3.2.6), we get

$$sAe^{st} + aAe^{st} = 0 \quad \text{or} \quad (s+a)Ae^{st} = 0 \tag{3.2.9}$$

which implies

$$s + a = 0 \quad \text{or} \quad s = -a \tag{3.2.10}$$

since A cannot be zero (or x_{tr} would be zero for all t causing a trivial solution), and e^{st} cannot be zero for all t regardless of s. Thus the transient solution is given by

$$x_{tr}(t) = Ae^{-at} \tag{3.2.11}$$

in which A is a constant yet to be determined.

We already know how to find the steady-state solution $x_{ss}(t)$ for dc or ac sources, and the principle of superposition can be used if we have more than one source. To find that part of the steady-state solution due to a dc source, deactivate or zero all the other sources, and replace inductors with short circuits, and capacitors with open circuits, and then solve the resulting circuit for the voltage or current of interest. To find that part of the steady-state solution due to an ac source, deactivate or zero all the other sources and use the phasor method of Section 3.1. The solution to Equation (3.2.3) or Equation (3.2.4) is then given by Equation (3.2.5) as the sum of the transient and steady-state solutions,

$$x(t) = x_{tr}(t) + x_{ss}(t) = Ae^{-at} + x_{ss}(t) \tag{3.2.12}$$

In order to evaluate A, let us apply Equation (3.2.12) for $t = 0^+$ which is immediately after $t = 0$,

$$x(0^+) = A + x_{ss}(0^+) \quad \text{or} \quad A = x(0^+) - x_{ss}(0^+) \tag{3.2.13}$$

where $x(0^+)$ is the value of $x(t)$ at the initial time and denotes the *initial condition* on $x(t)$, and $x_{ss}(0^+)$ would be found from $x_{ss}(t)$ corresponding to $t = 0$. Thus the total solution for $x(t)$ for all $t \geq 0$ is given by

$$x(t) = \underbrace{[x(0^+) - x_{ss}(0^+)]e^{-at}}_{\text{(transient response)}} + \underbrace{x_{ss}(t)}_{\text{(steady-state response)}} = \underbrace{[x(0^+)e^{-at}]}_{\text{(natural response)}} \underbrace{-x_{ss}(0^+)e^{-at} + x_{ss}(t)}_{\text{(forced response)}} \tag{3.2.14}$$

Note that the transient solution involving e^{-at} eventually goes to zero as time progresses (assuming a to be positive). At a time $t = 1/a$, the transient solution decays to $1/e$, or 37% of its initial value at $t = 0^+$. The reciprocal of a is known as the time constant with units of seconds,

$$\tau = 1/a \tag{3.2.15}$$

Thus Equation (3.2.14) can be rewritten as

$$x(t) = \left[x\left(0^+\right) - x_{ss}\left(0^+\right)\right]e^{-t/\tau} + x_{ss}(t) \tag{3.2.16}$$

which is an important result as a solution of Equation (3.2.3). Then we can easily write the solution of Equation (3.2.2) corresponding to the *RL* circuit of Figure 3.2.1, and that will involve $i(0^+)$, which is the same as $i_L(0^+)$, which is the value of the inductor current immediately after the switch is closed.

Since we know that the inductor current cannot change its value instantaneously, as otherwise the inductor voltage $v_L(t) = L\, di_L/dt$ would become infinite, it follows that the inductor current immediately after closing the switch, $i_L(0^+)$, must be the same as the inductor current just before closing the switch, $i_L(0^-)$,

$$i_L(0^+) = i_L(0^-) \tag{3.2.17}$$

In Figure 3.2.1 note that $i_L(0^-)$ is zero. While the inductor current satisfies Equation (3.2.17), the inductor voltage may change its value instantaneously, as we shall see in the following example.

EXAMPLE 3.2.1

Consider the *RL* circuit of Figure 3.2.1 with $R = 2\,\Omega$, $L = 5$ H, and $v(t) = V = 20$ V (a dc voltage source). Find the expressions for the inductor current $i_L(t)$ and the inductor voltage $v_L(t)$ for $t > 0$, and plot them.

Solution

It follows from Equation (3.2.16) that

$$i(t) = i_L(t) = \left[i_L\left(0^+\right) - i_{L,ss}\left(0^+\right)\right]e^{-t/\tau} + i_{L,ss}(t)$$

where the subscript ss denotes steady state, and $\tau = L/R$.

In our case $i_{L,ss} = V/R = 20/2 = 10$ A, and $i_L(0^-) = 0 = i_L(0^+)$. Thus we have

$$i_L(t) = (0 - 10)e^{-2t/5} + 10 = 10(1 - e^{-2t/5})\ \text{A}, \qquad \text{for } t > 0$$

Then the inductor voltage is obtained as

$$v_L(t) = L\frac{di_L}{dt} = +20e^{-2t/5}, \qquad \text{for } t > 0$$

$i_L(t)$ and $v_L(t)$ are plotted in Figure E3.2.1. The following points are worth noting.

1. $i_L(0^-) = i_L(0^+) = 0$; the inductor current cannot change instantaneously.

2. $v_L(0^-) = 0$ and $v_L(0^+) = 20$; the inductor voltage has changed instantaneously.

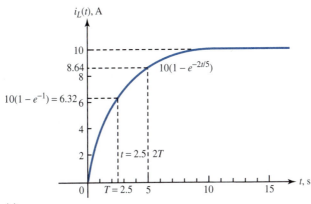

(a)

Figure E3.2.1 (a) Inductor current $i_L(t) = 10(1 - e^{-2t/5})$ A, for $t > 0$. (b) Inductor voltage $v_L(t) = 20e^{-2t/5}$ V, for $t > 0$.

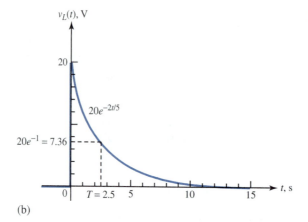

Figure E3.2.1 Continued

(b)

3. After one time constant has elapsed, the inductor current has risen to 63% of its final or steady-state value. After five time constants, it would reach 99% of its final value.

4. After one time constant, the inductor voltage has decayed to 37% of its initial value. The inductor voltage, in this example, eventually decays to zero as it should, since in the steady state the inductor behaves as a short circuit for dc sources.

The resistor voltage and resistor current, if needed, can be found as follows:

$$i_R(t) = i_L(t) = i(t) = 10\left(1 - e^{-2t/5}\right) \text{ A, for } t > 0$$

$$v_R(t) = R i_R(t) = 20\left(1 - e^{-2t/5}\right) \qquad \text{V, for } t > 0$$

Note that the KVL equation $v(t) = V = v_R(t) + v_L(t) = 20$ is satisfied.

EXAMPLE 3.2.2

Consider the RC circuit of Figure E3.2.2(a) with $R = 2\,\Omega$, $C = 5$ F, and $i(t) = I = 10$ A (a dc current source). Find the expressions for the capacitor voltage $v_C(t)$ and the capacitor current $i_C(t)$ for $t > 0$, and plot them.

Solution

The KCL equation at the upper node for $t > 0$ is given by

$$i_C + i_R = i(t), \qquad \text{or} \qquad C\frac{dv_C(t)}{dt} + \frac{v_C(t)}{R} = i(t) \qquad \text{or} \qquad \frac{dv_C(t)}{dt} + \frac{1}{RC}v_C(t) = \frac{i(t)}{C}$$

This equation is the same in form as Equation (3.2.3). The total solution, which is of the form of Equation (3.2.16), can be written as

$$v_C(t) = \left[v_C\left(0^+\right) - v_{C,ss}\left(0^+\right)\right]e^{-t/(RC)} + v_{C,ss}(t)$$

where RC is the time constant T, which is 10 s in this case. $v_{C,ss}(t) = RI = 20$ V in our case, since the source is dc and the capacitor current $i_C = C\left(dv_C/dt\right) = 0$ in the steady state for $t > 0$, denoting an open circuit. Then the solution is given by

$$v_C(t) = \left[v_C\left(0^+\right) - 20\right]e^{-t/10} + 20 \text{ V}, \qquad \text{for } t > 0$$

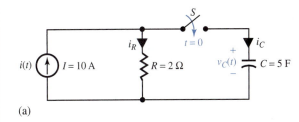

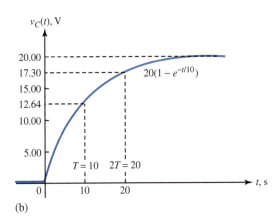

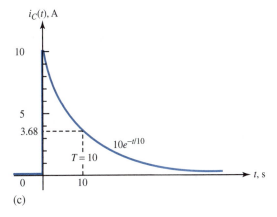

Figure E3.2.2 **(a)** *RC* circuit excited by $i(t) = I$. **(b)** Capacitor voltage $v_C(t) = 20(1 - e^{-t/10})$ V, for $t > 0$. **(c)** Capacitor current $i_C(t) = 10e^{-t/10}$ A, for $t > 0$.

Just as the inductor current cannot change instantaneously, the capacitor voltage cannot change instantaneously,

$$v_C(0^+) = v_C(0^-)$$

which happens to be zero in our case, as otherwise the capacitor current $i_C = C\,(dv_C/dt)$ would become infinite. Thus we have

$$v_C(t) = (0 - 20)\,e^{-t/10} + 20 = 20\left(1 - e^{-t/10}\right) \text{ V}, \qquad \text{for } t > 0$$

Then the capacitor current is obtained as

$$i_C(t) = C\frac{dv_C}{dt} = 10e^{-t/10} \text{ A}, \qquad \text{for } t > 0$$

$v_C(t)$ and $i_C(t)$ are plotted in Figures E3.2.2(b) and (c). The following points are noteworthy.

1. $v_C(0^+) = v_C(0^-)$; the capacitor voltage cannot change instantaneously.
2. $i_C(0^-) = 0$ and $i_C(0^+) = 10$ A; the capacitor current has changed instantaneously.
3. After one time constant has elapsed, the capacitor voltage has risen to 63% of its final (steady-state) value. After five time constants, it would reach 99% of its final value.
4. After one time constant the capacitor current has decayed to 37% of its initial value. The capacitor current, in this example, eventually decays to zero as it should, since in the steady state the capacitor behaves as an open circuit for dc sources.

The resistor voltage and current, if desired, can be found as follows:

$$v_R(t) = v_C(t) = 20\left(1 - e^{-t/10}\right) \text{ V}, \qquad \text{for } t > 0$$

$$i_R(t) = v_R(t)/R = 10\left(1 - e^{-t/10}\right) \text{ A}, \qquad \text{for } t > 0$$

Note that the KCL equation $i(t) = I = i_C(t) + i_R(t) = 10$ is satisfied. The charge transferred to the capacitor at steady-state is:

$$Q_{ss} = CV_{C,ss} = 5 \times 20 = 100\ C$$

Examples 3.2.1 and 3.2.2 are chosen with zero initial conditions. Let us now consider nonzero initial conditions for circuits still containing only one energy-storage element, L or C.

EXAMPLE 3.2.3

For the circuit of Figure E3.2.3(a), obtain the complete solution for the current $i_L(t)$ through the 5-H inductor and the voltage $v_x(t)$ across the 6-Ω resistor.

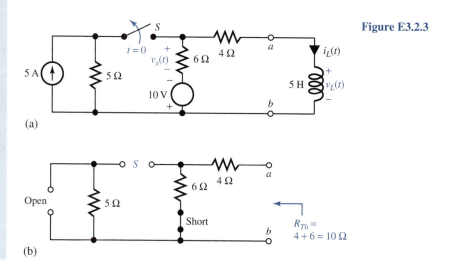

Figure E3.2.3

(a)

(b)

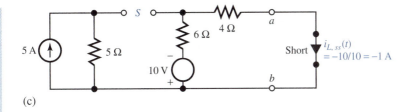

(c)

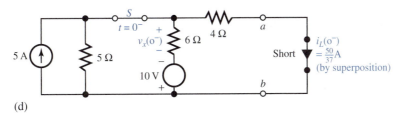

(d)

Figure E3.2.3 Continued

Solution

The Thévenin resistance seen by the inductor for $t > 0$ is found by considering the circuit for $t > 0$ while setting all ideal sources to zero, as shown in Figure E3.2.3(b).

The time constant of the inductor current is $\tau = L/R_{Th} = 5/10 = 0.5$ s. The steady-state value of the inductor current for $t > 0$, $i_{L,ss}(t)$, is found by replacing the inductor by a short circuit (since the sources are both dc) in the circuit for $t > 0$, as shown in Figure E3.2.3(c).

The initial current at $t = 0^-$, $i_L(0^-)$, is found from the circuit for $t < 0$ as the steady-state value of the inductor current for $t < 0$. Figure E3.2.3(d) is drawn for $t < 0$ by replacing the inductor with a short circuit (since both sources are dc).

One can solve for $i_L(0^-)$ by superposition to yield $i_L(0^-) = 50/37$ A $= i_L(0^+)$, by the continuity of inductor current. Then the solution for the inductor current for $t > 0$ can be written as

$$i_L(t) = \left[i_L(0^+) - i_{L,\,ss}(0^+)\right] e^{-R_{Th}t/L} + i_{L,ss}(t), \qquad \text{for } t > 0$$

In our example,

$$i_L(t) = \left(\frac{50}{37} + 1\right) e^{-2t} - 1 = \left(\frac{87}{37} e^{-2t} - 1\right) \text{ A}, \qquad \text{for } t > 0$$

In the circuit for $t = 0^-$ [Figure E3.2.3(d)], notice that $v_x(0^-)$ has a nonzero value, which can be evaluated by superposition as $v_x(0^-) = -220/37$ V.

For $t > 0$, $v_x(t) = -6i_L(t) = -(522/37) e^{-2t} + 6$. Note that $v_{x,\,ss} = 6$ V, and $v_x(0^+) = -522/37 + 6 = -300/37$ V, which shows an instantaneous change in v_x at $t = 0$.

EXAMPLE 3.2.4

Consider the circuit of Figure E3.2.4(a) and obtain the complete solution for the voltage $v_C(t)$ across the 5-F capacitor and the voltage $v_x(t)$ across the 5-Ω resistor.

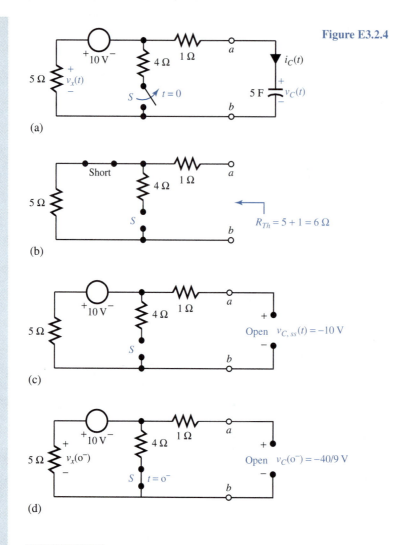

Figure E3.2.4

(a)

(b)

(c)

(d)

Solution

(Note that the procedure is similar to that of Example 3.2.3.) The Thévenin resistance seen by the capacitor for $t > 0$ is found by considering the circuit for $t > 0$ while setting all ideal sources to zero, as shown in Figure E3.2.4(b).

   The time constant of the capacitor voltage is $\tau = R_{Th}C = 6 \times 5 = 30$ s. The steady-state value of the capacitor voltage for $t > 0$, $v_{C,\,ss}(t)$, is found by replacing the capacitor by an open circuit (since the source is dc in the circuit for $t > 0$), as shown in Figure E3.2.4(c).

   The initial capacitor voltage at $t = 0^-$, $v_C(0^-)$, is found from the circuit for $t < 0$ as the steady-state value of the capacitor voltage for $t < 0$. Figure E3.2.4(d) is drawn for $t < 0$ by replacing the capacitor with an open circuit (since the source is dc).

   One can solve for $v_C(0^-)$ to yield $v_C(0^-) = -40/9$ V $= v_C(0^+)$, by the continuity of the capacitor voltage. Then the solution for the capacitor voltage for $t > 0$ can be written as

$$v_C(t) = \left[v_C\left(0^+\right) - v_{C,\,ss}\left(0^+\right)\right]e^{-t/(R_{Th}C)} + v_{C,\,ss}(t), \qquad \text{for } t > 0$$

In our example,

$$v_C(t) = \left(-\frac{40}{9} + 10\right)e^{-t/30} - 10 = \left(\frac{50}{9}e^{-t/30} - 10\right) \text{ V}, \qquad \text{for } t > 0$$

In the circuit of Figure E3.2.4(d) for $t = 0^-$, notice that $v_x(0^-) = 50/9$ V. For $t > 0$, $v_x(t) = -5i_C(t)$.

The capacitor current is found by

$$i_C(t) = C\frac{dv_C(t)}{dt} = (5)\left(-\frac{50}{9}\cdot\frac{1}{30}\right)e^{-t/30} = -\frac{25}{27}e^{-t/30} \text{ A}$$

Then $v_x(t) = -5i_C(t) = (125/27)e^{-t/30}$ A, for $t > 0$. Note that $v_x(0^+) = 125/27$ V, and $v_{x,\,ss} = 0$. The resistor voltage has changed value instantaneously at $t = 0$.

One should recognize that, in the absence of an infinite current, the voltage across a capacitor cannot change instantaneously: i.e., $v_C(0^-) = v_C(0^+)$. In the absence of an infinite voltage, the current in a inductor cannot change instantaneously: i.e., $i_L(0^-) = i_L(0^+)$. Note that an instantaneous change in inductor current or capacitor voltage must be accompanied by a change of stored energy in zero time, requiring an infinite power source. Also observe that the voltages across a resistor, $v_R(0^-)$ and $v_R(0^+)$, are in general not equal to each other, unless the equality condition is forced by $i_L(0^-)$ or $v_C(0^-)$. The voltage across a resistor can change instantaneously in its value.

So far we have considered circuits with only one energy-storage element, which are known as *first-order circuits,* characterized by first-order differential equations, regardless of how many resistors the circuit may contain. Now let us take up series LC and parallel LC circuits, both involving the two storage elements, as shown in Figures 3.2.2 and 3.2.3, which are known as *second-order circuits* characterized by second-order differential equations.

Referring to Figure 3.2.2, the KVL equation around the loop is

$$v_L(t) + v_C(t) + R_{Th}i(t) = v_{Th}(t) \qquad (3.2.18)$$

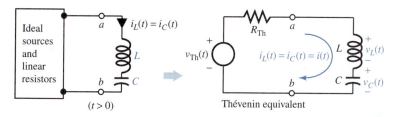

(t > 0)   Thévenin equivalent

Figure 3.2.2 Series LC case with ideal sources and linear resistors.

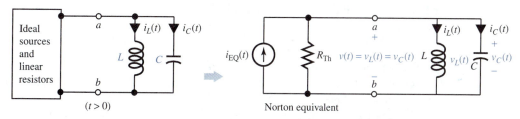

(t > 0)   Norton equivalent

Figure 3.2.3 Parallel LC case with ideal sources and linear resistors.

With $i_L(t) = i_C(t) = i(t)$, $v_L(t) = L \, di_L(t)/dt$, and $v_C(t) = \frac{1}{C} \int_{-\infty}^{t} i_C(\tau) \, d\tau$, one obtains the following equation in terms of the inductor current:

$$L \frac{di_L(t)}{dt} + \frac{1}{C} \int_{-\infty}^{\tau} i_L(\tau) \, d\tau + R_{\text{Th}} i_L(t) = v_{\text{Th}}(t) \tag{3.2.19}$$

Differentiating Equation (3.2.19) and dividing both sides by L, one gets

$$\frac{d^2 i_L(t)}{dt^2} + \frac{R_{\text{Th}}}{L} \frac{di_L(t)}{dt} + \frac{1}{LC} i_L(t) = \frac{1}{L} \frac{dv_{\text{Th}}(t)}{dt} \tag{3.2.20}$$

Referring to Figure 3.2.3, the KCL equation at node a is

$$i_L(t) + i_C(t) + \frac{v_C(t)}{R_{\text{Th}}} = i_{\text{EQ}}(t) \tag{3.2.21}$$

With $v_L(t) = v_C(t) = v(t)$, $i_C(t) = C \frac{dv_C(t)}{dt}$, and $i_L(t) = \frac{1}{L} \int_{-\infty}^{t} v_C(\tau) \, d\tau$, one obtains the following equation in terms of the capacitor voltage

$$C \frac{dv_C(t)}{dt} + \frac{1}{L} \int_{-\infty}^{t} v_C(\tau) \, d\tau + \frac{v_C(t)}{R_{\text{Th}}} = i_{\text{EQ}}(t) \tag{3.2.22}$$

Differentiating Equation (3.2.22) and dividing both sides by C, one gets

$$\frac{d^2 v_C(t)}{dt^2} + \frac{1}{C R_{\text{Th}}} \frac{dv_C(t)}{dt} + \frac{1}{LC} v_C(t) = \frac{1}{C} \frac{di_{\text{EQ}}(t)}{dt} \tag{3.2.23}$$

Equations (3.2.20) and (3.2.23) can be identified to be second-order, constant-coefficient, linear ordinary differential equations of the form

$$\frac{d^2 x(t)}{dt^2} + a \frac{dx(t)}{dt} + bx(t) = f(t) \tag{3.2.24}$$

Note for the series LC case,

$$a = \frac{R_{\text{Th}}}{L} \quad \text{and} \quad b = \frac{1}{LC} \tag{3.2.25}$$

and for the parallel LC case,

$$a = \frac{1}{R_{\text{Th}} C} \quad \text{and} \quad b = \frac{1}{LC} \tag{3.2.26}$$

Because of the linearity of Equation (3.2.24), the solution will consist of the sum of a transient solution $x_{tr}(t)$ and a steady-state solution $x_{ss}(t)$,

$$x(t) = x_{tr}(t) + x_{ss}(t) \tag{3.2.27}$$

where $x_{tr}(t)$ satisfies the homogeneous differential equation with $f(t) = 0$,

$$\frac{d^2 x_{tr}(t)}{dt^2} + a \frac{dx_{tr}(t)}{dt} + bx_{tr}(t) = 0 \tag{3.2.28}$$

and $x_{ss}(t)$ satisfies the following for a particular $f(t)$,

$$\frac{d^2 x_{ss}(t)}{dt^2} + a \frac{dx_{ss}(t)}{dt} + bx_{ss}(t) = f(t) \tag{3.2.29}$$

For obtaining the steady-state solution due to a dc source, recall replacing inductors with short circuits and capacitors with open circuits. For the steady-state solution due to a sinusoidal source, simple methods have been developed in Section 3.1.

Now for the transient solution, a possible form for $x_{tr}(t)$ in order to satisfy Equation (3.2.28) is the exponential function e^{st}. Thus let us try a solution of the form

$$x_{tr}(t) = Ae^{st} \tag{3.2.30}$$

Substituting Equation (3.2.30) in Equation (3.2.28), one gets

$$s^2 Ae^{st} + as Ae^{st} + bAe^{st} = 0 \quad \text{or} \quad \left(s^2 + as + b\right) Ae^{st} = 0 \tag{3.2.31}$$

which can be true only if

$$s^2 + as + b = 0 \tag{3.2.32}$$

Equation (3.2.32) is known as the characteristic equation of the differential equation. The two roots s_1 and s_2 of Equation (3.2.32) are given by

$$s_1, \ s_2 = -\frac{a}{2} \pm \frac{1}{2}\sqrt{a^2 - 4b} = -\frac{a}{2}\left(1 \pm \sqrt{1 - \frac{4b}{a^2}}\right) \tag{3.2.33}$$

Three possibilities arise:

Case 1: The two roots will be real and unequal (distinct), if $\left(\dfrac{4b}{a^2}\right) < 1$.

Case 2: The two roots will be real and equal, if $\left(\dfrac{4b}{a^2}\right) = 1$.

Case 3: The two roots will be complex conjugates of each other, if $\left(\dfrac{4b}{a^2}\right) > 1$.

Let us now investigate these cases in more detail.

Case 1: Roots Real and Unequal (Distinct)

For $4b/a^2 < 1$, from Equation (3.2.33) it is clear that both roots will be negative if a is positive. For realistic circuits with positive values of R_{Th}, L, and C, a will be positive, and hence both roots will be negative. With $s_1 = -\alpha_1$ and $s_2 = -\alpha_2$ (where α_1 and α_2 are positive numbers), the transient (natural) solution will be of the form

$$x_{tr}(t) = A_1 e^{-\alpha_1 t} + A_2 e^{-\alpha_2 t} \tag{3.2.34}$$

where A_1 and A_2 are constants to be determined later. Note that Equation (3.2.34) contains the sum of two decaying exponentials.

Case 2: Roots Real and Equal

For $4b = a^2$, the roots will be equal and negative, as one can see from Equation (3.2.33). With $s_1 = s_2 = -a/2 = -\alpha$ (where α is positive), let us look at the nature of the transient solution. For the case of real, repeated roots, not only $x_{tr}(t) = A_1 e^{-\alpha t}$ satisfies the differential equation, Equation (3.2.28), but also $x_{tr}(t) = A_2 t e^{-\alpha t}$ can be seen to satisfy Equation (3.2.28). By superposition, the form of the transient solution will be seen to be

$$x_{tr}(t) = A_1 e^{-\alpha t} + A_2 t e^{-\alpha t} = (A_1 + A_2 t) e^{-\alpha t} \tag{3.2.35}$$

which contains two constants (to be determined later), as it should for the solution of a second-order differential equation.

Case 3: Complex Conjugate Roots

For $(4b/a^2) > 1$, as seen from Equation (3.2.33), the roots will be complex conjugates of each other. With $s_1 = -\alpha + j\beta$ and $s_2 = -\alpha - j\beta$ (where α and β are positive numbers), the transient solution will be of the form

$$x_{tr}(t) = A_1 e^{(-\alpha+j\beta)t} + A_2 e^{(-\alpha-j\beta)t} = e^{-\alpha t}\left(A_1 e^{j\beta t} + A_2 e^{-j\beta t}\right) \qquad (3.2.36)$$

in which A_1 and A_2 will turn out to be complex conjugates of each other, since the transient solution must be real. An alternative, and more convenient, form of Equation (3.2.36) is given by

$$x_{tr}(t) = e^{-\alpha t}(C_1 \cos \beta t + C_2 \sin \beta t) \qquad (3.2.37)$$

in which C_1 and C_2 are real constants to be determined later. Another alternative form of Equation (3.2.37) can be shown to be

$$x_{tr}(t) = Ce^{-\alpha t} \sin (\beta t + \phi) \qquad (3.2.38)$$

Note that in all three cases [see Equations (3.2.34), (3.2.35), and (3.2.38)] with nonzero α's the transient solution will eventually decay to zero as time progresses. The transient solution is said to be *overdamped, critically damped,* and *underdamped,* corresponding to Cases 1, 2, and 3, respectively. Justification of the statement can be found from the solution of the following example.

EXAMPLE 3.2.5

Consider the circuit of Figure 3.2.2 for $t > 0$ with zero initial conditions, $v_{Th}(t) = 1$ V (dc), and $R_{Th} = 2\ \Omega$; $L = 1$ H. Determine the complete response for $v_C(t)$ for capacitance values of:

   (a) $(25/9)$ F

   (b) 1.0 F

   (c) 0.5 F

Solution

The KVL equation around the loop is given by

$$2i(t) + 1\left(\frac{di(t)}{dt}\right) + \frac{1}{C}\int_{-\infty}^{t} i(\tau)\ d\tau = 1.0$$

Since

$$v_C(t) = \frac{1}{C}\int_{-\infty}^{t} i(\tau)\ d\tau \qquad \text{and} \qquad i(t) = i_C(t) = C\frac{dv_C(t)}{dt}$$

the differential equation in terms of $v_C(t)$ is obtained as

$$2C\frac{dv_C(t)}{dt} + C\frac{d^2 v_C(t)}{dt^2} + v_C(t) = 1.0$$

The steady-state response $v_{C,\,ss}(t)$ due to a dc source is determined by replacing inductors with short circuits and capacitors with open circuits,

$$v_{C,\,ss}(t) = V_0 = 1.0 \text{ V}$$

[Note that $v_{C,\,ss}(t)$ is of the form of the excitation and is a constant V_0; substitution of V_0 for $v_{C,\,ss}(t)$ gives $0 + 0 + V_0 = 1.0$.]

The transient response is obtained from the force-free (homogeneous) equation, which is

$$2C\frac{dv_{C,\,tr}}{dt} + C\frac{d^2 v_{C,\,tr}(t)}{dt^2} + v_{C,\,tr}(t) = 0$$

Assuming $v_{C,\,tr}(t)$ to be of the form Ae^{st}, the values of s are determined from

$$Ae^{st}\left[2Cs + Cs^2 + 1\right] = 0 \qquad \text{or} \qquad Cs^2 + 2Cs + 1 = 0$$

and the roots are

$$s_1, s_2 = \frac{-2C \pm \sqrt{(2C)^2 - 4C}}{2C} = -1\left(1 \pm \sqrt{1 - \frac{1}{C}}\right)$$

(a) For $C = 25/9$ F, the values of s_1 and s_2 are

$$s_1 = -0.2 \qquad \text{and} \qquad s_2 = -1.8$$

The transient response is given by (see Case 1 with real and unequal roots)

$$v_{C,\,tr}(t) = A_1 e^{-0.2t} + A_1 e^{-1.8t} \text{ V}$$

and the total (complete) response is

$$v_C(t) = v_{C,\,ss}(t) + v_{C,\,tr}(t) \qquad \text{or} \qquad v_C(t) = 1.0 + A_1 e^{-0.2t} + A_2 e^{-1.8t} \text{ V}$$

where A_1 and A_2 are to be evaluated from the initial conditions,

$$v_C(0^-) = 0, \qquad i_L(0^-) = i(0^-) = 0$$

By the continuity principle, $v_C(0^+) = 0$ and $i_L(0^+) = 0$. Also, $i_L(0^+) = i_C(0^+)$ and $dv_C/dt\,(0^+) = i_C(0^+)/C = 0$. Expressing $dv_C(t)/dt$ as $dv_C(t)/dt = -0.2A_1 e^{-0.2t} - 1.8A_2 e^{-1.8t}$, evaluating $dv_C(0^+)/dt$ and $v_C(0^+)$ yields

$$v_C(0^+) = 0 = 1 + A_1 + A_2$$

$$\frac{dv_C}{dt}(0^+) = 0 = -0.2A_1 - 1.8A_2$$

Simultaneous solution yields $A_1 = -1.125$ and $A_2 = 0.125$. Hence the complete solution is

$$v_C(t) = 1.0 - 1.125e^{-0.2t} + 0.125e^{-1.8t} \text{ V}$$

(b) For $C = 1.0$ F, the values of s_1 and s_2 are both equal; $s_1 = s_2 = -1$. Corresponding to Case 2 with real and equal roots, the transient response is given by Equation (3.2.35),

$$v_{C,\,n}(t) = (A_1 + A_2 t)\,e^{-\alpha t} = (A_1 + A_2 t)\,e^{-t}$$

The complete response is $v_C(t) = 1.0 + (A_1 + A_2 t) e^{-t}$. Evaluating both v_C and dv_C/dt at $t = 0^+$, we have

$$v_C(0^+) = 0 = 1 + A_1$$

$$\frac{dv_C}{dt}(0^+) = 0 = A_2 - A_1$$

These equations result in $A_1 = A_2 = -1$, from which it follows that

$$v_C(t) = 1 - e^{-t} - te^{-t} = 1 - (t+1)e^{-t} \text{ V}$$

(c) For $C = 0.5$ F, the values of s_1 and s_2 are obtained as $s_1 = (-1 + j1)$ and $s_2 = (-1 - j1)$ (see Case 3 with complex conjugate roots). The transient response is of the form of Equation (3.2.38),

$$v_{C,n}(t) = Ae^{-\alpha t} \sin(\beta t + \phi)$$

where $\alpha = 1$ and $\beta = 1$ in our case. The complete solution is then $v_C(t) = 1.0 + Ae^{-t} \sin(\beta t + \phi)$ V. From the initial conditions $v_C(0^+) = 0$ and $dv_C/dt(0^+) = 0$, evaluating both at $t = 0^+$ yields

$$v_C(0^+) = 0 = 1 + A \sin \phi$$

$$\frac{dv_C}{dt}(0^+) = 0 = A(\cos \phi - \sin \phi)$$

Simultaneous solution yields $A = -\sqrt{2}$ and $\phi = \pi/4$. Thus the complete response is given by

$$v_C(t) = 1.0 - \sqrt{2}\, e^{-t} \sin\left(t + \frac{\pi}{4}\right) \text{ V}$$

The total responses obtained for the three cases are plotted in Figure E3.2.5. These cases are said to be overdamped for case (a), critically damped for case (b), and underdamped for case (c).

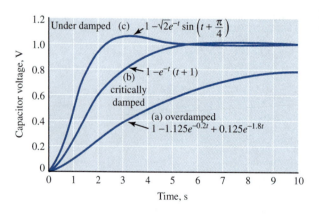

Figure E3.2.5 Total responses obtained.

A system that is overdamped [Figure E3.2.5(a)] responds slowly to any change in excitation. The critically damped system [Figure E3.2.5(b)] responds smoothly in the speediest fashion to approach the steady-state value. The underdamped system [Figure E3.2.5(c)] responds most quickly accompanied by overshoot, which makes the response exceed and oscillate about the steady-state value while it gradually approaches the steady-state value. Practical systems are generally designed to yield slightly underdamped response, restricting the overshoot to be less than 10%.

Let us next consider circuits with two energy-storage elements (L and C) and nonzero initial conditions.

EXAMPLE 3.2.6

Determine $i_L(t)$ and $v_C(t)$ for $t > 0$ in the circuit given in Figure E3.2.6(a).

Figure E3.2.6

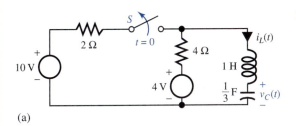

(a)

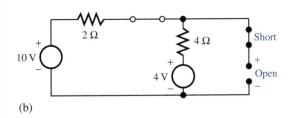 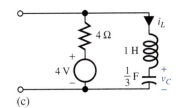

(b)

(c)

Solution

From the circuit for $t = 0^-$ drawn in Figure E3.2.6(b), we have

$$i_L(0^-) = 0 = i_L(0^+) \quad \text{and} \quad v_C(0^-) = \left(\frac{10-4}{6} \times 4\right) + 4 = 8 \text{ V} = v_C(0^+)$$

From the circuit for $t > 0$ drawn in Figure E3.2.6(c), we have $i_{L, \text{ss}} = 0$; $v_{C, \text{ss}} = 4$ V. With $R = 4 \, \Omega$, $L = 1$ H, and $C = \frac{1}{3}$ F, $s^2 + \frac{R}{L}s + \frac{1}{LC} = s^2 + 4s + 3 = 0$. With $s_1 = -1$ and $s_2 = -3$, $i_{L, \text{tr}}(t) = A_1 e^{-t} + A_2 e^{-3t}$; $v_{C, \text{tr}}(t) = B_1 e^{-t} + B_2 e^{-3t}$. The complete solutions are given by

$$i_L(t) = A_1 e^{-t} + A_2 e^{-3t} \quad \text{and} \quad v_C(t) = 4 + B_1 e^{-t} + B_2 e^{-3t}$$

Since

$$v_L(0^+) = L\frac{di_L}{dt}(0^+) = -A_1 - 3A_2 = 4 - 4i_L(0^+) - v_C(0^+) = 4 - 8 = -4$$

$$i_C(0^+) = C\frac{dv_C}{dt}(0^+) = \frac{1}{3}(-B_1 - 3B_2) = i_L(0^+) = 0$$

it follows that $A_1 + 3A_2 = 4$ and $B_1 + 3B_2 = 0$.

From $i_L(0^+) = 0$ and $v_C(0^+) = 8$ V, it follows that $A_1 + A_2 = 0$ and $B_1 + B_2 + 4 = 8$. Simultaneous solution yields $A_1 = -2$, $A_2 = 2$, $B_1 = 6$, and $B_2 = -2$. Thus we have, for $t > 0$,

$$i_L(t) = \left(-2e^{-t} + 2e^{-3t}\right) \text{ A}; \qquad v_C(t) = \left(4 + 6e^{-t} - 2e^{-3t}\right) \text{ V}$$

In order to represent the abrupt changes in excitation, encountered when a switch is opened or closed and in individual sequences of pulses, *singularity functions* are introduced. This type of representation leads to the *step* and *impulse* functions. The methods for evaluating the transient (natural) and steady-state (forced) components of the response can also be applied to these excitations.

The *unit-step* function, represented by $u(t)$, and defined by

$$u(t) = \begin{cases} 0, & t < 0 \\ 1, & t > 0 \end{cases} \tag{3.2.39}$$

is shown in Figure 3.2.4. The physical significance of the unit step can be associated with the turning on of something (which was previously zero) at $t = 0$. In general, a source that is applied at $t = 0$ is represented by

$$v(t) \quad \text{or} \quad i(t) = f(t)u(t) \tag{3.2.40}$$

A *delayed unit step* $u(t - T)$, shown in Figure 3.2.5, is defined by

$$u(t - T) = \begin{cases} 1, & t > T \\ 0, & t < T \end{cases} \tag{3.2.41}$$

A widely used excitation in communication, control, and computer systems is the *rectangular pulse*, whose waveform is shown in Figure 3.2.6 and whose mathematical expression is given by

$$f(t) = \begin{cases} 0, & -\infty < t < 0 \\ A, & 0 < t < T \\ 0, & T < t < \infty \end{cases} \tag{3.2.42}$$

Equation (3.2.42) can be expressed as the sum of two step functions,

$$f(t) = Au(t) - Au(t - T) \tag{3.2.43}$$

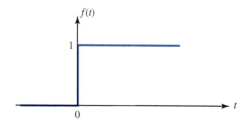

Figure 3.2.4 Unit-step function.

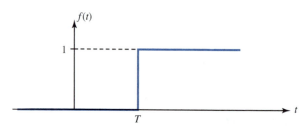

Figure 3.2.5 Delayed unit-step function.

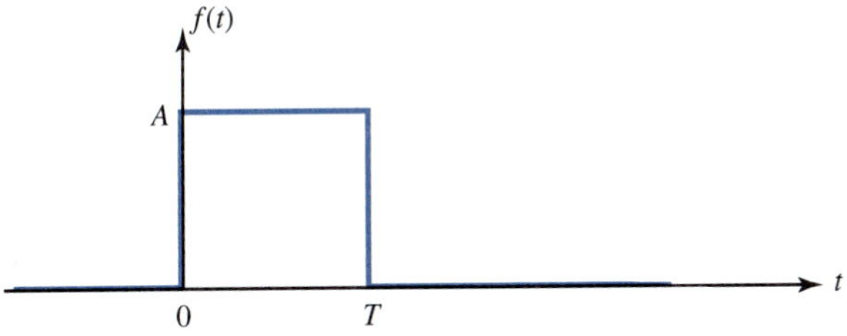

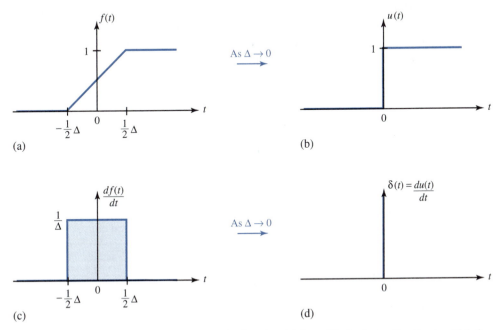

Figure 3.2.6 Rectangular pulse.

The student is encouraged to justify this statement by drawing graphically. For determining the response of a circuit to a pulse excitation, using the principle of superposition, the response to each component can be found and the circuit response obtained by summing the component responses. The decomposition of pulse-type waveforms into a number of step functions is commonly used.

In order to represent the effects of pulses of short duration and the responses they produce, the concept of *unit-impulse* function $\delta(t)$ is introduced,

$$\delta(t) = 0 \text{ for } t \neq 0, \qquad \text{and} \qquad \int_{-\infty}^{\infty} \delta(t) \, dt = \int_{0^-}^{0^+} \delta(t) \, dt = 1 \qquad (3.2.44)$$

Equation (3.2.44) indicates that the function is zero everywhere except at $t = 0$, and the area enclosed is unity. In order to satisfy this, $\delta(t)$ becomes infinite at $t = 0$. The graphical illustration of a unit-impulse function is shown in Figure 3.2.7. The interpretation of $i(t) = A\delta(t - T)$ is that $\delta(t - T)$ is zero everywhere except at $t = T$ (i.e., the current impulse occurs at $t = T$), and the area under $i(t)$ is A.

Figure 3.2.7 Graphic illustration of unit-impulse function. **(a)** A modified unit-step function in which the transition from zero to unity is linear over a Δ-second time interval. **(b)** Unit-step function: the graph of part (a) as $\Delta \to 0$. **(c)** The derivative of the modified unit-step function depicted in part (a) (area enclosed $= 1$). **(d)** Unit-impulse function: the graph of part (c) as $\Delta \to 0$ $[\int_{-\infty}^{\infty} \delta(t)dt = \int_{0^-}^{0^+} \delta(t)dt = 1]$

The form of the natural response due to an impulse can be found by the methods presented in this section. As for the initial capacitor voltages and inductor currents, notice that a current impulse applied to a capacitance provides an initial voltage of

$$v_C\left(0^+\right) = \frac{q}{C}$$

(3.2.45)

where q is the area under the current impulse. Similarly, a voltage impulse applied to an inductance provides an initial current of

$$i_L\left(0^+\right) = \frac{\lambda}{L}$$

(3.2.46)

where λ is the area under the voltage impulse. The impulse function allows the characterization of networks in terms of the system's natural response, which is dependent only on the network elements and their interconnections. Once the impulse response is known, the system response to nearly all excitation functions can be determined by techniques that are available, but not disclosed here in view of the scope of this text.

3.3　LAPLACE TRANSFORM

Many commonly encountered excitations can be represented by exponential functions. The differential equations describing the networks are transformed into algebraic equations with the use of exponentials. So far, methods have been developed to determine the forced response and the transient response in Sections 3.1 and 3.2. The *operational calculus* was developed by Oliver Heaviside (1850–1925) based on a collection of intuitive rules; the transformation is, however, named after Pierre Simon Laplace (1749–1827) because a complete mathematical development of Heaviside's methods has been found in the 1780 writings of Laplace. The Laplace transformation provides a systematic algebraic approach for determining the total network response, including the effect of initial conditions. The differential equations in the time domain are transformed into algebraic equations in the frequency domain. Frequency-domain quantities are manipulated to obtain the frequency-domain equivalent of the desired result. Then, by taking the inverse transform, the desired result in the time domain is obtained.

The single-sided Laplace transform of a function $f(t)$ is defined by

$$\mathcal{L}[f(t)] = F(s) = \int_0^\infty f(t)\, e^{-st}\, dt$$

(3.3.1)

where $f(t) = 0$ for $t < 0$, and s is a complex-frequency variable given by $s = \sigma + j\omega$. The frequency-domain function $F(s)$ is the Laplace transform of the time-domain function $f(t)$. When the integral of Equation (3.3.1) is less than infinity and converges, $f(t)$ is Laplace transformable. Note that for $\sigma > 0$, e^{-st} decreases rapidly, making the integral converge. The uniqueness of the Laplace transform leads to the concept of the *transform pairs*,

$$\mathcal{L}[f(t)] = F(s) \Leftrightarrow \mathcal{L}^{-1}[F(s)] = f(t)$$

(3.3.2)

which states that the *inverse Laplace transform* of $F(s)$ is $f(t)$. It should be noted that the Laplace transform is a linear operation such that

$$\mathcal{L}[Af_1(t) + Bf_2(t)] = AF_1(s) + BF_2(s)$$

(3.3.3)

in which A and B are independent of s and t, and $F_1(s)$ and $F_2(s)$ are the Laplace transforms of $f_1(t)$ and $f_2(t)$, respectively. Using the definition of Equation (3.3.1), Table 3.3.1 of Laplace transform pairs is developed for the most commonly encountered functions. Note that it is assumed that

$f(t) = 0$ for $t < 0$ and all $f(t)$ exist for $t \geq 0$. Also note that in Table 3.3.1, functions 8 through 20 can be considered as being multiplied by $u(t)$.

From Table 3.3.1 of Laplace transform pairs one can see that

$$\mathcal{L}\left[\frac{df(t)}{dt}\right] = sF(s) - f(0^+) \tag{3.3.4}$$

By recalling that $v_L = L\,di_L/dt$ and $i_C = C\,dv_C/dt$ and the principle of continuity of the inductor current and the capacitor voltage, the significance of the term $f(0^+)$ in Equation (3.3.4) is that the initial condition is automatically included in the Laplace transform of the derivative, and hence

TABLE 3.3.1 Laplace Transform Pairs

f(t) in time domain = $\mathcal{L}^{-1}$[F(s)] $\quad\Leftrightarrow$	F(s) in frequency domain = $\mathcal{L}$f[(t)]
(1) $\dfrac{df(t)}{dt}$	$sF(s) - f(0^+)$
(2) $\dfrac{d^2 f(t)}{dt^2}$	$s^2 F(s) - sf(0^+) - \dfrac{df}{dt}(0^+)$
(3) $\dfrac{d^n f(t)}{dt^n}$	$s^n F(s) - s^{n-1} f(0^+) - s^{n-2}\dfrac{df}{dt}(0^+) - \cdots - \dfrac{d^{n-1} f}{dt^{n-1}}(0^+)$
(4) $g(t) = \displaystyle\int_0^t f(t)\,dt$	$\dfrac{F(s)}{s} + \dfrac{g(0^+)}{s}$
(5) $\displaystyle\int_0^t f(\tau)g(t - \tau)\,d\tau$	$F(s)G(s)$
(6) $u(t)$, unit-step function	$\dfrac{1}{s}$
(7) $\delta(t)$, unit-impulse function	1
(8) t	$\dfrac{1}{s^2}$
(9) $\dfrac{t^{n-1}}{(n-1)!}$, n integer	$\dfrac{1}{s^n}$
(10) ε^{-at}	$\dfrac{1}{s + a}$
(11) $t\varepsilon^{-at}$	$\dfrac{1}{(s + a)^2}$
(12) $t^{n-1}\varepsilon^{-at}$	$\dfrac{(n-1)!}{(s + a)^n}$
(13) $\sin \omega t$	$\dfrac{\omega}{s^2 + \omega^2}$
(14) $\cos \omega t$	$\dfrac{s}{s^2 + \omega^2}$
(15) $\sin(\omega t + \theta)$	$\dfrac{s \sin \theta + \omega \cos \theta}{s^2 + \omega^2}$
(16) $\cos(\omega t + \theta)$	$\dfrac{s \cos \theta - \omega \sin \theta}{s^2 + \omega^2}$
(17) $\varepsilon^{-at} \sin \omega t$	$\dfrac{\omega}{(s + a)^2 + \omega^2}$
(18) $\varepsilon^{-at} \cos \omega t$	$\dfrac{s + a}{(s + a)^2 + \omega^2}$
(19) $t\varepsilon^{-at} \sin \omega t$	$\dfrac{2\omega(s + a)}{[(s + a)^2 + \omega^2]^2}$
(20) $t\varepsilon^{-at} \cos \omega t$	$\dfrac{(s + a)^2 - \omega^2}{[(s + a)^2 + \omega^2]^2}$

becomes an inherent part of the final total solution. For cases with zero initial condition, it can be seen that multiplication by s in the frequency domain corresponds to differentiation in the time domain, and dividing by s in the frequency domain corresponds to integration in the time domain.

Some other properties of the Laplace transform are listed in Table 3.3.2.

From the entries of Table 3.3.2, observe the time–frequency dualism regarding frequency differentiation, frequency integration, and frequency shifting. The initial-value and final-value theorems also display the dualism of the time and frequency domains. As seen later, these theorems will be effectively applied in the network solutions.

In order to use the tabulated transform pairs of Table 3.3.1, algebraic manipulations will become necessary to make $F(s)$ correspond to one of the tabulated entries. While the process may be simple in some cases, in other cases $F(s)$ may have to be rearranged in a systematic way as a sum of component functions whose inverse transforms are tabulated. A formalized approach to resolve $F(s)$ into a summation of simple factors is known as the method of *partial-fraction expansion* (Heaviside expansion theorem).

Let us consider a rational function (i.e., one that can be expressed as a ratio of two polynomials),

$$F(s) = \frac{N(s)}{D(s)} \tag{3.3.5}$$

where $N(s)$ denotes the numerator polynomial and $D(s)$ denotes the denominator polynomial. As a first step in the expansion of the quotient $N(s)/D(s)$, we check to see that the degree of the polynomial N is less than that of D. If this condition is not satisfied, divide the numerator by the denominator to obtain an expansion in the form

$$\frac{N(s)}{D(s)} = B_0 + B_1 s + B_2 s^2 + \cdots + B_{m-n} s^{m-n} + \frac{N_1(s)}{D(s)} \tag{3.3.6}$$

where m is the degree of the numerator and n is the degree of the denominator.

TABLE 3.3.2 Some Properties of the Laplace Transform

Property	Time Domain	Frequency Domain
Linearity	$af(t) \pm bg(t)$	$aF(s) \pm bG(s)$
Time delay of shift	$f(t-T)u(t-T)$	$e^{-sT}F(s)$
Periodic function $f(t) = f(t + nT)$	$f(t), \quad 0 \le t \le T$	$\dfrac{F(s)}{1 - e^{-Ts}}$ where $F(s) = \int_0^T f(t)e^{-st}\,dt$
Time scaling	$f(at)$	$\dfrac{1}{a}F\left(\dfrac{s}{a}\right)$
Frequency differentiation (multiplication by t)	$tf(t)$	$-\dfrac{dF(s)}{ds}$
Frequency integration (division by t)	$\dfrac{f(t)}{t}$	$\displaystyle\int_s^\infty F(s)\,ds$
Frequency shifting (exponential translation)	$f(t)e^{-at}$	$F(s+a)$
Initial-value theorem	$\lim\limits_{t \to 0} f(t) = f(0^+)$	$\lim\limits_{s \to \infty} sF(s)$
Final-value theorem	$\lim\limits_{t \to \infty} f(t) = f(\infty)$ where limit exists	$\lim\limits_{s \to 0} sF(s)$

The roots of the equation

$$N(s) = 0 \qquad (3.3.7)$$

are said to be the *zeros* of $F(s)$; and the roots of the equation

$$D(s) = 0 \qquad (3.3.8)$$

are said to be the *poles* of $F(s)$.

The new function $F_1(s)$ given by $N_1(s)/D(s)$ is such that the degree of the denominator polynomial is greater than that of the numerator. The denominator polynomial $D(s)$ is typically of the form

$$D(s) = a_n s^n + a_{n-1} s^{n-1} + \cdots + a_1 s + a_0 \qquad (3.3.9)$$

By dividing the numerator $N_1(s)$ and the denominator $D(s)$ by a_n, $F_1(s)$ may be rewritten as follows:

$$F_1(s) = \frac{N_2(s)}{D_1(s)} = \frac{N_2(s)}{s^n + \dfrac{a_{n-1}}{a_n} s^{n-1} + \cdots + \dfrac{a_0}{a_n}} \qquad (3.3.10)$$

The particular manner of evaluating the coefficients of the expansion is dependent upon the nature of the roots of $D_1(s)$ in Equation (3.3.10). We shall now discuss different cases of interest. The possible forms of the roots are (1) real and simple (or distinct) roots, (2) conjugate complex roots, and (3) multiple roots.

Real and Simple (or Distinct) Poles

If all the poles of $F_1(s)$ are of first order, Equation (3.3.10) may be written in terms of a partial-fraction expansion as

$$F_1(s) = \frac{N_2(s)}{(s - p_1)(s - p_2) \cdots (s - p_n)} \qquad (3.3.11)$$

or

$$F_1(s) = \frac{K_1}{s - p_1} + \frac{K_2}{s - p_2} + \cdots + \frac{K_n}{s - p_n} \qquad (3.3.12)$$

where $p_1, p_2, \ldots, p_n$ are distinct, and $K_1, K_2, \ldots, K_n$ are nonzero finite constants. For any k, the evaluation of the residue K_k of $F_1(s)$ corresponding to the pole $s = p_k$ is done by multiplying both sides by $(s - p_k)$ and letting $s \to p_k$,

$$K_k = \lim_{s \to p_k} [(s - p_k) F_1(s)] \qquad (3.3.13)$$

Equation (3.3.13) is valid for $k = 1, 2, \ldots, n$. Once the K's are determined in Equation (3.3.12), the inverse Laplace transform of each of the terms can be written easily in order to obtain the complete time solution.

Conjugate Complex Poles

It is possible that some of the poles of Equation (3.3.10) are complex. Since the coefficients a_k in Equation (3.3.9) are real, complex poles occur in complex conjugate pairs and will always be even in number. Let us consider a pair of complex poles for which case Equation (3.3.10) may be written as

$$F_1(s) = \frac{N_2(s)}{(s+a+jb)(s+a-jb)D_2(s)} \tag{3.3.14}$$

or

$$F_1(s) = \frac{K_1}{s+(a+jb)} + \frac{K_2}{s+(a-jb)} + \frac{N_3(s)}{D_2(s)} \tag{3.3.15}$$

The procedure for finding K_1 and K_2 is the same as outlined earlier for unrepeated linear factors or simple poles. Thus we have

$$K_1 = \lim_{s \to (-a-jb)} [(s+a+jb)F_1(s)] \tag{3.3.16}$$

and

$$K_2 = \lim_{s \to (-a+jb)} [(s+a-jb)F_1(s)] \tag{3.3.17}$$

It can be shown that K_1 and K_2 are conjugates of each other. The terms in the time function, $\mathcal{L}^{-1}[F(s)]$, due to the complex poles of $F_1(s)$, are then found easily.

Alternate Representation for Complex Poles

Complex poles can be combined to yield a quadratic term in the partial fraction expansion. The representation may best be illustrated by considering one real pole and two complex conjugate poles. Let us then consider

$$
\begin{aligned}
F_1(s) &= \frac{N_2(s)}{(s-p_1)(s+a+jb)(s+a-jb)} \\
&= \frac{N_2(s)}{(s-p_1)\left[(s+a)^2+b^2\right]} \\
&= \frac{N_2(s)}{(s-p_1)\left(s^2+As+B\right)}
\end{aligned}
\tag{3.3.18}
$$

which can be rewritten as

$$F_1(s) = \frac{K_1}{s-p_1} + \frac{K_2 s + K_3}{s^2+As+B} \tag{3.3.19}$$

Evaluating K_1 as before, one has

$$K_1 = \lim_{s \to p_1} [(s-p_1)F_1(s)] \tag{3.3.20}$$

It follows from Equations (3.3.18) and (3.3.19) that

$$N_2(s) = K_1\left(s^2+As+B\right) + (K_2 s+K_3)(s-p_1) \tag{3.3.21}$$

Since Equation (3.3.21) must hold for all values of s, the coefficients of various powers of s on both sides of the equality must be equal. These equations of equality are then solved to determine K_2 and K_3.

Even if many pairs of complex conjugate poles occur, this procedure may be used, remembering that the partial fraction for each complex conjugate pair will be of the form discussed.

Multiple Poles

Let us consider that $F_1(s)$ has all simple poles except, say, at $s = p_1$ which has a multiplicity m. Then one can write

$$F_1(s) = \frac{N_2(s)}{(s - p_1)^m (s - p_2) \cdots (s - p_n)} \tag{3.3.22}$$

The partial fraction expansion of $F_1(s)$ is given by

$$F_1(s) = \frac{K_{11}}{(s - p_1)^m} + \frac{K_{12}}{(s - p_1)^{m-1}} + \cdots + \frac{K_{1m}}{(s - p_1)}$$

$$+ \frac{K_2}{(s - p_2)} + \cdots + \frac{K_n}{(s - p_n)} \tag{3.3.23}$$

When a multiple root is involved, there will be as many coefficients associated with the multiple root as the order of multiplicity. For each simple pole p_k we have just one coefficient K_k, as before.

For simple poles one can proceed as discussed earlier and apply Equation (3.3.13) to calculate the residues K_k. To evaluate $K_{11}, K_{12}, \ldots, K_{1m}$ we multiply both sides of Equation (3.3.23) by $(s - p_1)^m$ to obtain

$$(s - p_1)^m F_1(s) = K_{11} + (s - p_1) K_{12} + \cdots$$

$$+ (s - p_1)^{m-1} K_{1m} + (s - p_1)^m$$

$$\times \left(\frac{K_2}{s - p_2} + \cdots + \frac{K_n}{s - p_n} \right) \tag{3.3.24}$$

The coefficient K_{11} may now be evaluated as

$$K_{11} = \lim_{s \to p_1} \left[(s - p_1)^m F_1(s) \right] \tag{3.3.25}$$

Next we differentiate Equation (3.3.24) with respect to s and let $s \to p_1$ in order to evaluate K_{12},

$$K_{12} = \lim_{s \to p_1} \left\{ \frac{d}{ds} \left[(s - p_1)^m F_1(s) \right] \right\} \tag{3.3.26}$$

The differentiation process can be continued to find the kth coefficient:

$$K_{1k} = \lim_{s \to p_1} \left\{ \frac{1}{(k-1)!} \frac{d^{k-1}}{ds^{k-1}} \left[(s - p_1)^m F_1(s) \right] \right\}, \qquad \text{for } k = 1, 2, \cdots, m \tag{3.3.27}$$

Note that $K_2, \ldots, K_n$ terms play no role in determining the coefficients $K_{11}, K_{12}, \ldots, K_{1m}$ because of the multiplying factor $(s - p_1)^m$ in Equation (3.3.24).

The alternate representation discussed for the case of complex poles may also be extended for multiple poles by combining the terms in Equation (3.3.23) corresponding to the multiple root. In an expansion of a quotient of polynomials by partial fractions, it may, in general, be necessary to use a combination of the rules given.

The denominator of $F_1(s)$ may not be known in the factored form in some cases, and it will then become necessary to find the roots of the denominator polynomial. If the order is higher than a quadratic and simple inspection (or some engineering approximation) does not help, one may have to take recourse to the computer. Computer programs for finding roots of a polynomial equation (using such methods as Newton–Raphson) are available these days in most computer libraries.

With the aid of theorems concerning Laplace transforms and the table of transforms, linear differential equations can be solved by the Laplace transform method. Transformations of the terms of the differential equation yields an algebraic equation in terms of the variable s. Thereafter, the solution of the differential equation is affected by simple algebraic manipulations in the s-domain. By inverting the transform of the solution from the s-domain, one can get back to the time domain. The response due to each term in the partial-fraction expansion is determined directly from the transform table. There is no need to perform any kind of integration. Because initial conditions are automatically incorporated into the Laplace transforms and the constants arising from the initial conditions are automatically evaluated, the resulting response expression yields directly the total solution. The flow diagram is illustrated in Figure 3.3.1.

Eliminating the need to write the circuit differential equations explicitly, *transformed networks*, which are networks converted directly from the time domain to the frequency domain, are used. For the three elements R, L, and C, the transformed network equivalents, using Table 3.3.1, are based on the Laplace transforms of their respective volt-ampere characteristics, as follows:

$$\mathcal{L}[v(t) = Ri(t)] \quad \rightarrow \quad V(s) = RI(s) \tag{3.3.28}$$

$$\mathcal{L}\left[v(t) = L\frac{di(t)}{dt}\right] \quad \rightarrow \quad V(s) = sLI(s) - Li(0^+) \tag{3.3.29}$$

$$\mathcal{L}\left[v(t) = \frac{1}{C}\int_{-\infty}^{t} i(t)\,dt\right] \quad \rightarrow \quad V(s) = \frac{1}{Cs}I(s) + \frac{v(0^+)}{s} \tag{3.3.30}$$

Note that $\int_{-\infty}^{0} i(t)\,dt = q(0^-)$ and $q(0^-)/C = v(0^-) = v(0^+)$ because of continuity of capacitor voltage. Figure 3.3.2 shows the time-domain networks and the transformed network equivalents in the frequency domain for elements R, L, and C. Notice the inclusion of the initial inductor current by means of the voltage term $[Li(0^+)]$ and the initial capacitor voltage by the voltage $[v(0^+)/s]$. Source conversion can be applied to obtain alternate transformed equivalent networks, as shown in Figure 3.3.2.

The following procedure is applied for the solution of network problems utilizing the transformed networks in the frequency domain.

1. Replace the time-domain current and voltage sources by the Laplace transforms of their time functions; similarly, replace the dependent-source representations. Transform the circuit elements into the frequency domain with the use of Figure 3.3.2.

2. For the resultant transformed network, write appropriate KVL and KCL equations using the mesh and nodal methods of analysis.

3. Solve algebraically for the desired network response in the frequency domain.

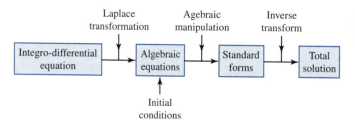

Figure 3.3.1 Flow diagram for Laplace transformation method of solving differential equations.

Time – Domain Network	Transformed Network in Frequency Domain

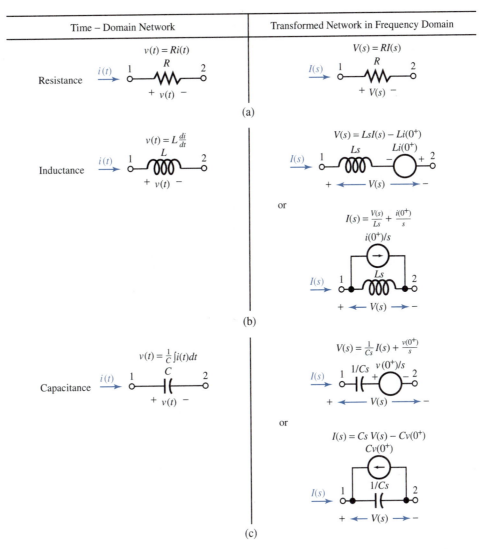

Figure 3.3.2 Time-domain networks and transformed network equivalents in frequency domain for R, L, and C.

4. By taking the inverse Laplace transform of the frequency-domain response, obtain the time-domain response.

Let us illustrate the use of this procedure with some examples.

EXAMPLE 3.3.1

Consider the circuit shown in Figure E3.3.1(a) in which the switch S has been in position 1 for a long time. Let the switch be changed instantaneously to position 2 at $t = 0$. Obtain $v(t)$ for $t \geq 0$ with the use of the Laplace transform method.

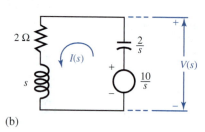

Figure E3.3.1

(a)

(b)

Solution

With switch S in position 1 for a long time,

$$v_C(0^-) = v_C(0^+) = 10 \text{ V}$$

$$i_L(0^-) = i_L(0^+) = 0$$

The transformed network in the frequency domain is shown in Figure E3.3.1(b).
The KVL equation is given by

$$I(s)\left(\frac{2}{s} + 2 + s\right) = \frac{10}{s}$$

or

$$I(s) = \frac{10}{s^2 + 2s + 2} \qquad \text{and} \qquad V(s) = I(s)(2 + s)$$

Thus,

$$V(s) = \frac{10(s+2)}{s^2 + 2s + 2} = \frac{K_1}{s + 1 - j1} + \frac{K_1^*}{s + 1 + j1}$$

where

$$K_1 = \frac{10(-1 + j1 + 2)}{(-1 + j1) + 1 + j1} = 5\sqrt{2}\, e^{-j\pi/4}$$

By taking the inverse Laplace transform, we get

$$v(t) = 10\sqrt{2}\, e^{-t} \cos\left(t - \frac{\pi}{4}\right) \text{ V}$$

EXAMPLE 3.3.2

Obtain $v(t)$ in the circuit of Figure E3.3.2(a) by using the Laplace transform method.

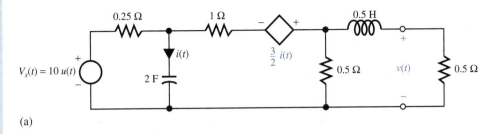

(a)

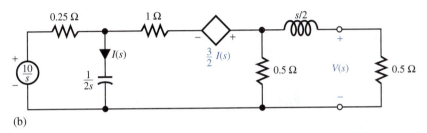

(b)

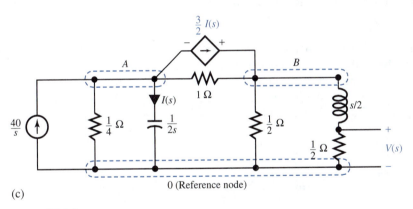

(c)

Figure E3.3.2

Solution

All initial values of inductor current and capacitor voltage are zero for $t < 0$ since no excitation exists. The transformed network is shown in Figure E3.3.2(b). Let us convert all voltage sources to current sources and use nodal analysis [Figure E3.3.2(c)].

The KCL equations are given by

$$V_A (4 + 2s + 1) - V_B (1) = \frac{40}{s} - \frac{3}{2} I_s$$

$$- V_A (1) + V_B \left(1 + 2 + \frac{1}{s/2 + 1/2} \right) = \frac{3}{2} I (s)$$

$$I (s) = 2s V_A$$

Substitution of $I(s)$ and solving for V_B gives

$$V_B = \frac{10\,(3s+1)\,(s+1)}{3s\,(s^2+3s+2)}$$

$$V_A = \frac{1/2}{1/2+s/2}V_B = \frac{1}{s+1}V_B = \frac{10(3s+1)}{3s\,(s+1)(s+2)} = \frac{K_1}{s} + \frac{K_2}{s+1} + \frac{K_3}{s+2}$$

where

$$K_1 = \frac{10\,(0+1)}{3\,(0+1)\,(0+2)} = \frac{5}{3}$$

$$K_2 = \frac{10\,(-3+1)}{3\,(-1)\,(-1+2)} = \frac{20}{3}$$

$$K_3 = \frac{10\,(-6+1)}{3\,(-2)\,(-2+1)} = -\frac{25}{3}$$

Therefore,

$$v\,(t) = \left(\frac{5}{3} + \frac{20}{3}e^{-t} - \frac{25}{3}e^{-2t}\right)u\,(t)\ \text{V}$$

Note that Thévenin's and Norton's network theorems are applicable in the frequency domain.

Network functions (also known as *system functions*) are defined as the ratio of the response to the excitation in the frequency domain, as illustrated in Figure 3.3.3. Driving-point impedances and admittances are network functions, as shown in Figure 3.3.4.

The concept of *transfer functions* is illustrated in Figure 3.3.5, in which the response measured at one pair of terminals is related to an excitation applied to another pair of terminals.

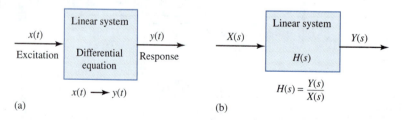

(a)

(b)

Figure 3.3.3 Linear system with no initial energy storage. (a) Time domain. (b) Frequency domain.

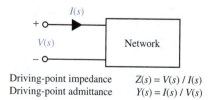

Figure 3.3.4 Network functions.

Driving-point impedance $\quad Z(s) = V(s)\,/\,I(s)$
Driving-point admittance $\quad Y(s) = I(s)\,/\,V(s)$

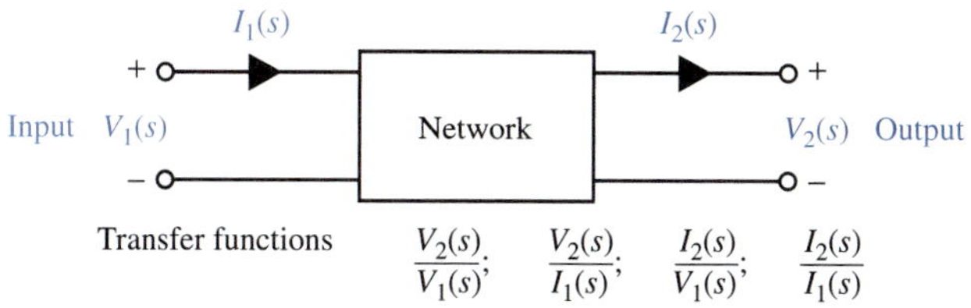
$I_1(s)$
$I_2(s)$
Input $V_1(s)$
Network
$V_2(s)$ Output
Transfer functions
$\frac{V_2(s)}{V_1(s)}$; $\frac{V_2(s)}{I_1(s)}$; $\frac{I_2(s)}{V_1(s)}$; $\frac{I_2(s)}{I_1(s)}$

Figure 3.3.5 Network transfer functions.

EXAMPLE 3.3.3

A network function is given by

$$H(s) = \frac{2(s+2)}{(s+1)(s+3)}$$

(a) For $x(t) = \delta(t)$, obtain $y(t)$.

(b) For $x(t) = u(t)$, obtain $y(t)$.

(c) For $x(t) = e^{-4t}$, obtain $y(t)$.

(d) Express the differential equation that relates $x(t)$ and $y(t)$.

Solution

(a) For $x(t) = \delta(t)$, $X(s) = 1$. Hence,

$$Y(s) = H(s)X(s) = H(s) = \frac{2(s+2)}{(s+1)(s+3)} = \frac{K_1}{s+1} + \frac{K_2}{s+3}$$

where

$$K_1 = \frac{2(-1+2)}{-1+3} = 1$$

$$K_2 = \frac{2(-3+2)}{-3+1} = 1$$

Thus,

$$y(t) = e^{-t} + e^{-3t}$$

Note that $\mathcal{L}^{-1}[H(s)]$ yields the natural response of the system.

(b) For $x(t) = u(t)$, $X(s) = 1/s$. Hence,

$$Y(s) = H(s)X(s) = H(s)\frac{1}{s} = \frac{2(s+2)}{(s+1)(s+3)}\frac{1}{s} = \frac{K_1}{s+1} + \frac{K_2}{s+3} + \frac{K_3}{s}$$

where

$$K_1 = \frac{2(-1+2)}{(-1+3)(-1)} = -1$$

$$K_2 = \frac{2(-3+2)}{(-3+1)(-3)} = -\frac{1}{3}$$

$$K_3 = \frac{2(0+2)}{(0+1)(0+3)} = \frac{4}{3}$$

Thus,

$$y(t) = -e^{-t} - \frac{1}{3}e^{-3t} + \frac{4}{3}$$

The first two terms on the right-hand side are the natural response arising from $H(s)$, whereas the last term is the forced response arising from $X(s)$. Note that K_3 is $H(s)|_{s=0}$, while $X(s) = 1/s$.

(c) For $x(t) = e^{-4t}$, $X(s) = 1/(s+4)$. Hence,

$$Y(s) = H(s)X(s) = H(s)\frac{1}{s+4} = \frac{2(s+2)}{(s+1)(s+3)(s+4)}$$

$$= \frac{K_1}{s+1} + \frac{K_2}{s+3} + \frac{K_3}{s+4}$$

where

$$K_1 = \frac{2(-1+2)}{(-1+3)(-1+4)} = \frac{1}{3}$$

$$K_2 = \frac{2(-3+2)}{(-3+1)(-3+4)} = 1$$

$$K_3 = \frac{2(-4+2)}{(-4+1)(-4+3)} = -\frac{4}{3}$$

Thus,

$$y(t) = \frac{1}{3}e^{-t} + e^{-3t} - \frac{4}{3}e^{-4t}$$

The first two terms on the right-hand side are the natural response arising from $H(s)$, whereas the last term is the forced response arising from $X(s)$. Note that K_3 is $H(s)|_{s=-4}$, whereas $X(s) = 1/(s+4)$.

(d) $Y(s) = H(s)X(s) = \dfrac{2(s+2)}{(s+1)(s+3)}X(s)$

or $(s+1)(s+3)Y(s) = 2(s+2)X(s)$, or $(s^2 + 4s + 3)Y(s) = (2s+4)X(s)$. Recognizing that multiplication by s in the frequency domain corresponds to differentiation in the time domain, the differential equation is expressed as

$$\frac{d^2y}{dt^2} + 4\frac{dy}{dt} + 3y = 2\frac{dx}{dt} + 4x$$

3.4 FREQUENCY RESPONSE

Earlier we examined the circuit response to sinusoids that have a fixed frequency. Now let us examine the response of a circuit to a sinusoidal source, called an *oscillator*, whose frequency can be varied. Known as the *frequency response*, it is often expressed as a network function, which is an output–input ratio. In order to visualize the changes in phase shift and amplitude as the frequency of the input signal changes, a graphical display of the frequency-response characteristics of networks is often employed.

A *filter* is a network used to select one range of frequencies, while rejecting all other frequencies. Let us now consider simple networks which function as filters. For the RC network shown in Figure 3.4.1, the voltage transfer function $\bar{V}_{out}/\bar{V}_{in}$ is given by

$$\bar{H}(j\omega) = \frac{\bar{V}_{out}}{\bar{V}_{in}} = \frac{1}{1+j\omega RC} = \frac{1}{\sqrt{1+(\omega RC)^2}} \angle - \tan^{-1}(\omega RC) = H(\omega) \angle \theta(\omega) \quad (3.4.1)$$

where $H(\omega)$ is the *amplitude ratio* and $\theta(\omega)$ is the *phase shift*. Defining the *cutoff frequency* as the frequency ω at which $H(\omega)$ is reduced to $1/\sqrt{2}$, or 0.707 of its maximum value (which is 1.0 at $\omega = 0$), it follows that

$$\omega_{CO} = \frac{1}{RC} \quad (3.4.2)$$

and the phase shift becomes $-45°$ at that cutoff frequency, as shown in Figure 3.4.2. The power delivered to the circuit, at this frequency, is one-half the maximum power. As a consequence, ω_{CO} is also called the *half-power point,* or half-power angular frequency. We see that this filter circuit of Figure 3.4.1 is indeed a *low-pass* filter because any input signal components at $\omega << \omega_{CO}$ are passed to the output with virtually unchanged amplitude and phase, whereas any components at $\omega >> \omega_{CO}$ have greatly reduced output amplitudes. ω_{CO} roughly divides the *passband* (region of transmitted frequencies) from the *stopband* (region of high attenuation). The range of transmitted frequencies is known as the *bandwidth*. Note that in Figure 3.4.1 the low-pass filtering properties come from the fact that the capacitor is an open circuit for $\omega = 0$ but becomes a short circuit for $\omega \to \infty$ and "shorts out" any high-frequency voltage components across the output.

The low-pass simple filter circuits are shown in Figure 3.4.3 and Figure 3.4.4 gives the high-pass simple filter circuits. Note the interchange of resistance and reactance, which converts a low-pass filter into a high-pass filter, or vice versa.

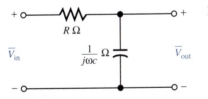

Figure 3.4.1 *RC* network.

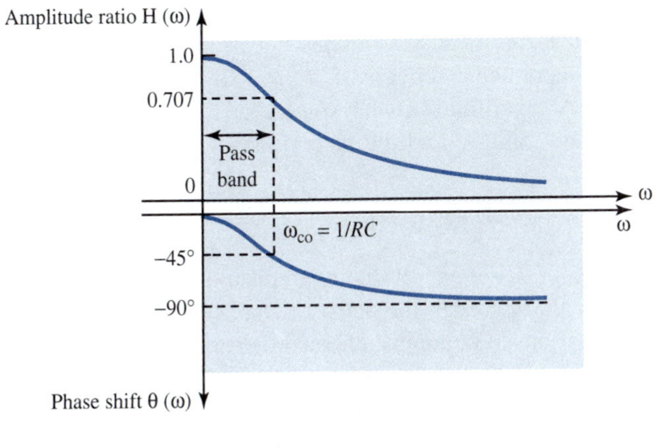

Figure 3.4.2 Frequency response of *RC* network of Figure 3.4.1.

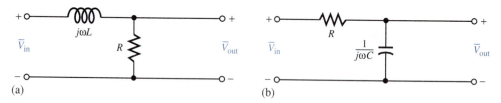

Figure 3.4.3 Low-pass simple filter circuits. (a) *RL* low-pass filter. (b) *RC* low-pass filter.

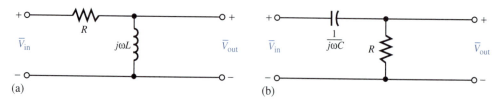

Figure 3.4.4 High-pass simple filter circuits (a) *RL* highpass filter. (b) *RC* high-pass filter.

The low-pass transfer function, given by Equation (3.4.1), can be written as

$$H(\omega) = \frac{1}{\sqrt{1 + (\omega/\omega_{CO})^2}}; \qquad \theta(\omega) = -\tan^{-1}\frac{\omega}{\omega_{CO}} \tag{3.4.3}$$

which also applies to the *RL* low-pass filter with $\omega_{CO} = R/L$.

The high-pass transfer function for either of the circuits of Figure 3.4.4 is given by

$$H(\omega) = \frac{(\omega/\omega_{CO})}{\sqrt{1 + (\omega/\omega_{CO})^2}}; \qquad \theta(\omega) = 90° - \tan^{-1}\frac{\omega}{\omega_{CO}} \tag{3.4.4}$$

where $\omega_{CO} = R/L$ or $1/RC$. The ideal response characteristics of low pass, high pass, bandpass, and band reject are shown in Figure 3.4.5.

Due to the greater availability and lower cost of capacitors and the undesirable resistance usually associated with inductors, virtually all simple filter designs employ capacitors rather than inductors, despite the similarity of *RL* and *RC* circuits. More sophisticated filters use two or more reactive elements in order to have essentially constant or flat $H(\omega)$ over the passband and narrow transition between the passband and the stopband.

One of the most common forms of display of the network function $\bar{H}(j\omega)$ is the *Bode diagram*. In this diagram the magnitude and phase of $\bar{H}(j\omega)$ are plotted separately as functions of the frequency variable ω. A logarithmic scale is used for the frequency variable in order to accommodate wide frequency ranges, and the magnitude of the network function $H(\omega)$ is expressed in *decibels* (dB) as

$$H(\text{dB}) = 20 \log_{10} H \tag{3.4.5}$$

The two portions of the Bode plot, $H(\omega)$ and $\theta(\omega)$, are graphed on semilog axes. The advantage of this technique is that rather than plotting the network characteristic point by point, one can employ straight-line approximations to obtain the characteristic quite efficiently.

Asymptotic Bode plots lead to simply drawn approximate characteristics which are quite adequate for many purposes and reduce the calculational complications greatly. The frequency response is found by substituting $s = j\omega$ in the network function $H(s)$. Let

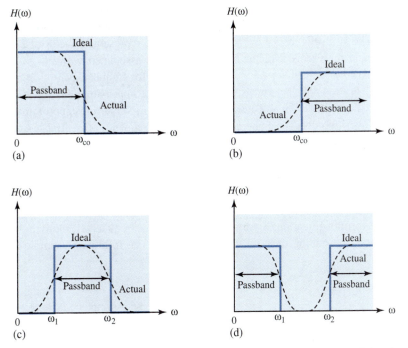

Figure 3.4.5 Ideal response characteristics. **(a)** Low pass. **(b)** High pass. **(c)** Band pass. **(d)** Band reject.

$$\bar{H}(j\omega) = k \frac{\left(1 + \frac{j\omega}{0_1}\right)\left(1 + \frac{j\omega}{0_2}\right)\cdots\left(1 + \frac{j\omega}{0_m}\right)}{\left(1 + \frac{j\omega}{p_1}\right)\left(1 + \frac{j\omega}{p_2}\right)\cdots\left(1 + \frac{j\omega}{p_n}\right)}$$ (3.4.6)

where $-0_1, -0_2, \ldots, -0_m$ are the zeros, and $-p_1, -p_2, \ldots, -p_n$ are the poles of the network function. Equation (3.4.6) is clearly the product of a constant and a group of terms having the form $(1 + j\omega/\omega_0)$ or $1/(1 + j\omega/\omega_0)$. Each of these terms can be considered as an individual phasor, and $\bar{H}(j\omega)$ has a magnitude given by the product of the individual magnitudes (or the sum of the individual terms expressed in dB) and an angle given by the sum of the individual angles. Thus the behavior of functions $(1 + j\omega/\omega_0)$ and $1/(1 + j\omega/\omega_0)$ is to be clearly understood for the construction of Bode plots.

Let us first consider $\bar{H}_1(j\omega) = 1 + j\omega/\omega_0$. For $\omega/\omega_0 << 1$ (i.e., low frequencies), the magnitude $H_1(\omega) \cong 1$, or $H_1(\omega)_{dB} = 20 \log 1 = 0$ dB. The *break frequency* is $\omega = \omega_0$. At high frequencies (i.e., $\omega/\omega_0 >> 1$), the function behaves as

$$H_1(\omega) \cong \frac{\omega}{\omega_0} \qquad \text{or} \qquad H_1(\omega)_{dB} = 20 \log \frac{\omega}{\omega_0}$$

at $\omega/\omega_0 = 1$,

$$H_1(\omega)_{dB} = 0 \text{ dB}$$

at $\omega/\omega_0 = 10$,

$$H_1(\omega)_{dB} = 20 \log 10 = 20 \text{ dB}$$

and at $\omega/\omega_0 = 100$,

$$H_1(\omega)_{dB} = 20 \log 100 = 40 \text{ dB}$$

Since factors of 10 are linear increments on the logarithmic frequency scale, the plot of 20 log (ω/ω_0) versus ω on a semilog paper is a straight line with a slope of +20 dB/decade or +6 dB/octave, as shown in Figure 3.4.6. Note that an octave represents a factor of 2 in frequency, whereas a decade represents a factor of 10. The angle associated with $\bar{H}_1(j\omega)$ is $\theta_1(\omega) = \tan^{-1}(\omega/\omega_0)$. At low frequencies ($\omega \leq 0.1\omega_0$), $\theta_1 \cong 0$; for frequencies $\omega \geq 10\omega_0$, $\theta_1 \cong 90°$; and at $\omega = \omega_0$, $\theta_1 = +45°$. This leads to the straight-line approximations for $\theta_1(\omega)$, as shown in Figure 3.4.6. For $0.1\omega_0 \leq \omega \leq 10\omega_0$, the slope of $\theta_1(\omega)$ versus ω is approximately given by +45°/decade (over two decades centered at the critical frequency) or +13.5°/octave. Thus the asymptotic Bode plot for $\bar{H}_1(j\omega) = 1 + j\omega/\omega_0$ is drawn in Figure 3.4.6. Figure 3.4.7 gives the asymptotic Bode plot for $\bar{H}_2(j\omega) = 1/(1 + j\omega/\omega_0)$ with a similar kind of reasoning. The student is encouraged to justify the plot. The dashed curves in Figures 3.4.6 and 3.4.7 indicate the exact magnitude and phase responses.

The process of developing asymptotic Bode plots is then one of expressing the network function in the form of Equation (3.4.6), locating the break frequencies, plotting the component asymptotic lines, and adding these to get the resultant. The following example illustrates the procedure.

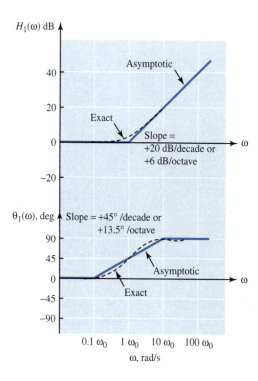

Figure 3.4.6 Asymptotic Bode plot for $\bar{H}_1(j\omega) = 1 + j\omega/\omega_0$.

EXAMPLE 3.4.1

Sketch the asymptotic Bode plot for

$$H(s) = \frac{10^4 (s + 50)}{s^2 + 510s + 5000}$$

Solution

$$H(s) = \frac{100(1 + s/50)}{(1 + s/10)(1 + s/500)} \quad \text{or} \quad \bar{H}(j\omega) = \frac{100(1 + j\omega/50)}{(1 + j\omega/10)(1 + j\omega/500)}$$

The break frequencies are 10 and 500 rad/s for the denominator and 50 rad/s for the numerator. The component straight-line segments are drawn as shown in Figure E3.4.1. Note that the effect of the constant multiplier 100 in $\bar{H}(j\omega)$ is to add a constant value, $20 \log 100 = 40$ dB, as marked in the Bode plot. The resultant asymptotic magnitude $H(\omega)$ and angle $\theta(\omega)$ characteristics are indicated by the dashed lines.

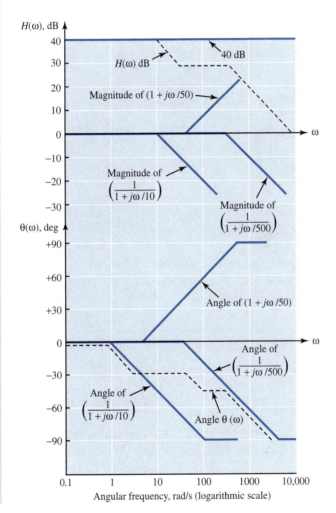

Figure E3.4.1 Asymptotic Bode plot.

As seen from Example 3.4.1, the frequency response in terms of the asymptotic Bode plot is obtained with far less computation than needed for the exact characteristics. With little additional effort, by correcting errors at a few frequencies within the asymptotic plot, one can get a sufficiently accurate result for most engineering purposes. While we have considered $\bar{H}(j\omega)$ of the type given by Equation (3.4.6) with only simple poles and zeros, two additional cases need further consideration:

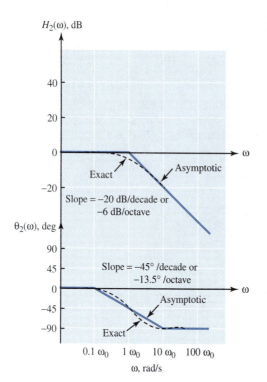

$H_2(\omega)$, dB

$\theta_2(\omega)$, deg

ω, rad/s

Figure 3.4.7 Asymptotic Bode plot for $\bar{H}_2(j\omega) = 1/(1 + j\omega/\omega_0)$.

Case 1: Zero or Pole at Origin

$\lim_{s\to0} H(s) = Ks^n$, where n can be any positive or negative integer. The magnitude characteristic for this factor has a slope of $20n$ dB/decade and passes through K dB at $\omega = 1$, while the component angle characteristic has a constant value of $n \times 90°$.

Case 2: Some Zeros or Poles Occur as Complex Conjugate Pairs

Let

$$H(s) = \left(1 + \frac{s}{\alpha + j\beta}\right)\left(1 + \frac{s}{\alpha - j\beta}\right) = 1 + 2\frac{\alpha}{\alpha^2 + \beta^2}s + \frac{s^2}{\alpha^2 + \beta^2}$$

With a break frequency of $\sqrt{\alpha^2 + \beta^2}$, the magnitude is characterized by a slope of 40 dB/decade for $\omega \geq \sqrt{\alpha^2 + \beta^2}$, and the angle characteristic shows a slope of 90°/decade in the range of $0.1\sqrt{\alpha^2 + \beta^2} \leq \omega \leq 10\sqrt{\alpha^2 + \beta^2}$. The asymptotic Bode plot for this case is shown in Figure 3.4.8.

Note that the angle and magnitude characteristics will have the opposite sign for the case of a complex conjugate pair of poles. It should also be pointed out that the straight-line quadratic representations are not generally good approximations (particularly in the one-decade region on either side of the break frequency), and a few points may be calculated to obtain the correct curves in the questionable frequency range.

For the series *RLC* circuit, the input admittance is given by

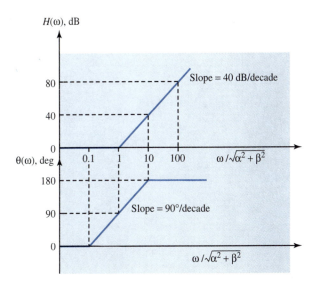

Figure 3.4.8 Asymptotic Bode plot for the quadratic $H(s) = 1 + 2\dfrac{\alpha}{\alpha^2 + \beta^2}s + \dfrac{s^2}{\alpha^2 + \beta^2}$.

$$\bar{Y}(j\omega) = \frac{1}{R + j\omega L + 1/j\omega C} = \frac{1}{R + j(\omega L - 1/\omega C)} \qquad (3.4.7)$$

At the *series resonant frequency* $\omega_0 = 1/\sqrt{LC}$,

$$\bar{Y}(j\omega_0) = \frac{1}{R} = Y_0$$

corresponds to maximum admittance (or minimum impedance). The ratio Y/Y_0 can be written as

$$\frac{\bar{Y}}{Y_0}(j\omega) = \frac{R}{R + j\left(\omega L - \dfrac{1}{\omega C}\right)} = \frac{1}{1 + j\left(\dfrac{L}{R}\right)\left(\omega - \dfrac{1}{\omega L C}\right)}$$

$$= \frac{1}{1 + j\dfrac{\omega_0 L}{R}\left(\dfrac{\omega}{\omega_0} - \dfrac{\omega_0}{\omega}\right)} \qquad (3.4.8)$$

Introducing a *quality factor* $Q_S = \omega_0 L/R$ and *per-unit source frequency deviation* $\delta = (\omega - \omega_0)/\omega_0$,

$$\frac{\bar{Y}}{Y_0}(j\omega) = \frac{1}{1 + jQ_S\left(\dfrac{\omega}{\omega_0} - \dfrac{\omega_0}{\omega}\right)} = \frac{1}{1 + j\delta Q_S\left(\dfrac{2 + \delta}{1 + \delta}\right)} \qquad (3.4.9)$$

For $\delta \ll 1$, i.e., for small frequency deviations around the resonant frequency, Equation (3.4.9) becomes, near resonance,

$$\frac{\bar{Y}}{Y_0}(j\omega) = \frac{1}{1 + j2\delta Q_S} \qquad (3.4.10)$$

which is the equation of the *universal resonance curve* plotted in Figure 3.4.9. The curve applies equally well to the parallel *GLC* circuit with $\bar{Z}/Z_0$ as the ordinate, when the value of $Q_p = \omega_0 C/G$. The bandwidth and the half-power points in a resonant circuit (either series or parallel) correspond to ω_0/Q and $2\delta Q = \pm 1$ when the magnitude Y/Y_0 or Z/Z_0 is 0.707.

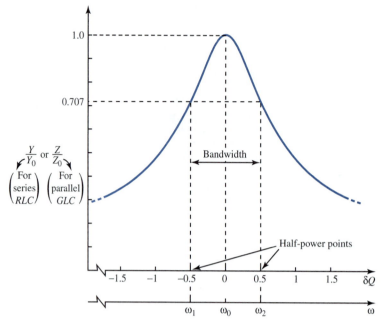

Figure 3.4.9 Universal resonance curve.

EXAMPLE 3.4.2

Consider the circuit shown in Figure E3.4.2 with $R \ll \omega L$. Find (a) the resonant angular frequency ω_0, (b) the quality factor Q, and (c) the maximum impedance Z_m. Comment on the applicability of the universal resonance curve.

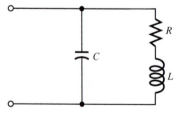

Figure E3.4.2 Circuit.

Solution

$$\bar{Z}(j\omega) = \frac{1}{j\omega C + 1/(R + j\omega L)} = \frac{R + j\omega L}{1 + j\omega RC - \omega^2 LC}$$

For $R \ll \omega L$,

$$\bar{Z}(j\omega) = \frac{j\omega L}{1 + j\omega RC - \omega^2 LC} = \frac{1}{(RC/L) + j(\omega C - 1/\omega L)}$$

$$\omega_0 = \frac{1}{\sqrt{LC}}; \qquad Q = \frac{\omega_0 L}{R}; \qquad Z_m = \frac{L}{RC} = RQ^2$$

The universal resonance curve and the conclusions obtained from it apply to this circuit, provided $\omega L \gg R$, or $Q \geq 10$.

Two-Port Network Parameters

A pair of terminals at which a signal may enter or leave a network is called a *port*, and a network having two such pairs of terminals, labeled generally input and output terminals, is known as a *two-port network*. A two-port network forms a very important building block in electronic systems, communication systems, automatic control systems, transmission and distribution systems, or in other systems in which an electric signal or electric energy enters the input terminals, is acted upon by the network, and leaves via the output terminals.

The *two-port network* (or two-terminal-pair network) shown in Figure 3.4.10 is described by port currents and voltages (i.e., four variables) related by a set of linear equations in terms of *two-port parameters*. Of the six possible parameter sets, the three used extensively in electronic circuit analysis are given next in the *frequency domain*.

1. Two-port admittance or y parameters (in siemens),

$$I_1 = y_{11} V_1 + y_{12} V_2 = y_i V_1 + y_r V_2 \tag{3.4.11}$$

$$I_2 = y_{21} V_1 + y_{22} V_2 = y_f V_1 + y_o V_2 \tag{3.4.12}$$

where

$$y_{11} = y_i = \frac{I_1}{V_1}\bigg|_{V_2=0} = \text{short-circuit input admittance}$$

$$y_{12} = y_r = \frac{I_1}{V_2}\bigg|_{V_1=0} = \text{short-circuit reverse transfer admittance}$$

$$y_{21} = y_f = \frac{I_2}{V_1}\bigg|_{V_2=0} = \text{short-circuit forward transfer admittance}$$

$$y_{22} = y_o = \frac{I_2}{V_2}\bigg|_{V_1=0} = \text{short-circuit output admittance}$$

2. Two-port impedance or z parameters (in ohms),

$$V_1 = z_{11} I_1 + z_{12} I_2 = z_i I_1 + z_r I_2 \tag{3.4.13}$$

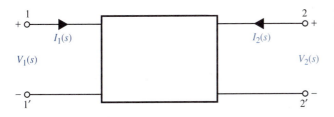

Figure 3.4.10 Two-port network (or two-terminal-pair network).

$$V_2 = z_{21}I_1 + z_{22}I_2 = z_f I_1 + z_o I_2 \tag{3.4.14}$$

where

$$z_{11} = z_i = \left.\frac{V_1}{I_1}\right|_{I_2=0} = \text{open-circuit input impedance}$$

$$z_{12} = z_r = \left.\frac{V_1}{I_2}\right|_{I_1=0} = \text{open-circuit reverse transfer impedance}$$

$$z_{21} = z_f = \left.\frac{V_2}{I_1}\right|_{I_2=0} = \text{open-circuit forward transfer impedance}$$

$$z_{22} = z_o = \left.\frac{V_2}{I_2}\right|_{I_1=0} = \text{open-circuit output impedance}$$

3. Two-port hybrid or h parameters,

$$V_1 = h_{11}I_1 + h_{12}V_2 = h_i I_1 + h_r V_2 \tag{3.4.15}$$
$$I_2 = h_{21}I_1 + h_{22}V_2 = h_f I_1 + h_o V_2 \tag{3.4.16}$$

where

$$h_{11} = h_i = \left.\frac{V_1}{I_1}\right|_{V_2=0} = \text{short-circuit input impedance (ohms)}$$

$$h_{12} = h_r = \left.\frac{V_1}{V_2}\right|_{I_1=0} = \text{open-circuit reverse voltage gain (dimensionless)}$$

$$h_{21} = h_f = \left.\frac{I_2}{I_1}\right|_{V_2=0} = \text{short-circuit forward current gain (dimensionless)}$$

$$h_{22} = h_o = \left.\frac{I_2}{V_2}\right|_{I_1=0} = \text{open-circuit output admittance (siemens)}$$

EXAMPLE 3.4.3

Consider Equations (3.4.11) through (3.4.16). Develop the y-parameter, z-parameter, and h-parameter equivalent circuits. Also express the z-parameters in terms of y-parameters.

Solution

Equations (3.4.11) and (3.4.12) can be expressed as

$$-I_1 + (y_{11} + y_{12})V_1 - y_{12}(V_1 - V_2) = 0$$

$$-I_2 + (y_{22} + y_{12})V_2 - y_{12}(V_2 - V_1) + (y_{21} - y_{12})V_1 = 0$$

The y-parameter equivalent circuit that satisfies these equations is shown in Figure E3.4.3(a).

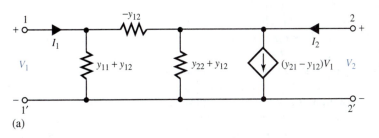

Figure E3.4.3 (a) y-parameter equivalent circuit. (b) z-parameter equivalent circuit. (c) h-parameter equivalent circuit.

(a)

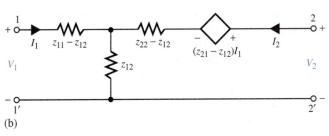

(b)

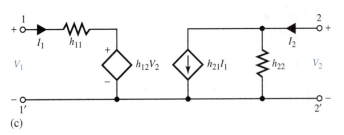

(c)

Equations (3.4.13) and (3.4.14) can be arranged as

$$V_1 = (z_{11} - z_{12})\, I_1 + z_{12}\,(I_1 + I_2)$$

$$V_2 = (z_{21} - z_{12})\, I_1 + (z_{22} - z_{12})\, I_2 + z_{12}\,(I_1 + I_2)$$

The z-parameter equivalent circuit that satisfies these equations is shown in Figure E3.4.3(b). The h-parameter equivalent circuit satisfying Equations (3.4.15) and (3.4.16) is shown in Figure E3.4.3(c).

Equations (3.4.11) and (3.4.12) can be solved simultaneously to yield

$$V_1 = \frac{y_{22}}{y_{11}y_{22} - y_{12}y_{21}} I_1 + \frac{-y_{12}}{y_{11}y_{22} - y_{12}y_{21}} I_2$$

$$V_2 = \frac{-y_{21}}{y_{11}y_{22} - y_{12}y_{21}} I_1 + \frac{y_{11}}{y_{11}y_{22} - y_{12}y_{21}} I_2$$

By comparing these with Equations (3.4.13) and (3.4.14), one gets

$$z_{11} = \frac{y_{22}}{y_{11}y_{22} - y_{12}y_{21}}; \qquad z_{12} = \frac{-y_{12}}{y_{11}y_{22} - y_{12}y_{21}}$$

$$z_{21} = \frac{-y_{21}}{y_{11}y_{22} - y_{12}y_{21}}; \qquad z_{22} = \frac{y_{11}}{y_{11}y_{22} - y_{12}y_{21}}$$

Block Diagrams

The mathematical relationships of control systems are usually represented by *block diagrams,* which show the role of various components of the system and the interaction of variables in it. It is common to use a block diagram in which each component in the system (or sometimes a group of components) is represented by a block. An entire system may, then, be represented by the interconnection of the blocks of the individual elements, so that their contributions to the overall performance of the system may be evaluated. The simple configuration shown in Figure 3.4.11 is actually the basic building block of a complex block diagram. In the case of linear systems, the input–output relationship is expressed as a transfer function, which is the ratio of the Laplace transform of the output to the Laplace transform of the input with initial conditions of the system set to zero. The arrows on the diagram imply that the block diagram has a unilateral property. In other words, signal can only pass in the direction of the arrows.

A box is the symbol for multiplication; the input quantity is multiplied by the function in the box to obtain the output. With circles indicating summing points (in an algebraic sense) and with boxes or blocks denoting multiplication, any linear mathematical expression may be represented by block-diagram notation, as in Figure 3.4.12 for the case of an elementary feedback control system.

The block diagrams of complex feedback control systems usually contain several feedback loops, and they may have to be simplified in order to evaluate an overall transfer function for the system. A few of the block diagram reduction manipulations are given in Table 3.4.1; no attempt is made here to cover all the possibilities.

Figure 3.4.11 Basic building block of a block diagram.

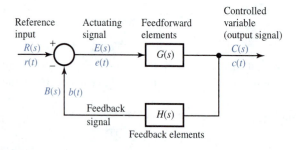

Figure 3.4.12 Block diagram of an elementary feedback control system.

$R(s)$ Reference input
$C(s)$ Output signal (controlled variable)
$B(s)$ Feedback signal $= H(s)C(s)$
$E(s)$ Actuating signal (error) $= [R(s) - B(s)]$
$G(s)$ Forward path transfer function or
open-loop transfer function $= C(s)/E(s)$
$M(s)$ Closed-loop transfer function $= C(s)/R(s) = G(s)/[1 + G(s)H(s)]$
$H(s)$ Feedback path transfer function
$G(s)H(s)$ Loop gain
$\dfrac{E(s)}{R(s)}$ = Error-response transfer function $\dfrac{1}{1 + G(s)H(s)}$

TABLE 3.4.1 Some of the Block Diagram Reduction Manipulations

Original Block Diagram	Manipulation	Modified Block Diagram
$R \to \boxed{G_1} \to \boxed{G_2} \to C$	Cascaded elements	$R \to \boxed{G_1 G_2} \to C$
$R \to \boxed{G_1} \to + \bigcirc \to$; $\to \boxed{G_2} \to$	Addition or subtraction (eliminating auxiliary forward path)	$R \to \boxed{G_1 \pm G_2} \to C$
$R \to \boxed{G} \to C$ (pickoff)	Shifting of pickoff point ahead of block	$R \to \boxed{G} \to C$; $\leftarrow \boxed{G}$
$R \to \boxed{G} \to C$ (pickoff)	Shifting of pickoff point behind block	$R \to \boxed{G} \to C$; $\leftarrow \boxed{1/G}$
$R \to \boxed{G} \to + \bigcirc \to E$; C	Shifting summing point ahead of block	$R + \bigcirc \to \boxed{G} \to E$; $\boxed{1/G} \leftarrow C$
$R \to + \bigcirc \to E \to \boxed{G} \to$; C	Shifting summing point behind block	$R \to \boxed{G} \to + \bigcirc \to E$; $\boxed{G} \leftarrow C$
$R \to + \bigcirc \to \boxed{G} \to C$; $\boxed{H}$	Removing H from feedback path	$R \to \boxed{1/H} \to + \bigcirc \to \boxed{H} \to \boxed{G} \to C$
$R \to + \bigcirc \to \boxed{G} \to C$; $\boxed{H}$	Eliminating feedback path	$R \to \boxed{\dfrac{G}{1 + GH}} \to C$

EXAMPLE 3.4.4

Feedback amplifiers are of great importance in electronic circuits. The block diagram of a class of feedback amplifier is shown in Figure E3.4.4(a). Determine the transfer function C/R for the block diagram.

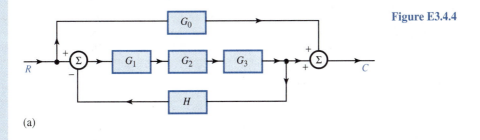

Figure E3.4.4

(a)

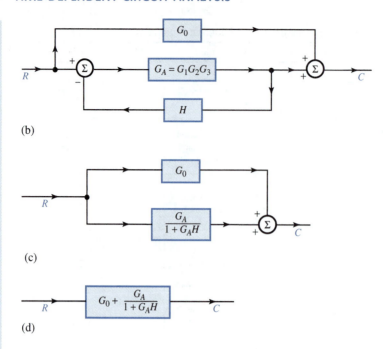

(b)

(c)

(d)

Solution

The three blocks in cascade are combined to yield $G_A = G_1G_2G_3$, as illustrated in Figure E3.4.4(b).

By eliminating the feedback path, the circuit of Figure E3.4.4(c) is obtained. By eliminating the auxiliary forward path, one gets the circuit of Figure E3.4.4(d). Thus,

$$\frac{C}{R} = G_0 + \frac{G_A}{1 + G_A H}$$

where $G_A = G_1G_2G_3$.

A linear system of equations can also be represented diagrammatically by a *signal-flow graph* (consisting of nodes and branches), which is used to describe a system schematically in terms of its constituent parts. However, in view of the scope of this text, the topic of signal-flow graph is not presented.

3.5 COMPUTER-AIDED CIRCUIT SIMULATION FOR TRANSIENT ANALYSIS, AC ANALYSIS, AND FREQUENCY RESPONSE USING PSPICE AND PROBE

Transient Analysis

PSpice is capable of performing transient circuit analysis, for which the request is given by the following statement:

- TRAN   TSTEP   TSTOP   TSTART   TMAX   UIC

where TSTEP is the interval between points printed, plotted, or reported to PROBE, TSTOP is the time value for which the analysis is stopped, TSTART is the beginning time value for which results are printed, plotted, or reported to PROBE, TMAX is the maximum time increment between computed values, and UIC is a keyword that causes PSpice to use initial conditions specified for inductors and capacitors in the element statements. In case of omission, TSTART defaults to zero and TMAX to (TSTOP − TSTART)/50. If UIC is omitted, PSpice computes initial conditions by assuming that the circuit is dc steady state prior to $t = 0$. Let us present an illustrative example.

EXAMPLE 3.5.1

Develop and execute a program to analyze the circuit shown in Figure E3.5.1(a), and use PROBE for plotting $v_C(t)$.

Figure E3.5.1

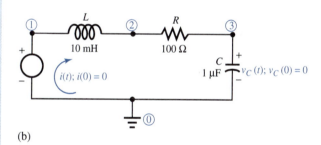

(a)

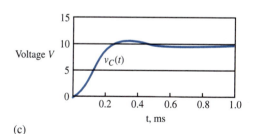

(b)

(c)

Solution

The program is as follows:

```
EXAMPLE E3.5.1
* THE CIRCUIT DIAGRAM SUITABLE FOR ANALYSIS IS GIVEN IN
FIGURE E3.5.1(b)
```

```
VS 1 0 10
L 1 2 10 MH IC = 0, INITIAL CURRENT IS ZERO
R 2 3 100
C 3 0 1 UF IC = 0; INITIAL VOLTAGE IS ZERO
* TRANSIENT ANALYSIS REQUEST
• TRAN   0.02 MS   IMS 0   0.02 MS   UIC
• PROBE
• END
```

A plot of $v_C(t)$ is shown in Figure E3.5.1(c).

Steady-State Sinusoidal Analysis with PSpice

PSpice can easily solve steady-state ac circuits for currents and voltages. Ac voltage sources, which must begin with the letter V, are specified by statements of the form

$$\text{VNAME} \quad \text{N+} \quad \text{N−} \quad \text{AC} \quad \text{VPEAK} \quad \text{PHASE}$$

where VNAME is the name of the voltage source, N+ is the node number of the positive reference, N− is the node number of the negative reference, AC indicates that this is an ac source, VPEAK is the peak value of the voltage, and PHASE is the phase angle of the source in degrees. Current sources, which must begin with the letter I, are specified in a similar manner, with the reference direction for the current source pointing from the first node number given in the statement to the second node number.

The ac analysis request is made with a statement of the form

$$\bullet \text{ AC } \quad \text{LIN} \quad \text{NPOINTS} \quad \text{STARTFREQ} \quad \text{STOPFREQ}$$

where • AC designates a request for frequency analysis, LIN defines a linear frequency sweep with NPOINTS specifying the number of points in the sweep starting at the STARTFREQ and finishing at the STOPFREQ. For example, to carry out a steady-state ac analysis for a frequency of 60 Hz, the analysis request would be

$$\bullet \text{ AC } \quad \text{LIN} \quad 1 \quad 60 \quad 60$$

Let us present an illustrative example.

EXAMPLE E3.5.2

Use PSpice to solve for the amplitude and phase of the current and the voltage across the inductor in the circuit shown in Figure E3.5.2(a).

Solution

The program is as follows:

```
* THE CIRCUIT DIAGRAM SUITABLE FOR ANALYSIS IS GIVEN IN
  FIGURE E3.5.2(b)
VS   1   0   AC   100   30
R    1   2   100
L    2   3   0.3 H
C    3   0   40 UF
```

```
• AC   LIN  1  79.58   79.58
* 79.58 IS THE SOURCE FREQUENCY IN HZ GIVEN BY 500/(2π)
• PRINT   AC   IM(R)   IP(R)   VM(2,3)   VP(2,3)
* IM(R) AND IP(R) DENOTE RESPECTIVELY MAGNITUDE AND PHASE OF THE
CURRENT THROUGH THE R ELEMENT
* VM (2,3) AND VP (2,3) DENOTE RESPECTIVELY MAGNITUDE AND PHASE OF
THE VOLTAGE BETWEEN NODES 2 AND 3 WITH THE POSITIVE REFERENCE AT
NODE 2
• END
```

After executing the program, the output file contains the following results:

FREQ	I(R)	IP(R)	V(2,3)	VP(2,3)
7.958E + 01	7.071E − 01	−1.500E + 01	1.061E + 02	7.500E + 01

Note that I(R) is the same as IM(R) directed from the first node number, 1 in this case, given for R to the second node number, 2 in this case; V(2,3) is the same as VM(2,3). Thus the phasor current is given by $\bar{I}_R = 0.7071\angle -15°$, and the voltage across the inductor is given by $\bar{V}_L = 106.1\angle 75°$. Note that the magnitudes are the peak values.

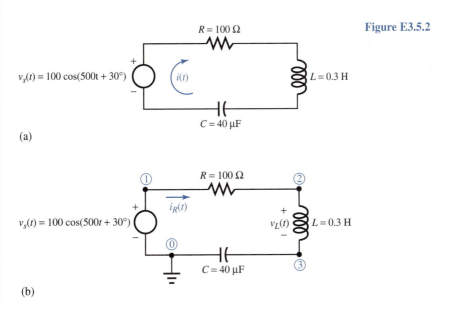

Figure E3.5.2

(a)

(b)

Analysis of Frequency Response with PSpice and PROBE

PSpice can readily accomplish the circuit analysis as a function of frequency, and PROBE can display Bode plots for magnitude and phase of network transfer functions. Besides the linear (LIN) frequency sweep that we used in steady-state sinusoidal analysis, with a similar syntax statement, a logarithmic frequency sweep, with points specifying the number of points per octave (an octave represents a twofold increase in frequency) or per decade (a decade represents a tenfold increase in frequency), can be specified. For example, the statement

• AC   DEC   20   10HZ   1MEGHZ

requests frequency analysis with 20 points per decade, which is usually a suitable value, starting at 10 Hz and ending at 1 MHz. Note that the frequencies would be uniformly spaced on a logarithmic frequency scale.

EXAMPLE E3.5.3

Develop a PSpice program and use PROBE to obtain Bode plots of the magnitude and phase of the transfer function $\bar{V}_{out}/\bar{V}_{in}$ for the high-pass filter circuit shown in Figure E3.5.3(a) along with node numbers. This filter is to pass components above 1 kHz and reject components below 1 kHz.

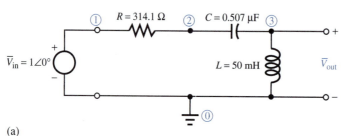

Figure E3.5.3 (a) High-pass filter circuit. (b) Magnitude Bode plot. (c) Phase plot.

(a)

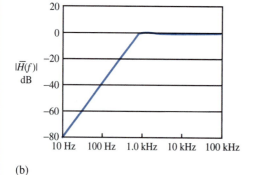

(b)

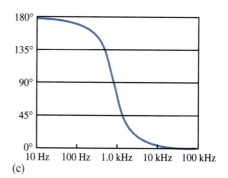

(c)

Solution

The transfer function $\bar{H}(f) = \bar{V}_{out}/\bar{V}_{in}$ is numerically equal to V(3) since the amplitude of $\bar{V}_{in}$ is given as 1 V. Thus to obtain a plot of the transfer-function magnitude in decibels, the PROBE menu commands can be used to request a plot of VDB (3). A Bode plot of the transfer function can be obtained by requesting a plot of VP (3). The program is as follows:

```
EXAMPLE E3.5.3
THE CIRCUIT DIAGRAM IS SHOWN IN FIGURE E3.5.3(a)
VIN 1 0 AC 1
R 1 2 314.1
C 2 3 0.507 UF
L 3 0 50 MH
```

- AC   DEC   20   10 HZ   100KHZ
* ANALYSIS EXTENDS FROM TWO DECADES BELOW TO TWO DECADES ABOVE
1 KHZ, THAT IS, FROM 10 HZ TO 100 KHZ.
- PROBE
* RESULTS ARE SAVED FOR DISPLAY BY THE PROBE PROGRAM
- END

Figures E3.5.3(b) and (c) display the magnitude and phase plots. Note that the filter gain is nearly 0 dB for frequencies above 1 kHz and falls at the rate of 40 dB per decade for frequencies below 1 kHz.

3.6 USE OF MATLAB IN COMPUTER-AIDED CIRCUIT SIMULATION

We shall illustrate the use of MATLAB by considering the following three examples.

EXAMPLE 3.6.1

This example is concerned with *steady-state sinusoidal analysis* by using mesh equations and MATLAB. Let us consider the circuit shown in Figure E3.6.1 and find the input impedance at the input interface terminals A–B and the ratio $\bar{V}_O/\bar{V}_S$.

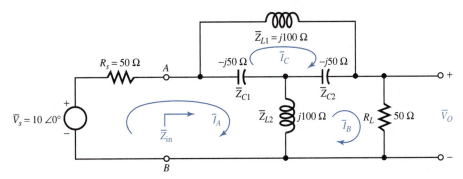

Figure E3.6.1

Solution

Note that

$$\bar{Z}_{\text{IN}} = \frac{\bar{V}_S}{\bar{I}_A} - R_S \quad \text{and} \quad \bar{K} = \frac{\bar{V}_O}{\bar{V}_S} = \frac{R_L \bar{I}_B}{\bar{V}_S}$$

The M-file and answers are as follows:

```
function example361
clc
% Circuit Parameters
RS = 50;
RL = 50;
VS = 10;
```

```
% Element parameters
ZL1 = 100j;
ZL2 = 100j;
ZC1 = -50j;
ZC2 = -50j;
% I = inverse (Z) *V
I = inv ([RS+ZC1+ZL2     -ZL2       -ZC1;
                 -ZL2   ZL2+ZC2+RL       -ZC2;
                 -ZC1      -ZC2    ZC1+ZC2+ZL1])  *  [VS 0 0]';
% Mesh Currents
IA = I(1)
IB = I(2)
IC = I(3)
% Answers
KI + RL*IB/VS
ZIN + VS/IA-RS


IA + 0.0100 - 0.0300i
IB = -0.0100 + 0.0300i
IC = 0 - 0.1000i
K = -0.0500 + 0.1500i
ZIN = 5.0000e+01 + 3.0000e+02i
```

EXAMPLE 3.6.2

Consider the circuit shown in Figure E3.6.2 in the t-domain as well as in the s-domain. Formulate the s-domain nodal equations and use MATLAB to solve for $V_A(s)$ and $V_B(s)$.

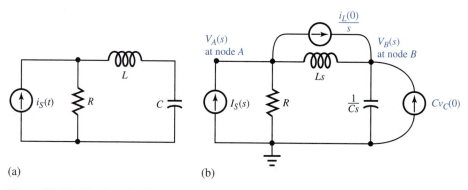

(a)   (b)

Figure E3.6.2 (a) t-domain. (b) s-domain.

Solution

At node A:
$$\frac{V_A(s)}{R} + \frac{V_A(s) - V_B(s)}{Ls} - I_S(s) + \frac{i_L(0)}{s} = 0$$

At node B: $$\frac{V_B(s)}{1/Cs} + \frac{V_B(s) - V_A(s)}{Ls} - \frac{i_L(0)}{s} - C\,v_C(0) = 0$$

Rearranging these equations, one gets

Node A: $$\left(G + \frac{1}{Ls}\right) V_A(s) - \left(\frac{1}{Ls}\right) v_B(s) = I_S(s) - \frac{i_L(0)}{s}$$

Node B: $$-\frac{1}{Ls} V_A(s) + \left(\frac{1}{Ls} + Cs\right) V_B(s) = C\,v_C(0) + \frac{i_L(0)}{s}$$

where $G = 1/R$. The M-file answers are as follows:

```
function example362
clc
syms s G L C IS iL vC
% Admittance Matrix
Y = [G+1/L/s − 1/L/s; −1/L/s C*s+1/L/s]
% Current Vector
I = [ IS − iL/s; iL/s+C*vC]
% Inverse of Y
inv(Y)
% Node Voltage Solutions
V = factor(inv(Y)*I)
Y =
[    G+1/L/s,    -1/L/s]
[    −1/L/s, C*s+1/L/s]
I =
[   IS − iL/s]
[   iL/s+C*vC]
ans =
[ (C*s^2*L+1)/(G*L*s^2*C+G+C*s),              1/(G*L*s^2*C+G+C*s)]
[            1/(G*L*s^2*C+G+C*s),     (G*L*s+1)/(G*L*s^2*C+G+C*s)]
V =
[ (C*s^2*L*IS-C*s*L*iL+IS+C*vC)/(G*L*s^2*C+G+C*s)]
[    (IS+G*L*iL+G*L*s*C*vC+C*vC)/(G*L*s^2*C+G+C*s)]
```

EXAMPLE 3.6.3

A *bandpass* filter design can be accomplished through the cascade connection shown in Figure E3.6.3, where the frequencies between the two cutoffs fall in the passband of both filters and are transmitted through the cascade connection, thereby producing the passband of the resulting bandpass filter.

Figure E3.6.3 Bandpass filter through cascade connection of high-pass and low-pass filters.

In this so-called Butterworth bandpass filter, the center frequency and the bandwidth of the bandpass filter are given by

$$\omega_0 = \sqrt{\omega_{C1}\omega_{C2}} \qquad \text{and} \qquad B = \omega_{C2} - \omega_{C1}$$

and

$$\frac{\omega_0}{B} = Q \simeq \sqrt{\frac{\omega_{C1}}{\omega_{C2}}} << 1$$

where $\omega_{CHP} = \omega_{C1}$ is the cutoff frequency of the high-pass filter and $\omega_{CLP} = \omega_{C2}$ is the cutoff frequency of the low-pass filter.

In order to design a bandpass filter with a passband gain of 0 dB and cutoff frequencies of $\omega_{C1} = 10$ rad/s and $\omega_{C2} = 50$ rad/s, with the stopband gains of less than -20 dB at 2 rad/s and 250 rad/s, one has chosen the following low-pass and high-pass transfer functions:

$$T_{LP}(s) = \frac{1}{(s/50)^2 + \sqrt{2}(s/50) + 1}$$

$$T_{HP}(s) = \frac{(s/10)^2}{(s/10)^2 + \sqrt{2}(s/10) + 1}$$

When circuits realizing these two functions are connected in cascade, the overall transfer function is given by

$$T_{BP}(s) = T_{HP}(s) \times T_{LP}(s)$$

Use MATLAB to illustrate the specified bandpass response with the center frequency $\omega_0 = \sqrt{10 \times 50} = 22.4$ rad/s, the bandwidth $B = 50 - 10 = 40$ rad/s, and the filter quality factor $Q = \omega_0/B = 0.56$. Note that Q being less than 1 indicates a broad-band response.

Solution

The M-file and answers are as follows:

```
function example363
clc
% 2nd order Butterworth low pass wc = 50
TLPs = tf(1,[1/50^2 sqrt (2)/50 1])
% 2nd order Butterworth high pass wc = 10
THPs = tf([1/10^2 0 0], [1/10^2 sqrt(2)/10 1])
% 4th order Butterworth bandpass wc1 = 10 wc2 = 50
TBPS = TLPs * THPs
bode (TBPs, {1,1e3});

Transfer function:
            1
-----------------------------
0.0004 s^2 + 0.02828 s + 1

Transfer function:
        0.01 s^2
-----------------------------
0.01 s^2 + 0.1414 s + 1
```

Transfer function:

$$\frac{0.01 \ s\char`^2}{\text{4e-06 } s\char`^4 + 0.0003394 \ s\char`^3 + 0.0144 \ s\char`^2 + 0.1697 \ s + 1}$$

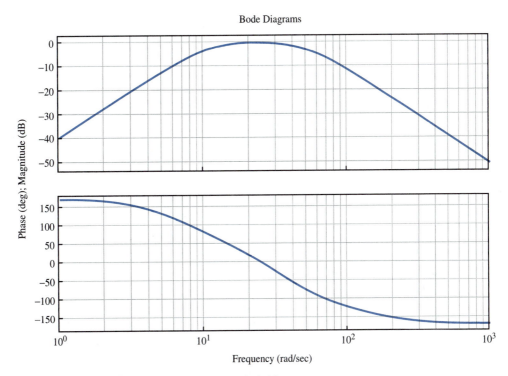

Figure E3.6.3 (Part Two) Bode plots for Example 3.6.3.

3.7 LEARNING OBJECTIVES

The *learning objectives* of this chapter are summarized here so that the student can check whether he or she has accomplished each of the following.

- Response to exponential excitations.
- Representation of sinusoidal signals by complex phasors.
- Impedance and admittance of common circuit elements in series–parallel combinations.
- Frequency-domain representations and transformed networks.
- Forced response to sinusoidal excitation.
- Phasor method of analysis and phasor diagrams.
- Real, reactive, and apparent powers in ac circuits.
- Fourier series (trigonometric and exponential forms).
- Transient and steady-state responses (natural and forced responses) of first- and second-order circuits.
- Accounting for the effect of nonzero initial conditions.

- Expressing pulses in terms of step and impulse functions.
- Laplace transformation method of solving differential equations.
- Time-domain networks and transformed networks in frequency domain for network analysis.
- Network transfer functions.
- Frequency response of networks.
- Low-pass and high-pass filter circuits.
- Bode diagrams and asymptotic Bode plots.
- Two-port network parameters.
- Block diagram representation of simple systems.
- Transient analysis, ac analysis, and frequency response using PSpice and PROBE.
- Use of MATLAB in circuit simulation.

3.8 PRACTICAL APPLICATION: A CASE STUDY

Automotive Ignition System

Ignition systems in automobiles have been designed as a straightforward application of electrical transients. Figure 3.8.1 shows a simplified ignition circuit for an internal-combustion engine. The primary inductance, current-limiting resistance, and capacitance form an underdamped series *RLC* circuit. Thus, when the points open, an oscillatory current flows through the primary, thereby inducing the required voltage in the secondary. The points form a switch, which opens and closes as the engine rotates, exactly at the instant when an ignition spark is needed by one of the cylinders.

During the period when the points are closed, current builds up rather slowly in the primary winding of the coil, which consists of a pair of mutually coupled inductors, as shown in Figure 3.8.1. The current is interrupted rapidly when the points open. The resulting large rate of change of current induces a large voltage across the coil secondary, which is connected to the appropriate spark plug by the distributor.

The resistance limits the current in case the engine stops with the points closed. Since the voltage across a capacitance cannot change instantaneously, the capacitor prevents the voltage

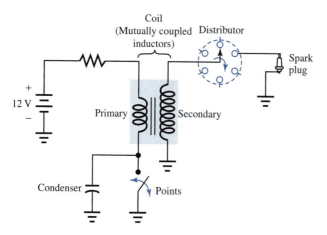

Figure 3.8.1 Simplified ignition circuit for an internal-combustion engine.

across the points from rising too rapidly when the points open. Otherwise arcing may occur across the points, which in turn may get burned and pitted.

Electrical transients in many practical systems can be analyzed by means of the techniques presented in this chapter.

PROBLEMS

3.1.1 (a) The concept of duality can be extended to nonelectric physical systems by means of *analogs*. For example, the mechanical system characteristics can be investigated by means of an equivalent electrical network. Consider Newton's second law, Hooke's law as applied to springs, and viscous friction law, and find the force–current analog as well as the force–voltage analog by identifying the analogous mathematical relations.

(b) Consider the analogy between electrical and hydraulic systems given in Table 1.5.2. Obtain the mass balance equation by equating the rate of change of fluid volume to the net difference between input and output flow. Identify this equation with that of an RC circuit excited by a current source $i(t) = I$.

3.1.2 Consider an RL series circuit excited by (a) $v(t) = 20e^{-2t}$ V, and (b) $v(t) = 20$ V. Determine the forced component of the voltage across the inductor for $R = 2\,\Omega$ and $L = 2$ H.

*3.1.3 Consider an RC parallel circuit excited by (a) $i(t) = 20e^{-2t}$ A, and (b) $i(t) = 20$ A. Find the forced component of the current through the capacitor for $R = 2\,\Omega$ and $C = 2$ F.

3.1.4 For the mechanical spring–mass–friction system shown in Figure P3.1.4, the differential equation relating the force $F(t)$ and the velocity $u(t)$ is given by

$$F(t) = M\frac{du}{dt} + Du + \frac{1}{C_m}\int u\,dt$$

where M is the mass, D is the friction, and C_m is the compliance (reciprocal of stiffness) of the spring. For $M = 20$ kg, $D = 4$ kg/s, and $C_m = 8$ N/m,

develop an electric equivalent network (a) using the force–current analog, and (b) using the force–voltage analog, and find $u(t)$ for $F(t) = 40e^{-t/4}$ N.

3.1.5 In an RLC series circuit excited by a voltage source $v(t)$, for $R = 10\,\Omega$, $L = 1$ H, and $C = 0.1$ F, determine $v(t)$ if the capacitor voltage $v_C(t) = 5e^{-10t}$ V.

3.1.6 In a GLC parallel circuit excited by a current source $i(t)$, for $G = 0.5$ S, $L = 3$ H, and $C = 0.5$ F, determine $i(t)$ if the inductor current $i_L(t) = 12e^{-0.5t}$.

3.1.7 Repeat Problem 3.1.6 for $i_L(t) = 2\cos t/3$ A.

*3.1.8 Repeat Problem 3.1.5 for $v_C(t) = 10\cos(2t - 30°)$ V.

3.1.9 An RL series circuit carries a current of $0.02\cos 5000t$ A. For $R = 100\,\Omega$ and $L = 20$ mH, find the impedance of the series combination and determine the voltage across the series combination. Sketch the phasor diagram showing all quantities involved.

3.1.10 The voltage across a parallel combination of a 100-Ω resistor and a 0.1-μF capacitor is $10\cos(5000t + 30°)$ V. Determine the admittance of the parallel combination and find the current from the supply source. Sketch the phasor diagram showing all quantities involved.

3.1.11 At the two terminals (A, B) of a one-port network, the voltage and the current are given to be $v(t) = 200\sqrt{2}\cos(377t + 60°)$ V and $i(t) = 10\sqrt{2}\cos(377t + 30°)$ A.

(a) Determine the average real power, reactive power, and volt-amperes absorbed by the network.

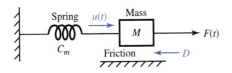

Figure P3.1.4

(b) Find the equivalent components of the network representing the effect at the terminals A–B.

(c) Find the Thévenin impedance (or the *driving-point impedance*) as seen from the terminals A–B.

3.1.12 Repeat Problem 3.1.11 if the 60-Hz voltage and the current at the terminals are given to be $V = 100$ V (rms), $I = 10$ A (rms), power factor = 0.8 leading.

*3.1.13 Use (a) mesh analysis and (b) nodal analysis to determine the current through the 4-Ω resistor of the circuit of Figure P3.1.13.

3.1.14 Use (a) mesh analysis and (b) nodal analysis to determine the voltage $\bar{V}$ at the terminals A–B of Figure P3.1.14.

3.1.15 (a) Obtain a Thévenin equivalent circuit at terminals A–B in the circuit of Problem 3.1.14.

(b) What impedance $\bar{Z}_L$, when connected to A–B, produces maximum power in $\bar{Z}_L$?

(c) Find the value of the maximum power in $\bar{Z}_L$.

3.1.16 In the circuit of Problem 3.1.13, find the Thévenin equivalent of the network as seen by the 4-Ω

resistor, and then determine the current in the 4-Ω resistor.

3.1.17 A 6.6-kV line feeds two loads connected in parallel. Load A draws 100 kW at 0.6 lagging power factor, and load B absorbs 100 kVA at 0.8 lagging power factor.

(a) For the combined load, calculate the real power, reactive power, volt-amperes, and line current drawn from the supply.

(b) If the power factor on the supply end is to be unity, determine the value of the capacitance to be placed across the load if the frequency of excitation is 60 Hz.

*3.1.18 Three loads in parallel are supplied by a single-phase 400-V, 60-Hz supply:

Load A:	10 kVA at 0.8 leading power factor
Load B:	15 kW at 0.6 lagging power factor
Load C:	5 kW at unity power factor

(a) Find the real power, reactive power, volt-amperes, and line current drawn from the supply by the combined load.

(b) If the supply line is to operate at 0.9 leading power factor, determine the value of the capacitance to be placed across the load.

Figure P3.1.13

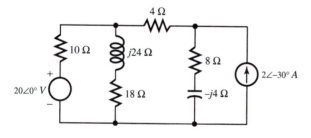

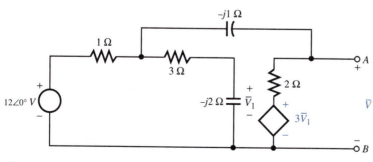

Figure P3.1.14

3.1.19 Determine the Fourier series for the periodic wave-forms given in Figure P3.1.19.

3.1.20 Find the exponential form of the Fourier series of the periodic signal given in Figure P3.1.20. Also determine the resulting series if $T = 2\tau$.

*3.1.21 The first four harmonics in the Fourier series of a current waveform given by

$$i(t) = \frac{2I_m}{\pi} \sin \frac{2\pi t}{T} - \frac{I_m}{\pi} \sin \frac{4\pi t}{T}$$

$$+ \frac{2I_m}{3\pi} \sin \frac{6\pi t}{T} - \frac{I_m}{2\pi} \sin \frac{8\pi t}{T}$$

where $I_m = 15$ mA and $T = 1$ ms. If such a current is applied to a parallel combination of $R = 5$ kΩ and $C = 0.1$ μF, determine the output waveform of the voltage $v_O(t)$ across the load terminals.

3.1.22 The full-wave rectified waveform, approximated by the first three terms of its Fourier series, is given by $v(\omega t) = V_m \sin(\omega t/2)$, for $0 \le \omega t \le 2\pi$, and

$$v(t) = \frac{2V_m}{\pi} - \frac{4V_m}{3\pi} \cos \omega t + \frac{4V_m}{15\pi} \cos 2\omega t$$

where $V_m = 100$ V and $\omega = 2\pi \times 120$ rad/s. If $v(t)$ is applied to the circuit shown in Figure P3.1.22, find the output voltage $v_O(t)$.

3.2.1 Determine $i(t)$ in the circuit of Figure P3.2.1 and sketch it.

*3.2.2 Obtain $i(t)$ in the circuit of Figure P3.2.2 and sketch it.

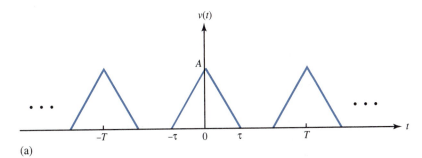

(a)

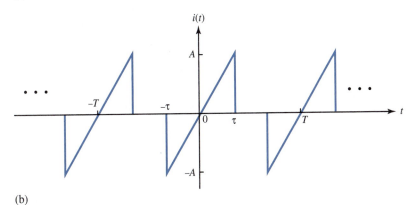

(b)

Figure P3.1.19

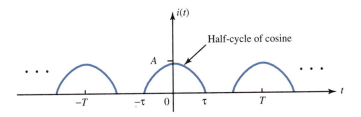

Figure P3.1.20

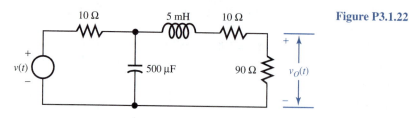

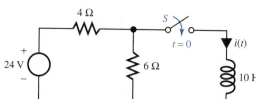

Figure P3.2.1

Figure P3.2.2

3.2.3 Determine and sketch $v(t)$ in the circuit of Figure P3.2.3.

3.2.4 Determine and sketch $v(t)$ in the circuit of Figure P3.2.4.

3.2.5 Obtain and sketch $i(t)$ in the circuit of Figure P3.2.5.

3.2.6 Obtain and sketch $i(t)$ in the circuit of Figure P3.2.6.

*3.2.7 Determine and sketch $i(t)$ in the circuit of Figure P3.2.7.

3.2.8 In the circuit of Figure P3.2.8,

$$i(t) = \begin{cases} 10 \text{ A}, & \text{for } t < 0 \\ 24e^{-t}, & \text{for } t \geq 0 \end{cases}$$

Find: $v(0^+)$, $i_L(0^+)$, $\dfrac{dv}{dt}(0^+)$, and $\dfrac{di_L}{dt}(0^+)$.

3.2.9 In the circuit of Figure P3.2.9,

$$v(t) = \begin{cases} 24 \text{ V}, & \text{for } t < 0 \\ 12 \cos t, & \text{for } t \geq 0 \end{cases}$$

Determine $v_C(0^+)$, $i_C(0^+)$, $\dfrac{dv_C}{dt}(0^+)$, and, $\dfrac{di_C}{dt}(0^+)$.

*3.2.10 In the circuit of Figure P3.2.10,

$$v(t) = \begin{cases} 0, & \text{for } t < 0 \\ 0.4t, & \text{for } t \geq 0 \end{cases}$$

Evaluate: $v_O(0^+)$ and $\dfrac{dv_O}{dt}(0^+)$.

3.2.11 Reconsider Problem 3.2.10 and obtain $v_O(t)$ for $t \geq 0$.

3.2.12 Consider the circuit of Figure P3.2.12. Determine and sketch $i_L(t)$ and $v_C(t)$ for capacitance values of

(a) $\frac{1}{6}$ F, (b) $\frac{1}{8}$ F, and (c) $\frac{1}{26}$ F. Note that the capacitance values are chosen here for calculational ease, even though they are too big and not typical.

3.2.13 For the circuit of Figure P3.2.13, determine and sketch $i_L(t)$ and $v_C(t)$ for inductance values of

(a) $\frac{3}{4}$ H, (b) $\frac{2}{3}$ H, and (c) $\frac{3}{17}$ H. Note that the inductance values are chosen here for calculational ease, even though they are too big and not typical.

*3.2.14 Consider the circuit of Figure P3.2.14 in which the switch S has been open for a long time and is closed at $t = 0$. Determine $v_C(t)$ for $t \geq 0$.

3.2.15 In the circuit of Figure P3.2.15, obtain $i_L(t)$ and $v_C(t)$.

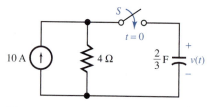

Figure P3.2.3

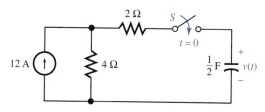

Figure P3.2.4

Figure P3.2.5

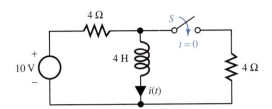

Figure P3.2.6

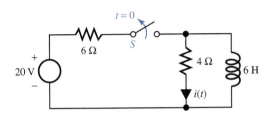

Figure P3.2.7

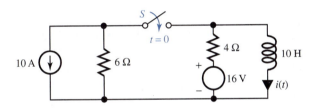

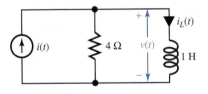

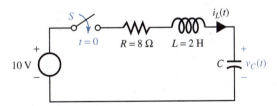

Figure P3.2.8

Figure P3.2.9

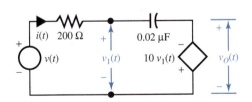

Figure P3.2.10

Figure P3.2.12

Figure P3.2.13

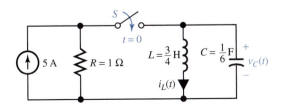

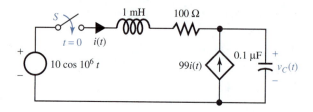

Figure P3.2.14

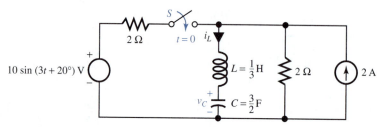

Figure P3.2.15

3.2.16 Determine $i_L(t)$ and $v_C(t)$ in the circuit of Figure P3.2.16.

3.2.17 Express the waveform of the staircase type shown in Figure P3.2.17 as a sum of step functions.

3.2.18 The voltage waveform of Figure P3.2.18 is applied to an *RLC* series circuit with $R = 2\ \Omega$, $L = 2$ H, and $C = 1$ F. Obtain $i(t)$ in the series circuit. (Note that the capacitor and inductor values are chosen here for calculational ease, although they are too big and not typical.)

3.2.19 (a) Let a unit impulse of current $i(t) = \delta(t)$ be applied to a parallel combination of $R = 3\ \Omega$ and $C = \frac{1}{2}$ F. Determine the voltage $v_C(t)$ across the capacitor.

  (b) Repeat (a) for $i(t) = \delta(t - 3)$.

   (Note that the capacitance value is chosen here for calculational ease, even though it is too big and not typical.)

3.2.20 (a) Let a unit impulse of voltage $v(t) = \delta(t)$ be applied to a series combination of $R = 20\ \Omega$ and $L = 10$ mH. Determine the current $i(t)$ in the series circuit.

  (b) Repeat (a) for $v(t) = \delta(t) + \delta(t - 3)$.

3.3.1 Determine the Laplace transform for each of the following functions from the basic definition of Equation (3.3.1).

  (a) $f_1(t) = u(t)$

  (b) $f_2(t) = e^{-at}$

  (c) $f_3(t) = \dfrac{df(t)}{dt}$, assuming $f(t)$ is transformable.

  (d) $f_4(t) = t$

  (e) $f_5(t) = \sin\ \omega t$

  (f) $f_6(t) = \cos(\omega t + \theta)$

  (g) $f_7(t) = te^{-at}$

  (h) $f_8(t) = \sinh\ t$

  (i) $f_9(t) = \cosh\ t$

  (j) $f_{10}(t) = 5f(t) + 2\dfrac{df(t)}{dt}$

3.3.2 Using the properties listed in Table 3.3.2, determine the Laplace transform of each of the following functions.

  (a) te^{-t}

  (b) $t^2 e^{-t}$

  (c) $te^{-2t} \sin 2t$

  (d) $\dfrac{1 - \cos\ t}{t}$

  (e) $\dfrac{e^{-2t} \sin 2t}{t}$

***3.3.3** Given the frequency-domain response of an *RL* circuit to be

$$I(s) = \frac{10}{2s + 5}$$

determine the initial value and the final value of the current by using the initial-value and final-value theorems given in Table 3.3.2.

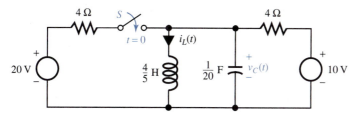

Figure P3.2.16

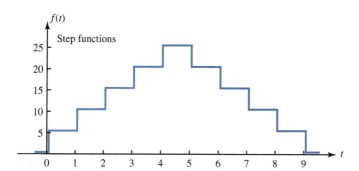

Figure P3.2.17

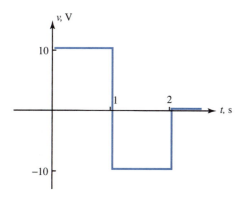

Figure P3.2.18

3.3.4 Determine the inverse Laplace transform of each
of the following functions.

(a) $\dfrac{6(s+3)}{s^2+2s+10}$

(b) $\dfrac{s+2}{(s+1)(s+3)(s+4)}$

(c) $\dfrac{8(s+1)}{s(s^2+2s+2)}$

(d) $\dfrac{s^2+4s+5}{s^2+3s+2}$

(e) $\dfrac{2(s+2)}{s(s+1)^3}$

(f) $\dfrac{s+1}{s^3+4s^2+3s}$

(g) $\dfrac{10(s^2+3s+2)}{3(s^2+2s+2)}$

(h) $\dfrac{(s+1)(s+2)}{s^2(s^2+2s+2)}$

(i) $\dfrac{4s}{(s^2+4)(s+2)}$

(j) $\dfrac{4s}{\left[(s+2)^2+4\right]^2}$

3.3.5 Solve the following differential equations (along
with the conditions) for $t \geq 0$.

(a) $\dfrac{d^2v}{dt^2}+5\dfrac{dv}{dt}+4v=10u\,(t)\,;\,v\left(0^+\right)=\dfrac{dv}{dt}$
$\left(0^+\right)=0$

(b) $3\dfrac{d^2i}{dt^2} + 7\dfrac{di}{dt} + 2i = 10\cos 2t;\ i\left(0^+\right) = 4\text{A};\ \dfrac{di}{dt}\left(0^+\right) = -4\text{A/s}$

(c) $\dfrac{d^2i}{dt^2} + 2\dfrac{di}{dt} + 2i = \sin t - e^{-2t};\ i\left(0^+\right) = 0\ ;\ \dfrac{di}{dt}\left(0^+\right) = 4\text{A/s}$

Identify the forced and natural response components in each case.

3.3.6 Determine the Laplace transform of the waveform shown in Figure P3.3.6.

3.3.7 In the circuit shown in Figure P3.3.7, the switch S has been open for a long time. At $t = 0$, the switch is closed. Find the currents $i_1(t)$ and $i_2(t)$ for $t \geq 0$ with the use of the Laplace transform method.

*3.3.8 Determine $v(t)$ and $i_L(t)$ in the circuit shown in Figure P3.3.8, given that $i(t) = 10te^{-t}u(t)$.

3.3.9 The switch S in the circuit of Figure P3.3.9 has been open for a long time before it is closed at $t = 0$. Determine $v_L(t)$ for $t \geq 0$.

3.3.10 Determine $v(t)$ in the circuit of Figure P3.3.10 if $i(t)$ is a pulse of amplitude 100 μA and duration 10 μs.

Figure P3.3.6

Figure P3.3.7

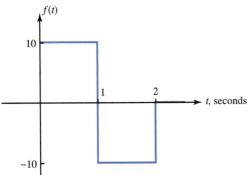

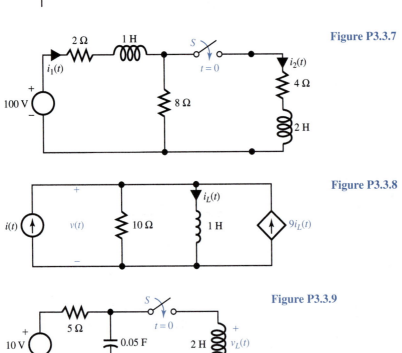

Figure P3.3.8

Figure P3.3.9

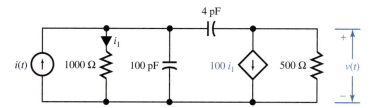

3.3.11 For the networks shown in Figure P3.3.11, determine the transfer function $G(s) = V_o(s)/V_i(s)$.

3.3.12 The response $v(t)$ of a linear system to a unit-step excitation $i(t)$ is given by $v(t) = (5 - 3e^{-t} + 2e^{-2t})\, u(t)$. Determine the transfer function $H(s) = V(s)/I(s)$.

*3.3.13 The response $y(t)$ of a linear system to a unit-step excitation is $y(t) = (4 - 10e^{-t} + 8e^{-2t})\, u(t)$.

    (a) Find the system function.

    (b) Find the frequency at which the forced response is zero.

3.3.14 The unit impulse response $h(t)$ of a linear system is $h(t) = 5e^{-t}\cos(2t - 30°)$. Determine $H(s)$.

3.3.15 The response $y(t)$ of a linear system to an excitation $x(t) = e^{-2t}\, u(t)$ is $y(t) = (t + 2)e^{-t}\, u(t)$. Find the transfer function.

3.3.16 A filter is a network employed to select one range of frequencies while rejecting all other frequencies. A basic building block often used in integrated-circuit filters is shown in Figure P3.3.16. Determine the following for the circuit:

    (a) Transfer function $V_o(s)/V_i(s)$.

    (b) Response $v_o(t)$ for $v_i(t) = u(t)$.

    (c) Driving-point impedance $V_i(s)/I_i(s)$.

3.4.1 For the circuits shown in Figure P3.4.1, sketch the frequency response (magnitude and phase) of $\bar{V}_{out}/\bar{V}_{in}$.

*3.4.2 Design the low-pass filter shown in Figure P3.4.2 (by determining L) to have a half-power frequency of 10 kHz.

3.4.3 Design the high-pass filter shown in Figure P3.4.3 (by determining C) to have a half-power frequency of 1 MHz.

3.4.4 Determine L and C of the bandpass filter circuit of Figure P3.4.4 to have a center frequency of 1 MHz and a bandwidth of 10 kHz. Also find the Q of the filter.

3.4.5 Determine L and C of the band reject filter circuit of Figure P3.4.5 to have a center frequency of 100

kHz and a bandwidth of 5 kHz. Also find the Q of the filter.

3.4.6 Determine the voltage transfer function of the low-pass filter circuit shown in Figure P3.4.6, and find the expression for ω_0.

*3.4.7 Determine the voltage transfer function of the high-pass filter circuit shown in Figure P3.4.7, and find the expression for ω_0.

3.4.8 Let a square-wave voltage source having an amplitude of 5 V, a frequency of 1 kHz, a pulse width of 0.5 ms, and an internal source resistance of 50 Ω be applied to a resistive load of 100 Ω. A filter (inductance L) is inserted between source and load in order to reduce all the high-frequency components above 5 kHz. Determine (a) L of the low-pass filter, (b) amplitude spectra of V_L and V_S, and (c) $v_L(t)$.

3.4.9 Sketch the asymptotic Bode diagrams for the following functions:

    (a) $H_1(s) = \dfrac{200}{(1+s)\left(1+\dfrac{s}{10}\right)}$

    (b) $H_2(s) = \dfrac{200s}{(1+s)\left(1+\dfrac{s}{10}\right)}$

    (c) $H_3(s) = \dfrac{200\left(1+\dfrac{s}{10}\right)}{(1+s)\left(1+\dfrac{s}{20}\right)}$

    (d) $H_4(s) = \dfrac{0.5\left(1+\dfrac{s}{10}\right)}{(1+s)^2\left(1+\dfrac{s}{40}\right)}$

    (e) $H_5(s) = \dfrac{0.5(1+s)^2}{s\left(1+\dfrac{s}{10}\right)\left(1+\dfrac{s}{50}\right)}$

    (f) $H_6(s) = \dfrac{20\left(1+\dfrac{s}{8}\right)}{(1+s)\left(1+\dfrac{s}{10}\right)^2\left(1+\dfrac{s}{40}\right)}$

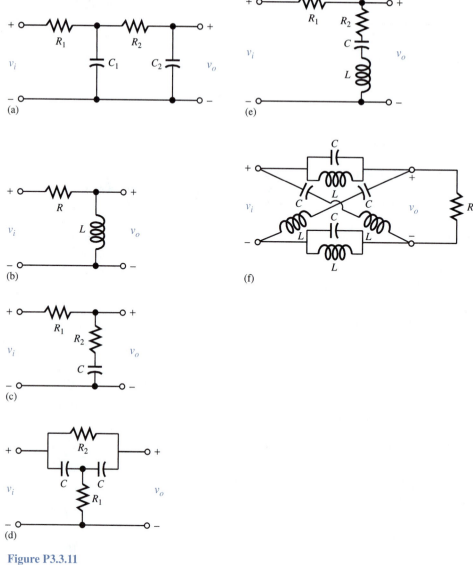

Figure P3.3.11

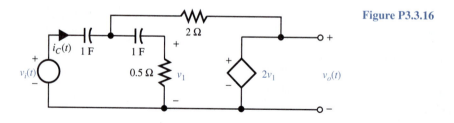

Figure P3.3.16

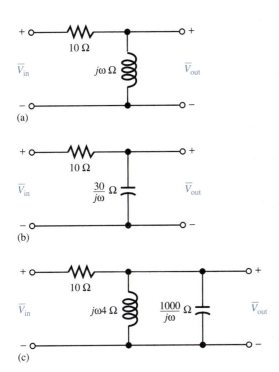

Figure P3.4.1

(a)

(b)

(c)

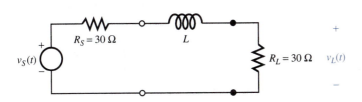

Figure P3.4.2

$R_S = 30\,\Omega$

L

$v_S(t)$

$R_L = 30\,\Omega$ $v_L(t)$

$+$

$-$

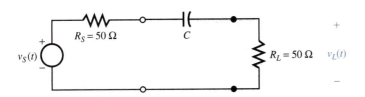

Figure P3.4.3

$R_S = 50\,\Omega$

C

$v_S(t)$

$R_L = 50\,\Omega$ $v_L(t)$

$+$

$-$

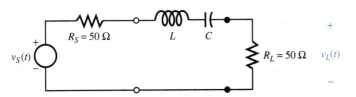

Figure P3.4.4

$R_S = 50\,\Omega$

L C

$v_S(t)$

$R_L = 50\,\Omega$ $v_L(t)$

$+$

$-$

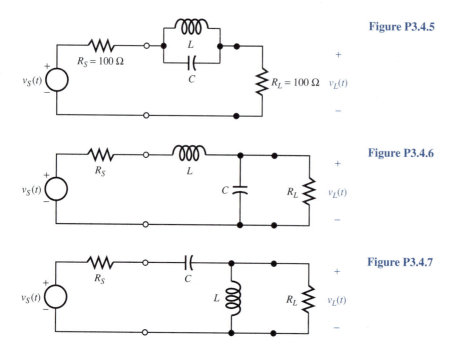

Figure P3.4.5

Figure P3.4.6

Figure P3.4.7

3.4.10 Reconsider Problem 3.4.9 and the corresponding asymptotic Bode plots.

(a) Find $\bar{H}_1$, $\bar{H}_2$, and $\bar{H}_3$ at $\omega = 5$ rad/s.

(b) At what angular frequency ω is the magnitude of $\bar{H}_4(j\omega)$ one-half of the magnitude of $\bar{H}_4(j5)$?

(c) Determine the angular frequency at which $H_6(\omega)$ is 0 dB and the angular frequency at which $\theta_6(\omega) = -180°$.

(d) Let $H_5(s) = V_2/V_1$. For $v_1(t) = 0.1 \cos 20t$, find the steady-state value of $v_2(t)$.

3.4.11 Sketch the idealized (asymptotic) Bode plot for the transfer function

$$\bar{H}(j\omega) = \frac{10(1 + j2\omega)}{(1 + j10\omega)(1 + j0.25\omega)}$$

Find the angular frequency at which $H(\omega)$ is 0 dB and the angular frequency at which $\theta(\omega) = -60°$.

*3.4.12 The loop gain of an elementary feedback control system (see Figure 3.4.12) is given by $G(s) \cdot H(s)$, which is $10/(1+s/2)(1+s/6)(1+s/50)$. Sketch the asymptotic Bode plot of the loop-gain function. *Gain margin* (GM) is defined by $[-20 \log |\bar{G}(\omega_\pi)$ $\bar{H}(\omega_\pi)|]$, which is the negative of the magnitude of the loop gain at $\omega = \omega_\pi$, ω_π represents the angular frequency at which the loop gain reaches a phase of $-\pi$. If ω_u represents the value of ω

where the loop gain has a magnitude of unity, then *phase margin* (PM) is defined by the phase of the loop gain at $\omega = \omega_u$ plus π. Evaluate GM and PM for this case from the asymptotic Bode plot.

3.4.13 Sketch the asymptotic Bode plots for the following loop-gain functions, and find the approximate values of gain and phase margins in each case. (For definitions of GM and PM, see Problem 3.4.12.)

(a) $G_1(s) H_1(s) =$

$$\frac{12(0.7 + s)}{(0.003 + s)(0.04 + s)(7 + s)}$$

(b) $G_2(s) H_2(s) = \dfrac{100(1 + s/3.9607)}{s(1 + 2s)(1 + s/39.607)}$

3.4.14 (a) For a series *RLC* resonant circuit, find an expression for the voltage across the resistance V_R and obtain the ratio V_R/V_S, where V_S is the applied voltage. Identify the expressions for the series resonant frequency and bandwidth.

(b) Determine the resonant frequency and bandwidth, given the voltage transfer function to be $10^3/(s^2 + 10^3 s + 10^{10})$.

3.4.15 A simple parallel resonant circuit with $L = 50 \, \mu H$ is used to perform the frequency selection. The circuit is to be tuned to the first station at a frequency of 1000 kHz. In order to minimize the interaction between signals, the tuning circuit response must attenuate the signal of the second station (with a frequency of 1020 kHz) by four times.

(a) Determine the values of G and C.

(b) Find the bandwidth of the circuit.

3.4.16 The twin-tee or notch network shown in Figure P3.4.16 is often used to obtain band-reject characteristics.

(a) Determine the transfer function V_2/I_1.

(b) Find the angular frequency at which the transfer function will be zero.

*3.4.17 A *gyrator* is sometimes used in integrated circuits to simulate inductances. Consider the circuit of Figure P3.4.17 consisting of a gyrator.

(a) Show that the impedance as seen to the right at terminals 1–1' can be represented by an inductor in series with a resistor.

(b) Determine the resonant frequency and bandwidth for the following values:

$$R = 10^5 \,\Omega; \, g = 10^{-3} \,\text{S}; \, C_1 = C_2 = 100 \,\text{pF}.$$

3.4.18 (a) Consider the capacitor-input filter circuit of Figure P3.4.18 and obtain the z-parameters for the circuit.

(b) Determine the transfer function V_2/V_1 when I_2 is zero.

3.4.19 The circuit shown in Figure P3.4.19 is the equivalent circuit of a field-effect transistor (FET) amplifier stage.

(a) Determine the y-parameters.

(b) For values of $\mu = g_m/g_d \gg 1$, obtain a y-parameter equivalent circuit.

3.4.20 For the circuit shown in Figure P3.4.20, obtain:

(a) z- and y-parameters;

(b) Transfer function I_2/I_1 when $V_2 = 0$.

3.4.21 For the capacitor input filter circuit of Problem 3.4.18, determine the h-parameters.

Figure P3.4.16

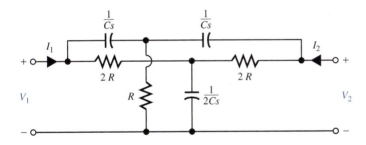

Figure P3.4.17

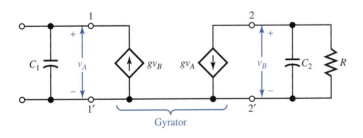

Figure P3.4.18

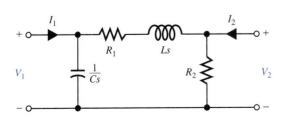

*3.4.22 Determine the *h*-parameters for the circuit shown in Figure P3.4.22 and obtain the transfer function V_2/V_1 when $I_2 = 0$.

3.4.23 A negative impedance converter circuit shown in Figure P3.4.23 is used in some applications where inductors cannot be utilized or where negative resistance is beneficial.

(a) Determine the *h*-parameters for the network.

(b) When the resistance R is connected across terminals 2–2′, find the impedance seen looking into terminals 1–1′.

3.4.24 Show that the block diagram of Figure P3.4.24 can be reduced to the form of Figure 3.4.12. Find $G_{eq}(s)$ and $H_{eq}(s)$.

3.4.25 Show that the block diagram of Figure P3.4.25(a) can be reduced to that of Figure P3.4.25(b).

Figure P3.4.19

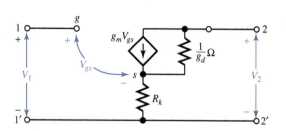

Figure P3.4.20

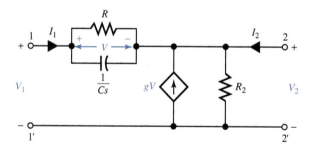

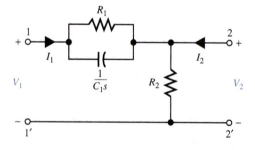

Figure P3.4.22

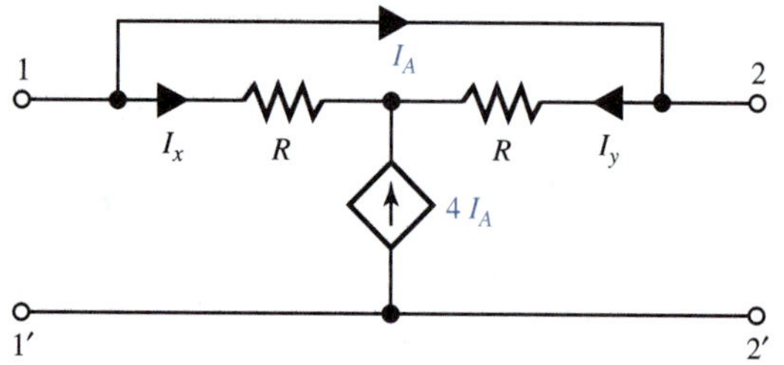

Figure P3.4.23

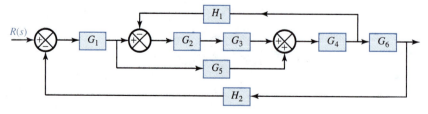

Figure P3.4.24

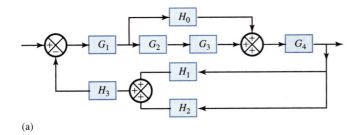

(a)

(b)

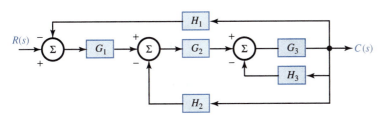

Figure P3.4.27

3.4.26 The block diagram of Figure P3.4.26 represents a multiloop control system.

(a) Determine the transfer function $C(s)/R(s)$.

(b) For $G_2 G_3 H_2 = 1$, evaluate $C(s)/R(s)$.

*3.4.27 Determine the transfer function $C(s)/R(s)$ of the nested-loop feedback system shown in Figure P3.4.27.

3.4.28 The equations for a two-port network are given by

$$V_1 = z_{11} I_1 + z_{12} I_2$$

$$0 = z_{21} I_1 + (z_{22} + Z_L) I_2$$

$$V_2 = -I_2 Z_L$$

(a) Satisfying the equations, develop a block diagram.

(b) Find the transfer function V_2/V_1.

3.5.1 In the circuit shown in Figure P3.5.1 the switch opens at $t = 0$. Develop and execute a PSpice program to solve for $v(t)$, and use PROBE to obtain a plot of $v(t)$.

Hint: Use the following analysis request:

• TRAN   0.1 MS   50 MS   0   0.1 MS   UIC

3.5.2 Obtain a plot of the current $i(t)$ in the circuit of Figure P3.5.2 by writing a PSpice program and using the following analysis request:

• TRAN   0.2 MS   80 MS   0   0.2 MS   UIC

Hint: Note that a sinusoidal voltage source is specified in PSpice by the statement

```
VSIN   NODEPLUS   NODEMINUS
  SIN (VDC   VPEAK   FREQ   TD
  DF   PHASE)
```

where VSIN is the name of the source, NODE-PLUS is the number identifying the positive node, NODEMINUS is the number identifying the negative node, VDC is the dc offset, VPEAK is the peak value of the ac component, FREQ is the frequency in hertz, TD is the time delay in seconds, DF is a damping factor, and PHASE is the phase angle in degrees. The voltage generated is then given by

$$v(t) = \text{VDC} + \text{VPEAK} \sin[\text{PHASE}],$$
$$\text{for } 0 < t < \text{TD}$$
$$= \text{VDC} + \text{VPEAK} \sin[2\pi$$
$$\text{FREQ}(t - \text{TD}) + \text{PHASE}]$$
$$\times \exp[-\text{DF}(t - \text{TD})],$$
$$\text{for TD} < t$$

Also note that DF = 0 for a constant-amplitude sinusoid.

*3.5.3 With the initial voltage across the capacitor being $v_C(0) = 10$ V, obtain a plot of $v_C(t)$ for the circuit of Figure P3.5.3 by writing a PSpice program. *Hint:* Use the following analysis request:

• TRAN   0.02   10   0   0.02   UIC

3.5.4 For the circuit shown in Figure P3.5.4, solve for $i(t)$ by writing and executing a PSpice program, and obtain a plot of $i(t)$ by using PROBE. Take the hint from Problem 3.5.2 for specifying the sinusoidal voltage source in the PSpice program. Also use the following analysis request:

• TRAN   0.1 MS   100 MS   0   0.1 MS   UIC

3.5.5 A dc source is connected to a series *RLC* circuit by a switch that closes at $t = 0$, as shown in Figure P3.5.5, with initial conditions. For the values of $R = 20$, 40, and 80 Ω, solve for $v_C(t)$ by writing and executing a PSpice program, and obtain plots by using PROBE. *Hint:* Use the following analysis request:

•TRAN   1US   2MS   0   1US   UIC

3.5.6 In the circuit of Figure P3.5.6, solve for phasors $\bar{V}_1$ and $\bar{V}_2$ with peak magnitudes by writing and executing a PSpice program.

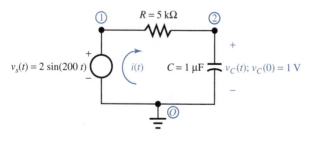

Figure P3.5.1

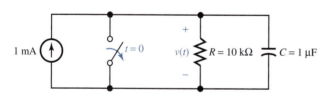

Figure P3.5.2

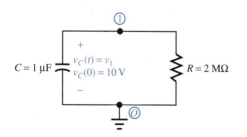

Figure P3.5.3

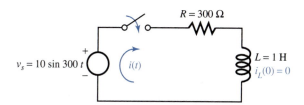

Figure P3.5.4

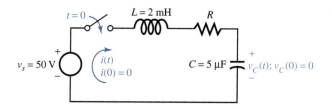

Figure P3.5.5

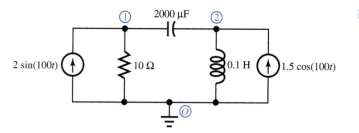

Figure P3.5.6

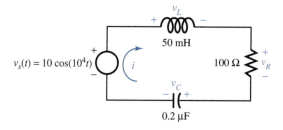

Figure P3.5.7

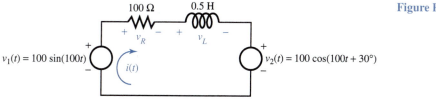

Figure P3.5.8

3.5.7 For the circuit shown in Figure P3.5.7, find the phasor values (with peak magnitudes) of $\bar{I}$, $\bar{V}_R$, $\bar{V}_L$, and $\bar{V}_C$ by using PSpice.

*3.5.8 Use PSpice to find the phasor $\bar{I}$ (with peak magnitude) for the circuit shown in Figure P3.5.8.

3.5.9 A low-pass filter circuit is shown in Figure P3.5.9.

Using a PSpice program and PROBE, obtain the Bode magnitude plot for the transfer function $\bar{H}(f) = \bar{V}_{out}/\bar{V}_{in}$ for the frequency range of 10 Hz to 100 kHz. Determine the fall-off rate (in dB/decade) of the magnitude at high frequencies, and also the half-power frequency.

3.5.10 A high-pass filter circuit is shown in Figure P3.5.10. Using a PSpice program and PROBE, obtain the Bode magnitude plot for the transfer function $\bar{H}(f) = \bar{V}_{out}/\bar{V}_{in}$ for frequency ranging from 10 Hz to 100 kHz. Determine the rate (in dB/decade) at which the magnitude falls off at low frequencies, and also the half-power frequency.

3.5.11 A bandpass filter circuit is shown in Figure P3.5.11. Develop a PSpice program and use PROBE to obtain a Bode magnitude plot for the transfer function $\bar{H}(f) = \bar{V}_{out}/\bar{V}_{in}$ for frequency ranging from 1 Hz to 1 MHz. At what rate (in dB/decade) does the magnitude fall off at low and high frequencies? Also determine the half-power frequencies.

3.6.1 A periodic sequence of exponential wave forms forms a pulse train whose first cycle is represented by

$$v(t) = [u(t) - u(t - T_0)]v_A e^{-t/T_C}$$

Use MATLAB to find V_{rms} of the pulse train for $v_A = 10$ V, $T_C = 2$ ms, and $T_0 = 5T_C$.

3.6.2 Use MATLAB to obtain the Laplace transform of the waveform

$$f(t) = [200te^{-25t} + 10e^{-50t}\ \sin(25t)]u(t)$$

which consists of a damped ramp and a damped sine. Also show the pole–zero plot of the transform $F(s)$.

3.6.3 With the use of MATLAB, find the expression for the waveform $f(t)$ corresponding to a transform

$F(s)$ with a zero at $s = -400$, a simple zero at $s = -1000$, a double pole at $s = j400$, a double pole at $s = -j400$, and a value at $s = 0$ of $F(0) = 2 \times 10^{-4}$. Plot $f(t)$.

3.6.4 Consider the circuit shown in Figure P3.6.4 in the time domain as well as in the s-domain. Its transfer function $V_2(s)/V_1(s)$ can be shown to be

$$T(s) = \frac{s/RC}{s^2 + s/RC + 1/LC}$$

$$= \frac{Ls/R}{LCs^2 + (Ls/R) + 1}$$

which is a second-order bandpass transfer function with a center frequency at $\omega_0 = 1/\sqrt{LC}$. Using MATLAB, evaluate the straight-line and actual gain response of the *RLC* circuit for the given values.

*3.6.5 Using MATLAB, plot the gain and phase response of the transfer function

$$T(s) = \frac{5000(s + 100)}{s^2 + 400s + (500)^2}$$

3.6.6 The dual situations of Figure E3.6.3 is shown in Figure P3.6.6, in which a high-pass and a low-pass filter are connected in parallel to produce a *bandstop* filter.
With $\omega_{CLP} = 10 \ll \omega_{CHP} = 50, \omega_0 = \sqrt{10 \times 50} = 22.4$ rad/s, and

$$T_{LP}(s) = \frac{1}{(s/10)^2 + \sqrt{2}(s/10) + 1}$$

$$T_{HP}(s) = \frac{(s/50)^2}{(s/50)^2 + \sqrt{2}(s/50) + 1}$$

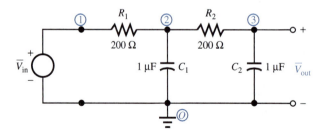

Figure P3.5.9

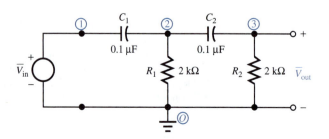

Figure P3.5.10

the overall transfer function is given by

$$T_{BS}(s) = T_{LP}(s) + T_{HP}(s)$$

Using MATLAB, plot the bandstop response.

3.6.7 An expression for a sawtooth wave over the internal $0 \leq t \leq T_0$ is given by $f(t) = At/T_0$. The student is encouraged to check the Fourier coefficients to be $a_0 = A/2$, $a_n = 0$ for all n, and $b_n = -A/(n\pi)$ for all n. The Fourier series for the sawtooth wave is then given by

$$f(t) = \frac{A}{2} + \sum_{n=1}^{\infty} \left(-\frac{A}{n\pi} \right) \sin \left(2\pi nt/T \right)$$

Using MATLAB, with $A = 10$ and $T_0 = 2$ ms, plot the truncated series representations of the waveform $f(5, t)$, which is the sum of the dc component plus the first 5 harmonics, and $f(10, t)$, which is the sum of the dc component plus the first 10 harmonics.

3.6.8 The steady-state circuit $i(t)$ in a series RL circuit due to a periodic sawtooth voltage is given by

$$i(t) = \frac{V_A}{2R} + \frac{V_A}{R} \sum_{n=1}^{\infty} \frac{1}{n\pi \sqrt{1 + (n\omega_0 L/R)^2}}$$

$$\cos(n\omega_0 t + 90° - \theta_n)$$

where $\theta_n = \tan^{-1}(n\omega_0 L/R)$. With the parameters $V_A = 25$ V, $T_0 = 5$ μs, $\omega_0 = 2\pi/T_0$, $L = 40$ μH, and $R = 50$ Ω, by using MATLAB, plot truncated Fourier series representations of $i(t)$ using the dc plus first 5 harmonics and the dc plus first 10 harmonics.

Hint:

$$I_0 = \frac{V_A}{2R}; \qquad I_n = \frac{V_A}{R} \frac{1}{n\pi \sqrt{1 + (n\omega_0 L R^{-1})^2}};$$

$$\theta_n = \tan^{-1}(n\omega_0 L/R)$$

$$f(k, t) = I_0 + \sum_{m=1}^{k} I_m \cos(m\omega_0 t + 0.5\pi - \theta_m)$$

Plot from $t = 0$ to $2T_0$ with a time-step interval of $T_0/400$.

Figure P3.5.11

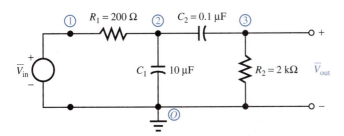

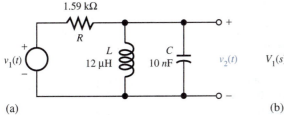

(a)   (b)

Figure P3.6.4 **(a)** t-domain. **(b)** s-domain.

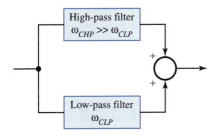

Figure P3.6.6 Bandstop filter through the parallel connection of high-pass and low-pass filters.

4 Three-Phase Circuits and Residential Wiring

We shall conclude Part 1 of this book on electric circuits with a treatment of three-phase circuits, an aspect of circuit theory important to the discussion of electric machines and to the bulk transfer of electric energy. After explaining the phase sequence of three-phase source voltages, balanced three-phase loads and power in three-phase circuits are considered. Then the elements of residential wiring, including grounding and safety considerations, are presented. The chapter ends with a case study of practical application.

The three-phase system is by far the most common polyphase system used for generation, transmission, and heavy power utilization of ac electric energy because of its economic and operating advantages. An ideal three-phase source generates three sinusoidal voltages of equal amplitudes displaced from each other by an angle of 120° in time. The voltages generated by the giant synchronous generators in power stations are practically sinusoidal with a frequency of 60 Hz in the United States, or 50 Hz in the United Kingdom and many other countries. Even though voltages and currents are sinusoidal, the power delivered to a balanced load is constant for a three-phase system. The three-phase scheme of power transmission offers the advantages of using the ac mode, constant power flow, and high power transfer capability.

4.1 THREE-PHASE SOURCE VOLTAGES AND PHASE SEQUENCE

The elementary three-phase, two-pole generator shown in Figure 4.1.1 has three identical stator coils (*aa'*, *bb'*, and *cc'*) of one or more turns, displaced by 120° in space from each other. The rotor carries a field winding excited by the dc supply through brushes and slip rings. When the rotor is driven at a constant speed, voltages of equal amplitude but different phase angle will be generated in the three phases in accordance with Faraday's law. Each of the three stator coils constitutes one phase of this single generator. If the field structure is so designed that the flux is distributed sinusoidally over the poles, the flux linking any phase will vary sinusoidally with time, and sinusoidal voltages will be induced in the three phases. These three induced voltage waves will be displaced by 120 electrical degrees in time because the stator phases are displaced by 120° in space. When the rotor is driven counterclockwise, Figure 4.1.2(a) shows the wave forms and Figure 4.1.2(b) depicts the corresponding phasors of the three voltages. The time origin and the reference axis are chosen on the basis of analytical convenience. In a balanced system, all three phase voltages are equal in magnitude but differ from each other in phase by 120°. The sequence of voltages in Figure 4.1.2(b), corresponding to that of Figure 4.1.2(a), is known as the *positive sequence (a–b–c)*. On the other hand, if the rotor is driven clockwise, then Figure 4.1.2(c) shows the corresponding phasor of the three voltages; the sequence of voltages in Figure 4.1.2(c) is known as *negative sequence (a–c–b)*. Notice that in positive sequence $\bar{E}_{bb'}$ lags $\bar{E}_{aa'}$ by 120°, and $\bar{E}_{cc'}$ lags $\bar{E}_{bb'}$ by 120°; in negative sequence, however, $\bar{E}_{cc'}$ lags $\bar{E}_{aa'}$ by 120°, and $\bar{E}_{bb'}$ lags $\bar{E}_{cc'}$ by 120°.

The stator phase windings may be connected in either wye (also known as star or symbolically represented as Y) or delta (also known as mesh or symbolically represented as Δ), as shown schematically in Figure 4.1.3. Almost all ac generators (otherwise known as alternators) have their stator phase windings connected in wye. By connecting together either all three primed terminals or all three unprimed terminals to form the *neutral* of the wye, a wye connection results. If a neutral conductor is brought out, the system is known as a *four-wire, three-phase system*; otherwise it is a *three-wire, three-phase system*. A delta connection is effected for the armature of the generator by connecting terminals a' to b, b' to c, and c' to a. The generator terminals A, B, C (and sometimes N for a wye connection) are brought out as shown in Figure 4.1.3. In the delta-connection, no neutral exists, and hence only a three-wire, three-phase system can be formed. Note that a phase is one of the three branch circuits making up a three-phase circuit. In a wye connection, a phase consists of those circuit elements connected between one line and neutral; in a delta circuit, a phase consists of those circuit elements connected between two lines.

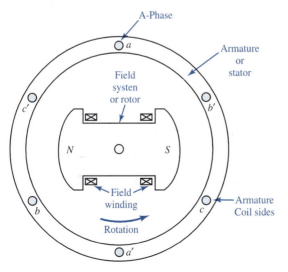

Figure 4.1.1 Elementary three-phase, two-pole ac generator.

Volts

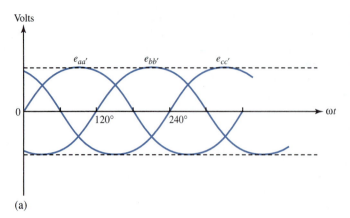

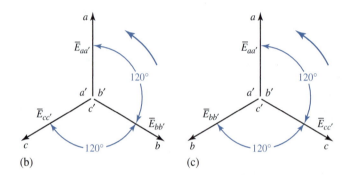

Figure 4.1.2 (a) Waveforms in time domain. (b) Positive-sequence (a–b–c) phasors. (c) Negative-sequence (a–c–b) phasors.

From the nature of the connections shown in Figure 4.1.3 it can be seen that the line-to-line voltages ($V_{L\text{-}L}$ or V_L) are equal to the phase voltage V_{ph} for the delta connection, and the line current is equal to the phase current for the wye connection. A balanced wye-connected three-phase source and its associated phasor diagram are shown in Figure 4.1.4, from which it can be seen that the line-to-line voltage is equal to $\sqrt{3}$ times the phase voltage (or the line-to-neutral voltage). The student should be able to reason on similar lines and conclude that, for the balanced delta-connected three-phase source, the line current will be equal to $\sqrt{3}$ times the phase current.

The notation using subscripts is such that V_{AB} is the potential at point A with respect to point B, I_{AB} is a current with positive flow from point A to point B, and I_A, I_B, and I_C are line currents with positive flow from the source to the load, as shown in Figure 4.1.5. The notation used is rather arbitrary. In some textbooks a different notation for the voltage is adopted such that the order of subscripts indicates the direction in which the voltage rise is taken. The student should be careful not to get confused, but try to be consistent with any conventions chosen. The rms values are usually chosen as magnitudes of the phasors for convenience. It is customary to use the letter symbol E for generated emf and V for terminal voltage. Sometimes the two are equal, but sometimes not. If we should neglect the existence of the generator winding impedance, the generated emf will be equal to the terminal voltage of the generator. Although the three-phase voltages are generated in one three-phase alternator, for analytical purposes this is modeled by three identical, interconnected, single-phase sources.

The one-line equivalent circuit of the balanced wye-connected three-phase source is shown in Figure 4.1.4(c). The line-to-neutral (otherwise known as phase) voltage is used; it may be taken

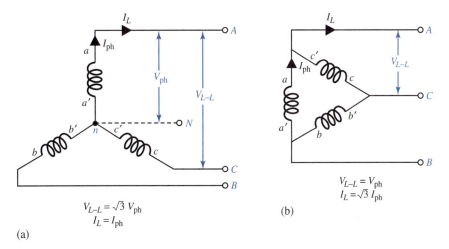

(a)

(b)

Figure 4.1.3 Schematic representation of generator windings. **(a)** Balanced wye connection. **(b)** Balanced delta connection.

as a reference with a phase angle of zero for convenience. This procedure yields the equivalent single-phase circuit in which all quantities correspond to those of one phase in the three-phase circuit. Except for the 120° phase displacements in the currents and voltages, the conditions in the other two phases are the same, and there is no need to investigate them individually. Line currents in the three-phase system are the same as in the single-phase circuit, and total three-phase real power, reactive power, and volt-amperes are three times the corresponding quantities in the single-phase circuit. Line-to-line voltages, in magnitude, can be obtained by multiplying voltages in the single-phase circuit by $\sqrt{3}$.

When a system of sources is so large that its voltage and frequency remain constant regardless of the power delivered or absorbed, it is known as an *infinite bus*. Such a bus has a voltage and a frequency that are unaffected by external disturbances. The infinite bus is treated as an ideal voltage source.

Phase Sequence

It is standard practice in the United States to designate the phase $A–B–C$ such that under balanced conditions the voltage and current in the A-phase lead in time the voltage and current in the B-phase by 120° and in the C-phase by 240°. This is known as *positive phase sequence A–B–C*. The phase sequence should be observed either from the waveforms in the time domain shown in Figure 4.1.2(a) or from the phasor diagrams shown in Figure 4.1.2(b) or 4.1.4(b), and not from space or schematic diagrams, such as Figures 4.1.3 and 4.1.4(a). If the rotation of the generator of Figure 4.1.1 is reversed, or if any two of the three leads from the armature (not counting the neutral) to the generator terminals are reversed, the phase sequence becomes $A–C–B$ (or $C–B–A$ or $B–A–C$), which is known as *negative phase sequence*.

Only the balanced three-phase sources are considered in this chapter. Selection of one voltage as the reference with a phase angle of zero determines the phase angle of all the other voltages in the system for a given phase sequence. As indicated before, the reference phasor is chosen arbitrarily for convenience. In Figure 4.1.4(b), $\bar{V}_{BC}$ is the reference phasor, and with the counterclockwise rotation (assumed positive) of all the phasors at the same frequency, the

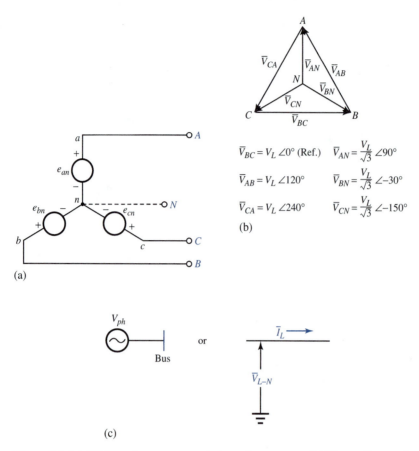

$$\bar{V}_{BC} = V_L \angle 0° \text{ (Ref.)} \qquad \bar{V}_{AN} = \frac{V_L}{\sqrt{3}} \angle 90°$$

$$\bar{V}_{AB} = V_L \angle 120° \qquad \bar{V}_{BN} = \frac{V_L}{\sqrt{3}} \angle -30°$$

$$\bar{V}_{CA} = V_L \angle 240° \qquad \bar{V}_{CN} = \frac{V_L}{\sqrt{3}} \angle -150°$$

(b)

Figure 4.1.4 **(a)** Balanced wye-connected, three-phase source. **(b)** Phasor diagram for three-phase source (sequence ABC). Note that such relations as $\bar{V}_{AB} = \bar{V}_{AN} + \bar{V}_{NB}$ are satisfied; also $\bar{V}_{AN} + \bar{V}_{BN} + \bar{V}_{CN} = 0$; $\bar{V}_{AB} + \bar{V}_{BC} + \bar{V}_{CA} = 0$. **(c)** Single-line equivalent circuit.

sequence can be seen to be A–B–C. Unless otherwise mentioned, the positive phase sequence is to be assumed.

4.2 BALANCED THREE-PHASE LOADS

Three-phase loads can be connected in either wye (also known as star or Y) or delta (otherwise known as mesh or Δ). If the load impedances in each of the three phases are the same in both magnitude and phase angle, the load is said to be balanced.

For the analysis of network problems, transformations for converting a delta-connected network to an equivalent wye-connected network and vice versa will be found to be useful. The relationships for interconversion of wye and delta networks are given in Figure 4.2.1. These are similar to those given in Section 2.4 for resistive network reduction. They can be obtained by imposing the condition of equivalence that the impedance between any two terminals for one network be equal to the corresponding impedance between the same terminals for the other network. The details are left as a desirable exercise for the student. For the balanced case, each wye impedance is one-third of each delta impedance; conversely, each delta impedance is three times each wye impedance.

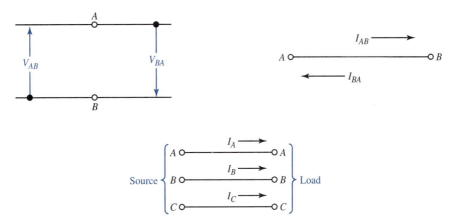

Figure 4.1.5 Notation using subscripts.

Balanced Wye-Connected Load

Let us consider a three-phase, four-wire 208-V supply system connected to a balanced wye-connected load with an impedance of $10\angle 20° \ \Omega$, as shown in Figure 4.2.2(a). We shall solve for the line currents and draw the corresponding phasor diagram.

Conventionally it is assumed that 208 V is the rms value of the line-to-line voltage of the supply system, and the phase sequence is positive, or A–B–C, unless mentioned otherwise. The magnitude of the line-to-neutral (or phase) voltages is given by $208/\sqrt{3}$, or 120 V. Selecting the line currents returning through the neutral conductor, as shown in Figure 4.2.2, we have

$$\bar{I}_A = \frac{\bar{V}_{AN}}{\bar{Z}} = \frac{(208/\sqrt{3})\angle 90°}{10\angle 20°} = 12\angle 70° \tag{4.2.1}$$

$$\bar{I}_B = \frac{\bar{V}_{BN}}{\bar{Z}} = \frac{(208/\sqrt{3})\angle - 30°}{10\angle 20°} = 12\angle -50° \tag{4.2.2}$$

$$\bar{I}_C = \frac{\bar{V}_{CN}}{\bar{Z}} = \frac{(208/\sqrt{3})\angle - 150°}{10\angle 20°} = 12\angle -170° \tag{4.2.3}$$

Note that $\bar{V}_{BC}$ has been chosen arbitrarily as the reference phasor, as in Figure 4.1.4(b). Assuming the direction of the neutral current toward the load as positive, we obtain

$$\bar{I}_N = - \left(\bar{I}_A + \bar{I}_B + \bar{I}_C \right)$$
$$= - (12\angle 70° + 12\angle - 50° + 12\angle 170°) = 0 \tag{4.2.4}$$

That is to say that the system neutral and the star point of the wye-connected load are at the same potential, even if they are not connected together electrically. It makes no difference whether they are interconnected or not.

Thus, for a *balanced* wye-connected load, the neutral current is always zero. The line currents and phase currents are equal in magnitude, and the line currents are in phase with the corresponding phase currents. The line-to-line voltages, in magnitude, are $\sqrt{3}$ times the phase voltages, and the phase voltages lag the corresponding line voltages by 30°.

The phasor diagram is drawn in Figure 4.2.2(b), from which it can be observed that the balanced line (or phase) currents lag the corresponding line-to-neutral voltages by the impedance

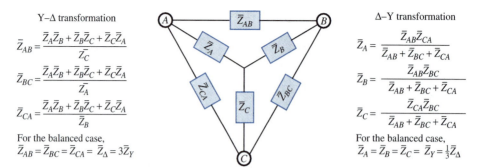

Y–Δ transformation

$$\bar{Z}_{AB} = \frac{\bar{Z}_A \bar{Z}_B + \bar{Z}_B \bar{Z}_C + \bar{Z}_C \bar{Z}_A}{\bar{Z}_C}$$

$$\bar{Z}_{BC} = \frac{\bar{Z}_A \bar{Z}_B + \bar{Z}_B \bar{Z}_C + \bar{Z}_C \bar{Z}_A}{\bar{Z}_A}$$

$$\bar{Z}_{CA} = \frac{\bar{Z}_A \bar{Z}_B + \bar{Z}_B \bar{Z}_C + \bar{Z}_C \bar{Z}_A}{\bar{Z}_B}$$

For the balanced case,
$$\bar{Z}_{AB} = \bar{Z}_{BC} = \bar{Z}_{CA} = \bar{Z}_\Delta = 3\bar{Z}_Y$$

Δ–Y transformation

$$\bar{Z}_A = \frac{\bar{Z}_{AB} \bar{Z}_{CA}}{\bar{Z}_{AB} + \bar{Z}_{BC} + \bar{Z}_{CA}}$$

$$\bar{Z}_B = \frac{\bar{Z}_{AB} \bar{Z}_{BC}}{\bar{Z}_{AB} + \bar{Z}_{BC} + \bar{Z}_{CA}}$$

$$\bar{Z}_C = \frac{\bar{Z}_{CA} \bar{Z}_{BC}}{\bar{Z}_{AB} + \bar{Z}_{BC} + \bar{Z}_{CA}}$$

For the balanced case,
$$\bar{Z}_A = \bar{Z}_B = \bar{Z}_C = \bar{Z}_Y = \tfrac{1}{3}\bar{Z}_\Delta$$

Figure 4.2.1 Wye–delta and delta–wye transformations.

angle (20° in our example). The load power factor is given by cos 20° for our problem, and it is said to be *lagging* in this case, as the impedance angle is positive and the phase current lags the corresponding phase voltage by that angle.

The problem can also be solved in a simpler way by making use of a single-line equivalent circuit, as shown in Figure 4.2.2.(c),

$$\bar{I}_L = \frac{\bar{V}_{L-N}}{\bar{Z}} = \frac{(208/\sqrt{3}) \angle 0°}{10 \angle 20°} = 12 \angle -20° \qquad (4.2.5)$$

in which $\bar{V}_{L-N}$ is chosen as the reference for convenience. The magnitude of the line current and the power factor angle are known; the negative sign associated with the angle indicates that the power factor is lagging. By knowing that the line (or phase) currents $\bar{I}_A$, $\bar{I}_B$, $\bar{I}_C$ lag their respective voltages $\bar{V}_{AN}$, $\bar{V}_{BN}$, and $\bar{V}_{CN}$ by 20°, the phase angles of various voltages and currents, if desired, can be obtained with respect to any chosen reference, such as $\bar{V}_{BC}$.

Balanced Delta-Connected Load

Next, let us consider the case of a balanced delta-connected load with impedance of $5\angle 45°$ Ω supplied by a three-phase, three-wire 100-V system, as shown in Figure 4.2.3(a). We shall determine the line currents and draw the corresponding phasor diagram.

With the assumed positive phase sequence (A–B–C) and with $\bar{V}_{BC}$ as the reference phasor, the line-to-line voltages $\bar{V}_{AB}$, $\bar{V}_{BC}$, and $\bar{V}_{CA}$ are shown in Figure 4.2.3. The rms value of the line-to-line voltages is 100 V for our example. Choosing the positive directions of the line and phase currents as in Figure 4.2.3(a), we have

$$\bar{I}_{AB} = \frac{\bar{V}_{AB}}{\bar{Z}} = \frac{100 \angle 120°}{5 \angle 45°} = 20 \angle 75° \qquad (4.2.6)$$

$$\bar{I}_{BC} = \frac{\bar{V}_{BC}}{\bar{Z}} = \frac{100 \angle 0°}{5 \angle 45°} = 20 \angle -45° \qquad (4.2.7)$$

$$\bar{I}_{CA} = \frac{\bar{V}_{CA}}{\bar{Z}} = \frac{100 \angle 240°}{5 \angle 45°} = 20 \angle 195° \qquad (4.2.8)$$

By the application of Kirchhoff's current law at each of the vertices of the delta-connected load, we obtain

$$\bar{I}_A = \bar{I}_{AB} + \bar{I}_{AC} = 20 \angle 75° - 20 \angle 195° = 34.64 \angle 45° \qquad (4.2.9)$$

$$\bar{I}_B = \bar{I}_{BA} + \bar{I}_{BC} = -20 \angle 75° + 20 \angle -45° = 34.64 \angle -75° \qquad (4.2.10)$$

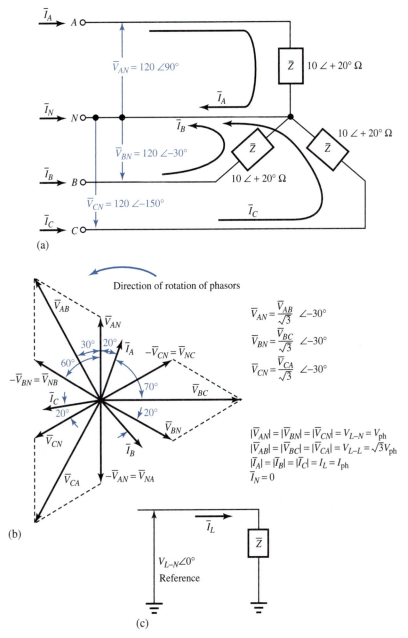

Figure 4.2.2 Balanced wye-connected load. **(a)** Connection diagram. **(b)** Phasor diagram. **(c)** Single-line equivalent circuit.

$$\bar{I}_C = \bar{I}_{CA} + \bar{I}_{CB} = 20\angle195° - 20\angle-45° = 34.64\angle165° \qquad (4.2.11)$$

The phasor diagram showing the line-to-line voltages, phase currents, and line currents is drawn in Figure 4.2.3(b). The load power factor is lagging, and is given by cos 45°.

For a *balanced* delta-connected load, the phase voltages and the line-to-line voltages are equal in magnitude, and the line voltages are in phase with the corresponding phase voltages.

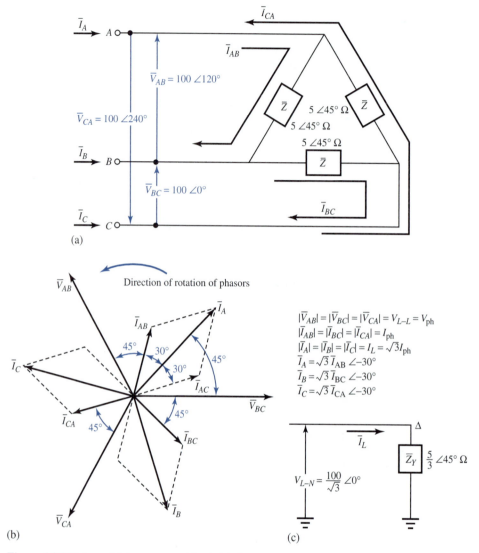

Figure 4.2.3 Balanced delta-connected load. **(a)** Connection diagram. **(b)** Phasor diagram. **(c)** Single-line equivalent circuit.

The line currents, in magnitude, are $\sqrt{3}$ times the phase currents, and the phase currents lead the corresponding line currents by 30°.

The preceding example can also be solved by the one-line equivalent method for which the delta-connected load is replaced by its equivalent wye-connected load. The single-line equivalent circuit is shown in Figure 4.2.3(c). The details are left as an exercise for the student.

Power in Balanced Three-Phase Circuits

The total power delivered by a three-phase source, or consumed by a three-phase load, is found simply by adding the power in each of the three phases. In a balanced circuit, however, this is the same as multiplying the average power in any one phase by 3, since the average power is the same for all phases. Thus one has

$$P = 3\, V_{ph} I_{ph}\, \cos\phi \qquad (4.2.12)$$

where V_{ph} and I_{ph} are the magnitudes of any phase voltage and phase current, $\cos\phi$ is the load power factor, and ϕ is the power factor angle between the phase voltage $\bar{V}_{ph}$ and the phase current $\bar{I}_{ph}$ corresponding to any phase. In view of the relationships between the line and phase quantities for balanced wye- or delta-connected loads, Equation (4.2.12) can be rewritten in terms of the line-to-line voltage and the line current for either wye- or delta-connected balanced loading as follows:

$$P = \sqrt{3}\, V_L I_L\, \cos\phi \qquad (4.2.13)$$

where V_L and I_L are the magnitudes of the line-to-line voltage and the line current. ϕ is still the load power factor angle as in Equation (4.2.12), namely, the angle between the phase voltage and the corresponding phase current.

In a balanced three-phase system, the sum of the three individually pulsating phase powers adds up to a constant, nonpulsating total power of magnitude three times the average real power in each phase. That is, in spite of the sinusoidal nature of the voltages and currents, the total instantaneous power delivered into the three-phase load is a constant, equal to the total average power. The real power P is expressed in watts when voltage and current are expressed in volts and amperes, respectively. You may recall that the instantaneous power in single-phase ac circuits absorbed by a pure inductor or capacitor is a double-frequency sinusoid with zero average value. The instantaneous power absorbed by a pure resistor has a nonzero average value plus a double-frequency term with zero average value. The instantaneous reactive power is alternately positive and negative, indicating the reversible flow of energy to and from the reactive component of the load. Its amplitude or maximum value is known as the reactive power.

The total reactive power Q (expressed as reactive volt-amperes, or VARs) and the volt-amperes for either wye- or delta-connected balanced loadings are given by

$$Q = 3\, V_{ph} I_{ph}\, \sin\phi \qquad (4.2.14)$$

or

$$Q = \sqrt{3}\, V_L I_L\, \sin\phi \qquad (4.2.15)$$

and

$$S = \left|\bar{S}\right| = \sqrt{P^2 + Q^2} = 3\, V_{ph} I_{ph} = \sqrt{3}\, V_L I_L \qquad (4.2.16)$$

where the complex power $\bar{S}$ is given by

$$\bar{S} = P + jQ \qquad (4.2.17)$$

In speaking of a three-phase system, unless otherwise specified, balanced conditions are assumed. The terms voltage, current, and power, unless otherwise identified, are conventionally understood to imply the line-to-line voltage (rms value), the line current (rms value), and the total power of all three phases. In general, the ratio of the real or average power P to the apparent power or the magnitude of the complex power S is the power factor, which happens to be $\cos\phi$ in the sinusoidal case.

EXAMPLE 4.2.1

(a) A wye-connected generator is to be designed to supply a 20-kV three-phase line. Find the terminal line-to-neutral voltage of each phase winding.

(b) If the windings of the generator of part (a) were delta-connected, determine the output line-to-line voltage.

(c) Let the 20-kV generator of part (a) supply a line current of 10 A at a lagging power factor of 0.8. Compute the kVA, kW, and kVAR supplied by the alternator.

Solution

(a) $V_{L-N} = V_{ph} = 20/\sqrt{3} = 11.547$ kV

(b) $V_L = V_{ph} = 11.547$ kV

(c) kVA $= \sqrt{3}\,(20)(10) = 346.4$

kW $= 346.4(0.8) = 277.12$

kVAR $= 346.4(0.6) = 207.84$

4.3 MEASUREMENT OF POWER

A *wattmeter* is an instrument with a potential coil and a current coil so arranged that its deflection is proportional to $VI \cos \theta$, where V is the voltage (rms value) applied across the potential coil, I is the current (rms value) passing through the current coil, and θ is the angle between $\bar{V}$ and $\bar{I}$. By inserting such a single-phase wattmeter to measure the average real power in each phase (with its current coil in series with one phase of the load and its potential coil across the phase of the load), the total real power in a three-phase system can be determined by the sum of the wattmeter readings. However, in practice, this may not be possible due to the nonaccessibility of either the neutral of the wye connection, or the individual phases of the delta connection. Hence it is more desirable to have a method for measuring the total real power drawn by a three-phase load while we have access to only three line terminals.

The three-phase power can be measured by three single-phase wattmeters having current coils in each line and potential coils connected across the given line and any common junction. Since this common junction is completely arbitrary, it may be placed on any one of the three lines, in which case the wattmeter connected in that line will indicate zero power because its potential coil has no voltage across it. Hence, that wattmeter may be dispensed with, and three-phase power can be measured by means of only two single-phase wattmeters having a common potential junction on any of the three lines in which there is no current coil. This is known as the *two-wattmeter method of measuring three-phase power.* In general, m-phase power can be measured by means of $m - 1$ wattmeters. The method is valid for both balanced and unbalanced circuits with either the load or the source unbalanced.

Figure 4.3.1 shows the connection diagram for the two-wattmeter method of measuring three-phase power. The total real power delivered to the load is given by the *algebraic sum* of the two wattmeter readings,

$$P = W_A + W_C \tag{4.3.1}$$

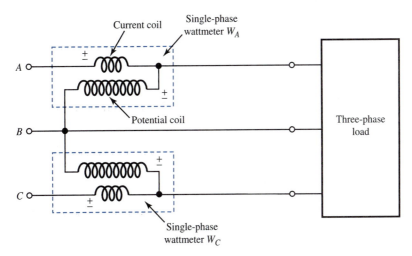

Figure 4.3.1 Connection diagram for two-wattmeter method of measuring three-phase power.

The significance of the algebraic sum will be realized in the paragraphs that follow. Two wattmeters can be connected with their current coils in any two lines, while their potential coils are connected to the third line, as shown in Figure 4.3.1. The wattmeter readings are given by

$$W_A = V_{AB} \cdot I_A \cdot \cos \theta_A \qquad (4.3.2)$$

where θ_A is the angle between the phasors $\bar{V}_{AB}$ and $\bar{I}_A$, and

$$W_C = V_{CB} \cdot I_C \cdot \cos \theta_C \qquad (4.3.3)$$

where θ_C is the angle between the phasors $\bar{V}_{CB}$ and $\bar{I}_C$.

The two-wattmeter method, when applied to the *balanced* loads, yields interesting results. Considering either balanced wye- or delta-connected loads, with the aid of the corresponding phasor diagrams drawn earlier for the phase sequence A–B–C (Figures 4.2.2. and 4.2.3), it can be seen that the angle between $\bar{V}_{AB}$ and $\bar{I}_A$ is $(30° + \phi)$ and that between $\bar{V}_{CB}$ and $\bar{I}_C$ is $(30 − \phi)$, where ϕ is the load power factor angle, or the angle associated with the load impedance. Thus, we have

$$W_A = V_L I_L \cos (30° + \phi) \qquad (4.3.4)$$

and

$$W_C = V_L I_L \cos (30° − \phi) \qquad (4.3.5)$$

where V_L and I_L are the magnitudes of the line-to-line voltage and line current, respectively. Simple manipulations yield

$$W_A + W_C = \sqrt{3} \, V_L I_L \cos \phi \qquad (4.3.6)$$

and

$$W_C − W_A = V_L I_L \sin \phi \qquad (4.3.7)$$

from which,

$$\tan \phi = \sqrt{3} \, \frac{W_C − W_A}{W_C + W_A} \qquad (4.3.8)$$

When the load power factor is unity, corresponding to a purely resistive load, both wattmeters will indicate the same wattage. In fact, both of them should read positive; if one of the wattmeters has a below-zero indication in the laboratory, an upscale deflection can be obtained by simply reversing the leads of either the current or the potential coil of the wattmeter. The sum of the wattmeter readings gives the total power absorbed by the load.

At zero power factor, corresponding to a purely reactive load, both wattmeters will again have the same wattage indication but with the opposite signs, so that their algebraic sum will yield zero power absorbed, as it should. The transition from a negative to a positive value occurs when the load power factor is 0.5 (i.e., ϕ is equal to 60°). At this power factor, one wattmeter reads zero while the other one reads the total real power delivered to the load.

For power factors (leading or lagging) greater than 0.5, both wattmeters read positive, and the sum of the two readings gives the total power. For a power factor less than 0.5 (leading or lagging), the smaller reading wattmeter should be given a negative sign and the total real power absorbed by the load (which has to be positive) is given by the difference between the two wattmeter readings. Figure 4.3.2 shows a plot of the load power factor versus the ratio W_l/W_h, where W_l and W_h are the lower and higher readings of the wattmeters, respectively.

Another method that is sometimes useful in a laboratory environment for determining whether the total power is the sum or difference of the two wattmeter readings is described here. To begin, make sure that both wattmeters have an upscale deflection. To perform the test, remove the lead of the potential coil of the lower reading wattmeter from the common line that has no current coil, and touch the lead to the line that has the current coil of the higher reading wattmeter. If the pointer of the lower reading wattmeter deflects upward, the two wattmeter readings should be added; if the pointer deflects in the below-zero direction, the wattage reading of the lower reading wattmeter should be subtracted from that of the higher reading wattmeter.

Given the two wattmeter readings from the two-wattmeter method used on a three-phase *balanced* load, it is possible to find the tangent of the phase impedance angle as $\sqrt{3}$ times the ratio of the difference between the two wattmeter readings and their sum, based on Equation (4.3.8). If one knows the system sequence and the lines in which the current coils of the wattmeters are located, the sign for the angle can be determined with the aid of the following expressions. For sequence A–B–C,

$$\tan \phi = \sqrt{3}\,\frac{W_C - W_A}{W_C + W_A} = \sqrt{3}\,\frac{W_A - W_B}{W_A + W_B} = \sqrt{3}\,\frac{W_B - W_C}{W_B + W_C} \qquad (4.3.9)$$

and for sequence C–B–A,

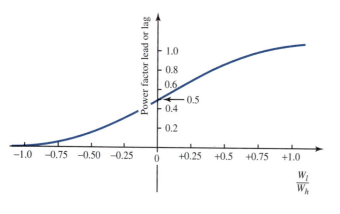

Figure 4.3.2 Plot of load power factor versus W_l/W_h.

$$\tan \phi = \sqrt{3} \, \frac{W_A - W_C}{W_A + W_C} = \sqrt{3} \, \frac{W_B - W_A}{W_B + W_A} = \sqrt{3} \, \frac{W_C - W_B}{W_C + W_B} \qquad (4.3.10)$$

The two-wattmeter method discussed here for measuring three-phase power makes use of single-phase wattmeters. It may be noted, however, that three-phase wattmeters are also available, which, when connected appropriately, indicate the total real power absorbed. The total reactive power associated with the three-phase *balanced* load is given by

$$Q = \sqrt{3} \, V_L I_L \, \sin \phi = \sqrt{3} \, (W_C - W_A) \qquad (4.3.11)$$

based on the two wattmeter readings of the two-wattmeter method.

With the generator action of the source assumed, $+P$ for the real power indicates that the source is supplying real power to the load; $+Q$ for the reactive power shows that the source is delivering inductive VARs while the current lags the voltage (i.e., the power factor is lagging); and $-Q$ for the reactive power indicates that the source is delivering capacitive VARs or absorbing inductive VARs, while the current leads the voltage (i.e., the power factor is leading).

EXAMPLE 4.3.1

Considering Figure 4.3.1, let balanced positive-sequence, three-phase voltages with $\bar{V}_{AB} = 100\sqrt{3}\angle 0°$ V (rms) be applied to terminals A, B, and C. The three-phase wye-connected balanced load consists of a per-phase impedance of $(10 + j10)\Omega$. Determine the wattmeter readings of W_A and W_C. Then find the total three-phase real and reactive powers delivered to the load. Based on the wattmeter readings of W_A and W_C, compute the load power factor and check the sign associated with the power factor angle.

Solution

$$\bar{V}_{AB} = 100\sqrt{3} \, \angle 0° \text{ V}; \qquad \bar{V}_{BC} = 100\sqrt{3} \, \angle - 120°; \qquad \bar{V}_{CA} = 100\sqrt{3} \, \angle 120°$$

$$\bar{V}_{AN} = 100\angle - 30° \text{ V}; \qquad \bar{V}_{BN} = 100\angle - 150°; \qquad \bar{V}_{CN} = 100\angle 90°$$

$$\bar{I}_A = \frac{\bar{V}_{AN}}{\bar{Z}} = \frac{100\angle - 30°}{10\sqrt{2} \, \angle 45°} = 5\sqrt{2} \, \angle - 75° \text{ A (rms)}$$

$$\bar{I}_C = \frac{\bar{V}_{CN}}{\bar{Z}} = \frac{100\angle 90°}{10\sqrt{2} \, \angle 45°} = 5\sqrt{2} \, \angle 45° \text{ A (rms)}$$

The load power factor angle $\phi = 45°$, and it is a case of lagging power factor with the inductive load.

$$W_A = V_{AB} I_A \, \cos(30° + \phi) = 100\sqrt{3} \left(5\sqrt{2}\right) \cos 75° = 317 \text{ W}$$

$$W_C = V_{CB} I_C \, \cos(30° - \phi) = 100\sqrt{3} \left(5\sqrt{2}\right) \cos 15° = 1183 \text{ W}$$

The total three-phase real power delivered to the load

$$W_A + W_C = 317 + 1183 = 1500 \text{ W}$$

which checks with

$$\sqrt{3} \, V_L I_L \, \cos \phi = \sqrt{3} \, 100\sqrt{3} \left(5\sqrt{2}\right) \cos 45° = 1500 \text{ W}$$

The total three-phase reactive power delivered to the load is

$$\sqrt{3}(W_C - W_A) = \sqrt{3}(1183 - 317) = 1500 \text{ VAR}$$

which checks with

$$\sqrt{3}V_L I_L \sin \phi = \sqrt{3}(100\sqrt{3})(5\sqrt{2}) \sin 45° = 1500 \text{ VAR}$$

For positive-sequence A–B–C,

$$\tan \emptyset = \sqrt{3}\left(\frac{W_C - W_A}{W_C + W_A}\right) = \sqrt{3}\left(\frac{1183 - 317}{1183 + 317}\right) = 1.0$$

or $\emptyset = 45°$ with a positive sign, implying thereby that it corresponds to an inductive or lagging load, which checks with the given load specification. The load power factor is given by

$$\cos \emptyset = 0.707$$

and is lagging in this case.

4.4 RESIDENTIAL WIRING AND SAFETY CONSIDERATIONS

Residential electric power service commonly consists of a three-wire ac system supplied by the local power company. A distribution transformer with a primary and two secondaries formed with a center tap on the secondary side, as illustrated in Figure 4.4.1, is located on a utility pole, from which three wires originate. Two of them are known as "hot" wires, while the third is called a neutral wire, which is connected to earth ground. In accordance with standard insulation color codes, "hot" wires are denoted by either B (black) or R (red), neutral by W (white), and ground (or uninsulated wire) by G (green). The functional difference between neutral and ground wires will be brought out later when considering circuit wiring.

Domestic loads consisting of 120-V appliances and lighting, divided nearly equally between the two 120-V (rms) secondaries, are connected from hot wires to neutral. Appliances such as electric ranges and water heaters are supplied with 240-V (rms) power from the series-connected secondaries, as shown in Figure 4.4.1.

Minimizing the power loss in the lines (known as I^2R *loss*) is important from the viewpoint of efficiency and reducing the amount of heat generated in the wiring for *safety* considerations. Since the power loss in the lines is directly related to the current required by the load, a lower line loss will be incurred with the 240-V wiring in delivering the necessary power to a load. For the lower voltage case, however, the size of the wires is increased, thereby reducing the wire resistance, in an effort to minimize line losses. Problem 4.4.2 deals with these considerations.

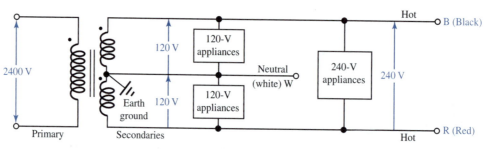

Figure 4.4.1 Three-winding distribution transformer providing dual-voltage ac supply.

The three-line cable coming out of the secondaries of the distribution transformer on the utility pole passes through the electric meter that measures energy consumption in kilowatt-hours and terminates at the *main panel*. Figure 4.4.2 shows a typical wiring arrangement for a residence. At the main panel, *circuit breakers* serve the joint role of disconnecting switches and overcurrent protection; the neutral is connected to a *busbar (bus)* and in turn to the local earth ground; the hot lines are connected to individual circuits for lighting and appliances, as illustrated in Figure 4.4.2. The circuit breaker labeled GFCI (*ground-fault circuit interruption*), used for safety primarily with outdoor circuits and in bathrooms, has additional features that will be described later. Note that every outgoing "hot" wire must be connected to a circuit breaker, whereas every neutral wire and ground wire must be tied directly to earth ground at the neutral busbar.

Today most homes have three-wire connections to their outlets, one of which is shown in Figure 4.4.3. The need for both ground and neutral connections needs to be explained, since the ground conductor may appear to be redundant, playing no role in the actual operation of a load that might be connected to the receptacle. From the viewpoint of *safety*, the ground connection is used to connect the metallic chassis of the appliance to earth ground. Without the ground conductor connected to the metal case of the appliance, as shown in Figure 4.4.4(a), the appliance chassis could be at any potential with respect to ground, possibly even at the "hot" wire's potential if a part of the "hot" wire were to lose some insulation and come in contact with the inside of the chassis. An unintended connection may occur because of the corrosion of insulation or a loose mechanical connection. Poorly grounded appliances can thus be a significant hazard by providing a path to ground through the body of a person touching the chassis with a hand. An undersized ground loop current limited by the body resistance may flow directly through the body to ground and could be quite harmful. Typically, the circuit breaker would not operate under such circumstances.

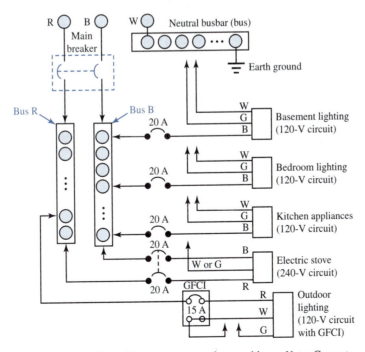

Figure 4.4.2 Typical wiring arrangement for a residence *Note:* Currents and voltages shown are rms magnitudes.

However, if the ground conductor is present and properly connected to the chassis of the appliance, as shown in Figure 4.4.4(b), the metal case will remain at ground potential or, at worst, a few volts from ground if a fault results in current through the ground wire. The "hot" conductor may be shorted to ground under fault conditions, in which case the circuit breaker would operate.

The body resistance of a normal person ranges from 500 kΩ down to 1 kΩ, depending upon whether the skin is dry or wet. Thus, a person with wet skin risks electrocution from ac voltages as low as 100 V. The amount of current is the key factor in electric shock, and Table 4.4.1 lists the effects of various levels of 60-Hz ac current on the human body. Note that the 100–300-mA range turns out to be the most dangerous.

The best possible shock protection is afforded by the *ground-fault circuit interrupter* (GFCI) shown in Figure 4.4.4(c). A sensing coil located around the "hot" and neutral wires in the GFCI detects the imbalance of currents between the neutral and the live conductor under fault conditions and opens the circuit in response when $|I_B - I_W| > 5$ mA. The GFCI may be located either at an outlet or at the main panel. Ground-fault interrupters are now required in branch circuits that serve outlets in areas such as bathrooms, basements, garages, and outdoor sites.

Various codes, such as the *National Electrical Code*, have been established to provide protection of personnel and property, while specifying requirements for the installation and maintenance of electrical systems. Only qualified and properly certified persons should undertake installation, alteration, or repair of electrical systems. Safety when working with electric power must always be a primary consideration. In addition to numerous deaths caused each year due to electrical accidents, fire damage that results from improper use of electric wiring and wiring faults amounts to millions of dollars per year.

Finally, Figure 4.4.5 illustrates how a device, such as a light, can be controlled independently from two different locations using single-pole double-throw (SPDT) switches, commonly known as three-way switches or staircase switches. When the "hot" wire is switched between two "travelers" at the first switch and from the travelers to the device at the second, a complete circuit is formed only when both switches are either up or down; flipping either switch opens the circuit. Note that neutral and ground wires are never switched.

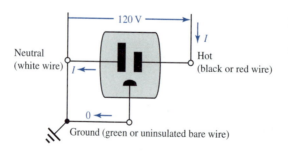

Figure 4.4.3 Three-wire outlet. *Note:* Currents are shown with rms magnitudes for normal-load operation.

TABLE 4.4.1 Effects of 60-Hz ac Current on Human Body

Range of Current	Physiological Effects
1–5 mA	Threshold of sensation
10–20 mA	Involuntary muscle contractions; pain; "can't let go" condition
20–100 mA	Breathing difficulties; severe pain
100–300 mA	Ventricular fibrilation; *possible death* with no intervention
300–500 mA	Respiratory paralysis (heartbeat may stop and may restart if shock is removed before death)
1–10 A	Severe burns; temporary heart contraction (not fatal unless vital organs are burned)

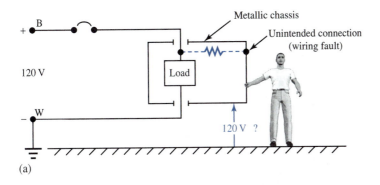

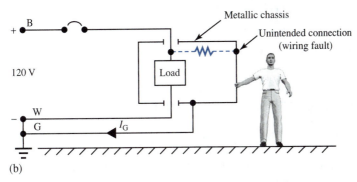

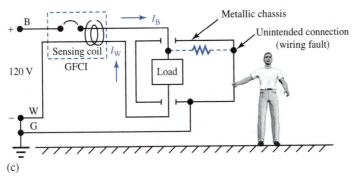

Figure 4.4.4 Appliance with wiring fault. *Note:* Currents and voltages shown are rms magnitudes. **(a)** Ungrounded chassis. **(b)** Grounded chassis. **(c)** Grounded chassis with GFCI in circuit.

4.5   LEARNING OBJECTIVES

The *learning objectives* of this chapter are summarized here, so that the student can check whether he or she has accomplished each of the following.

- Phase sequence of balanced three-phase voltages and currents.
- Wye–delta and delta–wye conversions for networks containing impedances.
- Circuit analysis with balanced three-phase source and balanced wye-connected load.
- Circuit analysis with balanced three-phase source and balanced delta-connected load.

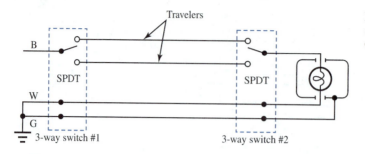

Figure 4.4.5 Three-way switch connections to operate a device.

- Real, reactive, and apparent powers in three-phase circuits.
- Measurement of three-phase power by two single-phase wattmeters.
- Basic notions of residential circuit wiring, including grounding and safety considerations.

4.6 PRACTICAL APPLICATION: A CASE STUDY

Physiological Effects of Current and Electrical Safety

Various effects that electric current may have on the human body and the current levels at which they occur are illustrated in Figure 4.6.1. Note that small currents can have serious effects. The ranges in Figure 4.6.1, however, represent variations among individuals in body size, condition, and tolerance to electric shock. The region of the body through which the current passes is crucial.

The current can easily be estimated by the application of Ohm's law,

$$I = V/R$$

where R is the resistance of the circuit of which a person's body might become a part. Most serious electric shocks occur where a person is in simultaneous contact with the so-called hot wire and the ground. Ground may be the dry or moist earth, the plumbing of a house, or even a concrete floor that is connected to the plumbing. The circuit resistance would include not only the body

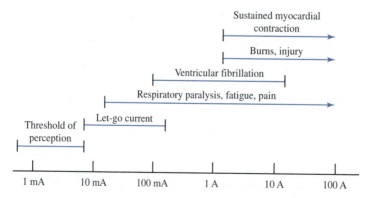

Figure 4.6.1 Physiological effects of electricity. (Adapted from J. G. Webster, *Medical Instrumentation, Application and Design,* Houghton Mifflin, 1978.)

TABLE 4.6.1 Resistance Estimates for Various Skin-Contact Conditions

Condition	Resistance	
	Dry	Wet
Finger touch	40 kΩ–1000 kΩ	4–15 kΩ
Hand holding wire	15–50 kΩ	3–6 kΩ
Finger–thumb grasp	10–30 kΩ	2–5 kΩ
Hand holding pliers	5–10 kΩ	1–3 kΩ
Palm touch	3–8 kΩ	1–2 kΩ
Hand around $1\frac{1}{2}$-in pipe (or drill handle)	1–3 kΩ	0.1–1.5 kΩ
Two hands around $1\frac{1}{2}$-in pipe	0.5–1.5 kΩ	250–750 Ω
Hand immersed		200–500 Ω
Foot immersed		100–300 Ω
Human body, internal excluding skin, 200–1000 Ω		

Adapted from R. Lee, "Electrical Safety in Industrial Plants," *IEEE Spectrum* (June 1971).

TABLE 4.6.2 Resistance Estimates for Equal Areas (130 cm^2) of Various Materials

Material	Resistance
Rubber gloves or soles	>20 MΩ
Dry concrete above grade	1–5 MΩ
Dry concrete on grade	0.2–1 MΩ
Leather sole, dry, including foot	0.1–0.5 MΩ
Leather sole, damp, including foot	5–20 kΩ
Wet concrete on grade	1–5 kΩ

Adapted from R. Lee, "Electrical Safety in Industrial Plants," *IEEE Spectrum* (June 1971).

resistance but also the resistance of the shoes and the resistance between the shoes and ground. Tables 4.6.1 and 4.6.2 give basic information for estimating the total resistance to ground.

Let us estimate the current caused when a person is standing on moist ground with leather-soled damp shoes and unwillingly grabs hold of 240-V wire with a wet palm. Referring to Tables 4.6.1 and 4.6.2 and taking the lowest values, one has 1 kΩ for the grasp, 200 Ω for the body, and 5 kΩ for the feet and shoes. Thus the largest current the victim might carry would be 240V/6.2 kΩ, or about 40 mA. From Figure 4.6.1 one can see that there is a good chance that the victim would be unable to release the grasp and unable to breathe. Thus a dangerous situation may exist.

When a person is experiencing electric shock, because the damaging effects are progressive, time becomes a critical factor. First, the source of electric energy should be removed from the shock victim, even before an emergency medical service is called in. CPR (cardiopulmonary resuscitation) may have to be administered by a trained person. The condition of ventricular fibrillation occurs, the heart loses its synchronized pumping action, and the blood circulation ceases. Sophisticated medical equipment may be needed to restore coherent heart pumping.

When working around electric power, the following general precautions are to be taken:

- Make sure that the power is off before working on any electric wiring or electric equipment.
- Wear rubber gloves and rubber-soled shoes, if possible.
- Avoid standing on a wet surface or on moist ground.
- Avoid working alone, as far as possible, around exposed electric power.

PROBLEMS

*4.1.1 The line-to-line voltage of a balanced wye-connected three-phase source is given as 100 V. Choose V_{AB} as the reference.

(a) For the phase sequence A–B–C, sketch the phasor diagram of the voltages and find the expressions for the phase voltages.

(b) Repeat part (a) for the phase sequence C–B–A.

4.2.1 A three-phase, three-wire 208-V system is connected to a balanced three-phase load. The line currents $\bar{I}_A$, $\bar{I}_B$, and $\bar{I}_C$ are given to be in phase with the line-to-line voltages $\bar{V}_{BC}$, $\bar{V}_{CA}$, and $\bar{V}_{AB}$, respectively. If the line current is measured to be 10 A, find the per-phase impedance of the load.

(a) If the load is wye-connected.

(b) If the load is delta-connected.

*4.2.2 Consider a three-phase 25-kVA, 440-V, 60-Hz alternator operating at full load (i.e., delivering its rated kVA) under balanced steady-state conditions. Find the magnitudes of the alternator line current and phase current.

(a) If the generator windings are wye-connected.

(b) If the generator windings are delta-connected.

4.2.3 Let the alternator of Problem 4.2.2 supply a line current of 20 A at a power factor of 0.8 lagging. Determine:

(a) The kVA supplied by the machine.

(b) The real power kW delivered by the generator.

(c) The reactive power kVAR delivered by the alternator.

4.2.4 A balanced wye-connected load with a per-phase impedance of $4 + j3\ \Omega$ is supplied by a 173-V, 60-Hz three-phase source.

(a) Find the line current, the power factor, the total volt-amperes, the real power, and the reactive power absorbed by the load.

(b) Sketch the phasor diagram showing all the voltages and currents, with $\bar{V}_{AB}$ as the reference.

(c) If the star point of the load is connected to the system neutral through an ammeter, what would the meter read?

4.2.5 A balanced delta-connected load with a per-phase impedance of $12 + j9\ \Omega$ is supplied by a 173-V, 60-Hz three-phase source.

(a) Determine the line current, the power factor, the total volt-amperes, the real power, and the reactive power absorbed by the load.

(b) Compare the results of Problem 4.2.5(a) with those obtained in Problem 4.2.4(a), and explain why they are the same.

(c) Sketch the phasor diagram showing all voltages and currents, with $\bar{V}_{AB}$ as the reference.

4.2.6 A 60-Hz, 440-V, three-phase system feeds two balanced wye-connected loads in parallel. One load has a per-phase impedance of $8 + j3\ \Omega$ and the other $4 - j1\ \Omega$. Compute the real power in kW delivered to (a) the inductive load, and (b) the capacitive load.

*4.2.7 Balanced wye-connected loads drawing 10 kW at 0.8 power factor lagging and 15 kW at 0.9 power factor leading are connected in parallel and supplied by a 60-Hz, 300-V, three-phase system. Find the line current delivered by the source.

4.2.8 A balanced delta-connected load with a per-phase impedance of $30 + j10\ \Omega$ is connected in parallel with a balanced wye-connected load with a per-phase impedance of $40 - j10\ \Omega$. This load combination is connected to a balanced three-phase supply through three conductors, each of which has a resistance of $0.4\ \Omega$. The line-to-line voltage at the terminals of the load combination is measured to be 2000 V. Determine the power in kW:

(a) Delivered to the delta-connected load.

(b) Delivered to the wye-connected load.

(c) Lost in the line conductors.

4.2.9 Three identical impedances of $30\angle 30°\ \Omega$ are connected in delta to a three-phase 173-V system by conductors that have impedances of $0.8 + j0.6\ \Omega$ each. Compute the magnitude of the line-to-line voltage at the terminals of the load.

4.2.10 Repeat Problem 4.2.9 for a delta-connected set of capacitors with a per-phase reactance of $-j60\ \Omega$ connected in parallel with the load, and compare the result with that obtained in Problem 4.2.9.

4.2.11 Two balanced, three-phase, wye-connected loads are in parallel across a balanced three-phase supply. Load 1 draws a current of 20 A at 0.8 power factor leading, and load 2 draws a current of 30 A at 0.8 power factor lagging. For the combined load compute:

(a) The current taken from the supply and the supply power factor.

(b) The total real power in kW supplied by the source, if the supply voltage is 400 V.

4.2.12 Two balanced, three-phase, wye-connected loads are in parallel across a balanced, three-phase supply. Load 1 draws 15 kVA at 0.8 power factor lagging, and load 2 draws 20 kVA at 0.6 power factor leading. Determine:

(a) The total real power supplied in kW.

(b) The total kVA supplied.

(c) The overall power factor of the combined load.

*4.2.13 Two balanced, wye-connected, three-phase loads are in parallel across a balanced, three-phase 60-Hz, 208-V supply. The first load takes 12 kW at 0.6 power factor lagging, and the second load takes 15 kVA at 0.8 power factor lagging.

(a) Draw the power triangle for each load and for the combination.

(b) What is the total kVA supplied?

(c) Find the supply power factor, and specify whether lagging or leading.

(d) Compute the magnitude of the current drawn by each load and drawn from the supply.

(e) If wye-connected capacitors are placed in parallel with the two loads in order to improve the supply power factor, determine the value of capacitance in farads in each phase needed to bring the overall power factor to unity.

4.2.14 A three-phase balanced load draws 100 kW at 0.8 power factor lagging. In order to improve the supply power factor to 0.95 leading, a synchronous motor drawing 50 kW is connected in parallel with the load. Compute the kVAR, kVA, and the power factor of the motor (specify whether lagging or leading).

4.2.15 (a) A balanced wye-connected load with per-phase impedance of $20 + j10 \, \Omega$ is connected to a balanced 415-V, three-phase supply through three conductors, each of which has a series impedance of $2 + j4 \, \Omega$. Find the line current, the voltage across the load, the power delivered to the load, and the power lost in the line conductors.

(b) Repeat the problem if the three per-phase load impedances of part (a) were connected in delta.

4.2.16 A balanced electrical industrial plant load of 9.8 MW with 0.8 lagging power factor is supplied by a three-phase 60-Hz system having a maximum rating (load-carrying capacity) of 660 A at 11 kV (line-to-line voltage).

(a) Determine the apparent power and the reactive power drawn by the load.

(b) Additional equipment, consisting of a load of 1.5 MW and 0.7 MVAR lagging, is to be installed in the plant. Compute the minimum rating in MVA of the power factor correction capacitor that must be installed if the rating of the line is not to be exceeded. Find the system power factor.

(c) If the capacitor, consisting of three equal sections, is connected in delta across the supply lines, calculate the capacitance required in each section.

4.2.17 Derive the relationships given in Figure 4.2.1 for the wye–delta and the delta–wye transformations.

4.2.18 Two three-phase generators are supplying to a common balanced three-phase load of 30 kW at 0.8 power factor lagging. The per-phase impedance of the lines connecting the generator G_1 to the load is $1.4 + j1.6 \, \Omega$, whereas that of the lines connecting the generator G_2 to the load is $0.8 + j1 \, \Omega$. If the generator G_1, operating at a terminal voltage of 800 V (line to line), supplies 15 kW at 0.8 power factor lagging, find the voltage at the load terminals, the terminal voltage of the generator G_2, and the real power, as well as the reactive power output of the generator G_2.

4.3.1 Determine the wattmeter readings when the two-wattmeter method is applied to Problem 4.2.4, and check the total power obtained.

4.3.2 When the two-wattmeter method for measuring three-phase power is used on a certain balanced load, readings of 1200 W and 400 W are obtained (without any reversals). Determine the delta-connected load impedances if the system voltage is 440 V. With the information given, is it possible to find whether the load impedance is capacitive or inductive in nature?

4.3.3 The two-wattmeter method for measuring three-phase power is applied on a balanced wye-connected load, as shown in Figure 4.3.2, and the readings are given by

$$W_C = 836W \quad \text{and} \quad W_A = 224 \text{ W}$$

If the system voltage is 100 V, find the per-phase impedance of the load. In this problem, is it possible to specify the capacitive or inductive nature of the impedance?

4.3.4 Two wattmeters are used, as shown in Figure 4.3.1, to measure the power absorbed by a balanced delta-connected load. Determine the total power in kW, the power factor, and the per-phase impedance of the load if the supply-system voltage is 120 V and the wattmeter readings are given by

(a) $W_A = -500$ W, and $W_C = 1300$ W.

(b) $W_A = 1300$ W, and $W_C = -500$ W.

*4.3.5 Referring to Problem 4.2.15(b), with a phase voltage as the reference phasor and with a positive-phase-sequence supply system, two single-phase wattmeters are used to measure the power delivered to the load. The wattmeter current coils are arranged as in Figure 4.3.1 to carry the currents I_A and I_C, whereas the potential coils share a common connection at the node B. Determine the readings of the two wattmeters W_A and W_C, and verify that the sum of these readings equals the total power in the load.

4.4.1 Referring to Figure 4.4.1, let $V_{BN} = V_{RN} = 120$ V rms magnitude, and $V_{BR} = 240$ V rms magnitude. Write down expressions for $v_{BN}(t)$, $v_{RN}(t)$, and $v_{BR}(t)$, and sketch them as a function of time.

*4.4.2 Consider a 240-V supply feeding a resistive load of 10 kW through wires having a total resistance of $R = 0.02$ Ω. For the same load, let a 120-V supply be used with a total wire resistance of $R/2 = 0.01$ Ω. Compute the I^2R loss in the lines for both cases and compare.

4.4.3 A person, while driving a car in a cyclone and waiting at an intersection, hears a thumping sound when a power line falls across the car and makes contact with the chassis. The power line voltage to ground is 2400 V. Instinctively he steps out onto the wet ground, while holding the door handle, to check out the external situation. Assuming his body resistance to be 10 kΩ and negligible auto-chassis resistance, comment on what might happen to the individual. Would it have been safer for him to remain in the car until some help arrived?

ELECTRONIC ANALOG AND DIGITAL SYSTEMS

5 Analog Building Blocks and Operational Amplifiers

Electronic systems usually process information in either analog or digital form. In order to process the two different kinds of signals, analog circuits and digital circuits have been devised. While almost all technology was of the analog type until around 1960, due to the advent of integrated circuits (ICs), digital technology has grown tremendously.

In analog systems, a signal voltage or current is made proportional to some physical quantity. Since voltages (or currents) can take on any values over a continuous range between some minimum and some maximum, analog systems are also known as *continuous-state* systems. These are to be distinguished from digital or *discrete-state* systems, in which only certain values of voltage (or current) are allowed.

Most circuits found in analog systems are linear circuits in which one voltage (or current) is meant to be linearly proportional to another. *Linear active circuits* are also known as *amplifiers*, which are the building blocks of linear systems with analog technology.

When describing and analyzing electric systems, which are often large and complex, it is very helpful to consider such large systems as being built from smaller units, called *building blocks*. These are then the subunits, which can be connected to form larger circuits or systems. More importantly, the building blocks can be described adequately by their simple *terminal properties*. Thus, with the building block point of view, one is not concerned with the interiors of the blocks, only with how they perform as seen from the outside.

The concept of a *model*, which is a collection of ideal linear circuit elements simulating approximately the behavior of a real circuit element or a building block under certain limitations, is utilized in order to predict the performance of electronic systems through the use of equations. In this chapter first models are developed for the amplifier block, then an ideal operational amplifier is presented, and later applications of operational amplifiers are discussed.

5.1 THE AMPLIFIER BLOCK

An amplifier can be modeled as a two-port device, that is, a box with two pairs of terminals designated as input and output, as shown in Figure 5.1.1 (a). The circuit model of the amplifier block shown in Figure 5.1.1 (b) is developed on the basis of the following considerations:

1. Since, for most amplifiers, the input current is proportional to the input voltage, the input terminals in the model are connected by a resistance R_i, known as the *input resistance* of the amplifier.

2. Since an amplifier delivers electric power (to a speaker, for example), the output current can be represented by its Thévenin-source model. The Thévenin resistance R_o is known as the *output resistance* and the Thévenin voltage is a *dependent voltage source* Av_{in}, where A is called the *open-circuit voltage amplification.*

Thus, the amplifier block is a linear circuit block in which the output is proportional to the input, and the amplifier is characterized by the three constants R_i, R_o, and A. The input and output resistances may be generalized to input and output impedances in ac systems. The advantage of the model is that all internal complexities are summarized in the three constants, thereby simplifying the analysis of electric systems with amplifiers. Power-supply connections are usually not shown in circuit diagrams since they would only clutter up the drawing. It is assumed, however, that there are always connections to some power source in order to bring in the power necessary to run the amplifier (as will be illustrated in Example 5.1.4).

EXAMPLE 5.1.1

Let the amplifier block be connected to a current source at the input terminals, as shown in Figure E5.1.1(a), and to a load resistance R_L at its output terminals. Find v_{out}.

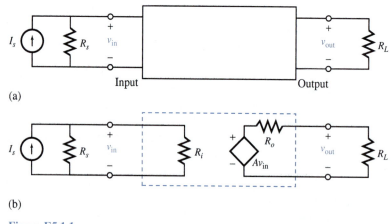

(a)

(b)

Figure E5.1.1

Solution

Using the model of Figure 5.1.1(b), we have the circuit configuration shown in Figure E5.1.1(b). Since R_i and R_S are connected in parallel,

$$v_{in} = I_S \frac{R_S R_i}{R_S + R_i}$$

Using the voltage-divider formula, one has

$$v_{out} = A v_{in} \frac{R_L}{R_o + R_L} = \frac{A R_L R_S R_i I_S}{(R_o + R_L)(R_S + R_i)}$$

Recall that A is the open-circuit voltage amplification. Let us consider the circuit shown in Figure 5.1.2 in order to explain *voltage amplification,* or *voltage gain.* In this circuit a signal voltage v_S is applied to the input of the amplifier block, whereas the output terminals are connected to a load resistance R_L. Let us evaluate the ratio of the voltage across the load to the signal voltage v_L/v_S, which is known as voltage gain G_V,

$$G_V = \frac{v_L}{v_S} = \frac{A v_{in} R_L}{(R_o + R_L) v_S} = \frac{A R_L}{(R_o + R_L)} \tag{5.1.1}$$

If there is no load, i.e., $R_L = \infty$, then it is easy to see that the voltage gain G_V will be equal to A; hence the justification to call A the open-circuit voltage amplification. The reduction in voltage gain due to the *effect of output loading* can be seen from Equation (5.1.1).

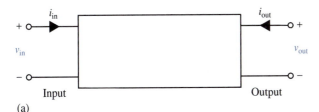

(a)

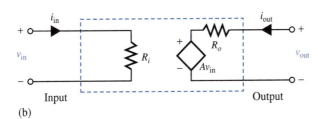

(b)

Figure 5.1.1 Amplifier block. (**a**) Two-port device. (**b**) Circuit model.

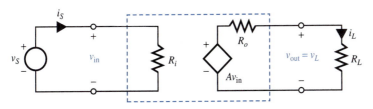

Figure 5.1.2 Circuit to explain voltage amplification or voltage gain.

By defining the *current gain* G_I to be the ratio of the current through R_L to the current through v_S, one gets

$$G_I = \frac{i_L}{i_S} = \frac{A v_{in} / (R_o + R_L)}{v_{in} / R_i} = \frac{A R_i}{R_o + R_L} \tag{5.1.2}$$

The *power gain* G_P, defined by the ratio of the power delivered to the load to the power given out by the signal source, is obtained as

$$G_P = \frac{v_L^2 / R_L}{v_S^2 / R_i} = \frac{G_V^2 R_i}{R_L} = \frac{A^2 R_L R_i}{(R_o + R_L)^2} = G_V G_I \tag{5.1.3}$$

Note that, for fixed values of R_o and R_i, G_P is maximized when R_L is chosen equal to R_o, and this corresponds to maximum power transfer to the load. One should also note that the added power emerging from the output comes from the power source that powers the amplifier, even though the power-supply connections are usually not shown on the circuit diagram, and that the existence of power gain does not violate the law of energy conservation.

EXAMPLE 5.1.2

The constants of an amplifier are given by $A = 1$, $R_i = 10,000\ \Omega$, and $R_o = 100\ \Omega$. It is driven by a Thévenin source with $v_{Th}(t) = V_O \cos \omega t$ and $R_{Th} = 20,000\ \Omega$. The amplifier output is connected to a 100-Ω load resistance. Find the power amplification, if it is defined as the ratio of the power delivered to the load to the maximum power available from the Thévenin source.

Solution

The corresponding circuit diagram is shown in Figure E5.1.2. The instantaneous power delivered to R_L is given by

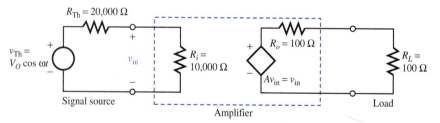

Figure E5.1.2 Circuit diagram.

$$P_L(t) = i_L^2(t) R_L = \left[\frac{A v_{in}(t)}{R_o + R_L} \right]^2 R_L$$

The time-averaged power P_L is obtained by representing $v_{in}(t)$ by its phasor $\bar{V}_{in}$,

$$P_L = \frac{1}{2} \left| \bar{V}_{in} \right|^2 \frac{A^2 R_L}{(R_o + R_L)^2}$$

Note that the time average of a product of sinusoids is given by

$$\text{Time average of } [v(t)\,v(t)] = \frac{1}{2}\text{Re}\left[\bar{V}\bar{V}^*\right] = \frac{|\bar{V}|^2}{2}$$

The maximum power is derived from a Thévenin source when it is loaded by a resistance equal to R_{Th}. The maximum power thus obtainable, known as *available power* P_{AVL}, is given by

$$P_{AVL} = \frac{|\bar{V}_{Th}|^2}{8R_{Th}}$$

In this circuit,

$$|\bar{V}_{in}|^2 = \frac{R_i^2}{(R_i + R_{Th})^2}\,|\bar{V}_{Th}|^2$$

Note that the peak value of the sinusoid is used here for the magnitude of the phasor in these expressions. Thus,

$$G_P = \frac{P_L}{P_{AVL}} = \frac{4A^2 R_L R_i^2 R_{Th}}{(R_o + R_L)^2\,(R_i + R_{Th})^2}$$

Substituting the values given, one gets

$$G_P = \frac{4\,(1)^2\,(100)\,10^8\,(20{,}000)}{(100 + 100)^2\,(10{,}000 + 20{,}000)^2} = \frac{200}{9} = 22.22$$

Note that although the amplifier's open-circuit voltage amplification A is only unity, substantial power gain is obtained in this case since the current gain is greater than unity and $G_P = G_V G_I$. In applications of microwave technology, the function of the amplifier is to magnify very small power to a measurable level with as little random noise as possible. The source signal may be from a radio telescope, and the available power may be determined by the power captured by the antenna dish.

EXAMPLE 5.1.3

Quite often an amplifier is used as a component of an amplifier circuit. Consider the amplifier circuit shown in Figure E5.1.3, which contains an amplifier block as an internal component. Find the input resistance R_i', the output resistance R_o', and the open-circuit voltage amplification A' of the larger circuit.

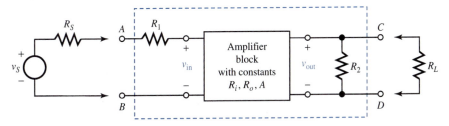

Figure E5.1.3

Solution

The input resistance looking into terminals A and B is

$$R_i' = R_1 + R_i$$

The output resistance is the same as the Thévenin resistance seen from terminals C and D. Turning off independent source v_S, v_{in} becomes zero, and as a consequence the dependent source in the amplifier block goes to zero. Looking to the left of terminals C and D, R_2 and R_o can be seen to be in parallel,

$$R_o' = \frac{R_2 R_o}{R_2 + R_o}$$

The open-circuit voltage gain of the larger circuit is

$$A' = \frac{v_{CD}}{v_{AB}}$$

With

$$v_{CD} = \frac{R_2}{R_o + R_2} A v_{in} \qquad \text{and} \qquad v_{in} = \frac{R_i}{R_i + R_1} v_{AB}$$

we have

$$A' = \frac{v_{CD}}{v_{AB}} = \frac{A R_2 R_i}{(R_o + R_2)(R_i + R_1)}$$

In further calculations, the dashed box can simply be replaced by the model with parameters R_i', R_o', and A'. The reader should note that in general R_i' may depend on R_L, and R_o' may depend on R_S, even though in this simple example they do not.

Two of the most important considerations that influence amplifier design are *power-handling capacity* and *frequency response*. In practice, the performance of an amplifier is limited by its ability to dissipate heat, known as its *power dissipation*. The electric power that is converted to heat in an amplifier can be calculated when the currents and voltages are known at all its terminals, including the power-supply terminals.

EXAMPLE 5.1.4

For the amplifier circuit shown in Figure E5.1.4 with $R_i \cong \infty$, $R_o \cong 0$, $A = 10$, $R_L = 100\ \Omega$, and $v_{in} = 1$ V, calculate the power dissipated in the amplifier if the voltage at the power-supply terminal V_{ps} is given to be 20 V and I_{ps} is assumed to be equal to I_L.

Figure E5.1.4

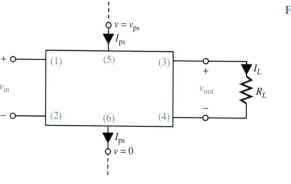

Solution

We have

$$P = \sum_{n=1}^{6} V_n I_n$$

where V_n is the voltage at terminal n and I_n is the current flowing *into* terminal n. Note that currents I_1 and I_2 are zero since $R_i = \infty$, $v_{\text{out}} = Av_{\text{in}} = 10(1) = 10$ V. The power supply maintains terminal 5 at $V_5 = V_{\text{ps}} = 20$ V and terminal 6 at $V_6 = 0$; $I_3 = -I_4$ and $I_5 = -I_6$. Then

$$P = V_3 I_3 + V_4 I_4 + V_5 I_5 + V_6 I_6 = V_3(-I_L) + V_4 I_L + V_5 I_{\text{ps}} - V_6 I_{\text{ps}}$$

$$= (V_4 - V_3) I_L + (V_{\text{ps}} I_{\text{ps}}) = -V_{\text{out}} I_L + V_{\text{ps}} I_{\text{ps}}$$

$$= (V_{\text{ps}} - V_{\text{out}}) I_{\text{ps}} = (V_{\text{ps}} - V_{\text{out}}) I_L$$

Noting that $I_L = V_{\text{out}}/R_L$, one has

$$P = (20 - 10)\left(\frac{10}{100}\right) = 1\text{W}$$

Since many amplifiers are designed with large values of R_i in order to keep the input power low, the approximation of $R_i \cong \infty$ is often justified. However, the assumption that $I_{\text{ps}} = I_L$ is not as justified because a certain amount of additional current (though kept small in order to minimize the waste) will pass directly through the amplifier from terminal 5 to terminal 6 without going through the load, making I_{ps} slightly larger than I_L.

The behavior of an amplifier always depends on the frequency of the sinusoidal signal in question. In general, the parameters of the amplifier model vary with the signal frequency. The variation of $|\bar{A}|$, as well as that of the phase angle θ_A, with frequency for an amplifier is known as its *frequency response*. The value of $|\bar{A}(f)|$ always drops off at sufficiently high frequencies. Since $\bar{A}(\omega) = \bar{V}_{\text{out}}/\bar{V}_{\text{in}}$, the phase difference θ_A between output and input sinusoids is also a function of frequency. However, at low frequencies θ_A is often zero, and therefore $\bar{A}$ can be regarded as real. Figure 5.1.3 shows variations of $|\bar{A}|$ and θ_A typical of audio, video, bandpass, and operational amplifiers. The input and output resistances should be generalized to impedances $Z_i(\omega)$ and $Z_o(\omega)$ when frequency effects are important. However, most commercial amplifiers are designed to make Z_i and Z_o real and constant over the useful frequency range of the amplifier block. If the frequency response of a block itself is known, the frequency response of a larger circuit containing the amplifier block can be found.

5.2   IDEAL OPERATIONAL AMPLIFIER

The operational amplifier, known also as *op amp*, consists of several transistors, diodes, capacitors, and resistors. It is available in integrated-circuit form for less than one U.S. dollar. Being inexpensive, compact, and versatile, operational amplifiers are used in a variety of simple circuits. Op-amp circuits, as we will see later, usually contain negative feedback.

Op-amp circuits themselves can be regarded as building blocks. These blocks are characterized by their input resistance, output resistance, and open-circuit voltage amplification. The symbol for the op amp is shown in Figure 5.2.1. Two terminals labeled $+$ and $-$ are available for inputs. The voltages of these terminals are labeled with respect to the common terminal,

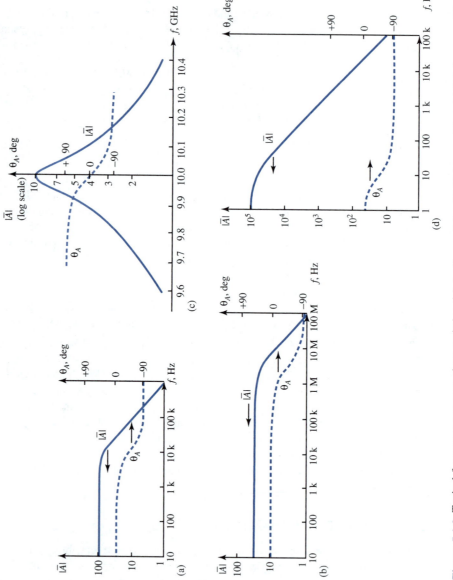

Figure 5.1.3 Typical frequency response characteristics. (**a**) Audio amplifier (**b**) Video amplifier. (**c**) Bandpass amplifier. (**d**) Operational amplifier. Note the separate scales for $|\bar{A}|$ and θ_A.

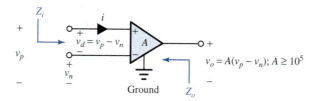

Figure 5.2.1 Operational amplifier.

denoted by a ground symbol. The output voltage is related to the difference between the two input voltages as

$$v_o = A(v_p - v_n) \tag{5.2.1}$$

where A is the open-loop voltage gain. Thus, the op amp is basically a form of differential amplifier, in which the difference $v_p - v_n$ is amplified. For an output voltage on the order of 12 V, the difference voltage v_d is on the order of 0.12 mV, or 120 μV. The input impedance Z_i is on the order of 1 MΩ, while the output impedance Z_o is on the order of 100 Ω to 1 kΩ.

The practical op-amp characteristics are approximated in the *ideal op amp* shown in Figure 5.2.2. Because of the high input impedance, very large gain, and the resulting small difference voltage v_d in practical op amps, the ideal op amp is approximated by the following two characteristics:

1. The input currents i_p and i_n are zero, $i_p = i_n = 0$.
2. The difference voltage v_d is zero, $v_d = 0$.

The principle of *virtual short circuit,* illustrated by the preceding, is utilized in analyzing circuits containing ideal op amps. The accompanying principle of negative feedback is explained later. The ideal op-amp technique is based on the approximations that $A \cong \infty$, $Z_i \cong \infty$, and $Z_o \cong 0$. The output voltage can always be found from the ideal op-amp technique in the usual op-amp feedback circuits, since the precise values of A, Z_i, and Z_o have negligible influence on the answer, as shown in Example 5.2.1.

The op amp is composed of a number of transistor stages on a single chip and provides the characteristics of a voltage-controlled voltage source. The so-called ideal op amp is characterized by infinite bandwidth. However, it is pertinent to mention that the bandwidth of an op amp without feedback is quite small. In commercially available op amps, while the open-loop voltage gain A is rather large (usually 10^5 or greater), the range of frequency for which this gain is achieved is limited. The asymptotic Bode diagram shown in Figure 5.2.3 illustrates the frequency characteristic for the open-loop voltage gain. For frequencies below f_h, the open-loop gain is a constant A_o; for frequencies beyond f_h, the open-loop gain decreases. The frequency f_h is known as the *open-loop bandwidth,* as it separates

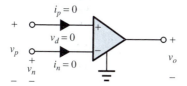

Figure 5.2.2 Ideal operational amplifier.

the constant region from the high-frequency band for which the open-loop gain decreases. f_h is also called the half-power frequency since the magnitude of A is 3 dB below its low-frequency value A_0. The frequency f_{GB} at which $A_{dB} = 0$, or the magnitude $A = 1$, is the *gain–bandwidth product* given by $A_0 f_h$. Typical values of f_h and f_{GB} are 10 Hz and 1 MHz, respectively. To overcome this frequency limitation, both *inverting* and *noninverting* op-amp stages (which are examples of feedback amplifiers) are frequently used. These will be discussed later in examples.

Feedback circuits have a desensitizing property, which implies that variations in the values of the op-amp parameters have little effect on the output of the circuit. "Feedback" implies that some of the output signal is fed back to be added to the input. Another property of feedback amplifiers is that the open-loop and closed-loop gain–bandwidth products are equal. Referring to Figure 5.2.4, which shows the open-loop and closed-loop frequency responses, it follows that

$$A_0 f_h = G_0 f_H \qquad (5.2.2)$$

where $A_0 f_h$ is the open-loop gain–bandwidth product, f_H is the closed-loop bandwidth, and G_0 is the *closed-loop gain*. The term closed loop signifies circuit performance when a path exists from output to input, whereas open loop implies the inherent characteristics of the op amp. Thus, for an op amp whose gain bandwidth product is 1 MHz with $A_0 = 10^5$ and $f_h = 10$ Hz, a stage with a closed-loop gain of 10 has a bandwidth $f_H = 100$ kHz, as shown in the asymptotic Bode diagram of Figure 5.2.4.

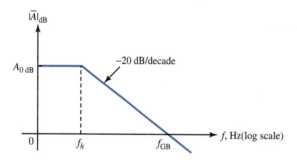

Figure 5.2.3 Asymptotic Bode diagram for open-loop voltage gain.

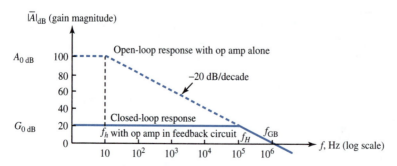

Figure 5.2.4 Typical frequency response of op-amp stage.

EXAMPLE 5.2.1

(a) Consider the circuit of the *inverting amplifier* shown in Figure E5.2.1(a), including an ideal op amp. Show that the voltage gain of the overall circuit v_o/v_i is independent of the op-amp parameters.

(b) Without considering the op-amp gain to be infinite, investigate the effect of a finite value of A, the open-circuit voltage amplification of the op amp, on the voltage gain of the overall circuit.

(c) Let $R_i = 2 \text{ k}\Omega$ and $R_f = 80 \text{ k}\Omega$ in Figure E5.2.1(a). Find the voltage gain of the overall circuit: (i) if the op-amp gain is infinite, and (ii) if the op amp gain is 100.

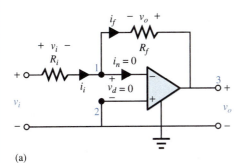

(a)

Figure E5.2.1 (a) Inverting amplifier. (b) Equivalent circuit.

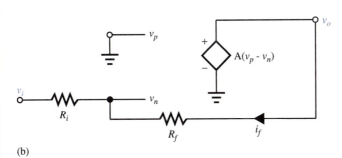

(b)

Solution

(a) Notice that a resistor R_i is placed in the − lead, whereas the + lead is connected to the common terminal (ground). Also, a feedback resistor R_f is connected between the − lead and the output lead. Noting that $i_n = 0$ and $i_i = i_f$, the KVL equation around the outside loop yields

$$v_i = R_i i_i + R_f i_f + v_o$$

Since $v_d = 0$, it follows that $i_i = v_i/R_i$. Thus, one gets

$$v_i = (R_i + R_f) \frac{v_i}{R_i} + v_o$$

Solving for the voltage gain, $v_o/v_i = -R_f/R_i$, where the negative sign implies that this is an *inverting amplifier*, i.e., the output voltage is 180° out of phase with the input voltage.

The voltage gain is independent of the op-amp parameters. By selecting the resistors to achieve the desired ratio, constructing a linear amplifier with a prescribed voltage gain is rather simple. It may further be noted that the output impedance is $Z_o = 0$, with zero output impedance of the ideal op amp; the input impedance is $Z_i = v_i/i_i = R_i$.

(b) The equivalent circuit of the inverting amplifier is shown in Figure E5.2.1(b). The current through the feedback resistor i_f is given by

$$i_f = \frac{v_o - v_i}{R_f + R_i}$$

The voltage at the inverting input v_n is

$$v_n = v_i + R_i i_f$$

Combining these two equations, one gets

$$v_n = v_i + R_i \left(\frac{v_o - v_i}{R_f + R_i} \right)$$

Since v_p is connected directly to ground, $v_p = 0$; the output voltage v_o is

$$v_o = -A v_n = -A \left[v_i + R_i \left(\frac{v_o - v_i}{R_f + R_i} \right) \right]$$

The voltage gain of the overall circuit is then given by

$$\frac{v_o}{v_i} = -\frac{1 - \left(\dfrac{R_i}{R_i + R_f} \right)}{\dfrac{1}{A} + \left(\dfrac{R_i}{R_i + R_f} \right)}$$

If A becomes infinitely large, $1/A \to 0$, and the preceding expression reduces to $-R_f/R_i$, obtained in part (a). Clearly as $A \to \infty$, the inverting terminal voltage $v_n = -v_o/A$ is going to be very small, practically on the order of microvolts. Then it may be assumed in the inverting amplifier that v_n is virtually zero, i.e., $v_n \cong 0$.

(c) (i) When A is infinitely large,

$$\frac{v_o}{v_i} = -\frac{R_f}{R_i} = -\frac{80}{2} = -40$$

(ii) When A has a finite value of 100,

$$\frac{v_o}{v_i} = -\frac{1 - \left(\dfrac{R_i}{R_i + R_f} \right)}{\dfrac{1}{A} + \left(\dfrac{R_i}{R_i + R_f} \right)} = -\frac{1 - \dfrac{2}{82}}{\dfrac{1}{100} + \dfrac{2}{82}} = -28.4$$

The principle of *negative feedback* is clearly illustrated in the operation of the inverting amplifier in Example 5.2.1, since negative feedback is used to keep the inverting terminal voltage as close as possible to the noninverting terminal voltage. One way of viewing negative feedback is to consider it as a self-balancing mechanism which allows the amplifier to preserve zero potential difference between its input terminals. The effect of the feedback connection from output to inverting input is then to force the voltage at the inverting input to be equal to that at the noninverting input. This is equivalent to stating that for an op amp with negative feedback, $v_n \cong v_p$. The analysis of the op amp is greatly simplified if one assumes that $i_n = 0$ and $v_n = v_p$. The voltage gain of the overall circuit of Figure E5.2.1(a) is called the *closed-loop gain*, because the presence of a feedback connection between the output and the input constitutes a closed loop as per the terminology used in the field of automatic control, which is presented in Section 16.2.

EXAMPLE 5.2.2

(a) Consider the circuit of the *noninverting amplifier* in Figure E5.2.2, including an ideal op amp. Obtain an expression for the voltage gain of the overall circuit.

(b) Let $R_i = 10 \text{ k}\Omega$ and $R_f = 240 \text{ k}\Omega$ in Figure E5.2.2. Find the voltage gain of the overall circuit.

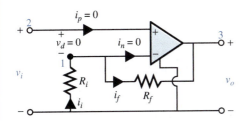

Figure E5.2.2 Noninverting amplifier.

Solution

(a) Note that the input is directly connected to the + terminal. Resistors R_i and R_f are connected as in the previous example. Since the input current $i_p = 0$, the input impedance is infinite. Because $i_n = 0$, $i_i = i_f$. The KVL equation around the loop containing R_i and R_f gives

$$v_o = -(R_i + R_f)i_i$$

Because of the virtual short circuit, $v_i = -R_i i_i$. Thus, for the case of a simple *noninverting amplifier*, one gets

$$\frac{v_o}{v_i} = 1 + \frac{R_f}{R_i}$$

(b) For the given values of R_i and R_f,

$$\frac{v_o}{v_i} = 1 + \frac{240}{10} = 25$$

5.3 PRACTICAL PROPERTIES OF OPERATIONAL AMPLIFIERS

To achieve voltage gain and consequently power gain, the op amp must be biased by a dc source. The biasing network is comprised of the power supply and the passive circuit elements surrounding the device that provide the correct dc levels at the terminals. Terminals at which the dc bias is to be connected are provided on the op-amp package with the actual biasing networks connected internally. The manufacturer specifies the permissible range of supply-voltage values and the corresponding op-amp characteristics.

Manufacturers add prefixes to type numbers such as 741 to indicate their own codings, even though the specifications are similar. For example,

μA 741   Fairchild
LM 741   National Semiconductors

An extra letter is sometimes added to indicate a temperature specification. For example,

μA 741 A   Military specification for guaranteed operation between −55 and +125°C
μA 741 C   Commercial specification for temperature range of 0 to 70°C

Practical op amps are composed of several amplifier stages, a typical structure of which is shown in Figure 5.3.1. The three building blocks are the input differential amplifier, the common emitter stage, and the emitter follower output stage. Some of the op-amp practical properties follow.

OPEN-LOOP VOLTAGE GAIN A

The op amp amplifies the difference v_d between the voltage on the noninverting (+) terminal and the inverting (−) terminal; see Figure 5.3.2. The term "open loop" implies that there is no external feedback connection between the output and either of the inputs. A is defined as the ratio of the change in output voltage to the change in differential input voltage, usually for a load resistance of no less than 2 Ωk. A typical value for A is 2×10^5.

INPUT RESISTANCE R_i

Input resistance is the open-loop incremental resistance looking into the two input terminals, and is typically 2 MΩ. Manufacturers sometimes quote the resistance between inputs and ground.

OUTPUT RESISTANCE R_o

The open-loop output resistance is usually between 50 and 500 Ω, with a typical value of 75 Ω for the 741. Thus one can see that Figure 5.3.3 is the representation of an op amp as a circuit element or block.

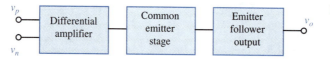

Figure 5.3.1 Typical op-amp stages.

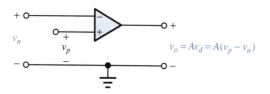

$$v_o = Av_d = A(v_p - v_n)$$

Figure 5.3.2 Op-amp symbol with signal voltages.

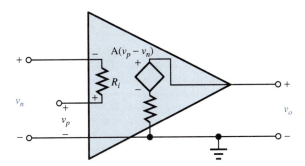

Figure 5.3.3 Op amp as a circuit component block.

Common-Mode Rejection Ratio (CMRR)

When there is a common-mode input voltage, i.e., when the input signals are equal and greater than zero, the output voltage of an ideal op amp is zero because v_d is equal to zero. In general, the common-mode input v_C is defined as $(v_p + v_n)/2$; and the difference signal $v_d = v_p - v_n$ is to be amplified. Common-mode gain A_C is defined as the ratio v_o/v_C when $v_d = 0$. The common-mode rejection ratio (CMRR) is defined by A/A_C, but is usually expressed in units of decibel (dB),

$$\text{CMRR} = 20 \ \log_{10} \frac{A}{A_C} \tag{5.3.1}$$

It is typically 90 dB, i.e., $A/A_C \cong 32{,}000$. The ideal op amp, however, has infinite CMRR.

To explain further, an op amp can be considered a special type of differential amplifier. The object of a differential amplifier, which is formally presented a little later, is to amplify "differences" in voltage between the two inputs, and to be unresponsive to voltage changes that appear simultaneously on both inputs. The *differential-mode input signal* is the difference between v_p and v_n; that is to say, $v_d = v_p - v_n$. The *common-mode input signal* is the average value of the two input signals; that is to say, $v_c = (v_1 + v_2)/2$. The output voltage of the amplifier v_o is given by $v_o = v_d A_d + v_c A_c$, where A_d (called simply A previously) is the differential-mode voltage gain and A_c is the common-mode voltage gain. Under ideal conditions, A_c is equal to zero and the differential amplifier completely rejects the common-mode signals. The departure from this ideal condition is a figure of merit for a differential amplifier and is measured by CMRR, which is the ratio of A_d to A_c. CMRR can thus be seen as a measure of an amplifier's ability to distinguish between differential-mode and common-mode signals. One of the practical advantages of a differential amplifier is its rejection of unwanted signals or noise.

INPUT OFFSET VOLTAGE V_{OS}

When both inputs are tied to ground, i.e., both differential-mode and common-mode inputs are zero, the output should be zero. In practice there will be mismatches in amplifier components, and if there is a mismatch in an input stage, the effect will be amplified, leading to a significant output voltage. The input offset voltage V_{OS} is the differential input voltage required to make the output zero, and is typically 1 mV. With some op amps, e.g., the 741, two voltage offset terminals are provided, and by connecting a potentiometer between them and taking the slider to the specified dc supply rail (or bus), as in Figure 5.3.4, the potentiometer can then be adjusted to zero the offset voltage.

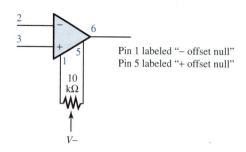

Pin 1 labeled "– offset null"
Pin 5 labeled "+ offset null"

Figure 5.3.4 Offset null terminals of 741 connected to 10-kΩ potentiometer and negative supply rail to zero the offset voltage.

INPUT BIAS CURRENT I_B

When operated at extremely low quiescent current values, the base bias currents will be low; but they do have to be taken into consideration. The input bias current I_B for an op amp is defined as the average of the two input currents with the inputs grounded (see Figure 5.3.5), i.e.,

$$I_B = \frac{I_{Bp} + I_{Bn}}{2} \tag{5.3.2}$$

A typical value for a μA 741 is 80 nA; for a μA 771 it is 50 pA. When the resistance of the source feeding the op amp is large enough, the input bias current may have an adverse effect. As shown in Figure 5.3.6, where the noninverting input is grounded and the inverting input is connected to a source of 1-MΩ resistance and of voltage v_S momentarily at 0 V, the bias current of 80 nA generates a voltage of -80 mV on the noninverting input.

INPUT OFFSET CURRENT I_{OS}

The adverse effect of the input bias current mentioned would be nullified if both inputs were connected to equal resistances (one of which could be a passive resistor) and if $I_{Bp} = I_{Bn}$. The differential input would then be zero. However, if the input bias currents are unequal, as shown in Figure 5.3.7, there will be a finite differential input voltage. It then becomes necessary to specify the difference between the bias currents, and this is the input offset current I_{OS},

$$I_{OS} = I_{Bp} - I_{Bn} \tag{5.3.3}$$

A typical value for I_{OS} is 20 nA. Both I_B and I_{OS} are usually measured with the output at 0 V; but in practice the output voltage has little effect.

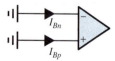

Figure 5.3.5 Bias currents of op amp.

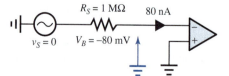

Figure 5.3.6 Illustration of input voltage generated by bias current.

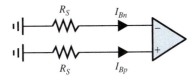

Figure 5.3.7 Illustration of unequal bias currents, which will cause a differential input voltage even with equal source resistances.

Power-Supply Rejection Ratio (PSRR)

An op amp's ability to disregard changes in power-supply voltage is measured by the power-supply rejection ratio (PSRR), which is specified by the change in offset voltage V_{OS} for a 1-V change in dc power supply and is usually expressed in μV/V. A typical value is 15 μV/V.

Maximum Differential Input Voltage

This is the maximum value of differential input voltage $v_p - v_n$ that can be applied without damaging the op amp.

Maximum Common-Mode Input Voltage

This is the maximum voltage that the two inputs can be raised above ground potential before the op amp becomes nonlinear.

Output Voltage Swing

Ideally this is equal to the difference between the two supply rail voltages, although in practice it is a few volts less.

Internal Frequency Compensation

Some op amps, such as the 741, have internal RC networks which are intentionally designed to reduce gain at high frequency. The result (shown in Figure 5.2.4) is that the open-loop gain begins to fall at a few hertz and then has a characteristic falling at 20 dB/decade or 6 dB/octave (i.e., gain $\propto 1/f$) until eventually the gain becomes unity (0 dB) at about 1 MHz.

External Frequency Compensation

For applications requiring a more extended high-frequency response, there are op amps with no internal compensation, and external frequency-compensation terminals are provided on these op amps (for example, the 709) so that the frequency response can be tailored to avoid instability without the heavy degrading of frequency response due to internal compensation. A typical frequency-compensation circuit for the μA 709 is illustrated in Figure 5.3.8. Op-amp manufacturers supply plots of open- and closed-loop responses for various values of components connected to the frequency-compensation terminals.

Slew Rate

Slew (or slewing) rate is a measure of how fast the output voltage can change. It is given by the maximum value of dv_o/dt, which is normally measured in response to a large input voltage step

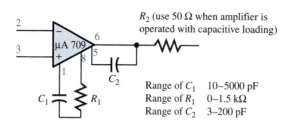

R_2 (use 50 Ω when amplifier is operated with capacitive loading)

Figure 5.3.8 Typical frequency-compensation circuit for Fairchild μA 709.

Range of C_1	10–5000 pF
Range of R_1	0–1.5 kΩ
Range of C_2	3–200 pF

and is therefore usually associated with low closed-loop voltage gain. For a 741 the slew rate is 0.5 V/μs at $A = 1$. For more recently developed op amps, the slew rate ranges from 5 to 100 V/μs. The effect of the slew rate in response to an input step voltage is shown in Figure 5.3.9. If one attempts to make the output voltage change faster than the slew rate, nonlinearity will be introduced. When specifying such output voltage requirements as rise time, output voltage, and frequency, it is necessary to choose an op amp with a slew rate that meets the specifications. With a sine-wave input, the slew rate limits a combination of maximum operating frequency and output voltage magnitude.

The slew rate occurs because at some stage in the amplifier a frequency-compensating capacitor will have to be charged, and the available limited charging current restricts the maximum rate of change of the capacitor voltage. With externally compensated op amps, such as the 709, slew rates will depend on the value of the compensating capacitors, which are in turn chosen on the basis of the closed-loop gain needed. The lower the gain, the higher the compensating capacitors, and hence the lower the slew rate. For a μA 709, the slew rate is 0.3 V/μs at $A = 1$ and 1.5 V/μs at $A = 10$.

Noise

This refers to the small, rapidly varying, random spurious signals generated by all electronic circuits. Noise places a limit on the smallness of signals that can be used. The subject of random signals and noise belongs to a branch of electrical engineering known as communication theory.

Stability

The amplifier is said to be stable when it performs its function reliably under all normal operating conditions. By definition, a system is stable if its response to an excitation that decays to

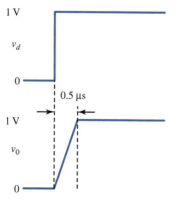

Figure 5.3.9 Effect of op-amp slew rate with step voltage input.

zero with time also decays to zero. Instability can come about in various ways, and it is a very common difficulty, which must be prevented. In almost all cases, op-amp circuits can be classified as feedback circuits. Feedback of an improper kind can lead to instability or oscillation of an op-amp circuit. In general, instability occurs when there is excessive phase shift in the op amp and feedback loop, so that negative feedback is changed to positive feedback. The frequency response of an op amp is usually designed to roll off smoothly at 20 dB decade, as mentioned earlier. This type of frequency response ensures stability for the more common op-amp circuits.

Frequency Response

As with all electronic circuits, op amps have limited frequency response. Because of the negative feedback of the circuit, the passband of an op-amp circuit is usually much larger than that of the op amp by itself. Typically with a given op amp, increasing the bandwidth of an op-amp circuit will decrease the voltage gain in the same proportion. Thus for a given op amp, the product of gain and bandwidth (i.e., the gain–bandwidth product) is a constant, as discussed earlier and shown in Figure 5.2.4.

Table 5.3.1 lists some representative op-amp parameters for different amplifier types, only to illustrate their characteristics. One can see from the table that a high-power op amp can deliver a large output current, but has relatively large offset parameters. On the other hand, a precision input amplifier has low offset and high gain, but at the price of bandwidth, slew rate, and output current. The general-purpose amplifier, as the name suggests, strikes a balance between these extremes.

TABLE 5.3.1 Typical Op-Amp Characteristics

Type	Input Offset Voltage (mV)	Input Offset Current (pA)	Voltage Gain (V/mV)	Gain–Bandwidth Product (MHz)	Slew Rate (V/μs)	Maximum Output Current (mA)
High Power	5	200	75	1	3	500
Wide band	3	200	15	30	30	50
High slew rate	5	100	4	50	400	50
Precision input	0.05	2	500	0.4	0.06	1
General purpose	2	50	100	1	3	10

EXAMPLE 5.3.1

In order to illustrate the insensitivity of feedback circuits to variations of the op-amp parameters, let us consider a simple feedback circuit using an op amp, as shown in Figure E5.3.1(a). Using the op-amp model of Figure 5.1.1, find the output voltage in terms of the input voltage under two sets of conditions:

(a) $A = 10^5$, $R_i = 10 \text{ k}\Omega$, $R_L = 1 \text{ k}\Omega$, $R_S = 1 \text{ k}\Omega$

(b) $A = 2 \times 10^5$, $R_i = 30 \text{ k}\Omega$, $R_L = 5 \text{ k}\Omega$, $R_S = 2 \text{ k}\Omega$

In order to simplify the calculations, let us assume that the op amp's output resistance R_o is zero.

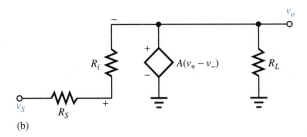

(a)

(b)

Solution

Replacing the op amp by its model, we have the circuit shown in Figure E5.3.1(b). The nodal equation at the node labeled + is

$$\frac{v_S - v_+}{R_S} + \frac{v_o - v_+}{R_i} = 0$$

Also,

$$v_o = A\,(v_+ - v_-) \qquad \text{and} \qquad v_- = v_o$$

Thus,

$$v_o = A\,(v_+ - v_o)$$

Solving, one gets

$$v_o = v_S \frac{A}{A+1} \frac{R_i}{R_i + [R_S/\,(1+A)]}$$

Evaluating v_o/v_S for the given two conditions, we have

(a) $v_o/v_S = 0.999989$

(b) $v_o/v_S = 0.999994$

Thus, we see that the output voltage (using negative feedback) is very nearly independent of R_i, R_o, and A, provided the op amp's parameters remain in their allowable ranges.

EXAMPLE 5.3.2

Find the maximum frequency of an output sine wave which can be produced at an amplitude of 1.5 V if the op-amp slew rate is 0.5 V/μs.

Solution

$$v_o = V_m \sin \omega t$$

$$\frac{dv_o}{dt} = \omega V_m \cos \omega t$$

$$\text{slew rate} = \text{maximum value of } \frac{dv_o}{dt} = \omega V_m = \frac{0.5}{10^{-6}}$$

Hence,

$$\omega = \frac{0.5}{1.5 \times 10^{-6}} \qquad \text{or} \qquad f = 53 \text{ kHz}$$

EXAMPLE 5.3.3

Consider the op-amp circuit shown in Figure E5.3.3 and obtain expressions for the open-loop voltage gain at (a) low and (b) high frequencies. Also determine relations for the 3-dB point frequency and the phase shift.

Solution

Using phasors and the voltage-divider rule,

$$\bar{V}_o = \left(\frac{X_C}{X_C + R_o} \right) A \bar{V}_d, \qquad \text{where } X_C = \frac{1}{j\omega C}$$

(a) At low frequencies, X_C becomes very large so that $X_C \gg R_o$. Then,

$$\bar{V}_o = A\bar{V}_d$$

The output signal is in phase with the input signal, since complex voltage gain $\bar{A}$ is the same as A.

(b) At higher frequencies, X_C becomes comparable to R_o. Let $\bar{A}_F$ be the complex open-loop voltage gain at frequency f. Then,

$$\bar{A}_F = \frac{1/j\omega C}{[R_o + (1/j\omega C)] A_o}$$

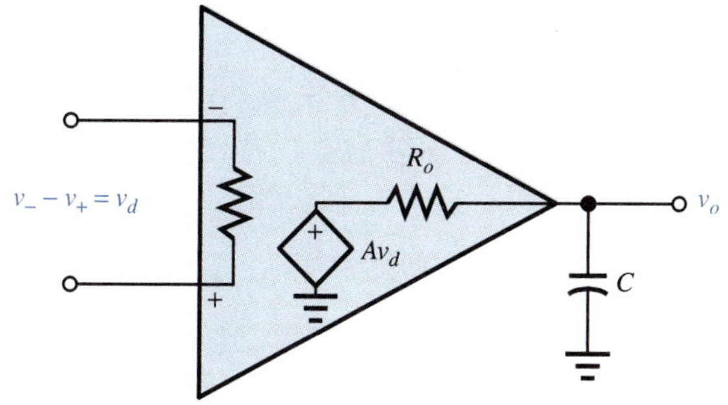

Figure E5.3.3

or

$$\bar{A}_F = \frac{A_o}{1 - j\omega R_o C}$$

$$= \frac{A_o(1 - j\omega R_o C)}{1 + (\omega R_o C)^2}$$

$$\left|\bar{A}_F\right| = A_F = \frac{A_o}{\sqrt{1 + (\omega R_o C)^2}}$$

At the 3-dB point, or half-power frequency f_h, $A_F = A_o/\sqrt{2}$, so that

$$\omega_h R_o C = 1 \qquad \text{or} \qquad f_h = \frac{1}{2\pi R_o C}$$

Denoting $\bar{A}_F = A_F e^{j\phi}$, where ϕ indicates that the output leads the input,

$$\tan \phi = -\omega R_o C$$

At very high frequencies, $\tan \phi \to -\infty$ and $\phi \to 3\pi/2$, so that the output leads the input by $3\pi/2$ or lags it by $\pi/2$. This is the largest phase shift possible.

5.4 APPLICATIONS OF OPERATIONAL AMPLIFIERS

An op amp along with a few external components (resistors and capacitors) is capable of performing many different operations—hence the name *operational*. The linear operations of integration, differentiation, addition, and subtraction that were needed in analog computers are some of the early applications. Other linear applications include instrumentation amplifiers, voltage-to-current and current-to-voltage converters, voltage followers, and active filters. Op amps are also utilized in nonlinear applications such as limiters, comparators, voltage regulators, signal rectifiers and detectors, logarithmic amplifiers, multipliers, and many digital circuits.

Negative feedback (which is said to be degenerative) is used to improve amplifier performance by sacrificing gain. The opposite situation of positive feedback (which is said to be regenerative) is also utilized, and gain is increased to the extent that an amplifier will produce an output signal with no input. This leads to sinusoidal oscillators and nonsinusoidal waveform generators.

This section concentrates on a limited range of op-amp applications.

Inverting Amplifier

One common op-amp circuit is shown in Figure 5.4.1. For the case of finite voltage gain A_o of an op amp that is otherwise ideal, the output voltage becomes

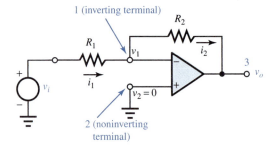

Figure 5.4.1 Inverting amplifier.

$$v_o = A_o \left(v_2 - v_1 \right) \tag{5.4.1}$$

Since $v_2 = 0$, because terminal 2 is grounded,

$$v_1 = -\frac{v_o}{A_o} \tag{5.4.2}$$

Thus one gets

$$i_1 = \frac{v_i - v_1}{R_1} = \frac{v_i + (v_o/A_o)}{R_1} \tag{5.4.3}$$

$$i_2 = \frac{v_1 - v_o}{R_2} = \frac{-(v_o/A_o) - v_o}{R_2} \tag{5.4.4}$$

Since no current is drawn by terminal 1 of the ideal op amp,

$$i_2 = i_1 \tag{5.4.5}$$

Hence,

$$\frac{v_o}{v_i} = -\frac{R_2}{R_1} \frac{1}{1 + [(R_1 + R_2)/A_o R_1]} \tag{5.4.6}$$

which is the expression for the circuit gain. Now, if $A_o \rightarrow \infty$, we have

$$\frac{v_o}{v_i} = -\frac{R_2}{R_1}, \qquad \text{if } A_o \rightarrow \infty \tag{5.4.7}$$

which is independent of the op amp's gain A_o (see Example 5.2.1). The sign inversion associated with Equation (5.4.7) makes the amplifier "inverting." Generally if A_o is at least 200 times the magnitude of R_2/R_1, then the actual gain given by Equation (5.4.6) will be within 1% of the ideal gain given by Equation (5.4.7). Assuming infinite gain of the op amp, it follows that

$$v_d = v_2 - v_1 = 0 \qquad \text{or} \qquad v_1 = v_2 \tag{5.4.8}$$

Since terminal 2 is at ground potential (zero) in Figure 5.4.1, terminal 1 is said to be a *virtual ground*. Notice that when $R_2 = R_1$, the inverting amplifier becomes a *voltage follower* with inverted sign and a gain magnitude of unity.

Noninverting Amplifier

It is called "noninverting" because there is no sign inversion. A typical circuit is shown in Figure 5.4.2. With a finite op-amp gain A_o,

$$v_2 - v_1 = \frac{v_o}{A_o} \qquad \text{or} \qquad v_1 = v_2 - \frac{v_o}{A_o} \tag{5.4.9}$$

Figure 5.4.2 Noninverting amplifier.

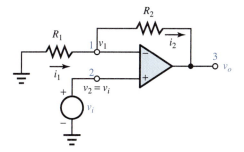

$$i_1 = \frac{-v_1}{R_1} = \frac{-v_2}{R_1} + \frac{v_o}{A_o R_1} \tag{5.4.10}$$

$$i_2 = \frac{v_1 - v_o}{R_2} = \frac{v_2}{R_2} - \frac{v_o}{R_2}\left(\frac{1}{A_o} + 1\right) \tag{5.4.11}$$

Since the ideal op amp draws no current,

$$i_1 = i_2 \tag{5.4.12}$$

Also, as seen from Figure 5.4.2,

$$v_2 = v_i$$

In equating Equations (5.4.10) and (5.4.11), one gets

$$\frac{v_o}{v_i} = \frac{v_o}{v_2} = \left(1 + \frac{R_2}{R_1}\right)\frac{1}{1 + [(R_1 + R_2)/A_o R_1]} \tag{5.4.13}$$

For $A_o \to \infty$, the ideal circuit gain is

$$\frac{v_o}{v_i} = 1 + \frac{R_2}{R_1} \tag{5.4.14}$$

(See Example 5.2.2.) Note that if $R_2 = 0$, for any nonzero value of R_1,

$$\frac{v_o}{v_i} = 1 \tag{5.4.15}$$

which is the gain of an ideal *voltage follower.* The same result applies if $R_1 \to \infty$ (i.e., open circuit) for any finite R_2.

Inverting Summing Amplifier

The circuit of Figure 5.4.1 is extended by adding other input points, as shown in Figure 5.4.3. Because of the virtual ground at terminal 1 (ideal op-amp assumption), the response of the circuit to any one input is not affected by the presence of the other $N - 1$ inputs. Based on Equation (5.4.7) one can write the total output as the sum of all responses,

$$v_o = -\left(\frac{R_f}{R_1}v_{i1} + \frac{R_f}{R_2}v_{i2} + \cdots + \frac{R_f}{R_N}v_{iN}\right) = -R_f \sum_{n=1}^{N} \frac{v_{in}}{R_n} \tag{5.4.16}$$

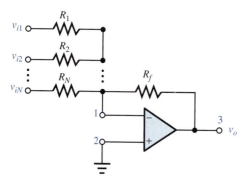

Figure 5.4.3 Inverting summing amplifier.

If all resistors R_n are the same, say $R_n = R_1$, for all n, then

$$v_o = -\frac{R_f}{R_1} \sum_{n=1}^{N} v_{in}, \qquad \text{if all } R_n = R_1 \qquad (5.4.17)$$

which corresponds to an inverting summing amplifier with gain.

Noninverting Summing Amplifier

The circuit of Figure 5.4.2 is generalized for multiple inputs, as shown in Figure 5.4.4. With the ideal op-amp assumption, based on Equation (5.4.14), the noninverting gain is

$$\frac{v_o}{v_2} = 1 + \frac{R_f}{R_d} \qquad (5.4.18)$$

By superposition, the output v_o is the sum of the responses taken individually. For input m,

$$v_2 = v_{im} \frac{[R_1 \| R_2 \| \cdots \| R_M]_{\text{without } R_m}}{R_m + [R_1 \| R_2 \| \cdots \| R_M]_{\text{without } R_m}} = \frac{v_{im}}{R_m} [R_1 \| R_2 \| \cdots \| R_M] \qquad (5.4.19)$$

where $[R_1 \| R_2 \| \cdots \| R_M]$ represents the resistance of all resistors $R_1, R_2, \ldots, R_M$ in parallel. Hence, due to input v_{im},

$$v_{om} = \left(1 + \frac{R_f}{R_d}\right) \frac{[R_1 \| R_2 \| \cdots \| R_M]}{R_m} v_{im} \qquad (5.4.20)$$

The total output voltage is

$$v_o = \sum_{m=1}^{M} v_{om} = \left(1 + \frac{R_f}{R_d}\right) [R_1 \| R_2 \| \cdots \| R_M] \sum_{m=1}^{M} \frac{v_{im}}{R_m} \qquad (5.4.21)$$

For the special case when all these resistors are equal, i.e., when $R_m = R$ for all m, then it follows

$$v_o = \left(1 + \frac{R_f}{R_d}\right) \frac{1}{M} \sum_{m=1}^{M} v_{im} \qquad (5.4.22)$$

which can be interpreted in the following ways:

1. A single-input noninverting amplifier gain $1 + R_f/R_d$ that has an input equal to the average of M inputs;

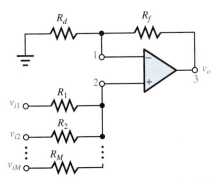

Figure 5.4.4 Noninverting summing amplifier.

2. An M-input noninverting summing amplifier for which the gain seen by each input is $(1 + R_f/R_d)/M$.

Current-to-Voltage Amplifier

The basic circuit is shown in Figure 5.4.5, which is similar to that of an inverting amplifier (Figure 5.4.1). The $-$ input is connected directly to a current source supplying current i_S. Node X is a virtual ground, so that

$$i_S = -\frac{v_o}{R_f} \tag{5.4.23}$$

Hence

$$v_o = -R_f i_S \tag{5.4.24}$$

which shows that the degree of amplification depends on the value of R_f.

Current-to-Current Amplifier

The circuit given in Figure 5.4.6 is to amplify a current fed to the input of the op amp. Applying KCL at Y,

$$i_1 + i_S = i_o \tag{5.4.25}$$

Since $i_1 = -v_Y/R_1$, Equation (5.4.25) becomes

$$i_S - \frac{v_Y}{R_1} = i_o \tag{5.4.26}$$

Because X is a virtual ground, $v_Y = -i_S R_f$. Hence Equation (5.4.26) becomes

$$i_S \left(1 + \frac{R_f}{R_1}\right) = i_o \tag{5.4.27}$$

The current gain is then given by

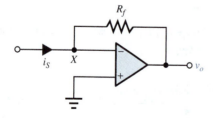

Figure 5.4.5 Current-to-voltage amplifier.

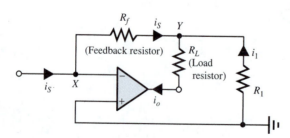

Figure 5.4.6 Current-to-current amplifier.

$$\frac{i_o}{i_S} = 1 + \frac{R_f}{R_1} \qquad (5.4.28)$$

Charge-to-Charge Amplifier

A circuit is shown in Figure 5.4.7 in which there is a capacitor C_1 in the $-$ input line and a capacitor C_f in the feedback loop. KCL at node X gives

$$\frac{dq_1}{dt} + \frac{dq_f}{dt} = 0 \qquad (5.4.29)$$

where q_1 and q_f are charges on the input and feedback capacitors. Thus,

$$q_1 = -q_f \qquad \text{or} \qquad C_1 v_i = -C_f v_o \qquad \text{or} \qquad \frac{v_o}{v_i} = -\frac{C_1}{C_f} \qquad (5.4.30)$$

Negative Impedance Converter

The op-amp circuit of Figure 5.4.8 causes a negative resistance R_{in} between the input terminal and ground. In the more general case, when R is replaced by an impedance Z, the circuit gives a negative impedance. Using ideal op-amp techniques, one has

$$v_1 = v_{in}; \qquad i_1 = \frac{v_1}{R_1} = \frac{v_{in}}{R_1} = i_2$$

$$v_o = i_2 (R_1 + R_2) = v_{in} \left(1 + \frac{R_2}{R_1} \right); \qquad i_3 = \frac{v_o - v_{in}}{R} = v_{in} \frac{R_2}{R R_1} = -i_{in}$$

so that

$$R_{in} = \frac{v_{in}}{i_{in}} = -R \frac{R_1}{R_2} \qquad (5.4.31)$$

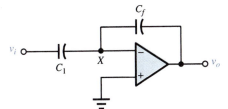

Figure 5.4.7 Charge-to-charge amplifier.

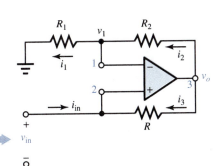

Figure 5.4.8 Negative impedance converter.

which is a negative resistance. The negative impedance converter is useful in transforming a voltage source into a current source by means of a voltage-to-current converter (see Problem 5.4.9). It has also been used in a class of active filters.

Differential Amplifier

Figure 5.4.9 shows a *weighted differencing amplifier*, where "weighted" refers to the fact that the output voltage has the form $v_o = w_a v_a + w_b v_b$, in which the weighting coefficients w_a and w_b are the voltage gains seen by the inputs. Since the network is linear, superposition can be applied for analysis. First, by grounding the input terminal of R_3, let us make $v_a = 0$. Except for the presence of R_3 in parallel with R_4, between terminal 2 and ground, the circuit is that of an inverting amplifier of gain $-R_2/R_1$ to the input v_b. Since no current flows at terminal 2, it follows that

$$w_b = -\frac{R_2}{R_1} \tag{5.4.32}$$

Next, by grounding the left terminal of R_1, let us make $v_b = 0$. With respect to the voltage on terminal 2, note that the circuit is that of a noninverting amplifier with gain $1 + R_2/R_1$. The voltage divider, consisting of R_3 and R_4, corresponds to a gain of $R_4/(R_3 + R_4)$ from the input to terminal 2. Thus the overall gain becomes

$$w_a = \left(1 + \frac{R_2}{R_1}\right) \frac{R_4}{R_3 + R_4} \tag{5.4.33}$$

Finally, the overall response is obtained by superposition,

$$v_o = w_a v_a + w_b v_b = \left(1 + \frac{R_2}{R_1}\right) \frac{R_4}{R_3 + R_4} v_a - \frac{R_2}{R_1} v_b$$

or

$$v_o = \frac{R_2}{R_1} \left[\frac{1 + (R_1/R_2)}{1 + (R_3/R_4)} v_a - v_b \right] \tag{5.4.34}$$

When $R_1/R_2 = R_3/R_4$,

$$v_o = \frac{R_2}{R_1} (v_a - v_b) \tag{5.4.35}$$

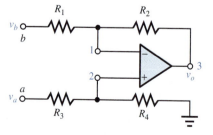

Figure 5.4.9 Weighted differencing amplifier.

which corresponds to the response of a *differential amplifier*. Usually resistor values are so chosen that $R_3 = R_1$ and $R_4 = R_2$ for some practical reasons. Improved versions of differential amplifiers are available commercially.

Integrators

Figure 5.4.10 shows a *noninverting integrator*, which can be seen to be a negative impedance converter added with a resistor and a capacitor. Noting that $v_o = 2v_1$ and $i_3 = v_1/R$, the total capacitor current is

$$i = i_{in} + i_3 = \frac{v_{in} - v_1}{R} + \frac{v_1}{R} = \frac{v_{in}}{R} \tag{5.4.36}$$

The capacitor voltage is given by

$$v_1(t) = \frac{1}{C} \int_{-\infty}^{t} i(\xi)\, d\xi = \frac{1}{RC} \int_{-\infty}^{t} v_{in}(\xi)\, d\xi \tag{5.4.37}$$

Thus,

$$v_o(t) = \frac{2}{RC} \int_{-\infty}^{t} v_{in}(\xi)\, d\xi \tag{5.4.38}$$

which shows that the circuit functions as an integrator.

Replacing R_2 in the inverting amplifier of Figure 5.4.1 by a capacitance C results in the somewhat simpler integrator circuit shown in Figure 5.4.11, known as an inverting integrator, or Miller integrator. With ideal op-amp techniques, $i_C = i_{in} = v_{in}/R$. The voltage across C is just v_o, so that

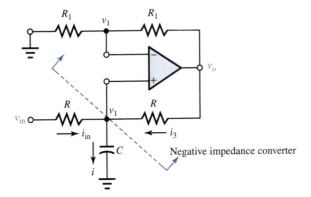

Figure 5.4.10 Noninverting integrator.

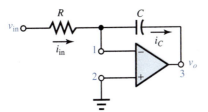

Figure 5.4.11 Inverting integrator (Miller integrator).

$$v_o(t) = -\frac{1}{C}\int_{-\infty}^{t} i_C(\xi)\,d\xi = -\frac{1}{RC}\int_{-\infty}^{t} v_{in}(\xi)\,d\xi \qquad (5.4.39)$$

which illustrates that the network behaves as an integrator with sign inversion.

Differentiator

Shown in Figure 5.4.12 is a differentiator which is obtained by replacing R_1 in the inverting amplifier of Figure 5.4.1 by a capacitor C. Assuming ideal op-amp characteristics, one has $i = i_C$ and $v_o = -Ri = -Ri_C$. But since $i_C(t) = C\,dv_{in}(t)/dt$, we get

$$v_o(t) = -RC\,\frac{dv_{in}(t)}{dt} \qquad (5.4.40)$$

which corresponds to a differentiator with a gain of $-RC$. In practice, however, differentiators are normally avoided because of high-frequency noise (which is accentuated due to a transfer function that increases with frequency) and stability problems (which make them oscillate).

Inductorless (Active) Filters

Filters (used to pass or eliminate certain frequency components of a signal) that are suitable for IC fabrication, but which do not contain inductors, are known as *active filters*. They have the op amp as a common component. Figure 5.4.13 shows the basic op-amp circuit with frequency-dependent impedances. The voltage gain or voltage transfer function of this circuit is given by

$$\frac{v_o}{v_i} = -\frac{Z_f}{Z_i} \qquad (5.4.41)$$

with $Z_i = R_i$ and $Z_f = R_f\|(1/j\omega C_f)$ (i.e., R_f and C_f in parallel). We have a *low-pass filter* with

$$\frac{v_o}{v_i} = -\frac{R_f}{R_i}\frac{1}{1 + j\omega R_f C_f} = -\frac{R_f}{R_i}\frac{1}{1 + (jf/f_l)} \qquad (5.4.42)$$

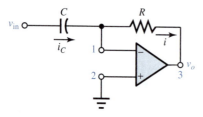

Figure 5.4.12 Differentiator.

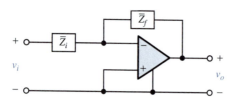

Figure 5.4.13 Basic op-amp active filter.

where

$$f_l = \frac{1}{2\pi R_f C_f} \tag{5.4.43}$$

The magnitude

$$\left|\frac{v_o}{v_i}\right| = \frac{R_f}{R_i} \frac{1}{\sqrt{1 + (f/f_l)^2}}$$

is sketched in Figure 5.4.14, from which one can see that at the half-power or 3-dB point of f_l, the dc value R_f/R_i goes down by a factor of $1/\sqrt{2}$ or 0.707.

With $Z_f = R_f$ and $Z_i = R_i + 1/j\omega C_i$ (i.e., R_i and C_i in series), we have a *high-pass filter* for which we obtain

$$\frac{v_o}{v_i} = -\frac{R_f}{R_i} \frac{jf/f_h}{1 + (jf/f_h)} \tag{5.4.44}$$

where

$$f_h = \frac{1}{2\pi R_i C_i} \tag{5.4.45}$$

and

$$\left|\frac{v_o}{v_i}\right| = \frac{R_f}{R_i} \frac{1}{\sqrt{1 + (f_h/f)^2}} \tag{5.4.46}$$

which is sketched in Figure 5.4.15. Note that the high-frequency gain is R_f/R_i, as in the low-filter case, and the half-power or 3-dB point is f_h.

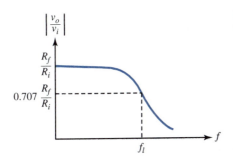

Figure 5.4.14 Frequency response of voltage-gain magnitude of a low-pass active filter.

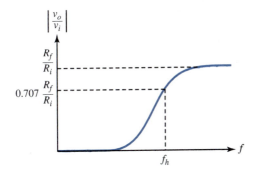

Figure 5.4.15 Frequency response of voltage-gain magnitude of a high-pass active filter.

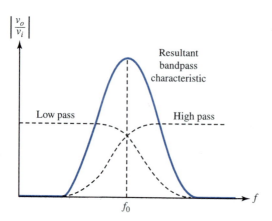

Figure 5.4.16 Bandpass filter seen as a combination of low-pass and high-pass filter characteristics.

By using a low-pass and a high-pass filter and overlapping their transfer functions, as shown in Figure 5.4.16, one can construct a *bandpass filter* with

$$Z_f = R_f \left\| \frac{1}{j\omega C_f} = \frac{R_f}{1 + j\omega R_f C_f} \right. \qquad \text{(i.e., } R_f \text{ and } C_f \text{ in parallel)}$$

and

$$Z_i = R_i + \frac{1}{j\omega C_i} \qquad \text{(i.e., } R_i \text{ and } C_i \text{ in series)}$$

in which case

$$\frac{v_o}{v_i} = -\frac{Z_f}{Z_i} = -\frac{R_f}{R_i} \frac{1}{\left(1 + j\omega R_f C_f\right)\left[1 + 1/\left(j\omega R_i C_i\right)\right]} \qquad (5.4.47)$$

Denoting

$$f_l = \frac{1}{2\pi R_i C_i} \qquad (5.4.48a)$$

and

$$f_h = \frac{1}{2\pi R_f C_f} \qquad (5.4.48b)$$

Equation (5.4.47) can be rewritten as

$$\frac{v_o}{v_i} = -\frac{R_f}{R_i} \frac{1}{(1 + f_l/f_h) + j\left(f/f_h - f_l/f\right)}$$

$$= -\frac{R_f}{R_i} \frac{f_h/(f_h + f_l)}{1 + j(f^2 - f_l f_h)/[f(f_l + f_h)]} \qquad (5.4.49)$$

whose magnitude is sketched in Figure 5.4.17. Notice that the magnitude response is a maximum with a value of R_f/R_i at $f_0 = \sqrt{f_l f_h}$, and if $f_l \ll f_h$, then the magnitude becomes ($0.707 R_f/R_i$ at f_l and f_h. The bandwidth (BW) may then be defined as

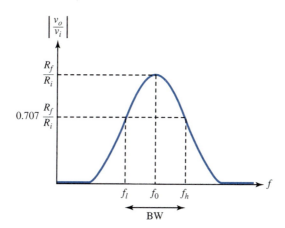

Figure 5.4.17 Frequency response of voltage-gain magnitude of a bandpass active filter.

$$BW = f_h - f_l, \qquad \text{for } f_l << f_h$$

Analog Computers

Although not used as much as the digital computer (which nowadays forms the basic tool for numerical analysis and the solution of algebraic as well as differential equations), the analog computer still retains some significant advantages over the digital computer. A physical system can be represented by a set of differential equations which can be modeled on an analog computer that uses continuously varying voltages to represent system variables. The differential equation is solved by the computer, whereas the modeled quantities are readily varied by adjusting passive components on the computer. The mathematical functions (integration, addition, scaling, and inversion) are provided by op amps.

The analog computer components are shown in Figure 5.4.18. Let the ordinary differential equation to be solved be

$$\frac{d^2 y(t)}{dt^2} + a_1 \frac{dy(t)}{dt} + a_2 y(t) = f(t) \tag{5.4.50}$$

which can be rearranged by isolating the highest derivative term as

$$\ddot{y} = -a_1 \dot{y} - a_2 y + f \tag{5.4.51}$$

subject to the initial conditions

$$y(0) = y_0 \tag{5.4.52a}$$

and

$$\left. \frac{dy}{dt} \right|_{t=0} = y_1 \tag{5.4.52b}$$

The *simulation* of the solution is accomplished by the connection diagram of Figure 5.4.19. Time scaling is done by redefining t as $t = \alpha\tau$, in which case the differential equation [Equation (5.4.50)] becomes

$$\frac{d^2 y(\tau)}{d\tau^2} = -\alpha a_1 \frac{dy(\tau)}{d\tau} - \alpha^2 a_2 y(\tau) + \alpha^2 f(\tau) \tag{5.4.53}$$

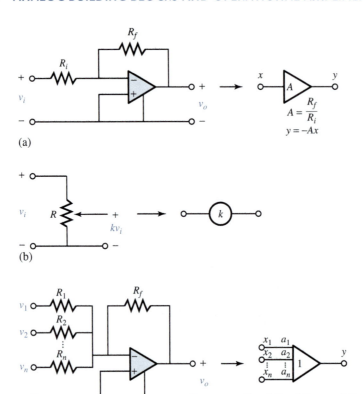

(a)

(b)

(c)

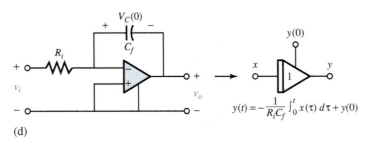

(d)

Figure 5.4.18 Analog computer components. **(a)** Basic op amp to provide integer multiplication. **(b)** Potentiometer for providing noninteger gains of less than 1. **(c)** Summer. [*Note:* $y = x_1 + x_2 + \cdots + x_n$, by choosing $R_f/R_1 = R_f/R_2 = \cdots = R_f/R_n = 1$.] **(d)** Integrator. [*Note:* C_f is charged to an initial value $V_C(0)$ to provide the initial condition on y; with $R_i C_f = 1$, $y = -\int_0^t x(\tau)\, d\tau + y(0)$.]

The coefficients may simply be scaled by the appropriate factors to change the solution time of the analog computer. For example, if $\alpha = 20$, then $t = 20$ s corresponds to $\tau = 1$ s, and thus the solution time may be reduced.

5.5   LEARNING OBJECTIVES

The *learning objectives* of this chapter are summarized here, so that the student can check whether he or she has accomplished each of the following.

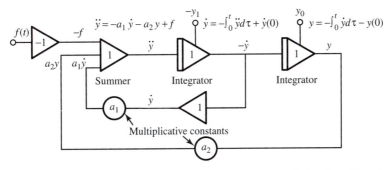

Figure 5.4.19 Connection diagram for analog computer simulation for solving the second-order differential equation $\ddot{y} = -a_1\dot{y} - a_2y + f$ subject to initial conditions $y(0) = y_0$ and $\dot{y}(0) = y_1$.

- Amplifier block as a two-port device and its circuit model.
- Ideal operational amplifier and its characteristics.
- Analysis and design of simple amplifier circuits including feedback.
- Properties of practical op-amps.
- Op-amp applications, including active filters.
- Understanding the operation of analog computers.

5.6 PRACTICAL APPLICATION: A CASE STUDY

Automotive Power-Assisted Steering System

In terms of negative feedback, there exists an analogy between the operational amplifier and the power-steering mechanism of an automobile. The hydraulic pump is analogous to the power supply in an op-amp circuit. The position of the booster-cylinder piston that is linked to the steering is analogous to the op-amp output signal; the mechanical linkage between the control valve and the booster-cylinder piston is analogous to the feedback circuit; the control-valve response to the difference between the input from the steering wheel and the position of the steering linkage is analogous to the op-amp response to its differential input signal. Thus the automotive power-assisted steering system is but an example of negative feedback in a mechanical sense.

Figure 5.6.1 illustrates in a simplified manner how a hydraulic pump driven by the engine continuously supplies pressure to a control valve which in turn supplies the fluid to the two sides of the booster cylinder. A negative feedback path is established from the booster cylinder through the mechanical linkage back to the control valve.

For straight steering, the pressure applied is equal on both sides of the cylinder and, as such, no turning force results. When the steering wheel is moved by the driver to turn the wheels in the desired direction, more pressure is applied to one side of the cylinder or the other.

A mechanical feedback arm from the steering linkage causes the valve to return to its neutral position as the wheels turn, thereby allowing the driver to make a gradual turn. As and when the steering wheel is turned, the wheels move a proportional amount rather than moving all the way to the extreme position.

Feedback control systems are discussed in Section 16.2 in more detail.

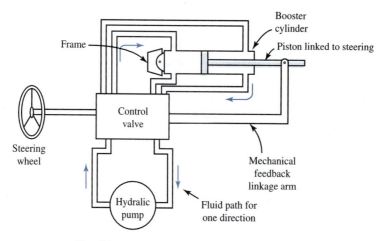

Figure 5.6.1 Simplified representation of an automotive power-assisted steering system.

PROBLEMS

5.1.1 (a) Determine the voltage at A in Figure P5.1.1.

(b) With $V_i = 10$ V, $R_1 = 10$ kΩ, $R_2 = 1000$ Ω, and $A = 100$, find the current i_2.

5.1.2 Consider the circuit configuration shown in Figure P5.1.2. Let $V_i = 1$ V, $R_1 = 1000$ Ω, $R_2 = 2000$ Ω, and $A = 3$.

(a) Compute i_2.

(b) If R_1 changed to 500 Ω, recompute i_2.

5.1.3 (a) Find v_{out} in the circuit shown in Figure P5.1.3.

(b) With $V_i = 2$ V, $R_1 = R_2 = 2.5$ kΩ, $R_3 = 5$ kΩ, and $A = 100$, find v_{out}.

5.1.4 Determine the Thévenin resistance viewed from terminals A–B of the circuit of Figure P5.1.4 by setting independent sources to be zero and applying a test voltage at terminals A–B.

*__5.1.5__ Show that any amplifier represented by the model of Figure 5.1.1 of the text can also be represented by the more general hybrid model of Figure P5.1.5. Evaluate the four parameters of the hybrid model (i.e., r_i, r_o, h_f, h_r) in terms of R_i, R_o, and A.

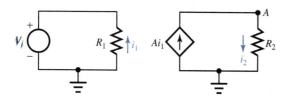

Figure P5.1.1

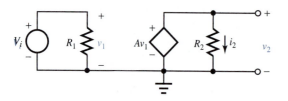

Figure P5.1.2

5.1.6 Consider Example 5.1.2. Let $A = 1$ and $R_i = R_o$. Show that no matter what values of R_{Th} and R_L are used, the power gain as defined in that example is less than unity, and find the maximum value it can have.

5.1.7 (a) Figure P5.1.7 shows a circuit containing an amplifier block. Find the open-circuit voltage amplification of the circuit.

(b) Let the output terminals be connected to a load resistance R_L. Find the input resistance R_i'.

5.1.8 (a) Consider the amplifier block in the circuit configuration of Figure P5.1.8. Find an expression for v_2/v_1 in terms of R_i, R_o, and A of the amplifier.

(b) Determine the output resistance of the circuit.

*5.1.9 (a) Two amplifiers, which can be represented by the model of Figure 5.1.1 of the text, are connected in *cascade* (that is, head to tail), as shown in Figure P5.1.9. Let their parameters be R_{i1}, R_{o1}, A_1, and R_{i2}, R_{o2}, A_2, respectively. Find v_x/v_1.

(b) Discuss how the answer is influenced by the ratio R_{i2}/R_{o1}, particularly when this ratio approaches zero or infinity.

5.1.10 Reconsider Problem 5.1.9. If we define a power gain G_P as the ratio of the power dissipated in R_L to the power produced by the source v_1, find an

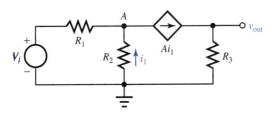

Figure P5.1.3

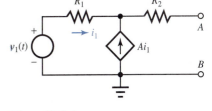

Figure P5.1.4

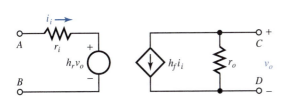

Figure P5.1.5

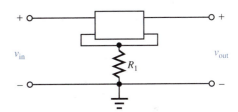

Figure P5.1.7

Figure P5.1.8

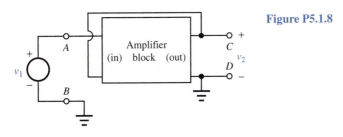

Figure P5.1.9

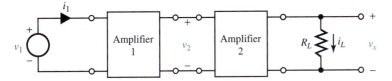

expression for G_P for the case when $R_{i1} = R_{i2} = R_i$, and $R_{o1} = R_{o2} = R_o$.

5.1.11 Consider Example 5.1.4. For what value of V_{out} is the power dissipation maximum?

5.1.12 In the circuit of Example 5.1.4 let $v_{in} = 1 - 0.5 \cos \omega t$. Determine the time-averaged power dissipated in the amplifier.

*5.1.13 In the circuit of Example 5.1.4, what is the smallest value of R_L so that no matter what value v_{out} takes in the range of 0 to 20 V, the power dissipation in the amplifier never exceeds 50 mW?

5.1.14 (a) An audio amplifier with $R_i = 10\,k\Omega$, $R_o = 0$, and $\bar{A}\,(\omega)$, as shown in Figure 5.1.13(a), is used in the circuit shown in Figure P5.1.14 with $R_S = 1\,k\Omega$, $R_L = 16\,\Omega$, and $C = 0.2\,\mu F$. Sketch $G_V(f)$ versus frequency if $G_V(f)$ is defined by $G_V\,(f) = |\bar{V}_L| / |\bar{V}_S|$.

    (b) Find f_{max} for this circuit if f_{max}, the maximum usable frequency, is defined as the frequency

at which G_V has dropped 6 dB below its maximum value.

5.2.1 Using the ideal op-amp technique, find the closed-loop voltage amplification A' for the circuit shown in Figure P5.2.1.

5.2.2 Find v_o in the circuit shown in Figure P5.2.2 by using the ideal op-amp technique.

5.2.3 In the circuit shown in Figure P5.2.3, use the ideal op-amp technique to find:

    (a) v_o as a function of v_i.

    (b) The voltage at A.

*5.2.4 Determine the open-circuit output voltage v_o of the system shown in Figure P5.2.4 as a function of the input voltage v_i.

5.2.5 Find the gain v_o/v_i for the circuit shown in Figure P5.2.5 and comment on the effects of R_3 and R_4 on the gain.

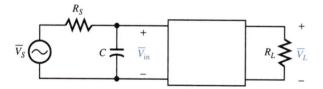

Figure P5.1.14

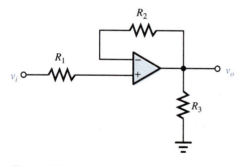

Figure P5.2.1

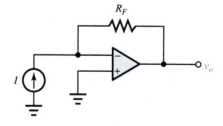

Figure P5.2.2

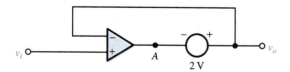

Figure P5.2.3

5.2.6 Determine v_o/v_i for the circuit shown in Figure P5.2.6 if the op amps are ideal.

5.2.7 Find v_o/v_i for the circuit shown in Figure P5.2.7 if the op amp is ideal.

5.2.8 Consider the inverting amplifier of Example 5.2.1. Let a voltmeter be connected (negative voltmeter terminal at terminal 3) between terminal 3 and ground to measure voltages accurately between 1 and 10 V. Let $R_i = 1$ kΩ and $v_i = 10$ V (dc). Determine the relation between voltage indication and R_f, noting that this circuit is an electronic ohmmeter capable of measuring the value of R_f.

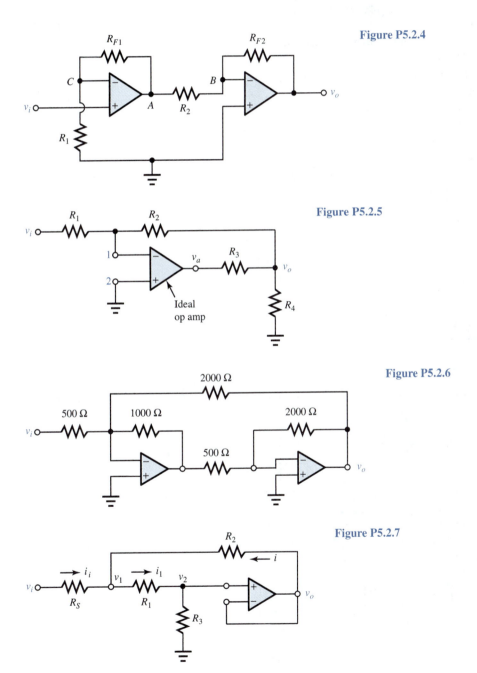

Figure P5.2.4

Figure P5.2.5

Figure P5.2.6

Figure P5.2.7

*5.2.9 Consider the noninverting amplifier of Example 5.2.2. Let $R_i = 1$ kΩ and $R_f = 2$ kΩ. Let the op amp be ideal, except that its output cannot exceed ±12 V at a current of ±10 mA.

(a) Find the minimum load resistor that can be added between terminal 3 and ground.

(b) Determine the largest allowable magnitude for v_i when using the minimum load resistance.

5.2.10 An op amp has an open-loop frequency response as shown in Figure 5.2.4.

(a) Find the approximate bandwidth of the circuit using this op amp:

(i) With a closed-loop voltage gain of 100;

(ii) With a voltage gain of 1000.

(b) Determine the gain–bandwidth product for this op amp.

5.2.11 Consider the generalized circuit shown in Figure P5.2.11, which contains two elements with impedances $\bar{Z}_1$ and $\bar{Z}_F$. Using the phasor techniques to study the response of op-amp circuits to sinusoidal signals, obtain a general expression for $\bar{A}' = \bar{V}_o/\bar{V}_i$ in terms of the op-amp parameters $\bar{A}$, R_i, and R_o and the impedances $\bar{Z}_1$ and $\bar{Z}_F$. In the limit, as $R_i \to \infty$, $R_o \to 0$, and $A \to \infty$, what is the effect on the result? Considering the high-frequency response, without the assumption of $A \to \infty$, find an expression for $\bar{A}'$.

5.3.1 Figure P5.3.1 gives the frequency-response graphs for a 709 op amp. Choose compensating components for the circuit (see Figure 5.3.8) to have a gain of 100 and a frequency response of up to 100 kHz.

5.3.2 Find the maximum amplitude of an output voltage sine wave that an op amp with a slew rate of 0.5 V/μs can deliver at $f = 100$ kHz.

*5.3.3 An op amp has a slew rate of 0.7 V/μs. Find the maximum amplitude of an undistorted output sine wave that the op amp can produce at a frequency of 50 kHZ. Also determine the maximum frequency of undistorted output that the op amp will produce at an amplitude of 3 V.

Figure P5.2.11

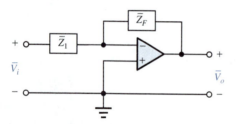

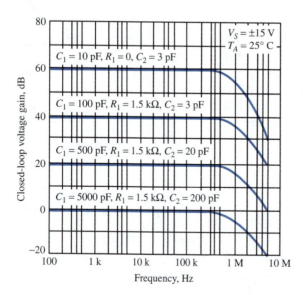

Figure P5.3.1 Frequency response for various parameters (709). *Source:* Fairchild.

5.3.4 In order to minimize output voltage offsets in practical op-amp circuits, one provides a dc path from each input terminal to ground, makes each input terminal see the same external resistance to ground, and uses external balancing circuits, if necessary, to null any remaining output offset voltage.

(a) Consider the input-offset voltage-nulling circuit for an inverting amplifier shown in Figure P5.3.4 (a). Let $R_1 = 1.5$ kΩ, $R_2 = 22$ kΩ, $R_6 = 100$ kΩ, $R_5 = 500$ kΩ, $R_4 = 200$ Ω, and $V = 12$ V. Find the range of input offset voltages that can be generated at terminal 2 of the op amp. Also find R_3 such that the input terminals see the same external resistance to ground.

(b) To examine the effects of input bias currents on the inverting amplifier, consider the circuit

shown in Figure P5.3.4(b). Show that it is desirable to choose R_3, which is equal to a parallel combination of R_1 and R_2. Compare the residual output voltage to what occurs if R_3 were zero.

(c) Reconsider the circuit of part (b). Let $R_1 = 5$ kΩ, and $R_2 = 70$ kΩ. Let the op-amp bias currents be $I_{b1} = 50$ nA and $I_{b2} = 60$ nA, but otherwise let the op amp be ideal. Determine the value of R_3 that should be used. Also, when the input signal is zero, find the residual output offset voltage.

5.3.5 A noninverting op-amp circuit and its closed-loop representation are given in Figure P5.3.5. Obtain an expression for the closed-loop transfer function $H(\omega) = Y(\omega)/X(\omega)$ and comment on how it behaves for large loop gain [i.e., when the product $H_1(\omega)H_2(\omega)$ is large].

Figure P5.3.4

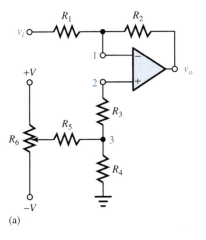

(a)

(b)

*5.4.1 An op amp has a finite gain of only 50, but is otherwise ideal. For the inverting-amplifier circuit of Figure 5.4.1, if $R_2 = 20$ kΩ, what value of R_1 would be needed to give a gain of -15? If the op amp's gain could be increased by 20%, find the new gain.

5.4.2 Consider an inverting-amplifier circuit with positive noninverting input, as shown in Figure P5.4.2 with values. Obtain v_o.

5.4.3 Let v_S in Problem 5.4.2 be a sine wave of peak voltage 3 V. Sketch v_o. Rework the problem with $R_2 = 40$ kΩ instead of 30 kΩ, and sketch v_o.

5.4.4 Consider the op-amp circuit of Figure P5.4.4. Sketch the waveforms of v_S and v_o, if v_S is a sinusoidal voltage source with a peak value of 2 V.

5.4.5 In the ideal inverting summing amplifier circuit (see Figure 5.4.3) with two inputs, for $R_f = 10$ kΩ, find R_1 and R_2 so that $v_o = -10v_{i1} - 5v_{i2}$.

5.4.6 Consider Figure 5.4.4 of the noninverting summing amplifier. Let $R_m = (m + 0.5)50$ Ω, $m = 1, 2, 3, 4$, and 5. Also let $R_d = 50$ Ω and $R_f = 3$ kΩ. Find v_o.

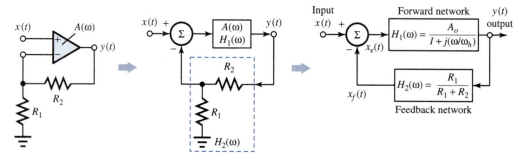

Figure P5.3.5

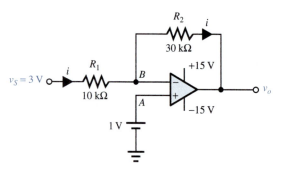

Figure P5.4.2

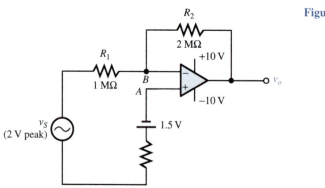

Figure P5.4.4

*5.4.7 In the noninverting summing amplifier of Figure 5.4.4, let $R_d = 1$ kΩ and $M = 6$. Find R_f so that
$$v_o = \sum_{m=1}^{6} v_{im}.$$

5.4.8 An inverting amplifier is designed with three inputs, v_1, v_2, and v_3, as shown in Figure P5.4.8. Determine the output voltage. Then indicate how the circuit may be modified to perform as a summer.

5.4.9 A negative impedance converter is used, as shown in Figure P5.4.9. Show that the load current i_L is given by v_{in}/R, which is independent of Z_L. Note that since the load sees a current source, the network is a *voltage-to-current converter.*

5.4.10 Determine how Equation (5.4.31) will be affected if the op amp of the negative impedance converter (Figure 5.4.8) has a finite gain A_0 but is otherwise ideal.

5.4.11 Find the input impedance Z_{in} for the *generalized impedance converter* circuit shown in Figure P5.4.11 if the op amps are ideal.

*5.4.12 Consider Figure 5.4.9 of the weighted differencing amplifier. Let $R_3 = R_1$ and $R_4 = R_2$. Determine the

input resistance between terminals a and b of the circuit.

5.4.13 In the circuit shown in Figure P5.4.13 with an ideal op amp, find v_o as a function of v_a and v_b.

5.4.14 For the low-pass filter configuration of Figure 5.4.13, with $R_i = R_f = 1$ MΩ, calculate C_f such that the 3-dB point is at 1 kHz.

5.4.15 For the high-pass filter configuration of Figure 5.4.13, with $v_o/v_i = 2.5$ at 10 MHz and $C_i = 100$ pF, determine R_i and R_f to yield a 3-dB point at 1 MHz.

5.4.16 Comment on the behavior of the circuit of Figure P5.4.16 at low and high frequencies.

*5.4.17 The input resistance of an ideal op amp with no feedback is infinite. Investigate the input resistance of an op amp with a feedback resistance R_F:

(a) When there is no resistance placed in the $-$ input line.

(b) When a resistance R_1 is placed in the $-$ input line.

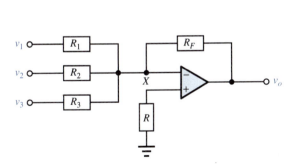

Figure P5.4.8

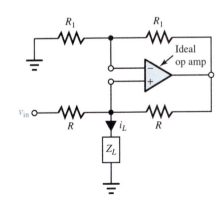

Figure P5.4.9

Figure P5.4.11

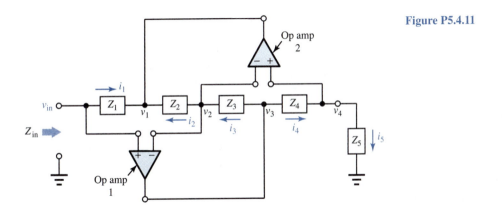

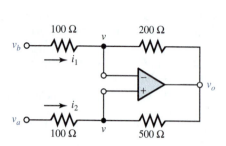

Figure P5.4.13

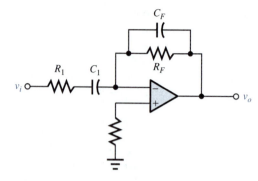

Figure P5.4.16

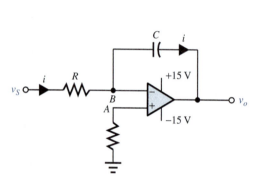

Figure P5.4.18

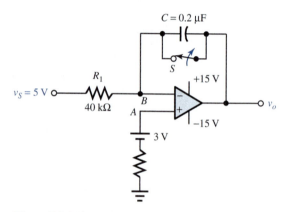

Figure P5.4.19

5.4.18 Consider the inverting integrator circuit shown in Figure P5.4.18. Let $C = 0.4\,\mu\text{F}$ and $R = 0.1\,\text{M}\Omega$. Sketch v_o for a period of 0.5 s after the application of a constant input of 2 V at the v_S terminal. Assume that C is discharged at the beginning of the operation.

5.4.19 An integrator with positive voltage on a noninverting input is shown in Figure P5.4.19. Sketch v_o for 60 ms after S has been opened.

5.4.20 Addition and integration can be combined by the summing integrator circuit shown in Figure P5.4.20. With the given component values and input waveforms, sketch v_o when S is opened at $t = 0$.

5.4.21 Refer to the noninverting amplifier circuit of Figure 5.4.2. Let $R_1 = 10\,\text{k}\Omega$, $R_2 = 30\,\text{k}\Omega$, and v_i be a sinusoidal source with a peak value of 1 V. Sketch the waveforms of v_i and v_o.

*5.4.22 Develop an analog computer simulation diagram to solve the differential equation

$$\frac{d^2 y\,(t)}{dt^2} + 12\frac{dy\,(t)}{dt} + 5y\,(t) = 10$$

with $y(0) = 2$ and $\dot{y}(0) = 0$.

5.4.23 If the solution to the differential equation of Problem 5.4.22 is to be obtained over $0 \le t \le 1$ ms, but one wants to expand this over an interval of 1 s, redraw the analog computer simulation diagram.

5.4.24 An integrator as shown in Figure 5.4.18(d) is to be designed to solve

$$\frac{dy\,(t)}{dt} + 2000y\,(t) = 0$$

with $y(0) = 5$. If R_i is chosen to be 10 kΩ, find C_f.

5.4.25 Determine I in the circuit shown in Figure P5.4.25.

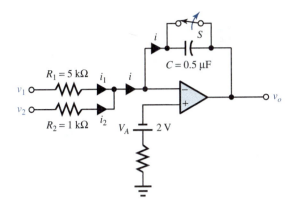

(a)

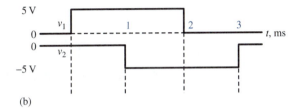

(b)

Figure P5.4.25

6 Digital Building Blocks and Computer Systems

Whereas a continuous change from one value to another is the essential characteristic of an analog signal and *continuous-state (analog) circuits* are used to process analog signals, signals represented by *discrete amplitudes* at *discrete times* are handled by *discrete-state (digital) circuits*. *Analog systems* process the information contained in the time function, which defines the signal, whereas *digital systems,* as the name implies, process digits, i.e., pulse trains, in which the information is carried in the pulse sequence rather than the amplitude–time characterization of the pulses. Figure 6.0.1 illustrates continuous signals, whereas Figure 6.0.2 depicts discrete signals.

Consider Figure 6.0.1(a) to be a voltage signal (as a function of a continuous-time variable t) representing a physical quantity, such as the output voltage of a phonograph cartridge. The discrete signal (which exists only at specific instances of time) of Figure 6.0.2(a) has the same amplitude at times $t = 0$, T_1, T_2, and T_3 as does the continuous signal of Figure 6.0.1(a). Figure 6.0.2(a) represents the *sampled-data signal* obtained by sampling the continuous signal at periodic intervals of time. The sequence of pulses in each time interval in Figure 6.0.2(b) is a numeric, or digital, representation of the corresponding voltage samples shown in Figure 6.0.2(a). Figure 6.0.1(b) may be considered as the signal, called a clock, which sets the timing sequence used in the generation of the pulses shown in Figure 6.0.2.

An *analog signal* is an electric signal whose value varies in analogy with a physical quantity such as temperature, force, or acceleration. Sampling of an analog signal makes it discrete in time. A *digital signal*, on the other hand, can only have a *finite number of discrete amplitudes* at any given time. Through a process known as *quantization*, which consists of rounding exact sample

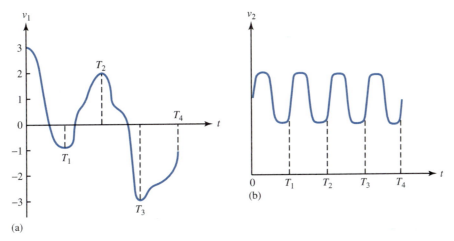

Figure 6.0.1 Continuous signals.

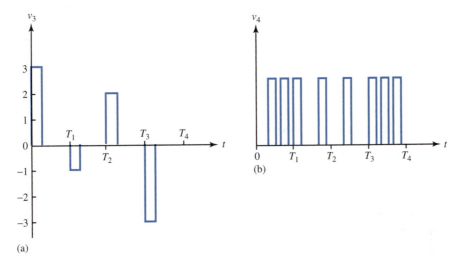

Figure 6.0.2 Discrete signals.

values to the nearest of a set of discrete amplitudes called quantum levels, a digitized signal can be obtained as discussed in more detail in Section 15.3.

The most common digital signals are *binary signals*. A binary signal is a signal that can take only one of two discrete values and is therefore characterized by transitions between two states. Figure 6.0.3 displays typical binary signals.

Voltmeters, for example, can be of either analog or digital (digital voltmeter, or DVM) variety. While the speedometer of a car is an analog device, the odometer (which records miles or kilometers) is digital because it records changes in units of one-tenth of a mile or kilometer. A toggle switch which is either on or off is digital, whereas the dimmer switch is analog because it allows the light intensity to be varied in a continuous way.

Digital electronic circuits have become increasingly important for several reasons. The present-day integrated-circuit (IC) technology allows the construction of an enormous number of

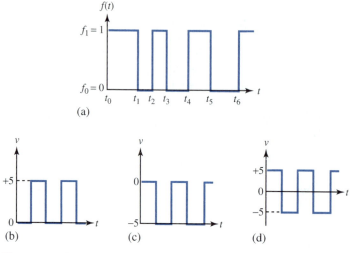

Figure 6.0.3 Typical binary signals.

transistors and diodes, as well as resistors and capacitors, on a very small chip (no larger than a pencil eraser). The variety of ICs available to process digital signals can contain a few to several thousand components on a single chip. A convenient method of digital IC classification, based on the number of components per chip, is as follows:

1. **Small-scale integration (SSI)**, containing fewer than 100 components
2. **Medium-scale integration (MSI)**, containing 100 to 1000 components
3. **Large-scale integration (LSI)**, containing 1000 to 10,000 components
4. **Very large-scale integration (VLSI)**, containing more than 10,000 components

Digital computers and other large digital systems use mostly LSI and MSI chips. The number of components per chip has increased from about 100 in the 1960s to 10^8 in the 1990s. Speed, power consumption, and the number of gates (switches) on the chip are three of the more important characteristics by which digital ICs are compared.

The general-purpose digital computer is the best known example of a digital system. Other examples include teletypewriters, word processors, dial-telephone switching exchanges, frequency counters, remote controls, and other peripheral equipment. Manipulation of finite, discrete elements of information is a characteristic of a digital system. Information in digital systems is represented by signals (currents or voltages) that take on a limited number of discrete values and are processed by devices that normally function only in a limited number of discrete states. A great majority of present digital devices are *binary* (i.e., have signals and states limited to two values) because of the lack of practical devices capable of performing reliably in more than two discrete states. Transistor circuitry, which can be constructed with extreme reliability, has two possible signal values, either on state or off state. With the advent of transistors, the computer industry flourished.

To start with, this chapter presents basic digital building blocks, which are usually designed to process binary signals. Then the digital system components are discussed. Later on an introduction to computer systems and networks is included. Thus, a basic foundation of digital systems is laid out for the reader.

6.1 DIGITAL BUILDING BLOCKS

The primary advantage of digital technology is the low cost, simplicity, and versatility of the digital building blocks. Because digital signals have a finite number of discrete amplitudes at any given time, distinctive digital building blocks are developed to process them. Digital systems are built by repeating a very few simple blocks. The approach becomes very powerful because the blocks (mass-produced in enormous numbers) are inexpensive in IC form and can be repeated thousands of times.

The two-state nature of digital technology makes the binary number system its natural tool. In a binary system there are only two digits, namely, 0 and 1. The vast majority of present digital computers use the binary system, which has two binary digits (*bits*), 0 and 1. Internal representation of information in a digital computer is in groups of bits. Besides the binary system (base 2), in the digital world, the most commonly used number systems are octal (base 8) and hexadecimal (base 16).

EXAMPLE 6.1.1

A signal stands for 0110 with positive logic in which low stands for 0 and high stands for 1. If negative logic (in which low stands for 1 and high stands for 0) is used, what digits are represented by the signal?

Solution

The signal sequence reads: low–high–high–low. With negative logic, the signal stands for 1001.

EXAMPLE 6.1.2

If the time occupied in transmitting each binary digit is 1 μs, find the rate of information transmission if 1 baud is equal to 1 bit per second (bit/s).

Solution

Since the time per digit is 1×10^{-6} s, one can transmit 10^6 digits per second. The information rate is then 1×10^6 bits, or 1 megabaud (M baud). Note that speeds of over 100 M baud are quite possible.

Number Systems

A number system, in general, is an ordered set of symbols (digits) with relationships defined for addition, subtraction, multiplication, and division. The base (radix) of the number system is the total number of digits in the system. For example, in our decimal system, the set of digits is {0, 1, 2, 3, 4, 5, 6, 7, 8, 9} and hence the base (radix) is ten (10); in the binary system, the set of digits (bits) is {0, 1} and hence the base or radix is two (2).

There are two possible ways of writing a number in a given system: positional notation and polynomial representation. For example, the number 2536.47 in our decimal system is represented in positional notation as $(2536.47)_{10}$, whereas in polynomial form it is $2 \times 10^3 + 5 \times 10^2 + 3 \times$

$10^1 + 6 \times 10^0 + 4 \times 10^{-1} + 7 \times 10^{-2}$. The radix or base is 10, whereas the most significant digit or bit (MSB) is 2, the least significant digit or bit (LSB) is 7; the number of integer bits (digits) is 4, and the number of fractional bits (digits) is 2.

The *binary number system* has a base of 2 with two distinct digits (bits), 1 and 0. A binary number is expressed as a string of 0s and 1s, and a binary point if a fraction exists. To convert the binary to the decimal system, the binary number is expressed in the polynomial form and the resulting polynomial is evaluated by using the decimal-system addition. For example,

$$(101101.101)^2 = 1 \times 2^5 + 0 \times 2^4 + 1 \times 2^3 + 1 \times 2^2 + 0 \times 2^1 + 1 \times 2^0 + 1 \times 2^{-1}$$
$$+ 0 \times 2^{-2} + 1 \times 2^{-3}$$
$$= 32 + 0 + 8 + 4 + 0 + 1 + 0.5 + 0 + 0.125 = (45.625)_{10}$$

For digital processing it is also often necessary to convert a decimal number into its equivalent binary number. This is accomplished by using the following steps.

1. Repeatedly *divide* the *integer part* of the decimal number by 2. Use the remainder after each division to form the equivalent binary number. Continue this process until a zero quotient is obtained. With the first remainder being the least significant bit, form the binary number by using the remainder after each division.

2. Repeatedly *multiply* the *decimal fraction* by 2. If 0 or 1 appears to the left of the decimal point of the product as a result of this multiplication, then add a 0 or 1 to the binary function. Continue this process until the fractional part of the product is zero or the desired number of binary bits is reached.

For example, the decimal number $(75)_{10}$ is converted into its binary equivalent:

Quotient		Remainder	
$75 \div 2 =$	37	1	LSB
$37 \div 2 =$	18	1	
$18 \div 2 =$	9	0	
$9 \div 2 =$	4	1	
$4 \div 2 =$	2	0	
$2 \div 2 =$	1	0	
$1 \div 2 =$	0	1	MSB
	Stop		

Thus, the binary number for $(75)_{10}$ is given by:

$$(1001011)_2$$

↑ MSB   ↑ LSB

Let us now convert the fractional decimal number $(0.4375)_{10}$ into its binary equivalent:

↓ MSB of binary fraction

$0.4375 \times 2 =$	0.8750	
$0.8750 \times 2 =$	1.7500	
$0.7500 \times 2 =$	1.5000	
$0.5000 \times 2 =$	1.0000	
	Stop	

↑ LSB of binary fraction

Thus, $(0.4375)_{10} = (.0111)_2$.

For representing binary data, octal and hexadecimal numbers are used. The *octal number system* is a base-8 system and therefore has 8 distinct digits {0, 1, 2, 3, 4, 5, 6, 7}. It is expressed as a string of any combination of the 8 digits. In order to convert octal to decimal, one follows the same procedure as for converting from binary to decimal, that is, by expressing the octal number in its polynomial form and evaluating that polynomial by using decimal-system addition.

For converting from decimal to octal form, one follows the same procedure as for conversion from decimal to binary; but instead of dividing by 2 for the integer part, one divides by 8 to obtain the octal equivalent. Also, instead of multiplying by 2 for the fractional part, one multiplies by 8 to obtain the fractional octal equivalent of the decimal fraction. However, it is more common to convert from binary to octal and vice versa. Binary to octal conversion is accomplished by grouping the binary number into groups of 3 bits each, starting from the binary point and proceeding to the right as well as to the left; each group is then replaced by its octal equivalent. For example, $(100101111011.01011)_2$ is arranged as 100 101 111 011 . 010 110 (note that a trailing 0 is added to complete the last group in the fractional part). Replacing each group by its octal "decimal" equivalent, one obtains $(4573.26)_8$. The conversion from octal to binary is done by replacing each octal digit with its 3-bit binary equivalent.

The *hexadecimal number system* is a base-16 system which has 16 distinct symbols in the set: $\{0, 1, 2, 3, 4, 5, 6, 7, 8, 9, A, B, C, D, E, F\}$, where A is equivalent to 10, B to 11, , and F to 15. A hexadecimal number is therefore expressed as a string of any combination of the 16 symbols. To convert from hexadecimal to decimal, for example,

$$(2AB)_{16} = 2 \times 16^2 + A \times 16^1 + B \times 16^0 = 2 \times 16^2 + 10 \times 16^1 + 11 \times 16^0 = (683)_{10}$$

$$(.F8)_{16} = F \times 16^{-1} + 8 \times 16^{-2} = 15 \times 16^{-1} + 8 \times 16^{-2} = (0.96875)_{10}$$

and

$$(2AB.F8)_{16} = (683.96875)_{10}$$

The conversion from binary to hexadecimal is accomplished by grouping the binary number into groups of 4 bits each, starting from the binary point and proceeding to the right as well as to the left. Each group is then replaced by its hexadecimal equivalent. For example,

$$(11101110100100.100111)_2 \Rightarrow 0011\ \ 1011\ \ 1010\ \ 0100\ .\ 1001\ \ 1100 \Rightarrow (3BA4.9C)_{16}$$

The conversion from hexadecimal to binary is achieved by reversing this process. Because the internal structures of most digital computers manipulate data in groups of 4-bit packets, the hexadecimal system is used for representing binary data.

Since there is a need for decimal-to-binary conversion at the input of a digital device and for binary-to-decimal conversion at the output of the device, from a user's point of view, the *binary-coded-decimal (BCD) number system* is developed for resolving the interface problem. In a BCD number each of the decimal digits is coded in binary using 4 bits. For example,

$$(163.25)_{10} = (10100011.01)_2 = (0001\ \ 0110\ \ 0011\ .\ 0010\ \ 0101)_{BCD}$$

Table 6.1.1 shows the first 20 numbers in the decimal, binary, octal, hexadecimal, and BCD systems.

Logic Blocks

Boolean algebra, using Boolean variables (which are binary variables that can assume only one of two distinct values, true or false), is a mathematical system with logic notation used to describe different interconnections of digital circuits. There are two basic types of digital building

TABLE 6.1.1 Representation of Numbers in Different Systems

Base 10 (Decimal)	Base 2 (Binary)	Base 8 (Octal)	Base 16 (Hexadecimal)	BCD
00	00000	00	00	0000 0000
01	00001	01	01	0000 0001
02	00010	02	02	0000 0010
03	00011	03	03	0000 0011
04	00100	04	04	0000 0100
05	00101	05	05	0000 0101
06	00110	06	06	0000 0110
07	00111	07	07	0000 0111
08	01000	10	08	0000 1000
09	01001	11	09	0000 1001
10	01010	12	0A	0001 0000
11	01011	13	0B	0001 0001
12	01100	14	0C	0001 0010
13	01101	15	0D	0001 0011
14	01110	16	0E	0001 0100
15	01111	17	0F	0001 0101
16	10000	20	10	0001 0110
17	10001	21	11	0001 0111
18	10010	22	12	0001 1000
19	10011	23	13	0001 1001

blocks: *logic (combinational) blocks* and *sequential blocks* (known as *flip-flops*). Logic blocks are electronic circuits that usually have several inputs and one output. The physical devices that perform the basic Boolean operations are known as logic gates. The basic logic gates are OR, AND, and NOT, the interconnections of which form a logic network (also known as combinational network), which does not generally contain memory devices. The Boolean function describing a combinational network can be easily derived by systematically progressing from the input(s) to the output on the logic gates.

The logic OR gate is an electronic circuit realization of the Boolean OR operation, which is represented by the symbol +. $F = A + B$, for example, is read as "F is equal to A OR B." The OR operation yields 1 if the value of *any* of its arguments is 1. Figure 6.1.1 shows a two-input OR gate along with its *truth table* (table of combinations for the gate) and an illustration of the OR operation. A three-input OR gate with $F = A + B + C$ would have a truth table with eight entries, which can be developed by the reader as an exercise.

The logic AND gate is an electronic circuit realization of the Boolean AND operation, which is represented by the symbol · or by the absence of an operator. Thus, for example, $F = A \cdot B$ (also written simply as $F = AB$) is read as "F is equal to A AND B." The AND operation yields 1 if and only if the values of all its arguments are 1s. Figure 6.1.2 shows a two-input AND gate, its truth table, and an illustration of the AND operation. A three-input AND gate with $F = A \cdot B \cdot C$ would have a truth table with $2^3 = 8$ possible entries, which can be developed by the reader as a desirable exercise.

The logic NOT gate is an electronic circuit realization of the Boolean NOT operation, which is represented by a bar over the variable. $F = \bar{A}$, for example, is read as "F equals NOT A." This operation is also known as the *complement* operation. Because the NOT operation is a unary operation and simply inverts a switching variable, the NOT gate has only one input and one output, and is also known as an *inverter*. Figure 6.1.3 shows a NOT gate and its truth table.

The AND operation followed by the NOT operation inverts the output coming out of the AND gate, and is known as NAND gate, which is typically shown in Figure 6.1.4 along with its

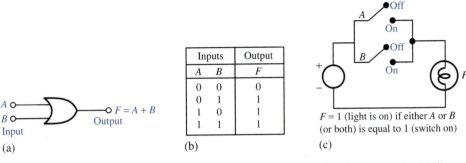

Inputs		Output
A	B	F
0	0	0
0	1	1
1	0	1
1	1	1

$A \circ$——$\circ F = A + B$
$B \circ$—— Output
Input

(a)   (b)   (c)

$F = 1$ (light is on) if either A or B (or both) is equal to 1 (switch on)

Figure 6.1.1 Two-input OR gate. **(a)** Symbol or physical representation. **(b)** Truth table. **(c)** Illustration of OR operation.

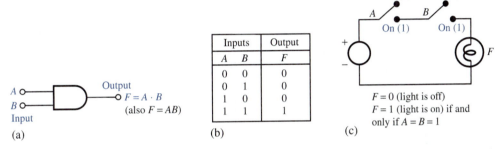

Inputs		Output
A	B	F
0	0	0
0	1	0
1	0	0
1	1	1

$A \circ$—— Output
$B \circ$—— $\circ F = A \cdot B$
Input   (also $F = AB$)

(a)   (b)   (c)

$F = 0$ (light is off)
$F = 1$ (light is on) if and only if $A = B = 1$

Figure 6.1.2 Two-input AND gate. **(a)** Symbol or physical representation. **(b)** Truth table. **(c)** Illustration of AND operation.

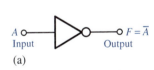

Input	Output
A	F
0	1
1	0

$A \circ$——$\circ F = \overline{A}$
Input   Output

(a)   (b)

Figure 6.1.3 NOT gate (complement). **(a)** Symbol or physical representation. **(b)** Truth table.

Figure 6.1.4 NAND gate. **(a)** Logic symbol. **(b)** Truth table.

Inputs			Output
A	B	C	F
0	0	0	1
0	0	1	1
0	1	0	1
0	1	1	1
1	0	0	1
1	0	1	1
1	1	0	1
1	1	1	0

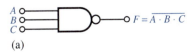

$A \circ$
$B \circ$——$\circ F = \overline{A \cdot B \cdot C}$
$C \circ$

(a)   (b)

truth table. Similarly, the OR operation followed by the NOT operation is known as the NOR operation. Figure 6.1.5 shows a typical NOR gate with its truth table.

Two other logic blocks that are sometimes used are the EXCLUSIVE OR (XOR) and EXCLUSIVE NOR (COINCIDENCE) gates. Their conventional symbols together with their truth tables are shown in Figures 6.1.6 and 6.1.7, respectively. It is also possible to synthesize these logic blocks by combining the basic OR, AND, as well as NOT logic gates.

The operations so far introduced are summarized in Table 6.1.2. Any Boolean function can be transformed from an algebraic expression into a combinational network by using the basic logic gates. It is always desirable, however, to implement a given Boolean function by using the *minimum* number of components, leading to a less complex and more economical network. Table 6.1.3 gives the basic Boolean identities for which the truth tables for the expressions on either side of the equals sign are the same. The reader should note that, in general, it is improper to draw a diagram in which the outputs of any two blocks are connected together. Two outputs can be combined by using them as two inputs to another gate.

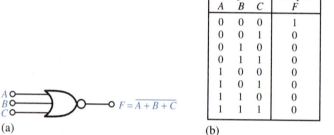

Inputs A	B	C	Output F
0	0	0	1
0	0	1	0
0	1	0	0
0	1	1	0
1	0	0	0
1	0	1	0
1	1	0	0
1	1	1	0

(a) $F = \overline{A + B + C}$ (b)

Figure 6.1.5 NOR gate. **(a)** Logic symbol. **(b)** Truth table.

Inputs A	B	Output F
0	0	0
0	1	1
1	0	1
1	1	0

(a) $F = A \oplus B$
$F = \overline{A} \cdot B + A \cdot \overline{B}$ (b)

Figure 6.1.6 EXCLUSIVE OR gate (XOR) (also known as binary comparator). **(a)** Logic symbol. **(b)** Truth table. NOTE: XOR operation yields 1 if the input A is 1 or B is 1, but yields 0 if both inputs A and B are 1 or 0.

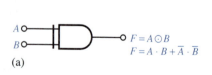

Inputs A	B	Output F
0	0	1
0	1	0
1	0	0
1	1	1

(a) $F = A \odot B$
$F = A \cdot B + \overline{A} \cdot \overline{B}$ (b)

Figure 6.1.7 EXCLUSIVE NOR gate (COINCIDENCE) (also known as XNOR or EQUIVALENT gate). **(a)** Logic symbol. **(b)** Truth table. NOTE: XNOR operation yields 1 if both inputs A and B have the same value of either 0 or 1, but yields 0 if the input A is 1 or B is 1. Hence it is a complement of XOR gate.

TABLE 6.1.2 Logic Symbols and Truth Tables

Symbol	Name	Notation	Truth Table
A ▷∘ F	COMPLEMENT	$F = \bar{A}$	A F 0 1 1 0
A B ⊃ F	OR	$F = A + B$	A B F 0 0 0 0 1 1 1 0 1 1 1 1
A B ⊃ F	AND	$F = AB$ or $A \cdot B$	A B F 0 0 0 0 1 0 1 0 0 1 1 1
A B ⊃∘ F	NOR	$F = \overline{A + B}$	A B F 0 0 1 0 1 0 1 0 0 1 1 0
A B ⊃∘ F	NAND	$F = \overline{AB}$	A B F 0 0 1 0 1 1 1 0 1 1 1 0
A B ⊃ F	EXCLUSIVE OR	$F = A \oplus B$	A B F 0 0 0 0 1 1 1 0 1 1 1 0
A B ⊃ F	COINCIDENCE	$F = A \odot B$	A B F 0 0 1 0 1 0 1 0 0 1 1 1

The last two entries in Table 6.1.3 are known as *DeMorgan's theorems*, whose importance stems from the fact that they offer a general technique for complementing Boolean expressions. They show that any logic function can be implemented using only OR and NOT gates, or using only AND and NOT gates. There exists a *duality* between AND and OR operations; any function can be realized by just one of the two basic operations, plus the complement operation. This gives rise to two families of logic functions: *sum of products* (SOP) and *product of sums* (POS). Any logic expression can be reduced to either one of these.

TABLE 6.1.3 Basic Boolean Identities

Identity	Comments
1. $X + 0 = X$	Identities 1–9 are basic to Boolean algebra
2. $X + 1 = 1$	
3. $X + X = X$	
4. $X + \bar{X} = 1$	
5. $X \cdot 0 = 0$	
6. $X \cdot 1 = X$	
7. $X \cdot X = X$	
8. $X \cdot \bar{X} = 0$	
9. $\bar{\bar{X}} = X$	
10. $X + Y = Y + X$	Commutative
11. $X \cdot Y = Y \cdot X$	Commutative
12. $X + (Y + Z) = (X + Y) + Z$	Associative
13. $X \cdot (Y \cdot Z) = (X \cdot Y) \cdot Z$	Associative
14. $X \cdot (Y + Z) = X \cdot Y + X \cdot Z$	Distributive
15. $X + Y \cdot Z = (X + Y) \cdot (X + Z)$	
16. $X + X \cdot Y = X$	Absorption
17. $X \cdot (X + Y) = X$	
18. $X \cdot Y + \bar{X} \cdot Z + Y \cdot Z = X \cdot Y + \bar{X} \cdot Z$	Consensus
19. $\overline{X + Y + Z} = \bar{X} \cdot \bar{Y} \cdot \bar{Z}$	DeMorgan
20. $\overline{X \cdot Y \cdot Z} = \bar{X} + \bar{Y} + \bar{Z}$	DeMorgan

EXAMPLE 6.1.3

For the switching function $F = A\left(\bar{A} + B\right)$, draw a corresponding set of logic blocks and write the truth table.

Solution

A suitable connection of logic blocks is shown in Figure E6.1.3. Using the intermediate variables, the truth table is as follows.

Figure E6.1.3

A	B	$\bar{A}$	$\bar{A} + B$	F
0	0	1	1	0
0	1	1	1	0
1	0	0	0	0
1	1	0	1	1

EXAMPLE 6.1.4

Derive the Boolean function for the combinational network shown in Figure E6.1.4(a).

Solution

The solution is given in Figure E6.1.4(b).

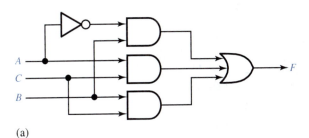

(a)

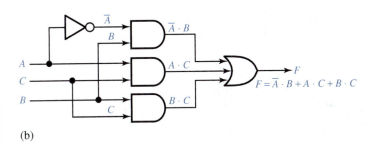

Figure E6.1.4

(b)

EXAMPLE 6.1.5

Prove the Boolean identity $A(B + C) = AB + AC$, which is distributive, by comparing the truth tables of both sides.

Solution

The truth tables with the necessary intermediate variables are as follows:

A	B	C	B + C	A(B + C)	AB	AC	AB + AC
0	0	0	0	0	0	0	0
0	0	1	1	0	0	0	0
0	1	0	1	0	0	0	0
0	1	1	1	0	0	0	0
1	0	0	0	0	0	0	0
1	0	1	1	1	0	1	1
1	1	0	1	1	1	0	1
1	1	1	1	1	1	1	1

↑_____ Same _____↑

Note: The direct representation of $AB + AC$ would require three gates (two ANDs and one OR). But two gates (one OR and one AND) can perform the same function, as seen from the identity.

EXAMPLE 6.1.6

(a) Express DeMorgan's laws given by

$$\overline{X + Y} = \overline{X} \cdot \overline{Y} \qquad \text{and} \qquad \overline{X \cdot Y} = \overline{X} + \overline{Y}$$

in terms of logic gates.

(b) Express the sum of products given by $(X \cdot Y) + (W \cdot Z)$ and the product of sums given by $(A + B) \cdot (C + D)$ in terms of logic gates.

Solution

(a) Figures E6.1.6(a) and (b) show the solution.

(b) Figures E6.1.6(c) and (d) show the solution.

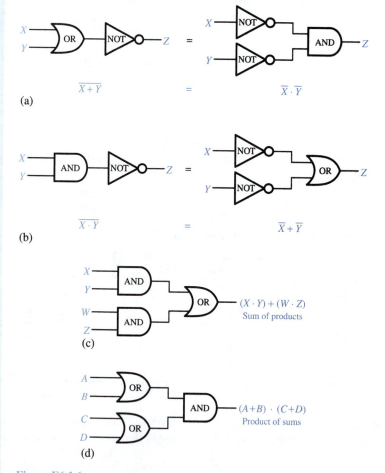

Figure E6.1.6

Thus far we have shown how to find the truth table for a given combination of logic blocks. The inverse process, that of finding an interconnection of blocks to produce a given truth table, is known as *logic synthesis*. Even though the resulting circuit is not necessarily minimal (thereby assuming a minimum number of components), the truth table can be realized by the two standard forms: *sum of products* (SOP) and *product of sums* (POS). SOP can be implemented by using two-level (AND–OR) networks, where each product term requires an AND gate (except for a term with a single variable) and the logic sum of these terms is obtained by using an OR gate with inputs from the AND gates or the single variables. POS can be implemented by using two-level (OR–AND) networks, where each term requires an OR gate (except for a term with a single variable) and the product of these terms is obtained by using an AND gate with inputs from the OR gates or the single variables.

Figure 6.1.8 shows four possible gates with an output of 1 only. For example, the truth table of Figure 6.1.9(a) can be realized by the interconnection of gates shown in Figure 6.1.9(b) with the resulting Boolean expression $F = \bar{A}B + AB$. The SOP or the POS method can be used to realize any truth table, but it is not efficient and minimal; usually other realizations with fewer blocks can be found. The realization can often be simplified by using the distributive law and the simple theorem $A + \bar{A} = 1$. For example, the expression $F = \bar{A}B + AB$ can be rewritten as $F = (\bar{A} + A) B$ by means of the distributive law, and can be simplified as $F = B$, since $\bar{A}+A = 1$. Finding minimal realizations with a minimum number of components is the challenge in *logic design*.

Addition of binary numbers is so common that a binary *half-adder* and a *full-adder* (FA) have become building blocks in their own right, and are available in IC form. The representation of the half-adder, which can add single-digit binary numbers, is shown in Figure 6.1.10(a) along with its truth table. *N* full adders are needed to add together two *N*-digit binary numbers. Figure 6.1.11 illustrates the addition of 4-digit binary numbers.

A ———⊐D— F
B ———

Output 1 only if $A = 1, B = 1$
(a)

A ———⊐D— F
B ——o

Output 1 only if $A = 1, B = 0$
(b)

Figure 6.1.8 Four possible gates with an output of 1 only.

A ——o⊐D— F
B ———

Output 1 only if $A = 0, B = 1$
(c)

A ——o⊐D— F
B ——o

Output 1 only if $A = 0, B = 0$
(d)

1st level   2nd level

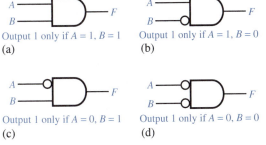

$F = \bar{A}B + AB$

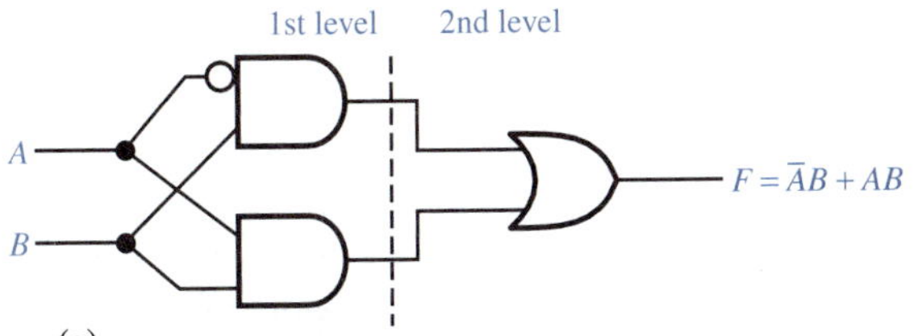

(a)

A	B	F
0	0	0
0	1	1
1	0	0
1	1	1

(b)

Figure 6.1.9 SOP realization of $F = \bar{A}B + AB$. **(a)** Truth table. **(b)** Interconnection of gates.

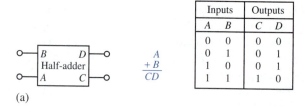

Inputs		Outputs	
A	B	C	D
0	0	0	0
0	1	0	1
1	0	0	1
1	1	1	0

(a)

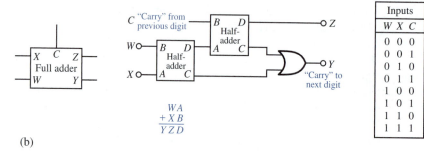

Inputs			Outputs	
W	X	C	Y	Z
0	0	0	0	0
0	0	1	0	1
0	1	0	0	1
0	1	1	1	0
1	0	0	0	1
1	0	1	1	0
1	1	0	1	0
1	1	1	1	1

(b)

Figure 6.1.10 (a) Half-adder and its truth table. (b) Full adder and its truth table.

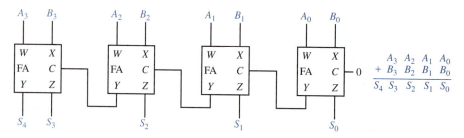

Figure 6.1.11 Addition of 4-digit binary numbers.

EXAMPLE 6.1.7

Refer to Figure 6.1.10(a) of the half-adder and its truth table for adding two single-digit binary numbers, A and B, to yield a two-digit number CD. Using the SOP method, develop a circuit to generate C and D.

Solution

In terms of the four possible gates (with an output of 1 only) shown in Figure 6.1.8, the circuit in Figure E6.1.7(a) can be drawn. Note that the resulting circuit will not be minimal, as is usually the case with the SOP method. The truth table can be realized with only three gates, using the circuit shown in Figure E6.1.7(b). The student is encouraged to verify by constructing its truth table.

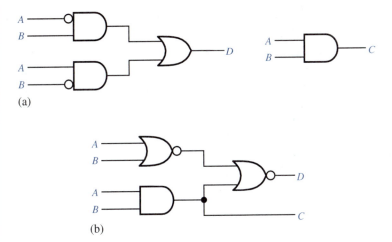

Figure E6.1.7

The basic logic gates described earlier are used to implement more advanced functions and are often combined to form logic blocks (or modules) which are available in compact IC packages. Although SSI packages were common basic units at one time, the trend now is to integration on an even larger scale. Entire digital systems, such as those to be discussed later, are now available in IC form, and those ICs in turn become blocks for building even larger systems. The ICs are mounted in packages known as DIPs (dual in-line packages), each with 8, 14, or more wires (pins) meant to be plugged into corresponding pin sockets. The actual gates, with the same logic diagrams, can be made with several different kinds of internal construction, giving rise to different *logic families.* In general blocks of one family are compatible with other blocks of the same family, but not with those of other families.

Karnaugh Maps and Logic Design

Even though a truth table uniquely represents a logic function, it is clear that the same function may appear in different algebraic forms. While the Boolean identities can be used for the simplification of a given algebraic form, it is desirable to have a systematic process that guarantees the minimum form with a minimum number of components. The map method, known as the *Karnaugh map* or, simply, the *K map*, which is a modified form of the truth table, provides a convenient procedure for obtaining a minimum SOP or POS of a Boolean expression. The K maps are usually restricted to up to five variables, since they become too cumbersome to manipulate for a larger number of variables. Before we get into the details of K maps, it is necessary to introduce *minterm* and *maxterm.*

When a product term contains each of the n variables of a function, it is called a *minterm.* For n variables, there are 2^n possible minterms. The ith minterm of n variables is denoted by m_i, where the subscript i, $0 \leq i \leq 2^n - 1$, represents the decimal equivalent of the binary number obtained when a variable in minterm is replaced by 1 and its complement is replaced by 0.

When a sum term contains each of the n variables of a function, it is called a *maxterm.* The ith maxterm of n variables, denoted by M_i, is the complement of the ith minterm m_i of the same n variables; that is to say, $M_i = \overline{m_i}$. Table 6.1.4 lists the eight possible maxterms and minterms of the three variables A, B, and C.

Any Boolean function can be expressed, algebraically, as a sum (OR) of minterms. An expression of this form is known as a *canonical sum of products.* Given a truth table for a logic function, its SOP form can be obtained by taking the sum of the minterms that correspond to a 1 in the output column of the table.

Any Boolean function can also be expressed, algebraically, as a product (AND) of maxterms. An expression of this form is known as a *canonical product of sums*. Given a truth table for a logic function, its POS form can be obtained by taking the product of the maxterms that correspond to a 0 (zero) in the output column of the table. Let us now consider an example to illustrate the use of minterms and maxterms.

TABLE 6.1.4 Minterms and Maxterms for Three Variables

A	B	C	i	Minterm m_i	Maxterm M_i
0	0	0	0	$\bar{A} \cdot \bar{B} \cdot \bar{C}$	$A + B + C$
0	0	1	1	$\bar{A} \cdot \bar{B} \cdot C$	$A + B + \bar{C}$
0	1	0	2	$\bar{A} \cdot B \cdot \bar{C}$	$A + \bar{B} + C$
0	1	1	3	$\bar{A} \cdot B \cdot C$	$A + \bar{B} + \bar{C}$
1	0	0	4	$A \cdot \bar{B} \cdot \bar{C}$	$\bar{A} + B + C$
1	0	1	5	$A \cdot \bar{B} \cdot C$	$\bar{A} + B + \bar{C}$
1	1	0	6	$A \cdot B \cdot \bar{C}$	$\bar{A} + \bar{B} + C$
1	1	1	7	$A \cdot B \cdot C$	$\bar{A} + \bar{B} + \bar{C}$

EXAMPLE 6.1.8

Given the truth table in Table E6.1.8 for an arbitrary Boolean function F, express F as a sum of minterms and a product of maxterms.

TABLE E6.1.8

m_i	A	B	C	F
m_0	0	0	0	0
m_1	0	0	1	1
m_2	0	1	0	0
m_3	0	1	1	1
m_4	1	0	0	0
m_5	1	0	1	0
m_6	1	1	0	1
m_7	1	1	1	1

Solution

Noting that F has an output of 1 that corresponds to minterms m_1, m_3, m_6, and m_7, F can be expressed as

$$F(A, B, C) = m_1 + m_3 + m_6 + m_7$$
$$= \bar{A} \cdot \bar{B} \cdot C + \bar{A} \cdot B \cdot C + A \cdot B \cdot \bar{C} + A \cdot B \cdot C$$

or in a compact form as

$$F(A, B, C) = \sum m_i(1, 3, 6, 7)$$

where $\sum m_i()$ means the sum of all the minterms whose subscript i is given inside the parentheses.

Noting that F has an output of 0 that corresponds to maxterms M_0, M_2, M_4, and M_5, F can therefore be expressed as

$$F(A, B, C) = M_0 \cdot M_2 \cdot M_4 \cdot M_5$$
$$= (A + B + C) \cdot (A + \bar{B} + C) \cdot (\bar{A} + B + C) \cdot (\bar{A} + B + \bar{C})$$

or in compact form as

$$F(A, B, C) = \prod M_i (0, \ 2, \ 4, \ 5)$$

where $\prod M_i(\ )$ means the product of all the maxterms whose subscript i is given inside the parentheses.

Now coming back to the K map, a K map is a diagram made up of cells (squares), one for each minterm of the function to be represented. An n-variable K map, representing an n-variable function, therefore has 2^n cells. Figures 6.1.12, 6.1.13, and 6.1.14 show the two-variable, three-variable, and four-variable K maps, respectively, in two different forms. Note particularly the code used in listing the row and column headings when more than one variable is needed.

The K map provides an immediate view of the values of the function in graphical form. Note that the arrangement of the cells in the K map is such that any two adjacent cells contain minterms that vary only in one variable. Consider the map to be continuously wrapping around itself, as if the top and bottom, and right and left, edges were touching each other. For an n-variable map, there will be n minterms adjacent to any given minterm. For example, in Figure 6.1.14, the cell corresponding to m_2 is adjacent to the cells corresponding to m_3, m_6, m_0, and m_{10}. It can be verified that any two adjacent cells contain minterms that vary only in one variable. The m_2 cell,

Figure 6.1.12 Two-variable K maps.

Figure 6.1.13 Three-variable K maps.

Figure 6.1.14 Four-variable K maps.

for example, has an assignment of $\bar{A} \cdot \bar{B} \cdot C \cdot \bar{D}$ ($= 0 \cdot 0 \cdot 1 \cdot 0$), whereas the m_3 cell has the assignment of $\bar{A} \cdot \bar{B} \cdot C \cdot D$ ($= 0 \cdot 0 \cdot 1 \cdot 1$), which differs only in the value of the variable D.

A function can be represented in a K map by simply entering 1s in the cells that correspond to the minterms of the function. This is best illustrated by an example.

EXAMPLE 6.1.9

Show the K-map representations of the following Boolean functions:

(a) $F(A, B, C) = \sum m_i (0, 2, 3, 5, 7)$

(b) $F(A, B, C, D) = \sum m_i (1, 3, 5, 6, 9, 10, 13, 14)$

Solution

Figure E6.1.9 shows the K-map representations of the functions.

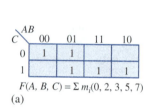

$F(A, B, C) = \Sigma\, m_i(0, 2, 3, 5, 7)$
(a)

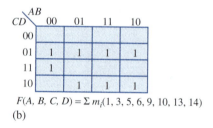

$F(A, B, C, D) = \Sigma\, m_i(1, 3, 5, 6, 9, 10, 13, 14)$
(b)

Figure E6.1.9

Simplification of an n-variable Boolean function by using a K map is achieved by grouping adjacent cells that contain 1s. The number of adjacent cells that may be grouped is always equal to 2^m, for $0 \leq m \leq n$, i.e., 1, 2, 4, 8, 16, 32, . . . cells. Such a group or set of 2^m cells is called a *subcube* and is expressed by the product term containing only the variables that are common to the adjacent cells. The larger the subcube is, the fewer variables are needed to express the product term. This statement can be justified by using Boolean identities. For example, in a four-variable K map, the following possible subcubes can be formed.

- 1-cell subcube (having one minterm) expressed by a product term containing four, or $n - m$, variables; note that $n = 4$ and $m = 0$ for this case.
- 2-cell subcube (having two minterms) expressed by a product term containing three, or $n - m$, variables; note that $n = 4$ and $m = 1$ for this case.
- 4-cell subcube (having four minterms) expressed by a product term containing two, or $n - m$, variables; note that $n = 4$ and $m = 2$ for this case.
- 8-cell subcube (having eight minterms) expressed by a product term containing one or $n - m$ variables; note that $n = 4$ and $m = 3$ for this case.
- 16-cell subcube (having sixteen minterms) expressed by the logic 1; that is, the function is always equal to 1.

Let us now present an example to illustrate the concepts related to subcubes.

EXAMPLE 6.1.10

Consider a Boolean function $F(A, B, C, D)$ defined by the K map of Figure E6.1.10(a). Form the possible subcubes and express them by their corresponding product terms.

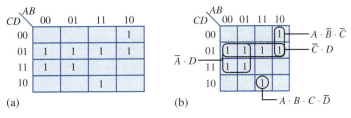

(a)   (b)

Figure E6.1.10 (a) K map. **(b)** K map with subcubes.

Solution

Figure E6.1.10(b) shows the K map of $F(A, B, C, D) = \sum m_i(1, 3, 5, 7, 8, 9, 13, 14)$ with subcubes.

Observe that the four-cell subcube consisting of minterms m_1, m_3, m_5, and m_7 can be expressed by the product term $(\bar{A} \cdot D)$, since $\bar{A}$ and D are the only variables common to the four minterms involved. The other subcubes shown in Figure E6.1.10(b) can be checked for their product-term expressions. Note that any cell may be included in as many subcubes as desired.

Once a Boolean function is represented in a K map and its different subcubes are formed, that Boolean function can be expressed as the logic SOP terms corresponding to the *minimum* set of subcubes that cover all its 1 cells. Obviously then, in forming a subcube, one should not select a subcube that is totally contained in another subcube. The product term representing the subcube containing the maximum possible number of adjacent 1 cells in the map is called a *prime implicant*. A prime implicant is known as an *essential prime implicant* if the subcube represented by the prime implicant contains at least one 1 cell that is not covered by any other subcube. If, on the other hand, *all* the 1 cells of a subcube of a prime implicant are covered by some other subcubes, then such a prime implicant is known as an *optional prime implicant*. The minimized Boolean expression is obtained by the logic sum (OR) of all the essential prime implicants and some other optional prime implicants that cover any remaining 1 cells that are not covered by the essential prime implicants.

The procedure for K-map simplification of an n-variable Boolean function may be summarized in the following steps.

1. Represent the function in an n-variable K map.

2. Mark all prime implicants that correspond to subcubes of the maximum adjacent 1 cells.

3. Determine the essential prime implicants.

4. Develop the minimum expression with the logic sum of the essential prime implicants.

5. In the K map, check the 1 cells that are covered by the subcubes expressed by the essential prime implicants.

6. No other terms are needed if all the 1 cells of the map are checked. Otherwise add to your expression the minimum number of optional prime implicants that correspond to the subcubes which include the unchecked 1 cells.

Let us now illustrate the application of this procedure in the following example.

EXAMPLE 6.1.11

Obtain a minimum Boolean expression for

$$F(A, B, C, D) = \sum m_i(1, 3, 4, 5, 6, 7, 10, 12)$$

Solution

1. The K map of F is shown in Figure E6.1.11(a).

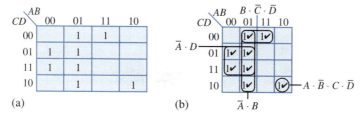

(a) (b)

Figure E6.1.11

2. The prime implicants, as depicted in Figure E6.1.11(b), are $(\bar{A} \cdot D)$, $(\bar{A} \cdot B)$, $(B \cdot \bar{C} \cdot \bar{D})$, and $(A \cdot \bar{B} \cdot C \cdot \bar{D})$.

3. All prime implicants happen to be essential, since each of the subcubes contains at least one 1 cell that is not covered by any other subcube.

4. $F(A, B, C, D) = \bar{A} \cdot D + \bar{A} \cdot B + B \cdot \bar{C} \cdot \bar{D} + A \cdot \bar{B} \cdot C \cdot \bar{D}$.

5. The 1 cells that are covered by the subcubes expressed by the essential prime implicants are checked, as shown in Figure E6.1.11(b).

6. Since all the 1 cells have been checked, it follows that

$$F(A, B, C, D) = \bar{A} \cdot D + \bar{A} \cdot B + B \cdot \bar{C} \cdot \bar{D} + A \cdot \bar{B} \cdot C \cdot \bar{D}$$

The minimum POS expression for a Boolean function can be obtained easily by using K maps in a procedure very similar to that used for the minimum SOP expression. By using the 0s of the map and obtaining prime implicants again by using the 0 cells, this can be done. Since the 0 cells represent the compliment of the function, complementing the minimum SOP expression for the 0s of the function yields the minimum POS form of the function. Even though the two forms (POS and SOP) are completely equivalent, logically, one of the two forms may lead to a realization involving a smaller number of gates.

In designing a digital system, one often encounters a situation where the output cannot be specified due to either physical or logical constraints. Figure 6.1.15 illustrates one such situation, where the input lines B and C to the black box of logic circuitry are physically connected together. For such a circuit, the input combination $A = 0$, $B = 1$, and $C = 0$ is not possible, and therefore the output corresponding to this combination cannot be specified. Such an output is hence known as a *don't-care output* because of a physical constraint. As another example, Figure 6.1.16 shows the logic circuitry in a black box with four input lines representing BCD data. For such a circuit, six of the input combinations (1010, 1011, 1100, 1101, 1110, and 1111) are not valid BCDs, and therefore the output of this circuitry cannot be specified because of a logic constraint. Such outputs are hence labeled as don't-care outputs. A don't-care output or condition is generally represented by the letter *d* in the truth table and the K map.

The *d*s can be treated either as 0s or 1s when the subcubes are formed in the K map, depending on which results in a greater simplification, thereby helping in the formation of the smallest number of maximal subcubes. Note that some or all of the *d*s can be treated as 0s or 1s. However, one should not form a subcube that contains only *d*s.

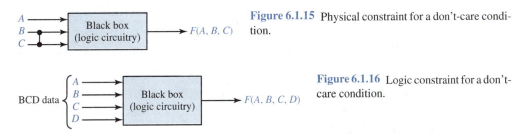

Figure 6.1.15 Physical constraint for a don't-care condition.

Figure 6.1.16 Logic constraint for a don't-care condition.

EXAMPLE 6.1.12

A Boolean function $F(A, B, C, D)$ is specified by the truth table of Figure E6.1.12 (a). Obtain: (a) a minimum SOP expression, and (b) a minimum POS expression.

Solution

The four-variable K map of F is shown in Figure E6.1.12(b).

(a) The prime implicants (with 1 cells) for F are marked in Figure E6.1.12(c). The minimum SOP expression is then given by

$$F = \bar{C} \cdot \bar{D} + B \cdot \bar{C}$$

Figure E6.1.12

A	B	C	D	F
0	0	0	0	1
0	0	0	1	0
0	0	1	0	0
0	0	1	1	0
0	1	0	0	1
0	1	0	1	1
0	1	1	0	0
0	1	1	1	0
1	0	0	0	1
1	0	0	1	0
1	0	1	0	d
1	0	1	1	d
1	1	0	0	d
1	1	0	1	d
1	1	1	0	d
1	1	1	1	d

(a)

CD \ AB	00	01	11	10
00	1	1	d	1
01	0	1	d	0
11	0	0	d	d
10	0	0	d	d

(b)

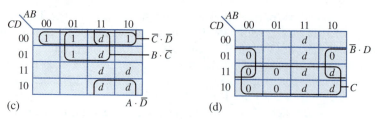

Figure E6.1.12 Continued

(b) The prime implicants (with 0 cells) for the complement of F are marked in Figure E6.1.12(d). The minimum POS expression is then given by

$$\bar{F} = C + \bar{B} \cdot D$$

Complementing both sides of this equation and using DeMorgan's rules, one obtains

$$F = \bar{C} \cdot (B + \bar{D})$$

Note in particular that in Figures E6.1.12(c) and (d) we did not form subcubes that covered only ds.

Sequential Blocks

Neglecting propagation delays, which are measures of how long it takes the output of a gate to respond to a transition at the input of the gate, the output of a logic block at a given time depends only on the inputs at that same time. The output of a *sequential* block, on the other hand, depends not only on the present inputs but also on inputs at earlier times. Sequential blocks have thus a kind of memory, and some of them are used as computer memories.

Most sequential blocks are of the kind known as *multivibrators*, which can be *monostable* (the switch remains in only one of its two positions), *bistable* (the switch will remain stable in either of its two positions), and *unstable* (the switch changes its position continuously as a kind of oscillator, being unstable in both of its two states). The most common sequential block is the *flip-flop*, which is a bistable circuit that remembers a single binary digit according to instructions. Flip-flops are the basic sequential building blocks. Various types of flip-flops exist, such as the SR flip-flop (SRFF), D flip-flop (or latch), and JK flip-flop (JKFF), which differ from one another in the way instructions for storing information are applied.

SR Flip-Flop (SRFF)

The symbol for the SRFF is shown in Figure 6.1.17(a), in which S stands for "set," R stands for "reset" on the input side, and there are two outputs, the normal output Q and the complementary output $\bar{Q}$. The operation of the SRFF can be understood by the following four basic rules.

1. If $S = 1$ and $R = 0$, then $Q = 1$ regardless of past history. This is known as the *set* condition.

2. If $S = 0$ and $R = 1$, then $Q = 0$ regardless of past history. This is known as the *reset* condition.

3. If $S = 0$ and $R = 0$, then Q does not change and stays at its previous value. This is a *highly stable* input condition.

4. The inputs $S = 1$ and $R = 1$ are not allowed (i.e., forbidden) because $Q\bar{Q} = 11$; $\bar{Q}$ is no longer complementary to Q. This is an unacceptable output state. Such a meaningless

instruction should not be used. Figure 6.1.17(b) summarizes the specification for an SRFF in terms of a truth table, in which Q_n is the state of the circuit before a clock pulse and Q_{n+1} is the state of the circuit following a clock pulse.

EXAMPLE 6.1.13

The inputs to an SRFF are shown in Figure E6.1.13. Determine the value of Q at times t_1, t_2, and t_3.

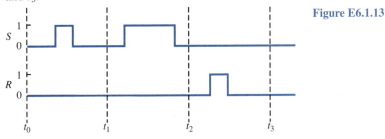

Figure E6.1.13

Solution

Notice that the value of Q at time t_0 is not given; however, it is not necessary to have this information. The first pulse of S sets the SRFF in the state $Q = 1$. Thus at $t = t_1$, $Q = 1$. While the second pulse of S tries again to set the SRFF, there will be no change since Q was already 1. Thus at $t = t_2$, $Q = 1$. The pulse of R then resets the SRFF and then at $t = t_3$, $Q = 0$.

Flip-flops can be constructed using combinations of logic blocks. The *realization* of an SRFF can be achieved from two NAND gates (plus two inverters), as shown in Figure 6.1.18. Because of the feedback in this circuit it is not possible to simply write its truth table, as we can do for a combinational circuit. However, a modified truth-table method can be used, in which we guess

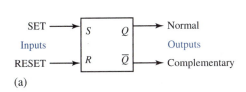

Inputs		Outputs	
S	R	Q_{n+1}	$\overline{Q}_{n+1}$
0	0	Q_n	$\overline{Q}_n$
0	1	0	1
1	0	1	0
1	1	Not allowed	

(a) (b)

Figure 6.1.17 SR flip-flop (SRFF) **(a)** Symbol. **(b)** Truth table.

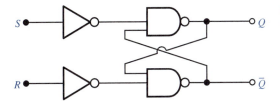

Figure 6.1.18 Realization of SRFF from two NAND gates.

an output and then go back to check it for self-consistency. This is given as a problem at the end of the chapter. Figure 6.1.18 is a realization of an SRFF in terms of a physical circuit that behaves similarly to the ideal device called SRFF which obeys the *SR* instruction rules. Thus, note that the circuit of Figure 6.1.18 cannot accurately be said to be an SRFF. While it is certainly possible to apply the "forbidden" input ($S = 1$ and $R = 1$) to the real circuit, the circuit is then not properly used as an SRFF.

Edge triggered SRFF symbols are illustrated in Figure 6.1.19. The triangle on the Ck (clock) input indicates that the flip-flop is triggered on the edge of a clock pulse. Positive-edge triggering is shown in Figure 6.1.19(a). Negative-edge triggering is indicated by the ring on the Ck input in Figure 6.1.19(b).

D FLIP-FLOP (LATCH OR DELAY ELEMENT)

The symbol for the clocked D flip-flop is shown in Figure 6.1.20(a), in which the two output terminals Q and $\bar{Q}$ behave just as in the SRFF, and the input terminals are D and Ck (clock). The term *clocked* flip-flop indicates that this device cannot change its state (i.e., Q cannot change) unless a specific change instruction is given through the clock (Ck) input. The value of Q after the change instruction is equal to the value of D at the time the change instruction is received. The value of Q before the change instruction does not matter. Figure 6.1.20(b) illustrates the values taken by Q after the change instruction for various inputs D and prior values of Q. While there are several variations of the device, in the rising-edge triggered flip-flop a change instruction is effected whenever the Ck input makes a change from 0 to 1. Note that only a positive-going transition of Ck is a change instruction, and a constant Ck input is not a change instruction.

Note that flip-flops have *propagation delay*, which means that there is a small delay (about 20 ns) between the change instruction and the time Q actually changes. The value of D that matters is its value when the change instruction is received, not its value at the later time when Q changes.

A convenient means of describing the series of transitions that occur as the signals set to the flip-flop input change is the *timing diagram*. A timing diagram depicts inputs and outputs (as a function of time) of the flip-flop (or any other logic device) showing the transitions that occur over time. The timing diagram thus provides a convenient visual representation of the evolution of the state of the flip-flop. However, the transitions can also be represented in tabular form.

Like logic blocks, flip-flops appear almost exclusively in IC form, and are more likely to be found in LSI and VLSI form. A very important application is in computer memories, in which a typical 256k RAM (random access memory) consists of about 256,000 flip-flops in a single IC.

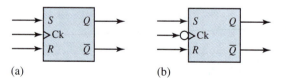

(a)　　　　　　(b)

Figure 6.1.19 Edge triggered SRFF symbols. **(a)** Triggered on positive edge of clock pulse. **(b)** Triggered on negative edge of clock pulse.

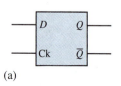

(a)

D	Q(before)	Q(after)
0	0	0
0	1	0
1	0	1
1	1	1

(b)

Figure 6.1.20 Clocked D flip-flop. **(a)** Symbol. **(b)** Q for various inputs D.

EXAMPLE 6.1.14

The positive-edge triggered D flip-flop is given the inputs shown in Figure E6.1.14(a), with a zero initial value of Q. Draw the timing diagram.

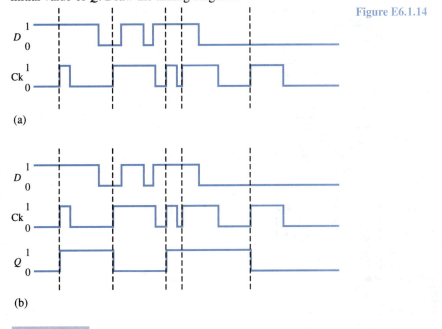

Figure E6.1.14

(a)

(b)

Solution

Q is illustrated in Figure E6.1.14(b) as a function of time.

In order to avoid the input condition $SR = 11$, an inverter is included, as shown in Figure 6.1.21. SR can be either 01 or 10. For $SR = 10$, the flip-flop *sets* (i.e., $Q\bar{Q} = 10$) whereas for $SR = 01$ it *resets* (i.e., $Q\bar{Q} = 01$). Thus, the output replicates the input state, but with a delay equal to the time for information to propagate through the flip-flop.

Most commercially available flip-flops also include two extra control input signals known as *preset* and *clear*. When activated, these input signals will set ($Q = 1$) or clear ($Q = 0$) the flip-flop, regardless of other input signals. Figure 6.1.22(a) shows the block diagram of a positive-edge triggered D flip-flop with preset and clear control signals; Figure 6.1.22(b) gives the functional truth table for this device. Note that the clock is the synchronizing element in a digital system. The particular device described here is said to be positive-edge triggered, or leading-edge triggered, since the final output of the flip-flop is set on a positive-going clock transition.

Figure 6.1.23 shows a realization of the D flip-flop using six NAND gates, in which the output part is very similar to the SRFF of Figure 6.1.18 and the other four gates are used to translate the D-type control instructions into a form that the SRFF can use.

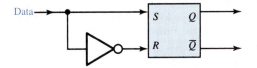

Figure 6.1.21 Flip-flop with inverter.

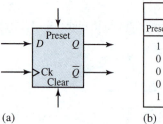

Inputs				Outputs	
Preset	Clear	Ck	D	Q	$\bar{Q}$
1	0	d	d	1	0
0	1	d	d	0	1
0	0	$\int$	0	0	1
0	0	$\int$	1	1	0
1	1	d	d	Not allowed	

Figure 6.1.22 Positive-edge triggered D flip-flop with preset and clear. **(a)** Block diagram. **(b)** Truth table. NOTE: d stands for don't-care condition. The symbol $\int$ indicates the occurance of a positive transition in the clock timing. The symbol > (knife edge) drawn next to the clock input in the block diagram indicates edge-triggered feature.

(a)

(b)

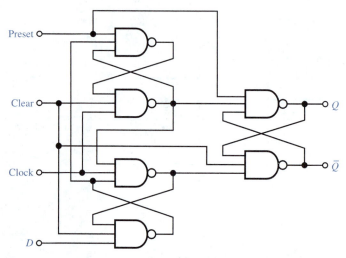

Figure 6.1.23 Realization of D flip-flop with six NAND gates.

JK FLIP-FLOP (JKFF)

The block diagram, truth table, and a practical realization of the JKFF are shown in Figure 6.1.24. The JKFF differs from an SRFF in that output Q is fed back to the K-gate input and $\bar{Q}$ to the J-gate input. Assuming $Q\bar{Q} = 01$, gate B is disabled by $Q = 0$ (i.e., $F = 1$). The only way to make the circuit change over is for gate A to be enabled by making $J = 1$ and $\bar{Q} = 1$ (which it is already). Then when Ck $= 1$, all inputs to gate A are 1 and E goes to zero, which makes $Q = 1$. With Q and F both equal to 1, $\bar{Q} = 0$, so the flip-flop has changed state. Note that the input condition $JK = 11$ is allowed, and in this condition, when the flip-flop is clocked, the output always changes state; thus it is said to *toggle*.

If the clock pulse is short enough to permit the flip-flop to change only once, the JKFF operates well. However, with modern high-speed ICs a *race* is more likely to occur, which is a condition in which two pulses are intended to arrive at a destination gate in some specific order, but due to each one racing through different paths in the logic with a different number of gates, the propagation delays stack up differently and the timing order is lost. This can be eliminated by introducing delays in the feedback paths between outputs (Q and $\bar{Q}$) and inputs (J and K). A better solution to the problem is the master–slave JKFF.

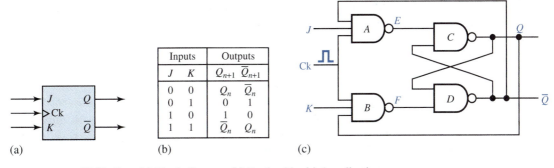

Inputs		Outputs	
J	K	Q_{n+1}	$\bar{Q}_{n+1}$
0	0	Q_n	$\bar{Q}_n$
0	1	0	1
1	0	1	0
1	1	$\bar{Q}_n$	Q_n

(a)　　　　　　　(b)　　　　　　　(c)

Figure 6.1.24 JK flip-flop. (**a**) Block diagram. (**b**) Truth table. (**c**) A realization.

MASTER–SLAVE JKFF

Figure 6.1.25 illustrates a master–slave JKFF, in which gates A, B, C, and D form the *master* flip-flop and T, U, V, and W form the *slave*. The output of the master–slave JKFF can be predicted for all combinations of J and K and for any duration of clock pulse. Thus it is the most versatile and universal type of flip-flop. SRFFs are also available in master–slave configuration.

Ck = 1 enables the master; $\bar{Ck}$ = 0 disables the slave. Let $Q\bar{Q} = 10$ and $JK = 11$ before the occurrence of a clock pulse. B is enabled by $Q = 1$ so that when the clock pulse arrives (i.e., Ck = 1), F goes to zero and $\bar{Q}_M$ to 1. Now with $E = 1$ and $\bar{Q}_M = 1$, Q_M goes to zero so that the master has been reset. However, the slave remains disabled until Ck goes to zero. Note that the slave, at this stage, is essentially an SRFF with inputs S and R equal to Q_M and $\bar{Q}_M$, respectively. Thus, when Ck goes to zero, $\bar{Ck}$ goes to 1 and the slave is now reset by its inputs $Q_M\bar{Q}_M = 01$. But the feedback to J and K cannot cause a race because, with Ck = 0, the master is disabled.

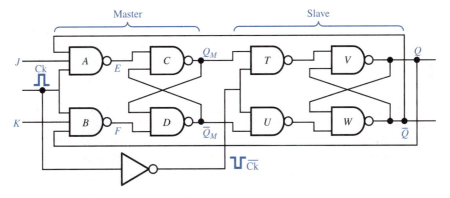

Figure 6.1.25 Master–slave JK flip-flop.

6.2 DIGITAL SYSTEM COMPONENTS

The basic combinational and sequential building blocks were introduced in Section 6.1. The reason that digital systems are so inexpensive and yet so powerful is that they consist of very large numbers of just a few building blocks, repeated in simple ways. In this section we shall deal with some common digital system components, such as decoders, encoders, multiplexers, registers, counters, digital-to-analog (D/A) and analog-to-digital (A/D) converters, memory, and display devices.

Decoders

An n-bit binary code is capable of encoding up to 2^n distinct elements of information. A decoder is a combinational network that decodes (converts) the n-bit binary-coded input to m outputs ($m \leq 2^n$). The block diagram of a 3-bit to 8-element decoder is shown in Figure 6.2.1(a), wherein the three inputs are decoded into eight outputs, one for each combination of the input variables. In the truth table shown in Figure 6.2.1(b), observe that for each input combination, there is only one output that is equal to 1 (i.e., each combination selects only one of the eight outputs). The logic diagram of the 3-to-8 decoder is shown in Figure 6.2.1(c).

Decoding is so common in digital design that decoders are commercially available as MSI (medium-scale integration) packages in the form of 2-to-4, 3-to-8, and 4-to-10 decoders. Integrated circuits for decoders are available in different forms.

Encoders

Encoding is the process of forming an encoded representation of a set of inputs, and it is the converse of the decoding operation. An encoder is a combinational network that generates an n-bit binary code that uniquely identifies the one out of m activated inputs ($0 \leq m \leq 2^n - 1$).

Figure 6.2.2(a) shows the block diagram of an 8-element to 3-bit encoder. The truth table is given in Figure 6.2.2(b). Notice that only one of the eight inputs is allowed to be activated at any given time. The logic diagram for the 8-to-3 encoder is shown in Figure 6.2.2(c).

Multiplexers

A *multiplexer* is a data selector, whereas a *demultiplexer* is a data distributor. A multiplexer is a combinational network that selects one of several possible input signals and directs that signal to a single output terminal. The selection of a particular input is controlled by a set of selection variables. A multiplexer with n selection variables can usually select one out of 2^n input signals.

Figure 6.2.3(a) shows the block diagram of a 4-to-1 multiplexer. The truth table is given in Figure 6.2.3(b). Notice that each of the four inputs (I_0, I_1, I_2, and I_3) is selected by S_1 and S_0, and directed to the output Q. In general, only the input whose *address* is given by the select lines is directed to the output. The logic diagram is shown in Figure 6.2.3(c). 2-, 4-, 8-, and 16-to-1 multiplexers are commercially available as MSI packages.

Registers

A register is a collection of flip flops (and some basic combinational gates to perform different binary arithmetic and logic operations), where each flip-flop is used to store 1 bit of information. Figure 6.2.4(a) shows the block diagram of a 4-bit *shift-right register* that uses D flip-flops. JKFFs and SRFFs are also used in shift-register construction. Observe in the timing diagram of Figure 6.2.4(b) that each successive clock pulse transfers (or shifts) the data bit from one flip-flop to the

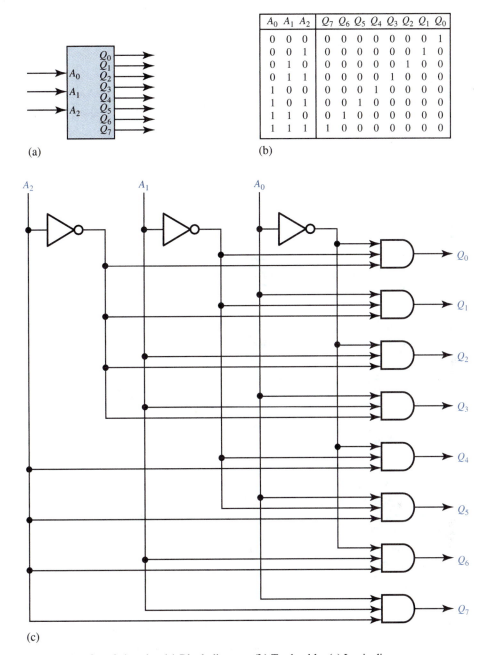

A_0	A_1	A_2	Q_7	Q_6	Q_5	Q_4	Q_3	Q_2	Q_1	Q_0
0	0	0	0	0	0	0	0	0	0	1
0	0	1	0	0	0	0	0	0	1	0
0	1	0	0	0	0	0	0	1	0	0
0	1	1	0	0	0	0	1	0	0	0
1	0	0	0	0	0	1	0	0	0	0
1	0	1	0	0	1	0	0	0	0	0
1	1	0	0	1	0	0	0	0	0	0
1	1	1	1	0	0	0	0	0	0	0

(a)   (b)

(c)

Figure 6.2.1 3-to-8 decoder. **(a)** Block diagram. **(b)** Truth table. **(c)** Logic diagram.

next one on the right. The waveforms also reveal that the data are read into the register *serially* and appear at the output in serial form. The shift register is then known as a *serial-in serial-out* (SISO) register.

PISO (parallel-in serial-out), SIPO (serial-in parallel-out), and PIPO (parallel-in parallel-out) registers are also often used to read in the input data and read out the output data in a

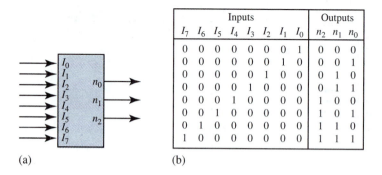

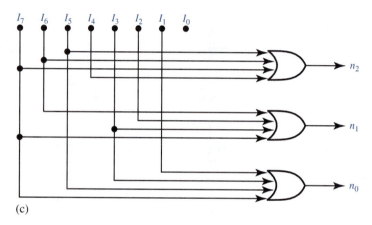

(c)

Figure 6.2.2 8-to-3 encoder. **(a)** Block diagram. **(b)** Truth table. **(c)** Logic diagram.

convenient way that is needed for the operations involved. Right-shifting registers are employed in multiplication algorithms, whereas left-shifting registers are utilized in division algorithms. Registers that are capable of shifting the data to the left or right are known as *bidirectional shift registers*. The register along with additional gates on a single chip forms an IC component known as the *universal register*, which usually includes the shift-left, shift-right, parallel-input, and no-change operations.

Counters

The shift register can be used as a counter because the data are shifted for each clock pulse. A counter is a register that goes through a predetermined sequence of states when input pulses are received. Besides, computers, timers, frequency meters, and various other digital devices contain counters for counting events.

There are what are known as *ripple (asynchronous)*, *synchronous*, and *ring* counters. In ripple counters, the output of each flip-flop activates the next flip-flop throughout the entire sequence of the counter's states. In a synchronous counter, on the other hand, all flip-flops are activated (triggered) simultaneously through a master clock connected to the clock inputs of all flip-flops. In a ring counter, as in a synchronous counter, all flip-flops are triggered

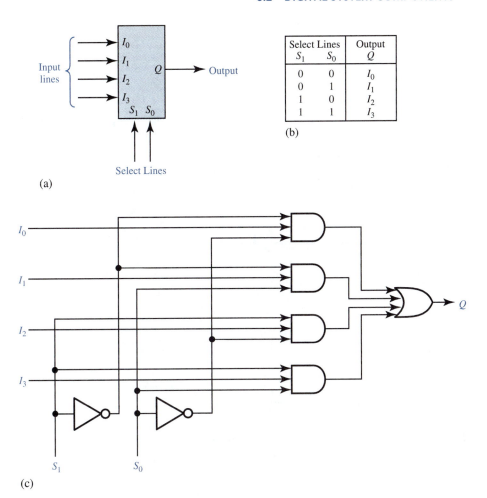

| Select Lines | | Output |
S_1	S_0	Q
0	0	I_0
0	1	I_1
1	0	I_2
1	1	I_3

(b)

(a)

(c)

Figure 6.2.3 4-to-1 multiplexer. **(a)** Block diagram. **(b)** Truth table. **(c)** Logic diagram.

simultaneously. However, the output of each flip-flop drives only an adjacent flip-flop. A single pulse propagates through the ring in a ring counter, whereas all remaining flip-flops are at the zero state.

Figure 6.2.5(a) shows a block diagram of a 3-bit ripple counter using JKFFs. Notice from the timing diagram shown in Figure 6.2.5(b) that the output Q_0 of the leftmost flip-flop will change its state at every clock pulse if the clear signal equals zero. The output Q_1, controlled by Q_0, will change its state every time Q_0 changes from 0 to 1. Similarly Q_2 is controlled by Q_1. Figure 6.2.5(c) shows the outputs for the first 8 clock pulses. Observe that a 3-bit counter will cycle through 8 states, 000 through 111. An n-bit ripple counter, in general, will cycle through 2^n states; it is known as a *divide-by-2^n counter* or *modulo-2^n binary counter.* Taking the outputs from $Q_2 \, Q_1 \, Q_0$, the counter becomes an *up-counter;* taking the outputs from $\bar{Q}_2 \, \bar{Q}_1 \, \bar{Q}_0$, the counter becomes a *down-counter,* which counts down from a preset number. Asynchronous ripple counters are available as MSI packages.

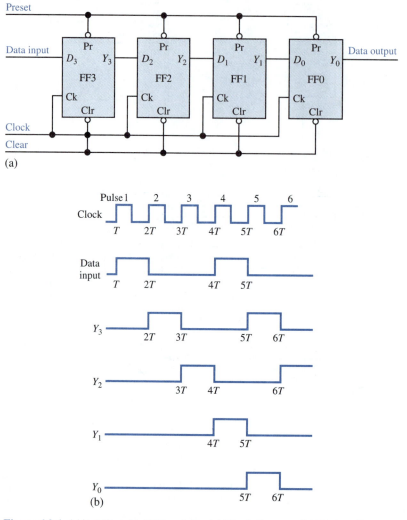

Figure 6.2.4 4-bit shift-right SISO register. (a) Block diagram. (b) Timing diagram.

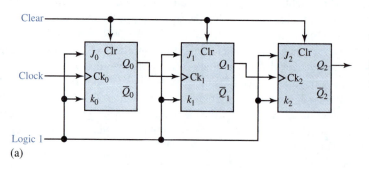

Figure 6.2.5 3-bit ripple counter. (a) Block diagram. (b) Timing diagram. (c) Outputs for the first 8 clock pulses.

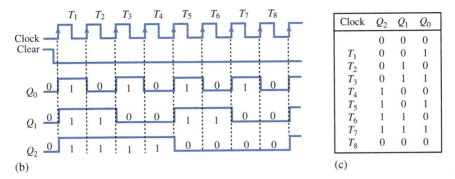

(b)

Clock	Q_2	Q_1	Q_0
	0	0	0
T_1	0	0	1
T_2	0	1	0
T_3	0	1	1
T_4	1	0	0
T_5	1	0	1
T_6	1	1	0
T_7	1	1	1
T_8	0	0	0

(c)

Figure 6.2.5 Continued

The slow speed of operation, caused by the long time required for changes in state to ripple through the flip-flops, is a disadvantage of ripple counters. This problem is overcome by using synchronous converters. However, additional control logic is needed to determine which flip-flops, if any, must change state, since flip-flops are triggered simultaneously. Figure 6.2.6 shows the logic diagram of a 3-bit binary synchronous converter using JKFFs. Synchronous counters can be designed to cycle through any sequence of states. Various n-bit synchronous converters are commercially available as MSI packages. Some counters are also programmable.

Figure 6.2.7(a) shows a 4-bit (modulo-4) ring counter using D flip-flops; its timing diagram is given in Figure 6.2.7(b). A modulo-n ring counter requires N flip-flops and no other gates, whereas modulo-N ripple and synchronous counters need only $\log_2 N$ flip-flops. However, ripple and synchronous counters generally use more components than ring counters.

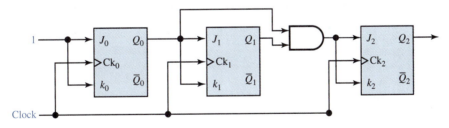

Figure 6.2.6 3-bit binary synchronous converter using JKFFs.

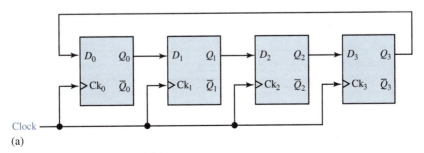

(a)

Figure 6.2.7 4-bit ring counter using D flip-flops. (a) Block diagram. (b) Timing diagram.

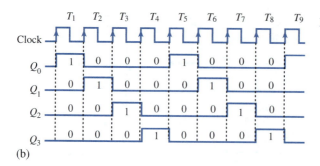

Figure 6.2.7 Continued

(b)

EXAMPLE 6.2.1

A table of minterms for three variables is as follows:

A	B	C	i	Minterm m_i
0	0	0	0	$\bar{A} \cdot \bar{B} \cdot \bar{C}$
0	0	1	1	$\bar{A} \cdot \bar{B} \cdot C$
0	1	0	2	$\bar{A} \cdot B \cdot \bar{C}$
0	1	1	3	$\bar{A} \cdot B \cdot C$
1	0	0	4	$A \cdot \bar{B} \cdot \bar{C}$
1	0	1	5	$A \cdot \bar{B} \cdot C$
1	1	0	6	$A \cdot B \cdot \bar{C}$
1	1	1	7	$A \cdot B \cdot C$

Implement the following Boolean functions by using one 3-to-8 decoder and three three-input OR gates:

$$F_1(A, B, C) = \sum m_i(1, 2, 3) = \bar{A} \cdot \bar{B} \cdot C + \bar{A} \cdot B \cdot \bar{C} + \bar{A} \cdot B \cdot C$$

$$F_2(A, B, C) = \sum m_i(2, 4, 6) = \bar{A} \cdot B \cdot \bar{C} + A \cdot \bar{B} \cdot \bar{C} + A \cdot B \cdot \bar{C}$$

$$F_3(A, B, C) = \sum m_i(3, 5, 7) = \bar{A} \cdot B \cdot C + A \cdot \bar{B} \cdot C + A \cdot B \cdot C$$

Solution

The implementation is shown in Figure E6.2.1.

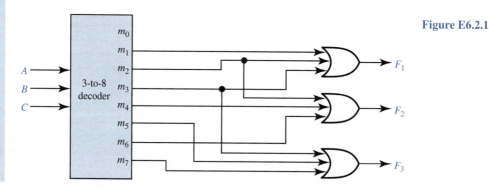

Figure E6.2.1

EXAMPLE 6.2.2

Show the logic diagram of an 8-to-1 multiplexer.

Solution

The logic diagram is depicted in Figure E6.2.2.

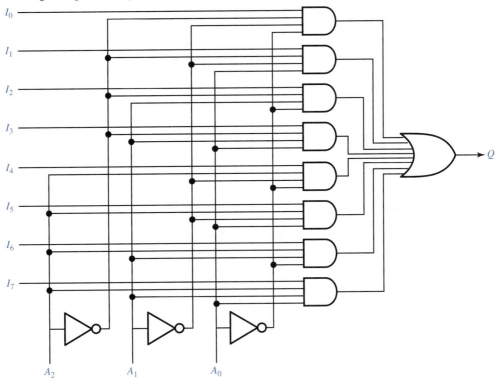

Figure E6.2.2

EXAMPLE 6.2.3

Given the block diagram for a 4-bit shift-left register shown in Figure E6.2.3(a), draw the output (Q_0, Q_1, Q_2, Q_3, and data out) as a function of time for the clock, clear, and data-in signals given in Figure E6.2.3(b).

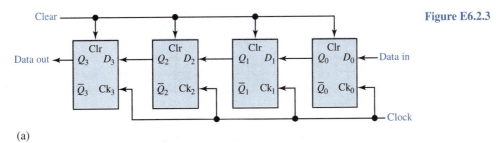

Figure E6.2.3

(a)

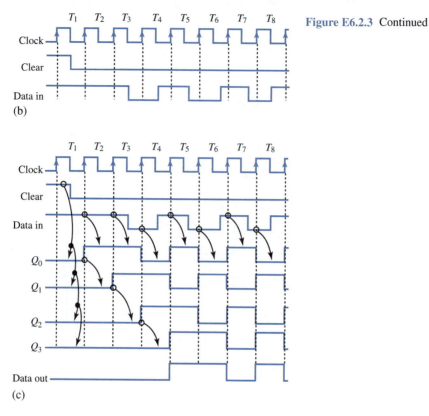

Figure E6.2.3 Continued

(b)

(c)

Solution

The output is shown in Figure E6.2.3(c).

EXAMPLE 6.2.4

The block diagram for a 3-bit ripple counter is shown in Figure E6.2.4(a). Obtain a *state table* for the number of pulses $N = 0$ to 8, and draw a *state diagram* to explain its operation.

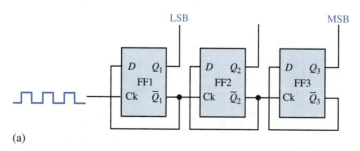

(a)

Figure E6.2.4 (a) Block diagram. (b) State table. (c) State diagram.

N	Q_{1N}	Q_{2N}	Q_{3N}	$(Q_3 Q_2 Q_1)_N$
0	0	0	0	000
1	1	0	0	001
2	0 ⟶ 1		0	010
3	1	1	0	011
4	0 ⟶ 0 ⟶ 1			100
5	1	0	1	101
6	0 ⟶ 1		1	110
7	1	1	1	111
8	0 ⟶ 0 ⟶ 0			000

Figure E6.2.4 Continued

(b)

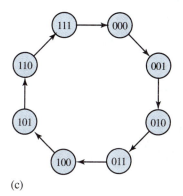

(c)

Solution

The state table and the state diagram are given in Figures E6.2.4(b) and (c). The horizontal arrows indicate the times when clock inputs are applied to FF2 and FF3. These times are located by noting that every time Q_1 makes a transition from 1 to 0, FF2 is clocked, and when Q_2 goes from 1 to 0, FF3 is clocked.

In the state diagram, the eight states of the system are indicated by the values of the three-digit binary number $Q_3 \, Q_2 \, Q_1$.

EXAMPLE 6.2.5

Given the block diagram of a synchronous counter shown in Figure E6.2.5(a), draw the timing diagram for the first input pulses, with Q_1, Q_2, and Q_3 initially at 0.

Solution

The timing diagram is shown in Figure E6.2.5(b). Corresponding to the rising transitions of the input, FF1 operates such that Q_1 can be drawn as a function of time, as shown in Figure E6.2.5(b). Because of the AND gate, FF2 can receive change instructions only when $Q_1 = 1$. Note that FF2 does not receive the first change instruction, but it does receive the second one. At the time of the second change instruction, Q_1 is still 1; it changes to 0 shortly afterward, but by that time FF2 has already been triggered. Similarly, FF3 is triggered only when Q_1 and Q_2 are both equal to 1. Thus, from the timing diagram, the successive states of $Q_3\ Q_2\ Q_1$ can be seen as 000, 001, 010, 011, 100, 101, 110, 111, 000, 001, . . . , as required for a counter.

Note that in a synchronous counter, all the flip-flops change at the same time (unlike in a ripple counter). The total delay is then the same as the propagation delay of a single flip-flop.

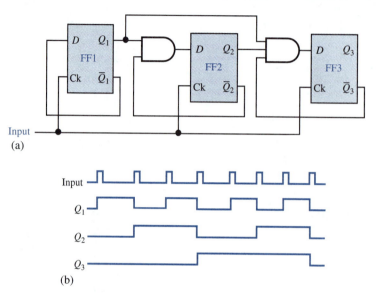

(a)

(b)

Figure E6.2.5 (a) Block diagram. **(b)** Timing diagram.

Digital-to-Analog (D/A) Converters

For the results of digital computations to be used in the analog world, it becomes necessary to convert the digital values to proportional analog values. Figure 6.2.8 shows the block diagram of a typical digital-to-analog (D/A) converter, which accepts an n-bit parallel digital code as input and provides an analog current or voltage as output. For an ideal D/A converter, the analog output for an n-bit binary code is given by

$$V_o = -V_{\text{ref}}(b_0 + b_1 \times 2^{-1} + b_2 \times 2^{-2} + \cdots + b_{n-1} \times 2^{-n+1}) \tag{6.2.1}$$

where

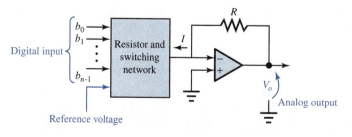

Figure 6.2.8 Block diagram of typical D/A converter.

V_0	analog output voltage
V_{ref}	reference analog input voltage
b_0	most significant bit of binary input code
b_{n-1}	least significant bit of binary input code

In order to provide current-to-voltage conversion and/or buffering, an op amp is used at the output. However, in some high-speed applications where a limited output voltage range is acceptable, a resistor, instead of an op amp, is used for the current-to-voltage conversion, thereby eliminating the delay associated with the op amp.

WEIGHTED-RESISTOR D/A CONVERTER

Figure 6.2.9 shows a 4-bit weighted-resistor D/A converter which includes a reference voltage source, a set of four electronically controlled switches, a set of four binary-weighted precision resistors, and an op amp. Each binary bit of digital input code controls its own switch. The switch closes with a bit value of 1, and the switch stays open with binary 0. The resistor connected to the most significant bit (MSB), b_0, has a value of R; b_1 is connected to $2R$, b_2 to $4R$, and b_3 to $8R$. Thus, each low-order bit is connected to a resistor that is higher by a factor of 2. For a 4-bit D/A converter, the binary input range is from 0000 to 1111.

An important design parameter of a D/A converter is the *resolution*, which is the smallest output voltage change, ΔV, which for an n-bit D/A converter is given by

$$\Delta V = \frac{V_{ref}}{2^{n-1}} \tag{6.2.2}$$

The range of resistor values becomes impractical for binary words longer than 4 bits. Also, the dynamic range of the op amp limits the selection of resistance values. To overcome these limitations, the R–$2R$ ladder D/A converter is developed.

R–2R LADDER D/A CONVERTER

Figure 6.2.10 shows a 4-bit R–$2R$ ladder D/A converter, which contains a reference voltage source, a set of four switches, two resistors per bit, and an op amp. The analog output voltage can be shown to be

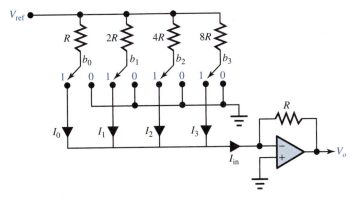

Figure 6.2.9 4-bit weighted-resistor D/A converter.

$$V_o = -V_{\text{ref}}(b_0 \times 2^{-1} + b_1 \times 2^{-2} + \cdots + b_{n-1} \times 2^{-n}) \qquad (6.2.3)$$

Because only two resistor values (R and $2R$) are used, the R–$2R$ ladder converter networks are relatively simple to manufacture, fast, practical, and reliable. The commercially available AD558, which is an 8-bit R-$2R$ D/A converter, is an example.

2^n-R D/A Converter

An n-bit 2^n–R D/A converter needs 2^n resistors of equal value R and ($2^{n+1} - 2$) analog switches. A 3-bit 2^n – R D/A converter is shown in Figure 6.2.11, which includes the eight resistors connected in series to form a voltage divider providing eight analog voltage levels, as well as 14 analog switches controlled by the digital input code such that each code creates a single path from the voltage divider to the converter output. A unit-gain amplifier is connected to the output in order to prevent loading of the voltage divider. 2^n–R D/A converters are economically manufactured as LSI packages in spite of the large number of components needed.

Analog-to-Digital (A/D) Converters

An A/D converter converts analog input signals into digital output data in many areas such as process control, aircraft control, and telemetry. Being the interface between analog systems and digital systems, it plays a key role in many industrial, commercial, and military systems. Several types of A/D converters exist: *counter-controlled, successive-approximation,* and *dual-ramp (dual-slope) converters.* One should understand its basic operation because analog comparators form the basis of A/D converters. Figure 6.2.12 shows the block diagram of an analog comparator. The commercially available LM311 is an example that is widely used by designers.

Counter-Controlled A/D Converter

Figure 6.2.13 shows the block diagram of a counter-controlled A/D converter. Resetting the binary counter to zero produces D/A output voltage $V_2 = 0$ and initiates the analog-to-digital conversion. When the analog input V_1 is larger than the DAC (D/A converter) output voltage, the comparator output will be high, thereby enabling the AND gate and incrementing the counter. V_2 is increased

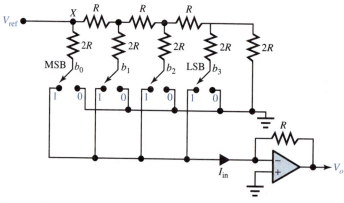

Figure 6.2.10 4-bit R–$2R$ ladder D/A converter.

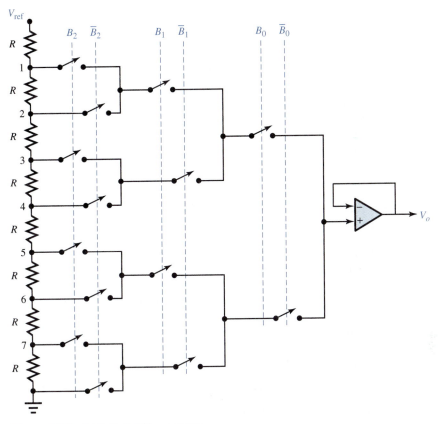

Figure 6.2.11 3-bit 2^n–R D/A converter.

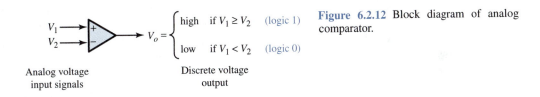

$$V_o = \begin{cases} \text{high} & \text{if } V_1 \geq V_2 \quad \text{(logic 1)} \\ \\ \text{low} & \text{if } V_1 < V_2 \quad \text{(logic 0)} \end{cases}$$

Figure 6.2.12 Block diagram of analog comparator.

Analog voltage input signals

Discrete voltage output

as the counter gets incremented; when V_2 is slightly greater than the analog input signal, the comparator signal becomes low, thereby causing the AND gate to stop the counter. The counter output at this point becomes the digital representation of the analog input signal. The relatively long conversion time needed to encode the analog input signal is the major disadvantage of this method.

SUCCESSIVE-APPROXIMATION A/D CONVERTER

This converter, shown in Figure 6.2.14, also contains a D/A converter, but the binary counter is replaced by a successive-approximation register (SAR), which makes the analog-to-digital conversion much faster. The SAR sets the MSB to 1 and all other bits to 0, after a start-of-conversion pulse. If the comparator indicates the D/A converter output to be larger than the signal

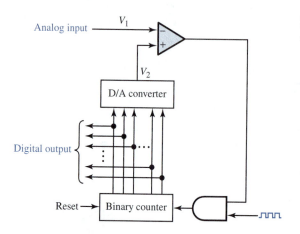

Figure 6.2.13 Block diagram of counter-controlled A/D converter.

to be converted, then the MSB is reset to 0 and the next bit is tried as the MSB. On the other hand, if the signal to be converted is larger than the D/A computer output, then the MSB remains 1. This procedure is repeated for each bit until the binary equivalent of the input analog signal is obtained at the end. This method requires only n clock periods, compared to the 2^n clock periods needed by the counter-controlled A/D converter, where n is the number of bits required to encode the analog signal. The National ADC 0844 is a popular 8-bit A/D converter based on the SAR.

DUAL-RAMP (DUAL-SLOPE) A/D CONVERTER

Figure 6.2.15(a) shows the block diagram of a dual-ramp (dual-slope) A/D converter. After a start-of-conversion pulse, the counter is cleared and the analog input V_{in} becomes the input of the ramp generator (integrator). When the output of the ramp generator V_o reaches zero, the counter starts to count. After a fixed amount of time T, as shown in Figure 6.2.15(b), the output of the ramp generator is proportional to the analog input signal. At the end of T, the reference voltage V_{ref} is selected, when the integrator gives out a ramp with a positive slope. As V_o increases, the counter is incremented until V_o reaches the comparator threshold voltage of 0 V, when the counter

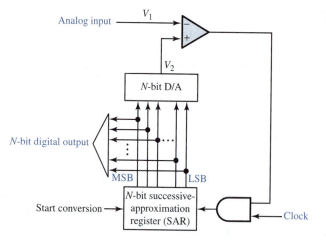

Figure 6.2.14 Block diagram of successive-approximation A/D converter.

stops being incremented again. The value of the counter becomes the binary code for the analog voltage V_{in}, since the number of clock pulses passing through the control logic gate for a time t is proportional to the analog signal V_{in}. Dual-ramp A/D converters can provide accuracy at low cost, even though the process is slow because a double clock pulse count is an inherent part of the process.

Memory

For a digital computer which stores both programs and data, memory can be divided into three types: *random-access memory, mass storage,* and *archival storage.* Random-access memory includes read-and-write memory (RAM), read-only memory (ROM), programmable read-only memory (PROM), and erasable programmable read-only memory (EPROM), in which any memory location can be accessed in about the same time. The time required to access data in a mass-storage device is relative to its location in the device. Mass storage, such as magnetic-disk memory, has a relatively large storage capacity and is lower in cost per bit than random-access memory. Archival storage, such as magnetic tape, is long-term storage with a very large capacity, but with a very slow access time, and may need user intervention for access by the system.

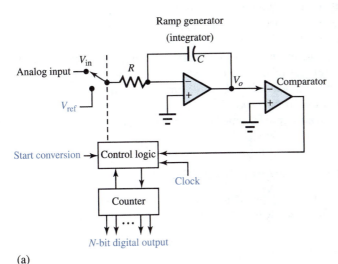

Figure 6.2.15 (a) Block diagram of dual-ramp (dual-slope) A/D converter. **(b)** Output of ramp generator in dual-ramp A/D converter.

(a)

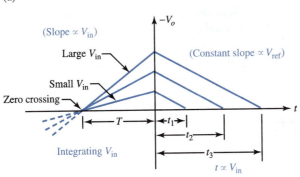

(b)

READ-AND-WRITE MEMORY (RAM)

Writing is the same as storing data into memory and reading is the same as retrieving the data later. RAM is said to be *volatile* because its contents are retained only as long as power is present. A RAM device is a collection of 2^n addressable storage locations, each of which contains k bits. Its block diagram is shown in Figure 6.2.16, in which each cell may be a flip-flop or a capacitor, and n address lines are decoded to select k cells. A *static* RAM, in which each cell is a flip-flop, is the read-and-write memory that retains its data so long as the power is applied, without any further action needed from the computer. Static RAM is used in microprocessor-based systems requiring small memory. Common static-RAM sizes are 2K × 8, 8K × 8, and 32K × 8, where K stands for $2^{10} = 1024$. A *dynamic* RAM, in which each cell is a capacitor (which leaks charges and therefore requires continuous refreshing from the computer to maintain its value), is the read-and-write memory that is used in large memory systems due to its lower cost and greater density. Common dynamic-RAM packages are available in 16K-bit, 64K-bit, 256K-bit, and 1M-bit (where M stands for mega = 2^{20}) sizes.

READ-ONLY MEMORY (ROM)

ROM is *nonvolatile* (because it maintains its contents even when its power is shut off) and is used to store data and programs that do not change during the operation of the system. The *mask-programmed* ROMs are read-only devices that are programmed for data storage during the manufacturing of the chip itself. These are generally less expensive devices for mass production. Character-font memory for laser printers is a good example. *Programmable* read-only memory (PROM) is a field-programmable memory that is fabricated by the manufacturer containing all 0s and is programmed irreversibly by the user by electrically changing appropriate 0s to 1s. PROMs are quite economical in small quantities. *Erasable* programmable read-only memory (EPROM) is nonvolatile and widely used in microprocessor systems for program storage. It can be erased by shining an ultraviolet light and reprogrammed if necessary. These are produced in low to moderate volumes.

ELECTRICALLY ERASABLE PROGRAMMABLE ROM (EEPROM)

EEPROM is used for remote-area applications. The device is provided with special pins which, when activated electrically, alter the rewriting of selected memory locations.

MAGNETIC STORAGE DEVICES

Magnetic disk memory is nonvolatile and provides large storage capabilities with moderate access times. The data are stored on one or more rigid aluminum circular disks coated with iron oxide. The

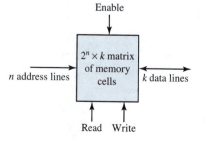

Figure 6.2.16 Block diagram of 2^n × k-bit RAM device.

most common disks have 11-in diameters and 200 tracks (concentric rings of data) per surface, numbered from 0 to 199, starting with the outside perimeter of the disk, with a typical track packing density of 4000 bits per inch. Disks are mounted on a common spindle, and all disks rotate at a typical speed of 3600 revolutions per minute (rpm). A typical disk has 17 sectors of fixed size per track and 512 bytes (1 byte = 8 bits) of information per sector. Any desired sector can be quickly accessed.

Floppy disks, also known as *flexible disks,* are the low-cost, medium-capacity, nonvolatile memory devices made of soft flexible mylar plastic with magnetically sensitive iron-oxide coating. The original 8-in standard floppy is no longer in popular use. The $5\frac{1}{4}$-in minifloppy has a disk and a disk jacket covering the mylar media for protection, along with a write-protect notch and index hole. The present-day minifloppy disks are either double-sided/double-density (DS/DD) with 9 sectors per track and 40 tracks per side or double-sided/quad-density (DS/QD) with 9 or 15 sectors per track and 80 tracks per side. In a DS/DD minifloppy disk, about 720 kbytes of data can be stored; whereas in a DS/QD mini-floppy disk, about 1–2 Mbytes of data can be stored.

The $3\frac{1}{2}$-in microfloppy disk, also known as a microdiskette, is enclosed in a rigid protective case and is provided with a write-protect notch. Microdiskettes are recorded in quad-density format with a capacity of 2 Mbytes; 4- to 16-Mbyte $3\frac{1}{2}$-in diskettes are being developed. Also, 2-in diskettes are introduced in electronic cameras and portable personal computers.

Magnetic tapes are ideal devices for storing vast quantities of information inexpensively. However, the access time is slow because the entire tape must be read sequentially. The most commonly used tapes are $\frac{1}{2}$-in wide, 2400 or 3600 ft long, and contained in a long $10\frac{1}{2}$-in reel. Tape densities of 200, 556, 800, 1600, 6250, and 12,500 bits per inch (BPI) are standard.

In addition to these magnetic storage devices, two newer types of secondary storage have come into use: *winchester disks* and *videodisks* (also known as *optical disks*). The former are sealed modules that contain both the disk and the read/write mechanism, requiring little maintenance and allowing higher-density recording. The latter have been introduced recently, with high reliability and durability and a storage capacity of 1 Gbyte of data (equivalent to almost 400,000 typewritten pages of information). A typical 14-in optical disk has 40,000 tracks and 25 sectors per track, with each sector holding up to 1 kbyte of information. While a write-once optical-disk drive is currently available, a read-and-write drive is being developed.

Display Devices

Display devices can be categorized as on/off indicators, numeric, alphanumeric, or graphical displays. They may also be classified as active and passive devices. Active display devices emit light, such as light-emitting diodes (LEDs), whereas passive display devices, such as liquid-crystal displays (LCDs), reflect or absorb light.

LIGHT-EMITTING DIODE (LED)

This is a reliable, rugged, and inexpensive semiconductor display device requiring about 10 mA of current flow for full illumination. An LED is shown in Figure 6.2.17. It is available in red, yellow, or green color, in common sizes of T-1 and T-$1\frac{3}{4}$, where the number after the T denotes the diameter of the lamp in units of eighths of an inch. Two-color LEDs are also available.

LIQUID-CRYSTAL DISPLAY (LCD)

This display needs only microwatts of power (over a thousand times less than a LED) and is used in such devices as electronic wristwatches. With the application of an electric field the molecules

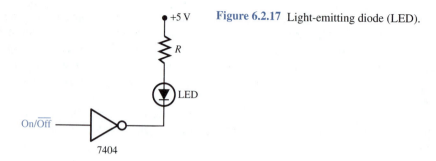

Figure 6.2.17 Light-emitting diode (LED).

of the liquid-crystal material are straightened out, absorbing the light, and the display appears black. With no electric field applied, the display appears as a silver mirror because the light is reflected.

Segment Displays

Seven-segment displays are the most commonly used numeric display devices, while 10- and 16-segment (starburst) display devices are also available. Figure 6.2.18(a) depicts a common-cathode seven-segment LED display; Figure 6.2.18(b) shows its internal structure, consisting of a single LED for each of the segments; and Figure 6.2.18(c) displays the digits by an appropriate combination of lighted segments. The 16-segment display shown in Figure 6.2.18(d) is commonly used for alphanumeric data.

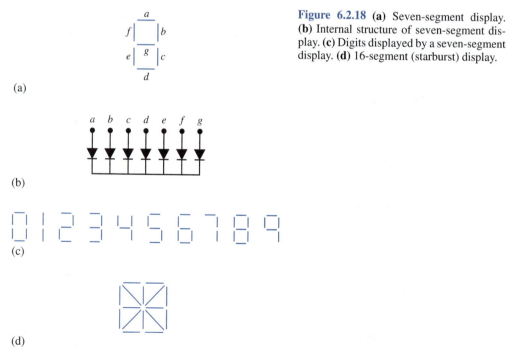

Figure 6.2.18 **(a)** Seven-segment display. **(b)** Internal structure of seven-segment display. **(c)** Digits displayed by a seven-segment display. **(d)** 16-segment (starburst) display.

(a)

(b)

(c)

(d)

CATHODE-RAY TUBE (CRT)

While the CRT display is one of the oldest, it still remains one of the most popular display technologies. The *raster-scanned* display devices work similarly to commercial television sets, whereas *graphics* display devices use different technologies to achieve extremely high resolution. A CRT video signal has only two levels, with a 0 level causing a dark spot and a 1 level causing a bright spot. The appropriate combination of 1s and 0s displays data on a CRT screen, with each character displayed by dot-matrix displays, typically in 5×7 or 7×9 display fonts.

EXAMPLE 6.2.6

For the 4-bit D/A converter of Figure 6.2.9 with $V_{ref} = -5$ V, determine the range of analog output voltage and the smallest increment.

Solution

The binary input range is from 0000 to 1111 for a 4-bit D/A converter. From Equation (6.2.1) it follows that the range of the analog output voltage is from 0 V to

$$5 \times (1 + 1 \times 2^{-1} + 1 \times 2^{-2} + 1 \times 2^{-3}) = 5 \times 1.875 = 9.375 \text{ V}$$

The smallest increment is given by $5 \times 1 \times 2^{-3} = 5/8 = 0.625$ V.

EXAMPLE 6.2.7

For the 3-bit 2^n–R D/A converter of Figure 6.2.11, calculate the analog output voltage when the input is (a) 100, and (b) 010.

Solution

a. For the binary input 100, switches controlled by B_0, $\bar{B}_1$, and $\bar{B}_2$ will be closed. A path is then produced between the output V_o and point 4, where the voltage is equal to $V_{ref}/2$. The analog output voltage is therefore $V_{ref}/2$.

b. For the binary input 010, switches controlled by $\bar{B}_0$, B_1, and $\bar{B}_2$ will be closed. A path exists between V_o and point 6, where the voltage is equal to $V_{ref}/4$. The analog output voltage is thus $V_{ref}/4$.

EXAMPLE 6.2.8

The speed of an 8-bit A/D converter is limited by the counter, which has a maximum speed of 4×10^7 counts per second. Estimate the maximum number of A/D conversions per second that can be achieved.

Solution

The rate of the clock will be constant, independent of the analog input. It must be slow enough to allow the counter to count up to the highest possible input voltage. This will require $255(= 2^7 + 2^6 + 2^5 + 2^4 + 2^3 + 2^2 + 2^1 + 2^0)$ counts, which will take $255/(40 \times 10^6) = 6.375 \ \mu s$. Thus, the process can be repeated $10^6/6.375 = 156,863$ times per second.

EXAMPLE 6.2.9

If it is desired to store English-language writing with 1 byte representing each letter, find the minimum number of bits per byte that could be used.

Solution

For the 26 letters in the English alphabet, we must have $2^N \geq 26$, where N is the number of bits per byte. N being an integer, the smallest possible value for N is 5. One can represent the letter A by 00000, B by 00001, and so on up to Z by 11001; the remaining six combinations could be used for representing punctuation marks or spaces. This is, in fact, the method by which letters are represented in teletype systems.

If one wants to store capital English-language letters also, there will be 52 letters instead of 26; in such a case 6-bit bytes would have to be used.

6.3 COMPUTER SYSTEMS

Digital computers, in general, are automatic machines that accept data and instructions, perform predefined operations very quickly on the data, and have the results available to the user in various forms. They can be classified as *microcomputers, minicomputers, mainframes,* and *supercomputers.*

Microcomputers have become a common part of everyday life. The cost ranging from a few hundred to about ten thousand dollars, today's 16- and 32-bit microcomputer systems are also dedicated for real-time applications in a distributed system. A *microprocessor* is an LSI device, which is a realization of the computer *central-processor unit* (CPU) in IC form. The microprocessor is the CPU of the microcomputer system.

Minicomputers, developed in the early 1960s, are high-performance, general-purpose multi-user computers. These are also designed for real-time dedicated applications. The PDP-11 series from Digital Equipment Corporation (DEC) have been the most prominent 16-bit minicomputers, and are now obsolete. The 32-bit minicomputers, known as *superminis,* were developed in the 1970s, the most prominent one being the VAX 8600 from DEC, which was capable of executing about 5 million instructions per second (MIPS). New VAX lines with larger MIPS have been developed since.

Mainframes, capable of executing in excess of 53 MIPS, are high-performance, general-purpose computers supporting very large databases, ranging in price from one to ten million dollars. These are used by many universities, large businesses, and government agencies, and are supplied mainly by IBM. Examples include IBM 360, CDC 7600 of Control Data Corporation, and Texas Instrument Advanced Scientific Computer (TI–ASC).

Supercomputers, capable of executing in excess of one billion floating-point operations per second (FLOPS), are very powerful, extremely high-performance computers for applications that are beyond the reach of the mainframes, and cost more than ten million dollars. These are used for weather prediction, image processing, and nuclear-energy studies that require high-precision processing of ordered data achieved by a speed advantage due to parallel processors. Cyber 205, Cray X-MP, and Cray 2 are some examples of supercomputers. In the 1980s, supercomputing centers were developed at six American universities for high performance computing. By 1990 it was possible to build chips with a million components; semiconductor memories became standard in all computers; widespread use of computer networks and workstations had occurred. Explosive growth of wide area networking took place with transmission rates of 45 million bits per second.

Organization

There are two principal components: *hardware* and *software*. The former refers to physical components such as memory unit (MU), arithmetic and logic unit (ALU), control unit (CU), input/output (I/O) devices, etc.; whereas the latter refers to the *programs* (collections of ordered instructions) that direct the hardware operations.

Figure 6.3.1 illustrates the basic organization of a digital computer. The MU stores both data and programs that are currently processed and executed. The ALU processes the data obtained from the MU and/or input devices, and puts the processed data back into the MU and/or output devices. The CU coordinates the operations of the MU, ALU, and I/O units. While retrieving instructions from programs resident in the MU, the CU decodes these instructions and directs the ALU to perform the corresponding processing; it also oversees the I/O operations. Input devices may consist of card readers, keyboards, magnetic tape readers, and A/D converters; output devices may consist of line printers, plotters, and D/A converters. Devices such as terminals and magnetic disk drives have both input and output capabilities. For communicating with the external world and for storing large quantities of data, a variety of I/O devices are used.

Software may be classified as *system software* and *user software*. The former refers to the collection of programs provided by the computer system for the creation and execution of the user programs, whereas the latter refers to the programs generated by various computer users for solving their specific problems.

A *program* generally consists of a set of instructions and data specifying the solution of a particular problem. Programs (and data) expressed in the binary system (using 0s and 1s) are known as *machine-language* programs. Writing programs in this form, which demands

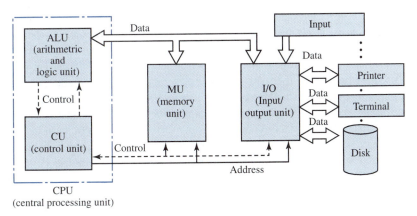

Figure 6.3.1 Typical organization of a computer.

a detailed knowledge of the computer structure, is rather tedious and error-prone. *Assembly-language* programming is developed by using symbolic names, known as *mnemonics,* and matching machine-language instructions on a more or less one-for-one basis. An *assembler* is then used to translate assembly-language programs into their equivalent machine-language programs. Because both assembly-language and machine-language programmings are specific to a particular computer, *high-level languages* (HLL) such as Fortran, Pascal, Basic, LISP, and C have been developed such that programs written by using them could be run on virtually any computer. Also, these are problem-oriented languages, which allow the user to write programs in forms that are as close as possible to the human-oriented languages.

An *interpreter* translates each high-level-language statement into its equivalent set of machine-language instructions, which are then executed right away. Interpretive languages such as Basic are very inefficient for programs with loops (repetitive instructions). The inefficiency is corrected by a *compiler*, which translates the complete high-level language into machine language. Once the whole program is compiled, it can be executed as many times as desired without any need for recompilation. Examples of compiler languages are Fortran, Pascal, and C.

An operating system, such as DOS, VMS, or UNIX, consisting of a set of system programs, performs resource management and human-to-machine translation, supporting a given computer architecture. Operations such as starting and stopping the execution, as well as selecting a specific compiler or assembler for translating a given program into machine language, are taken care of by the operating system, which is unique for a given microcomputer, minicomputer, or mainframe.

Architecture

Replacing the ALU and CU (i.e., CPU) of Figure 6.3.1 by a microprocessor, and storing instructions and data in the same memory, one arrives at a stored-program computer or a microcomputer. A *bus,* which is a set of wires carrying address, data, and control signals, is employed for interconnecting the major components of a microcomputer system. The address lines are unidirectional signals that specify the address of a memory location of an I/O device. With a typical 24-bit address bus, the microprocessor can access 2^{24} (over 16 million) memory locations. Memory is generally organized in blocks of 8, 16, or 32 bits. The data bus is a bidirectional bus, varying in size from 8 to 32 bits, which carries data between the CPU, MU, and I/O units. The control bus provides signals to synchronize the memory and I/O operations, select either memory or an I/O device, and request either the read or the write operation from the device selected.

While there are virtually countless variations in microprocessor circuit configurations, the system architecture of a typical microprocessor is shown in Figure 6.3.2. The *arithmetic logic unit* (ALU) accepts data from the data bus, processes the data as per program-storage instructions and/or external control signals, and feeds the results into temporary storage, from which external control and actuator control functions can be performed. The *accumulators* are parallel storage registers used for processing the work in progress, temporarily storing addresses and data, and housekeeping functions. The *stacks* provide temporary data storage in a sequential order and are of use during the execution of subroutines. A *subroutine* is a group of instructions that appears only once in the program code, but can be executed from different points in the program. The *program counter* is a register/counter that holds the address of the memory location containing the next instruction to be executed. The *status register* contains condition-code bits or *flags* (set to logic 1 or logic 0, depending on the result of the previous instruction) that are used to make decisions and redirect the program flow. The *control unit* (CU), which consists of the timing and data-routing circuits, decodes the instruction being processed and properly establishes data paths among the various elements of the microprocessor. Interconnections may take the form of gates that the control section enables or disables according to the program instructions. That is to say, programming at the machine-language level amounts to wiring with software instead of hard-wired connections.

Microprocessors have instruction sets ranging from 20 to several hundred instructions, known as *microprograms,* which are stored in ROM to initiate the microprocessor routines. The instructions generally consist of a series of arithmetic and logic type operations, and also include directions for fetching and transferring data. Microprocessors are classified by word size in bits, such as 1-, 4-, 8-, and 16-bit microprocessors; generally speaking, the larger the word size, the more powerful the processor. Three popular 8-bit microprocessors (μp) are Intel 8085, Zilog Z80, and Motorola MC6800; the 16-bit microprocessors dominating the market are Intel 8086 and 80286, Motorola MC68000, and Zilog Z8000; the powerful 32-bit microprocessors at the very high end of the market started with Intel 80386, Intel 80486, Motorola MC68020, and National N532032. Still more powerful Intel Pentium processors such as Pentium II and III were introduced in the 1990's, and even these are going to be replaced soon by Intel Itanium Processors.

A microprocessor system bus consists of three physical buses: the address bus, the data bus, and the control bus. The types of circuits connected to microprocessor buses are registers, accumulators, or buffer circuits between the bus and the external memory or I/O devices. Multiplexing is usually used, which is to connect each register to a bus on a time-shared basis, only when it is being read. An operating bus is used to transfer various internal operations and commands. An interface bus, such as the IEEE-488 (developed in 1975 by the Institute of Electrical and Electronics Engineers) or the GPIB (general-purpose interface bus) provides the

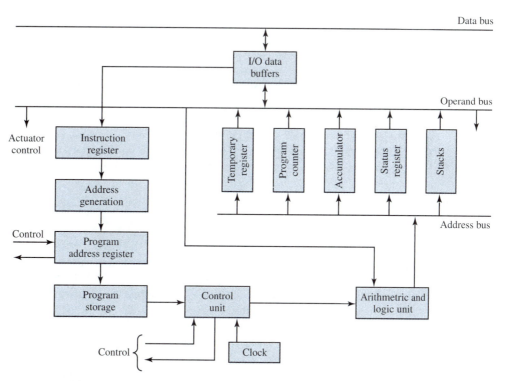

Figure 6.3.2 Block diagram of typical microprocessor system architecture.

means of communicating between the computer and the outside world with external devices such as oscilloscopes, data collection devices, and display devices. Most present-day *data acquisition* systems are designed to be compatible with microprocessors for processing the measured data. A typical data acquisition and processing system is illustrated in Figure 6.3.3.

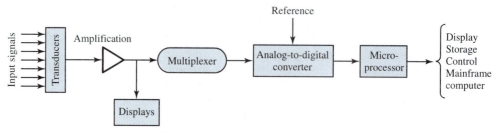

Figure 6.3.3 Block diagram of typical data acquisition and processing system.

6.4   COMPUTER NETWORKS

Computer communications networks are the outcome of a combination of computers and telecommunication products. An interconnected group of independent computers and peripheral devices that communicate with each other for the purpose of sharing software and hardware resources is known as a computer communications network. A *local-area network* (LAN) is any physical network technology that operates at high speed (usually tens of Mbits per second through several Gbits per second) over short distances (up to a few thousand meters). Examples include Ethernet and proNET-10. The stations, also known as *nodes*, within a LAN are physically linked with each other through twisted pairs of copper wires, coaxial cables, or fiber-optic cables. A *wide-area network* (WAN, also known as *long-haul network*) is any physical network that spans large geographic distances, usually operates at slower speeds, and has significantly higher delays than a LAN. Stations within a WAN communicate with each other through standard telephone lines, dedicated telephone lines, line-of-sight microwave systems, or fiber-optic links. *Public-data network* (PDN) is network service offered by a common carrier, such as Telnet, Tymnet, and Dataphone Digital Services of AT&T. ARPANET (Advanced Research Projects Agency of the U.S. Department of Defense) is an example of a private communications network. CYBERNET is an example of a remote-access communications network that provides access to huge databases for their users in different countries. The *Internet* is a collection of networks and gateways (special-purpose dedicated computers attached to various networks routing packets of information from one to the other), including the Milnet (military network) and NSFNET (National Science Foundation network), that function as a single, cooperative virtual network providing universal connectivity and application-level services such as the *electronic mail*. The Internet reaches many universities, government research labs, and military installations in several countries.

The future scope of the Internet, along with the World Wide Web (born in 1990), and the commercialization of the Internet are bound to grow exponentially. In 1991, the U.S. Congress passed the High Performance Computing Act to establish the National Research and Education Network (NREN), which allows the electronic transfer of the entire Encyclopedia Britannica in one second. Computer networks worldwide will feature 3–D animated graphics, radio and cellular phone-links to portable computers, as well as fax, voice, and high-definition television. While the Web is fast becoming a part of our everyday lives, the real Internet of the future may bear very little resemblance to today's plans.

Network Architecture

A *protocol* is a formal description of message formats and the rules two or more machines must follow to exchange those messages. Because TCP (transmission control protocol) and IP (Internet protocol) are the two most fundamental protocols, the entire protocol suite that is used by the Internet is often referred to as TCP/IP. X.25 is the CCITT (Consultative Committee on International Telephony and Telegraphy) standard protocol employed by Telnet, and is most popular in Europe. Ethernet utilizes CSMA/CD (carrier sense multiple access with collision detection) protocol technology.

Computer network architecture refers to the convention used to define how the different protocols of the system interact with each other to support the end users. The most common network architecture model is the open-systems interconnections (OSI). Figure 6.4.1 shows the ISO (International Standards Organization) seven-layer model for an OSI. Although not all layers need be implemented, the more layers that are used, the more functionality and reliability are built into the system. Starting from the bottom layer, the functions of the layers are as follows.

1. *Physical*—Defines the type of medium, the transmission method, and the transmission rates available for the network; provides the means for transferring data across the interconnection channel and controlling its use.

2. *Data Link*—Defines how the network medium is accessed, which protocols are used,

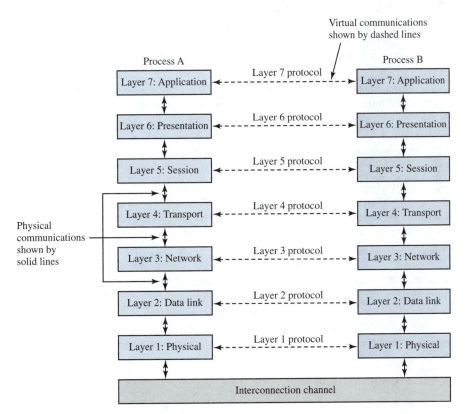

Figure 6.4.1 ISO 7-layer model for an OSI.

the packeting/framing methods, and the virtual circuit/connection services; responsible for the transfer of data across the link; provides for the detection and correction of data-transmission errors.

3. *Network*—Standardizes the way in which addressing is accomplished between linked networks; performs networking functions and internetworking.

4. *Transport*—Handles the task of reliable message delivery and flow control between applications on different stations; provides source-to-destination data integrity.

5. *Session*—Establishes two-way communication between applications running on different stations on the network; provides the user interface into the transport layer.

6. *Presentation*—Translates data formats so that computers with different "languages" can communicate; provides the syntax (rules) of representation of data between devices.

7. *Application*—Interfaces directly with the application programs running on the stations; provides services such as file access and transfer, peer-to-peer communication among applications, and resource sharing; provides support to process end users' applications such as electronic mail, database management, and file management.

Note that the interconnection channel is not a part of the OSI specification.

Network Topology

This deals with the geometrical arrangement of *nodes* (endpoints consisting of physical devices such as terminals, printers, PCs, and mainframes) interconnected by *links* (transmission channels). Network topologies may be classified as bus topology (multidrop topology), star topology, ring topology, tree topology, and distributed (mesh or hybrid) topology, as illustrated in Figure 6.4.2. Bus topology is used predominantly by LANs, whereas star topology is commonly used by private-branch exchange (PBX) systems. Ring topology may have centralized control (with one node as the controller) or decentralized control (with all nodes having equal status). Tree topology is used in most of the remote-access networks, whereas distributed topology is common in public and modern communications networks. A fully distributed network allows every set of nodes to communicate directly with every other set through a single link and provides an alternative route between nodes.

The Internet is physically a collection of packet switching networks interconnected by gateways along with protocols that allow them to function logically as a single, large, virtual network. Gateways (often called IP routers) route packets to other gateways until they can be delivered to the final destination directly across one physical network. Figure 6.4.3 shows the structure of physical networks and gateways that provide interconnection. Gateways do not provide direct connections among all pairs of networks. The TCP/IP is designated to provide a universal interconnection among machines, independent of the particular network to which they are attached. Besides gateways that interconnect physical networks, as shown in Figure 6.4.3, Internet access software is needed on each host (any end-user computer system that connects to a network) to allow application programs to use the Internet as if it were a single, real physical network. Hosts may range in size from personal computers to supercomputers.

Transmission Media

These, also known as *physical channels*, can be either *bounded* or *unbounded*. Bounded media, in which signals representing data are confined to the physical media, are twisted pairs of wires,

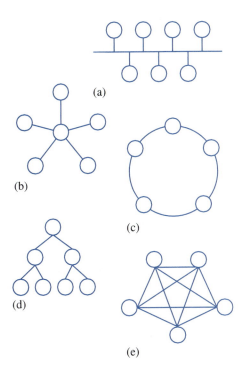

Figure 6.4.2 Network topologies. **(a)** Bus (multidrop) topology. **(b)** Star topology. **(c)** Ring topology. **(d)** Tree topology. **(e)** Distributed (mesh or hybrid) topology.

(a)

(b)

(c)

(d)

(e)

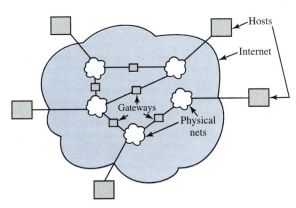

Hosts

Internet

Gateways

Physical nets

Figure 6.4.3 Structure of typical TCP/IP Internet.

coaxial cables, and optical-fiber cables, used in most LANs. Unbounded media, such as the atmosphere, the ocean, and outer space in which the transmission is wireless, use infrared radiation, lasers, microwave radiation, radio waves, and satellites. Data are transmitted from one node to another through various transmission media in computer communications networks.

Twisted pairs are used in low-performance and low-cost applications with a data rate of about 1 Mbit per second (Mbps) for a transmission distance of about 1 km. *Baseband coaxial cables* used for digital transmission are usually 50-Ω cables with a data rate of about 10 Mbps over a distance of about 2 km. *Broad-band coaxial cables* used for analog transmission (cable TV) are usually 75-Ω cables with a data rate of about 500 Mbps over a distance of about 10 km. The lighter and cheaper *fiber-optic cables* support data transmission of about 1 Gbps over a distance of about 100 km.

Data Transmission and Modems

Data can be transferred between two stations in either *serial* or *parallel* transmission. Parallel data transmission, in which a group of bits moves over several lines at the same time, is used when the two stations are close to each other (usually within a few meters), as in a computer–printer configuration. Serial data transmission, in which a stream of bits moves one by one over a single line, is used over a long distance. Serial data transfer can be either *asynchronous* or *synchronous*. Asynchronous data communication is most commonly applied in low-speed terminals and small computers. Large-scale integration (LSI) devices known as UARTs (universal asynchronous receivers/transmitters) are commercially available for asynchronous data transfer. Synchronous data communication is used for transferring large amounts of data at high speed. USARTs (universal synchronous/asynchronous receivers/transmitters) are commercially available LSI devices.

Frequency-division multiplexing (FDM) is a technique for data transmission widely used in telephone, radio, and cable TV systems in which the transmission frequency spectrum (i.e., bandwidth) is divided into smaller bands known as subchannels.

Data transmission between two stations can be achieved in either *simplex, half-duplex,* or *full-duplex* mode. In a simplex mode, mainly used in radio and TV broadcasts, information travels only in one direction. This mode is rarely used in data communications. In half-duplex mode, used by radio communications, information may travel in both directions, but only in one direction at a time. The transmitter becomes the receiver and vice versa. In a full-duplex mode, information may travel in both directions simultaneously. This mode, used in telephone systems, adopts two different carrier frequencies.

A *modem* (modulator/demodulator) is an electronic device that takes digital data as a serial stream of bits and produces a modulated carrier signal as an output. That is to say, the digital signals are converted to an analog form with a relatively narrow bandwidth. The carrier signal is then transmitted over the telephone line to a similar modem at the receiving end, where the carrier signal is demodulated back into its original serial stream of bits, as shown in Figure 6.4.4. The serial digital data to be transmitted are modulated, filtered, and amplified for analog transmission; the analog data received at the receiving end are amplified, filtered, and demodulated to produce serial digital signals.

There are four different types of modems: *half-duplex, full-duplex, synchronous,* and *asynchronous.* With half-duplex modems data can be transmitted in only one direction at a time. Full-duplex modems transmit data in both directions at the same time; one modem is designated as the originating modem and the other as the answering modem, while transmitting and receiving data are done at different frequencies. Asynchronous modems are low-data-rate modems transmitting serial data at a rate of about 1800 bits per second (bps). Synchronous modems are high-data-rate modems transmitting serial data at a rate of about 10,800 bps.

Modems can also be classified as *voice-band* or *wide-band* modems. Voice-band modems are low-to-high speed modems designed for use on dial-up, voice-grade, standard telephone lines up to

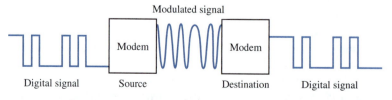

Modulated signal

Digital signal   Source   Destination   Digital signal

Figure 6.4.4 Input/output signals of a modem.

a rate of about 10,800 bps. Microprocessor-controlled modems are known as *smart* modems, such as the Hayes modem, manufactured by Hayes Microcomputer Products. The portable *acoustic-coupler* device, which is a different type of voice-band modem, is a low-speed modem with a rate of about 600 bps that is connected acoustically to a standard telephone. Wide-band modems are very high-speed modems with rates of 19,200 bps and above, designed for use with dedicated telephone lines. These are currently used mostly on private communications systems.

6.5 LEARNING OBJECTIVES

The *learning objectives* of this chapter are summarized here, so that the student can check whether he or she has accomplished each of the following.

- Understanding continuous, discrete, and binary digital signals.
- Performing operations with binary, octal, hexadecimal, and binary-coded-decimal number systems.
- Logic symbols and truth tables for various gates.
- Designing simple combinational logic circuits using logic gates.
- Using Karnaugh maps to realize logic expressions.
- Sketching timing diagrams for sequential circuits based on flip-flops.
- Digital system components such as decoders, encoders, multiplexers, registers, counters, D/A and A/D converters, memory, and display devices.
- Basic understanding of computer systems.
- Basic notions about computer networks.

6.6 PRACTICAL APPLICATION: A CASE STUDY

Microcomputer-Controlled Breadmaking Machine

Figure 6.6.1 shows a simplified schematic diagram of a microcomputer-controlled breadmaking machine. A microcomputer along with its timing circuit, keypad, and display unit controls the heating resistor, fan motor, and bread-ingredient mixing motor by means of digitally activated switches. An analog temperature sensor, through an A/D converter, provides the status of temperature to the microcomputer. A digital timer circuit counts down, showing the time remaining in the process.

The control programs are stored in ROM and determine when and how long the machine should mix the ingredients added to the bread pan, when and how long the heating resistor should be turned on or off for various parts of the cycle, and when and how long the fan should be on to cool the loaf after baking is finished. The parameters such as light, medium, or dark bread crust are entered through the keypad into RAM.

According to the programs stored and the parameters entered, the machine initially mixes the ingredients for several minutes. The heating resistor is turned on to warm the yeast, causing the dough to rise while a temperature of about 90°F is maintained. The time remaining and the temperature are continually checked until the baked loaf is cooled, and the finished bread is finally ready in about 4 hours.

Microprocessors and computers in various forms are used extensively in household appliances, automobiles, and industrial equipment.

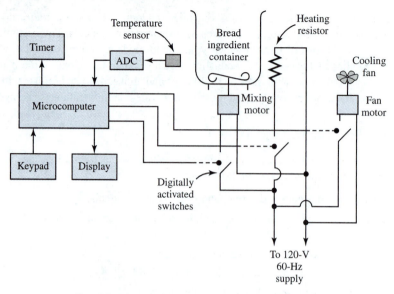

Figure 6.6.1 Simplified schematic diagram of a microcomputer-controlled breadmaking machine.

PROBLEMS

6.1.1 Convert the following binary numbers into decimal numbers:

(a) $(10110)_2$

(b) $(101100)_2$

(c) $(11010101)_2$

(d) $(11101.101)_2$

(e) $(.00101)_2$

6.1.2 Convert the following decimal numbers into binary numbers:

(a) $(255)_{10}$

(b) $(999)_{10}$

(c) $(1066)_{10}$

(d) $(0.375)_{10}$

(e) $(1259.00125)_{10}$

*6.1.3 Convert the following octal numbers into decimal numbers:

(a) $(257)_8$

(b) $(367)_8$

(c) $(0.321)_8$

(d) $(367.240)_8$

(e) $(2103.45)_8$

6.1.4 Convert the following decimal numbers into octal numbers:

(a) $(175)_{10}$

(b) $(247)_{10}$

(c) $(65,535)_{10}$

(d) $(0.125)_{10}$

(e) $(379.25)_{10}$

6.1.5 Convert the following octal numbers into binary numbers:

(a) $(3425)_8$

(b) $(3651)_8$

(c) $(.214)_8$

(d) $(4125.016)_8$

(e) $(4573.26)_8$

6.1.6 Convert the following binary numbers into octal numbers:

(a) $(011\ 100\ 010\ 101)_2$

(b) $(1011010)_2$

(c) $(.110101)_2$

(d) $(100101111011.01011)_2$

(e) $(1110110111.1011)_2$

6.1.7 Convert the following hexadecimal numbers into decimal numbers:

(a) $(6B)_{16}$

(b) $(1F4)_{16}$

(c) $(C59)_{16}$

(d) $(256.72)_{16}$

(e) $(.0E3)_{16}$

*6.1.8 Convert the following decimal numbers into hexadecimal numbers:

(a) $(97)_{10}$

(b) $(864)_{10}$

(c) $(5321)_{10}$

(d) $(0.00125)_{10}$

(e) $(449.375)_{10}$

6.1.9 Convert the following binary numbers into hexadecimal numbers:

(a) $(11101110100100.100111)_2$

(b) $(1011011101)_2$

(c) $(.11101)_2$

(d) $(1101110001.11011110)_2$

(e) $(.0000110111000101)_2$

6.1.10 Convert the following hexadecimal numbers into binary numbers:

(a) $(2ABF5)_{16}$

(b) $(3BA4.9C)_{16}$

(c) $(0.0DC5)_{16}$

(d) $(15CE.FB3)_{16}$

(e) $(2AB.F8)_{16}$

6.1.11 Convert the following decimal numbers into BCD numbers:

(a) $(567)_{10}$

(b) $(1978)_{10}$

(c) $(163.25)_{10}$

(d) $(0.659)_{10}$

(e) $(2153.436)_{10}$

6.1.12 Convert the following BCD numbers into decimal numbers:

(a) $(010101100111)_{BCD}$

(b) $(.011001011001)_{BCD}$

(c) $(.100110000100)_{BCD}$

(d) $(10010010.00000001)_{BCD}$

(e) $(0010000101010011.010000110110)_{BCD}$

*6.1.13 Obtain the truth table for the logic block shown in Figure P6.1.13.

6.1.14 Find the output function Y for the logic circuits of Figure P6.1.14 (a) and (b).

6.1.15 An AOI (AND-OR-INVERT) gate is shown in Figure P6.1.15 with its two possible realizations. Obtain the output function Y and show that the two circuits are equivalent.

6.1.16 Figure P6.1.16(a) shows the seven-segment array that is widely used to form the decimal digits 0 to 9 in LED displays, as indicated in Figure P6.1.16(b). Let the inputs be the four binary digits used to represent the decimal digits 0 to 9. (a) For the segment Y_1, (b) For the segment Y_2, and (c) For the segment Y_3 to be turned on, develop the logic expression and find one possible logic circuit to realize the output in each case.

6.1.17 The DeMorgan's theorems suggest that the basic logic operations can be realized by use of inverters and NAND gates only. For the circuits shown in Figure P6.1.17, find the truth table, the type of gate realized, and the expression for the logic output, in each case.

6.1.18 For the NOR and inverter realizations shown in Figure P6.1.18, find the truth table, the type of gate realized, and the expression for the logic output, in each case.

*6.1.19 Obtain the Boolean expressions for the logic circuits shown in Figure P6.1.19.

6.1.20 Draw the logic diagram for the following Boolean expressions (without any simplification).

(a) $Y = AB + \bar{B}C$

(b) $Y = (A + B)(\bar{A} + C)$

(c) $Y = A \cdot B + \bar{B} \cdot C + A \cdot B \cdot D + A \cdot C \cdot D$

(d) $Y = (\bar{A} + B) \cdot \overline{(A + \bar{C})} \cdot (B + C)$

6.1.21 Obtain the Boolean expressions for the logic circuits shown in Figure P6.1.21.

6.1.22 Using Boolean identities, simplify the following:

(a) $Y = A + \bar{A} \cdot B$

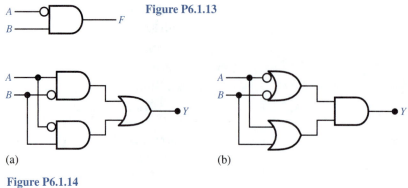

Figure P6.1.13

(a)

(b)

Figure P6.1.14

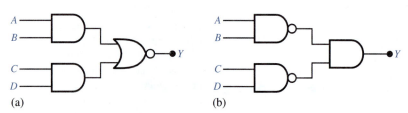

(a)

(b)

Figure P6.1.15

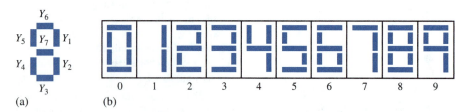

(a)

(b)

Figure P6.1.16

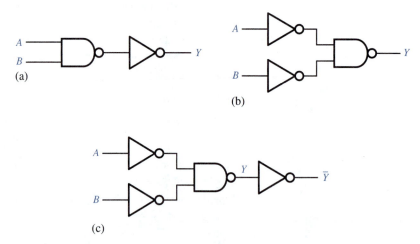

(a)

(b)

(c)

Figure P6.1.17

Figure P6.1.18

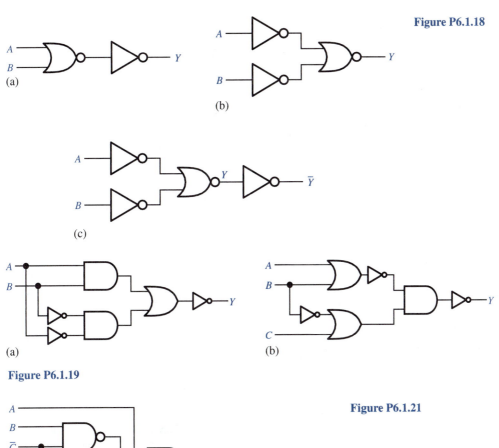

(a)

(b)

(c)

Figure P6.1.19

(a)

(b)

Figure P6.1.21

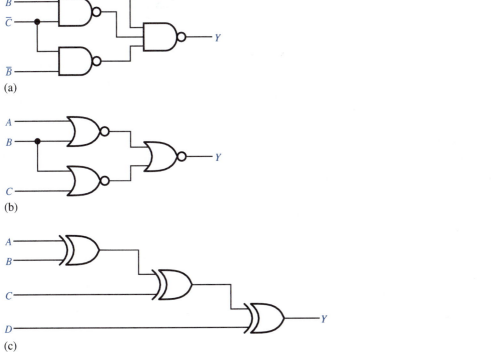

(a)

(b)

(c)

(b) $Y = A \cdot B + \bar{B} \cdot C + A \cdot C \cdot D + A \cdot B \cdot D$

(c) $Y = \overline{(\bar{A} + B + C)} \cdot \overline{(\bar{A} + B + C)} \cdot \bar{C}$

(d) $Y = B \cdot C + \bar{B} \cdot \bar{C} + A \cdot \bar{B} \cdot C$

6.1.23 The truth table for $F(A, B, C) = \sum m_i(2, 3, 4, 5)$ is as follows:

A	B	C	F
0	0	0	0
0	0	1	0
0	1	0	1
0	1	1	1
1	0	0	1
1	0	1	1
1	1	0	0
1	1	1	0

(a) Express F in a canonical sum-of-products form.

(b) Minimize F in an SOP form, and obtain a possible realization.

*6.1.24 The truth table for $F(A, B, C) = \prod M_i(0, 1, 6, 7)$ is as follows:

A	B	C	F
0	0	0	0
0	0	1	0
0	1	0	1
0	1	1	1
1	0	0	1
1	0	1	1
1	1	0	0
1	1	1	0

(a) Express F in a canonical product-of-sums form.

(b) Minimize F in a POS form and obtain a possible realization.

6.1.25 Using K maps, simplify the following Boolean expressions:

(a) $F = A \cdot \bar{B} + A \cdot B$

(b) $F = A \cdot C + C \cdot D + B \cdot C \cdot D$

(c) $F = A \cdot B \cdot \bar{C} + B \cdot C + A \cdot B \cdot D + B \cdot C \cdot D$

6.1.26 Simplify the following Boolean functions into their minimum SOP form, by using K maps.

(a) $F(A, B) = \sum m_i(0, 1, 3)$

(b) $F(A, B, C) = \sum m_i(0, 2, 3, 4, 5, 6)$

(c) $F(A, B, C, D) = \sum m_i(0, 4, 5, 6, 7, 12, 14)$

(d) $F(A, B) = \prod M_i(0, 2)$

(e) $F(A, B, C) = \prod M_i(0, 6)$

(f) $F(A, B, C, D) = \prod M_i(1, 3, 11, 14, 15)$

6.1.27 Given the following truth table, design the logic circuit with the use of a K map by using only two-input gates.

A	B	C	D	y
0	0	0	0	1
0	0	0	1	1
0	0	1	0	1
0	0	1	1	0
0	1	0	0	0
0	1	0	1	1
0	1	1	0	0
0	1	1	1	0
1	0	0	0	1
1	0	0	1	1
1	0	1	0	0
1	0	1	1	1
1	1	0	0	0
1	1	0	1	1
1	1	1	0	0
1	1	1	1	1

6.1.28 For the logic circuit of Figure P6.1.28, construct a truth table and obtain the minimum SOP expression.

*6.1.29 Simplify the logic circuit of Figure P6.1.29 by reducing the number of gates to a minimum.

6.1.30 Given the following truth table:

(a) Realize the function f by a K map using 0s.

(b) Realize the function f by a K map using 1s.

x	y	z	f
0	0	0	0
0	0	1	1
0	1	0	1
0	1	1	1
1	0	0	1
1	0	1	1
1	1	0	0
1	1	1	0

6.1.31 Realize the logic function defined by the following truth table as the simplest POS form.

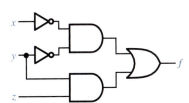

Figure P6.1.28

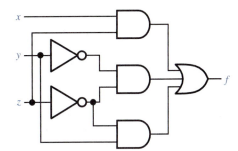

Figure P6.1.29

x	y	z	f
0	0	0	1
0	0	1	0
0	1	0	1
0	1	1	0
1	0	0	1
1	0	1	0
1	1	0	0
1	1	1	0

6.1.32 With the use of a K map, simplify the following Boolean expressions and draw the logic diagram.

(a) $F_1 = A \cdot B + \bar{B} \cdot C + A \cdot B \cdot D + A \cdot C \cdot D$

(b) $F_2 = (X + Y) \cdot (\bar{X} + Z) \cdot (Y + \bar{Z})$

(c) $F_3 = \overline{\overline{A \cdot C} + B \cdot \bar{C} + A \cdot B \cdot C}$

(d) $F_4 = \overline{(\bar{X} + Y) \cdot (\overline{X + \bar{Z}}) \cdot (Y + Z)}$

6.1.33 The K map of a logic function is shown in Figure P6.1.33.

(a) Obtain a POS expression and its corresponding realization.

(b) For the purpose of comparison, obtain the corresponding SOP circuit, and comment on the number of gates needed.

*6.1.34 Given the K map of a logic function as shown in Figure P6.1.34, in which ds denote don't-care conditions, obtain the SOP expression.

6.1.35 The K map of a logic function is shown in Figure P6.1.35, in which ds denote don't-care conditions. Obtain the SOP expressions.

6.1.36 Obtain a minimum two-level NAND–NAND realization for the following Boolean expressions.

(a) $F(A, B, C) = \sum m_i(1, 6) + \sum d_i(2, 4, 5)$

(b) $F(A, B, C, D) = \sum m_i(0, 4, 5, 7, 13) + \sum d_i(2, 6, 8, 10, 11)$

Note that $\sum d_i(\ )$ denotes the sum of minterms corresponding to don't-care outputs.

6.1.37 (a) Show the equivalent NOR realizations of the basic NOT, OR, and AND gates.

(b) Show the equivalent NAND realization of the basic NOT, AND, and OR gates.

6.1.38 Using a minimum number of NAND gates, realize the following Boolean expression: $F(A, B, C) = \sum m_i(0, 3, 4, 5, 7)$.

6.1.39 Figure P6.1.39 shows a *full adder* with the idea of adding C_i to the partial sum S', which is the same logic process as addition with pencil and paper.

(a) Draw the truth table for the full adder.

(b) Add decimal numbers $A = 7$ and $B = 3$ in binary form, showing the values of $A, B, S',$

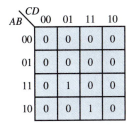

AB\CD	00	01	11	10
00	0	0	0	0
01	0	0	0	0
11	0	1	0	0
10	0	0	1	0

Figure P6.1.33

AB\CD	00	01	11	10
00	0	1	1	1
01	d	1	0	d
11	1	0	0	d
10	0	0	1	d

Figure P6.1.34

AB\CD	00	01	11	10
00	1	0	1	0
01	0	0	0	1
11	0	d	d	d
10	0	1	d	d

Figure P6.1.35

C'_o, C_i, C''_o, and S for each pair of bits.

*6.1.40 Show that the circuit of Figure P6.1.40 is a NOR-gate realization of a flip-flop.

6.1.41 Consider a 1-bit version of the *digital comparator* shown in Figure P6.1.41. Note that the operation of this circuit is such that whichever output is 1 gives the desired magnitude comparison.

    (a) Using NAND and INVERTER gates only, determine the number of gates required.

    (b) Using NOR and INVERTER gates only, determine the number of gates required.

    (c) Which realization requires the least number of gates?

6.1.42 Draw the logic diagram of an *SR* latch using only NAND gates, and obtain the truth table for that implementation.

6.1.43 Complete the timing diagram of Figure P6.1.43 of an *SR* latch.

6.1.44 You are to construct a modified truth table for the circuit realization of the SRFF shown in Figure 6.1.18. As indicated in the text, you guess an output and then go back to check it for self-consistency. In order to get you started, a part of the table is given:

	S	R	Q (guess)	Q̄ (guess)	Q	Q̄
(a)	0	0	1	0		
(b)	0	0	0	1		
(c)	1	0	0	1		
(d)	1	0	1	0		
(e)	0	1	0	1		
(f)	1	1	1	1		

For each row in the table, determine Q and $\bar{Q}$, and comment on the result.

*6.1.45 Let the circuit of Figure 6.1.18 be given the inputs shown in Figure P6.1.45, with the initial value of Q being 1. Find the *timing diagram* (i.e., Q as a function of time).

6.1.46 Consider the case of a positive-edge triggered D flip-flop with preset and clear, as shown in Figure 6.1.22 of the text. With the input signals shown in Figure P6.1.46, find the timing diagram (i.e., plot Q and $\bar{Q}$ as a function of time).

6.1.47 (a) Draw the logic diagram of the enabled *D* latch using only NAND gates.

    (b) Complete the timing diagram of Figure P6.1.47(a) of the *D* latch whose block diagram and truth table are given in Figure P6.1.47(b).

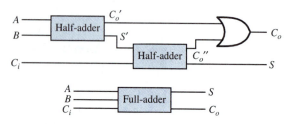

Figure P6.1.39

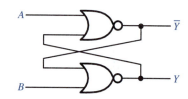

Figure P6.1.40

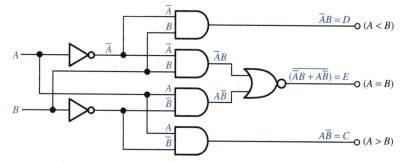

Figure P6.1.41

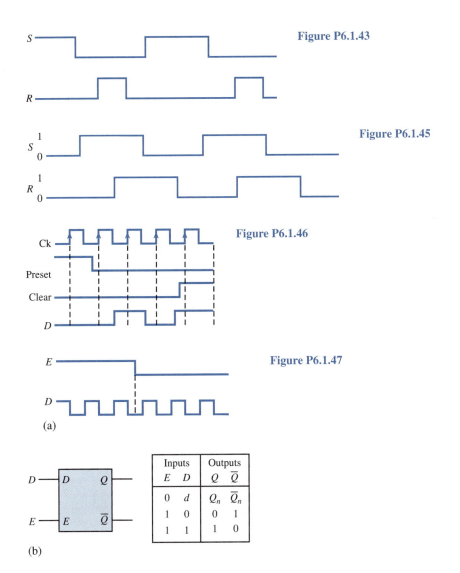

Figure P6.1.43

Figure P6.1.45

Figure P6.1.46

Figure P6.1.47

(a)

(b)

Inputs		Outputs	
E	D	Q	$\bar{Q}$
0	d	Q_n	$\bar{Q}_n$
1	0	0	1
1	1	1	0

6.1.48 A JK flip-flop is shown in Figure P6.1.48(a).

(a) Modify it to operate like the D flip-flop of Figure P6.1.48(b).

(b) Modify the JK flip-flop to operate like the T flip-flop of Figure P6.1.48(c).

6.1.49 When the J and K inputs of a JKFF are tied to logic 1, this device is known as a divide-by-2 counter. Complete the timing diagram shown in Figure P6.1.49 for this counter.

*6.1.50 An interesting application of the SRFF is as a *buffer* in overcoming *contact bounce* in mechanical switches. These switches, of the *toggle* type, may be used to change the logic state in a circuit.

However, they suffer from a major problem in that their contacts do not close immediately but continue to make and break for some time after. To avoid such an undesirable state (because it causes the logic state of the circuit to fluctuate), an SRFF is placed between the switch and the circuit, as shown in Figure P6.1.50. Explain the operation as a buffer.

6.1.51 J and K are the external inputs to the JKFF shown in Figure P6.1.51. Note that gates 1 and 2 are enabled only when the clock pulse is high. Consider the four cases of operation and explain what happens.

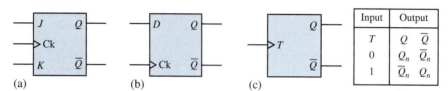

Figure P6.1.48

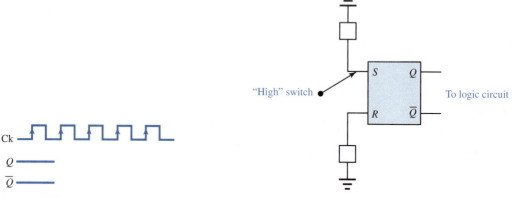

Ck

Q

$\overline{Q}$

Figure P6.1.49

Figure P6.1.50

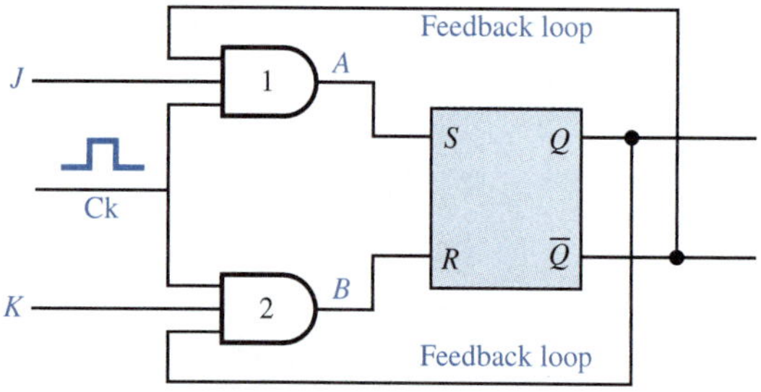

Figure P6.1.51

(a) $JK = 00$

(b) $JK = 10$

(c) $JK = 01$

(d) $JK = 11$

6.1.52 Figure P6.1.52 shows the master–slave JKFF. Assuming that the output changes on the falling edge of the clock pulse (i.e., when the clock pulse goes from high to low), discuss the operation of the flip-flop, and obtain a table indicating the state of normal output Q after the passage of one clock pulse for various combinations of the inputs JK.

6.2.1 A table for the direct 3-bit binary decoding is given. Show a block diagram for a 3-to-8 decoder and suggest a method for its implementation.

A	B	C	Output = 1
0	0	0	F_0
0	0	1	F_1
0	1	0	F_2
0	1	1	F_3
1	0	0	F_4
1	0	1	F_5
1	1	0	F_6
1	1	1	F_7

*6.2.2 (a) *Excess-3 code* is a 4-bit binary code for the 10 decimal digits and is found useful in digital computer arithmetic. Each combination is found by adding 3 to the decimal number being coded and translating the result into direct

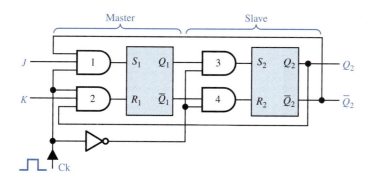

Figure P6.1.52

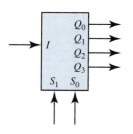

Figure P6.2.9

S_1	S_0	I	Q_3	Q_2	Q_1	Q_0
0	0	0	0	0	0	0
0	0	1	0	0	0	1
0	1	0	0	0	0	0
0	1	1	0	0	1	0
1	0	0	0	0	0	0
1	0	1	0	1	0	0
1	1	0	0	0	0	0
1	1	1	1	0	0	0

binary form. Set up a table for the excess-3 code for the 10 decimal digits.

(b) Set up a table for a 4-to-10 line excess-3 decoding.

6.2.3 A common requirement is conversion from one digital code to another. Develop a table of the BCD code and the excess-3 code [see Problem 6.2.2(a)] to be derived from it, for the decimal digits 0 to 9. Show a block diagram for a BCD to excess-3 code converter.

6.2.4 Draw a block diagram for a 2-to-4 decoder. Obtain the truth table, and develop a logic diagram.

6.2.5 Illustrate BCD-to-decimal decoding with a 4-to-16 decoder, and draw the corresponding truth table.

6.2.6 Based on the 8421 BCD code for decimal digits 0 through 9, develop a block diagram for a BCD encoder and its implementation scheme.

*6.2.7 Implement the following Boolean functions by employing 8-to-1 multiplexers (see Example 6.2.1 in the text for a table of *minterms*).

(a) $F_1(A, B, C) = \sum m_i(0, 2, 4, 6)$

(b) $F_2(A, B, C) = \sum m_i(1, 3, 7)$

6.2.8 Using two 8-to-1 multiplexers and one 2-to-1 multiplexer, show how a 16-to-1 multiplexer can be obtained in the form of a block diagram.

6.2.9 Given the block diagram and the truth table of a *demultiplexer,* as shown in Figure P6.2.9, obtain its implementation.

6.2.10 Use a 4-to-1 multiplexer to simulate the following:

(a) NAND logic function.

(b) EXCLUSIVE-OR logic function.

(c) $\sum m_i(1, 2, 4)$.

6.2.11 Show how a 16-to-1 multiplexer can be used to implement the logic function described by the following truth table.

A	B	C	D	Q
0	0	0	0	1
0	0	0	1	1
0	0	1	0	1
0	0	1	1	0
0	1	0	0	1
0	1	0	1	0
0	1	1	0	1
0	1	1	1	1
1	0	0	0	0
1	0	0	1	0
1	0	1	0	1
1	0	1	1	0
1	1	0	0	0
1	1	0	1	1
1	1	1	0	1
1	1	1	1	0

*6.2.12 Show an arrangement for multiplexing 64-to-1 by using four 16-to-1 multiplexers and one 4-to-1 multiplexer.

6.2.13 Sketch the output waveforms for the register of Figure 6.2.4(a) in the text if JKFFs are used in place of D flip-flops.

6.2.14 Show a block diagram of a 4-bit, parallel-input shift-right register and briefly explain its operation.

6.2.15 Draw the timing diagram of Example 6.2.3 for a 4-bit shift-right register.

6.2.16 Let the content of the register of Example 6.2.3 be initially 0111. With data in being 101101, what is the content of the register after six clock pulses?

6.2.17 A shift register can be used as a binary (a) divide-by-2, and (b) multiply-by-2 counter. Explain.

*6.2.18 Show a block diagram of a 4-bit shift-right register using JKFFs.

6.2.19 Obtain a block diagram of a shift-left/right register using D flip-flops.

6.2.20 Design a 4-bit universal shift register.

6.2.21 (a) Show a block diagram of an SRFF connected to store 1 bit.

(b) Using 4 SRFFs obtain the block diagram for an SISO shift register.

(c) See what can be done to convert the SISO device to SIPO.

*6.2.22 Draw a block diagram of a 4-bit PIPO register and briefly describe its operation.

6.2.23 Taking parallel data from a computer to be fed out over a single transmission line needs a PISO device. Develop a block diagram for such a shift register and briefly explain its operation.

6.2.24 Give a block diagram for a modulo-5 binary ripple counter using JKFFs and draw its timing diagram.

6.2.25 (a) For a JKFF with $JK = 11$, the output changes on every clock pulse. The change will be co-incident with the clock pulse trailing edge and the flip-flop is said to toggle, when $T = 1$, for the T flip-flop. Show JKFF connected as a T flip-flop and its timing diagram.

(b) Using T flip-flops, show the block diagram for a 3-bit ripple counter and its input and output waveforms.

6.2.26 Sketch the timing diagram for a 4-bit ripple counter which uses T flip-flops. (See Problem 6.2.25.)

*6.2.27 Counting to moduli other than 2^n is a frequent requirement, the most common being to count through the binary-coded decimal (BCD) 8421 sequence. All that is required is a four-stage counter which, having counted from 0000 to 1001 (i.e., decimal 0 to 9; ten states), resets to 0000 on the next clock pulse. Develop a block diagram of an *asynchronous decade counter* and show its timing diagram.

6.2.28 Consider the synchronous counter shown in Figure 6.2.6 of the text.

(a) Draw its timing diagram.

(b) Show the implementation of the same synchronous counter using D flip-flops.

(c) Draw the timing diagram for part (b).

6.2.29 Consider a *series-carry synchronous counter* with T flip-flops shown in Figure P6.2.29 in which the AND gates carry forward the transitions of the flip-flops, thereby improving the speed. Sketch the output waveform for the synchronous counter.

6.2.30 Figure P6.2.30 shows the *mod-8 counter* which counts from 0_{10} to 7_{10} before resetting. Explain the operation of the counter and sketch the timing diagram.

6.2.31 Counters are used to realize various dividers in the schematic representation of the digital clock shown in Figure P6.2.31. The blocks labeled "logic array" are logic gate combinations required to activate the corresponding segments in order to display the digits.

(a) Check to see that the six outputs (Y_0 through Y_5) display the number of hours, minutes, and seconds.

(b) If the date is also to be displayed, suggest additional circuitry.

*6.2.32 Determine the bits required for a D/A converter to detect 1-V change when $V_{\text{ref}} = 15$ V.

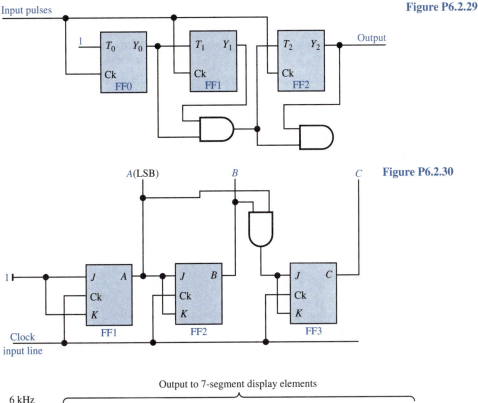

Figure P6.2.30

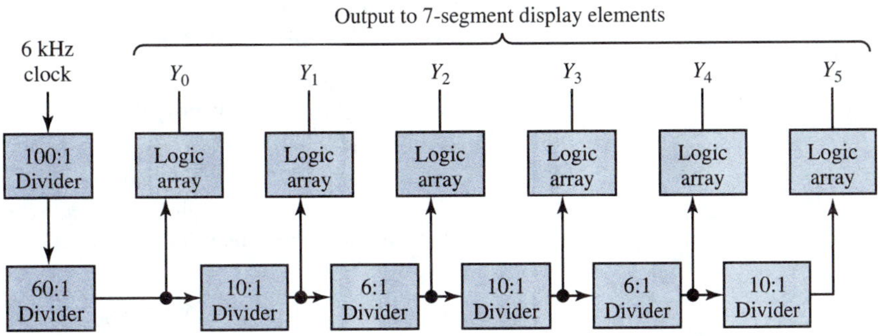

Figure P6.2.31

6.2.33 For the 4-bit D/A converter of Figure 6.2.9, calculate:

   (a) The maximum analog output voltage.

   (b) The minimum analog output voltage.

   (c) The smallest detectable analog output voltage when $V_{ref} = -10$ V.

6.2.34 For the 4-bit weighted-resistor D/A converter shown in Figure 6.2.9, prepare a table showing decimal, binary equivalent, and current I_{in} in per unit, where 1 pu $= V_{ref}/R$. Also sketch the analog

output waveform, i.e., I_{in} as a function of digital binary input.

6.2.35 For a 6-bit weighted-resistor D/A converter, if R is the resistor connected to the MSB, find the other resistor values needed, and calculate the maximum analog output voltage, the minimum analog output voltage, and the smallest detectable analog output voltage if $V_{ref} = -15$ V.

6.2.36 Analyze the 2-bit R–$2R$ ladder-network D/A converter, and corresponding to binary 01, 10, and 11, obtain the equivalent circuits and determine the analog output voltage as a fraction of the reference voltage.

*6.2.37 Consider the 4-bit R–$2R$ ladder D/A converter with $V_{ref} = -10$ V. Determine the analog output voltage when the binary input code is 1100. Also, find what reference voltage is to be used in order to obtain the corresponding decimal output voltage.

6.2.38 For a 10-bit R–$2R$ ladder-network D/A converter with an MSB resistor value of 10 kΩ, what is the value of the LSB resistor?

6.2.39 What is the basic difference between the weighted-resistor and the R–$2R$ ladder D/A converters?

6.2.40 (a) Design a 6-bit R–$2R$ ladder D/A converter.

(b) For $V_{ref} = 10$ V, find the maximum output voltage.

(c) Determine the output voltage increment.

(d) If the output voltage is to indicate increments of 0.1 V, find the bits that must be used.

6.2.41 What is the basic difference between counter-controlled and successive-approximation A/D converters?

*6.2.42 Consider the dual-slope A/D converter of Figure 6.2.15.

(a) Calculate the total charge on the integrator due to the input voltage V_{in} during the signal integration time T.

(b) Obtain an expression for the discharge time t_d in terms of V_{in}, V_{ref}, and T.

6.2.43 An 8-bit A/D converter is driven by a 1-MHz clock. Estimate the maximum conversion time if:

(a) It is a counter-controlled A/D converter.

(b) It is a successive-approximation A/D converter.

6.2.44 How many 500-page books can be stored on a 2400-ft, 1600-BPI magnetic tape if a typewritten page contains about 2500 bytes?

6.2.45 Suppose a ROM holds a total of 8192 bits.

(a) How many bits long would the individual addresses have to be?

(b) If the bits are organized into 8-bit memory words or bytes, how many words would there be, and how many bits long would the addresses have to be?

(c) How is such a ROM described?

(d) If each location requires its own word line emanating from a decoder AND gate, how many gates would the decoder for 1K-byte ROM have to contain?

(e) Develop a *two-dimensional addressing* system using a 6-to-64 decoder, a 64-word × 128-bit matrix, and 16-input multiplexers. How many gates would such a system require?

6.2.46 Show the schematic arrangement for: (a) one-dimensional addressing, and (b) two-dimensional addressing (see Problem 6.2.45), if a 32-kbit ROM is used to provide an 8-bit output word.

6.2.47 Repeat Problem 6.2.46 if a 64-kbit ROM is to provide a 16-bit output word.

*6.2.48 Sketch a typical circuit for a 2-input, 4-output decoder.

6.2.49 Digital watches display time by turning on a certain combination of the seven-segment display device.

(a) Show a typical seven-segment display.

(b) Develop a truth table for turning on the segments. The truth table should have inputs W, X, Y, and Z to represent the binary equivalents of the decimal integers, and outputs S_0, $S_1, \ldots, S_7$.

(c) Develop a typical circuit for one segment, S_0.

(d) Show a schematic diagram of the seven-segment decoder/driver block (available in IC form).

6.2.50 Develop a schematic diagram of a system in which the D/A converter of Figure 6.2.13 can be employed in a digital voltmeter.

7 Semiconductor Devices

Turning our attention to the internal structure of integrated-circuit (IC) building blocks, we encounter a new family of circuit elements known as *semiconductor devices*, which include diodes and transistors of various kinds. These circuit elements are nonlinear in their *i–v* characteristics. Nonlinearity complicates circuit analysis and calls for new methods of attack. However, the semiconductor devices are extremely important for the *electronic circuits*.

Active circuits contain not only passive circuit elements (such as resistors, capacitors, or inductors) but also active elements such as transistors. All of the analog and digital blocks discussed in Chapters 5 and 6 are active circuits, and transistors are essential for their internal construction.

We shall present in this chapter the most important semiconductor devices, such as diodes, bipolar junction transistors, and field-effect transistors. Chapters 8 and 9 will deal with their applications in analog and digital circuits.

7.1 SEMICONDUCTORS

Semiconductors are crystalline solid materials whose resistivities have values between those of conductors and insulators. Conductivity ranges from about 10^{-6} to about 10^5 S/m. Silicon is by far the most important semiconductor material used today. The conductivity of pure silicon is about

4.35×10^{-4} S/m. Good electrical characteristics and feasible fabrication technology have been requisites for the prevalence of silicon technology. Compound semiconductors, such as gallium arsenide, are being developed for microwave and photonic applications, while germanium is used for a few special purposes.

When unbound negatively charged electrons move through a crystal, as shown in Figure 7.1.1(a), electrical conduction in semiconductors can take place with a direction of current opposite to the direction of movement of the electrons. When a bound electron that should be present in the valence bond is missing, the vacancy that arises is known as a *hole*. Holes are positively charged particles with a charge equal in magnitude to that of the electron. Mobile positively charged holes can also give rise to a current, as shown in Figure 7.1.1(b), with the direction of current in the same direction as the movement of holes. Both holes and unbound electrons are known as charge carriers, or simply *carriers*.

Pure semiconductors, known as *intrinsic semiconductors,* have very few charge carriers and may hence be classified as almost insulators or very poor electrical conductors. However, by adding (through a process known as *doping*) tiny controlled amounts of impurities (such as boron, gallium, indium, antimony, phosphorus, or arsenic), a semiconductor can be made to contain a desired number of either holes or free electrons and is then known as *extrinsic* (impure) material. A *p-type semiconductor* contains primarily holes, whereas an *n-type semiconductor* contains primarily free electrons. While holes are the *majority carriers* in a *p*-type material, it is possible to inject electrons artificially into *p*-type material, in which case they become excess *minority carriers*. Minority carriers do play a vital role in certain devices. The doping substance is called an *acceptor* when the extrinsic semiconductor is the *p*-type with holes forming the majority carriers and electrons forming the minority carriers. The doping substance is known as the *donor* when the extrinsic semiconductor is the *n*-type with free electrons forming the majority carriers and holes forming the minority carriers. Both *p*- and *n*-type semiconductors are vitally important in solid-state device technology. Diodes, transistors, and other devices depend on the characteristics of a *pn*-junction formed when the two materials are joined together as a single crystal.

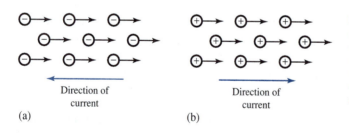

Figure 7.1.1 Electrical conduction. (**a**) Electrons moving from left to right give rise to a current directed from right to left. (**b**) Holes moving from left to right give rise to a current directed from left to right.

7.2 DIODES

A single *pn*-junction with appropriate contacts for connecting the junction to external circuits is called a semiconductor *pn-junction diode.* The fundamental building block upon which all semiconductor devices are based is the *pn*-junction. The most common two-terminal nonlinear resistor is the semiconductor diode, whose symbol is shown in Figure 7.2.1(a). The terminal voltage and current are denoted by v_D and i_D, respectively. The physical structure of the *pn*-junction is shown in Figure 7.2.1(b). The junction is made by doping the two sides of the crystal with different impurities. Figure 7.2.1(c) shows the *volt–ampere curve* (or *static characteristic*) of the ideal (or perfect) diode. Note that when v_D is zero, i_D is not and vice versa, a condition corresponding to a switch. A diode acts like a switch that closes to allow current flow in the *forward direction*, but opens to prevent current flow in the *reverse direction*. The diode thus acts like a unilateral circuit element providing an on–off characteristic.

The physical operation of the junction can be described in terms of the charge-flow processes. Usually there is a greater concentration of holes in the *p*-region than in the *n*-region; similarly, the electron concentration in the *n*-region is greater than that in the *p*-region. The differences in concentration establish a potential gradient across the junction, resulting in a *diffusion* of carriers, as indicated in Figure 7.2.2(a). Holes diffuse from the *p*-region to the *n*-region, electrons from the *n*-region to the *p*-region. The result of the diffusion is to produce immobile ions of opposite charge on each side of the junction, as shown in Figure 7.2.2(b), and cause a *depletion region* (or *space-charge region*) in which no mobile carriers exist.

The immobile ions (or space charge), being of opposite polarity on each side of the junction, establish an electric field because of which a potential barrier is formed and *drift current* is produced. The drift current causes holes to move from the *n*- to the *p*-region and electrons to move from the *p*- to the *n*- region, as shown in Figure 7.2.2(c). In equilibrium and with no external circuit, the drift and diffusion components of current are equal and oppositely directed. The potential barrier established across the depletion region prohibits the flow of carriers across the junction without the application of energy from an external source.

pn-Junction under Bias

Let an external source be connected between the *p*- and *n*-regions, as shown in Figure 7.2.3(a). Figure 7.2.3(b) shows the circuit representation of the *pn*-junction or diode, and its external circuit. The voltage source V, called the *bias*, either decreases or increases the potential barrier, thereby controlling the flow of carriers across the junction. With $V = 0$, the barrier is unaffected and the *pn*-junction has zero current. Positive values of V, known as *forward biasing,* decrease the potential barrier, thereby increasing the number of electrons and holes diffusing across the junction. The increased diffusion results in a net current, called the forward current, from the *p*- to the *n*-region. With increased V, the forward current further increases rapidly because the barrier is reduced even further. Making V negative (*reverse biasing*), on the other hand, increases the potential barrier and reduces the number of carriers diffusing across the boundary. The drift component produced by the electric field from the *n*- to the *p*-region causes a small current, called the *reverse current* (or *saturation current*) I_S. The magnitude of the saturation current depends on the doping levels in the *p*- and *n*-type materials and on the physical size of the

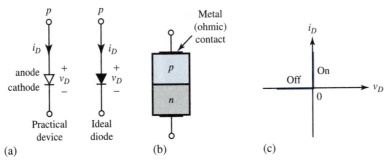

Figure 7.2.1 *pn*-junction. **(a)** Circuit symbol for *pn*-junction diode. **(b)** Physical structure. **(c)** Volt–ampere characteristic of an ideal (or perfect) diode.

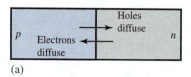

(a)

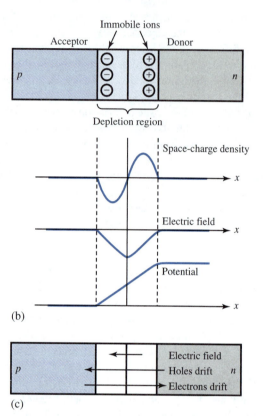

(b)

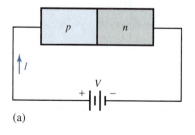

(c)

Figure 7.2.2 *pn*-junction diode with no external voltage source. (a) Hole and electron diffusion. (b) Depletion region. (c) Drift of electrons and holes.

junction. Increasing the reverse bias, however, does not affect the reverse current significantly until breakdown occurs.

The static characteristic of a junction diode is shown in Figure 7.2.4(a), which describes the dc behavior of the junction and relates the diode current I and the bias voltage V. Such a characteristic is analytically expressed by the *Boltzmann diode equation*

$$I = I_S(e^{V/\eta V_T} - 1) \tag{7.2.1}$$

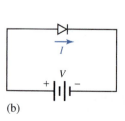

(a)   (b)

Figure 7.2.3 *pn*-junction under bias. (a) Physical structure. (b) Semiconductor diode symbol.

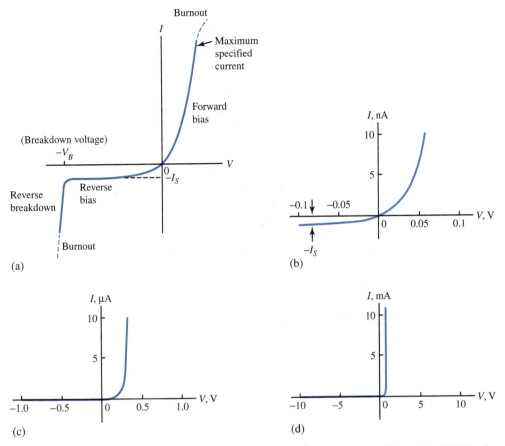

Figure 7.2.4 Typical static volt–ampere characteristic (dc behavior) of a *pn*-junction diode. **(a)** Showing reverse breakdown. **(b)**, **(c)**, **(d)** Omitting reverse breakdown (plotted on different scales).

in which η depends on the semiconductor used (2 for germanium and nearly 1 for silicon), and V_T is the thermal voltage given by

$$V_T = \frac{kT}{q} = \frac{T}{11,600} \qquad (7.2.2)$$

where k is Boltzmann's constant ($= 1.381 \times 10^{-23}$ J/K), q is the magnitude of the electronic charge ($= 1.602 \times 10^{-19}$ C), and T is the junction temperature in kelvins (K $= °$C $+ 273.15$). At room temperature ($T = 293$ K), V_T is about 0.025 V, or 25 mV. Using $\eta = 1$, Equation (7.2.1) is expressed by

$$I = I_S(e^{40V} - 1) \qquad (7.2.3)$$

or by the following, observing that $e^4 \gg 1$ and $e^{-4} \ll 1$,

$$I = \begin{cases} I_S e^{40V}, & V > 0.1 \text{ V} \\ -I_S, & V < -0.1 \text{ V} \end{cases} \qquad (7.2.4)$$

which brings out the difference between the forward-bias and reverse-bias behavior. The reverse saturation current is typically in the range of a few nanoamperes (10^{-9} A). In view of the exponential factor in Equation (7.2.3), the apparent shape of the I–V curve depends critically upon the scale of the voltage and current axes. Figures 7.2.4(b), (c), and (d) illustrate this point, taking $I_S = 1$ nA $= 10^{-9}$ A. A comparison of Figure 7.2.4(d) with Figure 7.2.1(c) suggests that one can use the ideal diode as a *model* for a semiconductor diode whenever the forward voltage drop and the reverse current of the semiconductor diode are unimportant.

Based on the ability of the junction to dissipate power in the form of heat, the *maximum forward current* rating is specified. Based on the maximum electric field that can exist in the depletion region, the *peak inverse voltage* (maximum instantaneous value of the reverse-bias voltage) rating is specified.

The most apparent difference between a real diode and the ideal diode is the nonzero voltage drop when a real diode conducts in the forward direction. The finite voltage drop across the diode is accounted for by V_{on}, known as the *offset* or *turn-on* or *cut-in* or *threshold* voltage, as shown in the alternate representation of the junction diode in Figure 7.2.5(a). Typical values of V_{on} are 0.6 to 0.7 V for silicon devices and 0.2 to 0.3 V for germanium devices.

A closer approximation to the actual diode volt–ampere characteristic than that in Figure 7.2.5(a) is depicted in Figure 7.2.5(b), which includes the effect of the *forward (dynamic) resistance* R_f, whose value is the reciprocal of the slope of the straight-line portion of the approximate characteristic beyond the threshold voltage V_{on}.

As an extension of the diode model of Figure 7.2.5(b), to allow for more realistic volt–ampere characteristic slopes, the diode's *reverse resistance* R_r for $v < V_{on}$ is included in the model of Figure 7.2.6.

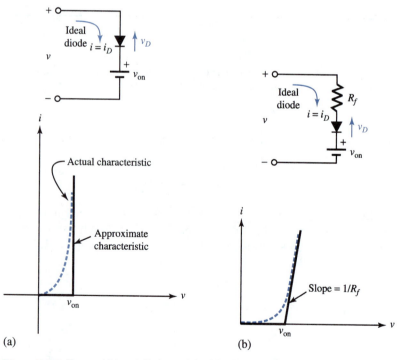

(a)

(b)

Figure 7.2.5 Forward-biased diode models. **(a)** With threshold voltage V_{on}. **(b)** With threshold voltage V_{on} and forward resistance R_f.

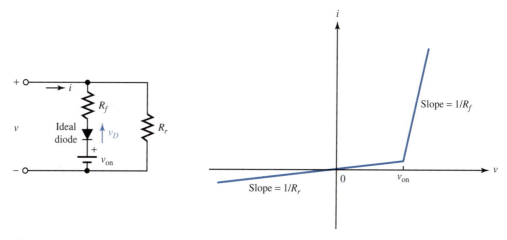

Figure 7.2.6 Piecewise-linear model of a diode, including the threshold voltage V_{on}, forward resistance R_f, and reverse resistance R_r.

Two types of capacitors are associated with a *pn*-diode: the *junction capacitance* C_J (also known as *depletion capacitance* or *space-charge capacitance*), which is dominant for a reverse-bias diode; and the *diffusion capacitance* C_D, which is most significant for the forward-bias condition and is usually negligible for a reverse-biased diode. For applications where the diode capacitance is important, the small-signal equivalent circuit under back (reverse)-biased operation includes R_r in parallel with C_J, and the parallel combination of R_f, C_J, and C_D for forward-biased operation.

Elementary Diode Circuits

Semiconductor diodes are used in a wide variety of applications. Their usage abounds in communication systems (limiters, gates, clippers, mixers), computers (clamps, clippers, logic gates), television (clamps, limiters, phase detectors), radar (power detectors, phase detectors, gain-control circuits, parametric amplifiers), and radio (mixers, automatic gain-control circuits, message detectors). Several simple diode circuits are presented in this section to serve only as examples.

In solving circuit problems with a nonlinear element, such as a diode, a useful technique employed in many cases is a graphical approach. After plotting a nonlinear characteristic of the element, such as the volt–ampere characteristic of a diode, one can superimpose a plot of the circuit response (excluding the nonlinear element), which is the equation of a straight line in the i–v plane given by the loop equation for the network. Such an equation of the straight line is known as the *load line* equation. The intersection of the load line with the characteristic of the nonlinear element in the i–v plane determines the *quiescent (operating) point*, which is the desired solution. Example 7.2.1 illustrates the procedure.

EXAMPLE 7.2.1

Determine whether the diode (considered to be *ideal*) in the circuit of Figure E7.2.1(a) is conducting.

Solution

For determining the condition of the ideal diode, let us initially assume that it does not conduct, and let us replace it with an open circuit, as shown in Figure E7.2.1(b). The voltage across the 10-Ω resistor can be calculated as 8 V by the voltage-divider rule. Then, applying KVL around the right-hand loop, we get

$$8 = v_D + 10 \qquad \text{or} \qquad v_D = -2 \text{ V}$$

That is to say, the diode is not conducting since $v_D < 0$. This result is consistent with the initial assumption, and therefore the diode does not conduct.

The student is encouraged to reverse the initial assumption by presuming that the diode is conducting, and show the same result as obtained in the preceding.

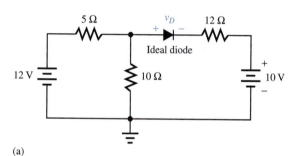

Figure E7.2.1

(a)

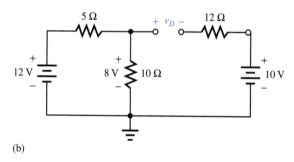

(b)

EXAMPLE 7.2.2

Use the offset diode model with a threshold voltage of 0.6 V to determine the value of v_1 for which the diode D will first conduct in the circuit of Figure E7.2.2(a).

Solution

Figure E7.2.2(b) shows the circuit with the diode replaced by its circuit model. When v_1 is zero or negative, it is safe to assume that the diode is off. Assuming the diode to be initially off, no current flows in the diode circuit. Then, applying KVL to each of the loops, we get

$$v_1 = v_D + 0.6 + 2 \qquad \text{and} \qquad v_0 = 2$$

Since $v_D = v_1 - 2.6$, the condition for the diode to conduct is $v_1 > 2.6$ V.

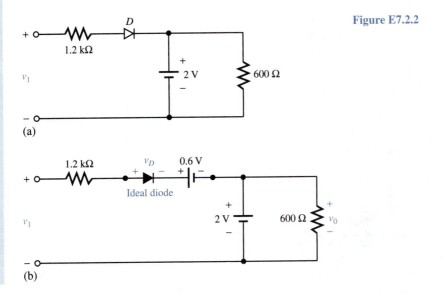

EXAMPLE 7.2.3

We shall demonstrate load-line analysis to find the diode current and voltage, and then compute the total power output of the battery source in the circuit of Figure E7.2.3(a), given the diode i–v characteristic shown in Figure E7.2.3(b).

Solution

The Thévenin equivalent circuit as seen by the diode is shown in Figure E7.2.3(c). The load-line equation, obtained by the KVL, is the equation of a line with slope $-1/R_{Th}$ and ordinate intercept given by V_{Th}/R_{Th},

$$i_D = -\frac{1}{R_{Th}}v_D + \frac{1}{R_{Th}}V_{Th}$$

Superposition of the load line and the diode i–v curve is shown in Figure E7.2.3(d). From the sketch we see that the load line intersects the diode curve at approximately 0.67 V and 27.5 mA, given by the Q point (quiescent or operating point). The voltage across the 10-Ω resistor of Figure E7.2.3(a) is then given by

$$V_{10\Omega} = 40I_Q + V_Q = 1.77 \text{ V}$$

The current through the 10-Ω resistor is thus 0.177 A, and the total amount out of the source is therefore given by $0.177 + 0.0275 = 0.2045$ A. The total power supplied to the circuit by the battery source is then

$$12 \times 0.2045 = 2.454 \text{ W}$$

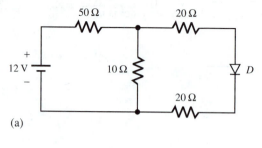

(a)

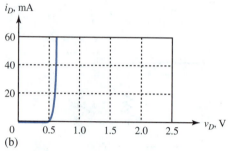

(b)

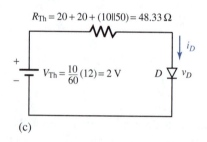

(c)

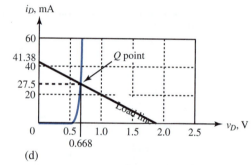

(d)

Figure E7.2.3

EXAMPLE 7.2.4

Consider the circuit of Figure E7.2.4(a) with $v_S(t) = 10 \cos \omega t$. Use the piecewise-linear model of the diode with a threshold voltage of 0.6 V and a forward resistance of 0.5 Ω to determine the rectified load voltage v_L.

Solution

Figure E7.2.4(b) shows the circuit with the diode replaced by its piecewise-linear model. Applying KVL,

$$v_S = v_1 + v_2 + v_D + 0.6 + v_L \qquad \text{or} \qquad v_D = v_S - v_1 - v_2 - 0.6 - v_L$$

The diode is off corresponding to the negative half-cycle of the source voltage. Thus no current flows in the series circuit; the voltages v_1, v_2, and v_L are all zero. So when the diode is not conducting, the following KVL holds:

$$v_D = v_S - 0.6$$

When $v_D \geq 0$ or $v_S \geq 0.6$ V, the diode conducts. Once the diode conducts, the expression for the load voltage can be obtained by the voltage divider rule, by considering that the ideal diode behaves like a short circuit. The complete expression for the load voltage is therefore given by

$$v_L = \frac{10}{10 + 1 + 0.5}(v_S - 0.6) = 8.7 \cos \omega t - 0.52, \qquad \text{for } v_S \geq 0.6 \text{ V}$$

and $v_L = 0$ for $v_S < 0.6$ V. The source and load voltages are sketched in Figure E7.2.4(c).

Figure E7.2.4

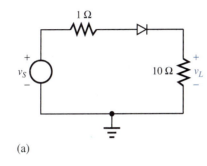

(a)

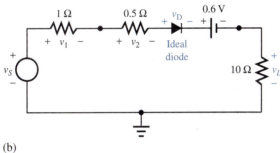

(b)

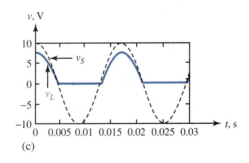

(c)

EXAMPLE 7.2.5

Consider a forward-biased diode with a load resistance. Let the static volt–ampere characteristic of the diode be given by Equations (7.2.1) and (7.2.2), and typically represented by Figure 7.2.4.

(a) For a dc bias voltage V_B, obtain the load-line equation and the operating (quiescent) point

(I_Q, V_Q) by graphical analysis. Extend the graphical analysis for different values of (i) load resistance, and (ii) supply voltage.

(b) If, in addition to the constant potential V_B, an alternating or time-varying potential $v_S(t) = \sqrt{2}\, V_S \sin \omega t$ is impressed across the circuit, discuss the dynamic (ac) characteristics of the diode in terms of (i) small-signal current and voltage waveforms, and (ii) large-signal current and voltage waveforms.

Solution

(a) The circuit of a forward-biased diode with a load resistance R_L is shown in Figure E7.2.5(a). The KVL equation yields

$$V_B = I R_L + V \qquad \text{or} \qquad I = \frac{V_B - V}{R_L}$$

which is the load-line equation. The device equation (Boltzmann diode equation) and the load-line equation involve two variables, I and V, whose values must satisfy both equations simultaneously. As seen from Figure E7.2.5(b), Q is the only condition satisfying the restrictions imposed by both the diode and the external circuit. The intersection Q of the two curves is called the quiescent or operating point, indicated by the diode current I_Q and the diode voltage V_Q.

Extension of the graphical analysis for different values of load resistance and different values of supply voltage is shown in Figures E7.2.5(c) and (d).

(b) The diode circuit with dc and ac sources is shown in Figure E7.2.5(e). The total instantaneous voltage impressed across the circuit is given by

$$v_t = V_B + \sqrt{2}\, V_S \sin \omega t \qquad \text{and} \qquad v_t = v + i R_L$$

The maximum and minimum values of v_t are $(V_B + \sqrt{2}\, V_S)$ and $(V_B - \sqrt{2}\, V_S)$, corresponding to the values of $\sin \omega t$ equal to $+1$ and -1, respectively. The small-signal current and voltage waveforms, for values of V_S much less than V_B, are shown in Figure E7.2.5(f). The large-signal current and voltage waveforms, for values of V_S comparable to those of V_B, are shown in Figure E7.2.5(g).

The motion of the load line traces the shaded area of the characteristics. The line segment $Q_1 Q_2$ is the locus of the position of the operating point Q. It is clear from the figures that the waveforms of the diode voltage and current are functions of time. A point-by-point method should be used for plotting the waveforms. For the small-signal case, the diode can be considered to behave linearly and the segment $Q_1 Q_2$ is approximated by a straight line. For the large-signal case, on the other hand, the behavior is nonlinear. The time-varying portion of the response is not directly proportional to $v_S(t)$, and a simple superposition of the direct and alternating responses does not apply.

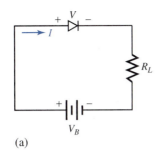

(a)

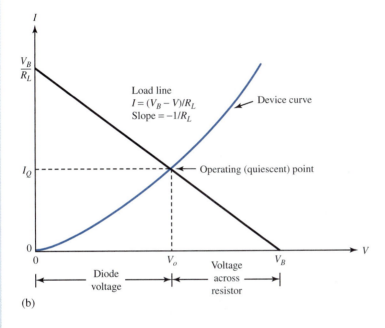

Load line
$I = (V_B - V)/R_L$
Slope $= -1/R_L$

Device curve

I_Q

Operating (quiescent) point

V_o

V_B

Diode voltage

Voltage across resistor

(b)

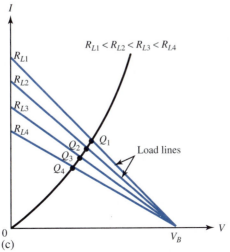

$R_{L1} < R_{L2} < R_{L3} < R_{L4}$

R_{L1}
R_{L2}
R_{L3}
R_{L4}

Q_1
Q_2
Q_3
Q_4

Load lines

V_B

(c)

Figure E7.2.5 (a) Circuit of forward-biased diode. (b) Graphical analysis of forward-biased diode with load resistance. (c) Graphical analysis for different values of load resistance. (d) Graphical analysis for different values of supply voltage. (e) Diode circuit with dc and ac sources. (f) Small-signal current and voltage waveforms. (g) Large-signal current and voltage waveforms.

Figure E7.2.5 Continued

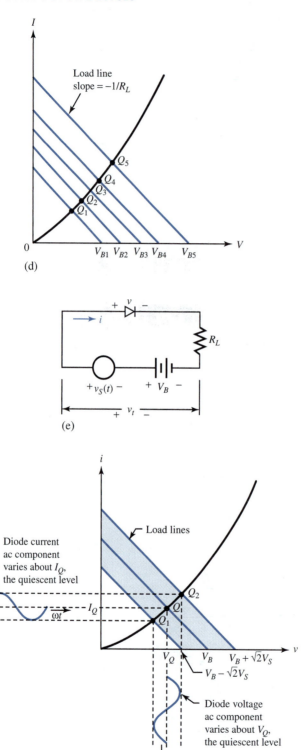

(d)

(e)

(f)

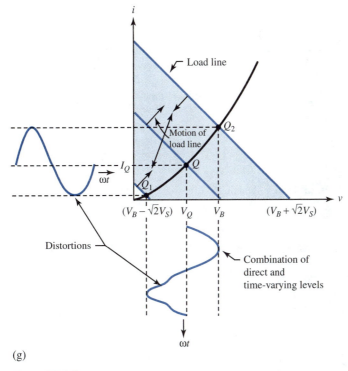

(g)

Figure E7.2.5 Continued

Zener Diodes

Most diodes are not intended to be operated in the reverse breakdown region [see Figure 7.2.4(a)]. Diodes designed expressly to operate in the breakdown region are called *zener diodes*. A nearly constant voltage in the breakdown region is obtained for a large range of reverse current through the control of semiconductor processes. The principal operating region for a zener diode is the negative of that for a regular diode in terms of both voltage and current. Zener diodes are employed in circuits for establishing *reference voltages* and for maintaining a constant voltage for a load in regulator circuits. Figure 7.2.7 shows the device symbol along with the linearized i–v curve and the circuit model.

As seen from the i–v characteristic, a zener diode approximates an ideal diode in the forward region. However, when the reverse bias exceeds the zener voltage V_Z, the diode starts to conduct in the reverse direction and acts like a small reverse resistance R_Z in series with a battery V_Z. Zener diodes are available with values of V_Z in the range of 2 to 200 V. The circuit model of Figure 7.2.7(c) incorporates two ideal diodes, D_f and D_r, to reflect the forward and reverse characteristics of the zener diode. The i–v curve thus has two *breakpoints*, one for each ideal diode, and three straight-line segments.

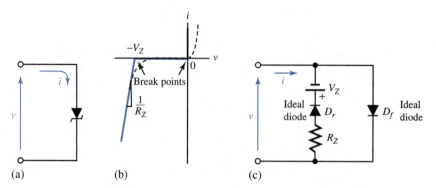

Figure 7.2.7 Zener diode. **(a)** Device symbol. **(b)** Linearized i–v curve. **(c)** Circuit model.

EXAMPLE 7.2.6

Consider a simple zener voltage regulator with the circuit diagram shown in Figure E7.2.6(a).

Figure E7.2.6

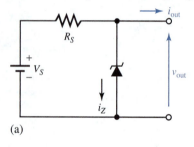

(a)

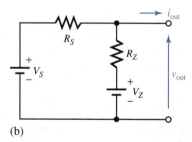

(b)

(a) For a small reverse resistance $R_Z \ll R_S$ and $V_S - R_S i_{out} > V_Z$, show that $v_{out} \cong V_Z$.

(b) For values of $V_S = 25$ V, $R_S = 100 \ \Omega$, $V_Z = 20$ V, and $R_Z = 4 \ \Omega$, find:

   (i) v_{out} for $i_{out} = 0$ and $i_{out} = 50$ mA.

   (ii) The corresponding values of the reverse current i_Z through the zener diode.

Solution

(a) When $V_S - R_S i_{out} > V_Z$, the zener diode will be in reverse breakdown. The forward diode D_f in our model of Figure 7.2.7(c) will be off while the reverse diode D_r is on. The equivalent circuit is then given by Figure E7.2.6(b).

    Straightforward circuit analysis yields

$$v_{out} = \frac{R_S}{R_S + R_Z}\left(V_Z + \frac{R_Z}{R_S}V_S - R_Z i_{out}\right)$$

For $R_Z \ll R_S$, $R_Z |V_S/R_S - i_{out}| \ll V_Z$, in which case

$$v_{out} \cong V_Z$$

Thus, the zener diode regulates v_{out} by holding it at the fixed zener voltage V_Z, in spite of the possible variations of V_S or i_{out}.

(b) For $i_{out} = 0$,

$$v_{out} = \frac{100}{100 + 4}\left[20 + \frac{4}{100}(25)\right] = \frac{100}{104} \times 21 = 20.19\ V$$

we have $20.19 - 20 = 4i_Z$, or $i_Z = 0.19/4 = 47.5$ mA.

For $i_{out} = 50$ mA,

$$v_{out} = \frac{100}{104}\left[20 + \frac{4}{100}(25) - (4 \times 0.05)\right] = \frac{100}{104} \times 20.8 = 20\ V$$

we have $20 - 20 = 4i_Z$, or $i_Z = 0$.

Breakpoint Analysis

When a circuit consists of two or more ideal diodes, it will have several distinct operating conditions resulting from the off and on states of the diodes. A systematic way of finding those operating conditions is the method of *breakpoint analysis*. For a two-terminal network containing resistors, sources, and N ideal diodes, and driven by a source voltage v, the i–v characteristic will in general consist of $N + 1$ straight-line segments with N breakpoints. The i–v curve can be constructed by following these steps:

1. For $v \to \infty$, determine the states of all diodes, and write i in terms of v; do the same for $v \to -\infty$.
2. With one diode to be at its breakpoint (i.e., having zero voltage drop and zero current), find the resulting values of i and v at the terminals; do the same for each of the other diodes.
3. Plot the i–v breakpoints obtained from step 2; connect them with straight lines and add the end lines found in step 1.

Note that in step 2, if two or more diodes are simultaneously at breakpoint conditions, the numbers of breakpoints and line segments of the i–v curve are correspondingly reduced.

EXAMPLE 7.2.7

Determine the i–v characteristic of the network shown in Figure E7.2.7(a) by the use of breakpoint analysis.

Solution

For $v \to \infty$, D_1 will be forward-biased while D_2 will be reverse-biased (because $v_1 > 10$ V). Hence, with D_1 on and D_2 off, $v = 2i - 12 + 4i$, or $i = v/6 + 2$.

For $v \to -\infty$, $i = 0$ since D_1 will be off and D_2 on. With D_1 at its breakpoint, the circuit is drawn in Figure E7.2.7(b). It follows that $i = 0$ and $v_1 = v + 12$; but one does not know the value of v_1 and the state of D_2. If one assumes $v_1 > 10$ V, then D_2 will be off and there is no source for the current $i_1 = v_1/4$. Hence one concludes that D_2 must be on and $v_1 = 10$ V. Then the corresponding i–v breakpoint is at $i = 0$ and $v = -2$ V.

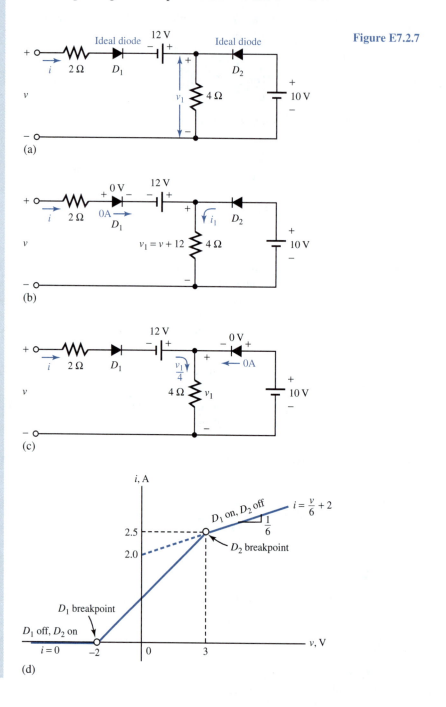

Figure E7.2.7

With D_2 at its breakpoint, the circuit is drawn in Figure E7.2.7(c). It follows that $v_1 = 10$ V and D_1 must be on to carry $i_1 = v_1/4 = 2.5$ A. Thus, the second breakpoint is at $i = 2.5$ A and $v = 3$ V (because $v = 2i - 12 + v_1$).

The complete i–v characteristic based on our results is shown in Figure E7.2.7(d). It can be seen that both D_1 and D_2 will be on over the middle region $-2 < v < 3$.

Rectifier Circuits

A simple *half-wave rectifier* using an ideal diode is shown in Figure 7.2.8(a). The sinusoidal source voltage v_S is shown in Figure 7.2.8(b). During the positive half-cycle of the source, the ideal diode is forward-biased and closed so that the source voltage is directly connected across the load. During the negative half-cycle of the source, the ideal diode is reverse-biased so that the source voltage is disconnected from the load and the load voltage as well as the load current are zero. The load voltage and current are of one polarity and hence said to be rectified. The output current through the load resistance is shown in Figure 7.2.8(c).

In order to smooth out the pulsations (i.e., to eliminate the higher frequency harmonics) of the rectified current, a filter capacitor may be placed across the load resistor, as shown in Figure 7.2.9(a). As the source voltage initially increases positively, the diode is forward-biased since the load voltage is zero and the source is directly connected across the load. Once the source reaches its maximum value V_S and begins to decrease, while the load voltage and the capacitor voltage are momentarily maintained at V_S, the diode becomes reverse-biased and hence open-circuited. The capacitor then discharges over time interval t_2 through R_L until the source voltage $v_S(t)$ has increased to a value equal to the load voltage. Since the source voltage at this point in time exceeds the capacitor voltage, the diode becomes once again forward-biased and hence closed. The capacitor once again gets charged to V_S. The output current of the rectifier with the filter capacitor is shown in Figure 7.2.9(b), and the circuit configurations while the capacitor gets charged and discharged are shown in Figure 7.2.9(c). The smoothing effect of the filter can be improved by increasing the time constant CR_L so that the discharge rate is slowed and the output current more closely resembles a true dc current.

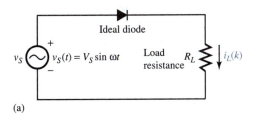

(a)

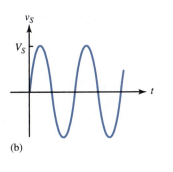

(b)

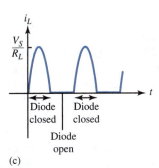

(c)

Figure 7.2.8 Simple half-wave rectifier. **(a)** Circuit with ideal diode. **(b)** Input source voltage. **(c)** Output current through load resistance.

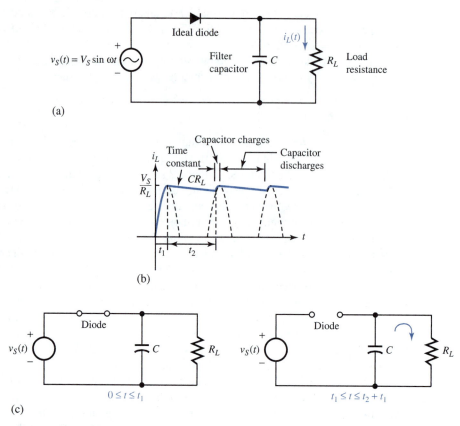

(a)

(b)

(c)

Figure 7.2.9 Rectifier with filter capacitor. **(a)** Circuit. **(b)** Output current of rectifier with filter capacitor. **(c)** Circuit configurations while capacitor gets charged and discharged.

The *full-wave rectifier* using ideal diodes is shown in Figure 7.2.10(a). Figure 7.2.10(b) shows circuit configurations for positive and negative half-cycles of the input source voltage $v_S (= V_S \sin \omega t)$, and Figure 7.2.10(c) shows the rectified output voltage across the load resistance R_L. The full-wave rectification can be accomplished by using either a center-tapped transformer with two diodes or a bridge rectifier circuit with four diodes.

7.3 BIPOLAR JUNCTION TRANSISTORS

The family of bipolar junction transistors has two members: the *npn* BJT and the *pnp* BJT. Both types contain semiconductor junctions which operate with bipolar internal currents consisting of holes and electrons. These are illustrated in Figure 7.3.1 along with their circuit symbols. The emitter in the circuit symbol is identified by the lead having the arrowhead. The arrow points in the direction of conventional emitter current flow when the base–emitter junction is forward biased. A transistor can operate in three modes: *cutoff, saturation,* and *active.* In the active mode, for an *npn* BJT, the base–emitter junction (BEJ) is forward-biased by a voltage v_{BE}, while the collector–base junction (CBJ) is reverse-biased by a voltage v_{CB}. Thus for an *npn* BJT, as shown in Figure 7.3.1(a), i_B and i_C are positive quantities such that $i_B + i_C = i_E$. For a *pnp* BJT on the other hand, in the active region, the base–emitter and collector–base voltages are negative, and

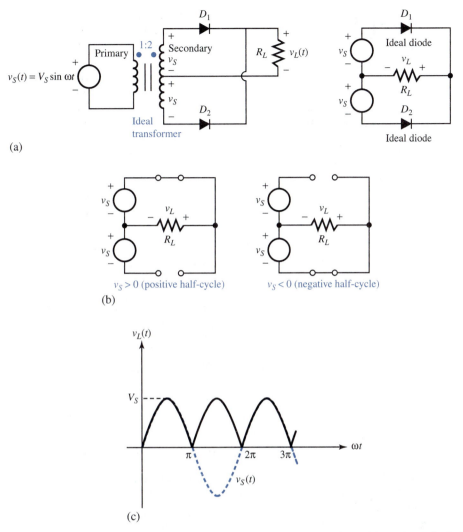

Figure 7.2.10 Full-wave rectifier. (a) Circuit. (b) Circuit configurations for positive and negative half-cycles. (c) Rectified output voltage.

currents i_B, i_C, and i_E are all negative quantities such that $i_E = i_B + i_C$, that is to say, the bias voltages as well as current directions are reversed compared to those of an *npn* BJT.

In an *npn* BJT, current flow is due to majority carriers at the forward-biased BEJ. While the electrons diffuse into the base from the emitter and holes flow from the base to the emitter, the electron flow is by far the more dominant part of the emitter current since the emitter is more heavily doped than the base. Electrons become minority carriers in the base region, and these are quickly accelerated into the collector by action of the reverse bias on the CBJ because the base is very thin. While the electrons are going through the base region, however, some are removed by recombination with majority-carrier holes. The number lost through recombination is only 5% of the total or less. Due to the usual minority-carrier drift current at a reverse-biased *pn* junction, a small current flow, on the order of a few microamperes, denoted by I_{CBO} (collector current when

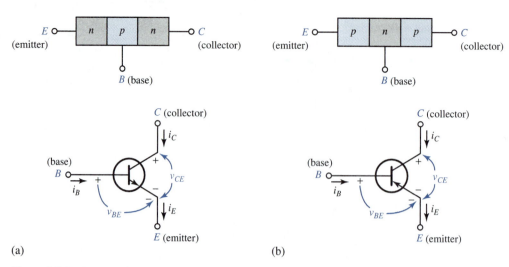

Figure 7.3.1 Bipolar junction transistors. **(a)** *npn* BJT structure and circuit symbol. **(b)** *pnp* BJT structure and circuit symbol.

emitter is open-circuited), called *reverse saturation current,* results. BJTs biased in the active region are shown in Figure 7.3.2.

It can be shown that the currents in a BJT are approximately given by

$$i_E = I_{SE} \, e^{V_{BE}/V_T} = i_B + i_C = \frac{1}{\alpha}I_C - \frac{1}{\alpha}I_{CBO} \tag{7.3.1}$$

$$i_C = i_E + \alpha I_{CBO} \tag{7.3.2}$$

$$i_B = (1 - \alpha)i_E - I_{CBO} = \frac{1-\alpha}{\alpha}i_C - \frac{1}{\alpha}I_{CBO} \tag{7.3.3}$$

where I_{SE} is the reverse saturation current of the BEJ, I_{CBO} is the reverse saturation current of the CBJ, α (known as *common-base current gain* or *forward-current transfer ratio,* typically ranging from about 0.9 to 0.998) is the fraction of i_E that contributes to the collector current, and $V_T = kT/q$ is the thermal voltage (which is the voltage equivalent of temperature, having a value of 25.861×10^{-3} V when $T = 300$ K). Note that the symbol h_{FB} is also used in place of α.

Another important BJT parameter is the *common-emitter current gain,* denoted by β (also symbolized by h_{FE}), which is given by

$$\beta = \frac{\alpha}{1-\alpha} \qquad \left(\text{or} \qquad \alpha = \frac{\beta}{1+\beta} \right) \tag{7.3.4}$$

which ranges typically from about 9 to 500, being very sensitive to changes in α. In terms of β, one can write

$$i_C = \beta i_B + \frac{I_{CBO}}{1-\alpha} = \beta i_B + (\beta + 1)I_{CBO} = \beta i_B + I_{CEO} \tag{7.3.5}$$

where $I_{CEO} = (\beta + 1)I_{CBO}$ is the *collector cutoff current* when the base is open-circuited (i.e., $i_B = 0$).

Figure 7.3.3 illustrates common-base static curves for a typical *npn* silicon BJT. In a common-emitter configuration in which transistors are most commonly used, where the input is to the base and the output is from the collector, the input and output characteristics are shown in Figure 7.3.4.

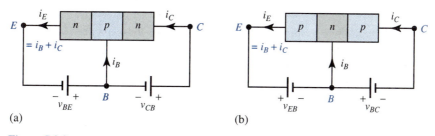

(a)

(b)

Figure 7.3.2 BJTs biased in the active region. **(a)** *npn* BJT (i_B, i_C, and i_E are positive). **(b)** *pnp* BJT (i_B, i_C, and i_E are negative).

With varying but positive base current, as seen from Figure 7.3.4(a), v_{BE} stays nearly constant at the *junction threshold voltage* V_γ, which is about 0.7 V for a typical silicon BJT.

The *Early effect* and the *Early voltage* $-V_A$ (whose magnitude is on the order of 50 to 100 V) for a typical *npn* BJT are illustrated in Figure 7.3.5, in which the linear curves are extrapolated back to the v_{CE}-axis to meet at a point $-V_A$. The Early effect causes the nonzero slope and is due to the fact that increasing v_{BE} makes the width of the depletion region of the CBJ larger, thereby reducing the effective width of the base. I_{SE} in Equation (7.3.1) is inversely proportional to the base width; so i_C increases according to Equation (7.3.2). The increase in i_C can be accounted for by adding a factor to I_{SE} and modifying Equation (7.3.2) such that αi_E is replaced by $\alpha i_E (1 + v_{CE}/V_A)$.

The common-emitter collector characteristics for a typical *pnp* BJT are shown in Figure 7.3.6.

A *small-signal equivalent circuit* of a BJT that applies to both *npn* and *pnp* transistors and is valid at lower frequencies (i.e., ignoring capacitance effects) is given in Figure 7.3.7, where the notation is as follows:

$$\Delta i_C = g_m \Delta v_{BE} + \frac{\Delta v_{CE}}{r_o} \tag{7.3.6}$$

in which

$$\text{Transconductance } g_m = \frac{\partial i_C}{\partial v_{BE}}\bigg|_Q = \frac{I_{CQ}}{V_T} \tag{7.3.7}$$

$$\text{Reciprocal of output resistance } \frac{1}{r_o} = \frac{\partial i_C}{\partial v_{CE}}\bigg|_Q = \frac{I_{CQ}}{V_A} \tag{7.3.8}$$

The derivatives are evaluated at the quiescent or operating point Q at which the transistor is biased to a particular set of static dc currents and voltages. Notice the dependence of i_C on both v_{BE} and v_{CE}. Considering a small base-current change Δi_B occurring due to Δv_{BE}, one can define

$$r_\pi = \frac{\Delta v_{BE}}{\Delta i_B}\bigg|_Q = \frac{\Delta i_C}{\Delta i_B}\frac{\Delta v_{BE}}{\Delta i_C}\bigg|_Q \cong \frac{\partial i_C}{\partial i_B}\bigg|_Q \frac{1}{g_m} = \frac{\beta}{g_m} \tag{7.3.9}$$

and

$$v_\pi = \Delta v_{BE} = r_\pi \Delta i_B \tag{7.3.10}$$

The *large-signal models* of a BJT for the active, saturated, and cutoff states are given in Figure 7.3.8. Note that in Figure 7.3.8(a) $i_E \cong \beta i_B = i_C$ if $\beta \gg 1$. In Figure 7.3.8(b) the collector battery may be replaced by a short circuit when the small value of V_{sat} can be neglected. In Figure

7.3.8(c) I_{CEO} may often be ignored at room temperature, in which case the model reduces to an open circuit at all three terminals. Representative values for a silicon BJT at room temperature are V_γ (junction threshold voltage) = 0.7 V, V_{sat} = 0.2 V, and I_{CEO} = 0.001 mA. The one BJT parameter that must be specified is the common-emitter current gain β, because it is subject to considerable variation.

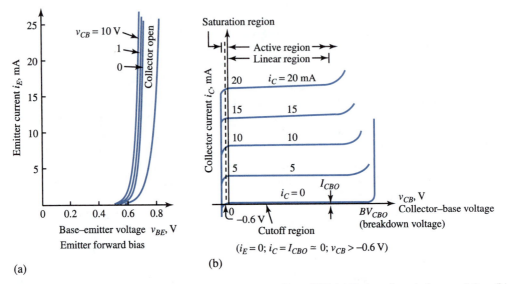

(a)

(b)

Figure 7.3.3 Common-base static curves for typical *npn* silicon BJT. (a) Emitter (input) characteristics. (b) Collector (output) characteristics.

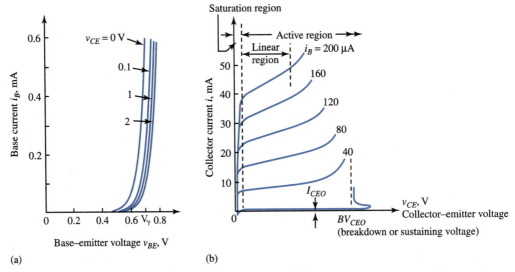

(a)

(b)

Figure 7.3.4 Common-emitter static curves for typical *npn* silicon BJT. (a) Input characteristics. (b) Output characteristics.

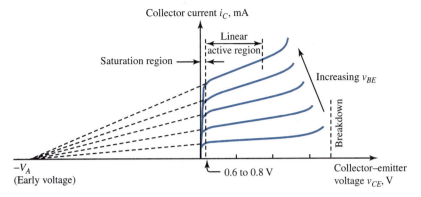

Figure 7.3.5 Early effect and Early voltage of typical *npn* silicon BJT.

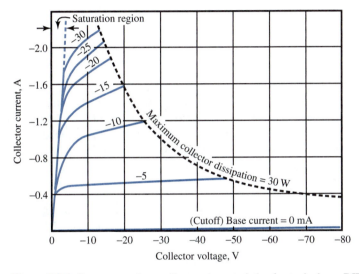

Figure 7.3.6 Common-emitter collector characteristics for typical *pnp* BJT.

Despite the structural similarities, a *pnp* BJT has smaller current gain than a comparable *npn* BJT because holes are less mobile than electrons. Most applications of *pnp* BJTs involve pairing them with *npn* BJTs to take advantage of complementary operation. The large-signal models of Figure 7.3.8 also hold for *pnp* BJTs if all voltages, currents, and battery polarities are reversed. BJTs can provide the circuit properties of a controlled source or a switch.

Figure 7.3.7 Small-signal equivalent circuit of BJT.

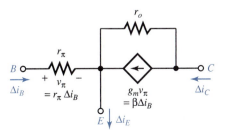

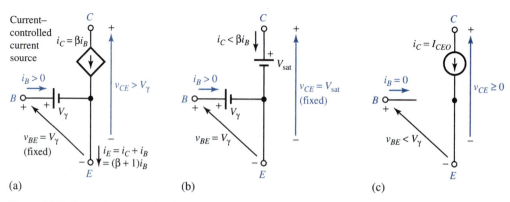

Figure 7.3.8 Large-signal models of *npn* BJT. **(a)** Linear circuit model in idealized active state. **(b)** Idealized saturated state. **(c)** Idealized cutoff state.

EXAMPLE 7.3.1

Consider the common-emitter BJT circuit shown in Figure E7.3.1(a). The static characteristics of the *npn* silicon BJT are given in Figure E7.3.1(b) along with the load line. Calculate i_B for $v_S = 1$ V and 2 V. Then estimate the corresponding values of v_{CE} and i_C from the load line, and compute the voltage amplification $A_v = \Delta v_{CE}/\Delta v_S$ and the current amplification $A_i = \Delta i_C/\Delta i_B$.

Solution

$i_B = 0$ for $v_S < V_\gamma$ and $i_B = (v_S - V_\gamma)/R_B$ for $v_S > V$. With varying but positive base current, v_{BE} stays nearly constant at the junction threshold voltage V_γ, which is 0.7 V for a silicon BJT [see Figure 7.3.4(a)].

$$\text{Then, } I_{BQ1} = \frac{v_{S1} - 0.7}{R_B} = \frac{1 - 0.7}{20,000} = 15 \ \mu A, \text{ for } v_{S1} = 0.7 \text{ V}$$

Corresponding to 15-μA interpolated static curve and load line [see Fig. E7.3.1(b)], we get $v_{CE1} = 9.4$ V, and $i_{C1} = 1.3$ mA;

$$\text{for } v_{S2} = 2 \text{ V}, \ I_{BQ2} = \frac{2 - 0.7}{20,000} = \frac{1.3}{20,000} = 65 \ \mu A$$

Figure E7.3.1 (a) Circuit. **(b)** Static curves and load line.

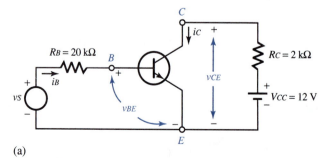

(a)

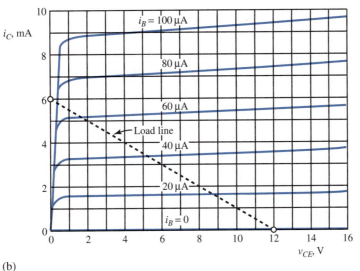

Figure E7.3.1 Continued

(b)

Corresponding to 65-μA interpolated static curve and load line [see Fig. E7.3.1(b)], we get $v_{CE2} = 1$ V and $i_{C2} = 5.5$ mA. Hence,

$$A_v = \frac{\Delta v_{CE}}{\Delta v_S} = \frac{1 - 9.4}{2 - 1} = -8.4$$

and

$$A_i = \frac{\Delta i_C}{\Delta i_B} = \frac{(5.5 - 1.3)10^{-3}}{(65 - 15)10^{-6}} = \frac{4.2}{50} \times 10^3 = 84$$

EXAMPLE 7.3.2

Given that a BJT has $\beta = 60$, an operating point defined by $I_{CQ} = 2.5$ mA, and an Early voltage $V_A = 50$ V. Find the small-signal equivalent circuit parameters g_m, r_o, and r_π.

Solution

$$g_m = \frac{I_{CQ}}{V_T} = \frac{2.5 \times 10^{-3}}{25.681 \times 10^{-3}} = 97.35 \times 10^{-3} \text{ S}$$

$$r_o \cong \frac{V_A}{I_{CQ}} = \frac{50}{2.5 \times 10^{-3}} = 20 \text{ k}\Omega$$

$$r_\pi \cong \frac{\beta}{g_m} = \frac{60}{97.35 \times 10^{-3}} = 616 \text{ } \Omega$$

EXAMPLE 7.3.3

Considering the circuit shown in Figure E7.3.3(a), find the state of operation and operating point if the BJT has $\beta = 80$ and other typical values of a silicon BJT at room temperature.

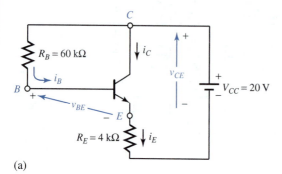

Figure E7.3.3

(a)

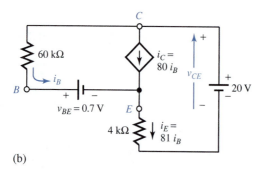

(b)

Solution

Let us check the state of operation through some preliminary calculations. Application of KVL yields

$$v_{BE} = v_{CE} - R_B i_B = V_{CC} - R_E i_E - R_B i_B$$

If we assume the saturated state, then $v_{CE} = V_{sat}$ and $i_B > 0$, so that

$$v_{BE} = V_{sat} - R_B i_B < 0.2$$

which is in violation of the saturation condition: $v_{BE} = V_\gamma = 0.7$ V.

If we assume the cutoff state, then $i_B = 0$ and $i_E = i_C = I_{CEO}$, so that

$$v_{BE} = V_{CC} - R_E I_{CEO} \cong 20 \text{ V}$$

which is in violation of the cutoff condition: $v_{BE} < V_\gamma$.

Having thus eliminated saturation and cutoff, the active-state model is substituted, as shown in Figure E7.3.3(b).

The outer loop equation gives

$$20 - 60i_B - 0.7 - 4 \times 81i_B = 0$$

where i_B is the base current in mA. Solving,

$$i_B = \frac{19.3}{384} = 0.05 \text{ mA}$$

$$i_C = 80 i_B = 4 \text{ mA}$$

Hence,

$$v_{CE} = 20 - 4 \times 81 i_B = 3.8 \text{ V}$$

which does satisfy the active-state condition: $v_{CE} > V_\gamma$.

7.4 FIELD-EFFECT TRANSISTORS

Field-effect transistors (FETs) may be classified as JFETs (junction field-effect transistors), *depletion* MOSFETs (metal-oxide-semiconductor field-effect transistors), and *enhancement* MOSFETs. Each of these classifications features a semiconductor channel of either *n*-type or *p*-type, whose conduction is controlled by a field effect. Consequently, all FETs behave in a similar fashion. The FET classification is illustrated in Figure 7.4.1 along with the corresponding circuit symbols. FETs have the useful property that very little current flows through their input (gate) terminals.

Junction FETs (JFETs)

The JFET is a three-terminal, voltage-controlled current device, whereas the BJT is principally a three-terminal, current-controlled current device. The advantages associated with JFETs are much higher input resistance (on the order of 10^7 to 10^{10} Ω), lower noise, easier fabrication, and in some cases even the ability to handle higher currents and powers. The disadvantages, on the other hand, are slower speeds in switching circuits and smaller bandwidth for a given gain in an amplifier.

Figure 7.4.2 shows the *n*-channel JFET and the *p*-channel JFET along with their circuit symbols. The JFET, which is a three-terminal device, consists of a single junction embedded in a semiconductor sample. When the base semiconductor forming the channel is of a *n*-type material, the device is known as an *n*-channel JFET; otherwise it is a *p*-channel JFET when the channel is

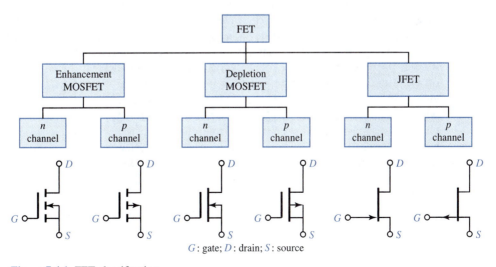

G: gate; D: drain; S: source

Figure 7.4.1 FET classification.

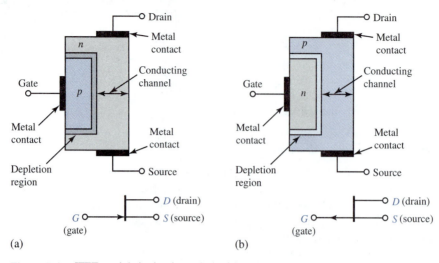

Figure 7.4.2 JFETs and their circuit symbols. **(a)** *n*-channel JFET. **(b)** *p*-channel JFET.

formed of a *p*-type semiconductor. The functions of source, drain, and gate are analogous to the emitter, collector, and base of the BJT. The gate provides the means to control the flow of charges between source and drain.

The junction in the JFET is reverse-biased for normal operation. No gate current flows because of the reverse bias and all carriers flow from source to drain. The corresponding drain current is dependent on the resistance of the channel and the drain-to-source voltage v_{DS}. As v_{DS} is increased for a given value of v_{GS}, the junction is more heavily reverse-biased, when the depletion region extends further into the conducting channel. Increasing v_{DS} will ultimately block or *pinch off* the conducting channel. After the pinch-off, the drain current i_D will be constant, independent of v_{DS}. Changing v_{GS} (gate-to-source voltage) controls where pinch-off occurs and what the value of drain current is.

It is the active region beyond pinch-off that is useful for the controlled-source operation, since only changes in v_{GS} will produce changes in i_D. Figure 7.4.3 illustrates the JFET characteristics. Part (a) shows the idealized *static characteristics* with two regions separated by the dashed line, indicating the *ohmic (controlled-resistance or triode) region* and the *active (controlled-source) region* beyond pinch-off. Note that i_D is initially proportional to v_{DS} in the ohmic region where the JFET behaves much like a voltage-variable resistance; i_D depends on v_{GS} for a given value of v_{DS} in the active region. In a practical JFET, however, the curves of i_D versus v_{DS} are not entirely flat in the active region but tend to increase slightly with v_{DS}, as shown in Figure 7.4.3(b); when extended, these curves tend to intersect at a point of $-V_A$ on the v_{DS} axis. Another useful characteristic indicating the strength of the controlled source is the *transfer characteristic*, relating the drain current i_D to the degree of the negative bias v_{GS} applied between gate and source; a cutoff region exists, indicated by the *pinch-off voltage* $-V_P$, for which no drain current flows, because both v_{GS} and v_{DS} act to eliminate the conducting channel completely.

Mathematically, the drain current in the active controlled-source region is approximately given by [see Figure 7.4.3(a)]:

$$i_D = I_{DSS} \left(1 + \frac{v_{GS}}{V_P} \right)^2 \tag{7.4.1}$$

where I_{DSS}, known as the *drain–source saturation current*, represents the value of i_D when $v_{GS} = 0$.

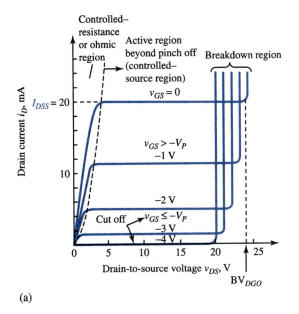

(a)

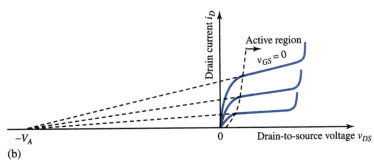

(b)

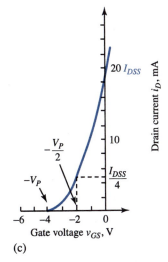

(c)

Figure 7.4.3 JFET characteristics. **(a)** Idealized static characteristics. **(b)** Practical static characteristics. **(c)** Transfer characteristic.

Also shown in Figure 7.4.3(a) is a *breakdown voltage,* denoted by BV_{DGO}, at which breakdown in the drain–gate junction occurs in the channel near the drain. For most JFETs, BV_{DGO} ranges from about 20 to 50 V. The dependence of i_D on v_{DS}, as shown in Figure 7.4.3(b), can be accounted for by applying a first-order correction to Equation (7.4.1),

$$i_D = I_{DSS} \left(1 + \frac{v_{GS}}{V_P}\right)^2 \left(1 + \frac{v_{DS}}{V_A}\right) \tag{7.4.2}$$

A small-signal equivalent circuit (valid at low frequencies where capacitances can be neglected) can now be developed based on Equation (7.4.2). Denoting the dc values at the operating point by V_{GSQ}, I_{DQ}, and V_{DSQ}, the small changes that occur can be expressed by

$$\Delta i_D = \left.\frac{\partial i_D}{\partial v_{GS}}\right|_Q \Delta v_{GS} + \left.\frac{\partial i_D}{\partial v_{DS}}\right|_Q \Delta v_{DS} = g_m \Delta v_{GS} + \frac{1}{r_o}\Delta v_{DS} \tag{7.4.3}$$

where

$$g_m = \left.\frac{\partial i_D}{\partial v_{GS}}\right|_Q = 2 I_{DSS} \left(1 + \frac{v_{GS}}{V_P}\right)\left(1 + \frac{v_{DS}}{V_A}\right)\left(\frac{1}{V_P}\right)\Bigg|_Q$$

$$= \frac{2 I_{DSS}}{V_P} \left[\left(\frac{I_{DQ}}{I_{DSS}}\right)\left(1 + \frac{V_{DSQ}}{V_A}\right)\right]^{1/2} \cong \left(\frac{2}{V_P}\right)(I_{DSS}I_{DQ})^{1/2} \tag{7.4.4}$$

and

$$\frac{1}{r_o} = \left.\frac{\partial i_D}{\partial v_{DS}}\right|_Q = I_{DSS}\left(1 + \frac{v_{GS}}{V_P}\right)^2 \left(\frac{1}{V_A}\right)\Bigg|_Q = \frac{I_{DQ}/V_A}{1 + (V_{DSQ}/V_A)} \cong \frac{I_{DQ}}{V_A} \tag{7.4.5}$$

The small-signal equivalent circuit based on Equation (7.4.3) is shown in Figure 7.4.4.

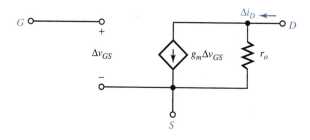

Figure 7.4.4 JFET small-signal equivalent circuit for low frequencies.

EXAMPLE 7.4.1

Measurements made on the self-biased *n*-channel JFET shown in Figure E7.4.1 are $V_{GS} = -1$ V, $I_D = 4$ mA; $V_{GS} = -0.5$ V, $I_D = 6.25$ mA; and $V_{DD} = 15$ V.

(a) Determine V_P and I_{DSS}.

(b) Find R_D and R_S so that $I_{DQ} = 4$ mA and $V_{DS} = 4$ V.

Solution

(a) From Equation (7.4.1),

$$4 \times 10^{-3} = I_{DSS}\left(1 + \frac{-1}{V_P}\right)^2$$

and

$$6.25 \times 10^{-3} = I_{DSS}\left(1 + \frac{-0.5}{V_P}\right)^2$$

Simultaneous solution yields

$$I_{DSS} = 9 \text{ mA} \qquad \text{and} \qquad V_P = 3 \text{ V}$$

(b) For $I_{DQ} = 4$ mA, $V_{GS} = -1$ V,

$$R_S = \frac{1}{4 \times 10^{-3}} = 250 \ \Omega$$

The KVL equation for the drain loop is

$$- V_{DD} + I_{DQ}R_D + V_{DS} + I_{DQ}R_S = 0$$

or

$$- 15 + 4 \times 10^{-3}R_D + 4 + 4 \times 10^{-3} \times 250 = 0$$

or

$$R_D = 2.5 \text{ k}\Omega$$

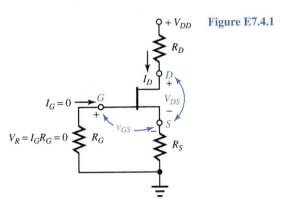

Figure E7.4.1

MOSFETs

The metal-oxide-semiconductor construction leads to the name MOSFET, which is also known as insulated-gate FET or IGFET. One type of construction results in the *depletion* MOSFET, the other in the *enhancement* MOSFET. The names are derived from the way in which channels are formed and operated. Both *n*-channel and *p*-channel MOSFETs are available in either type.

The input resistance of a MOSFET is even higher than that of the JFET (typically on the order of 10^{10} to 10^{15} Ω) because of the insulating layer of the gate. As in JFETs, the conductive gate current is negligibly small in most applications. The insulating oxide layer can, however, be damaged easily due to buildup of static charges. While the MOSFET devices are often shipped with leads conductively tied together to neutralize static charges, users too must be careful in handling MOSFETs to prevent damage due to static electricity. The MOSFETs are used primarily in digital electronic circuits. They can also provide controlled-source characteristics, which are utilized in amplifier circuits.

ENHANCEMENT MOSFETS

Figure 7.4.5 illustrates the cross-sectional structure of an *n*-channel enhancement MOSFET and its symbol showing as a *normally off* device when used for switching purposes. When the gate-to-source voltage $v_{GS} > 0$, an electric field is established pushing holes in the substrate away from the gate and drawing mobile electrons toward it, as shown in Figure 7.4.6(a). When v_{GS} exceeds the *threshold voltage* V_T of the MOSFET, an *n-type channel* is formed along the gate and a *depletion region* separates the channel from the rest of the substrate, as shown in Figure 7.4.6(b). With $v_{GS} > V_T$ and $v_{DS} > 0$, electrons are injected into the channel from the heavily doped n^+ source region and collected at the n^+ drain region, thereby forming drain-to-source current i_D, as shown in Figure 7.4.6(b). Note that none of the electrons comes from the *p*-type portion of the substrate, which now forms a reverse-biased junction with the *n*-type channel. As the gate voltage increases above V_T, the electric field increases the channel depth and *enhances* conduction. For a fixed v_{GS} and small v_{DS}, the channel has uniform depth d, acting like a resistance connected between the drain and source terminals. The MOSFET is then said to be operating in the *ohmic state*.

With a fixed $v_{GS} > V_T$, increasing v_{DS} will reduce the gate-to-drain voltage $v_{GD}(= v_{GS} - v_{DS})$, thereby reducing the field strength and channel depth at the drain end of the substrate. When $v_{DS} > (v_{GS} - V_T)$, i.e., $v_{GD} < V_T$, a *pinched-down* condition occurs when the electron flow is limited due to the narrowed neck of the channel, as shown in Figure 7.4.6(c). The MOSFET is then said to be operating in a *constant-current state*, when i_D is essentially constant, independent of v_{DS}.

Figure 7.4.7 illustrates the MOSFET behavior explained so far. When $v_{GS} \leq V_T$, however, the field is insufficient to form a channel so that the i_D–v_{DS} curve for the normally off state is simply a horizontal line at $i_D = 0$. The *drain breakdown voltage* BV_{DS} ranges between 20 and

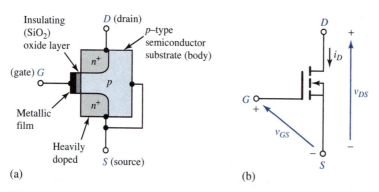

(a)   (b)

Figure 7.4.5 *n*-channel enhancement MOSFET. **(a)** Cross-sectional structure. **(b)** Symbol.

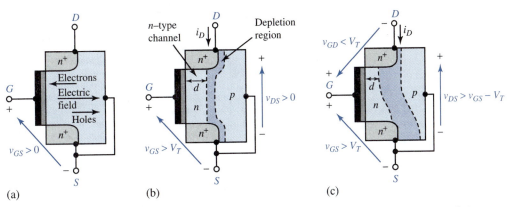

Figure 7.4.6 Internal physical picture in an *n*-channel enhancement MOSFET. **(a)** Movement of electrons and holes due to electric field. **(b)** Formation of *n*-type channel. **(c)** Pinched-down channel.

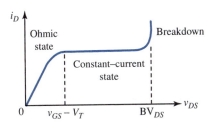

Figure 7.4.7 Idealized i_D–v_{DS} curve of an *n*-channel enhancement MOSFET with fixed $v_{GS} > V_T$.

50 V, at which the drain current abruptly increases and may damage the MOSFET due to heat if operation is continued. The *gate breakdown voltage,* at about 50 V, may also cause a sudden and permanent rupture of the oxide layer.

Figure 7.4.8 shows the characteristics of a typical *n*-channel enhancement MOSFET. In the ohmic region where $v_{GS} > V_T$ and $v_{DS} < v_{GS} - V_T$, the drain current is given by

$$i_D = K \left[2(v_{GS} - V_T)v_{DS} - v_{DS}^2 \right] \tag{7.4.6}$$

where K is a constant given by I_{DSS}/V_T^2 having the unit of A/V², and I_{DSS} is the value of i_D when $v_{GS} = 2V_T$. The boundary between ohmic and active regions occurs when $v_{DS} = v_{GS} - V_T$. For $v_{GS} > V_T$ and $v_{DS} \geq v_{GS} - V_T$, in the active region, the drain current is ideally constant and given by

$$i_D = K(v_{GS} - V_T)^2 \tag{7.4.7}$$

To account for the effect of v_{DS} on i_D, however, a factor is added,

$$i_D = K(v_{GS} - V_T)^2 \left(1 + \frac{v_{DS}}{V_A} \right) \tag{7.4.8}$$

where V_A is a constant that is in the range of 30 to 200 V. The *n*-channel enhancement MOSFET with characteristics depicted in Figure 7.4.8 has typical values of $V_T = 4$ V, $V_A = 200$ V, and $K = 0.4$ mA/V².

The small-signal equivalent circuit for low frequencies is of the same form as Figure 7.4.4 for a JFET, with g_m and r_o evaluated from the equations

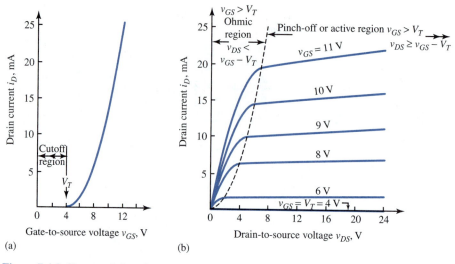

Figure 7.4.8 Characteristics of an n-channel enhancement MOSFET. (a) Transfer characteristic. (b) Static characteristics.

$$g_m = \left. \frac{\partial i_D}{\partial v_{GS}} \right|_Q = 2K(v_{GS} - V_T)\left(1 + \frac{v_{DS}}{V_A}\right)\Bigg|_Q \cong 2\sqrt{K I_{DQ}} \qquad (7.4.9)$$

and

$$r_o = \left(\left.\frac{\partial i_D}{\partial v_{DS}}\right|_Q\right)^{-1} = \left.\frac{V_A}{K(v_{GS} - V_T)^2}\right|_Q \cong \frac{V_A}{I_{DQ}} \qquad (7.4.10)$$

in which all definitions are the same as those used previously for the JFET.

EXAMPLE 7.4.2

Consider the basic MOSFET circuit shown in Figure E7.4.2 with variable gate voltage. The MOSFET is given to have very large V_A, $V_T = 4$ V, and $I_{DSS} = 8$ mA. Determine i_D and v_{DS} for $v_{GS} = 1$, 5, and 9 V.

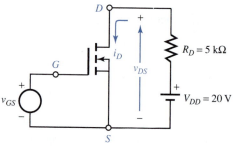

Figure E7.4.2

Solution

(a) For $v_{GS} = 1$ V: Since it is less than V_T, the MOSFET is in the cutoff region so that $i_D = 0$, which corresponds to the normally off state of the MOSFET,

$$v_{DS} = V_{DD} = 20 \text{ V}$$

(b) For $v_{GS} = 5$ V: The MOSFET operates in the active region,

$$i_D = \frac{I_{DSS}}{V_T^2}(v_{GS} - V_T)^2 = I_{DSS}\left(\frac{v_{GS}}{V_T} - 1\right)^2 = 8 \times 10^{-3}\left(\frac{5}{4} - 1\right)^2 = 0.5 \text{ mA}$$

$$v_{DS} = V_{DD} - R_D i_D = 20 - (5 \times 10^3)(0.5 \times 10^{-3}) = 17.5 \text{ V}$$

The active MOSFET behaves like a nonlinear voltage-controlled current source.

(c) For $v_{GS} = 9$ V: Since it is greater than $2V_T$ so that i_D increases while v_{DS} decreases, the MOSFET is presumably in the linear ohmic state when

$$|v_{DS}| \le \frac{1}{4}(v_{GS} - V_T) \qquad \text{and} \qquad v_{GS} > V_T$$

The theory for this case predicts that $i_D \cong v_{DS}/R_{DS}$, where R_{DS} is the equivalent drain-to-source resistance given by

$$R_{DS} = \frac{V_T^2}{2I_{DSS}(v_{GS} - V_T)}$$

Here the MOSFET acts as a voltage-controlled resistor. For our example,

$$R_{DS} = \frac{4^2}{2(8 \times 10^{-3})(9 - 4)} = 200 \ \Omega$$

Since this resistance appears in series with R_D,

$$i_D = \frac{V_{DD}}{R_D + R_{DS}} = \frac{20}{5000 + 200} = 3.85 \text{ mA}$$

and

$$v_{DS} = R_{DS} i_D = 200 \times 3.85 \times 10^{-3} = 0.77 \text{ V}$$

which is less than $1/4(9 - 4)$ V.

Figure 7.4.9 shows a *p*-channel enhancement MOSFET which differs from an *n*-channel device in that the doping types are interchanged. Channel conduction now requires negative gate-to-source voltage or positive source-to-gate voltage. With $v_{SG} > V_T$ and $v_{SD} > 0$, i_D flows from source to drain, as shown in Figure 7.4.9(a). The *p*-channel and *n*-channel MOSFETs are *complementary* transistors, having the same general characteristics but opposite current direction and voltage polarities. By replacing v_{GS} and v_{DS} with v_{SG} and v_{SD}, respectively, the equations of the *n*-channel MOSFET apply to the *p*-channel MOSFET. However, a *p*-type channel does not conduct as well as an *n*-type channel of the same size because holes are less mobile than electrons. Consequently, smaller values of I_{DSS} are typical of the *p*-channel MOSFETs.

DEPLETION MOSFETS

Figure 7.4.10 illustrates depletion MOSFETs and their symbols. Because the channel is built in, no field effect is required for conduction between the drain and the source. The depletion

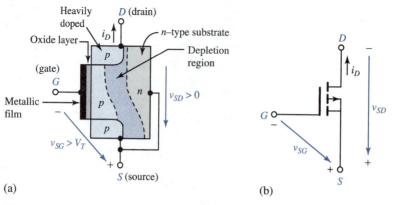

Figure 7.4.9 *p*-channel enhancement MOSFET. **(a)** Cross-sectional structure. **(b)** Symbol.

MOSFETs, like JFETs, are *normally on* transistors, in which the field effect reduces conduction by depleting the built-in channel. Figure 7.4.11(a) shows the formation of depletion regions due to electron-hole recombinations with a negative gate voltage. With $v_{GS} \leq -V_P$, where V_P is the *pinch-off voltage*, the depletion regions completely block the channel, making $i_D = 0$, as shown in Figure 7.4.11(b), which corresponds to the cutoff condition. With $v_{GS} > -V_P$ and $v_{GD} < -V_P$, so that $v_{DS} > v_{GS} + V_P$, the channel becomes partially blocked or pinched down when the device operates in its active state.

Figure 7.4.12 shows the characteristics of a typical *n*-channel depletion MOSFET. With $V_P = 3$ V, $i_D = 0$ for $v_{GS} \leq -3$ V. If -3 V $< v_{GS} \leq 0$, the device operates in the depletion mode; if $v_{GS} > 0$, it operates in the enhancement mode. The equations describing the drain current are of the same form as for the JFET.

In the ohmic region, when $v_{DS} < v_{GS} + V_P$,

$$i_D = I_{DSS} \left[2 \left(1 + \frac{v_{GS}}{V_P} \right) \left(\frac{v_{DS}}{V_P} \right) - \left(\frac{v_{DS}}{V_P} \right)^2 \right] \tag{7.4.11}$$

In the active region, when $v_{DS} \geq v_{GS} + V_P$,

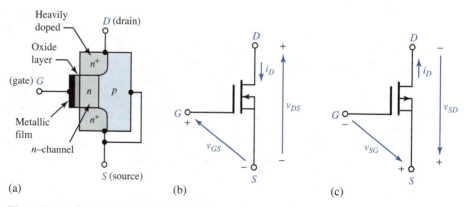

Figure 7.4.10 Depletion MOSFETs. **(a)** Structure of *n*-channel depletion MOSFET. **(b)** Symbol of *n*-channel depletion MOSFET. **(c)** Symbol of *p*-channel depletion MOSFET.

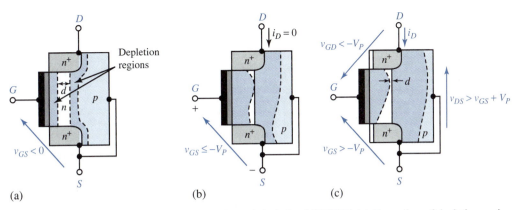

Figure 7.4.11 Internal physical picture in *n*-channel depletion MOSFET. **(a)** Formation of depletion regions. **(b)** Cutoff condition. **(c)** Active state.

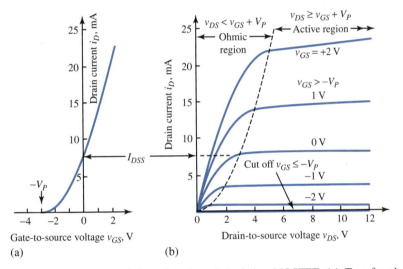

Figure 7.4.12 Characteristics of *n*-channel depletion MOSFET. **(a)** Transfer characteristic. **(b)** Static characteristics.

$$i_D = I_{DSS} \left(1 + \frac{v_{GS}}{V_P} \right)^2 \left(1 + \frac{v_{DS}}{V_A} \right) \tag{7.4.12}$$

where V_A and I_{DSS} are positive constants, and the factor $(1 + v_{DS}/V_A)$ is added to account approximately for the nonzero slope of the i_D–v_{DS} curves of a practical device, as was done in Equation (7.4.2). The small-signal equivalent circuit for low frequencies is of the same form as Figure 7.4.4 for a JFET.

EXAMPLE 7.4.3

An *n*-channel depletion MOSFET, for which $I_{DSS} = 7$ mA and $V_P = 4$ V, is said to be operating in the ohmic region with drain current $i_D = 1$ mA when $v_{DS} = 0.8$ V. Neglecting the effect of v_{DS} on i_D, find v_{GS} and check to make sure the operation is in the ohmic region.

Applying Equation (7.4.11):

$$v_{GS} = V_P \left[\left\{ \frac{i_D}{I_{DSS}} + \left(\frac{v_{DS}}{V_P} \right)^2 \right\} \left(\frac{V_P}{2v_{DS}} \right) - 1 \right]$$

$$= 4 \left[\left\{ \frac{1}{7} + \left(\frac{0.8}{4} \right)^2 \right\} \left(\frac{4}{1.6} \right) - 1 \right] = -2.17 \text{ V}$$

Check: $v_{DS} = 0.8 \text{ V} < v_{GS} + V_P = -2.17 + 4 = 1.83 \text{ V}$. The ohmic-region operation is verified.

While there is no difference in the general shape of the characteristics between the depletion and enhancement MOSFETs, the practical distinction is the gate voltage range. In particular, a depletion MOSFET can be in the active region when $v_{GS} = 0$, whereas an enhancement MOSFET must have $v_{GS} > V_T > 0$.

While a JFET behaves much like a depletion MOSFET, there are several minor differences between JFETs and depletion MOSFETs. First, with $v_{GS} < 0$, the junction in the JFET carries a reverse saturation gate current $i_C \cong -I_{GSS}$, which is quite small (on the order of 1 nA) and can usually be neglected. Second, any positive gate voltage above about 0.6 V would forward-bias the junction in the JFET, resulting in a large forward gate current. Thus, enhancement-mode operation is not possible with JFETs. On the side of advantages for JFETs, the channel in a JFET has greater conduction than the channel in a MOSFET of the same size, and the static characteristic curves are more nearly horizontal in the active region. Also, JFETs do not generally suffer permanent damage from excessive gate voltage, whereas MOSFETs would be destroyed.

The transistor is operated within its linear zone and acts like a controlled source in electronic amplifiers. It is also used in instrumentation systems as an active device. In digital computers or other electronic switching systems, a transistor effectively becomes a switch when operated at the extremes of its nonlinear mode.

7.5 INTEGRATED CIRCUITS

For the fabrication of semiconductor circuits, there are three distinct technologies employed:

1. *Discrete-component technology,* in which each circuit element is an individual component and circuit construction is completed by interconnecting the various components.

2. *Monolithic technology,* in which all the parts (such as transistors, resistors, capacitors, and diodes) needed for a complete circuit (such as an amplifier circuit) are constructed at the same time from one silicon wafer (which is typically 5 mils or 0.005 inch in thickness).

3. *Hybrid technology,* a combination of the preceding two technologies, in which various circuit components constructed on individual chips are connected so that the hybrid IC resembles a discrete circuit packaged into a single, small case.

Integrated circuits (ICs), in which several transistors, resistors, wires, and even other components are all fabricated in a single chip of semiconductor, are ideal building blocks for electronic systems. Space, weight, cost, and reliability considerations gave much impetus for the development of ICs. The ability to place circuit elements closer on an IC chip helps in extending the frequency range of the devices. Whereas the IC technology involves the use of only solid-state devices, resistors, and capacitors, the elimination of inductors is necessitated by the fact that typical semiconductors do not exhibit the magnetic properties needed to realize practical inductance values.

ICs are made by *microfabrication* technologies. The low cost of IC production is a result of *planar processing* in which fabrication begins with a very flat disc of silicon wafer, 5 to 10 cm in diameter and only 0.5 mm thick. The small electronic structures to be built on it are then produced photographically. The technique is known as *photolithography*, in which a photosensitive lacquer (known as *photoresist*), which has the property of hardening when struck by light, is used. The fabrication method requires a series of masks, photoetching, and diffusions.

MOSFET chips generally utilize either a *p*-channel or an *n*-channel device; hence, these chips are known as PMOS and NMOS, respectively. Alternatively, both *p*-channel and *n*-channel devices are used to form compound devices, in which case they are known as complementary MOS (CMOS). Whereas the CMOS has the advantage of low power consumption, only a smaller number of devices can be placed on the chip. MOS technologies are popularly used in computer circuits due to their higher packing densities. Bipolar technologies, however, are used in high-speed applications because they respond more quickly. The device fabrication methods are too involved to be presented in this introductory text.

Small-scale integration (SSI) is used typically for a 20-component op amp, whereas *large-scale integration* (LSI) puts an entire microprocessor, typically with 10,000 components, on a single chip. The chief benefits from integrating many components on an IC are low cost, small size, high reliability, and matched characteristics. Of the many IC packaging technologies, the most popular is the *dual-in-line package* (DIP), which consists of a rectangular plastic or ceramic case enclosing the IC, with protruding pin terminals. While an op amp is commonly supplied in an 8-pin DIP for insertion into some larger circuit, a microprocessor may have a 40- to 64-pin DIP to accommodate the many external connections needed for an LSI chip.

7.6 LEARNING OBJECTIVES

The *learning objectives* of this chapter are summarized here so that the student can check whether he or she has accomplished each of the following.

- Understanding of electrical conduction in semiconductor materials.
- *i–v* characteristics of a semiconductor diode (or of a *pn*-junction).
- Diode modeling and analysis of elementary diode circuits.
- Zener diode, its circuit model, and simple applications.
- Breakpoint analysis of a circuit containing two or more ideal diodes.
- Half-wave and full-wave rectifier circuits.
- Basic operation of bipolar junction transistors, along with their input and output characteristics.
- Small-signal equivalent circuits and large-signal models of BJT.
- Recognizing the more common BJT configurations, and determining the voltage and current gains.
- Basic operation of JFETs and their characteristics.
- JFET small-signal equivalent circuit (for low frequencies) and its applications for simple circuit configurations.
- Basic operation of MOSFETs (enhancement and depletion types) and their characteristics.

- MOSFET small-signal equivalent circuit (for low frequencies) and its application for simple circuit configurations.
- Basic notions of integrated circuits.

7.7 PRACTICAL APPLICATION: A CASE STUDY

Electronic Photo Flash

A simplified schematic diagram of the electric circuit of an electronic photo flash typically used on cameras is shown in Figure 7.7.1. By passing a high current through the flash tube, a bright flash of light is to be produced while the camera shutter is open. The power level is quite high, as much as 1 kW. The total energy delivered, however, is only on the order of 1 j, since the flash lasts less than 1 ms.

The electronic switch alternates between opening and closing approximately 10,000 times per second. Energy is delivered by the battery over a period of several seconds and stored in the capacitor. The stored energy is extracted from the capacitor whenever needed. The battery source causes the current in the inductor to build up while the electronic switch is closed. Recall that the current in an inductor cannot change instantaneously. When the switch opens, the inductor forces current to go through the diode in one direction, charging the capacitor. Thus the diode allows the capacitor to be charged whenever the electronic switch is open, and prevents the flow in the other direction when the electronic switch is closed. The voltage on the capacitor eventually reaches several hundred volts. When the camera shutter is opened, another switch is closed, enabling the capacitor to discharge through the flash tube. Nowadays several practical electronic circuits employ diodes, BJTs, FETs, and integrated circuits.

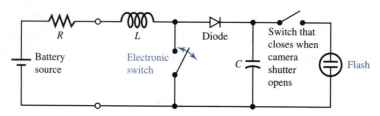

Figure 7.7.1 Simplified schematic diagram of the electric circuit of an electronic photoflash.

PROBLEMS

7.2.1 Explain the action of a pn-junction with bias. Consider both the forward bias and the reverse bias, and use sketches wherever possible.

7.2.2 Assuming the diode to obey $I = I_S(e^{V/0.026} - 1)$, calculate the ratio V/I for an ideal diode with $I_S = 10^{-13}$ A for the applied voltages of $-2, -0.5, 0.3, 0.5, 0.7, 1.0$, and 1.5 V, in order to illustrate

that a diode is definitely not a resistor with a constant ratio of V/I.

*7.2.3 A semiconductor diode with $I_S = 10\mu$A and a 1-kΩ resistor in series is forward-biased with a voltage source to yield a current of 30 mA. Find the source voltage if the diode I–V equation is given by $I = I_S(e^{40V} - 1)$. Also find the source voltage that would yield $I = -8\mu A$.

7.2.4 A silicon diode is forward-biased with $V = 0.5$ V at a temperature of 293 K. If the diode current is 10 mA, calculate the saturation current of the diode.

7.2.5 With $V = 50$ mV, a certain diode at room temperature is found to have $I = 16~\mu A$ and satisfies $I = I_S(e^{40V} - 1)$. Find the corresponding diffusion current.

7.2.6 A diode is connected in series with a voltage source of 5 V and a resistance of 1 kΩ. The diode's saturation current is given to be 10^{-12} A and the I–V curve is shown in Figure P7.2.6. Find the current through the diode in the circuit by graphical analysis.

7.2.7 For the circuit in Figure P7.2.7(a), determine the current i, given the i–v curve of the diode shown in Figure P7.2.7(b).

7.2.8 A diode with the i–v characteristic shown in Figure P7.2.8 is used in series with a voltage source of 5 V (forward bias) and a load resistance of 1 kΩ.

   (a) Determine the current and the voltage in the load resistance.

   (b) Find the power dissipated by the diode.

   (c) Compute the load current for different load resistance values of 2, 5, 0.5, and 0.2 kΩ.

*7.2.9 The diode of Problem 7.2.8 is connected in series with a forward-bias voltage of 10 V and a load resistance of 2 kΩ.

   (a) Determine the load voltage and current.

   (b) Calculate the power dissipated by the diode.

   (c) Find the load current for different values of supply voltage of 2.5, 5.0, 15.0, and 20.0 V.

7.2.10 Consider a reverse-biased diode with a source voltage V_B in series with a load resistance R_L. Write the KVL equation for the circuit.

7.2.11 Two identical junction diodes whose volt–ampere relation is given by Equation (7.2.1) in which $I_S = 0.1~\mu A$, $V_T = 25$ mV, and $\eta = 2$, are connected as shown in Figure P7.2.11. Determine the current in the circuit and the voltage across each diode.

7.2.12 Consider the diode of Problem 7.2.6 with $V_{on} = 0.7$ V and the model of Figure 7.2.5(a). Evaluate the effect of V_{on} on the answer.

*7.2.13 Consider the model of Figure 7.2.5(a). In the circuit of Figure P7.2.13, the diode is given to have $V_{on} = 0.7$ V. Find i_1 and i_2 in the circuit.

7.2.14 For the circuit shown in Figure P7.2.14(a), determine the diode current and voltage and the power delivered by the voltage source. The diode characteristic is given in Figure P7.2.14(b).

7.2.15 Let the diode of Problem 7.2.14, with its given v–i curve, be connected in a circuit with an operating point of $V_d = 0.6$ V and $I_d = 2$ mA. If the diode is to be represented by the model of Figure 7.2.5(b), determine R_f and V_{on}.

7.2.16 Consider the circuit shown in Figure P7.2.16. Determine the current in the diode by assuming:

   (a) The diode is ideal.

   (b) The diode is to be represented by the model of Figure 7.2.5(a) with $V_{on} = 0.6$ V, and

Figure P7.2.6

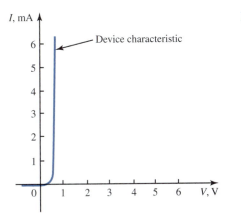

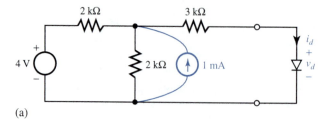

Figure P7.2.7

(a)

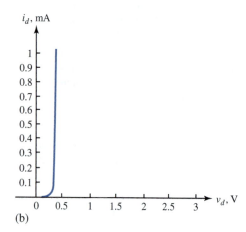

(b)

Figure P7.2.8

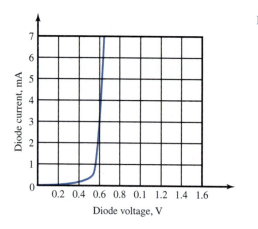

Diode voltage, V

Figure P7.2.11

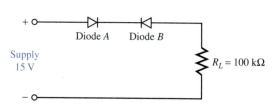

Diode A   Diode B

Supply
15 V

$R_L = 100$ kΩ

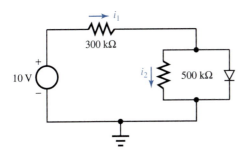

Figure P7.2.13

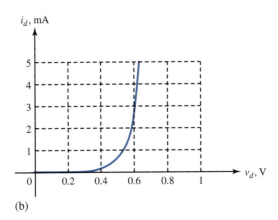

Figure P7.2.14

(a)

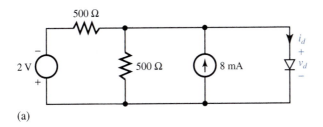

(b)

(c) The diode is to be represented by the model of Figure 7.2.5(b) with $V_{on} = 0.6$ V and $R_f = 20 \, \Omega$.

7.2.17 Sketch the output waveform of $v_o(t)$ in the circuit shown in Figure P7.2.17 for the interval $0 \le t \le$ 10 ms.

*7.2.18 Consider the circuit shown in Figure P7.2.18 with ideal diodes in order to approximate a two-terminal nonlinear resistor whose v–i curve satisfies $i = 0.001v^2$ in a piecewise-linear fashion.

(a) Find the values of R_1, R_2, and R_3.

(b) Suppose that both diodes have $R_f = 10 \, \Omega$ and $V_{on} = 0.5$ V. Find the revised values of the resistors and voltage sources (v_1 and v_2) in order to accomplish the same objective.

7.2.19 Consider the small-signal operation of a diode as represented in the model of Figure 7.2.5(b) and the v–i curve given in Problem 7.2.14. Using the circuit shown in Figure P7.2.19, develop an approximate equation for the diode current.

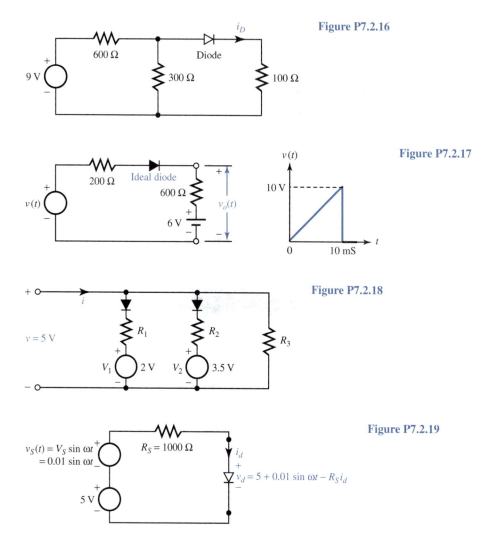

Figure P7.2.16

Figure P7.2.17

Figure P7.2.18

Figure P7.2.19

7.2.20 (a) In the circuit shown in Figure P7.2.20, the zener diode (with zero zener resistance) operates in its reverse breakdown region while the voltage across it is held constant at V_Z and the load current is held constant at V_Z/R_L, as the source voltage varies within the limits $V_{S,min} < V_S < V_{S,max}$. Find I_{max} and I_{min} for the corresponding R_{min} and R_{max}, respectively.

(b) Assuming $R_S = 0$ and the source voltage to vary between 120 and 75 V, for a load resistance of 1000 Ω, determine the maximum value of the regulator resistor R if it is desired to maintain the load voltage at 60 V. Also find the required power rating of the zener.

7.2.21 For the zener diode regulator of Figure P7.2.20, assuming that V_S varies between 40 and 60 V, with $R_S = 100$ Ω and $R_L = 1$ kΩ, select a zener diode and its regulator resistor such that V_L is maintained at 30 V. You may assume zero zener resistance.

7.2.22 Choose R and find the smallest load resistance allowed in Figure P7.2.20 when $V_Z = 12$ V and the source is 25 V $\pm$ 20% with $R_S = 0$. Assume a maximum desired diode current of 20 mA and a minimum of 1 mA.

*7.2.23 Two zener diodes are connected as shown in Figure P7.2.23. For each diode $V_Z = 5$ V. Reverse saturation currents are 2 μA for D_1 and 4 μA for D_2. Calculate v_1 and v_2: (a) when $V_S = 4$ V, and (b) if V_S is raised to 8 V.

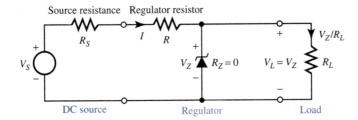

Figure P7.2.20

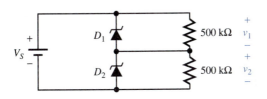

Figure P7.2.23

7.2.24 Consider the circuit of Figure P7.2.20 with $V_S = 94$ V, $V_Z = 12$ V, $R = 820$ Ω, $R_L = 220$ Ω, $R_S = 0$, and $R_Z = 25$ Ω. Assume the reverse saturation current of the zener diode to be zero.

(a) Find the load voltage, current, and power.

(b) Calculate the power dissipated in R and in the diode.

7.2.25 For the circuit of Example 7.2.7 let the direction of D_2 be reversed. Find the i–v curve.

*7.2.26 Consider the periodic pulsating dc voltage produced by a half-wave rectifier. Find the Fourier series representation and the average dc value.

7.2.27 For the half-wave rectifier of Figure 7.2.8(a), let the diode characteristic be the one given in Figure P7.2.14(b) instead of being an ideal one. For $V_S = 2$ V and $R_L = 500$ Ω, sketch $v_L(t)$.

7.2.28 Consider the circuit of Figure 7.2.9(a) with $V_S = 10$ V, $\omega = 2\pi \times 10^3$ rad/s, $C = 10$ μF, and $R_L = 1000$ Ω. Sketch $v_L(t)$ and find the minimum value of $v_L(t)$ at any time after steady-state operation has been achieved.

7.2.29 For the rectifier circuit of Figure 7.2.9(a), sketch the load current for $C = 50$ μF, $R = 1$ kΩ, and $v_S(t) = 165 \sin 377t$ V.

7.2.30 Consider the bridge rectifier shown in Figure P7.2.30. Describe its action as a full-wave rectifier, assuming the diodes to be ideal.

7.2.31 Consider a simple limiter circuit using ideal diodes, as shown in Figure P7.2.31. Analyze its action to restrict the variation of voltage within certain limits.

7.3.1 A transistor has a base current $i_B = 25$ μA, $\alpha = 0.985$, and negligible I_{CBO}. Find β, i_E, and i_C.

7.3.2 A particular BJT has a nominal value of α 0.99. Calculate the nominal β. If α can easily change $\pm 1\%$, compute the percentage changes that can occur in β.

*7.3.3 A silicon BJT has an emitter current of 5 mA at 300 K when the BEJ is forward-biased by $v_{BE} = 0.7$ V. Find the reverse saturation current of the BEJ. Neglecting I_{CBO}, calculate i_C, β, and i_B if $\alpha = 0.99$.

7.3.4 The parameters of a BJT are given by $\alpha = 0.98$, $I_{CBO} = 90$ nA, and $i_C = 7.5$ mA. Find β, i_B, and i_E.

7.3.5 For a BJT with $v_{BE} = 0.7$ V, $I_{CBO} = 4$ nA, $i_E = 1$ mA, and $i_C = 0.9$ mA, evaluate α, i_B, i_{SE}, and β.

7.3.6 Consider the circuit of Figure P7.3.6 in which the silicon BJT has $\beta = 85$ and other typical values at room temperature. I_{CBO} may be neglected.

(a) Compute i_B, i_C, and i_E, and check whether the transistor is in the active mode of operation.

(b) Check what happens if β is reduced by 10%.

(c) Check what happens if β is increased by 20%.

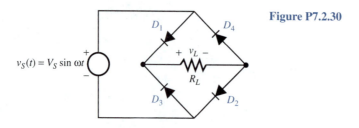

Figure P7.2.30

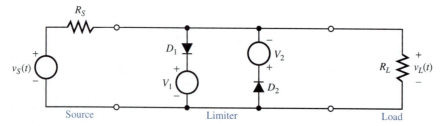

Figure P7.2.31

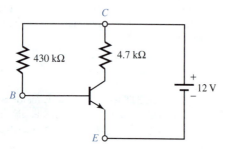

Figure P7.3.6

*7.3.7 Using the small-signal equivalent circuit of a BJT with $g_m = 0.03$ S, $\beta = 75$, and $V_A = 65$ V, a load resistor R_L is connected from the collector to the emitter, as shown in Figure P7.3.7. The transistor is biased to have a dc collector current of 6 mA.

(a) Calculate Δv_L due to the small change Δv_{BE}.

(b) Find the corresponding change Δi_B in the base current.

7.3.8 The common-emitter configuration shown in Figure P7.3.8(a) for a *pnp* BJT is most frequently used because the base current exerts a greater control on the collector current than does the emitter current. The idealized collector characteristics of the *pnp* transistor are shown in Figure P7.3.8(b) along with the load line.

The main power source V_{CC} in conjunction with the base-bias source I_{BB} is used to establish the operating point Q in Figure P7.3.8(b). Let the controlling signal be $i_b = \sqrt{2}\, I_b \sin \omega t$. Sketch the sinusoidal variations of collector current i_C, collector voltage v_C, and base current i_B, superimposed on the direct values I_{CQ}, V_{CQ}, and I_{BQ}, respectively, and calculate the current gain corresponding to a change in base current of 10 mA (peak value).

7.3.9 (a) A simple circuit using an *npn* BJT containing only one supply is shown in Figure P7.3.9(a). Outline a procedure for determining the operating point Q. The collector characteristics of the transistor are given in Figure P7.3.9(b).

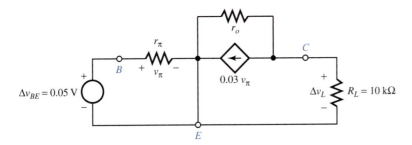

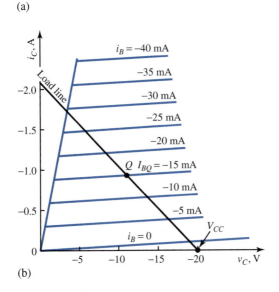

(a)

(b)

(b) For $V_{CC} = 18$ V, if the operating Q point is at a collector voltage of 10 V and a collector current of 16 mA, determine the R_C and R_B needed to establish the operating point.

7.3.10 The two-transistor combination known as Darlington pair or Darlington compound transistor is often used as a single three-terminal device, as shown in Figure P7.3.10. Assuming the transistors to be identical, neglecting I_{CBO} of each BJT, find α_C and β_C of the combination.

*7.3.11 The circuit of Figure P7.3.11 uses a *pnp* BJT whose characteristics are shown in Figure 7.3.6. The parameter values are $R_C = 30 \ \Omega$, $R_B = 6 \ k\Omega$, $V_{CC} = 60$ V, and $V_{BE} = -0.7$ V.

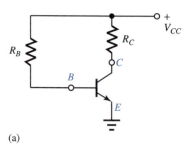

(a)

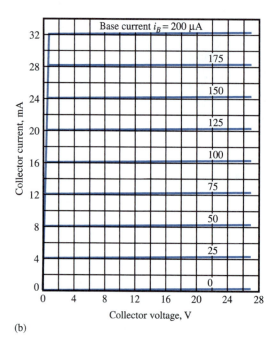

(b)

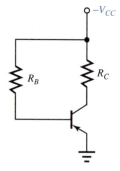

Figure P7.3.10

Figure P7.3.11

(a) Find I_C and V_{CE} at the operating point.

(b) Determine the power supplied by the V_{CC} source.

7.3.12 In the circuit of Figure P7.3.12 the transistor has $\beta = 99$ and $V_{BE} = 0.6$ V. For $V_{CC} = 10$ V, $R_F = 200$ kΩ, and $R_C = 2.7$ kΩ, determine the operating point values of V_{CE} and I_C.

7.3.13 If the circuit of Example 7.3.1 is to switch from cutoff to saturation, find the condition on v_S, given that the transistor has $\beta = 100$.

7.3.14 Consider the circuit of Example 7.3.3. For $R_E = 0$ and $\beta = 50$, find i_E.

7.3.15 If the BJT in the circuit of Example 7.3.1 has $\beta = 150$, find i_C and v_{CE} when: (a) $i_B = 20\ \mu A$, and (b) $i_B = 60\ \mu A$. Specify the state of the BJT in each case.

7.3.16 The circuit shown in Figure P7.3.16 has a *pnp* BJT turned upside down. Find R_B when $v_{EC} = 4$ V and $\beta = 25$.

7.3.17 Reconsider the circuit of Figure P7.3.16. With $\beta = 25$, find the condition on R_B such that i_C has the largest possible value.

7.4.1 Consider JFET characteristics shown in Figures 7.4.3 (a) and (c).

(a) Write down the conditions for the operation to take place in the active region.

(b) Obtain an expression for the drain current i_D in the active region, and for the value of i_D for the boundary between the ohmic and active regions.

(c) Find the conditions for the linear ohmic operation and the equivalent drain-to-source resistance.

(d) Express the condition for operation in the cutoff region.

7.4.2 The JFET with parameters $V_P = 6$ V and $I_{DSS} = 18$ mA is used in the circuit shown in Figure P7.4.2 with a positive supply voltage. Find v_{GS}, i_D, and v_{DS}. Note that the gate current is negligible for the arrangement shown.

*7.4.3 Consider the circuit of Figure P7.4.2 with the same JFET parameters. Let R_S be not specified. Determine v_{GS}, v_{DS}, and R_S for active operation at $i_D = 2$ mA.

7.4.4 A JFET with $I_{DSS} = 32$ mA and $V_P = 5$ V is biased to produce $i_D = 27$ mA at $v_{DS} = 4$ V. Find the region in which the device is operating.

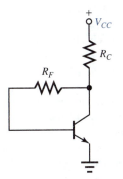

Figure P7.3.12

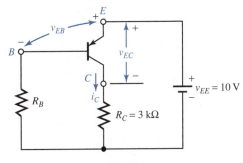

Figure P7.3.16

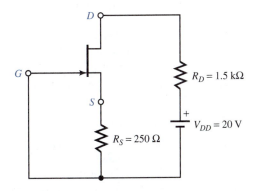

Figure P7.4.2

7.4.5 For a p-channel JFET in its active region, specify the polarities of voltages and the directions of conventional currents.

7.4.6 Consider the common-source JFET circuit shown in Figure P7.4.6 with fixed bias. Sketch the sinusoidal variations of drain current, drain voltage, and gate voltage superimposed on the direct values at the operating point. Assume reasonable common-source drain characteristics.

7.4.7 A self-biased n-channel JFET used in the circuit of Example 7.4.1 has the characteristics given in Figure P7.4.7 and a supply voltage $V_{DD} = 36$ V, $R_S = 1$ kΩ, and $R_D = 9$ kΩ. Determine the operating point and the values of V_{GSQ}, I_{DQ}, and V_{DSQ}.

*7.4.8 In the p-channel version of the circuit of Example 7.4.1, with the JFET having $V_P = 4$ V and $I_{DSS} = -5$ mA, find R_D and R_S to establish $I_{DQ} = -2$ mA and $V_{DSQ} = -4$ V when $V_{DD} = -12$ V.

7.4.9 An n-channel JFET in the circuit configuration shown in Example 7.4.1 is operating at $I_{DQ} = 6$ mA and $V_{GSQ} = -1$ V when $V_{DS} = 5$ V. Determine R_S and V_{DD} if: (a) $R_D = 2$ kΩ, and (b) $R_D = 4$ kΩ.

7.4.10 An n-channel JFET is given to have $V_P = 3$ V and $I_{DSS} = 6$ mA.

(a) Find the smallest value of v_{DS} when $v_{GS} = -2$ V if the operation is to be in the active region.

(b) Determine the corresponding i_D for the smallest v_{DS}.

7.4.11 Given that a silicon n-channel JFET has $V_P = 5$ V and $I_{DSS} = 12$ mA, check whether the device is operating in the ohmic or active region when $v_{GS} = -3.2$ V and $i_D = 0.5$ mA.

7.4.12 For an n-channel JFET with $V_A = 350$ V, $I_{DSS} = 10$ mA, and $V_P = 3$ V, find V_{DS} that will cause $i_D = 11$ mA when $v_{GS} = 0$.

7.4.13 An n-channel JFET with $V_A = 300$ V, $V_P = 2$ V, and $I_{DSS} = 10$ mA is to be operated in the active mode. Determine i_D when $v_{DS} = 10$ V and $v_{GS} = -0.5$ V.

7.4.14 Sketch g_m versus v_{GS} for a JFET with $I_{DSS} = 10$ mA, $V_P = 3$ V, $V_A = 100$ V, and $v_{DS} = 10$ V. See what happens if $V_A \to \infty$. Also sketch r_o versus v_{GS}.

7.4.15 The drain current of a JFET in the ohmic region is approximated by

$$i_D = I_{DSS}\left[2\left(1 + \frac{v_{GS}}{V_P}\right)\left(\frac{v_{DS}}{V_P}\right) - \left(\frac{v_{DS}}{V_P}\right)^2\right]$$

Assuming small v_{DS}, find the channel resistance r_{DS} for $v_{GS} = -2$ V if the JFET's parameters are $I_{DSS} = 25$ mA and $V_P = 3$ V.

7.4.16 An n-channel enhancement MOSFET operates in the active region with very large V_A, $v_{GS} = 6$ V, $V_T = 4$ V, and $i_D = 1$ mA. Calculate K.

7.4.17 Consider the MOSFET circuit with variable voltage shown in Example 7.4.2, with $R_D = 2$ kΩ and $V_{DD} = 12$ V. The static characteristics of the n-channel enhancement MOSFET are given in Figure P7.4.17.

(a) Draw the load line and find the operating point if $v_{GS} = 4$ V.

(b) Sketch the resulting transfer curves (i.e., i_D and v_{DS} as a function of v_{GS}) showing cutoff, active, and saturation regions.

(c) For relatively undistorted amplification, the MOSFET circuit must be restricted to signal variations within the active region. Let $v_{GS}(t) = 4 + 0.2 \sin \omega t$ V. Sketch $i_D(t)$ and $v_{DS}(t)$, and estimate the resulting voltage amplification A_v.

(d) Let $v_{GS}(t) = 6 \sin \omega t$, where ω is slow enough to satisfy the static condition. Sketch $i_D(t)$ and $v_{DS}(t)$ obtained from the transfer curves. Comment on the action of the MOSFET in the switching circuit.

7.4.18 Find idealized expressions for the active and ohmic states and sketch the universal characteristics of an n-channel enhancement MOSFET operated below breakdown.

7.4.19 Let the circuit in Example 7.4.2 have $V_{DD} = 12$ V and $R_D = 2$ kΩ, and let the MOSFET have very large V_A, $V_T = 2.5$ V, and $I_{DSS} = 8.3$ mA.

(a) Determine v_{GS} and v_{DS} when $i_D = 4$ mA.

(b) Determine i_D and v_{DS} when the gate voltage is 6 V.

7.4.20 Consider the MOSFET connected as a two-terminal device, as shown in Figure P7.4.20. Discuss its states of operation.

7.4.21 Find the parameter values V_T and I_{DSS} for a p-channel MOSFET with $i_D = 0$ when $v_{GS} \leq -3$ V, and $i_D = 5$ mA when $v_{GS} = v_{DS} = -8$ V. You may neglect the effect of v_{DS} on i_D.

7.4.22 In a depletion MOSFET for which $V_P = 3$ V and $I_{DSS} = 11$ mA, the drain current is 3 mA when v_{DS} is set at the largest value that will maintain ohmic-region operation. Find v_{GS} if V_A is very large.

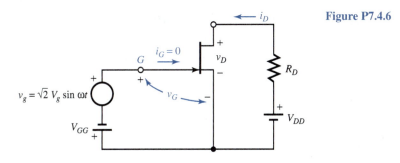

Figure P7.4.6

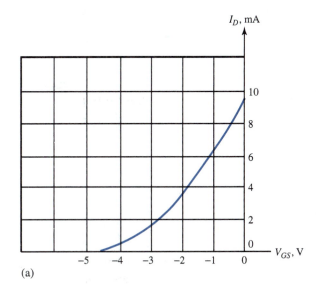

(a)

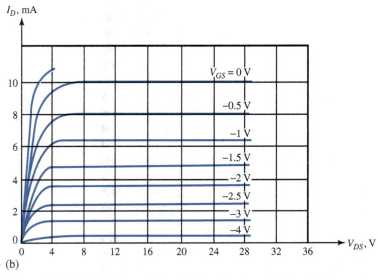

(b)

Figure P7.4.7 JFET characteristics. (a) Transfer characteristic. (b) Output characteristics.

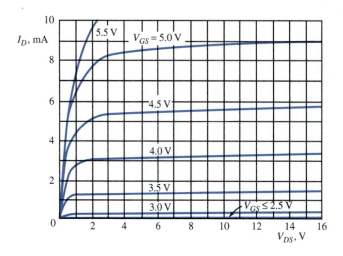

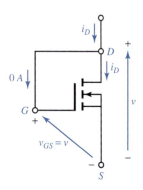

Figure P7.4.20

*7.4.23 A depletion MOSFET is given to have large V_A, $V_P = 2.8$ V, $I_{DSS} = 4.3$ mA, $v_{DS} = 4.5$ V, and $v_{GS} = 1.2$ V.

(a) Is the MOSFET operating in the active region?

(b) Find i_D.

(c) Comment on whether the device is operating in the depletion mode or in the enhancement mode.

8 Transistor Amplifiers

Amplifiers are circuits that produce an output signal which is larger than, but proportional to, an input signal. The input and output signals can be both voltages or currents, or one or the other, as in voltage-in current-out and current-in voltage-out amplifiers. The amplifier gain is just the network's transfer function, which is the ratio of output-to-input complex signals in the frequency domain as found by complex analysis. Amplifiers find extensive use in instrumentation applications. Sometimes, amplifiers are used for reasons other than gain alone. An amplifier may be designed to have high input impedance so that it does not affect the output of a sensor while at the same time giving a low output impedance so that it can drive large currents into its load, such as a lamp or heating element. In some other applications, an amplifier with a low input impedance might be desirable.

The first step in designing or analyzing any amplifier is to consider the biasing. The biasing network consists of the power supply and the passive circuit elements surrounding the transistor that provide the correct dc levels at the terminals. This is known as setting the Q point (quiescent or operating point) with no signal applied. A good bias circuit must not only establish the correct dc levels, but must maintain them in spite of changes in temperature, variations in transistor characteristics, or any other sources of variation.

Thus, a biasing signal (current or voltage), in the absence of any other signals, places the transistor at an operating or quiescent point of its i–v characteristics. Time-varying signals are usually superimposed on dc biasing signals. Small variations of voltage and current about the operating point are known as small-signal voltages and currents. While small-signal variations are just a fraction of the power-supply voltage, the large-scale excursions of a power amplifier may be comparable to the supply voltage. This chapter is devoted to the study of small-signal amplifiers, in which the relationships between small-signal variables are linear. Graphical solutions including the transistor's general nonlinearity are not considered in this text.

A transistor model, having the same number of terminals as the transistor, is a collection of ideal linear elements designed to approximate the relationships between the transistor small-signal variables. While the small-signal model cannot be used to obtain information about biasing, the ac device model considered in this chapter deals only with the response of the circuit to small signals about the operating point. The transistor model is then substituted for the transistor in the circuit in order to analyze an amplifier circuit. Under the assumption of small-signal linear operation, the technique of superposition can be used effectively to simplify the analysis.

Amplifier circuits can be treated conveniently as building blocks when analyzing larger systems. The amplifier block may be represented by a simple small-signal model. A multistage amplifier is a system obtained by connecting several amplifier blocks in sequence or cascade, in which the individual amplifier blocks are called stages. Input stages are designed to accept signals coming from various sources; intermediate stages provide most of the amplification; output stages drive various loads. Most of these stages fall in the category of small-signal amplifiers.

The subject of amplifier frequency response has to do with the behavior of an amplifier as a function of signal frequency. Circuit capacitances and effects internal to the transistors impose limits on the frequency response of an amplifier. Minimization of capacitive effects is a topic of great interest in circuit design.

After discussing biasing the BJTs and FETs to establish the operating point, BJT and FET amplifiers are analyzed, and the frequency response of amplifiers is looked into. Advantages of negative feedback in amplifier circuits are also mentioned.

8.1 BIASING THE BJT

A simple method of biasing the BJT is shown in Figure 8.1.1. While no general biasing procedure that will work in all cases can be outlined, a reasonable approach is to assign $\frac{1}{2} V_{CC}$ as the drop across the transistor, $\frac{3}{8} V_{CC}$ as the drop across R_C to allow an adequate ac voltage swing capability in the collector circuit, and $\frac{1}{8} V_{CC}$ as the drop across R_E, so that R_E is about $\frac{1}{3} R_C$. With a specified supply voltage V_{CC}, biasing consists mainly of selecting values for V_{CEQ}, I_{CQ}, and $I_{BQ} = I_{CQ}/\beta$, which define the operating Q point. One can then select

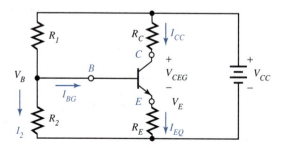

Figure 8.1.1 Method of biasing the BJT.

$$R_C = \frac{3V_{CC}}{8I_{CQ}} \tag{8.1.1}$$

$$R_E = \frac{V_{CC}}{8(I_{CQ} + I_{BQ})} = \frac{V_{CC}\beta}{8(1+\beta)I_{CQ}} \tag{8.1.2}$$

Noting that $V_B = V_E + V_{BE}$ in Figure 8.1.1, or $V_B = (V_{CC}/8) + 0.7$ for silicon, and selecting $I_2 = 5I_{BQ}$, it follows then

$$R_2 \cong \frac{0.7 + (V_{CC}/8)}{5I_{BQ}} \tag{8.1.3}$$

$$R_1 \cong \frac{V_{CC} - V_B}{6I_{BQ}} = \frac{(7\,V_{CC}/8) - 0.7}{6I_{BQ}} \tag{8.1.4}$$

EXAMPLE 8.1.1

Apply the rule-of-thumb dc design presented in this section for a silicon *npn* BJT with $\beta = 70$ when the operating Q point is defined by $I_{CQ} = 15$ mA and $I_{BQ} = 0.3$ mA, with a dc supply voltage $V_{CC} = 12$ V, and find the resistor values of R_C, R_E, R_1, and R_2.

Solution

Applying Equations (8.1.1) through (8.1.4), we get

$$R_C = \frac{3(12)}{8(15 \times 10^{-3})} = 300\ \Omega$$

$$R_E = \frac{12(70)}{8(71)(15 \times 10^{-3})} = 98.6\ \Omega$$

$$R_1 \cong \frac{[7(12)/8] - 0.7}{6(0.3 \times 10^{-3})} = 5444\ \Omega$$

$$R_2 \cong \frac{0.7 + (12/8)}{5(0.3 \times 10^{-3})} = 1467\ \Omega$$

8.2 BIASING THE FET

Let us first consider biasing the JFET and then go on to biasing the MOSFET.

Biasing JFET

A practical method of biasing a JFET is shown in Figure 8.2.1. Neglecting the gate current, which is usually very small for a JFET, we have

$$V_G = \frac{V_{DD} R_2}{R_1 + R_2} \tag{8.2.1}$$

The transfer characteristic of the JFET [see Figure 7.4.3(c)], neglecting the effect of v_{DS} on i_D, is given by

$$i_D = I_{DSS} \left(1 + \frac{v_{GS}}{V_P}\right)^2 \tag{8.2.2}$$

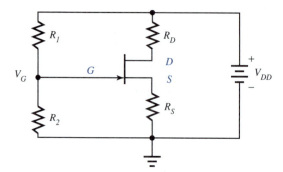

Figure 8.2.1 Method of biasing the JFET.

From Figure 8.2.1, applying the KVL to the loop containing R_2 and R_S, the load-line equation is

$$i_D = \frac{V_G - v_{GS}}{R_S} \qquad (8.2.3)$$

The operating point Q is the intersection of the load line with the transfer characteristic, from which I_{DQ} and V_{GSQ} can be read. While no systematic bias-point design procedure to serve all applications exists, a simple procedure that will serve a good number of problems is outlined here. In establishing the operating point Q, compromising between high stability and high gain, one may choose

$$I_{DQ} = \frac{I_{DSS}}{3} \qquad (8.2.4)$$

Substituting this into Equation (8.2.2), we get

$$V_{GSQ} = \left(\frac{1 - \sqrt{3}}{\sqrt{3}}\right) V_P \cong -0.423 V_P \qquad (8.2.5)$$

Next, select V_G to yield a reasonably low slope to the load line so that drain-current changes are small,

$$V_G = 1.5 V_P \qquad (8.2.6)$$

The voltage drop across R_S is then given by

$$V_G - V_{GSQ} \cong 1.5 V_P + 0.423 V_P = 1.923 V_P \qquad (8.2.7)$$

so that

$$R_S = \frac{1.923 V_P}{I_{DQ}} = \frac{1.923 V_P}{I_{DSS}/3} = \frac{5.768 V_P}{I_{DSS}} \qquad (8.2.8)$$

Choosing R_2 arbitrarily as

$$R_2 = 100 R_S \qquad (8.2.9)$$

to maintain large resistance across the gate, R_1 can be found,

$$R_1 = \frac{R_2(V_{DD} - V_G)}{V_G} = 100 R_S \left(\frac{V_{DD}}{1.5 V_P} - 1\right) \qquad (8.2.10)$$

Next, to find R_D, choosing the transistor's drop to be equal to that across R_D plus the pinch-off voltage necessary to maintain the active-mode operation,

$$V_{DD} - I_{DQ} R_S = V_P + 2 I_{DQ} R_D \qquad (8.2.11)$$

or

$$R_D = \frac{V_{DD} - I_{DQ}R_S - V_P}{2I_{DQ}} = \frac{3(V_{DD} - 2.923V_P)}{2I_{DSS}} \qquad (8.2.12)$$

Note that the source voltage V_{DD} must be larger than the minimum necessary to maintain adequate voltage swings for ac signals. A value of $4.923V_P$ is considered to be a reasonable minimum in order to allow peak collector voltage changes of $\pm V_P$.

EXAMPLE 8.2.1

Consider and obtain the values for R_S, R_2, R_1, and R_D. Apply the rule-of-thumb dc design procedure outlined in this section for a JFET with $V_P = 3$ V, $I_{DSS} = 20$ mA, and a source voltage $V_{DD} = 24$ V.

Solution

Applying Equations (8.2.8) through (8.2.10),

$$R_S = \frac{5.768V_P}{I_{DSS}} = \frac{5.768(3)}{20 \times 10^{-3}} = 865.2 \ \Omega$$

$$R_2 = 100R_S = 86,520 \ \Omega$$

$$R_1 = 100R_S \left(\frac{V_{DD}}{1.5V_P} - 1 \right) = 86,520 \left(\frac{24}{1.5 \times 3} - 1 \right) = 374,920 \ \Omega$$

From Equation (8.2.12),

$$R_D = \frac{3(V_{DD} - 2.923V_P)}{2I_{DSS}} = \frac{3[24 - 2.923(3)]}{2 \times 20 \times 10^{-3}} = 1142.3 \ \Omega$$

Note that $V_{DD} = 24$ V is greater than $4.923V_P = 14.77$ V; the device is biased above the reasonable minimum.

The dc design procedure may have to be adjusted after the ac design in some cases because of signal values, and refined further to suit the available components and power-supply voltages. In any case, one should ensure that any transistor maximum rating is not exceeded.

Biasing Depletion MOSFETs

Recognizing that v_{GS} can exceed zero in a MOSFET and hence the operating point may be placed at a higher I_{DQ} than in a JFET circuit, the same procedure as that outlined for dc biasing of JFETs can be applied directly to the biasing of depletion-type MOSFETs.

Biasing Enhancement MOSFETs

A biasing method for an n-channel enhancement-type MOSFET is illustrated in Figure 8.2.2. Since the gate draws no current, V_G is given by

$$V_G = \frac{V_{DD}R_2}{R_1 + R_2} \qquad (8.2.13)$$

Note that R_S in Figure 8.2.2 is only to provide operating-point stability and not to establish a quiescent point since its voltage drop is not of the correct polarity. Neglecting the effect of v_{DS} on i_D, the drain current for the active mode is given by [see Equations (7.4.2) and (7.4.3)]

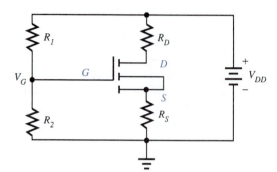

Figure 8.2.2 Biasing an n-channel enhancement MOSFET.

$$i_D = K(v_{GS} - V_T)^2 \qquad (8.2.14)$$

in which V_T and K are specified based on the transfer characteristic [Figure 7.4.8(a)] in the manufacturer's data sheets.

The load-line equation is

$$i_D = \frac{V_G - v_{GS}}{R_S} \qquad (8.2.15)$$

The required gate–source voltage V_{GSQ} at the operating point is then

$$V_{GSQ} = V_T + \sqrt{\frac{I_{DQ}}{K}} \qquad (8.2.16)$$

V_{DSQ} can then be selected to yield a desired operating point on the I_D–V_{DS} static characteristics of the device. It follows then

$$R_S + R_D = \frac{V_{DD} - V_{DSQ}}{I_{DQ}} \qquad (8.2.17)$$

By trading off ac gain (larger R_D) versus dc stability (larger R_S), R_D and R_S need to be chosen. Once R_S is chosen, then V_G is set by

$$V_G = V_{GSQ} + I_{DQ} R_S \qquad (8.2.18)$$

Finally, R_1 and R_2 can be selected arbitrarily to yield V_G while keeping both large enough to maintain a large gate impedance. The outlined approach is best illustrated by an example.

EXAMPLE 8.2.2

Given an n-channel enhancement MOSFET having $V_T = 4$ V, $K = 0.15$ A/V^2, $I_{DQ} = 0.5$ A, $V_{DSQ} = 10$ V, and $V_{DD} = 20$ V. Using the dc design approach outlined in this section, determine V_{GSQ}, V_G, R_D, R_S, R_1, and R_2.

Solution

Applying Equation (8.2.16),

$$V_{GSQ} = V_T + \sqrt{\frac{I_{DQ}}{K}} = 4 + \sqrt{\frac{0.5}{0.15}} = 5.826 \text{ V}$$

Noting the supply voltage to be 20 V and the drop across the transistor 10 V, about 10 V can be allowed for ac swing across R_D. Allowing some drop of, say, 3 V across R_S for dc stability,

$$R_D = \frac{7}{0.5} = 14 \ \Omega$$

$$R_S = \frac{4}{0.5} = 8 \ \Omega$$

From Equation (8.2.18),

$$V_G = V_{GSQ} + I_{DQ}R_S = 5.826 + 4 = 9.826 \text{ V}$$

Selecting arbitrarily $R_2 = 10{,}000 \ \Omega$ to maintain a large gate impedance,

$$R_1 = \frac{R_2(V_{DD} - V_G)}{V_G} = \frac{10^4(20 - 9.826)}{9.826} = 10.354 \text{ k}\Omega$$

One should also check to ensure that the voltage, current, and power ratings of the device are not exceeded.

Biasing methods using resistors have been presented for the sake of simplicity and ease of understanding. However, nowadays biasing techniques for modern amplifiers utilize transistors.

8.3 BJT AMPLIFIERS

The purpose of electronic amplifiers is essentially to increase the amplitude and power of a signal so that either useful work is done or information processing is realized. The output signal power being greater than the input signal power, the additional power is supplied by the bias supply. Thus, the amplifier action is one of energy conversion in which the bias power is converted to signal power within the device.

A single-stage amplifier is one in which there is only one amplifying element. By combining several single-stage amplifier circuits, a multistage amplifier is produced. Audio amplifiers are designed to amplify signals in the frequency range of 30 to 15,000 Hz perceptible to the human ear. A video amplifier is designed to amplify the signal frequencies needed for television imaging (see Chapter 15). An amplifier, in general, is then made up of a cascade of several stages. A stage usually consists of an elementary amplifier, which normally has only one transistor. The cascade is formed by making the output of the first stage as the input of the second stage, the output of the second stage as the input of the third stage, and so on. This section is devoted to the study of three basic forms of amplifier stages which use a BJT.

Common-Emitter (CE) Configuration

The emitter part of a circuit being common to both the input and the output portions, Figure 8.3.1(a) illustrates a common-emitter (CE) BJT amplifier. The resistors R_1, R_2, R_C, and R_E are primarily set by biasing. The input ac source is represented by its Thévenin equivalent. The amplified output ac voltage v_L appears across the load resistor R_L, which could represent the input resistance of the next stage in a cascade. Capacitors C_B, C_C, and C_E are so chosen that they represent short circuits at the lowest frequency of interest. C_E would be made large enough so that $1/\omega C_E$ is small relative to R_E in parallel with the impedance looking into the emitter at the smallest ω of interest. Similarly, the reactances of C_C and C_B would be chosen small relative to the resistances in their parts of the circuit. Capacitors C_B and C_C appear as short circuits to the ac signals, but block the dc voltages and currents out of one part of the circuit from coupling with

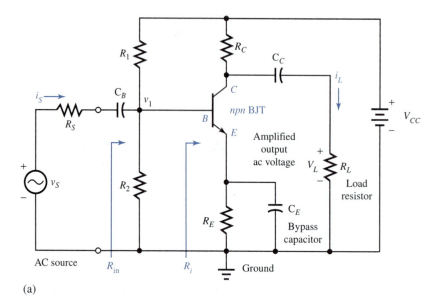

(a)

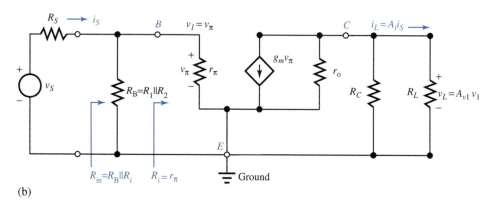

(b)

Figure 8.3.1 Common-emitter (CE) BJT amplifier. **(a)** Circuit. **(b)** Small-signal ac equivalent circuit.

another part. Capacitor C_E, known as the bypass capacitor, bypasses the ac current around R_E so that no significant ac voltage is generated across R_E, and helps to increase the gain.

The small-signal ac equivalent circuit is shown in Figure 8.3.1(b), in which the small-signal model of Figure 7.3.7 for the transistor is used. While omitting the details of analysis and summarizing the results, we have

$$R_B = R_1 \| R_2 = \frac{R_1 R_2}{R_1 + R_2} \tag{8.3.1}$$

$$R_i = \left[\begin{array}{c} \text{resistance between} \\ \text{transistor base and ground} \\ \text{(as seen looking into base)} \end{array} \right] = r_\pi \tag{8.3.2}$$

$$v_1 = v_\pi = \frac{v_S R_{\text{in}}}{R_S + R_{\text{in}}} \tag{8.3.3}$$

where

$$R_{in} = \begin{bmatrix} \text{total input resistance} \\ \text{seen by source} \end{bmatrix} = R_B \| R_i = \frac{R_B r_\pi}{R_B + r_\pi} \qquad (8.3.4)$$

Voltage and current gains are given by

$$A_{v_1} = \frac{v_L}{v_1} = \frac{-g_m R_L [r_o \| R_C]}{R_L + [r_o \| R_C]} \qquad (8.3.5)$$

$$A_i = \frac{i_L}{i_S} = \frac{-g_m (r_o \| R_C)(r_\pi \| R_B)}{R_L + (r_o \| R_C)} \qquad (8.3.6)$$

The CE configuration yields signal inversion since the gains can be seen to be negative.

EXAMPLE 8.3.1

Consider the transistor biased in Example 8.1.1. Given that $R_L = 500\ \Omega$ and $V_A = 75$ V for the transistor, determine the ac voltage and current gains.

Solution

Taking $V_T = 25.861 \times 10^{-3}$ V, as indicated in Section 7.3, and applying Equation (7.3.7),

$$g_m = \frac{I_{CQ}}{V_T} = \frac{15 \times 10^{-3}}{25.861 \times 10^{-3}} = 0.58\ \text{S}$$

By using Equation (7.3.8),

$$r_o = \frac{V_A}{I_{CQ}} = \frac{75}{15 \times 10^{-3}} = 5000\ \Omega$$

Taking $R_C = 300\ \Omega$ from the solution of Example 8.1.1,

$$r_o \| R_C = 5000 \| 300 = \frac{5000(300)}{5000 + 300} = 283\ \Omega$$

Applying Equation (8.3.5), one gets

$$A_{v_1} = \frac{-0.58(500)(283)}{(500 + 283)} = -104.8$$

From Equation (7.3.9),

$$r_\pi = \frac{\beta}{g_m} = \frac{70}{0.58} = 120.7\ \Omega$$

and

$$R_{in} = R_1 \| R_2 \| r_\pi = \frac{1}{(1/R_1) + (1/R_2) + (1/r_\pi)} = \frac{1}{(1/5444) + (1/1467) + (1/120.7)} = 109.3\ \Omega$$

where values for R_1 and R_2 are taken from the solution of Example 8.1.1. From Equation (8.3.6),

$$A_i = \frac{-0.58(283)(109.3)}{500 + 283} = -22.9$$

Common-Collector (CC) Configuration

A common-collector (CC) amplifier is also known as an emitter follower (or a voltage follower) due to the fact that the output voltage "follows" the input by being approximately equal to the input voltage. The amplifier is shown in Figure 8.3.2(a), in which the collector forms a common terminal between the input and output circuits, and resistors R_1, R_2, and R_E are determined by biasing. Capacitors C_B and C_E are chosen large enough to appear as short circuits at the lowest frequency of interest in the input signal v_S. The output voltage v_L is taken across the load resistor R_L. The small-signal ac equivalent circuit of the amplifier is shown in Figure 8.3.2(b), whose analysis yields the following results:

$$R_B = R_1 \| R_2 = \frac{R_1 R_2}{R_1 + R_2} \qquad (8.3.7)$$

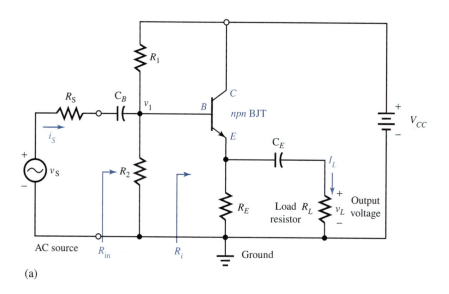

(a)

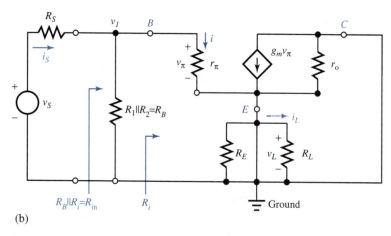

(b)

Figure 8.3.2 Common-collector (CC) BJT amplifier. **(a)** Circuit. **(b)** Small-signal ac equivalent circuit.

$$R_{\text{in}} = R_B \| R_i = \frac{R_B R_i}{R_B + R_i} \qquad (8.3.8)$$

in which

$$R_i = r_\pi + R_W(1 + g_m r_\pi) \qquad (8.3.9)$$

where

$$R_W = r_o \| R_E \| R_L \qquad (8.3.10)$$

The voltage and current gains are given by

$$A_{v_1} = \frac{v_L}{v_1} = \frac{R_W(1 + g_m r_\pi)}{r_\pi + R_W(1 + g_m r_\pi)} \qquad (8.3.11)$$

$$A_i = \frac{i_L}{i_S} = \frac{v_L R_{\text{in}}}{v_1 R_L} = \frac{R_{\text{in}}}{R_L} A_{v_1} = \frac{R_{\text{in}} R_W(1 + g_m r_\pi)}{R_L[r_\pi + R_W(1 + g_m r_\pi)]} \qquad (8.3.12)$$

r_o is generally large enough so that the following results hold. For $r_o \to \infty$,

$$R_i \cong r_\pi + (1 + g_m r_\pi)(R_E \| R_L) \qquad (8.3.13)$$

$$A_{v_1} \cong \frac{(1 + g_m r_\pi)(R_E \| R_L)}{r_\pi + (1 + g_m r_\pi)(R_E \| R_L)} \cong 1 \qquad (8.3.14)$$

$$A_i \cong \frac{(1 + g_m r_\pi)(R_E \| R_L) R_B}{R_L[r_\pi + R_B + (1 + g_m r_\pi)(R_E \| R_L)]} \qquad (8.3.15)$$

Note that the voltage gain of the CC amplifier is about unity, but never exceeds unity. The current gain, on the other hand, is large since $R_{\text{in}} \gg R_L$ typically.

Common-Base (CB) Configuration

The common-base (CB) amplifier is shown in Figure 8.3.3(a), in which the base forms the common terminal between the input and output circuits, and resistors $R_1, R_2, R_C,$ and R_E are selected through biasing. Capacitors $C_B, C_C,$ and C_E are chosen large enough to act as short circuits at the lowest frequency of interest in the input signal v_S. The output voltage v_L is taken from the collector across resistor R_L. The small-signal ac equivalent circuit of the amplifier is shown in Figure 8.3.3(b), whose analysis yields the following results:

$$R_H = R_C \| R_L = \frac{R_C R_L}{R_C + R_L} \qquad (8.3.16)$$

$$R_i = \frac{r_\pi(r_o + R_H)}{r_\pi + R_H + r_o(1 + g_m r_\pi)} \qquad (8.3.17)$$

$$R_{\text{in}} = R_E \| R_i = \frac{R_E R_i}{R_E + R_i} = \frac{(r_\pi \| R_E)(r_o + R_H)}{r_o + R_H + (r_\pi \| R_E)(1 + g_m r_o)} \qquad (8.3.18)$$

The voltage and current gains are given by

$$A_{v_1} = \frac{v_L}{v_1} = \frac{v_L}{-v_\pi} = \frac{R_H(1 + g_m r_o)}{R_H + r_o} \qquad (8.3.19)$$

$$A_i = \frac{R_{\text{in}}}{R_L} A_{v_1} = \frac{R_H(r_\pi \| R_E)(1 + g_m r_o)}{R_L[r_o + R_H + (r_\pi \| R_E)(1 + g_m r_o)]} \qquad (8.3.20)$$

For $r_o \to \infty$,

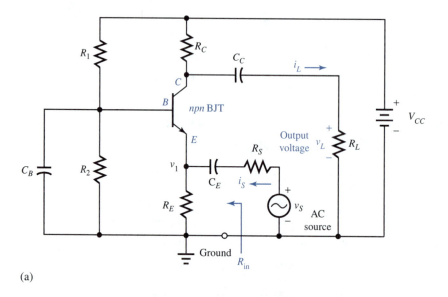

(a)

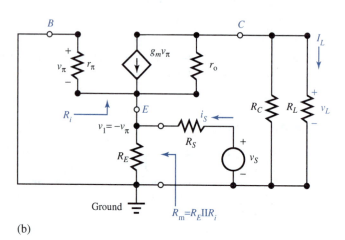

(b)

Figure 8.3.3 Common-base (CB) BJT amplifier. **(a)** Circuit. **(b)** Small-signal ac equivalent circuit.

$$R_i \cong \frac{r_\pi}{1 + g_m r_\pi} \tag{8.3.21}$$

$$R_{in} \cong \frac{(r_\pi \| R_E)}{1 + g_m (r_\pi \| R_E)} \tag{8.3.22}$$

$$A_{v_1} \cong g_m (R_C \| R_L) \tag{8.3.23}$$

$$A_i \cong \frac{g_m (R_C \| R_L)(r_\pi \| R_E)}{R_L[1 + g_m (r_\pi \| R_E)]} \tag{8.3.24}$$

Because of the low input resistance, the CB amplifier is often used as a current amplifier which accepts an input current into its low impedance and provides an output current into a high impedance. That is why a CB amplifier is sometimes referred to as a current follower.

On comparing the CE, CC, and CB configurations one can come up with the following observations:

1. R_i and R_{in} are largest for the CC configuration, smallest for the CB, and in between those extremes for the CE.
2. While there is a sign inversion in the voltage-gain expression with the CE amplifier, about the same magnitude of gain (which can be greater than unity) results for the CE and CB configurations. For the CC amplifier, however, the voltage gain cannot exceed unity.
3. Both CE and CB configurations can yield large current-gain magnitudes; the CB amplifier has a current gain less than unity.

8.4 FET AMPLIFIERS

Just like the BJT amplifiers, FET amplifiers are constructed in common-source (CS, analogous to CE), common-drain (CD, analogous to CC), and common-gate (CG, analogous to CB) configuration. First we shall consider JFET amplifiers, then we show how the results are modified with MOSFETs.

Common-Source (CS) JFET Amplifier

Figure 8.4.1(a) shows a CS JFET amplifier in which resistors R_1, R_2, R_D, and R_{SS} are selected by the bias design, and capacitors C_G, C_D, and C_S are chosen to be large enough that they act as short circuits at the lowest frequency of interest in the input signal v_S. Figure 8.4.1(b) gives its small-signal equivalent circuit. Noting that the input impedance of a JFET is very large, we have

$$R_i \cong \infty \tag{8.4.1}$$

$$R_{in} = R_1 \| R_2 = \frac{R_1 R_2}{R_1 + R_2} \tag{8.4.2}$$

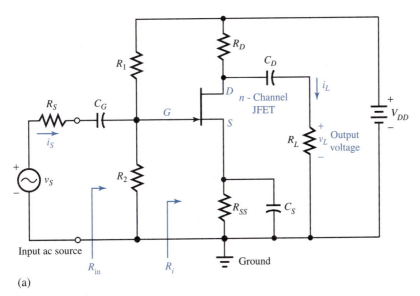

(a)

Figure 8.4.1 Common-source (CS) JFET amplifier. **(a)** Circuit. **(b)** Small-signal ac equivalent circuit.

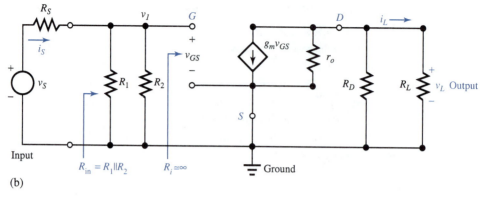

$R_{\text{in}} = R_1 \| R_2$ $R_i = \infty$ Ground

Input

(b)

Figure 8.4.1 Continued

Further analysis yields

$$A_{v_1} = \frac{-g_m r_o R_F}{r_o + R_F} \tag{8.4.3}$$

$$A_i = \frac{-g_m r_o R_F R_{\text{in}}}{R_L(r_o + R_F)} = \frac{R_{\text{in}}}{R_L} A_{v_1} \tag{8.4.4}$$

where

$$R_F = R_D \| R_L = \frac{R_D R_L}{R_D + R_L} \tag{8.4.5}$$

The CS JFET amplifier is capable of large voltage and current gains.

EXAMPLE 8.4.1

A JFET for which $V_A = 80$ V, $V_P = 4$ V, and $I_{DSS} = 10$ mA has a quiescent drain current of 3 mA when used as a common-source amplifier for which $R_D = R_{SS} = 1$ kΩ and $R_L = 3$ kΩ. For the case of fully bypassed R_{SS}, find the amplifier's voltage gain A_{v_1}. Also determine the current gain A_i if $R_1 = 300$ kΩ and $R_2 = 100$ kΩ.

Solution

From Equations (7.4.5) and (7.4.4),

$$r_o = \frac{V_A}{I_{DQ}} = \frac{80}{3 \times 10^{-3}} = 26{,}666.7 \ \Omega$$

$$g_m = \frac{2}{V_P}\sqrt{I_{DSS}I_{DQ}} = \frac{2}{4}\sqrt{10 \times 10^{-3} \times 3 \times 10^{-3}} = 2.7386 \times 10^{-3} \ \text{S}$$

$$R_F = R_D \| R_L = \frac{1}{(1/1000) + (1/3000)} = 750 \ \Omega$$

$$A_{v_1} = \frac{-g_m r_o R_F}{r_o + R_F} = \frac{-2.7386(10^{-3})(26{,}666.7)750}{26{,}666.7 + 750} \cong -2$$

$$R_{in} = R_1 \| R_2 = \frac{300(100)}{400} = 75 \text{ k}\Omega$$

$$A_i = \frac{R_{in}}{R_L} A_{v_1} = \frac{75}{3}(-2) = -50$$

Common-Drain (CD) JFET Amplifier

Figure 8.4.2(a) shows a CD JFET amplifier in which resistors R_1, R_2, and R_{SS} are selected by the bias design, and capacitors C_G and C_S are chosen to be large enough to act as short circuits at frequencies in the band of interest (known as the midband). Figure 8.4.2(b) gives its small-signal ac equivalent circuit. Analysis of this circuit yields

$$R_i \cong \infty \qquad (8.4.6)$$

$$R_{in} = R_1 \| R_2 = \frac{R_1 R_2}{R_1 + R_2} \qquad (8.4.7)$$

$$A_{v_1} = \frac{v_L}{v_1} = \frac{g_m r_o (R_{SS} \| R_L)}{r_o + (R_{SS} \| R_L)(1 + g_m r_o)} \qquad (8.4.8)$$

$$A_i = \frac{i_L}{i_S} = \frac{g_m r_o (R_{SS} \| R_L)(R_1 \| R_2)}{R_L [r_o + (R_{SS} \| R_L)(1 + g_m r_o)]} \qquad (8.4.9)$$

In many cases $A_{v_1} \cong 1$ and $v_L = v_1$; that is to say the load voltage "follows" the input. Hence a CD amplifier is often known as a *source follower*, which becomes an excellent buffer to couple a high-resistance source to a low-resistance load with nearly no loss in signal voltage. The current gain, however, can be very large, leading to significant power gain.

Common-Gate (CG) JFET Amplifier

Figure 8.4.3(a) shows a CG JFET amplifier in which resistors R_1, R_2, R_D, and R_{SS} are selected by the bias design, and capacitors C_D and C_S are chosen to be large enough to act as short circuits at

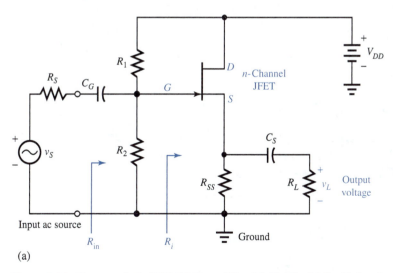

(a)

Figure 8.4.2 Common-drain (CD) JFET amplifier. **(a)** Circuit. **(b)** Small-signal ac equivalent circuit.

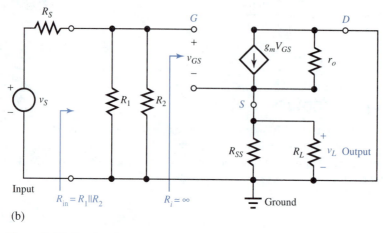

(b)

Figure 8.4.2 Continued

the lowest frequency of interest in the input signal v_S. Figure 8.4.3(b) shows its small-signal ac equivalent circuit, whose analysis yields the following results:

$$R_{in} = \frac{R_{SS}(r_o + R_F)}{r_o + R_F + R_{SS}(1 + g_m r_o)} \tag{8.4.10}$$

where

$$R_F = R_D \| R_L = \frac{R_D R_L}{R_D + R_L} \tag{8.4.11}$$

$$A_{v_1} = \frac{v_L}{v_1} = \frac{R_F(1 + g_m r_o)}{r_o + R_F} \tag{8.4.12}$$

$$A_i = \frac{i_L}{i_s} = \frac{R_F(1 + g_m r_o)R_{SS}}{R_L[r_o + R_F + R_{SS}(1 + g_m r_o)]} \tag{8.4.13}$$

With practical values, quite often r_o is rather large and $g_m R_{SS} \gg 1$ so that $R_{in} \cong 1/g_m$ and $A_i \cong R_F/R_L = R_D/(R_D + R_L)$, which turns out to be less than unity.

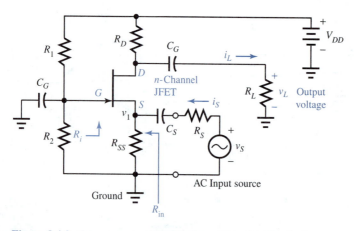

Figure 8.4.3 Common-gate (CG) JFET amplifier. **(a)** Circuit. **(b)** Small-signal ac equivalent circuit.

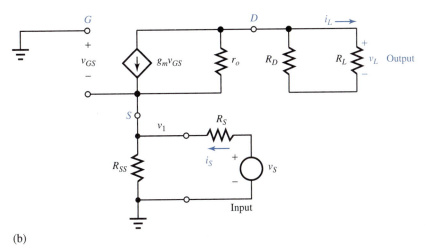

(b)

Figure 8.4.3 Continued

On comparing the CS, CD, and CG configurations one can make the following observations:

1. For CS and CD configurations $R_{in} = R_1 \| R_2$, which can be selected to be large during the bias design. For the CG configuration, however, R_{in} is not very large, on the order of a few hundred ohms.

2. For the CD configuration the voltage gain is generally less than unity or near unity, while it can exceed unity in the other configurations. The voltage gain of the CG configuration is slightly larger than that of the CS configuration.

3. For the CG configuration the current gain cannot be larger than unity. But for the CS and CD configurations it can be large by the choice of R_{in}.

MOSFET Amplifiers

Because the same small-signal equivalent circuits apply to both the JFET and the MOSFETs, all the equations developed for the JFET amplifiers hold good for the MOSFET amplifiers, so long as g_m and r_o are computed properly. For the depletion MOSFET the equations for g_m and r_o, given by Equations (7.4.4) and (7.4.5), are the same as for the JFET. For the enhancement MOSFET, they are given by Equations (7.4.9) and (7.4.10).

8.5   FREQUENCY RESPONSE OF AMPLIFIERS

All amplifiers exhibit variations of performance as the signal frequency is changed. The frequency response of an amplifier may be defined as the functional dependence of output amplitude and phase upon frequency, for all frequencies. Invariably there is a maximum frequency above which amplification does not occur. Depending on the design of the circuit, there may also be a lower frequency limit below which amplification disappears. The range of frequencies over which significant amplification of the magnitude is obtained is known as the *passband*, which is bounded by the upper cutoff frequency and the lower cutoff frequency (if one exists). *Dc-coupled* amplifiers, in which the stages are coupled together for all frequencies down to zero, have no low-frequency limit. On the other hand, circuits containing coupling capacitors that couple stages together for

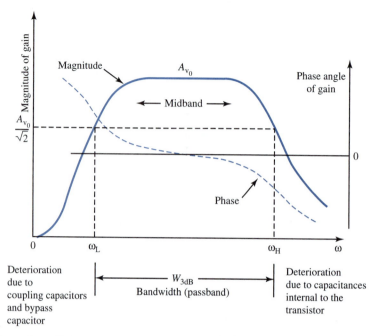

Figure 8.5.1 Typical frequency response of a voltage amplifier.

ac while isolating them for dc are limited in their low-frequency response. The high-frequency deterioration of the voltage gain is due to the effect of capacitances that are internal to the transistor.

Figure 8.5.1 shows a typical frequency response of a voltage amplifier. While most often the magnitude of the gain is discussed (because it defines the frequency range of useful gain), the phase response becomes important for transient calculations. The *midband* region is the range of frequencies where gain is nearly constant. Amplifiers are normally considered to operate in this useful midband region. The *bandwidth* of an amplifier is customarily defined as the band between two frequencies, denoted by ω_H and ω_L, which corresponds to the gain falling to 3 dB below the midband constant gain. Thus,

$$W_{3dB} = \omega_H - \omega_L \tag{8.5.1}$$

which is shown in Figure 8.5.1. When both ω_L and ω_H have significant values, the amplifier is known as a bandpass amplifier. A narrow-bandpass amplifier is one in which W_{3dB} is small relative to the center frequency of the midband region. In a low-pass unit (dc amplifier) there is no low 3-dB frequency.

The general problems of analyzing any given amplifier to determine ω_L or ω_H, and of designing amplifiers with specific values of ω_L and ω_H, are extremely complex and well beyond the scope of this text. Specific amplifier cases can, however, be considered in order to gain a sense of what is involved in determining the frequency response. Toward that end let us consider the CS JFET amplifier shown in Figure 8.4.1(a). The small-signal ac equivalent circuit for low frequencies near ω_L is depicted in Figure 8.5.2(a), that for high frequencies near ω_H in Figure 8.5.2(b).

Assuming r_o to be large for convenience, the gain of the low-frequency circuit of Figure 8.5.2(a) can be found to be

$$A_v = \frac{v_L}{v_s} = \frac{A_{v_0}(j\omega)^2(\omega_Z + j\omega)}{(\omega_{L_1} + j\omega)(\omega_{L_2} + j\omega)(\omega_{L_3} + j\omega)} \tag{8.5.2}$$

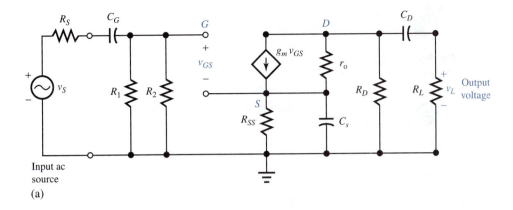

(a)

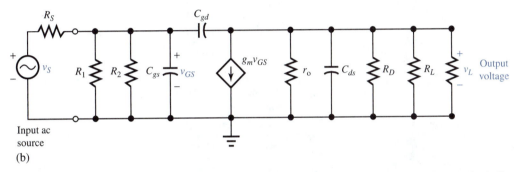

(b)

Figure 8.5.2 Small-signal equivalent circuits for CS JFET amplifier. **(a)** For low frequencies near ω_L (including coupling capacitors and bypass capacitor). **(b)** For high frequencies near ω_H (including parasitic capacitances of transistor).

where

$$A_{v_0} = \frac{-g_m R_G R_D R_L}{(R_S + R_G)(R_D + R_L)} \tag{8.5.3}$$

$$R_G = R_1 \| R_2 = \frac{R_1 R_2}{R_1 + R_2} \tag{8.5.4}$$

$$\omega_Z = \frac{1}{R_{SS} C_S} \tag{8.5.5}$$

$$\omega_{L_1} = \frac{1}{R_{L_1} C_S} \tag{8.5.6}$$

$$\omega_{L_2} = \frac{1}{R_{L_2} C_D} \tag{8.5.7}$$

$$\omega_{L_3} = \frac{1}{R_{L_3} C_G} \tag{8.5.8}$$

$$R_{L_1} = \frac{R_{SS}}{1 + g_m R_{SS}} \tag{8.5.9}$$

$$R_{L_2} = R_D + R_L \tag{8.5.10}$$

$$R_{L_3} = R_S + R_G \tag{8.5.11}$$

The frequencies ω_Z, ω_{L_1}, ω_{L_2}, and ω_{L_3} are known as break frequencies, at which the behavior of $|A_v|$ changes. For ω above all break frequencies $A_v \cong A_{v_0}$, which is the midband value of the gain. The largest break frequency is taken to be ω_L, the low 3-dB frequency.

The amplifier gain at higher frequencies near ω_H is found from the analysis of Figure 8.5.2(b). Neglecting the second-order term in ω^2 in the denominator, the gain can be found to be

$$A_v = \frac{v_L}{v_s} \cong \frac{A_{v_0}\omega_H(\omega_Z - j\omega)}{\omega_Z(\omega_H + j\omega)} \tag{8.5.12}$$

where

$$A_{v_0} = \frac{-g_m R_G R_{DL}}{R_S + R_G} \tag{8.5.13}$$

$$R_{DL} = r_o \| R_D \| R_L = \frac{r_o R_D R_L}{r_o R_D + r_o R_L + R_D R_L} \tag{8.5.14}$$

$$R_G = R_1 \| R_2 = \frac{R_1 R_2}{R_1 + R_2} \tag{8.5.15}$$

$$\omega_Z = \frac{g_m}{C_{gd}} \tag{8.5.16}$$

$$\omega_H = \frac{1}{R_A[C_{gs} + C_{gd}(1 + g_m R_{DL} + R_{DL}/R_A)]} \tag{8.5.17}$$

$$R_A = R_S \| R_G = \frac{R_S R_G}{R_S + R_G} \tag{8.5.18}$$

An illustrative example of both low- and high-frequency designs is worked out next.

EXAMPLE 8.5.1

Consider a CS JFET amplifier with the following parameters: $R_1 = 350\,\text{k}\Omega$; $R_2 = 100\,\text{k}\Omega$, $R_{SS} = 1200\,\Omega$, $R_D = 900\,\Omega$, $R_L = 1000\,\Omega$, $R_S = 2000\,\Omega$, $r_o = 15\,\text{k}\Omega$, $g_m = 6 \times 10^{-3}$ S, $C_{gs} = 3$ pF, $C_{gd} = 1$ pF, and $\omega_L = 2\pi \times 100$ rad/s. Discuss the low- and high-frequency designs.

Solution

For the low-frequency design,

$$R_G = R_1 \| R_2 = \frac{R_1 R_2}{R_1 + R_2} = \frac{350(100)}{450} = 77.78\,\text{k}\Omega$$

$$R_{L_1} = \frac{R_{SS}}{1 + g_m R_{SS}} = \frac{1200}{1 + (6 \times 10^{-3} \times 1200)} = \frac{1200}{8.2} = 146.34\,\Omega$$

$$R_{L_2} = R_D + R_L = 900 + 1000 = 1900\,\Omega$$

$$R_{L_3} = R_S + R_G = 2000 + 77,780 = 79,780\,\Omega$$

Noting that R_{L_1} is the smallest, from Equation (8.5.6),

$$C_S = \frac{1}{R_{L_1}\omega_L} = \frac{1}{146.34(2\pi \times 100)} = 10.87\,\mu\text{F}$$

From Equations (8.5.7) and (8.5.8),

$$C_D = \frac{1}{R_{L_2}\omega_{L_2}} = \frac{1}{R_{L_2}\omega_L/10} = \frac{10}{1900(200\pi)} = 8.37 \ \mu\text{F}$$

where ω_{L_2} is chosen to be $\omega_L/10$ so that the break frequency of the capacitor is at least 10 times smaller than ω_L.

$$C_G = \frac{1}{R_{L_3}\omega_{L_3}} = \frac{1}{R_{L_3}\omega_L/10} = \frac{10}{79,780(200\pi)} = 0.2 \ \mu\text{F}$$

where ω_{L_3} is chosen to be $\omega_L/10$ so that the break frequency of the capacitor is at least 10 times smaller than ω_L.

In order to determine ω_H, we first find

$$R_A = R_S \| R_G = \frac{2000(77,780)}{79,780} = 1969.9 \ \Omega$$

$$R_D \| R_L = \frac{900(1000)}{1900} = 473.68 \ \Omega$$

$$R_{DL} = r_o \| R_D \| R_L = \frac{15,000(473.68)}{15,473.68} = 459.2 \ \Omega$$

From Equation (8.5.17),

$$\omega_H = \frac{10^{12}}{1950[3 + 1 + (6 \times 10^{-3})(459.2) + 459.2/1950]} = 73.26 \times 10^6 \ \text{rad/s}$$

or

$$\frac{\omega_H}{2\pi} = 11.655 \ \text{MHz}$$

Midband gain from Equation (8.5.3) is

$$A_{v_0} = \frac{-(6 \times 10^{-3})(77,780)(900)(1000)}{(2000 + 77,780)(900 + 1000)} = -2.77$$

Amplifiers with Feedback

Almost all practical amplifier circuits include some form of negative feedback. The advantages gained with feedback may include the following:

- Less sensitivity to transistor parameter variations.
- Improved linearity of the output signal by reducing the effect of nonlinear distortion.
- Better low- and high-frequency response.
- Reduction of input or output loading effects.

The gain is willingly sacrificed in exchange for the benefits of negative feedback. Compensating networks are often employed to ensure stability. A voltage amplifier with negative feedback is shown by the generalized block diagram in Figure 8.5.3. The negative feedback creates a self-correcting amplification system that compensates for the shortcomings of the amplifying unit. The gain with feedback can be seen to be

$$A_f = \frac{v_{\text{out}}}{v_{\text{in}}} = \frac{A}{1 + AB} \qquad (8.5.19)$$

Also noting that $v_d = v_{\text{out}}/A$, it follows

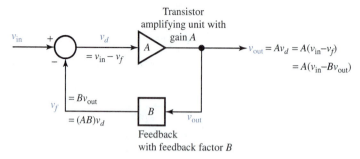

Figure 8.5.3 Voltage amplifier with negative feedback.

$$\frac{v_d}{v_{in}} = \frac{1}{1 + AB} \tag{8.5.20}$$

The product AB is known as the loop gain because $v_f = B(Av_d) = (AB)v_d$. In order to obtain negative feedback, the loop gain must be positive. With $AB > 0$ in Equation (8.5.19) it follows that $|A_f| < |A|$. Thus, a negative feedback amplifier has always less gain than the amplifying unit itself. With $AB >> 1$, A_f remains essentially constant, thereby reducing the effect of transistor parameter variations in the amplifying unit. If the feedback unit has $|B| < 1$, one can still obtain useful amplification since $|A_f| \cong 1/|B| > 1$.

The amplifiers discussed in Sections 8.3 and 8.4 as well as in this section are small-signal linear stages that use capacitance coupling. These are also known as *RC amplifiers* since their circuits need only resistors and capacitors. There are of course other transistor amplifiers that do not have the small-signal limitation and are also not limited to the use of resistors and capacitors in their circuits. For larger input signals the linear dependence of the current on the signal level may not be maintained.

8.6 LEARNING OBJECTIVES

The *learning objectives* of this chapter are summarized here so that the student can check whether he or she has accomplished each of the following.

- Biasing the BJTs and FETs.
- BJT and FET amplifier configurations, and their small-signal equivalent circuits, based on which voltage and current gains are computed.
- Frequency response of amplifiers, and low- and high-frequency designs.
- Advantages of negative feedback in amplifier circuits.

8.7 PRACTICAL APPLICATION: A CASE STUDY

Mechatronics: Electronics Integrated with Mechanical Systems

Electronic circuits have become so intimately integrated with mechanical systems that a new term, *mechatronics*, has been coined for the combination. Various subjects related to mechatronics are being developed at a number of universities in the United States, supported by major industrial sponsors. Too many engineers, at present, are not well equipped and trained to design mechatronic products. Interdisciplinary experts who can blend many different technologies harmoniously are rather rare.

Let us think for a moment about various electronic and electric systems in an automobile of today. Only a few of those are listed here:

- *Body electronics:* Airbags, security and keyless entry, memory seats and mirrors
- *Vehicle control:* Antilock brakes, traction control, electronic navigation, adaptable suspension systems
- *Power train:* Engine and transmission, cruise control, electronic ignition, four-wheel drive
- *Instrumentation:* Analog/digital dash, computerized performance evaluation and maintenance scheduling, tire inflation sensors
- *Communications and entertainment:* cellular phone, AM/FM radio, CD/tape player, digital radio
- *Alternative propulsion systems* such as electric vehicles, advanced batteries, and hybrid vehicles are being developed. Fiber-optics in communications and *electrooptics*, replacing the conventional wire harness, are already in practice.

In recent years, the conventional electric ignition system has been replaced by *electronic ignition*. Mechanically operated switches, or the so-called points, have been replaced by bipolar junction transistors (BJTs). The advantages of transistorized ignition systems over the conventional mechanical ones are their greater reliability, durability, and ease of control. The transistor cycles between saturation (state in which it behaves as a closed switch) and cutoff (state in which it behaves as an open switch). The ignition spark is produced as a result of rapidly switching off current through the coil. Modern engine control systems employ electric sensors to determine operating conditions, electronic circuits to process the sensor signals, and special-purpose computers to compute the optimum ignition timing.

Automotive electronics has made tremendous advances, and still continues to be one of the most dominating topics of interest to automotive engineers. Certainly, today's mechanical engineers have to be familiar with electronic circuit capabilities and limitations while they try to integrate them with mechanical design and material science.

PROBLEMS

8.1.1 A silicon *npn* BJT is biased by the method shown in Figure 8.1.1, with $R_E = 240\ \Omega$, $R_2 = 3000\ \Omega$, and $V_{CC} = 24$ V. The operating point corresponds to $V_{BEQ} = 0.8$ V, $I_{BQ} = 110\ \mu$A, $V_{CEQ} = 14$ V, and $I_{CQ} = 11$ mA. Determine R_C and R_1.

8.1.2 By using the rule-of-thumb procedure indicated in the text, find values of all resistors in the bias method of Figure 8.1.1 with $V_{CC} = 10$ V for a silicon *npn* BJT for which $\beta = 100$ and the operating point is given by $I_{CQ} = 5$ mA.

8.1.3 A fixed-bias method is illustrated in Figure P8.1.3. Assuming I_{CBO} to be small compared to I_{BQ} and I_{CQ}, find R_B such that the operating point corre-

sponds to $I_{CQ} = 14$ mA, $V_{CEQ} = 7$ V, when $V_{CC} = 12$ V and the silicon BJT has a nominal $\beta = 70$. Also determine R_C.

*8.1.4 Consider the collector–base biasing method shown in Figure P8.1.4. With the same data as in Problem 8.1.3, find R_B and R_C.

8.1.5 A BJT is biased by the method shown in Figure P8.1.5. If $V_{BEQ} = 0.7$ V, $\beta = 100$, $V_{CEQ} = 10$ V, and $I_{CQ} = 5$ mA, find I_1, I_2, and I_{EQ}.

8.2.1 For the method of biasing a JFET shown in Figure 8.2.1 of the text, use the design procedure outlined there to find V_G, R_S, R_D, R_2, and R_1, given that

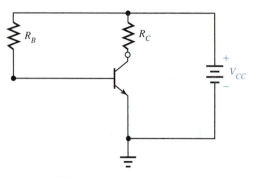

Figure P8.1.3

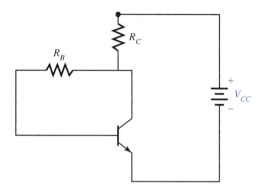

Figure P8.1.4

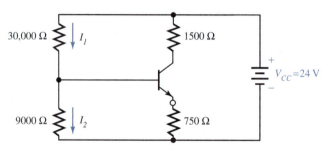

Figure P8.1.5

$V_{DD} = 24$ V and for the JFET $V_P = 3.5$ V and $I_{DSS} = 5$ mA.

8.2.2 An n-channel JFET having $V_P = 3.5$ V and $I_{DSS} = 5$ mA is biased by the circuit of Figure 8.2.1 with $V_{DD} = 28$ V, $R_S = 3000$ Ω, and $R_2 = 100$ kΩ. If the operating point is given by $I_{DQ} = 2$ mA and $V_{DSQ} = 12$ V, determine V_G, R_D, and R_1.

*8.2.3 A self-biasing method for the JFET is shown in Figure P8.2.3. The JFET has $V_P = 3$ V and $I_{DSS} = 24$ mA. It is to operate at an active-region Q point that is given by $I_{DQ} = 5$ mA and $V_{DSQ} = 8$ V. Determine R_S and R_D so that the desired Q point is achieved with $V_{DD} = 16$ V.

8.2.4 An n-channel depletion MOSFET that has $V_P = 3$ V and $I_{DSS} = 3$ mA (when $V_{DS} = 10$ V) is biased by the circuit of Figure 8.2.1 with $V_{DD} = 20$ V and $R_2 = 1$MΩ. If the operating point is given by $I_{DQ} = 3$ mA and $V_{DSQ} = 10$ V, and 40% of the voltage drop across R_D and R_S is across R_S, determine R_S, R_D, V_G, and R_1.

8.2.5 Given that an n-channel enhancement MOSFET has $V_T = 1.5$ V, $K = 5$ mA/V^2, $I_{DQ} = 10$ mA, $V_{DSQ} = 15$ V, and a maximum continuous power dissipation of 0.3 W. For the biasing circuit of Figure 8.2.2 with $V_{DD} = 24$ V, $R_2 = 1$ MΩ, and the voltage across $R_S = 3$ V, find R_S, R_D, V_G, and R_1. Also find the continuous power dissipated in the MOSFET.

*8.3.1 By analyzing the small-signal ac equivalent circuit of the CE BJT amplifier shown in Figure 8.3.1(b), show that Equation (8.3.6) is true.

8.3.2 The input resistance of a CE amplifier can be increased at the expense of reduced voltage and current gains by leaving a portion of the emitter resistance unbypassed. Consider the CE amplifier shown in Figure P8.3.2 having an unbypassed emitter resistance R_{E_1}. Assuming for simplicity that $r_o \to \infty$, draw the ac equivalent circuit for small signals and discuss the effects on input resistance and voltage and current gains.

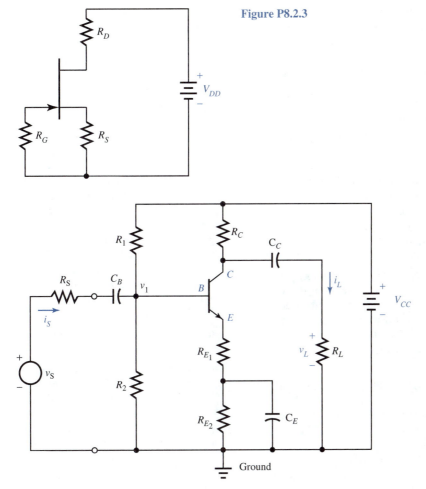

Figure P8.2.3

Figure P8.3.2 CE amplifier having an unbypassed emitter resistance R_{E_1}.

8.3.3 Consider the CE BJT amplifier of Figure 8.3.1(a) with $R_1 = 1600$ Ω, $R_2 = 400$ Ω, $R_C = 70$ Ω, $R_E = 20$ Ω, and $R_L = 150$ Ω. The transistor has $\beta = 70$, $V_A = 50$ V, and $I_{CQ} = 80$ mA when $V_{CC} = 15$ V. Compute the amplifier's voltage and current gains. Take V_T to be 25.861 mV.

8.3.4 Let the bypass capacitor C_E be removed in Problem 8.3.3. How would the gains be altered?

8.3.5 Let half of R_E be unbypassed in Problem 8.3.3. How would the gains be altered?

*8.3.6 The CE BJT amplifier of Figure 8.3.1(a) has the following parameters: $R_1 = 28,000$ Ω, $R_2 = 8000$ Ω, $R_C = 1400$ Ω, $R_E = 700$ Ω, $R_L = 2000$ Ω, $V_{CC} = 24$ V, and $I_{CQ} = 5$ mA when $\beta = 100$ and $V_A = 100$ V. Calculate the voltage and current gains of the amplifier.

8.3.7 Let 200 Ω of R_E be not bypassed in Problem 8.3.6. How would the gains be altered?

8.3.8 Determine voltage and current gains for the CE BJT amplifier shown in Figure 8.3.1(a) with the following parameters: $R_1 = 30,000$ Ω, $R_2 = 9000$ Ω, $R_C = 750$ Ω, $R_E = 250$ Ω, $R_L = 1000$ Ω, $V_{CC} = 9$ V; and $I_{CQ} = 5$ mA when $\beta = 150$ and $V_A = 50$ V.

8.3.9 Assuming that the bypass capacitor C_E is removed in Problem 8.3.8, find the voltage and current gains.

*8.3.10 Find an expression for the power gain A_P for a common-collector stage if the power gain of an amplifier stage is defined as the ratio of the load power $v_L i_L$ to the power delivered by the source $v_1 i_s$. (See Figure 8.3.1 for notation.)

8.3.11 Starting from Equations (8.3.9), (8.3.11), and (8.3.12), show that equations (8.3.13) through (8.3.15) hold for $r_o \to \infty$.

8.3.12 Consider the CC BJT amplifier circuit shown in Figure 8.3.2(a) with $R_1 = 150\,k\Omega$, $R_2 = 150\,k\Omega$, $R_E = 2\,k\Omega$, $R_L = 3\,k\Omega$, and $V_{CC} = 10\,V$. The BJT has $\beta = 100$, $I_{CQ} = 1.2\,mA$, and $V_A = 75\,V$. Determine R_i, R_{in}, A_{v_1}, and A_i.

8.3.13 Develop a formula for the power gain A_P for a common-collector stage if the power gain of an amplifier stage is defined as the ratio of the load power $v_L i_L$ to the power delivered by the stage $v_1 i_s$. (See Figure 8.3.2 for notation.)

8.3.14 In the CC amplifier stage of Figure 8.3.2(a), let $R_1 = 32\,k\Omega$, $R_2 = 22\,k\Omega$, $R_E = 400\,\Omega$, $R_L = 250\,\Omega$, and $V_{CC} = 9\,V$. Given that $V_A = 70\,V$, $\beta = 50$, and $I_{CQ} = 4\,mA$ for the BJT, find R_i, R_{in}, A_{v_1}, and A_i.

*8.3.15 Obtain an expression for the power gain A_P for a common-base stage if the power gain of an amplifier stage is defined as the ratio of the load power $v_L i_L$ to the power delivered by the source $v_1 i_s$. (See Figure 8.3.3 for notation.)

8.3.16 Show that Equation (8.3.17) holds for the circuit of Figure 8.3.3(b).

8.3.17 Consider the CB BJT amplifier circuit shown in Figure 8.3.3(a), with $R_C = 1\,k\Omega$, $R_L = 6\,k\Omega$, and $R_{in} = 20\,\Omega$. The transistor parameters are given by $\beta = 60$, $V_A = 70\,V$, and $g_m = 0.03\,S$. Find R_i, R_E, A_{v_1}, and A_i.

8.3.18 Consider the CE BJT amplifier circuit shown in Figure P8.3.18. In order to make it into a common-base amplifier, terminals a and b are connected together, and capacitor C_E is disconnected from ground and used to couple a signal from a source to the emitter. Determine R_i, R_{in}, A_{v_1}, and A_i.

8.4.1 For the CS JFET amplifier circuit of Figure 8.4.1(a), $R_D = 2\,k\Omega$ and $R_L = 3\,k\Omega$. The JFET with $r_o = 15\,k\Omega$ has a voltage gain $A_{v_1} = -4.5$ when the entire source resistance is bypassed. Find g_m.

8.4.2 Considering the CS JFET amplifier circuit of Figure 8.4.1(a), a portion of the source resistance R_{SS} in the JFET CS stage is sometimes left unbypassed. Let $R_{SS} = R_{SS_1} + R_{SS_2}$ in which R_{SS_1} is that portion of R_{SS} that is not bypassed and R_{SS_2} is that part of R_{SS} that is bypassed. Find expressions for the amplifier voltage gain A_{v_1} and the current gain A_i. Comment on the results, particularly with reference to Equations (8.4.3) and (8.4.4).

8.4.3 The common-source amplifier of Figure 8.4.1(a) with $R_L = 300\,\Omega$, $R_D = 150\,\Omega$, and $R_{SS} = 100\,\Omega$ (fully bypassed) has a JFET with $V_A = 80\,V$, $r_o = 2\,k\Omega$, and $V_P = 4\,V$.

(a) Compute g_m and I_{DSS} if the voltage gain is -2.8.

(b) If the voltage gain is to be reduced to -1.4 by leaving part of R_{SS} unbypassed, find R_{SS_1}, which is that portion of R_{SS} that is not bypassed.

*8.4.4 Consider Figure 8.4.1 (a) of a CS JFET amplifier with $R_1 = 330\,k\Omega$, $R_2 = 110\,k\Omega$, $R_D = 1\,k\Omega$, $R_{SS} = 1\,k\Omega$, $R_L = 1\,k\Omega$, $I_{DQ} = 6\,mA$, $V_A = 90\,V$, $V_P = 4\,V$, and $I_{DSS} = 20\,mA$. Compute R_{in}, A_{v_1}, and A_i.

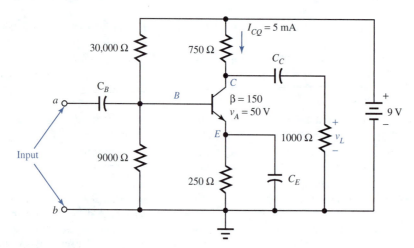

Figure P8.3.18

8.4.5 For the CD JFET amplifier of Figure 8.4.2(a), show that Equations (8.4.8) and (8.4.9) can be rearranged as follows:

$$A_{v_1} = \frac{g_m(r_o \| R_{SS} \| R_L)}{1 + g_m(r_o \| R_{SS} \| R_L)}$$

$$A_i = \frac{g_m(r_o \| R_{SS} \| R_L)(R_1 \| R_2)}{R_L[1 + g_m(r_o \| R_{SS} \| R_L)]}$$

8.4.6 In the source-follower circuit of Figure 8.4.2(a) the voltage gain is to be 0.7 when $R_{SS} = 300 \ \Omega$ and a JFET with $r_o = 5 \ k\Omega$ and $g_m = 0.025$ S is used. Find the required load resistance.

8.4.7 Obtain an expression for the output resistance seen by the load when looking back into the amplifier between source and ground for the CD JFET source-follower circuit of Figure 8.4.2(a).

8.4.8 If the circuit of Example 8.4.1 is converted to a CG amplifier having the same component values, compute the amplifier voltage gain A_{v_1}, current gain A_i, and R_{in}.

*8.4.9 Obtain an expression for R_i for the CG JFET amplifier circuit of Figure 8.4.3(a).

8.4.10 Consider the CG JFET amplifier circuit of Figure 8.4.3(a) with $R_L = 15k\Omega$, $R_D = 7.5 \ k\Omega$, $R_{SS} = 5 \ k\Omega$, $r_o = 100 \ k\Omega$, and $g_m = 5 \times 10^{-3}$ S. Evaluate A_{v_1}, A_i, and R_{in}.

8.5.1 If the voltage gain of an amplifier is given by

$$A_v = \frac{20 j\omega}{(120\pi + j\omega)[1 + j\omega/(5\pi \times 10^{-4})]}$$

find ω_H, ω_L, and the midband gain.

8.5.2 Consider a CS JFET amplifier (Figure 8.5.2) with $R_1 = 520 \ k\Omega$, $R_2 = 140 \ \Omega$, $R_D = 1 \ k\Omega$, and $R_{SS} = 1.4 \ k\Omega$. The FET parameters are $g_m = 5 \ mS$, $r_o = 20k\Omega$, and $C_{gs} = C_{gd} = 2 \ pF$. The source resistance R_S is 500 Ω and the load resistance R_L is 2 kΩ. For the low-frequency 3-dB angular frequency to be 40π rad/s, compute C_S, C_D, C_G, A_{v_0}, and ω_H.

8.5.3 Determine R_L, C_{gs}, and A_{v_0} of a CS JFET amplifier (Figure 8.5.2), given that $R_S = 1 \ k\Omega$ which can be considered to be much less than $(R_1 \| R_2)$, $R_L = R_D$, $\omega_L \ll \omega_H$, $\omega_{3dB} = 25 \times 10^6$ rad/s, and $\omega_Z/\omega_H = 75$. The transistor parameters are $g_m = 5 \ mS$, $r_o \gg (R_D \| R_L)$, and $C_{gs} = C_{gd}$.

8.5.4 (a) Let the nonlinear input–output relationship of a transistor amplifying unit be given by $v_{out} = 50(v_d + 3v_d^2)$. Plot the transfer characteristic and check for what range of v_d the undistorted output results.

(b) Let a negative feedback loop be added such that $v_d = v_{in} - B v_{out}$. For values of $B = 0.03$ and 0.08, investigate how the negative feedback reduces nonlinear distortion, but reduces gain.

*8.5.5 (a) Consider the voltage amplifier with negative feedback shown in Figure 8.5.3. For $A = -100$, find the feedback factor B in order to get $A_f = -20$.

(b) With the value of B found in part (a), determine the resulting range of A_f if the transistor parameter variations cause A to vary between -50 and -200.

8.5.6 (a) Consider a BJT in the common-emitter configuration. The equations that describe the behavior of the transistor in terms of the hybrid h parameters are given by

$$v_B = h_{ie}i_B + h_{re}v_C$$

$$i_C = h_{fe}i_B + h_{oe}v_C$$

Develop the h-parameter equivalent circuit of the transistor in the common-emitter mode.

(b) Setting h_{re} and h_{oe} equal to zero, obtain the h-parameter approximate equivalent circuit of the common-emitter transistor.

(c) Figure P8.5.6(a) shows the small-signal equivalent circuit of a transistor amplifier in the common-emitter mode. Find expressions for gains $A_{I_1} = I_L/I_B$, $A_{I_2} = I_L/I_S$, $A_{V_1} = V_C/V_B$, and $A_{V_2} = V_C/V_S$.

(d) Now consider the single-stage transistor amplifier shown in Figure P8.5.6(b). The parameters of the 2N104 transistor are $h_{ie} = 1.67 \ k\Omega$, $h_{fe} = 44$, and $1/h_{oe} = 150 \ k\Omega$. Evaluate the performance of the amplifier by computing current gain, voltage gain, power gain, input resistance as seen by the signal source, and output resistance appearing at the output terminals.

(e) Next consider the two-stage transistor amplifier depicted in Figure P8.5.6(c). Develop the h-parameter equivalent circuit of the two-stage amplifier and evaluate its performance.

(f) Then consider the linear model of the common-emitter BJT amplifier shown in Figure P8.5.6(d) that is applicable to the midband frequency range. Find an expression for the current gain $A_{Im} = I_2/I_S$ that remains invariant with frequency. (Note the subscript m

denotes a reference to the midband frequency range.)

(g) The linear model of a common-emitter BJT amplifier shown in Figure P8.5.6(e) is applicable at low frequencies. Obtain an expression for the current gain $A_{Il} = I_2/I_S$ in terms of A_{Im} and ω_l, where ω_l corresponds to $0.707A_{Im}$ on the frequency response curve, and ω_l identifies the lower end of the useful bandwidth of the amplifier.

(h) The hybrid-π equivalent circuit for computing the high-frequency performance of a BJT

RC-coupled amplifier is shown in Figure P8.5.6(f). Analyze the circuit (by representing in terms of isolated input and output sections) to find $A_{Ih} = I_2/I_S$ in terms of A_{Im} and ω_h, where ω_h corresponds to $0.707A_{Im}$ on the frequency response curve, and denotes highband frequencies.

(i) Sketch the frequency response curve of an *RC*-coupled amplifier by plotting the current gain as a function of frequency on semilogarithmic paper and showing the bandwidth.

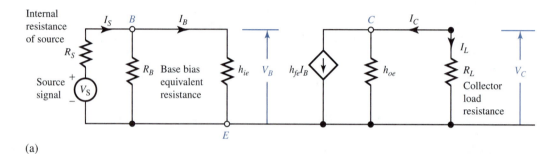

(a)

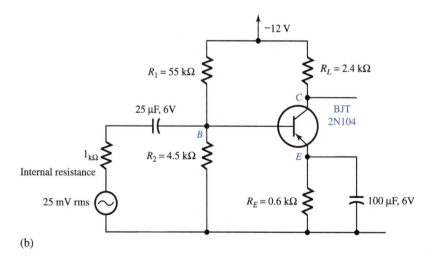

(b)

Figure P8.5.6

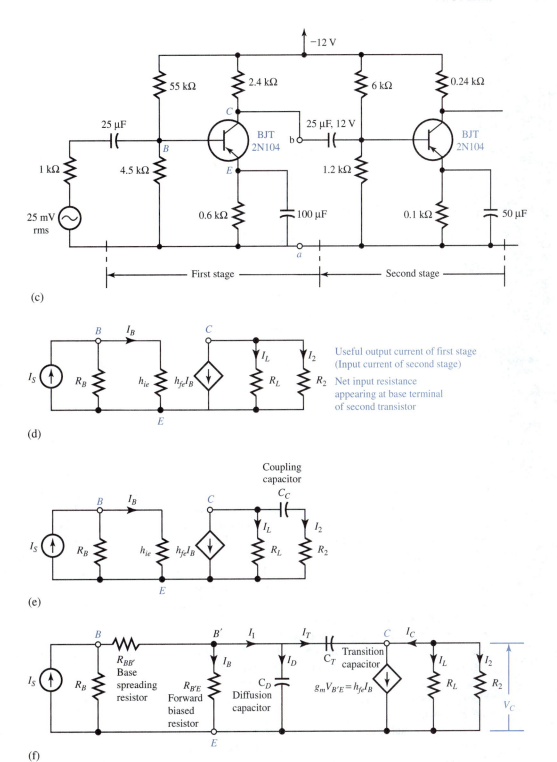

(c)

(d)

(e)

(f)

Figure P8.5.6 Continued

9 Digital Circuits

The use of nonlinear devices (BJT and FET) in constructing linear amplifiers has been studied in Chapter 8. Although these devices are inherently nonlinear, their operation was confined to the linear portions of their characteristics in order to produce linear amplification of a signal. When these devices are operated in the nonlinear regions of their characteristics, the primary use lies in electronic switches for use in computers and other digital systems. Digital electronic circuits are becoming of increasing importance. Revolutionary advances in integrated-circuit (IC) technology have made it possible to place a large number of switches (gates) on a chip of very small size. Besides the advantage of utilizing less space, another significant advantage of digital systems over analog systems is their inherent immunity to noise and interference.

Digital building blocks (made up of transistor circuits) and computer systems have been presented in Chapter 6. Digital circuits are almost always purchased as ready-made IC building blocks. However, users of digital hardware must decide which technology is best suited for each case because competing technologies known as *logic families* are available. Digital circuit design is of course of great interest to engineers in the IC industry for designing and building digital blocks. The users of digital technology will certainly benefit from knowing what is inside the blocks they use.

A fundamental circuit of digital logic is the transistor switch, or inverter, which is considered in Section 9.1. Basic logic circuits are developed in Section 9.2 using bipolar technology. Section 9.3 deals with other logic families, particularly the FET-based family known as CMOS (complementary metal-oxide semiconductor) technology.

9.1 TRANSISTOR SWITCHES

The basic element of logic circuits is the transistor switch, a simplified model of which is shown in Figure 9.1.1. The control signal is the input voltage v_{in}, which must lie in either the "low" range or the "high" range for a digital circuit. When v_{in} is low, the switch may take either the open or the closed position; when v_{in} is high, the switch takes the other position. Looking at Figure 9.1.1, closing the switch makes v_{out} zero, while opening the switch yields an output near V_{CC}, assuming that not much current flows through the output terminal. If V_{CC} is chosen to lie in the high range and $V = 0$ is inside the low range, then allowed values of v_{in} control the switch and give rise to allowed values of v_{out}. A common special circuit in which a high input yields a low output, and vice versa, is known as an *inverter*, which performs the complement operation.

A typical inverter circuit using a bipolar transistor is shown in Figure 9.1.2(a), and the operation of the BJT switch is illustrated in Figure 9.1.2(b). The KVL around the collector–emitter loop is given by

$$v_{CE} = V_{CC} - R_C i_C \tag{9.1.1}$$

from which the load line may be drawn on the collector–emitter characteristic. The KVL around the base–emitter loop is given by

$$v_{BE} = V_i - R_B i_B \tag{9.1.2}$$

and the corresponding load line may be drawn on the base–emitter characteristic.

If the input is at zero volts, that is logic 0, then the base current is zero, and the operating point is at ①, as shown in Figure 9.1.2(b). The transistor is then said to be *cut off*, or simply *off*, when only a very small value of collector current ($I_{C \, cutoff} \cong I_{CEO}$) flows. Thus, in the cutoff state the output voltage is given by

$$V_o = V_{CE \, cutoff} \cong V_{CC} - R_C I_{CEO} \cong V_{CC} = 5V \qquad \text{(i.e., logic 1 level)} \tag{9.1.3}$$

If the input V_i changes to +5 V (logic 1 level), the base current is given by

$$I_{B \, sat} = \frac{V_i - v_{BE}}{R_B} \tag{9.1.4}$$

where $v_{BE} \cong 0.7$ V, which is the *threshold voltage* V_T. Supposing that R_B is chosen so as to drive the transistor into saturation such that $I_{B \, sat} = 50 \, \mu A$, the operating point switches to point ② on the collector–emitter characteristic, when the transistor is said to be *saturated*, or simply *on*. In this saturated state $V_{CE \, sat} = V_{sat}$, which is typically 0.2 to 0.3 V, depending on i_B. The collector current in saturation is given by

$$I_{C \, sat} = \frac{V_{CC} - V_{sat}}{R_C} \cong \frac{V_{CC}}{R_C} \tag{9.1.5}$$

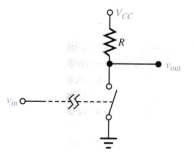

Figure 9.1.1 Model of transistor switch.

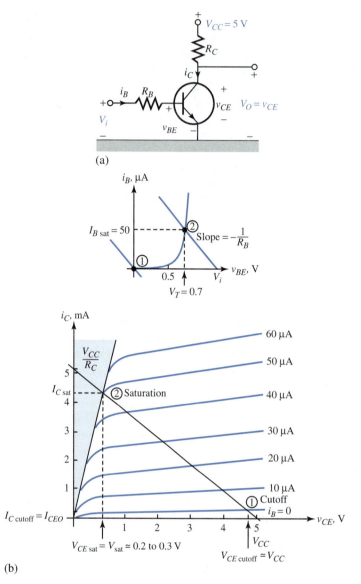

Figure 9.1.2 BJT inverter switch. **(a)** Circuit with *npn* switching transistor.
(b) Typical operation using load lines.

Thus, the transistor behaves like an ideal switch, as shown in Figure 9.1.3. It can be shown that saturation will occur when

$$\frac{V_i - V_T}{R_B} > \frac{V_{CC} - V_{sat}}{\beta \, R_C} \qquad \text{or} \qquad V_i > (V_{CC} - V_{sat})\frac{R_B}{\beta \, R_C} + V_T \qquad (9.1.6)$$

The power dissipated in the transistor $p = v_{CE}i_C + v_{BE}i_B \cong v_{CE}i_C$ is very small or approximately zero in either cutoff or saturation. It should be noted, however, that power is expended in switching from one state to the other, going through the linear or active region.

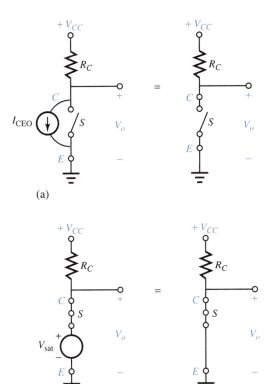

(a)

(b)

Figure 9.1.3 BJT switch. (a) Off (cutoff) position. (b) On (saturation) position.

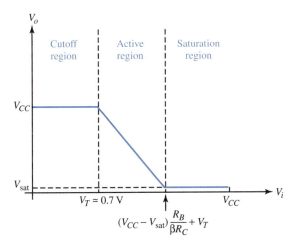

Figure 9.1.4 Transfer characteristic of BJT switch.

Figure 9.1.4 shows the transfer characteristic of a BJT switch relating V_i and V_o, whereas Figure 9.1.5 depicts a plot of a typical input-voltage waveform and the resulting output-voltage waveform, illustrating the essential inherent inversion property of the switch, the waveform distortion, propagation delays, and rise and fall times of the BJT switch. Note the following time parameters.

• *Propagation delay* t_d is a certain amount of delay for a change to occur. It is defined here as the time to change from 0 to 10% of the final value. (Sometimes propagation delay is defined as the time between the 50% level of V_i and the 50% level of V_o.) When V_i abruptly changes from 0 to 5 V, note that the output voltage and collector current do not initially change during the propagation delay time.

• *Rise time* t_r is the time needed to change from 10 to 90% of the final level.

• *Propagation delay* t_s occurs when the input pulse returns to 0 V (i.e., logic level 0). This is a result of the time required to remove charge stored in the base region before the transistor begins to switch out of saturation and is usually longer than t_d.

• *Fall time* t_f is the time required for the output voltage and the collector current to change state. That is the time required to switch through the active region from saturation to cutoff.

These times are influenced by the various capacitances inherent in the transistor and other stray capacitances. In turn, the design of a digital switch is influenced by these propagation delays, and the rise and fall times, which are grouped under the general category of *switching speed*.

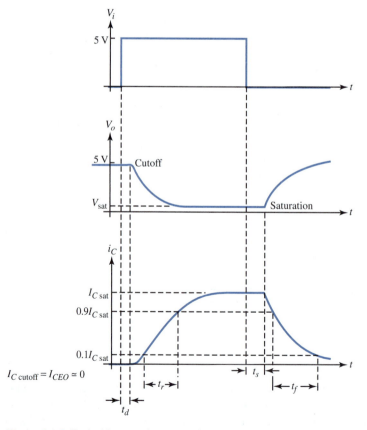

Figure 9.1.5 Typical input-voltage waveform and resulting output-voltage waveform.

EXAMPLE 9.1.1

For the circuit of Figure 9.1.2, given that $V_{CC} = 5$ V, $R_C = 1$ kΩ, $\beta = 100$, and the high range is 4 to 5 V, choose R_B such that any high input will saturate the transistor with the base overdriven by a factor of at least 5. Assume V_{sat} to be 0.2 V.

Solution

Since $i_C \cong \beta i_B + I_{CEO}$, one must have (neglecting I_{CEO}) $I_{B\,\text{sat}} > I_{C\,\text{sat}}/\beta$. But

$$I_{C\,\text{sat}} = \frac{V_{CC} - V_{\text{sat}}}{R_C}$$

Hence,

$$I_{B\,\text{sat}} > \frac{V_{CC} - V_{\text{sat}}}{\beta R_C}$$

With a factor of 5 included for the desired overdrive, one has

$$i_B = 5\frac{V_{CC} - V_{\text{sat}}}{\beta R_C} = 5\frac{5 - 0.2}{100(1000)} = 240 \ \mu A$$

When v_i is in the high range, the emitter–base junction is forward-biased. The base current that flows is given approximately by

$$i_B = \frac{v_i - 0.7}{R_B} \quad \text{or} \quad R_B = \frac{v_i - 0.7}{i_B}$$

Setting $v_i = 4$ V, which is the lowest value in the high range,

$$R_B = \frac{4 - 0.7}{2.4 \times 10^{-4}} = 13{,}750 \ \Omega$$

One chooses the closest standard resistance smaller than 13,750 Ω, since making R_B smaller increases the overdrive and hence improves the margin of safety.

9.2 DTL AND TTL LOGIC CIRCUITS

Bipolar transistors were the first solid-state switching devices commonly used to implement digital logic circuits in the 1950s and 1960s. These circuits used diodes at the input of the gate for logic operation followed by a transistor (BJT) output device for signal inversion. One of the bipolar logic families that emerged was called DTL (*diode–transistor logic*). The TTL (*transistor–transistor logic*) soon replaced DTL and then became the principal bipolar technology for the next two decades. It is still often used today. In TTL circuits, the diodes used in DTL at the gate input are replaced with a multiemitter transistor for increased performance. Primarily by reducing the size of the transistors and other components, speed and performance improvements have been made in TTL circuits.

Both DTL and TTL are called *saturating* logic families, because the BJTs in the circuit are biased into the saturated region to achieve the effect of a closed switch. The inherent slow switching speed is a major difficulty with saturating logic because saturated BJTs store significant charge and switch rather slowly. Schottky TTL is a nonsaturating logic family that was later developed to achieve higher speed performance by preventing the transistors from saturating. Another bipolar nonsaturating logic family is ECL (*emitter-coupled logic*), which has BJTs that remain biased in the active region. These circuits consume more power, are less dense, but are extremely fast.

Using the transistor switch, logic gates can be constructed to perform the basic logic functions such as AND, OR, NAND, NOR, and NOT. Since the individual gates are available in the form of small packages (such as the dual-in-line package, or DIP), it is generally not necessary to design individual gates in order to design an overall digital system. However, the designer needs to observe fan-out restrictions (i.e., the maximum number of gates that may be driven by the device), fan-in restrictions (i.e., the maximum number of gates that may drive the device), propagation delays, proper supply voltage to the unit, and proper connections to perform the intended logic function. Gates of the same logic family can be interconnected since they have the same logic voltage levels, impedance characteristics, and switching times. Some of the logic families are discussed in this section.

DTL (Diode Transistor Logic) Gate

A circuit realization of a NAND gate will now be developed. By connecting NAND gates together in various ways, one can synthesize other gates and flip-flops. Thus in principle, a single NAND gate circuit, repeated many times, would be sufficient to build up digital systems.

One possible NAND gate circuit is shown in Figure 9.2.1, in which it can have as many inputs as desired (indicated by the dashed-line input C), and typical values of $V_{CC} = 5$ V, $R_A = 2$ kΩ, $R_C = 5$ kΩ, and $\beta = 50$. We shall now consider the two inputs A and B that are quite adequate for our discussion, with the high range defined to be 4 to 5 V and the low range defined to be 0 to 0.5 V.

While the input voltages v_A and v_B are constrained to lie inside the high or low range of the allowed voltage ranges, the resulting output voltage v_F is supposed to be inside the high or low range as well. For different combinations of v_A and v_B, we need to find v_F. Since the circuit consists of no less than five nonlinear circuit elements ($D_A, D_B, D_1, D_2,$ and T_1), an approximation technique is used for analysis, while taking the voltage across a current-carrying forward-biased *pn* junction to be 0.7 V. The reader may have realized that it is not obvious at the start which diodes are forward-biased and which are reverse-biased. Hence a guessing procedure is used in which a guess is made and checked for self-consistency.

Let us start by letting $v_A = v_B = 0$, in which case a probable current path is from V_{CC} down through R_A and through inputs A and B. Since this guess implies current flow through D_A and D_B in the forward direction, we may guess that the voltage at X is 0.7 V. There is also a current path from X down to ground through D_1 and D_2 and the base–emitter junction of the transistor. Even though the sign of the guessed voltage is correct to forward-bias these three junctions, its magnitude of 0.7 V is insufficient since 2.1 V (or 3×0.7) is needed to make current flow through this path. Then $i_B = 0$; the transistor is cut off; and the output voltage $v_F = V_{CC} = 5$ V, assuming that

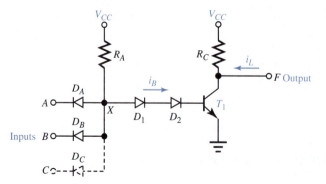

Figure 9.2.1 DTL NAND gate.

no load current flows through the output terminal. Thus in this case, D_A and D_B are conducting, whereas D_1, D_2, and T_1 are not. This is illustrated by the first line in Table 9.2.1.

For the inputs indicated on the other lines of Table 9.2.1 the reader can reason out the other columns shown. In terms of the high and low ranges, the operation of the gate is illustrated in Table 9.2.2. The truth table is given in Table 9.2.3 using the positive logic in which high is indicated by 1 and low is indicated by 0. One can now see that the circuit does function as a NAND gate.

TABLE 9.2.1 Inputs and Outputs for NAND Gate of Figure 9.2.1

Volts			Conduction			Volts
v_A	v_B	v_X	D_A	D_B	$D_1D_2T_1$	v_F
0	0	0.7	YES	YES	NO	5
0	0.2	0.7	YES	NO	NO	5
0.3	4.6	1.0	YES	NO	NO	5
4.6	0.3	1.0	NO	YES	NO	5
4.8	4.1	2.1	NO	NO	YES	0.2 $(= V_{CE\,sat})$

TABLE 9.2.2 Operation of NAND gate of Figure 9.2.1

v_A	v_B	v_F
LOW	LOW	HIGH
LOW	HIGH	HIGH
HIGH	LOW	HIGH
HIGH	HIGH	LOW

TABLE 9.2.3 Truth Table using positive logic

A	B	F
0	0	1
0	1	1
1	0	1
1	1	0

EXAMPLE 9.2.1

What logic function does the circuit of Figure 9.2.1 perform if negative logic is used?

Solution

Using negative logic, Table 9.2.2 can be translated into the following truth table. This can be seen to be the logic function NOR.

A	B	F
1	1	0
1	0	0
0	1	0
0	0	1

The DTL NAND gate circuit shown in Figure 9.2.1 and others closely related to it belong to the DTL logic family. Since all circuits in the same family have the same high and low ranges, they can be interconnected to build up digital systems. Although we have described the DTL gate as our first example for our convenience and easy understanding, the DTL technology is almost obsolete and is replaced by TTL.

TTL NAND Gate

The TTL logic family uses only transistors instead of a combination of transistors and diodes. TTL has the same high and low ranges as DTL. TTL is the most popular BJT logic family. A primary reason for the popularity of TTL over DTL is its higher speed. TTL has a typical fan-out of 10 or more, small propagation delays on the order of 2 to 10 ns, and a power consumption of about 2 mW. By contrast, DTL has a typical fan-out of 8 to 10, a propagation delay on the order of 30 to 90 ns, and a power consumption of around 15 mW. The noise margins and fan-out of DTL are generally better than those of RTL (resistor–transistor logic; see Problem 9.2.8), but the switching speeds are about the same. The emitter-coupled logic (ECL) family is also available with increased switching times, although not discussed here in any detail. The name arises from the common attachment of the emitters of the input transistors. The propagation delays are on the order of 1 ns, but the power consumption is quite high (on the order of 25 mW per gate). Low power consumption combined with relatively small noise margins (< 0.3 V) has made TTL a popular choice and, in particular, the high-speed Schottky diode TTL gates.

A typical TTL NAND gate is shown in Figure 9.2.2. Transistor T_1 is a multiple-emitter transistor (*npn* BJT) which acts as an AND gate. Replacing the base–emitter and base–collector junctions with diodes, T_1 can be represented as shown in Figure 9.2.3.

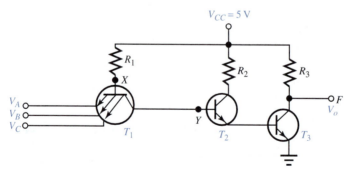

Figure 9.2.2 TTL NAND gate.

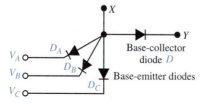

Figure 9.2.3 Diode replacement of three-input transistor T_1 of Figure 9.2.2.

EXAMPLE 9.2.2

Considering the TTL NAND gate circuit of Figure 9.2.2, with one or more inputs low, show that the output will be high.

Solution

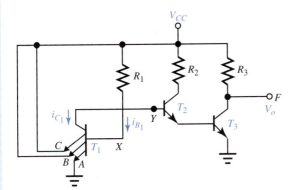

Figure E9.2.2 TTL NAND gate with one input low and the other two high.

Let one of the emitters, say A, be grounded and therefore let V_A be low while V_B and V_C are high. The situation is then as shown in Figure E9.2.2. There is a current path from V_{CC} down through R_1 and emitter A to ground. With $v_X = 0.7$ V a base current $i_{B_1} \cong (V_{CC} - 0.7)/R_1$ will be flowing in transistor T_1. Note, however, that i_{C_1} must be flowing out of the base of transistor T_2. That being the wrong direction for the base current of an *npn* transistor, it must represent the reverse current through one of the junctions in T_2. Reverse currents being very small, $i_{B_1} \gg i_{C_1}/\beta$. Hence T_1 will be saturated, and its collector-to-emitter A voltage will be $V_{CE\,sat}$, which is about 0.2 V. Thus, the voltage at Y will be about 0.2 V, which is much less than the 1.4 V needed to forward-bias the emitter junctions of T_2 and T_3. Thus T_3 is cut off and output V_o at F is high.

The TTL family is large and widely used. TTL circuits can switch fairly quickly allowing data rates on the order of 10 to 40 Mbits/s (mega stands for 2^{20}). They also have good output current capability. However, they consume too much power and space to be used in LSI. Hence, TTL is used primarily in SSI and MSI, which are the less densely packed ICs. MOS (metal-oxide semiconductor) technology, which has low power consumption and high packing density, but low output current capability, is widely used in LSI.

9.3 CMOS AND OTHER LOGIC FAMILIES

Both DTL and TTL are based on the saturating BJT inverter. The transistor acts as a switch that connects or disconnects the collector and emitter. The switch is closed when sufficient base current is applied to saturate the transistor. While the BJT technologies have dominated during the 1970s, logic families based on MOSFET technology are now more widely used because of the advantages of fewer fabrication steps and generally lower power consumption. After PMOS and NMOS technologies, CMOS emerged as the dominant MOS technology and remains so today.

Let us consider the simple MOSFET inverter with resistive load shown in Figure 9.3.1(a), quite similar in principle to the BJT inverter, although the circuit is rather impractical, as we shall see later. The load line for $R_D = 23$ kΩ and $V_{DD} = 7$ V is shown along with the transistor I–V characteristics in Figure 9.3.1(b). On finding v_{out} ($= v_{DS}$) for different values of v_{in} ($= v_{GS}$), the

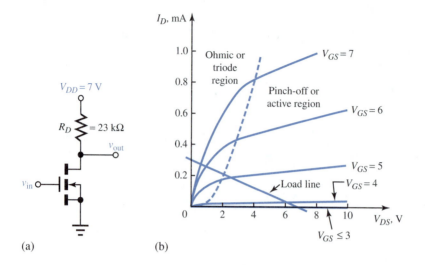

(a) (b)

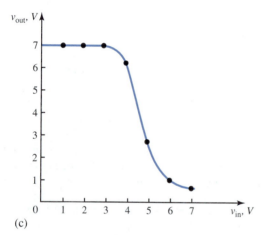

(c)

Figure 9.3.1 MOSFET inverter with resistive load. **(a)** Circuit with *n*-channel enhancement MOSFET and a resistor. **(b)** Transistor *I–V* characteristics. **(c)** Voltage-transfer characteristic for (a).

voltage-transfer characteristic (v_{out} versus v_{in}) is plotted in Figure 9.3.1(c). By choosing the low range to be 0 to 3 V (i.e., less than the threshold voltage) and the high range to be 5 to 7 V, we can see that any input voltage in the low range gives an output of 7 V (high) and inputs in the high range give outputs in the low range. Thus, the circuit is seen to be an inverter.

While the circuit of Figure 9.3.1(a) functions correctly as an inverter, in order to maintain low current consumption, large values of R_D are needed, but they are undesirable in ICs because they occupy too much space. In order to increase the number of circuits per IC, R_D is usually replaced by a second MOS transistor, which is known as an *active load*. When R_D is replaced by an *n*-channel MOSFET, it results in the logic family known as NMOS, which is widely used in VLSI circuits such as memories and microprocessors. NMOS logic circuits are more compact and, as such, more of them can be put on each chip. Also, MOS fabrication is simpler than bipolar fabrication; with fewer defects, production costs are less. As mentioned before, the main disadvantage of MOS technology, as compared to TTL, is low output current capacity.

FET switches offer significant advantages over BJT switches. An FET switch does not draw current from a previous stage because the gate current of an FET is practically zero. Consequently, no significant power-loading effects are present, whereas BJTs, in contrast, do load down previous stages. Another advantage of FETs is that their logic voltage levels (typically, $V_{DD} = 15$ V) are higher than those of the BJTs (typically, $V_{CC} = 5$ V). Thus, the FET logic circuits tend to tolerate more noise than the comparable BJT logic circuits. However, their switching speeds tend to be somewhat smaller than those for BJTs in view of the larger inherent capacitances of the FETs.

MOSFETs are preferred over JFETs for digital integrated circuits. Either PMOS (p-channel metal-oxide semiconductor) or NMOS (n-channel metal-oxide semiconductor) logic circuits can be constructed. Requiring only about 15% of the chip area of a BJT, MOSFETs offer very high packing densities.

The usage of a p-channel MOSFET as the active load for an n-channel MOSFET leads to a logic family known as complementary-symmetry MOS, or CMOS. CMOS technology has significant advantages and has become most popular. A basic CMOS inverter is shown in Figure 9.3.2(a), in which both a p-channel and an n-channel enhancement MOSFET are used as a symmetrical pair, with each acting as load for the other.

When the input v_{in} is low, the gate–source voltage of the NMOS is less than the threshold voltage and is cut off. The gate–source voltage of the PMOS, on the other hand, is $-V_{SS}$, where V_{SS}, the supply voltage, is greater than the threshold voltage, V_T. Then the PMOS is on, and the supply voltage appears at the output v_{out}. When the input v_{in} is high ($v_{\text{in}} \cong V_{SS}$), the PMOS turns off; the NMOS turns on. Then V_{SS} appears across the drain–source terminals of the PMOS and the output v_{out} drops to zero. Thus, the circuit functions as an inverter. Figure 9.3.2(b) shows the simplified circuit model of the CMOS inverter by depicting each transistor as either a short or an open circuit, depending on its state.

Note that when the output is in the high state, the PMOS is on with the NMOS being off; when the output is in the low state, the NMOS is on with the PMOS being off. Virtually no current is drawn from the power supply in either case. The CMOS has the advantage that it uses no power except when it is actually switching. This property of virtually no power consumption, coupled with the small chip surface needed, makes the CMOS very favorable for miniature and low-power applications such as wristwatches and calculators. Because of the poor switching speeds (compared to TTL), they are applied to low- to medium-speed devices. The principal disadvantage of the CMOS is a more complex fabrication procedure than that of the NMOS, leading to more defects and higher cost.

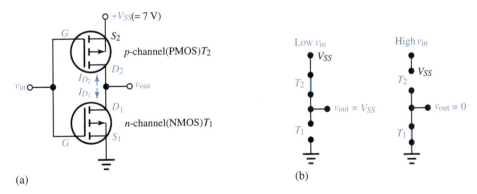

(a)   (b)

Figure 9.3.2 CMOS inverter and circuit model.

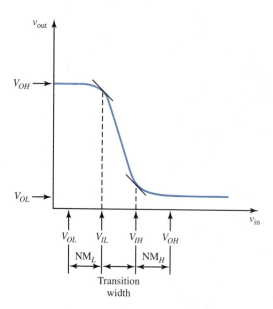

Figure 9.3.3 Typical voltage-transfer characteristic of a CMOS logic circuit, showing noise margins.

The switching action in the CMOS circuit is rather sharp, as compared with that of Figure 9.3.1(c). A general voltage-transfer characteristic shown in Figure 9.3.3 illustrates this feature. A very small change in v_{in} is sufficient to produce a relatively large change in v_{out}. The voltages V_{OH} and V_{OL}, indicated in Figure 9.3.3, are, respectively, the nominal high and low output voltages of the circuit. V_{IL} and V_{IH} are the input voltages at which $|dV_{out}/dV_{in}| = 1$. These are seen to be the boundaries of the low and high ranges. The region between V_{IL} and V_{IH} is known as the *transition region,* and the *transition width* is given by $V_{IH} - V_{IL}$. The lower and upper *noise margins* are given by $V_{IL} - V_{OL} \equiv NM_L$ and $V_{OH} - V_{IH} \equiv NM_H$, respectively. The noise margins indicate the largest random noise voltages that can be added to v_{in} (when it is low) or subtracted from v_{in} (when it is high), without yielding an input in the forbidden transition region. A typical voltage-transfer characteristic and drain current I_{D_1} versus input voltage v_{in} for the CMOS inverter are shown in Figure 9.3.4. Note that when v_{in} is inside either the low or the high states (and not in the transition region), the drain current $I_{D_1} = -I_{D_2} \cong 0$. The maximum value of I_{D_1} that occurs during transition is indeed very small.

EXAMPLE 9.3.1

For the CMOS inverter with characteristics shown in Figure 9.3.4, determine the noise margins.

Solution

From Figure 9.3.4(a), $V_{OH} = 7$ V and $V_{OL} = 0$. From the location of the approximate points where the absolute value of the slope is unity, $V_{IL} = 3.1$ V and $V_{IH} = 3.9$ V. Thus, noise margins are given by

$$NM_L = V_{IL} - V_{OL} \cong 3.1 - 0 = 3.1 \text{ V}$$
$$NM_H = V_{OH} - V_{IH} \cong 7 - 3.9 = 3.1 \text{ V}$$

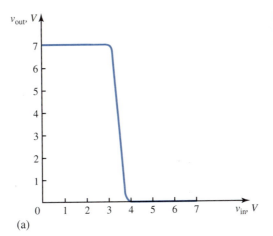

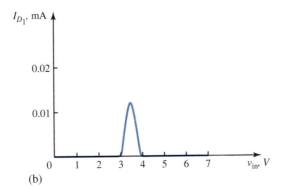

CMOS NAND and NOR gates with two inputs are shown in Figure 9.3.5. For the NAND gate of Figure 9.3.5(a), if at least one input is in the low state, the associated PMOS device will be on and the NMOS device will be off, thereby yielding a high output state. If, on the other hand, both V_A and V_B are high, both PMOS devices will be off while both NMOS devices will be on, thereby giving a low output V_{out}. Thus, the CMOS device of Figure 9.3.5(a) functions as a NAND gate.

For the NOR gate of Figure 9.3.5(b), when both inputs V_A and V_B are low, both NMOS devices will be off with both PMOS devices on, and the output V_{out} is high around V_{SS}. If, on the other hand, one of the inputs goes to be high, the associated PMOS device turns off and the corresponding NMOS device turns on, thereby yielding an output V_{out} to be low. Thus, the CMOS device of Figure 9.3.5(b) functions as a NOR gate. Note that neither gate in Figure 9.3.5 draws virtually any power-supply current, so that there is virtually no power consumption.

Another CMOS circuit that is conveniently built into CMOS technology is known as the *transmission gate*, which is not strictly speaking a logic circuit. It is a switch controlled by a logic input. In its circuit shown in Figure 9.3.6, the control signal C and its complement $\bar{C}$ determine whether or not the input is connected to the output. When C is low (and $\bar{C}$ is high), neither gate can induce a channel and the circuit acts as a high resistance between input and output. The transmission gate is then effectively an open circuit. On the other hand, when C is high (and $\bar{C}$ is low), the transmission gate provides a low-resistance path between input and output for all

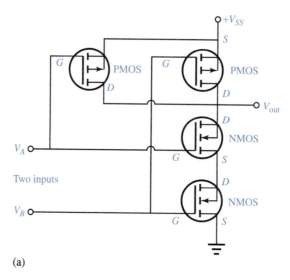

(a)

Figure 9.3.5 CMOS gates with two inputs. (a) NAND gate. (b) NOR gate.

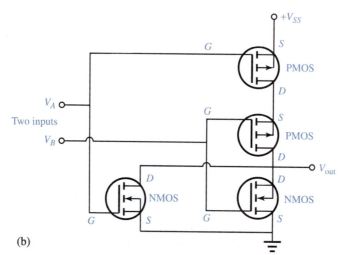

(b)

allowed values of v_{in}. The reader is encouraged to reason out these statements. Since the gate conducts well in either direction when the switch is on, the circuit is known to be *reciprocal*, in which case the labels "input" and "output" can be interchanged.

In addition to TTL, NMOS, and CMOS, several other logic families have been developed to suit various purposes. Emitter-coupled logic (ECL) is a high-speed bipolar technology with high power consumption. Transistors operate in the active mode and do not saturate in ECL; propagation delays per gate are as short as 1 ns. Besides high power consumption, the logic swing (which is the difference between high and low voltage levels) is small, being on the order of 1 V, which makes the ECL vulnerable to random noise voltages, thereby leading to errors.

The integrated-injection logic (IIL or I^2L) is a compact, low-power bipolar technology competitive with MOS technology for LSI. This technology offers packing densities ten times larger than TTL: 100 to 200 gates per square millimeter as compared with probably 10 to 20 for TTL. Power consumption and delay time have a kind of trade-off relationship in IIL technology.

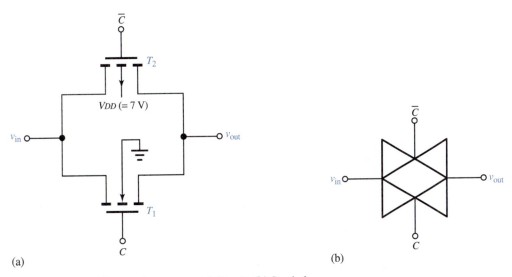

(a)

(b)

Figure 9.3.6 CMOS transmission gate. **(a)** Circuit. **(b)** Symbol.

As with ECL, the logic swing is small, thereby implying vulnerability to noise. While MOS can be made compatible with TTL, IIL voltage ranges are incompatible with other families. The IIL is used inside LSI blocks, with interfacing circuits provided at the inputs and outputs to make the blocks externally compatible with TTL.

A particular choice of a logic technology involves a compromise of many practical characteristics. The *power-delay product* (PDP) is the approximate energy consumed by a gate every time its output is switched, and it forms a general figure of merit. PDP is the product of P_{av}, the average dc power consumed by the gate, and T_D, the gate's propagation delay. It is the energy required to effect a single change, since T_D is the time required for a single change of logic state, and is a measure of the electrical efficiency of the switch.

For TTL and ECL, PDP is on the order of 100 pJ; for NMOS it is around 10 pJ. Slower versions of I²L can operate at 1 pJ per change. The average power consumption in CMOS is linearly proportional to the data rate. CMOS, run at maximum speed, has comparable PDP to that of NMOS. By reducing logic swing and/or device capacitance, the PDP can be reduced. Improvements in PDP can also be achieved by making conventional circuits physically smaller with consequent increased package density. Microfabrication technologies have gradually reduced the *minimum feature size*, which refers to the typical minimum dimension used in an IC.

Improved performance is also achieved with substantially different kinds of technology, such as cryogenic *Josephon digital technology*, which makes use of superconductors and can offer 0.04-ns delay times and a PDP of about 2×10^{-4} pJ.

9.4 LEARNING OBJECTIVES

The *learning objectives* of this chapter are summarized here, so that the student can check whether he or she has accomplished each of the following.

- BJT inverter switch, its characteristics, and its operation.
- Analysis of DTL and TTL logic circuits.
- MOSFET inverter and its characteristics.

- Advantages of PMOS and NMOS logic circuits.
- CMOS inverter switch and its characteristics.
- CMOS gates and CMOS technology.
- Notions of other logic families.

9.5 PRACTICAL APPLICATION: A CASE STUDY

Cardiac Pacemaker, a Biomedical Engineering Application

When blockage occurs to the biological signals that stimulate the heart to beat, the application of electric pacemaker pulses forcing the heart to beat at a higher rate is very helpful. A demand pacemaker is particularly helpful to the patient with partial blockage, in which case an electric pulse to the heart muscle is only applied when a beat does not occur within a predetermined time interval.

The pacemaker circuitry along with a lithium battery are enclosed in a metal case, which is implanted under the skin on the patient's chest. The mental case and the tip of the *catheter* (a wire enclosed in an insulating tube) form the electrical terminals of the pacemaker. A simplified block diagram of a typical demand pacemaker is shown in Figure 9.5.1.

The input amplifier amplifies the natural heart signals, which have a small amplitude on the order of 1 mV. Filtering eliminates certain frequency components so that heartbeats can be better detected. A comparative circuit is then used to compare the amplified and filtered signal with a threshold value and detect either a natural heartbeat or an output pacing pulse. This detection decision is passed on to the counting and timing circuitry through an AND gate. The second input to the AND gate comes from the counter circuit, such that input signals for 0.4 s after the start of a natural or forced beat are ignored.

The timing functions are performed by counting the output cycles of a timing oscillator, which generates a square wave with a precise period of 0.1 s. The digital signals produced by the counter are passed on to a digital comparator, which compares them with signals from a reference count generator and decides when the pulse generator is to deliver an output pulse of a specified amplitude and duration.

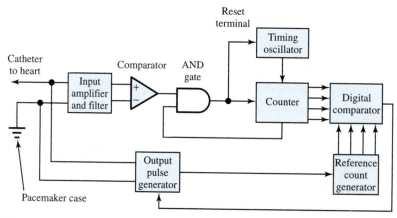

Figure 9.5.1 Simplified block diagram of demand cardiac pacemaker (power source connections not shown).

Extremely low power consumption and high reliability become important criteria for all pacemaker designs. While electrical engineers can come up with better circuit designs, mechanical and chemical engineers have to select better materials to produce the case and the catheter, and above all, the physicians have to provide the pacemaker specifications. Thus, doctors and engineers have to work in teams in order to develop better biomedical products.

PROBLEMS

9.1.1 Show that saturation will occur when Equation (9.1.6) is satisfied.

*9.1.2 Consider a BJT switch connected to the next stage, as shown in Figure P9.1.2, in which i_{out} is likely to be negative when v_{out} is high. Assume $V_{CC} = 5$ V, $R_C = 1k\Omega$, and the high range to be 4 to 5 V. Find the largest $|i_{out}|$ that can be tolerated.

9.1.3 The transistor switch of Figure 9.1.2(a) is to be designed to operate in saturation and in cutoff when the pulse signal shown in Figure P9.1.3 is applied to the input. Assume an ideal transistor with $\beta = 100$, $V_T = 0.7$ V, $V_{sat} = 0.2$ V, and $I_{CEO} = 0.1$ mA. Letting the supply voltage $V_{CC} = 5$ V and $R_C = 500$ Ω, determine the minimum value of R_B, and sketch the output-voltage waveform.

9.1.4 For the BJT switch described in Figure 9.1.2(a), let $V_{CC} = 5$ V, $V_T = 0.7$ V, $V_{sat} = 0.2$ V, and $\beta = 25$. If V_l switches between 0 and 5 V and $i_B \leq 0.1$ mA, find the minimum values of R_B and R_C for proper operation.

9.1.5 Sketch the transfer characteristic for the BJT switch described in Figure 9.1.2(a), given that $V_{CC} = 5$ V, $V_{sat} = 0.2$ V, $V_T = 0.7$ V, $R_C = 500$ Ω, $R_B = 10$ kΩ, and $\beta = 100$.

9.1.6 The transistor switch of Figure 9.1.2(a) with $R_B = 10$ kΩ and $R_C = 750$ Ω employs a BJT which

has the characteristics shown in Figure P9.1.6(a). Sketch V_o as a function of time for the input signal given in Figure P9.1.6(b).

9.2.1 Considering Table 9.2.1, the first line has been reasoned out in the text. Justify the other four lines.

*9.2.2 With $v_A = v_B = 0$ in Figure 9.2.1, show that a guess $v_X = 2.1$ V would lead to a contradiction, and hence cannot be correct.

9.2.3 Consider the circuit of Figure 9.2.1 with $v_A = 0.4$ V and $v_B = 0.3$ V. Find v_X and v_F.

9.2.4 With $v_A = 0.2$ V and $v_B = 4.5$ V in Figure 9.2.1, justify why $v_X = 0.7$ V will be an incorrect guess.

9.2.5 Considering Figure 9.2.1 of the DTL NAND gate circuit, inquire as to why D_1 and D_2 are used in the circuit. (*Hint:* Consider the third line of Table 9.2.1.)

9.2.6 Consider the DTL gate circuit shown in Figure P9.2.6. Assume diodes with $V_T = 0.7$ V and the transistor to have $\beta = 35$, $V_T = 0.7$ V, $I_{CEO} = 0$, and $V_{sat} = 0.2$ V.

(a) For inputs $V_A = 5$ V and $V_B = 0$ V, determine V_o.

(b) For inputs $V_A = V_B = 0$ V, determine V_o.

(c) Is this a configuration of a DTL NOR gate?

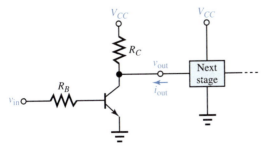

Figure P9.1.2

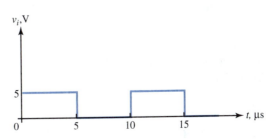

Figure P9.1.3

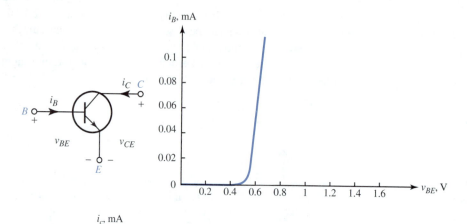

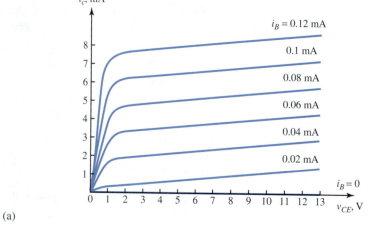

(a)

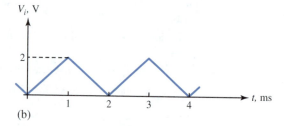

(b)

Figure P9.1.6

9.2.7 For the DTL gate shown in Figure P9.2.6, assume ideal diodes, $V_T = 0.7$ V, $V_{sat} = 0.2$ V, $R_B = 10$ kΩ, $R_C = 500$ Ω, and $V_{CC} = 5$ V. Assuming $\beta = 20$ for the BJT and using reasonable approximations, sketch the output V_o for the input signals of Figure P9.2.7.

*9.2.8 A gate using resistor-transistor logic (RTL) is shown in Figure P9.2.8. Justify that this gate performs the NOR function.

9.2.9 Study the DTL gate circuit shown in Figure P9.2.9 and state whether it behaves like a NOR gate or a NAND gate.

Figure P9.2.6

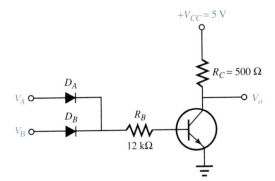

Figure P9.2.7

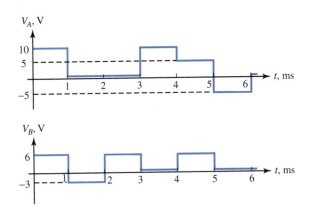

9.2.10 Considering the TTL NAND gate circuit of Figure 9.2.2, with all inputs high, show that the output will be low.

9.2.11 Consider the TTL gate circuit of Figure 9.2.2. If $V_A = 0.1$ V, $V_B = 0.2$ V, and $V_C = 0.3$ V, determine the approximate values of V_X, V_Y, and V_o.

9.2.12 A TTL NAND gate with a multiple-emitter *npn* BJT (which acts as an AND gate) is shown in Figure P9.2.12.

   (a) With all inputs in the high state, show that the output V_o will be in the low state.

   (b) With at least one input, say V_A, being in the low state, show that the output V_o will be in the high state.

9.2.13 Discuss the significance of R_4, T_4, T_3, and the diode between T_4 and T_3 in the TTL NAND gate circuit of Figure P9.2.12.

*9.2.14 For the TTL NAND gate circuit of Figure P9.2.12, assuming that the inputs vary between 0 and 5 V and $V_{CC} = 5$ V, determine the maximum value of R_B to saturate T_2 if $i_{C\ \text{sat}} = 3.8$ mA.

9.3.1 Consider the *n*-channel JFET switch shown in Figure P9.3.1(a) with the characteristics shown in Figure P9.3.1(b).

   (a) Explain its operation.

   (b) Draw the circuit of a depletion MOSFET switch.

   (c) For the input shown in Figure P9.3.1(c), sketch the output as a function of t.

9.3.2 For the JFET switch shown in Figure P9.3.1(a) with $R_D = 3$kΩ and $V_{DD} = 12$ V, sketch the output voltage as a function of t, if the input voltage is as shown in Figure P9.3.2(a) and the JFET characteristics are as given in Figure P9.3.2(b).

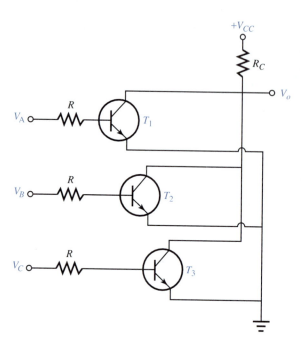

Figure P9.2.8

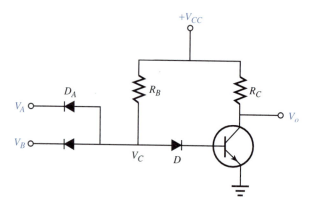

Figure P9.2.9 DTL gate circuit.

9.3.3 FET logic circuits can be used with positive pulses by using enhancement-mode n-channel MOSFETs as switches. Figure P9.3.3(a) shows the circuit and Figure P9.3.3(b) the MOSFET characteristics. For the input shown in Figure P9.3.3(c), sketch the output as a function of time.

*9.3.4 The complementary-symmetry MOSFET (CMOS) switch shown in Figure 9.3.2 has MOSFETs with $V_T = 5$ V and $V_{sat} = 1$ V. If $V_{SS} = 20$ V and v_{in} as shown in Figure P9.3.4, sketch the output voltage as a function of time.

9.3.5 A typical CMOS inverter is shown in Figure

P9.3.5(a). The n-channel MOSFET T_1 has the characteristics shown in Figure P9.3.5(b). T_2 has identical characteristics except for the changes of sign appropriate to a p-channel device.

(a) Outline a graphical procedure in order to find the operating point of the CMOS inverter by superposing the I–V characteristics of T_1 and T_2.

(b) Sketch the resulting voltage-transfer characteristics (v_{out} versus v_{in}) and drain current versus input voltage v_{in} for the CMOS inverter.

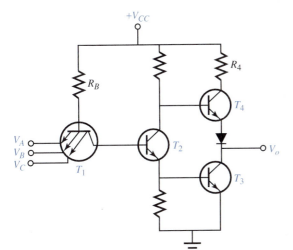

Figure P9.2.12 TTL NAND gate.

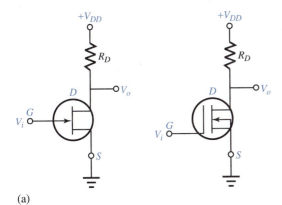

Figure P9.3.1 (a) *n*-channel JFET switch. (b) Characteristics (transfer characteristics; terminal characteristics and load line). (c) Input.

(a)

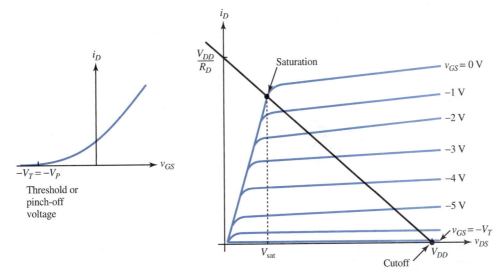

(b)

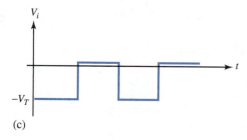

(c)

Figure 9.3.1 Continued

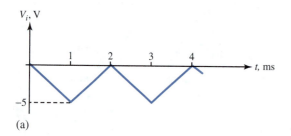

(a)

Figure P9.3.2 (a) Input. (b) JFET characteristics.

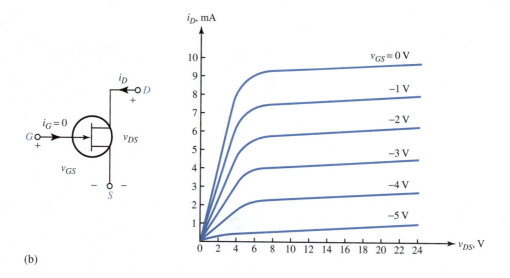

(b)

9.3.6 Explain the principle of operation of the CMOS transmission gate shown in Figure 9.3.6.

9.3.7 Consider the CMOS NAND gate shown in Figure

P9.3.7. Explain its operation and the approximate behavior of transistors in CMOS logic.

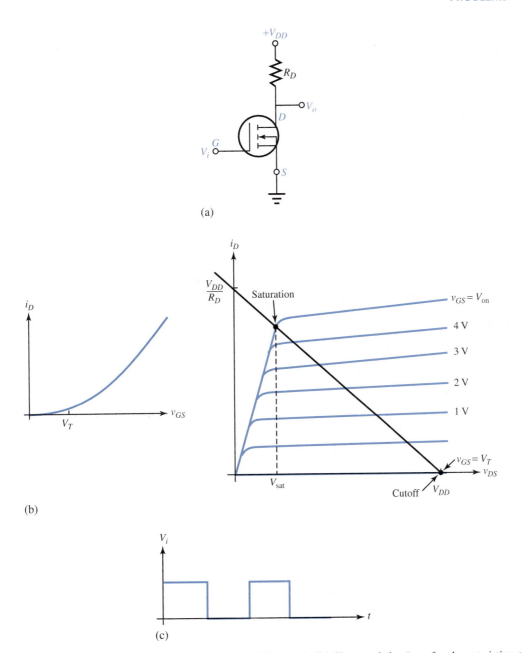

Figure P9.3.3 (a) *n*-channel enhancement MOSFET switch. (b) Characteristics (transfer characteristics; terminal characteristics and load line). (c) Input.

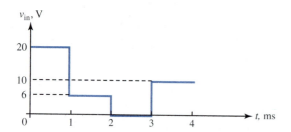

Figure P9.3.4 **Figure P9.3.4** Input voltage as a function of time.

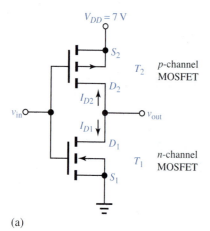

(a)

Figure P9.3.5 (a) Typical CMOS inverter. (b) Characteristics of T_1.

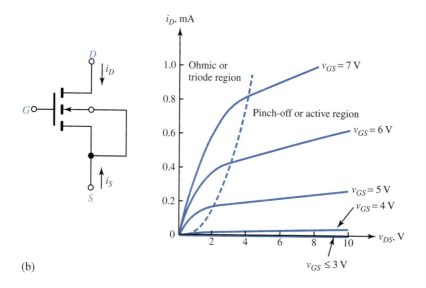

(b)

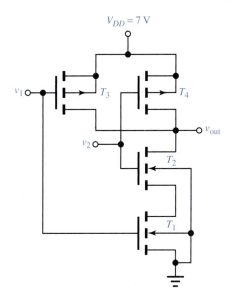

$V_{DD} = 7$ V

v_1

T_3

T_4

v_{out}

v_2

T_2

T_1

Figure P9.3.7 CMOS NAND gate.

ENERGY SYSTEMS

10 AC Power Systems

Electric power is indispensable for any modern society. Our use and demand for electric power grows annually at the rate of 2 to 3%; our need nearly doubles about every 25 to 35 years. We, in the United States, have become so accustomed to reliable and accessible electric power that we virtually take it for granted.

Even though energy appears in many different forms, the vast majority of all energy delivered from one point to another across the country is handled by ac power systems. Transporting electric energy most efficiently from place to place becomes extremely important. Electric power generation, transmission, distribution, and utilization are the features of any practical power system.

In the United States power is generated by many generating stations interconnected in an overall network, known as the *power grid,* which spans the entire country. The bulk of the system is privately owned, whereas a part of the network is owned federally and a part municipally. Some utility companies in a given geographical area operate as *power pools* for reasons of economy and reliability.

The principal source of energy comes from the burning of fossil fuels such as coal and oil to generate steam, which drives steam turbines, which in turn drive electric generators. Other important energy sources are hydroelectric and nuclear. In the former, electric generators are driven by waterwheel (hydraulic) turbines near natural or human-made waterfalls; in the latter, nuclear reactions generate heat to drive the steam-turbine–generator chain. There are also other less widely used sources of energy such as geothermal sources, wind, sun, and tides.

In this chapter, to start with, an introduction to power systems will be presented. Then we analyze three-phase systems, a topic that was covered in part in Chapter 4. Finally, the components of an ac power network are identified, and topics related to power transmission and distribution are introduced.

10.1 INTRODUCTION TO POWER SYSTEMS

Thomas A. Edison's work in 1878 on the electric light led to the concept of a centrally located power station with distributed electric power for lighting in a surrounding area. The opening of the historic Pearl Street Station in New York City on September 4, 1882, with dc generators (dynamos) driven by steam engines, marked the beginning of the electric utility industry. Edison's dc systems expanded with the development of three-wire 220-V dc systems. But as transmission distances and loads continued to grow, voltage problems were encountered. With the advent of William Stanley's development of a commercially practical transformer in 1885, alternating current became more attractive than direct current because of the ability to transmit power at high voltage with corresponding lower current and lower line-voltage drops. The first single-phase ac line (21 km at 4 kV) in the United States operated in 1889 between Oregon City and Portland.

Nikola Tesla's work in 1888 on electric machines made evident the advantages of polyphase over single-phase systems. The first three-phase line (12 km at 2.3 kV) in the United States became operational in California during 1893. The three-phase induction motor conceived by Tesla became the workhorse of the industry.

Most electric energy has been generated by steam-powered (accounting for about 85% of U.S. generation) and by water-powered, or hydro, turbine plants (accounting for about 10% of U.S. generation). Gas turbines are also used for short periods to meet peak loads. Steam plants are fueled primarily by coal, gas, oil, and uranium. While coal is the most widely used fuel in the United States, due to its abundance, nuclear units of 1280-MW steam-turbine capacity are in service today. However, rising construction costs, licensing delays, and public concerns have stopped the growth of nuclear capacity in the United States.

Other types of electric power generation are also prevalent, accounting for about 1% of U.S. generation. These include wind-turbine generators, solar-cell arrays, tidal power plants, and geothermal power plants, wherein energy in the form of steam or hot water is extracted from the earth's upper crust. Substantial research now under way shows nuclear fusion energy to be the most promising technology for producing safe, pollution-free, and economical electric energy in this century and beyond, since the needed fuel (deuterium) consumed in a nuclear fusion reaction is present in seawater abundantly.

Today the two standard frequencies for the generation, transmission, and distribution of electric power in the world are 60 Hz (in the United States, Canada, Japan, and Brazil) and 50 Hz (in Europe, the former Soviet Republics, South America except Brazil, India, and also Japan). Relatively speaking, the 60-Hz power-system apparatus is generally smaller in size and lighter in weight than the corresponding 50-Hz equipment with the same ratings. On the other hand, transmission lines and transformers have lower reactances at 50 Hz than at 60 Hz.

Along with increases in load growth, there have been continuing increases in the size of generating units and in steam temperatures and pressure, leading to savings in fuel costs and overall operating costs. Ac transmission voltages in the United States have also been rising steadily: 115, 138, 161, 230, 345, 500, and now 765 kV. Ultrahigh voltages (UHV) above 1000 kV are now being studied. Some of the reasons for increased transmission voltages are:

• Increases in transmission distance and capacity

• Smaller line-voltage drops

• Reduced line losses

- Reduced right-of-way requirements
- Lower capital and operating costs.

In association with ac transmission, there have been other significant developments:

- Suspension insulator
- High-speed relay system
- High-speed, extra-high-voltage (EHV) circuit breakers
- EHV surge arrester to protect from lightning strokes and other surges
- Communications via power-line carrier, microwave, and fiber optics
- Energy control centers with supervisory control and data acquisition (SCADA) and with automatic generation control (AGC)
- Extensive use of microprocessors for various tasks.

Along with ac transmission in the United States, there have been modern high-voltage dc (HVDC) transmission lines: the ±400-kV, 1360-km Pacific Intertie line between Oregon and California in 1970 as well as four other HVDC lines up to 400 kV and five back-to-back ac–dc links as of 1991. A total of 30 HVDC lines up to 533 kV are in place worldwide. For an HVDC line interconnected with an ac system, solid-state converters at both ends of the dc line are needed to operate as rectifiers and inverters. Studies in the United States have shown that overhead HVDC transmission is economical for transmission distances longer than about 600 km. Also, HVDC links seem to improve the overall system stability.

An interconnected system (in contrast with isolated systems) has many advantages:

- Better maintenance of continuity of service
- Increase in reliability and improved economy
- Reduction of reserve requirements
- Scheduling power transfers taking advantage of energy-cost differences in respective areas, load diversity, and seasonal conditions
- Shared ownership of larger and more efficient generating units.

Some of the disadvantages of interconnected operations are:

- Increased fault currents during short circuits
- Occasional domino effect leading to a regional blackout (such as the one that occurred in 1965 in the northern United States) due to an initial disturbance in some part of the interconnected grid system.

Present and Future Trends

According to the Edison Electric Institute, electricity's share of U.S. primary energy was almost 36% in 1989, and it is likely to reach 46% by the year 2010. The growth rate in the use of electric energy in the United States is projected to increase by about 2.4% per year for the near future, in spite of conservation practices, more efficient use of electricity, and a slackening population growth. One should also be aware of large growth of power systems internationally.

Because of the large amount of U.S. coal reserves, there is the continuing shift away from the use of gas and oil and toward increasing use of coal. Unless the construction time and cost per kW can be reduced significantly, no new nuclear units will be commissioned. Also, safety concerns seem to demand inherently safe reactor designs with standardized, modular construction of nuclear units. Since the major U.S. hydroelectric sites (except in Alaska) have been fully developed, one can foresee a trend for continuing percentage decline in hydroelectric energy generation.

By the year 2000, the total U.S. generating capacity has reached 817 GW (1 GW = 1000 MW) and continues to grow. Current lead times of about a decade for the construction and licensing of large coal-fired units may cause insufficient reserve margins in some regions of the United States.

As of 1989, U.S. transmission systems consisted of about 146,600 circuit-miles of high-voltage transmission. During the 1990s, additions have totalled up to 13,350 circuit-miles, which include 230-kV, 345-kV, and 500-kV lines. Because of the right-of-way costs, the possibility of six-phase transmission (instead of the current three-phase transmission) is being looked into.

U.S. distribution-network construction is expected to increase over the next decade. The older 2.4-, 4.1-, and 5-kV distribution systems are being converted to 12 or 15 kV. Higher distribution voltages such as 25 and 34.5 kV are also contemplated.

Recently some concern has surfaced about the effect of electromagnetic waves on the human and animal environment. The result of this remains to be seen.

Computers in Power Systems

The control and stability of any electric power system is indeed extremely important, in particular when a system is expected to maintain uninterrupted continuity of service within set limits of frequency and voltage, and to guarantee reliability of the system. Digital computers and microprocessors, along with highly developed software programs, have made their way into planning, designing, operating, and maintaining complex interconnected power systems. A large volume of network data must also be acquired and accurately processed. Digital computer programs in power-system engineering include power-flow, stability, short-circuit, and transients programs.

For a network under steady-state operating conditions, power-flow programs compute the voltage magnitudes, phase angles, and transmission-line power flows. Today's computers are capable of handling networks with more than 2000 buses (nodes) and 2500 transmission lines in less than 1 minute power-flow solutions. Interactive power-flow programs have also been developed along with CRT displays.

Stability programs are used to analyze power systems under various disturbances. Short-circuit programs compute fault currents and voltages under various fault conditions. These, in turn, will help in circuit-breaker selection, relay coordination, and overall system protection.

Transients programs yield the magnitudes and shapes of transient overvoltages and currents that may result from lightning strikes and other surges on the system. Based on the results of such studies, insulation coordination and surge-arrester selection are configured.

Planning and Research

Energy research worldwide is assuming top priority, in particular because of economic, environmental, and resource constraints. DOE (Department of Energy) was established in the United States in 1977. A major private utility sponsored energy research organization, EPRI (Electric Power Research Institute) in Palo Alto, California, has been in existence since 1972. In addition, large utility companies, such as AEP (American Electric Power), have their own research programs.

The major goals for the future are:

- New primary resources (such as nuclear fusion and solar energy) for electric bulk power generation
- Development of better means of generation (such as superconducting generators) and transmission (such as six-phase)
- Emphasis on energy conservation (better utilization of electricity with less waste)
- Electric-energy storage facilities such as pumped storage, compressed gas, storage batteries, and superconducting magnetic coils.

10.2 SINGLE- AND THREE-PHASE SYSTEMS

It would be very helpful if the reader would review Section 3.1 on sinusoidal steady-state phasor analysis and Chapter 4 on three-phase circuits before studying this section.

Ac power has significant practical advantages over dc power in generation, transmission, and distribution. One major drawback of the single-phase circuit is the oscillatory nature of the instantaneous power flow $p(t)$ as seen in Equation (3.1.36). The consequent shaft vibration and noise in single-phase machinery are rather undesirable. A three-phase circuit, on the other hand, under balanced conditions has constant, nonpulsating (time invariant), instantaneous power, as seen from Equation (4.2.13); the pulsating strain on generating and load equipment is eliminated. Also for power transmission, a balanced three-phase system delivers more watts per kilogram of conductor than an equivalent single-phase system. For these reasons, almost all bulk electric power generation and consumption take place in three-phase systems.

The majority of three-phase systems are four-wire, wye-connected systems, in which a grounded neutral conductor is used. Some three-phase systems such as delta-connected and three-wire wye-connected systems do not have a neutral conductor. Because the neutral current is nearly zero under normal operating conditions, neutral conductors for transmission lines are typically smaller in size and current-carrying capacity than the phase conductors. Thus, the cost of a neutral conductor is substantially less than that of a phase conductor. The capital and operating costs of three-phase transmission and distribution systems, with or without neutral conductors, are comparatively much less than those of separate single-phase systems.

Ratings of three-phase equipment, such as generators, motors, transformers, and transmission lines, are usually given as total three-phase real power in MW, or as total three-phase apparent power in MVA, and as line-to-line voltage in kV.

Power

The essential concepts related to power have been presented in Sections 3.1 and 4.2. However, for better clarity and understanding, those concepts are revisited in a slightly different form.

The *complex power* $\bar{S}$ in a single-phase system is the complex sum of the *real* (P) and *reactive* (Q) power, expressed as follows:

$$\bar{S} = P + jQ = \bar{V}\,\bar{I}^* = \bar{I}\,\bar{I}^*\bar{Z} = I^2 Z\angle\theta_Z = VI\angle\theta_V - \theta_I \tag{10.2.1}$$

where θ_V is the angle associated with $\bar{V}$ (with respect to any chosen reference), θ_I is the angle associated with $\bar{I}$ (with respect to the same reference chosen), θ_Z is the impedance angle, $\bar{V}$ is the rms voltage phasor, $\bar{I}$ is the rms current phasor, and $\bar{Z} = R \pm jX$ is the complex impedance with a magnitude of Z. Note that * stands for complex conjugate. If the voltage phasor $\bar{V}$ itself is taken to be the reference, $\theta_V = 0$ and θ_I may be replaced by θ without any subscript.

The *power factor* PF is given by the ratio of the real power P (expressed in watts) to the apparent power $S = \sqrt{P^2 + Q^2}$ expressed in volt-amperes,

$$PF = P/S = \cos\theta \qquad (10.2.2)$$

where $\theta = \tan^{-1} Q/P$. Inductive loads cause current to lag voltage and are referred to as lagging power factor loads. Conversely, capacitive loads cause current to lead voltage and are referred to as leading power factor loads.

For a case with lagging power factor, Figure 10.2.1(a) shows voltage and current phasors; Figure 10.2.1(b) depicts the real, reactive, and apparent powers; and Figure 10.2.1(c) gives the power triangle. The corresponding diagrams with leading power factor are shown in Figure 10.2.2. Loads on the electric power system are generally inductive, which will cause the phase current to lag the corresponding applied phase voltage. The real power component represents the components of voltage and current that are in phase, whereas the reactive power component represents the components of voltage and current that are in quadrature (that is, 90° out of phase).

The convention used for positive power flow is described with the help of Figure 10.2.3, in which Figure 10.2.3 (a) applies to a generator (source), and Figure 10.2.3 (b) applies to a load (sink).

The power expressions for the three-phase case, in terms of the line quantities, are (see Section 3.2)

$$\bar{S}_{3\phi} = P_{3\phi} + jQ_{3\phi} \qquad (10.2.3)$$

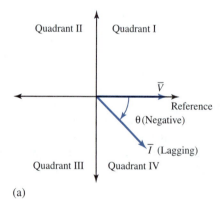

(a)

Figure 10.2.1 Lagging power factor. **(a)** Voltage and current phasors. **(b)** Real, reactive, and apparent powers. **(c)** Power triangle.

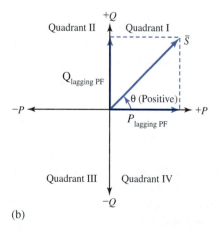

(b)

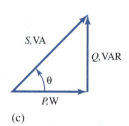

(c)

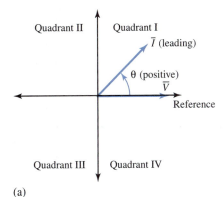

Figure 10.2.2 Leading power factor. **(a)** Voltage and current phasors. **(b)** Real, reactive, and apparent powers. **(c)** Power triangle.

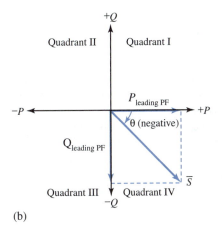

(b)

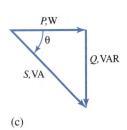

(c)

where

$$P_{3\phi} = \sqrt{3}\, V_L I_L \cos\theta_P \tag{10.2.4}$$

$$Q_{3\phi} = \sqrt{3}\, V_L I_L \sin\theta_P \tag{10.2.5}$$

$$S_{3\phi} = \sqrt{3}\, V_L I_L = \sqrt{P_{3\phi}^2 + Q_{3\phi}^2} \tag{10.2.6}$$

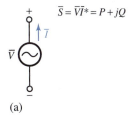

(a)

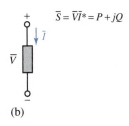

(b)

Figure 10.2.3 Convention for positive power flow. **(a)** Generator case. If P is positive, then real power is delivered. If Q is positive, then reactive power is delivered. If P is negative, then real power is absorbed. If Q is negative, then reactive power is absorbed. **(b)** Load case. If P is positive, then real power is absorbed. If Q is positive, then reactive power is absorbed. If P is negative, then real power is delivered. If Q is negative, then reactive power is delivered.

in which θ_P is the phase angle between the voltage and the current of any particular phase, and $\cos \theta_P$ is the power factor.

Under most normal operating conditions, the various components of the three-phase system are characterized by complete *phase symmetry*. If such phase symmetry is assured throughout the power system, it is desirable to simplify the analytical efforts to a great extent by the use of *per-phase analysis*. Also recall that three-phase systems are most often represented by *single-line (one-line) diagrams*.

The *energy E* associated with the instantaneous power over a period of time T seconds is given by

$$E = PT \tag{10.2.7}$$

where E is the energy in joules that is transferred during the interval, and P is the average value of the real-power component in joules per second. Note that the reactive-power component does not contribute to the energy that is dissipated in the load. The energy associated with the reactive power component is transferred between the electric fields (which result from the application of the sinusoidal voltage between the phase conductors and ground) and the magnetic fields (which result from the flow of sinusoidal current through the phase conductors).

Many industrial loads have lagging power factors. Electric utilities may assess penalties for the delivery of reactive power when the power factor of the customer's load is below a minimum level, such as 90%. Capacitors are often used in conjunction with such loads for the purpose of *power factor correction or improvement*. An appropriate capacitor connected in parallel with an inductive load cancels out the reactive power and the combined load may have unity power factor, thereby minimizing the current drawn from the source. For problems and examples on this subject, one may also refer to Sections 3.1 and 4.2.

EXAMPLE 10.2.1

A 60-Hz, three-phase motor draws 25 kVA at 0.707 lagging power factor from a 220-V source. It is desired to improve the power factor to 0.9 lagging by connecting a capacitor bank across the terminals of the motor.

(a) Determine the line current before and after the addition of the capacitor bank.

(b) Specify the required kVA (kVAR) rating of the capacitor bank. Also sketch a power triangle depicting the power factor correction by using capacitors.

(c) If the motor and the capacitor bank are wye-connected in parallel, find the capacitance per phase of the capacitor bank assuming that it is balanced.

(d) How would the result of part (c) change if the motor and the capacitor bank were to be delta-connected in parallel?

Solution

The real and reactive powers of the load (motor) are:

$$P_M = 25 \times 0.707 = 17.68 \text{ kW}$$
$$Q_M = 25 \sin(\cos^{-1} 0.707) = 17.68 \text{ kVAR}$$

(a) The line current of the motor, before the addition of the capacitor bank, is

$$I_M = \frac{S_{3\phi}}{\sqrt{3}\ V_L} = \frac{25,000}{\sqrt{3} \times 220} = 65.6\text{A}$$

After the addition, the new value of line current is

$$I_{\text{corr}} = \frac{P_M}{\sqrt{3}\ V_L \text{PF}_{\text{corr}}} = \frac{17,680}{\sqrt{3} \times 220 \times 0.9} = 51.55\text{A}$$

(b) The corrected new value of reactive power is

$$Q_{\text{corr}} = P_M \tan(\cos^{-1} 0.9) = 17.68 \tan 25.8° = 8.56 \text{ kVAR}$$

The kVA (kVAR) rating of the capacitor bank needed to improve the power factor from 0.707 to 0.9 lagging is then found as

$$Q_{\text{cap}} = Q_{\text{corr}} - Q_M = 8.56 - 17.68 = -9.12 \text{ kVAR}$$

The power triangle is shown in Figure E10.2.1.

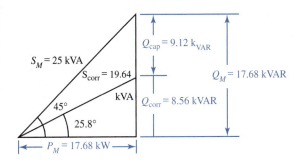

Figure E10.2.1 Power triangle

(c) Per-phase capacitive kVAR $= 9.12/3 = 3.04$ kVAR. With wye connection,

$$\frac{V_{LN}^2}{X_C} = \frac{\left(220/\sqrt{3}\right)^2}{X_C} = 3.04 \times 10^3$$

or

$$X_C = \frac{\left(220/\sqrt{3}\right)^2}{3.04 \times 10^3} = 5.31 \ \Omega$$

Hence,

$$C = \frac{1}{2\pi \times 60 \times 5.31} = 500 \ \mu\text{F}$$

(d) With delta connection,

$$\frac{V_L^2}{X_C} = \frac{220^2}{X_C} = 3.04 \times 10^3$$

or

$$X_C = \frac{220^2}{3.04 \times 10^3} = 15.92 \ \Omega$$

Hence,

$$C = \frac{1}{2\pi \times 60 \times 15.92} = 166.6 \ \mu F$$

Note that $Z_\Delta = 3Z_Y$ (see Figure 4.2.1), or $C_\Delta = (\frac{1}{3})C_Y$, as it should be for a balanced case.

10.3   POWER TRANSMISSION AND DISTRIBUTION

The structure of a power system can be divided into generation (G), transmission, (T), and distribution (D) facilities, as shown in Figure 10.3.1. An ac three-phase generating system provides the electric energy; this energy is transported over a transmission network, designed to carry power at high or extrahigh voltages over long distances from generators to bulk power substations and major load points; the subtransmission network is a medium to high-voltage network whose purpose is to transport power over shorter distances from bulk power substations to distribution substations. The transmission and subtransmission systems are meshed networks with multiple-path structure so that more than one path exists from one point to another to increase the reliability of the transmission system.

The transmission system, in general, consists of overhead transmission lines (on transmission towers), transformers to step up or step down voltage levels, substations, and various protective devices such as circuit breakers, relays, and communication and control mechanisms.

Figure 10.3.2 shows a typical electric-power distribution system and its components. Below the subtransmission level, starting with the distribution substation, the distribution system usually consists of distribution transformers, primary distribution lines or main feeders, lateral

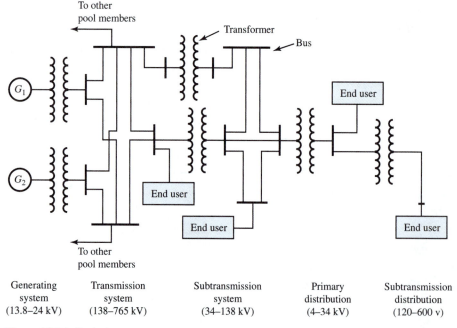

| Generating system (13.8–24 kV) | Transmission system (138–765 kV) | Subtransmission system (34–138 kV) | Primary distribution (4–34 kV) | Subtransmission distribution (120–600 v) |

Figure 10.3.1 Typical power-system structure.

feeders, distribution transformers, secondary distribution circuits, and customers' connections with metering. Depending on the size of their power demand, customers may be connected to the transmission system, the subtransmission system, or the primary or secondary distribution circuit.

In central business districts of large urban areas, the primary distribution circuits consist of underground cables which are used to interconnect the distribution transformers in an electric network. With this exception, the primary system is most often *radial*. However, for additional reliability and backup capability, a *loop-radial* configuration is frequently used. The main feeder is looped through the load area and brought back to the substation, and the two ends of the loop are connected to the substation by two separate circuit breakers. For normal operation, selected sectionalizing switches are opened so as to form a radial configuration. Under fault conditions, the faulted section is isolated and the rest of the loop is used to supply the unaffected customers.

Most residences and small buildings are supplied with power by means of *single-phase, three-wire* service, as illustrated in Figure 10.3.3. A distribution transformer is located on a power pole or underground, near the residential customer. Inside residences, the 220-V supply, being available between the two "hot" wires, is used for major appliances such as dryers, ranges, and ovens. The 110-V loads up to 20 to 40 A are connected between the ground wire and either "hot" wire, while nearly balancing the loads on the two "hot" wires. Each of these circuits, protected by its own fuse or circuit breaker, supplies lighting loads and/or convenience outlets. In the wiring of residential and commercial buildings, safety considerations are of paramount importance, the principal hazards being fire and electric shock. Residential wiring is also discussed in Section 4.4.

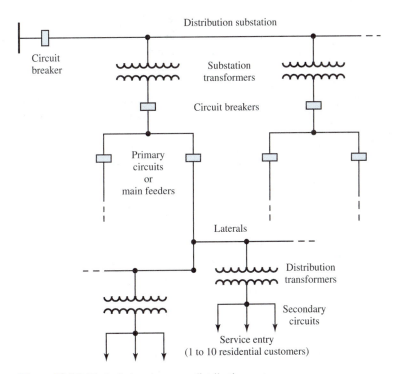

Figure 10.3.2 Typical electric power distribution system.

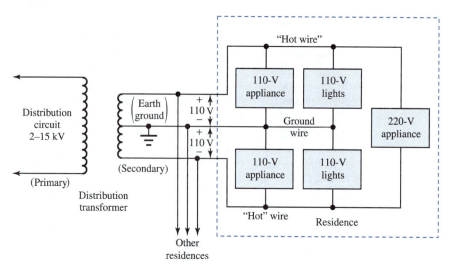

Figure 10.3.3 Single-phase, three-wire residential wiring circuit.

Power-System Loads

Figure 10.3.4 represents a *one-line (single-line) diagram* of a part of a typical three-phase power system. Notice the symbols that are used for generators, transformers, buses, lines, and loads. Recall that a bus is a nodal point.

Let us now consider the addition of a *load bus* to the operating power system. Service classifications assigned by the electric utilities include residential, commercial, light industrial, and heavy industrial loads, as well as municipal electric company loads. Let us prepare a load-bus specification, which is a summary of the service requirements that must be provided by the electric utility. Referring to the simple model of Figure 10.3.5, as an *example,* let the total new load connected to the system be 220 MVA at 0.8 power factor lagging; let the service be provided at the subtransmission level from a radial 115-kV circuit; let the 115-kV transmission line have a per-phase series impedance of $3 + j8\,\Omega$.

The load-bus data (specifications), transmission line data, and source-bus data are given in Tables 10.3.1, 10.3.2, and 10.3.3, respectively. The reader is encouraged to work out the details.

Looking at the load-bus data, the amount of reactive power that is needed to provide 100% compensation is 120 MVAR. Let us then add a three-phase shunt capacitor bank with a nominal voltage rating of 115 kV and a reactive power rating of 120 MVAR. The per-phase reactance of the bank can be computed as $X = V^2/Q = 115^2/120 = 110.2\,\Omega$. The load-bus data with power factor correction are given in Table 10.3.4.

The *transmission-system efficiency* is defined as the ratio of the real power delivered to the receiving-end bus to the real power transferred from the sending-end bus. This efficiency, which is a measure of the real-power loss in the transmission line, comes out as 94.7% without power factor correction, and 96.5% with power factor correction.

The *transmission-line voltage regulation* (TLVR) is the ratio of the per-phase voltage drop between the sending-end and receiving-end buses to the receiving-end per-phase voltage (or nominal system per-phase voltage). It can also be expressed as

$$\%\text{TLVR} = \frac{V_{RNL} - V_{RFL}}{V_{RFL}} \times 100 \qquad (10.3.1)$$

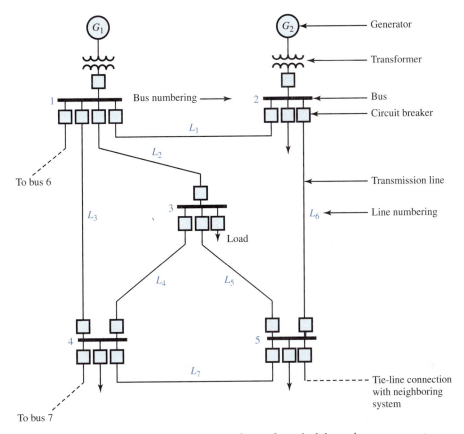

Figure 10.3.4 One-line (single-line) diagram of part of a typical three-phase power system.

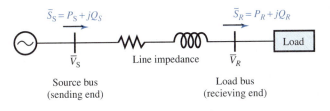

Figure 10.3.5 Simple model of part of an operating power system.

$\bar{S}_S = P_S + jQ_S$

$\bar{S}_R = P_R + jQ_R$

Load

$\bar{V}_S$ Line impedance $\bar{V}_R$

Source bus
(sending end)

Load bus
(recieving end)

where subscript R denotes receiving end, NL indicates no-load, and FL stands for full-load. The TLVR for our example can be seen to be 12.8% prior to power factor improvement, and 10.2% after power factor improvement.

The two components of electric service are *demand* and *energy*. Demand is the maximum level of real power which the electric utility must supply to satisfy the load requirements of its customers. Energy is the cumulative use of electric power over a period of time. The demand component of the electric-rate structure represents the capital investment needed by the utility to provide the generation, transmission, and distribution facilities in order to meet the maximum customer demand. The energy component represents the operating costs, which include fuel and maintenance that must be provided to meet the demand requirements over a period of time.

The *load factor* is the ratio of the actual energy usage to the rated maximum energy usage over a given period. A low load factor is indicative of a substantial period during which the capacity

TABLE 10.3.1 Load-Bus Data

Quantity	Symbol	Value
Voltage	V	115 kV
Current	I	1 kA
Apparent power	S	220 MVA
Real power	P	160 MW
Reactive power	Q	120 MVAR
Power factor	$\cos\theta$	0.8 lagging
Phase angle	θ	36.9°
Load impedance	Z	66.1 Ω
Load resistance	R	52.9 Ω
Load reactance	X_L	39.7 Ω

TABLE 10.3.2 Transmission-Line Data

Quantity	Symbol	Value
Resistance	R	3 Ω
Inductive reactance	X_L	8 Ω
Series impedance	Z	8.5 Ω
Series impedance angle	θ	69.4°
Real power loss	P_{loss}	9 MW
Reactive power loss	Q_{loss}	24 MVAR
Voltage drop	V_{drop}	8.5 kV

TABLE 10.3.3 Source-Bus Data

Quantity	Load Bus	Line	Source Bus
Voltage	115	8.5	127.7 kV
Real power	160	9	169 MW
Reactive power	120	24	144 MVAR
Apparent power	200	—	222 MVA
Angle	36.9°	—	40.4°

TABLE 10.3.4 Load-Bus Data with
Power-Factor Correction

Quantity	Symbol	Value
Voltage	V	115 kV
Current	I	0.8 kA
Apparent power	S	160 MVA
Real power	P	160 MW
Reactive power	Q	0 MVAR
Power factor	$\cos\theta$	1
Phase angle	θ	0°
Load impedance	Z	82.7 Ω
Load resistance	R	82.7 Ω
Load reactance	X_L	0 Ω

of the system is underutilized. Utilities often define load periods in terms of *on-peak* and *off-peak* hours. In order to level the demand by diverting a portion of the energy usage from the on-peak to the off-peak periods, economic incentives (such as lower electric rates for the sale of off-peak energy) are generally offered.

EXAMPLE 10.3.1

A three-phase, 34.5-kV, 60-Hz, 40-km transmission line has a per-phase series impedance of $0.2 + j0.5 \, \Omega/\text{km}$. The load at the receiving end absorbs 10 MVA at 33 kV. Calculate the following:

(a) Sending-end voltage at 0.9 PF lagging.

(b) Sending-end voltage at 0.9 PF leading.

(c) Transmission system efficiency and transmission-line voltage regulation corresponding to cases (a) and (b).

Solution

The per-phase model of the transmission line is shown in Figure E10.3.1.

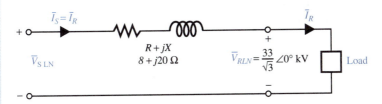

Figure E10.3.1 Per-phase model of transmission line (with only series impedance).

$$\bar{V}_R = \frac{33,000}{\sqrt{3}} \angle 0° = 19,052\angle 0° \, \text{V}$$

(a)
$$\bar{I}_R = \frac{10 \times 10^6}{\sqrt{3} \times 33 \times 10^3} \angle - \cos^{-1} 0.9 = 175\angle - 25.8° \, \text{A}$$

$$\bar{V}_{S\,LN} = 19,052\angle 0° + (175\angle - 25.8°)(8 + j20) = 21,983\angle 6.6°V_{LN}$$

$$V_{S\,LL} = 21.983\sqrt{3} = 38.1 \, \text{kV (line - to - line)}$$

(b)
$$\bar{I}_R = \frac{10 \times 10^6}{\sqrt{3} \times 33 \times 10^3} \angle + \cos^{-1} 0.9 = 175\angle + 25.8° \, \text{A}$$

$$\bar{V}_{S\,LN} = 19,052\angle 0° + (175\angle + 25.8°)(8 + j20) = 19,162\angle 11.3° \, \text{V}_{LN}$$

$$V_{S\,LL} = 19.162\sqrt{3} = 33.2 \, \text{kV (line - to - line)}$$

(c) (i) 0.9 PF lagging:

$$P_R = 10 \times 0.9 = 9 \, \text{MW}$$

$$P_S = \sqrt{3} \times 38.1 \times 0.175 \cos(25.8 + 6.6)° = 9.75 \, \text{MW}$$

$$\eta = \frac{9}{9.75} \times 100 = 92.3\%$$

$$\text{TLVR} = \frac{V_{R\text{ NL}} - V_{R\text{ FL}}}{V_{R\text{ FL}}} \times 100 = \frac{38.1 - 33}{33} \times 100 = 15.45\%$$

Note: If there were no load, the sending-end voltage would appear at the receiving end.

(ii) 0.9 PF leading:

$$P_R = 10 \times 0.9 = 9 \text{ MW}$$

$$P_S = \sqrt{3} \times 33.2 \times 0.175 \cos(25.8 - 11.3)° = 9.74 \text{ MW}$$

$$\eta = \frac{9}{9.74} \times 100 = 92.4\%$$

$$\text{TLVR} = \frac{33.2 - 33}{33} \times 100 = 0.61\%$$

10.4 LEARNING OBJECTIVES

The *learning objectives* of this chapter are summarized here, so that the student can check whether he or she has accomplished each of the following.

- Past, present, and future trends of power systems in the United States.
- Three-phase power calculations and power factor improvement.
- Basic notions of power transmission and distribution.
- Working with load-bus data, transmission-line data, and source-bus data.
- Simple transmission-line model and its circuit calculations.

10.5 PRACTICAL APPLICATION: A CASE STUDY

The Great Blackout of 1965

In the United States, electric utilities grew first as isolated systems. Gradually, however, neighboring utilities began to interconnect, which allowed utility companies to draw upon each others' generation reserves during time of need and to schedule power transfers that take advantage of energy-cost differences. Although overall system reliability and economy have improved dramatically through interconnection, there is a remote possibility that an initial disturbance may lead to instability and a regional blackout.

The worst power failure in history occurred on Tuesday evening, November 9, 1965. At the height of the rush hour, complete darkness descended, and practically everything came to a standstill over an area of 83,000 square miles. About 30 million people were affected in eight states and part of Canada. The system involved was the Canadian–United States eastern power complex (CANUSE).

Many guesses were offered as to the cause of the blackout, but most were far from correct. At first, little hope was held out for pinpointing the exact reason for the failure within a short time. However, on November 15 it was announced that the cause had been found. The failure was attributed to a faulty relay at Sir Adam Beck Plant no. 2 at Queenston, Ontario (see Figure

R. McCaw, "The Great Blackout," *Power Engineering*, Dec. 1965.

10.5.1). This plant is part of the Ontario Hydro-Electric System on the Niagara River. Six lines run into Ontario from the Beck plant, and the line controlled by the faulty relay was carrying 300 MW. Failure of the relay dumped this load onto the other five lines. Even though these lines were not overloaded, all five tripped out.

Total power flow on these lines had been 1600 MW into Ontario, including 500 MW being imported from the Power Authority of the State of New York. All this power was suddenly dumped on the New York system. The resultant surge knocked out the Power Authority's main east–west transmission line and shut down seven units that had been feeding the northeastern grid. The resulting drain on systems to the south and east caused the whole system to collapse.

New York City, for example, had been drawing about 300 MW from the network just before the failure. Loss of the upstate plants caused a sudden reversal of flow and placed a heavy drain on the City generators. The load was much greater than the plants still in service could supply, and the result was a complete collapse. Automatic equipment shut down the units to protect them from damage.

After the total failure the individual systems started up in sections, and most power was restored by a little after midnight. However, Manhattan, with the greatest concentration of load, was not fully restored unitl after six o'clock the following morning.

Only a rare combination of faults, however, will result in a cascading of tripouts and a complete shutdown over an entire region. In order to avoid such large blackouts, stronger grids have been planned and various techniques developed to operate large interconnected networks in parallel with a high degree of operating stability, and with increasing dependence on automated

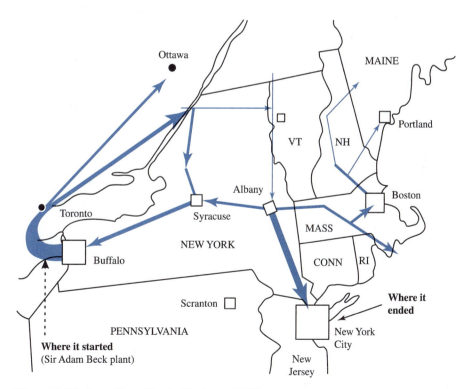

Figure 10.5.1 Area affected by the blackout of 1965.

controls. Pumped storage plants, which represent a large reserve that can be put into operation on short notice, can help the situation.

The great blackout of 1965 had indeed been a sobering experience and possibly a needed warning to reexamine some of our policies and plans. The warning will not have been wasted if we learn from it how to build a more reliable grid to suppply all the power that is needed in the years ahead. After all, our whole economy is almost completely dependent on electric power.

PROBLEMS

10.2.1 Two ideal voltage sources are connected to each other through a feeder with impedance $\bar{Z} = 1.5 + j6 \ \Omega$, as shown in Figure P10.2.1. Let $\bar{E}_1 = 120\angle 0°$ V and $\bar{E}_2 = 110\angle 45°$.

(a) Determine the real power of each machine, and state whether the machine is supplying or absorbing real power.

(b) Compute the reactive power of each machine, and state whether each machine is delivering or receiving reactive power.

(c) Find the real and reactive power associated with the feeder impedance, and state whether it is supplied or absorbed.

*10.2.2 If the impedance between the voltage sources in Problem 10.2.1 is changed to $\bar{Z} = 1.5 - j6 \ \Omega$, redo parts (a) through (c) of Problem 10.2.1.

10.2.3 A single-phase industrial plant consists of two loads in parallel:

$$P_1 = 48\text{kW} \qquad \text{PF}_1 = 0.60 \text{ lagging}$$
$$P_2 = 24\text{kW} \qquad \text{PF}_2 = 0.96 \text{ leading}$$

It is operated from a 500-V, 60-Hz source. An additional capacitor C is added in parallel to improve the plant's overall power factor to unity. Determine the value of C in μF. Also sketch the power triangles before and after the addition of C, and find the overall current drawn from the supply.

10.2.4 A 230-V, single-phase, 60-Hz source supplies two loads in parallel. One draws 10 kVA at a

lagging power factor of 0.80 and the other draws 6 kW at a lagging power factor of 0.90. Compute the source current.

10.2.5 A three-phase power system consists of a wye-connected ideal generator, supplying a wye-connected balanced load through a three-phase feeder. The load has $\bar{Z}_L = 20\angle 30°\Omega/$ phase, and the feeder has an impedance $\bar{Z}_{\text{fdr}} = 1.5\angle 75° \ \Omega/\text{phase}$. If the terminal voltage of the load is 4.16 kV, determine:

(a) The terminal voltage of the generator.

(b) The line current supplied by the generator.

10.2.6 A 345-kV, 60-Hz, three-phase transmission line delivers 600 MVA at 0.866 power factor lagging to a three-phase load connected to its receiving-end terminals. Assuming that the voltage at the receiving end is 345 kV and the load is wye-connected, find the following:

(a) Complex load impedance per phase $\bar{Z}_L/\text{ph}$.

(b) Line and phase currents.

(c) Real and reactive powers per phase.

(d) Total three-phase real and reactive powers.

*10.2.7 A three-phase load, connected to a 440-V bus, draws 120 kW at a power factor of 0.85 lagging. In parallel with this load is a three-phase capacitor bank that is rated 50 kVAR . Determine the resultant line current and power factor.

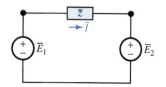

Figure P10.2.1

10.2.8 A balanced three-phase, wye-connected, 2400-V, 60-Hz source supplies two balanced wye-connected loads in parallel. The first draws 15 kVA at 0.8 power factor lagging, and the second needs 20 kW at 0.9 power factor leading. Compute the following:

(a) Current supplied by the source.

(b) Total real and reactive power drawn by the combined load.

(c) Overall power factor.

10.2.9 A three-phase, 60-Hz substation bus supplies two wye-connected loads that are connected in parallel through a three-phase feeder that has a per-phase impedance of $0.5 + j2$ Ω. Load 1 draws 50 kW at 0.866 lagging PF, and load 2 draws 36 kVA at 0.9 leading PF. If the line-to-line voltage at the load terminals is 460 V, find the following:

(a) Total line current flowing through the feeder.

(b) Per-phase impedance of each load.

(c) Line-to-line voltage at the substation bus.

(d) Total real and reactive power delivered by the bus.

10.2.10 A balanced delta-connected load has a per-phase impedance of $45\angle60°$ Ω. It is connected to a three-phase, 208-V, 60-Hz supply by a three-phase feeder that has a per-phase impedance of $1.2 + j1.6$ Ω.

(a) Determine the line-to-line voltage at the load terminals.

(b) If a delta-connected capacitor bank, with a per-phase reactance of 60Ω, is connected in parallel with the load at its terminals, compute the resulting line-to-line voltage at the load terminals.

*10.3.1 Consider a lossless transmission line with only a series reactance X, as shown in Figure P10.3.1.

(a) Find an expression for the real power transfer capacity of the transmission system.

(b) What is P_{max} (the theoretical steady-state limit of a lossless line), which is the maximum power that the line can deliver?

(c) How could the same expressions be used for a three-phase transmission line?

10.3.2 A 20-km, 34.5-kV, 60-Hz, three-phase transmission line has a per-phase series impedance of $\bar{Z} = 0.19 + j0.34$ Ω/km. The load at the receiving end absorbs 10 MVA at 33 kV. Calculate:

(a) The sending-end voltage for a load power factor of 0.9 lagging.

(b) The sending-end voltage for a load power factor of 0.9 leading.

(c) The transmission system efficiency for cases (a) and (b).

(d) The transmission-line voltage regulation (TLVR) for cases (a) and (b).

10.3.3 It is sometimes convenient to represent a transmission line by a two-port network, as shown in Figure P10.3.3. The relations between sending-end and receiving-end quantities are given by

$$\bar{V}_S = \bar{A}\,\bar{V}_R + \bar{B}\,\bar{I}_R \text{ and } \bar{I}_S = \bar{C}\,\bar{V}_R + \bar{D}\,\bar{I}_R,$$

in which the generally complex parameters $\bar{A}$, $\bar{B}$, $\bar{C}$, and $\bar{D}$ depend on the transmission-line models. For the model that includes only the series impedance $\bar{Z}$ of the transmission line, find the parameters $\bar{A}$, $\bar{B}$, $\bar{C}$, and $\bar{D}$ and specify their units. Also evaluate $(\bar{A}\,\bar{D} - \bar{B}\,\bar{C})$.

10.3.4 For a transmission-line model that includes only the series impedance $\bar{Z}$, sketch phasor diagrams for:

(a) Lagging power factor load.

(b) Leading power factor load.

10.3.5 In terms of the parameters $\bar{A}$, $\bar{B}$, $\bar{C}$, and $\bar{D}$ introduced in Problem 10.3.3, find an expression for $\bar{V}_R$ at no load in terms of $\bar{V}_S$ and the parameters.

*10.3.6 Consider an upgrade of a three-phase transmission system in which the operating line-to-line voltage is doubled, and the phase or line currents are reduced to one-half the previous value, for the same level of apparent power transfer. Discuss the consequent effects on the real and reactive power losses, and on the voltage drop across the series impedance of the transmission system.

10.3.7 Justify the entries made in Tables 10.3.1, 10.3.2, and 10.3.3 for load-bus data, transmission-line data, and source-bus data, respectively, for the example considered in the text.

10.3.8 Justify the entries made in Table 10.3.4 for load-bus data with power factor correction, and discuss the effects of power factor correction on P_{loss}, Q_{loss}, and V_{drop} of the transmission line.

10.3.9 Check the figures given in the text (for the example considered) regarding the transmission-system efficiencies and TLVR for the two cases, with and without power factor correction.

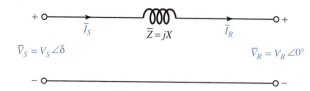

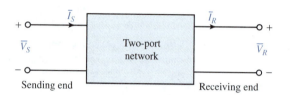

Figure P10.3.1 Simple model of loss-less transmission line.

Figure P10.3.3

Sending end   Receiving end

10.3.10 On a per-phase basis, let $v = \sqrt{2}\, V \cos \omega t$ and $i = \sqrt{2}\, I \cos (\omega t - \theta)$.

(a) Express the instantaneous power $s(t)$ in terms of real power P and reactive power Q.

(b) Now consider the energy E associated with the instantaneous power and show that $E = Pt$, where t is the duration of the time interval in seconds.

10.3.11 A 60-Hz, three-phase transmission line has a total per-phase series impedance of $35 + j140$ Ω. If it delivers 40 MW at 220 kV and 0.9 power factor lagging, find:

(a) The voltage, current, and power factor at the sending end of the line.

(b) The voltage regulation and efficiency of the line.

10.3.12 A three-phase, 60-Hz transmission line has a total series impedance of $22.86\angle 62.3°$ Ω per phase. It delivers 2.5 MW at 13.8 kV to a load connected to its receiving end. Compute the sending-end voltage, current, real power, and reactive power for the following conditions:

(a) 0.8 power factor lagging.

(b) Unity power factor.

(c) 0.9 power factor leading.

11 Magnetic Circuits and Transformers

A basic understanding of electromagnetism is essential to the study of electrical engineering because it is the key to the operation of many electric apparatus found in industry and at home. Electric systems that generate, convert, or control huge amounts of energy almost always involve devices whose operation depends on magnetic phenomena. Thus, we focus our attention here on magnetic aspects as related to electric-power engineering. In this chapter we shall proceed from introductory magnetic-circuit concepts to relatively simple magnetic structures and then to a consideration of transformers.

The magnetic material determines the size of the equipment, its capabilities, and the limitations on its performance. The transformer, although not an energy-conversion device, is an important auxiliary in the transfer and conversion of electric energy. Practically all transformers and electric machinery utilize magnetic material for shaping the magnetic fields which act as the medium for transferring and converting energy. Transformers are found in various applications, such as radio, television, and electric power transmission and distribution circuits. Other devices, such as circuit breakers, automatic switches, and relays, also require the presence of a confined magnetic field for their proper operation.

11.1 MAGNETIC MATERIALS

For magnetic material media, the *magnetic flux density B*, expressed in tesla (T) or Wb/m², and the *field intensity H*, expressed in A/m or ampere-turns per meter (At/m), are related through the relationship

$$B = \mu H \qquad (11.1.1)$$

where μ stands for the *permeability* of the material expressed in henrys per meter (H/m). The free-space permeability μ_0 is a constant given by $4\pi \times 10^{-7}$ H/m in the SI system of units. The same value holds good for air as well as for any nonmagnetic material. For a linear magnetic material which exhibits a straight-line relationship between B and H, the permeability is a constant given by the slope of the linear *B–H* characteristic, and it is related to the free-space permeability as

$$\mu = \mu_r \mu_0 \qquad (11.1.2)$$

where μ_r is the relative permeability, which is a dimensionless constant. If the *B–H* characteristic is nonlinear, as with a number of common magnetic materials, then the permeability varies as a function of the magnetic induction. The variation of B with H is depicted by the saturation curve of Figure 11.1.1, in which the slope of the curve clearly depends upon the operating flux density, as classified for convenience into regions I, II, and III.

There are several material classifications with their distinguishing characteristics. *Ferromagnetic* materials, for which $\mu_r \gg 1$, exhibit a high degree of magnetizability and are generally subdivided into *hard* and *soft* materials. Soft ferromagnetic materials include most of the steels and iron, whereas hard ferromagnetic materials include the *permanent-magnet* materials such as alnicos and alloys of cobalt with a rare-earth element such as samarium. *Ferrimagnetic* materials are ferrites composed of iron oxides, subdivided into hard and soft categories. *Ferrofluids* (magnetic fluids with iron-oxide particles suspended) and *amorphous* magnetic (soft ferromagnetic) materials were also developed later. Typical magnetic characteristics of some core materials are shown in Figure 11.1.2.

Core (Iron) Losses

Iron-core losses are usually divided into two components: *hysteresis loss* and *eddy-current loss*. The former is proportional to the area enclosed by the *hysteresis loop*, shown in Figure 11.1.3,

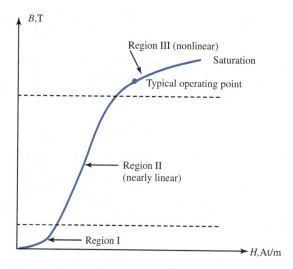

Figure 11.1.1 Typical magnetization characteristic showing three regions.

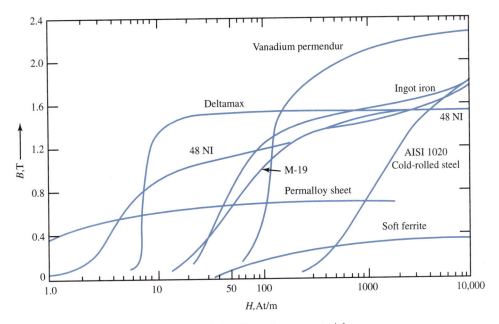

Figure 11.1.2 Typical magnetic characteristics of certain core materials.

which is a characteristic of a magnetic core material. The area of the loop represents the heat-energy loss during one cycle in a unit cube of the core material. The hysteresis loss per cycle $P_{h\ cycle}$ in a core of volume V, possessing a uniform flux density B throughout its volume, is given by

$$P_{h\ cycle} = V(\text{area of hysteresis loop}) = V \oint H\ dB \qquad (11.1.3)$$

The hysteresis loss per second is approximated empirically by

$$P_{h\ second} = k_h V f B_m^{1.5\ \text{to}\ 2.5}\ W \qquad (11.1.4)$$

where k_h is a proportionality constant dependent on the characteristics of iron, f is the frequency of excitation in hertz, and B_m is the maximum value of the core flux density. This loss component of the core loss can be reduced by choosing a core of electrical steel that has a narrow hysteresis loop.

Square-loop magnetic materials (such as ferrites and permalloy) that have a nearly rectangular hysteresis loop are used in switching circuits, as storage elements in computers, and in special types of transformers in electronic circuits.

Another feature of an ac-operated magnetic system is the eddy-current loss, which is the loss due to the eddy currents induced in the core material. The eddy-current loss is empirically approximated as

$$P_e = k_e V f^2 \tau^2 B_m^2\ W \qquad (11.1.5)$$

where k_e is a constant dependent on the characteristics of iron, V is the volume of iron, f is the frequency, τ is the lamination thickness (usually a stack of thin laminations makes up the core), and B_m is the maximum core flux density. By choosing very thin laminations (making τ smaller), the eddy-current loss can be reduced.

The laminations (or thin sheets), insulated from each other by a thin coat of varnish, are oriented parallel to the direction of flux, as shown in Figure 11.1.4. Laminating a core generally results in an increase in the overall volume. The ratio of the volume actually occupied by the

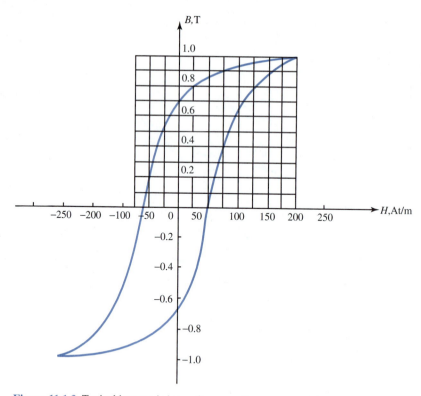

Figure 11.1.3 Typical hysteresis loop of a magnetic-core material.

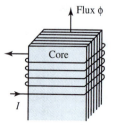

Figure 11.1.4 Part of a laminated core.

magnetic material to the total (gross) volume of the core is known as the *stacking factor.* The thinner the lamination thickness, the lower the stacking factor. This factor usually ranges between 0.5 and 0.95.

EXAMPLE 11.1.1

(a) Estimate the hysteresis loss at 60 Hz for a toroidal (doughnut-shaped) core of 300-mm mean diameter and a square cross section of 50 mm by 50 mm. The symmetrical hysteresis loop for the electric sheet steel (of which the torus is made) is given in Figure 11.1.3.

(b) Now suppose that all the linear dimensions of the core are doubled. How will the hysteresis loss differ?

(c) Next, suppose that the torus (which was originally laminated for reducing the eddy-current losses) is redesigned so that it has half the original lamination thickness. Assume the stacking factor to be unity in both cases. What would be the effect of such a change in design on the hysteresis loss?

(d) Suppose that the toroidal core of part (a) is to be used on 50-Hz supply. Estimate the change in hysteresis loss if no other conditions of operation are changed.

Solution

(a) Referring to Figure 11.1.3, the area of each square represents

$$(0.1 \text{ T}) \times (25 \text{ A/m}) = 2.5\frac{\text{Wb}}{\text{m}^2} \times \frac{\text{A}}{\text{m}} = 2.5\frac{\text{V} \times \text{s} \times \text{A}}{\text{m}^3} = 2.5 \text{ J/m}^3$$

By counting, the number of squares in the upper half of the loop is found to be 43; the area of the hysteresis loop is given by

$$2 \times 43 \times 2.5 = 215 \text{ J/m}^3$$

The toroidal volume is

$$\pi \times 0.3 \times 0.05^2 = 2.36 \times 10^{-3} \text{m}^3$$

The hysteresis loss in the torus is then given by

$$2.36 \times 10^{-3} \times 60 \times 215 = 30.44 \text{ W}$$

(b) When all the linear dimensions of the core are doubled, its volume will be eight times the previous volume. Hence the new core hysteresis loss will be

$$8 \times 30.44 = 243.52 \text{ W}$$

(c) The lamination thickness has no bearing on the hysteresis loss; it changes only the eddy-current loss. Hence, the hysteresis loss remains unchanged.

(d) Since the hysteresis loss is directly proportional to the frequency, the loss on the 50 Hz supply will be

$$30.44 \times \frac{50}{60} = 25.37 \text{ W}$$

11.2 MAGNETIC CIRCUITS

A magnetic circuit provides a path for magnetic flux, just as an electric circuit provides a path for electric current. Magnetic circuits are integral parts of transformers, electric machines, and many other common devices such as doorbells. Analogous to voltage (electric potential difference) in the electric circuit, in a magnetic circuit we have the *magnetomotive force* (mmf), or the *magnetic potential difference* which produces a magnetic field, and which has units of amperes or ampere-turns. The two sources of mmf in magnetic circuits are the electric current and the permanent magnet (which stores energy and is capable of maintaining a magnetic field with no expenditure

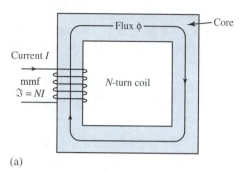

Figure 11.2.1 Simple magnetic circuit. (a) Mmf and flux. (b) Leakage flux and fringing flux.

(a)

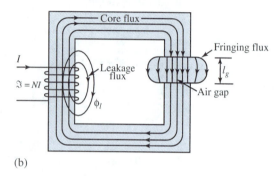

(b)

of power). The current source is commonly a coil of N turns, carrying a current I known as the *exciting current*; the mmf is then said to be NI At.

Figure 11.2.1(a) shows, schematically, a simple magnetic circuit with an mmf $\Im$ $(= NI)$ and magnetic flux ϕ. Note that the *right-hand rule* gives the direction of flux for the chosen direction of current. The concept of a magnetic circuit is useful in estimating the mmf (excitation ampere-turns) needed for simple electromagnetic structures, or in finding approximate flux and flux densities produced by coils wound on ferromagnetic cores. Magnetic circuit analysis follows the procedures that are used for simple dc electric circuit analysis.

Calculations of excitation are usually based on *Ampere's law*, given by

$$\oint \bar{H} \, \overline{dl} = \text{ampere-turns enclosed} \tag{11.2.1}$$

where $H(|\bar{H}| = B/\mu)$ is the magnetic field intensity along the path of the flux. If the magnetic field strength is approximately constant ($H = H_C$) along the closed flux path, and l_C is the average (mean) length of the magnetic path in the core, Equation (11.2.1) can be simplified as

$$\Im = NI = H_C l_C \tag{11.2.2}$$

Analogous to Ohm's law for dc electric circuits, we have this relation for magnetic circuits,

$$\text{flux } \phi = \frac{\text{mmf } \Im}{\text{reluctance } \Re} = \frac{\Im}{l/\mu A} \tag{11.2.3}$$

where μ is the permeability, A is the cross-sectional area perpendicular to the direction of l, and l stands for the corresponding portion of the length of the magnetic circuit along the flux path. Based on the analogy between magnetic circuits and dc resistive circuits, Table 11.2.1 summarizes the corresponding quantities. Further, the laws of resistances in series and parallel also hold for reluctances.

TABLE 11.2.1 Analogy with dc Resistive Circuits

DC Resistive Circuit	Magnetic Circuit
Current I (A)	Flux ϕ (Wb)
Voltage V (V)	Magnetomotive force (mmf) $\Im$ (At)
Resistance $R = l/\sigma A$ (Ω)	Reluctance $\Re = l/\mu A$ (H^{-1})
Conductivity σ (S/m)	Permeability μ (H/m)
Conductance $G = 1/R$ (S)	Permeance $\mathbf{P} = 1/\Re$ (H)
$I = V/R$	$\phi = \Im/\Re$
Current density $J = I/A$ (A/m^2)	Flux density $B = \phi/A$ (Wb/m^2)

Analogous to KVL in electrical circuits, Ampere's law applied to the analysis of a magnetic circuit leads to the statement that the algebraic sum of the magnetic potentials around any closed path is zero. Series, parallel, and series–parallel magnetic circuits can be analyzed by means of their corresponding electric-circuit analogs. All methods of analysis that are valid for dc resistive circuits can be effectively utilized in an analogous manner.

The following differences exist between a dc resistive circuit and a magnetic circuit

- Reluctance $\Re$ is not an energy-loss component like a resistance R (which leads to an I^2R loss). Energy must be supplied continuously when a direct current is established and maintained in an electric circuit; but a similar situation does not prevail in the case of a magnetic circuit, in which a flux is established and maintained constant.

- Magnetic fluxes take *leakage* paths [as ϕ_l in Figure 11.2.1(b)]; but electric currents flowing through resistive networks do not.

- *Fringing* or bulging of flux lines [shown in Figure 11.2.1(b)] occurs in the *air gaps* of magnetic circuits; but such fringing of currents does not occur in electric circuits. Note that fringing increases with the length of the air gap and increases the effective area of the air gap.

- There are no magnetic insulators similar to the electrical insulators.

In the case of ferromagnetic systems containing air gaps, a useful approximation for making quick estimates is to consider the ferromagnetic material to have infinite permeability. The relative permeability of iron is considered so high that practically all the ampere-turns of the winding are consumed in the air gaps alone.

Calculating the mmf for simple magnetic circuits is rather straightforward, as shown in the following examples. However, it is not so simple to determine the flux or flux density when the mmf is given, because of the nonlinear characteristic of the ferromagnetic material.

EXAMPLE 11.2.1

Consider the magnetic circuit of Figure 11.2.1(b) with an air gap, while neglecting leakage flux. Correct for fringing by adding the length of the air gap $l_g = 0.1$ mm to each of the other two dimensions of the core cross section $A_C = 2.5$cm $\times$ 2.5cm. The mean length of the magnetic path in the core l_C is given to be 10 cm. The core is made of 0.15- mm-thick laminations of M-19 material whose magnetization characteristic is given in Figure 11.1.2. Assume the stacking factor to be 0.9. Determine the current in the exciting winding, which has 100 turns and produces a core flux of 0.625 mWb.

The net cross-sectional area of the core is $2.5^2 \times 10^{-4} \times 0.9 = 0.5625 \times 10^{-3}$ m^2. Note that the stacking factor of 0.9 is applied for the laminated core. It does not, however, apply for the air-gap portion,

$$B_C = \frac{\phi}{\text{net area}} = \frac{0.625 \times 10^{-3}}{0.5625 \times 10^{-3}} = 1.11 \text{ T}$$

The corresponding H_C from Figure 11.1.2 for M-19 is 130 A/m. Hence,

$$\Im_C = H_C l_C = 130 \times 0.1 = 13 \text{ At}$$

The cross-sectional area of the air gap, corrected for fringing, is given by

$$A_g = (2.5 + 0.01)(2.5 + 0.01)10^{-4} = 0.63 \times 10^{-3} \text{ m}^2$$

$$B_g = \frac{\phi}{A_g} = \frac{0.625 \times 10^{-3}}{0.63 \times 10^{-3}} = 0.99 \text{ T}$$

$$H_g = \frac{B_g}{\mu_0} = \frac{0.99}{4\pi \times 10^{-7}} = 0.788 \times 10^6 \text{ A/m}$$

Hence,

$$\Im_g = H_g l_g = 0.788 \times 10^6 \times 0.1 \times 10^{-3} = 78.8 \text{ At}$$

For the entire magnetic circuit (see Figure E11.2.1)

$$NI = \Im_{\text{TOTAL}} = \Im_C + \Im_g = 13 + 78.8 = 91.8 \text{ At}$$

Thus, the coil current

$$I = \frac{NI}{N} = \frac{91.8}{100} = 0.92 \text{ A}$$

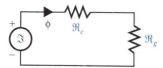

Figure E11.2.1 Equivalent magnetic circuit.

EXAMPLE 11.2.2

In the magnetic circuit shown in Figure E11.2.2(a) the coil of 500 turns carries a current of 4 A. The air-gap lengths are $g_1 = g_2 = 0.25$ cm and $g_3 = 0.4$ cm. The cross-sectional areas are related such that $A_1 = A_2 = 0.5A_3$. The permeability of iron may be assumed to be infinite. Determine the flux densities B_1, B_2, and B_3 in the gaps g_1, g_2, and g_3, respectively. Neglect leakage and fringing.

Noting that the reluctance of the iron is negligible, the equivalent magnetic circuit is shown in Figure E11.2.2(b).

Given $\Im = NI = 500 \times 4 = 2000$ At, $g_1 = g_2 = 0.25$ cm, and $g_3 = 0.4$ cm; $A_1 = A_2 = 0.5A_3$; $\mu_1 = \mu_2 = \mu_3 = \mu_0$; $\phi_3 = \phi_1 + \phi_2$; and $H_1 g_1 + H_3 g_3 = H_2 g_2 + H_3 g_3 = \Im = 2000$, we have

$$H_1 g_1 = H_2 g_2 \quad \text{or} \quad H_1 = H_2$$

since $g_1 = g_2$. Thus, $B_1 = B_2$, since $\mu_1 = \mu_2 = \mu_0$. Because $A_1 = A_2$, it follows that $\phi_1 = \phi_2$ and $\phi_3 = 2\phi_1$. But $B_3 = \phi_3/2A_1 = B_1$. Thus,

$$B_1 = B_2 = B_3; \quad H_1 = H_2 = H_3$$

We had $H_1 g_1 + H_3 g_3 = 2000$, which is rewritten as

$$\frac{B_1}{\mu_0}(0.25 \times 10^{-2}) + \frac{B_1}{\mu_0}(0.4 \times 10^{-2}) = 2000$$

or,

$$B_1 = \frac{2000 \times 4\pi \times 10^{-7}}{0.65 \times 10^{-2}} = 0.387 \text{ T} = B_2 = B_3$$

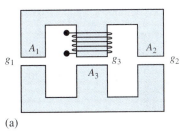

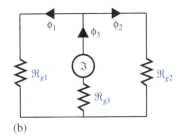

(a)   (b)

Figure E11.2.2

A simple magnetic structure, similar to those examined in the previous examples, finds common application in the so-called *variable-reluctance position sensor*, which, in turn, finds widespread application in a variety of configurations for the measurement of linear and angular position and velocity.

For *magnetic circuits with ac excitation*, the concepts of inductance and energy storage come into play along with *Faraday's law* of induction. Those concepts have been presented in some detail in Section 1.2. The reader is encouraged to review that section, as that background will be helpful in solving some of the problems related to this section.

11.3 TRANSFORMER EQUIVALENT CIRCUITS

It would be appropriate for the reader to review the material on transformers in Section 1.2. Transformers come in various sizes, from very small, weighing only a few ounces, to very large, weighing hundreds of tons. The ratings of transformers cover a very wide range. Whereas transformers applied with electronic circuits and systems usually have ratings of 300 VA or less, power-system transformers used in transmission and distribution systems have the highest volt-ampere ratings (a few kVA to several MVA) as well as the highest continuous voltage ratings. Instrument transformers (potential transformers and current transformers) with very small volt-ampere ratings are used in instruments for sensing voltages or currents in power systems.

Transformers may be classified by their frequency range: power transformers, which usually operate at a fixed frequency; audio and ultrahigh-frequency transformers; wide-band and narrow-band frequency transformers; and pulse transformers. Transformers employed in supplying power to electronic systems are generally known as power transformers. In power-system applications, however, the term *power transformer* denotes transformers used to transmit power in ratings larger than those associated with distribution transformers, usually more than 500 kVA at voltage levels of 67 kV and above.

Conventional transformers have two windings, but others (known as *autotransformers*) have only one winding, and still others (known as *multiwinding transformers*) have more than two windings. Transformers used in polyphase circuits are known as *polyphase transformers*. In the most popular three-phase system, the most common connections are the wye (star or Y) and the delta (mesh or Δ) connections.

Figure 11.3.1 shows, schematically, a transformer having two windings with N_1 and N_2 turns, respectively, on a common magnetic circuit. The transformer is said to be *ideal* when:

- Its core is infinitely permeable.
- Its core is lossless.
- It has no leakage fluxes.
- Its windings have no losses.

The mutual flux linking the N_1-turn and N_2-turn windings is ϕ. Due to a finite rate of change of ϕ, according to Faraday's law of induction, emf's e_1 and e_2 are induced in the primary and secondary windings, respectively. Thus, we have

$$e_1 = N_1 \frac{d\phi}{dt} ; \qquad e_2 = N_2 \frac{d\phi}{dt} ; \qquad \frac{e_1}{e_2} = \frac{N_1}{N_2} \qquad (11.3.1)$$

The polarity of the induced voltage is such as to produce a current which opposes the flux change, according to Lenz's law. For the ideal transformer, since $e_1 = v_1$ and $e_2 = v_2$, it follows that

$$\frac{v_1}{v_2} = \frac{e_1}{e_2} = \frac{N_1}{N_2} = a; \qquad \frac{V_1}{V_2} = \frac{E_1}{E_2} = \frac{N_1}{N_2} = a \qquad (11.3.2)$$

where V_1, V_2, E_1, and E_2 are the rms values of v_1, v_2, e_1, and e_2, respectively, and a stands for the *turns ratio*. When a passive external load circuit is connected to the secondary winding terminals, the terminal voltage v_2 will cause a current i_2 to flow, as shown in Figure 11.3.1. Further, we have

$$v_1 i_1 = v_2 i_2; \qquad \frac{i_1}{i_2} = \frac{v_2}{v_1} = \frac{N_2}{N_1} = \frac{1}{a}; \qquad \frac{I_1}{I_2} = \frac{V_2}{V_1} = \frac{N_2}{N_1} = \frac{1}{a} \qquad (11.3.3)$$

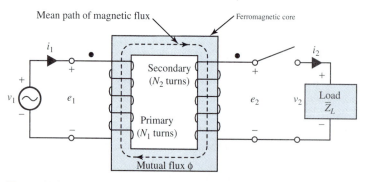

Figure 11.3.1 Schematic of two-winding transformer.

in which I_1 and I_2 are the rms value of i_1 and i_2, respectively.

If the flux varies sinusoidally, such that

$$\phi = \phi_{max} \sin \omega t \qquad (11.3.4)$$

where ϕ_{max} is the maximum value of the flux and $\omega = 2\pi f$, f being the frequency, then the induced voltages are given by

$$e_1 = \omega N_1 \phi_{max} \cos \omega t; \qquad e_2 = \omega N_2 \phi_{max} \cos \omega t \qquad (11.3.5)$$

Note that the induced emf leads the flux by 90°. The rms values of the induced emf's are given by

$$E_1 = \frac{\omega N_1 \phi_{max}}{\sqrt{2}} = 4.44 f N_1 \phi_{max}; \qquad E_2 = \frac{\omega N_2 \phi_{max}}{\sqrt{2}} = 4.44 f N_2 \phi_{max} \qquad (11.3.6)$$

Equation (11.3.6) is known as the *emf equation*.

From Equations (11.3.2) and (11.3.3) it can be shown that if an impedance $\bar{Z}_L$ is connected to the secondary, the impedance $\bar{Z}_1$ seen at the primary is given by

$$\bar{Z}_1 = a^2 \bar{Z}_L = \left(\frac{N_1}{N_2}\right)^2 \bar{Z}_L \qquad (11.3.7)$$

Major applications of transformers are in voltage, current, and impedance transformations, and in providing isolation while eliminating direct connections between electric circuits.

A *nonideal* (or *practical*) transformer, in contrast to an ideal transformer, has the following characteristics, which have to be accounted for:

- Core losses (hysteresis and eddy-current losses)

- Resistive ($I^2 R$) losses in its primary and secondary windings

- Finite permeability of the core requiring a finite mmf for its magnetization

- Leakage fluxes (associated with the primary and secondary windings) that do not link both windings simultaneously.

Now our goal is to develop an *equivalent circuit* of a practical transformer by including the nonideal effects. First, let us consider the simple magnetic circuit of Figure 11.2.1(a), excited by an ac mmf, and come up with its equivalent circuit. With no coil resistance and no core loss, but with a finite constant permeability of the core, the magnetic circuit along with the coil can be represented just by an inductance L_m or, equivalently, by an inductive reactance $X_m = \omega L_m$, when the coil is excited by a sinusoidal ac voltage of frequency $f = \omega/2\pi$. This reactance is known as the *magnetizing reactance*. Thus, Figure 11.3.2(a) shows X_m (or impedance $\bar{Z} = jX_m$) across which the terminal voltage with an rms value of V_1 (equal to the induced voltage E_1) is applied.

Next, in order to include the core losses, since these depend directly upon the level of flux density and hence the voltage V_1, a resistance R_C is added in parallel to jX_m, as shown in Figure 11.3.2(b). Then the resistance R_1 of the coil itself and the *leakage reactance* X_1, representing the effect of leakage flux associated with the coil, are included in Figure 11.3.2(c) as a series impedance given by $R_1 + jX_1$.

Finally, Figure 11.3.3 shows the equivalent circuit of a nonideal transformer as a combination of an ideal transformer and of the nonideal effects of the primary winding, the core, and the secondary winding. Note that the effects of distributed capacitances across and between the windings are neglected here. The following notation is used:

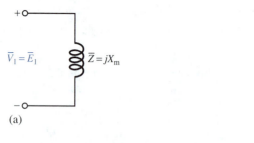

(a)

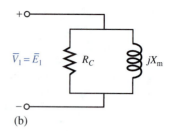

(b)

Figure 11.3.2 Equivalent circuits of an iron core excited by an ac mmf.

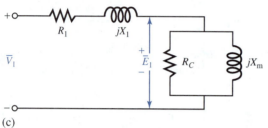

(c)

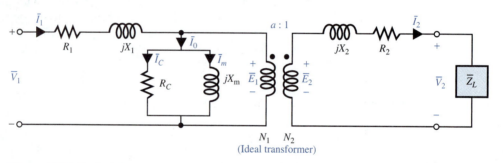

$N_1 \quad N_2$
(Ideal transformer)

Figure 11.3.3 Equivalent circuit of a two-winding iron-core transformer as a combination of an ideal transformer and the nonideal effects.

a	Turns ratio N_1/N_2
E_1, E_2	Induced voltage in primary and secondary
V_1, V_2	Terminal voltage of primary and secondary
I_1, I_2	Input and output current
I_0	No-load primary exciting current
R_1, R_2	Winding resistance of primary and secondary
X_1, X_2	Leakage reactance of primary and secondary
I_m, X_m	Magnetizing current and reactance (referred to primary)
I_C, R_C	Current and resistance accounting for core losses (referred to primary)
$\bar{Y}_o$	Exciting admittance (referred to primary), which is the admittance of the parallel combination of X_m and R_C
$\bar{Z}_L$	Load impedance $= \bar{V}_2/\bar{I}_2$

Note that phasor notation and *rms* values for all voltages and currents are used.

It is generally more convenient to have the equivalent circuit entirely referred to either primary or secondary by using the ideal-transformer relationships [Equations (11.3.2), (11.3.3), and (11.3.7)], thereby eliminating the need for the ideal transformer to appear in the equivalent circuit. Figure 11.3.4 shows such circuits, which are very useful for determining the transformer characteristics.

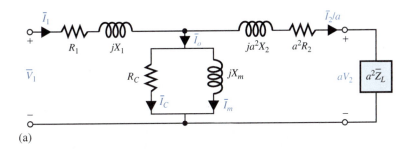

(a)

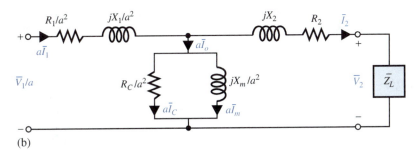

(b)

Figure 11.3.4 Equivalent circuits of a transformer. **(a)** Referred to primary. **(b)** Referred to secondary.

EXAMPLE 11.3.1

A single-phase, 50-kVA, 2400:240-V, 60-Hz distribution transformer has the following parameters:

Resistance of the 2400-V winding $R_1 = 0.75$
Resistance of the 240-V winding $R_2 = 0.0075$
Leakage reactance of the 2400-V winding $X_1 = 1 \ \Omega$
Leakage reactance of the 240-V winding $X_2 = 0.01 \ \Omega$
Exciting admittance on the 240-V side $= 0.003 - j0.02$ S

Draw the equivalent circuits referred to the high-voltage side and to the low-voltage side. Label the impedances numerically.

Solution

(a) The equivalent circuit referred to the high-voltage side is shown in Figure E11.3.1 (a). The quantities, referred to the high-voltage side from the low-voltage side, are calculated as

$$R'_2 = a^2 R_2 = \left(\frac{2400}{240}\right)^2 (0.0075) = 0.75 \ \Omega$$

$$X'_2 = a^2 X_2 = \left(\frac{2400}{240}\right)^2 (0.01) = 1.0 \ \Omega$$

Note that the exciting admittance on the 240-V side is given. The exciting branch conductance and susceptance referred to the high-voltage side are given by

$$\frac{1}{a^2}(0.003) \quad \text{or} \quad \frac{1}{100} \times 0.003 = 0.03 \times 10^{-3} \ \text{S}$$

and

$$\frac{1}{a^2}(0.02) \quad \text{or} \quad \frac{1}{100} \times 0.02 = 0.2 \times 10^{-3} \ \text{S}$$

(b) The equivalent circuit referred to the low-voltage side is given in Figure E11.3.1 (b). Note the following points.

(i) The voltages specified on the nameplate of a transformer yield the turns ratio directly. The turns ratio in this problem is 2400:240, or 10:1.

(ii) Since admittance is the reciprocal of impedance, the reciprocal of the referring factor for impedance must be used when referring admittance from one side to the other.

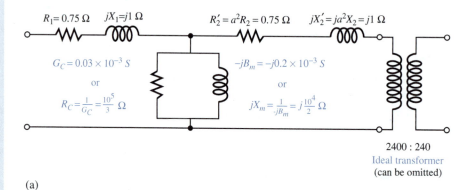

(a)

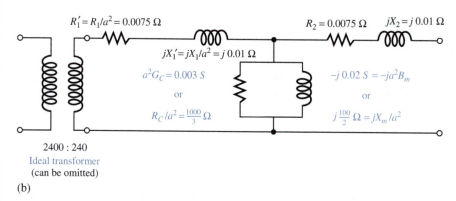

(b)

Figure E11.3.1 Equivalent circuit. **(a)** Referred to high-voltage side. **(b)** Referred to low-voltage side.

The equivalent circuits shown in Figure 11.3.4 are often known as transformer *T-circuits* in which winding capacitances have been neglected. Other modifications and simplifications of this basic T-circuit are used in practice. Approximate circuits (referred to the primary) commonly used for the constant-frequency power-system transformer analysis are shown in Figure 11.3.5. By moving the parallel combination of R_C and jX_m from the middle to the left, as shown in Figure 11.3.5(a), computational labor can be reduced greatly with minimal error. The series impedance $R_1 + jX_1$ can be combined with $a^2R_2 + ja^2X_2$ to form an equivalent series impedance, $\bar{Z}_{eq} = R_{eq} + jX_{eq}$. Further simplification is gained by neglecting the exciting current altogether, as shown in Figure 11.3.5(b), which represents the transformer by its equivalent series impedance. When R_{eq} is small compared to X_{eq}, as in the case of large power-system transformers, R_{eq} may frequently be neglected for certain system studies. The transformer is then modeled by its equivalent series reactance X_{eq} only, as shown in Figure 11.3.5(c). The student should have no difficulty in drawing these approximate equivalent circuits referred to the secondary.

The modeling of a circuit or system consisting of a transformer depends on the frequency range of operation. For variable-frequency transformers, the high-frequency-range equivalent circuit with capacitances is usually considered, even though it is not further pursued here.

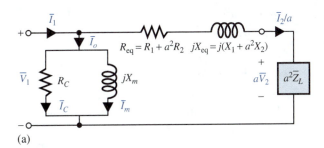

(a)

Figure 11.3.5 Approximate transformer equivalent circuits referred to primary.

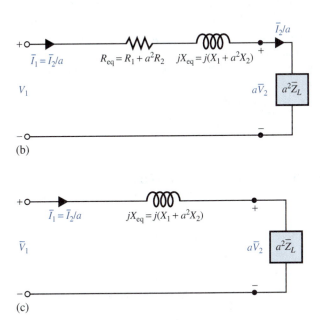

(b)

(c)

11.4 TRANSFORMER PERFORMANCE

The characteristics of most interest to power engineers are *voltage regulation* and *efficiency*. The voltage regulation of a transformer is a measure of the change in the magnitude of the secondary voltage as the current changes from no load to full load while the primary voltage is held fixed. This is defined as

$$\% \text{ voltage regulation} = \frac{V_{2\,NL} - V_{2\,FL}}{V_{2\,FL}} \times 100 \tag{11.4.1a}$$

which may also be expressed as

$$\frac{V_1 - aV_2}{aV_2} \times 100 \tag{11.4.1b}$$

or

$$\frac{(V_1/a) - V_2}{V_2} \times 100 \tag{11.4.1c}$$

The *power efficiency* (generally known as *efficiency*) of a transformer is defined as

$$\eta = \frac{\text{real power output}}{\text{real power input}} \times 100\% \tag{11.4.2a}$$

$$= \frac{\text{real power output}}{\text{real power output} + \text{losses}} \times 100\% \tag{11.4.2b}$$

$$= \frac{\text{real power output}}{\text{real power output} + \text{core loss} + \text{copper } (I^2R) \text{ loss}} \times 100\% \tag{11.4.2c}$$

in which the I^2R loss is obviously load-dependent, whereas the core loss is considered to be constant and independent of the load. It can be shown that the core loss should be equal to the copper loss for maximum operating efficiency at a given load.

The rated core loss can be approximated by the power input in the *open-circuit test* in which one winding is open-circuited and rated voltage at rated frequency is applied to the other winding. The rated copper loss (at full load) can be approximated by the power input in the *short-circuit test*, in which one winding is short-circuited and a reduced voltage at rated frequency is applied to the other winding such that the rated current results.

The *energy efficiency* generally taken over a 24-hour period, known as *all-day efficiency*, is also of interest. It is given by

$$\eta_{AD} = \frac{\text{energy output over 24 hours}}{\text{energy input over 24 hours}} \times 100\% \tag{11.4.3}$$

EXAMPLE 11.4.1

The transformer of Example 11.3.1 is supplying full load (i.e., rated load of 50 kVA) at a rated secondary voltage of 240 V and 0.8 power factor lagging. Neglecting the exciting current of the transformer,

(a) Determine the percent voltage regulation of the transformer.

(b) Sketch the corresponding phasor diagram.

(c) If the transformer is used as a step-down transformer at the load end of a feeder whose impedance is $0.5 + j2.0 \ \Omega$, find the voltage V_S and the power factor at the sending end of the feeder.

Solution

(a) The equivalent circuit of the transformer, referred to the high-voltage (primary) side, neglecting the exciting current of the transformer, is shown below in Figure E11.4.1(a). Note that the voltage at the load terminals referred to the high-voltage side is $240 \times 10 = 2400$ V. Further, the load current corresponding to the rated (full) load condition is $50 \times 10^3 / 2400 = 20.8$ A, referred to the high-voltage side. With a lagging power factor of 0.8,

$$\bar{I}_1 = \bar{I}_2/a = 20.8\angle - \cos^{-1} 0.8 = 20.8\angle - 36.9°$$

Using KVL,

$$\bar{V}_1 = 2400\angle 0° + (20.8\angle - 36.9°)(1.5 + j2.0) = 2450\angle 0.34°$$

If there is no load, the load-terminal voltage will be 2450 V. Therefore, from Equation (11.4.1b),

$$\% \text{ voltage regulation} = \frac{V_1 - aV_2}{aV_2} \times 100 = \frac{2450 - 2400}{2400} \times 100 = 2.08\%$$

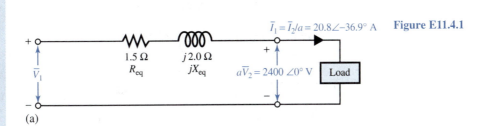

(a)

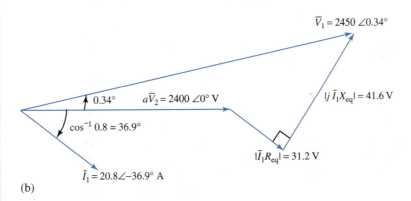

(b)

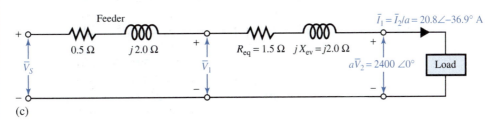

(c)

(b) The corresponding phasor diagram is shown in Figure E11.4.1(b).

(c) The equivalent circuit of the transformer, along with the feeder impedance, referred to the high-voltage side, neglecting the exciting current of the transformer, is shown in Figure E11.4.1(c). The total series impedance is $(0.5 + j2.0) + (1.5 + j2.0) = 2 + j4 \ \Omega$. Using KVL,

$$\bar{V}_S = 2400\angle 0° + (20.8\angle - 36.9°)(2 + j4) = 2483.5\angle 0.96° \ V$$

The voltage at the sending end is then 2483.5 V. The power factor at the sending end is given by $\cos(36.9 + 0.96)° = 0.79$ lagging.

EXAMPLE 11.4.2

Compute the efficiency of the transformer of Example 11.3.1 corresponding to (a) full load, 0.8 power factor lagging, and (b) one-half load, 0.6 power factor lagging, given that the input power P_{oc} in the open-circuit conducted at rated voltage is 173 W and the input power P_{sc} in the short-circuit test conducted at rated current is 650 W.

Solution

(a) Corresponding to full load, 0.8 power factor lagging,

$$\text{Output} = 50,000 \times 0.8 = 40,000 \ W$$

The $I^2 R$ loss (or copper loss) at rated (full) load equals the real power measured in the short-circuit test at *rated current*,

$$\text{Copper loss} = I^2_{HV} R_{eq \ HV} = P_{sc} = 650 \ W$$

where the subscript HV refers to the high-voltage side. The core loss, measured at rated voltage, is

$$\text{Core loss} = P_{oc} = 173 \ W$$

Then

$$\text{Total losses at full load} = 650 + 173 = 823 \ W$$

$$\text{Input} = 40,000 + 823 = 40,823 \ W$$

The full-load efficiency of η at 0.8 power factor is then given by

$$\eta = \frac{\text{output}}{\text{input}} \times 100 = \frac{40,000}{40,823} \times 100 = 98\%$$

(b) Corresponding to one-half rated load, 0.6 power factor lagging,

$$\text{Output} = \frac{1}{2} \times 50,000 \times 0.6 = 15,000 \ W$$

$$\text{Copper loss} = \frac{1}{4} \times 650 = 162.5 \ W$$

Note that the current at one-half rated load is half of the full-load current, and that the copper loss is one-quarter of that at rated current value.

$$\text{Core loss} = P_{oc} = 173 \text{ W}$$

which is considered to be unaffected by the load, as long as the secondary terminal voltage is at its rated value. Then

$$\text{Total losses at one-half rated load} = 162.5 + 173 = 335.5 \text{ W}$$

$$\text{Input} = 15,000 + 335.5 = 15,335.5 \text{ W}$$

The efficiency at one-half rated load and 0.6 power factor is then given by

$$\eta = \frac{15,000}{15,335.5} \times 100 = 97.8\%$$

EXAMPLE 11.4.3

The distribution transformer of Example 11.3.1 is supplying a load at 240 V and 0.8 power factor lagging. The open-circuit and short-circuit test data are given in Example 11.4.2.

(a) Determine the fraction of full load at which the maximum efficiency of the transformer occurs, and compute the efficiency at that load.

(b) The load cycle of the transformer operating at a constant 0.8 lagging power factor is 90% full load for 8 hours, 50% (half) full load for 12 hours, and no load for 4 hours. Compute the all-day energy efficiency of the transformer.

Solution

(a) For maximum efficiency to occur at a certain load, the copper loss at that load should be equal to the core loss. So,

$$k^2 P_{sc} = P_{oc}$$

where k is the fraction of the full-load rating at which the maximum efficiency occurs. Therefore,

$$k = \sqrt{\frac{P_{oc}}{P_{sc}}} = \sqrt{\frac{173}{650}} = 0.516$$

The output power corresponding to this condition is

$$50,000 \times 0.516 \times 0.8 = 20,640 \text{ W}$$

where 0.8 is the power factor given in the problem statement. Also, core loss = copper loss = 173 W. The maximum efficiency is then given by

$$\eta_{max} = \frac{\text{output}}{\text{output} + \text{losses}} = \frac{20,640}{20,640 + 173 + 173} = \frac{20,640}{20,986} = 0.9835, \text{ or } 98.35\%$$

(b) During 24 hours,

$$\text{energy output} = (8 \times 0.9 \times 50 \times 0.8) + (12 \times 0.5 \times 50 \times 0.8) = 528 \text{ kWh}$$

$$\text{core loss} = 24 \times 0.173 = 4.15 \text{ kWh}$$

$$\text{copper loss} = (8 \times 0.9^2 \times 0.65) + (12 \times 0.5^2 \times 0.65) = 6.16 \text{ kWh}$$

The all-day (or energy) efficiency of the transformer is given by

$$\eta_{AD} = \frac{528}{528 + 4.15 + 6.16} = 0.9808, \text{ or } 98.08\%$$

11.5 THREE-PHASE TRANSFORMERS

As we have seen in Chapter 10, three-phase transformers are used quite extensively in power systems between generators and transmission systems, between transmission and subtransmission systems, and between subtransmission and distribution systems. Most commercial and industrial loads require three-phase transformers to transform the three-phase distribution voltage to the ultimate utilization level.

Transformation in three-phase systems can be accomplished in either of two ways: (1) connecting three identical single-phase transformers to form a *three-phase bank* (each one will carry one-third of the total three-phase load under balanced conditions); or (2) a three-phase transformer manufactured for a given rating. A three-phase transformer, compared to a bank of three single-phase transformers, for a given rating will weigh less, cost less, require less floor space, and have somewhat higher efficiency.

The windings of either core-type or shell-type three-phase transformers may be connected in either wye or delta. Four possible combinations of connections for the three-phase, two-winding transformers are Y–Δ, Δ–Y, Δ–Δ, and Y–Y. These are shown in Figure 11.5.1 with the transformers assumed to be ideal. The windings on the left are the primaries, those on the right are the secondaries, and a primary winding of the transformer is linked magnetically with the secondary winding drawn parallel to it. With the per-phase primary-to-secondary turns ratio ($N_1/N_2 = a$), the resultant voltages and currents for balanced applied line-to-line voltages V and line currents I are marked in the figure.

As in the case of three-phase circuits under balanced conditions, only one phase needs to be considered for circuit computations, because the conditions in the other two phases are the same except for the phase displacements associated with a three-phase system. It is usually convenient to carry out the analysis on a per-phase-of-Y (i.e., line-to-neutral) basis, and in such a case the transformer series impedance can then be added in series with the transmission-line series impedance. In dealing with Y–Δ and Δ–Y connections, all quantities can be referred to the Y-connected side. For Δ–Δ connections, it is convenient to replace the Δ-connected series impedances of the transformers with equivalent Y-connected impedances by using the relation

$$Z_{\text{per phase of Y}} = \frac{1}{3} Z_{\text{per phase of }\Delta} \tag{11.5.1}$$

It can be shown that to transfer the ohmic value of impedance from the voltage level on one side of a three-phase transformer to the voltage level on the other side, the multiplying factor is the square of the ratio of line-to-line voltages, regardless of whether the transformer connection is Y–Y, Δ–Y, or Δ–Δ. In some cases, where Y–Y transformation is utilized in particular, it is quite common to incorporate a third winding, known as *tertiary winding,* connected in delta. Such multiwinding transformers with three or more windings are not considered in this text.

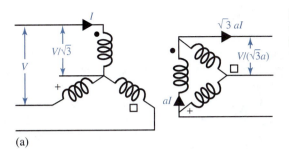

Figure 11.5.1 Wye–delta, delta–wye, delta–delta, and wye–wye transformer connections in three-phase systems.

(a)

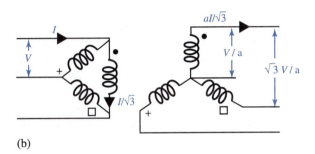

(b)

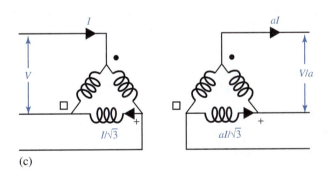

(c)

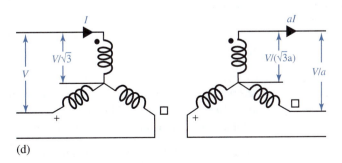

(d)

11.6 AUTOTRANSFORMERS

In contrast to a two-winding transformer, the autotransformer is a single-winding transformer having a tap brought out at an intermediate point. Thus, as shown in Figure 11.6.1, a–c is the single N_1-turn winding wound on a laminated core, and b is the intermediate point where the tap is brought out such that b–c has N_2 turns. The autotransformer may generally be used as either a step-up or a step-down operation. Considering the step-down arrangement, as shown in Figure 11.6.1, let the primary applied voltage be V_1, resulting in a magnetizing current and a core flux ϕ_m. Voltage drops in the windings, exciting current, and small phase-angular differences are usually neglected for the analysis. Then it follows that

$$\frac{V_1}{V_2} = \frac{I_2}{I_1} = \frac{N_1}{N_2} = a \qquad (11.6.1)$$

in which $a > 1$ for step-down, $a < 1$ for step-up transformers.

The input apparent power is $S_1 = V_1 I_1$, while the output apparent power is given by $S_2 = V_2 I_2$. The apparent power transformed by electromagnetic induction (or transformer action) is $S_{ind} = V_2 I_3 = (V_1 - V_2) I_2$. The output transferred by electrical conduction (because of the direct electrical connection between primary and secondary windings) is given by $S_{cond} = V_2 I_2 - V_2 I_3 = V_2 I_1$.

For the same output the autotransformer is smaller in size, weighing much less than a two-winding transformer, and has higher efficiency. An important disadvantage of the autotransformer is the direct copper connection (i.e., no electrical isolation) between the high- and low-voltage sides. A type of autotransformer commonly found in laboratories is the variable-ratio autotransformer, in which the tapped point b (shown in Figure 11.6.1) is movable. It is known as the *variac* (variable ac). Although here, for the sake of simplicity, we have considered only the single-phase autotransformer, three-phase autotransformers, which are available in practice, can be modeled on a per-phase basis and also analyzed just as the single-phase case.

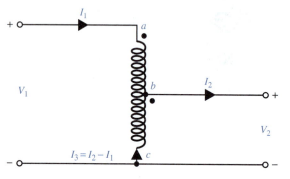

Figure 11.6.1 Single-phase step-down autotransformer.

EXAMPLE 11.6.1

The single-phase, 50-kVA, 2400:240-V, 60-Hz, two-winding distribution transformer of Example 11.3.1 is connected as a step-up autotransformer, as shown in Figure E11.6.1. Assume that the 240-V winding is provided with sufficient insulation to withstand a voltage of 2640 V to ground.

(a) Find V_H, V_X, I_H, I_C, and I_X corresponding to rated (full-load) conditions.

(b) Determine the kVA rating as an autotransformer, and find how much of that is the conducted kVA.

(c) Based on the data given for the two-winding transformer in Example 11.4.2, compute the efficiency of the autotransformer corresponding to full load, 0.8 power factor lagging, and compare it with the efficiency calculated for the two-winding transformer in part (a) of Example 11.4.2.

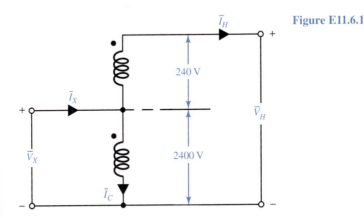

Figure E11.6.1

(a) The two windings are connected in series so that the polarities are additive. Neglecting the leakage-impedance voltage drops,

$$V_H = 2400 + 240 = 2640 \text{ V}$$

$$V_X = 2400 \text{ V}$$

The full-load rated current of the 240-V winding, based on the rating of 50 kVA as a two-winding transformer, is $50{,}000/240 = 208.33$ A. Since the 240-V winding is in series with the high-voltage circuit, the full-load current of this winding is the rated current on the high-voltage side of the autotransformer,

$$I_H = 208.33 \text{ A}$$

Neglecting the exciting current, the mmf produced by the 2400-V winding must be equal and opposite to that of the 240-V winding,

$$I_C = 208.33 \left(\frac{240}{2400} \right) = 20.83 \text{ A}$$

in the direction shown in the figure. Then the current on the low-voltage side of the autotransformer is given by

$$I_X = I_H + I_C = 208.33 + 20.33 = 229.16 \text{ A}$$

(b) The kVA rating as an autotransformer is

$$\frac{V_H I_H}{1000} = \frac{2640 \times 208.33}{1000} = 550 \text{ kVA}$$

or

$$\frac{V_X I_X}{1000} = \frac{2400 \times 229.16}{1000} = 550 \text{ kVA}$$

which is 11 times that of the two-winding transformer. The transformer must boost the current I_H of 208.33 A through a potential rise of 240 V. Thus the kVA transformed by electromagnetic induction is given by

$$\frac{240 \times 208.33}{1000} = 50 \text{ kVA}$$

The remaining 500 kVA is the conducted kVA.

(c) With the currents and voltages shown for the autotransformer connections, the losses at full load will be 823 W, the same as in Example 11.4.2. However, the output as an autotransformer at 0.8 power factor is given by

$$550 \times 1000 \times 0.8 = 440,000 \text{ W}$$

The efficiency of the autotransformer is then calculated as

$$\eta = \frac{440,000}{440,823} = 0.9981, \text{ or } 99.81\%$$

whereas that of the two-winding transformer was calculated as 0.98 in Example 11.4.2. Because the only losses are those due to transforming 50 kVA, higher efficiency results for the autotransformer configuration compared to that of the two-winding transformer.

11.7 LEARNING OBJECTIVES

The *learning objectives* of this chapter are summarized here, so that the student can check whether he or she has accomplished each of the following.

- Analysis and design of simple magnetic circuits.
- Equivalent circuits of transformers.
- Predicting transformer performance from equivalent circuits.
- Basic notions of three-phase transformers.
- Elementary autotransformer calculations and performance.

11.8 PRACTICAL APPLICATION: A CASE STUDY

Magnetic Bearings for Space Technology

The high magnetic energy stored in rare-earth-cobalt permanent magnets allows the design of lightweight motors and magnetic bearings for high-speed rotors. Magnetic bearings are not subject to wear, and with the ability to operate under high vacuum conditions, they appear to be ideal for applications requiring high rotational speeds, such as 100,000 r/min. Important applications are for turbomolecular pumps, laser scanners, centrifuges, momentum rings for satellite stabilizations, and other uses in space technology.

Because of the absence of mechanical contact and lubricating fluids, magnetic bearings can be operated in high vacuum at higher speeds with extremely low friction, low noise, and longer operating life. With no risk of contamination by oil or gas, and with less heat dissipation, it is possible to ascertain clean, stable, and accurate operating conditions with reliability and repeatability. In space technology the magnetic bearings are used successfully in reaction, momentum, and energy wheels, helium pumps, and telescope pointing. Terrestrial applications include scanners, high-vacuum pumps, beam choppers in high vacuum, energy storage wheels, and accurate smooth rotating machines. The magnetic bearings in a reaction wheel are shown in Figure 11.8.1.

The design of magnetic bearings involves the calculation of magnetic forces and stiffness as part of designing an electromechanical servo system. In earlier days, the magnetic force of a suspension block was calculated approximately, with reasonable accuracy, by assuming simple straight flux paths. However, higher magnetic flux densities are increasingly used for reducing the weight and size of magnetic bearings, particularly in the case of a single-axis servoed magnetic bearing which utilizes fringing rings. In such cases the nonlinear characteristics of the ferromagnetic materials become quite significant; analytical techniques fail to yield sufficiently accurate results. Hence it becomes essential to take recourse to numerical analysis of the nonlinear magnetic fields with the aid of a high-speed digital computer in order to determine more accurately the flux distribution corresponding to various conditions of operation, compute leakage, and evaluate forces at the air gap, so as to optimize the design of nonlinear magnetic bearings.

The strength of the electromagnet and/or the permanent magnet can easily be changed to observe their effect on the leakage as well as the flux-density distribution, particularly at the air-gap level and fringing rings. The number of fringing rings and their location may also be easily changed in order to evaluate their effects on the forces at the air-gap level.

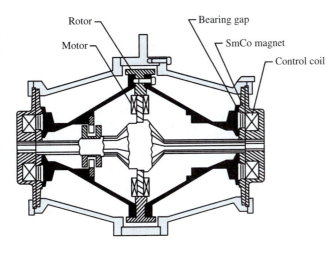

Figure 11.8.1 Magnetic bearings in a reaction wheel.

PROBLEMS

11.1.1 From the magnetic material characteristics shown in Figure 11.1.2, estimate the relative permeability μ_r at a flux density of 1 T for M-19

and AISI 1020 materials.

11.1.2 Determine the units for the area of the hysteresis loop of a ferromagnetic material.

11.1.3 In plotting a hysteresis loop the following scales are used: 1 cm = 400 At/m and 1 cm = 0.3 T. The area of the loop for a certain magnetic material is found to be 6.2 cm². Calculate the hysteresis loss in joules per cycle for the specimen tested if the volume is 400 cm³.

11.1.4 A sample of iron having a volume of 20 cm³ is subjected to a magnetizing force varying sinusoidally at a frequency of 400 Hz. The area of the hysteresis loop is found to be 80 cm² with the flux density plotted in Wb/m² and the magnetizing force in At/m. The scale factors used are 1 cm = 0.03 T and 1 cm = 200 At/m. Find the hysteresis loss in watts.

11.1.5 The flux in a magnetic core is alternating sinusoidally at a frequency of 500 Hz. The maximum flux density is 1 T. The eddy-current loss then amounts to 15 W. Compute the eddy-current loss in this core when the frequency is 750 Hz and the maximum flux density is 0.8 T.

*11.1.6 The total core loss for a specimen of magnetic sheet steel is found to be 1800 W at 60 Hz. When the supply frequency is increased to 90 Hz, while keeping the flux density constant, the total core loss is found to be 3000 W. Determine the hysteresis and eddy-current losses separately at both frequencies.

11.1.7 A magnetic circuit is found to have an ac hysteresis loss of 10 W when the peak current is $I_m = 2$ A. Assuming the exponent of B_m to be 1.5 in Equation (11.1.4), estimate P_h for $I_m = 0.5$ A and 8 A.

11.1.8 Ac measurements with constant voltage amplitude reveal that the total core loss of a certain magnetic circuit is 10 W at $f = 50$ Hz, and 13 W at $f = 60$ Hz. Find the total core loss if the frequency is increased to 400 Hz.

*11.2.1 Consider the magnetic circuit of Figure 11.2.1(a). Let the cross-sectional area A_C of the core, be

16 cm², the average length of the magnetic path in the core l_C be 40 cm, the number of turns N of the excitation coil be 100 turns, and the relative permeability μ_r of the core be 50,000. For a magnetic flux density of 1.5 T in the core, determine:

(a) The flux ϕ

(b) Total flux linkage $\lambda(= N\phi)$.

(c) The current required through the coil.

11.2.2 Now suppose an air gap 0.1 mm long is cut in the right leg of the core of Figure 11.2.1(a), making the magnetic circuit look like that of Figure 11.2.1(b). Neglect leakage and fringing. With the new core configuration, repeat Problem 11.2.1 for the same dimensions and values given in that problem. See what a difference that small air gap can make!

11.2.3 A toroid with a circular cross section is shown in Figure P11.2.3. It is made from cast steel with a relative permeability of 2500. The magnetic flux density in the core is 1.25 T measured at the mean diameter of the toroid.

(a) Find the current that must be supplied to the coil.

(b) Calculate the magnetic flux in the core.

(c) Now suppose a 10-mm air gap is cut across the toroid. Determine the current that must be supplied to the coil to produce the same value of magnetic flux density as in part (a). You may neglect leakage and fringing.

11.2.4 For the magnetic circuit shown in Figure P11.2.4, neglecting leakage and fringing, determine the mmf of the exciting coil required to produce a flux density of 1.6 T in the air gap. The material is M-19. The dimensions are $l_{m1} = 60$ cm, $A_{m1} = 24$ cm², $l_{m2} = 10$ cm, $A_{m2} = 16$ cm², $l_g = 0.1$ cm, and $A_g = 16$ cm².

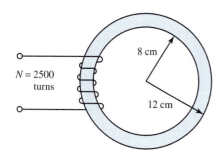

Figure P11.2.3 Toroid with circular cross section.

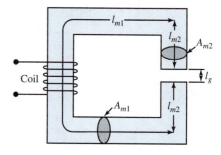

Figure P11.2.4

11.2.5 Consider the toroid shown in Figure P11.2.5(a) made up of three ferromagnetic materials. Material a is nickel–iron alloy having a mean arc length l_a of 0.6 m; material b is medium silicon steel having a mean arc length l_b of 0.4 m; material c is cast steel having a mean arc length l_c of 0.2 m. Each of the materials has a cross-sectional area of 0.002 m^2. Find the mmf needed to establish a magnetic flux of $\phi = 1.2$ mWb by making use of the magnetization curves given in

Figure P11.2.5(b).

11.2.6 The configuration of a magnetic circuit is given in Figure P11.2.6. Assume the permeability of the ferromagnetic material to be $\mu = 1000\mu_0$. Neglect leakage, but correct for fringing by increasing each linear dimension of the cross-sectional area by the length of the air gap. The magnetic material has a square cross-sectional area of 4 cm^2. Find the air-gap flux density, and the magnetic field intensity in the air gap.

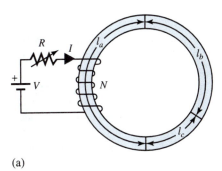

(a)

Figure P11.2.5 **(a)** Toroid. **(b)** Magnetization curves of typical ferromagnetic materials.

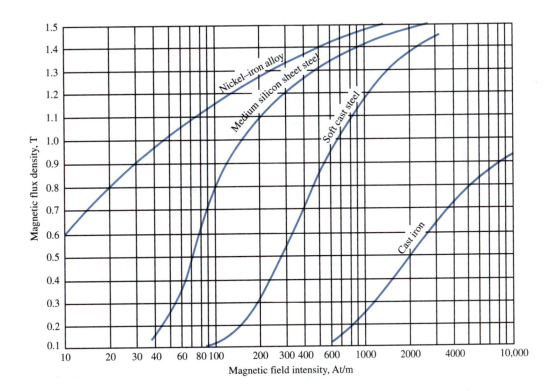

Nickel–iron alloy

Medium silicon sheet steel

Soft cast steel

Cast iron

Magnetic flux density, T

Magnetic field intensity, At/m

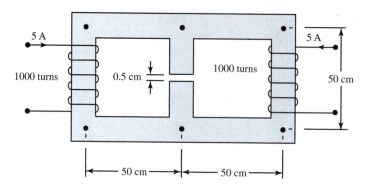

Figure P11.2.6

*11.2.7 Consider the magnetic circuit shown in Figure P11.2.7. Assume the relative permeability of the magnetic material to be 1000 and the cross-sectional area to be the same throughout. Determine the current needed in the coil to produce a flux density of 1 T in the center limb, if the excitation coil has 500 turns.

11.2.8 In the magnetic circuit shown in Figure P11.2.8 the center leg has the same cross-sectional area as each of the outer legs. The coil has 400 turns. The permeability of iron may be assumed to be infinite. If the air-gap flux density in the left leg is 1.2 T, find:

(a) The flux density in the air gap of the right leg.

(b) The flux density in the center leg.

(c) The current needed in the coil.

11.2.9 Figure P11.2.9 shows the cross section of a rectangular iron core with two air gaps g_1 and g_2. The ferromagnetic iron can be assumed to have infinite permeability. The coil has 500 turns.

(a) Gaps g_1 and g_2 are each equal to 0.1 cm, and a current of 1.83 A flows through the

winding. Compute the flux densities in the two air gaps, B_{g1} in the center gap g_1, and B_{g2} in the end gap g_2.

(b) Let gap g_2 now be closed by inserting an iron piece of the correct size and infinite permeability so that only the center gap $g_1 = 0.1$ cm remains. If a flux density of 1.25 T is needed in gap g_1, find the current that is needed in the winding.

11.2.10 Consider the magnetic circuit in Figure P11.2.10, in which all parts have the same cross section. The coil has 200 turns and carries a current of 5 A. The air gaps are $g_1 = 0.4$ cm and $g_2 = 0.5$ cm. Assuming the core has infinite permeability, compute the flux density in tesla in (a) gap g_1, (b) gap g_2, and (c) the left limb.

11.2.11 The magnetic circuit shown in Figure P11.2.11 has an iron core which can be considered to be infinitely permeable. The core dimensions are $A_C = 20$ cm², $g = 2$ mm, and $l_C = 100$ cm. The coil has 500 turns and draws a current of 4 A from the source. Magnetic leakage and fringing may be neglected. Calculate the following:

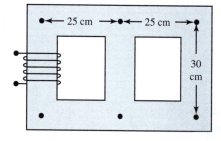

Figure P11.2.7

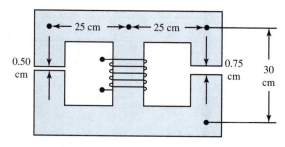

Figure P11.2.8

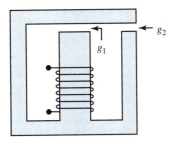

Figure P11.2.9

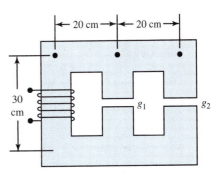

Figure P11.2.10

Figure P11.2.11

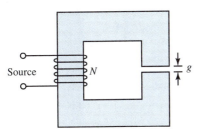

(a) Total magnetic flux.

(b) Flux linkages of the coil.

(c) Coil inductance.

(d) Total stored magnetic energy.

*11.2.12 Repeat Problem 11.2.11 accounting for the core's relative permeability of 2000.

11.2.13 Reconsider Problem 11.2.3. Calculate the coil inductance L and the total stored magnetic energy before and after the air gap is cut.

11.3.1 A 10-kVA, 4800:240-V, 60-Hz, single-phase transformer has an equivalent series impedance of $120 + j300$ Ω referred to the primary high-voltage side. The exciting current of the transformer may be neglected.

(a) Find the equivalent series impedance referred to the secondary low-voltage side.

(b) Calculate the voltage at the primary high-voltage terminals if the secondary supplies rated secondary current at 230 V and unity power factor.

*11.3.2 A 25-kVA, 2300:230-V, 60-Hz, single-phase transformer has the following parameters: resistance of high-voltage winding 1.5 Ω, resistance of low-voltage winding 0.015 Ω, leakage reactance of high-voltage winding 2.4 Ω, and leakage reactance of low-voltage winding 0.024 Ω. Compute the following:

(a) Equivalent series impedance referred to the high-voltage side.

(b) Equivalent series impedance referred to the low-voltage side.

(c) Terminal voltage on the high-voltage side when the transformer is delivering full load at 230 V and 0.866 power factor lagging.

11.3.3 A 100-kVA, 2300:230-V, 60-Hz, single-phase transformer has the following parameters: $R_1 = 0.30$ Ω, $R_2 = 0.003$ Ω, $R_{C1} = 4.5$ kΩ, $X_1 = 0.65$ Ω, $X_2 = 0.0065$ Ω, and $X_{m1} = 1.0$ kΩ, where subscripts 1 and 2 refer to high-voltage and low-voltage sides, respectively. Set up the equivalent T-circuit of the transformer and determine the input current, input voltage, input power, and power factor when the transformer is delivering 75 kW at 230 V and 0.85 lagging power factor.

11.3.4 A 150-kVA, 2400:240-V, 60-Hz, single-phase transformer has the following parameters: $R_1 = 0.2$ Ω, $R_2 = 0.002$ Ω, $X_1 = 0.45$ Ω, $X_2 = 0.0045$ Ω, $R_C = 10$ kΩ, and $X_m = 1.55$ kΩ, where the notation is that of Figure 11.3.4. Form the equivalent T-circuit of the transformer referred to the high-voltage side, and calculate the supply voltage on the high-voltage side when the transformer supplies rated (full) load at 240 V and 0.8 lagging power factor.

11.3.5 For a single-phase, 60-Hz transformer rated 500

kVA, 2400:480 V, following the notation of Figure 11.3.4, the equivalent circuit impedances in ohms are $R_1 = 0.06$, $R_2 = 0.003$, $R_C = 2000$, and $X_1 = 0.3$, $X_2 = 0.012$, $X_m = 500$. The load connected across the low-voltage terminals draws rated current at 0.8 lagging power factor with rated voltage at the terminals.

(a) Calculate the high-voltage winding terminal voltage, current, and power factor.

(b) Determine the transformer series equivalent impedance for the high-voltage and low-voltage sides, neglecting the exciting current of the transformer.

(c) Considering the T-equivalent circuit, find the Thévenin equivalent impedance of the transformer under load as seen from the primary high-voltage terminals.

11.3.6 A 20-kVA, 2200:220-V, 60-Hz, single-phase transformer has these parameters:

Resistance of the 2200-V winding $R_1 = 2.50\ \Omega$
Resistance of the 220-V winding $R_2 = 0.03\ \Omega$
Leakage reactance of the 2200-V winding $X_1 = 0.1\ \Omega$
Leakage reactance of the 220-V winding $X_2 = 0.1\ \Omega$
Magnetizing reactance on the 2200-V side $X_m = 25,000\ \Omega$

(a) Draw the equivalent circuits of the transformer referred to the high-voltage and low-voltage sides. Label impedances numerically in ohms.

(b) The transformer is supplying 15 kVA at 220 V and a lagging power factor of 0.85. Determine the required voltage at the high-voltage terminals of the transformer.

11.4.1 These data were obtained from tests carried out on a 10-kVA, 2300:230-V, 60-Hz distribution transformer:

• Open-circuit test, with low-voltage winding excited: applied voltage 230 V, current 0.45 A, input power 70 W

• Short-circuit test, with high-voltage winding excited: applied voltage 120 V, current 4.35 A, input power 224 W

(a) Compute the efficiency of the transformer when it is delivering full load at 230 V and 0.85 power factor lagging.

(b) Find the efficiency of the transformer when it is delivering 7.5 kVA at 230 V and 0.85 power factor lagging.

(c) Determine the fraction of rating at which the transformer efficiency is a maximum, and calculate the efficiency corresponding to that load if the transformer is delivering the load at 230 V and a power factor of 0.85.

(d) The transformer is operating at a constant load power factor of 0.85 on this load cycle: 0.85 full load for 8 hours, 0.60 full load for 12 hours, and no load for 4 hours. Compute the transformer's all-day (or energy) efficiency.

(e) If the transformer is supplying full load at 230 V and 0.8 lagging power factor, determine the voltage regulation of the transformer. Also, find the power factor at the high-voltage terminals.

*11.4.2 A 3-kVA, 220:110-V, 60-Hz, single-phase transformer yields these test data:

• Open-circuit test: 200 V, 1.4 A, 50 W

• Short-circuit test: 4.5 V, 13.64 A, 30 W

Determine the efficiency when the transformer delivers a load of 2 kVA at 0.85 power factor lagging.

11.4.3 A 75-kVA, 230/115-V, 60-Hz transformer was tested with these results:

• Open-circuit test: 115 V, 16.3 A, 750 W

• Short-circuit test: 9.5 V, 326 A, 1200 W

Determine:

(a) The equivalent impedance in high-voltage terms.

(b) The voltage regulation at rated load, 0.8 power factor lagging.

(c) The efficiency at rated load, 0.8 power factor lagging, and at half-load, unity power factor.

(d) The maximum efficiency and the current at which it occurs.

11.4.4 A 300-kVA transformer has a core loss of 1.5 kW and a full-load copper loss of 4.5 kW.

(a) Calculate its efficiency corresponding to 25, 50, 75, 100, and 125% loads at unity power factor.

(b) Repeat the efficiency calculations for the 25% load at power factors of 0.8 and 0.6.

(c) Determine the fraction of load for which the efficiency is a maximum, and calculate the corresponding efficiencies for the power factors of unity, 0.8, and 0.6.

11.4.5 Consider the solution of Example 11.4.1. By means of a phasor diagram, determine the load power factor for which the regulation is maximum (i.e., the poorest), and find the corresponding regulation.

11.4.6 A 10-kVA, 200:400-V, single-phase transformer gave these test results:

- Open-circuit test (LV winding supplied): 200 V, 3.2 A, 450 W

- Short-circuit test (HV winding supplied): 38 V, 25 A, 600 W

Compute the efficiency when the transformer delivers half its rated kVA at 0.85 power factor lagging.

*11.4.7 Find the percent voltage regulation and the efficiency of the transformer for the following cases:

(a) Problem 11.3.1(b).

(b) Problem 11.3.2(c).

11.4.8 The transformer of Problem 11.3.3 is delivering a full load of 100 kVA at a secondary load voltage of 230 V. Neglect the exciting current of the transformer and determine the voltage regulation if:

(a) The load power factor is 0.8 lagging.

(b) The load power factor is 0.8 leading.

11.4.9 A 25-kVA, 2400/240-V, 60-Hz, single-phase transformer has an equivalent series impedance of $3.45 + j5.75\ \Omega$ referred to the primary high-voltage side. The core loss is 120 W. When the transformer is delivering rated kVA to a load at rated secondary voltage and a 0.85 lagging power factor, find the percent voltage regulation and the efficiency of the transformer.

11.4.10 A 25-kVA, 2200:220-V, 60-Hz, single-phase transformer has an equivalent series impedance of $3.5 + j4.0\ \Omega$ referred to the primary high-voltage side.

(a) Determine the highest value of voltage regulation for full-load output at rated secondary terminal voltage.

(b) At what power factor does it occur?

(c) Sketch the corresponding phasor diagram.

11.4.11 The following data were obtained on a 25-kVA, 2400:240-V, 60-Hz, single-phase distribution transformer:

- Open-circuit test with meters on LV side: 240 V, 3.2 A, 165 W

- Short-circuit test with meters on HV side: 55 V, 10.4 A, 375 W

Compute the worst voltage regulation and the power factor at which it occurs, when the transformer is delivering rated output at rated secondary terminal voltage of 240 V. Sketch the corresponding phasor diagram.

*11.4.12 The efficiency of a 400-kVA, single-phase, 60-Hz transformer is 98.77% when delivering full-load current at 0.8 power factor, and 99.13% with half-rated current at unity power factor. Calculate:

(a) The iron loss.

(b) The full-load copper loss.

(c) The efficiency at 3/4 load, 0.9 power factor.

11.4.13 A transformer has its maximum efficiency of 0.9800 when it delivers 15 kVA at unity power factor. During the day it is loaded as follows:

12 hours	2 kW at power factor 0.5
6 hours	12 kW at power factor 0.8
6 hours	18 kW at power factor 0.9

Determine the all-day efficiency.

11.4.14 A single-phase, 3-kVA, 220:110-V, 60-Hz transformer has a high-voltage winding resistance of $0.3\ \Omega$, a low-voltage winding resistance of $0.06\ \Omega$, a leakage reactance of $0.8\ \Omega$ on its high-voltage side, and a leakage reactance of $0.2\ \Omega$ on its low-voltage side. The core loss at rated voltage is 45 W, and the copper loss at rated load is 100 W. Neglect the exciting current of the transformer. Find the per-unit voltage regulation when the transformer is supplying full load at 110 V and 0.9 lagging power factor.

11.4.15 The transformer of Problem 11.3.4 operates on the following load cycle:

12 hours	full load, 0.8 power factor lagging
4 hours	no load
8 hours	one-half full load, unity power factor

Compute the all-day energy efficiency.

11.4.16 A 75-kVA transformer has an iron loss of 1 kW and a full-load copper loss of 1 kW. If the transformer operates on the following load cycle, determine the all-day efficiency:

8 hours	full load at unity power factor
8 hours	no load
8 hours	one-half full load at unity power factor

*11.4.17 A 10-kVA transformer is known to have an iron loss of 150 W and a full-load copper loss of 250 W. If the transformer has the following load cycle, compute its all-day efficiency:

4 hours	full load at 0.8 power factor
8 hours	75% full load at 1.0 power factor
12 hours	50% full load at 0.6 power factor

11.4.18 Show polarity markings for a single-phase transformer for (a) subtractive polarity, and (b) additive polarity.

11.4.19 Two single-phase transformers, each rated 2400: 120-V, are to be interconnected for (a) 4800:240-V operation, and (b) 2400:120-V operation. Draw circuit diagrams and show polarity markings.

11.4.20 Two 1150:115-V transformers are to be interconnected for (a) 2300:230-V operation, and (b) 1150:230-V operation. Show the interconnections and appropriate polarity markings.

11.5.1 A one-line diagram of a three-phase distribution system is given in Figure P11.5.1. Determine the line-to-line voltage at the sending end of the high-voltage feeder when the transformer delivers rated load at 240 V (line-to-line) and 0.8 lagging power factor. Neglect the exciting current of the transformer.

*11.5.2 A 5-MVA, 66:13.2 kV, three-phase transformer supplies a three-phase resistive load of 4500 kW at 13.2 kV. What is the load resistance in ohms as measured from line to neutral on the high-voltage side of the transformer, if it is:

(a) Connected in Y–Y?

(b) Reconnected in Y–Δ, with the same high-tension voltage supplied and the same load resistors connected?

11.5.3 A three-phase transformer bank consisting of three 10-kVA, 2300:230-V, 60-Hz, single-phase transformers connected in Y–Δ is used to step down the voltage. The loads are connected to the transformers by means of a common three-phase low-voltage feeder whose series impedance is $0.005 + j0.01$ Ω per phase. The transformers themselves are supplied by means of a three-phase high-voltage feeder whose series impedance is $0.5 + j5.0$ Ω per phase. The equivalent series impedance of the single-phase transformer referred to the low-voltage side is $0.12 + j0.24$ Ω. The star point on the primary side of the transformer bank is grounded. The load consists of a heating load of 2 kW per phase and a three-phase induction-motor load of 20 kVA with a lagging power factor of 0.8, supplied at 230 V line-to-line.

(a) Draw a one-line diagram of this three-phase distribution system.

(b) Neglecting the exciting current of the transformer bank, draw the per-phase equivalent circuit of the distribution system.

(c) Determine the line current and the line-to-line voltage at the sending end of the high-voltage feeder.

11.5.4 A single-line diagram of a three-phase transformer bank connected to a load is given in Figure P11.5.4. Find the magnitudes of the line-to-line voltages, line currents, phase voltages, and phase currents on either side of the transformer bank. Determine the primary to secondary ratio of the line-to-line voltages and the line currents.

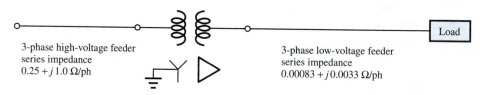

3-phase high-voltage feeder
series impedance
$0.25 + j1.0$ Ω/ph

3-phase low-voltage feeder
series impedance
$0.00083 + j0.0033$ Ω/ph

Load

3 single-phase transformers
each rated 50 kVA, 2400:240 V
identical with that of Example 11.4.1

Figure P11.5.1

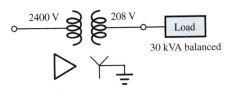

Figure P11.5.4

2400 V   208 V   Load
30 kVA balanced

3 single-phase transformers
each rated 10 kVA, 2400:120 V, 60 Hz

11.5.5 Three single-phase 100-kVA, 2400:240-V, 60-Hz transformers (each of which has an equivalent series impedance of $0.045 + j0.16$ Ω referred to its low-voltage side) are connected to form a three-phase, 4160:240-V transformer bank, which in turn is connected to a three-phase feeder with an impedance of $0.5 + j1.5$ Ω/phase. When the three-phase transformer bank delivers 250 kW at 240 V and 0.866 lagging power factor, determine:

  (a) The transformer winding currents.

  (b) The sending-end voltage (line to line) at the source.

11.5.6 Three single-phase, 10-kVA, 2400/120-V, 60-Hz transformers are connected to form a three-phase, 4160/208-V transformer bank. Each of the single-phase transformers has an equivalent series impedance of $10 + j25$ Ω referred to the high-voltage side. The transformer bank is said to deliver 27 kW at 208 V and 0.9 power factor leading.

  (a) Draw a schematic diagram of transformer connections, and develop a per-phase equivalent circuit.

  (b) Determine the primary current, primary voltage, and power factor.

  (c) Compute the voltage regulation.

*11.5.7 A three-phase, 600-kVA, 2300:230-V, Y–Y transformer bank has an iron loss of 4400 W and a full-load copper loss of 7600 W. Find the efficiency of the transformer for 70% full load at 230 V and 0.85 power factor.

11.5.8 Three identical single-phase transformers are to be connected to form a three-phase bank rated at 300 MVA, 230:34.5 kV. For the following configurations, determine the voltage, current, and kVA ratings of each single-phase transformer: (a) Δ–Δ, (b) Y–Δ, (c) Y–Y, (d) Δ–Y.

11.6.1 A single-phase, 10-kVA, 2300:230-V, 60-Hz, two-winding distribution transformer is connected as an autotransformer to step up the voltage from 2300 V to 2530 V.

  (a) Draw a schematic diagram of the arrangement showing all the voltages and currents while delivering full load.

  (b) Find the permissible kVA rating of the autotransformer if the winding currents are not to exceed those for full-load operation as a two-winding transformer. How much of that is transformed by electromagnetic induction?

  (c) Based on the data given for the two-winding transformer in Problem 11.4.1, compute the efficiency of the autotransformer corresponding to full load and 0.8 lagging power factor. Comment on why the efficiency of the autotransformer is higher than that of the two-winding transformer.

11.6.2 A two-winding, single-phase transformer rated 3 kVA, 220:110 V, 60 Hz is connected as an autotransformer to transform a line input voltage of 330 V to a line output voltage of 110 V and to deliver a load of 2 kW at 0.8 lagging power factor. Draw the schematic diagram of the arrangement, label all the currents and voltages, and calculate all the quantities involved.

11.6.3 A two-winding, 15-kVA, 2300:115-V, 60-Hz, single-phase transformer, which is known to have a core loss of 75 W and a copper loss of 250 W, is connected as an autotransformer to step up 2300 V to 2415 V. With a load of 0.8 power factor lagging, what kVA load can be supplied without exceeding the current rating of any winding? Determine the efficiency at this load.

11.6.4 A single-phase, two-winding, 10-kVA, 440:110-V, 60-Hz transformer is to be connected as an autotransformer to supply a load at 550 V from a 440-V supply. Draw a schematic diagram of the connections and determine: (a) The maximum kVA rating as an autotransformer. (b) The maximum apparent power transferred by conduction. (c) The maximum apparent power transferred by electromagnetic induction.

*11.6.5 A 15-kVA, 2200:220-V, two-winding, single-phase transformer is connected as an autotransformer to step up voltage from 220 V to 2420 V. Without exceeding the rated current of any winding, determine the kVA rating of the autotransformer, the kVA transformed by transformer action, and the kVA conducted.

11.6.6 A 5-kVA, 480:120-V, two-winding, 60-Hz, single-phase transformer has an efficiency of 95% while delivering rated load at rated voltage and 0.8 power factor lagging. This transformer is to be connected as an autotransformer to step down a 600-V source to 480 V.

(a) Draw a schematic connection diagram as an autotransformer.

(b) Find the maximum kVA rating as an autotransformer.

(c) Compute the efficiency as an autotransformer delivering its rated kVA at 480 V and 0.8 power factor lagging.

12 Electromechanics

12.1 BASIC PRINCIPLES OF ELECTROMECHANICAL ENERGY CONVERSION

Energy available in many forms is often converted to and from electrical form because electric energy can be transmitted and controlled simply, reliably, and efficiently. Among the energy-conversion devices, electromechanical energy converters are the most important. Electromechanical energy conversion involves the interchange of energy between an electric system and a mechanical system, while using magnetic field as a means of conversion.

Devices that convert control signals from one form to another are known as *transducers*, most of which have an output signal in the form of electric energy. For example, a potentiometer is used to convert a mechanical position to an electric voltage, a tachometer generator converts a velocity into a voltage (dc or ac), and a pressure transducer indicates a pressure drop (or rise) in terms of a corresponding drop (or rise) in electric potential. Thus, a transducer translates the command signal appropriately into an electrical form usable by the system, and forms an important part of control systems. Electromechanical transducers form a link between electric and mechanical systems.

When the energy is converted from electrical to mechanical form, the device is displaying *motor action*. A *generator action* involves converting mechanical energy into electric energy. The electromechanical energy-conversion process can be expressed as

Electric energy $\quad\quad\quad$ $\overset{\text{Motor}}{\rightarrow}$ $\quad\quad\quad$ Mechanical energy
$\underset{\text{Generator}}{\leftarrow}$

Electromechanical energy converters, simply known as electric machines, embody three essential features: (1) an electric system, (2) a mechanical system, and (3) a coupling field. Figure 12.1.1 is a schematic representation of an ideal electric machine (or a lossless electromechanical device) for which the following relations hold:

Electric input (or output) energy $vi\,\Delta t$ = mechanical output (or input) energy $T\omega_m\Delta t$

or

Electric input (or output) power vi = mechanical output (or input) power $T\omega_m$ $\quad$ (12.1.1)

where v and i are the voltage and the current associated with the electrical port, and T and ω_m are the torque and the angular rotational velocity associated with the mechanical port.

The principle of *conservation of energy* may be stated as follows: Energy can be neither created nor destroyed, even though, within an isolated system, energy may be converted from one form to another form, and transferred from an energy source to an energy sink. The total energy in the system is constant.

A practical electromechanical system with losses can be represented by adding on the lossy portion of the electric system and the lossy portion of the mechanical system modeled externally. Excluding all types of dissipation and losses makes the energy-conversion part to be *lossless* or *conservative* with a coupling field.

Both electric and magnetic fields store energy, from which useful mechanical forces and torques can be derived. With a normal working electric field intensity of about 3×10^6 V/m, the stored electric-energy density is on the order of

$$\frac{1}{2}\varepsilon_0 E^2 = \frac{1}{2}\frac{10^{-9}}{36\pi}(3 \times 10^6)^2 \cong 40 \text{ J/m}^3$$

where ε_0 is the permittivity of free space, given by $10^{-6}/36\pi$ or 8.854×10^{-12} F/m, and E is the electric field intensity. This corresponds to a force density of 40 N/m². The stored magnetic energy density in air, on the other hand, with a normal working magnetic flux density of about 1.6 T, comes to

$$\frac{1}{2}\frac{B^2}{\mu_0} = \frac{1}{2}\frac{1.6^2}{4\pi \times 10^{-7}} \cong 1 \times 10^6 \text{ J/m}^3$$

where μ_0 is the permeability of free space, and B is the magnetic flux density. As this is nearly 25,000 times as much as for the electric field, almost all industrial electric machines are magnetic

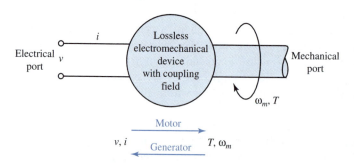

Figure 12.1.1 Schematic representation of a lossless electromechanical device.

in principle and are magnetic-field devices.

Three basic principles associated with all electromagnetic devices are (1) *induction,* (2) *interaction,* and (3) *alignment.*

Induction

The essentials for producing an emf by magnetic means are electric and magnetic circuits, mutually interlinked. Figure 12.1.2(a) shows a load (or sink or motor) convention with the induced emf (or back emf) e directed in opposition to the positive current, in which case Faraday's law of induction,

$$e = +\frac{d\lambda}{dt} = +N\frac{d\phi}{dt} \qquad (12.1.2a)$$

applies, where λ is the flux linkages, N is the number of turns, and ϕ is the flux. On the other hand, the induced emf e (or generated emf) acting in the direction of positive current, as shown in Figure 12.1.2(b) with a source (or generator) convention, satisfies Faraday's law of induction,

$$e = -\frac{d\lambda}{dt} = -N\frac{d\phi}{dt} \qquad (12.1.2b)$$

The change in flux linkage in a coil may occur in one of the following three ways:

1. The coil remaining stationary with respect to the flux, the flux varies in magnitude with time. Since no motion is involved, no energy conversion takes place. Equation (12.1.2a) gives the *transformer emf* (or the *pulsational emf*) as in the case of a transformer (see Chapter 11), in which a time-varying flux linking a stationary coil yields a time-varying voltage.

2. The flux remaining constant, the coil moves through it. A conductor or a coil moving through a magnetic field will have an induced voltage, known as the *motional emf* (or *speed emf*), given by

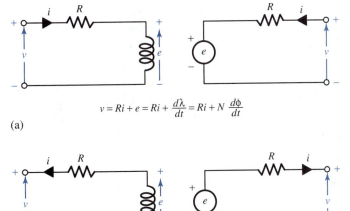

$$v = Ri + e = Ri + \frac{d\lambda}{dt} = Ri + N\frac{d\phi}{dt}$$

(a)

$$v = e - iR = -\frac{d\lambda}{dt} - iR = -N\frac{d\phi}{dt} - iR$$

(b)

Figure 12.1.2 Circuit conventions. **(a)** Load (or sink or motor) convention. Note that the power *absorbed* by the circuit is positive when v and i are positive, and when current flows in the direction of voltage drop. **(b)** Source (or generator) convention. Note that the power *delivered* by this circuit to the external circuit is positive when v and i are positive, and when current flows out of the positive terminal.

$$\text{Motional emf } e = BlU \qquad (12.1.3)$$

which is often called the *cutting-of-flux* equation, where B is the flux density of a non-time-varying, uniform magnetic field, l is the length of the conductor, U is the velocity of the conductor, and $\bar{B}$, $\bar{l}$, and $\bar{U}$ are mutually perpendicular in their directions. If the motion is rotary in nature, it is also known as *rotational voltage*. The direction for the motional emf can be worked out from the *right-hand rule:* if the thumb, first, and second fingers of the right hand are extended so that they are mutually perpendicular to each other, and if the thumb represents the direction of $\bar{U}$ and the first finger the direction of $\bar{B}$, the second finger will then represent the direction of the emf along $\bar{l}$. This is depicted in Figure 12.1.3(a).

The generation of motional emf is further illustrated by a simple example, as shown in Figure 12.1.3(b), where a single-turn coil formed by the moving (or sliding) conductor (moving with velocity U), the two conducting rails, and the voltmeter are situated in a magnetic field of flux density B. The conductor moving with a velocity U, in a direction at right angles to both B and l, sweeps the area lU in 1 second. The flux per unit time in this area is BlU, which is also the flux linkage per unit time with the single-turn coil. Thus, the induced emf e is simply given by BlU. The motional emf (or speed emf) is always associated with the conversion of energy between the mechanical and electrical forms.

3. The coil may move through a time-varying flux; that is to say, both changes (1) and (2) may occur together. Usually one of the two phenomena is so predominant in a given device that the other may be neglected for the purposes of analysis.

Because magnetic poles occur in pairs (north and south) and the movement of a conductor through a natural north–south sequence induces an emf that changes direction in accordance with the magnetic polarity (i.e., an alternating emf), the devices are inherently ac machines.

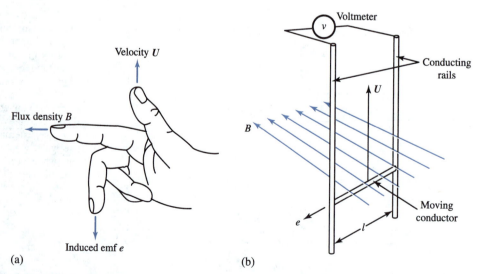

Figure 12.1.3 Generation of motional emf. **(a)** Right-hand rule. **(b)** Simple example.

Interaction

Current-carrying conductors, when placed in magnetic fields, experience mechanical force. Considering only the effect of the magnetic field, the *Lorentz force equation* gives the force F as

$$F = BlI \qquad (12.1.4)$$

when a current-carrying conductor of length l is located in a uniform magnetic field of flux density B, and the direction of the current in the conductor is perpendicular to the direction of the magnetic field. The direction of the force is orthogonal (perpendicular) to the directions of both the current-carrying conductor and the magnetic field. Equation (12.1.4) is often used in electric machine analysis.

The principle of interaction is illustrated in Figure 12.1.4, in which $\bar{B}$ is the flux density, $\bar{I}$ the current, and $\bar{F}$ the force. Shown in Figure 12.1.4(a) is the flux density $\bar{B}$ of an undisturbed uniform field, on which an additional field is imposed due to the introduction of a current-carrying conductor. For the case in which the current is directed into and perpendicular to the plane of the paper, the resultant flux distribution is depicted in Figure 12.1.4(b). It can be seen that in the neighborhood of the conductor the resultant flux density is greater than B on one side and less than B on the other side. The direction of the mechanical force developed is such that it tends to restore the field to its original undisturbed and uniform configuration. Figure 12.1.4(c) shows the conditions corresponding to the current being in the opposite direction to that of Figure 12.1.4(b). The force is always in such a direction that the energy stored in the magnetic field is minimized. Figure 12.1.5 shows a one-turn coil in a magnetic field and illustrates how torque is produced by forces caused by the interaction between current-carrying conductors and magnetic fields.

Alignment

Pieces of highly permeable material, such as iron, situated in ambient medium of low permeability, such as air, in which a magnetic field is established, experience mechanical forces that tend to align them with the field direction in such a way that the reluctance of the system is minimized (or the inductance of the system is maximized). Figure 12.1.6 illustrates this principle of alignment

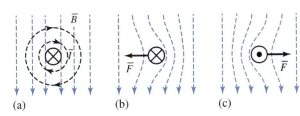

(a)   (b)   (c)

Figure 12.1.4 Principle of interaction.

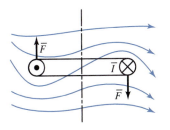

Figure 12.1.5 Torque produced by forces caused by interaction of current-carrying conductors and magnetic fields.

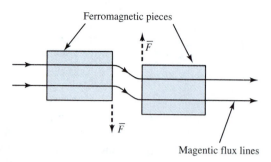

Ferromagnetic pieces

Magentic flux lines

Figure 12.1.6 Principle of alignment.

and shows the direction of forces. The force is always in a direction that reduces the net magnetic reluctance and shortens the magnetic flux path.

A mechanical force is exerted on ferromagnetic material, tending to align it with or bring it into the position of the densest part of the magnetic field. This force is the familiar attraction of a magnet for pieces of iron in its field. In magnetic circuits, for example, definite forces are exerted on the iron at the air–iron boundary. Energy changes associated with a differential displacement of the iron cause the mechanical force. This force is the essential operating mechanism of many electromagnetic devices, such as lifting magnets, magnetic clutches, chucks, brakes, switches (known as contactors), and relays. Solenoid-operated (solenoid being another name for the operating coil) valves are common elements in piping systems. Actuators used in control systems operate due to the mechanical force or torque converted from the electric, pneumatic, or hydraulic inputs.

In motor and generator action, the magnetic fields tend to line up, pole to pole. When their complete alignment is prevented by the need to furnish torque to a mechanical shaft load, motor action results when electric to mechanical energy conversion takes place. On the other hand, when the alignment is prevented by the application of a mechanical torque to the rotor from a source of mechanical energy, generator action results when mechanical to electric energy conversion takes place.

EXAMPLE 12.1.1

A magnetic crane used for lifting weights can be modeled and analyzed as a simple magnetic circuit, as shown in Figure E12.1.1. Its configuration consists of two distinct pieces of the same magnetic material with two air gaps. Obtain an expression for the total pulling force on the bar in terms of the flux density B in the air-gap region and the cross-sectional area A perpendicular to the plane of paper, while making reasonable approximations.

Solution

Reasonable approximations include infinite permeability of the magnetic material, as well as neglecting leakage and fringing. The magnetic energy stored in an incremental volume of the air-gap region is given by

$$dW_m = \frac{1}{2} B \left(\frac{B}{\mu_0} \right) dv$$

where B and $H \, (= B/\mu_0)$ are in the same direction. Working with one pole of the magnetic circuit, in view of the symmetry, the incremental change in volume is $dv = A \, dg$. Hence,

$$dW_m = \frac{1}{2}\left(\frac{B^2}{\mu_0}\right) A\, dg$$

The definition of work gives us

$$dW = F\, dg$$

where F is the pulling force *per pole* on the bar. While a magnetic pull is exerted upon the bar, an energy dW equal to the magnetic energy dW_m stored in the magnetic field is expended. Thus,

$$dW_m = dW$$

or

$$\frac{1}{2}\left(\frac{B^2}{\mu_0}\right) A\, dg = F\, dg$$

which yields the pulling force per pole on the bar as

$$F = \frac{1}{2}\left(\frac{B^2}{\mu_0}\right) A$$

The total pulling on the bar is then given by

$$F_{\text{total}} = 2F = \left(\frac{B^2}{\mu_0}\right) A$$

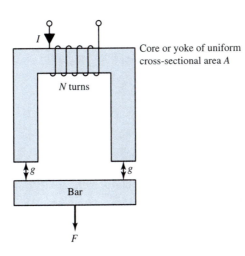

Figure E12.1.1 Simple magnetic circuit for weight lifting.

Core or yoke of uniform cross-sectional area A

N turns

g

g

Bar

F

I

EXAMPLE 12.1.2

Consider the arrangement shown in Figure E12.1.2. A conductor bar of length l is free to move along a pair of conducting rails. The bar is driven by an external force at a constant velocity of U m/s. A constant uniform magnetic field $\bar{B}$ is present, pointing into the paper of the book page. Neglect the resistance of the bar and rails, as well as the friction between the bar and the rails.

(a) Determine the expression for the motional voltage across terminals 1 and 2. Is terminal 1 positive with respect to terminal 2?

(b) If an electric load resistance R is connected across the terminals, what are the current and power dissipated in the load resistance? Show the direction of current on the figure.

(c) Find the magnetic-field force exerted on the moving bar, and the mechanical power required to move the bar. How is the principle of energy conservation satisfied?

(d) Since the moving bar is not accelerating, the net force on the bar must be equal to zero. How can you justify this?

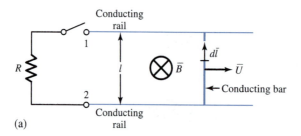

Figure E12.1.2

(a)

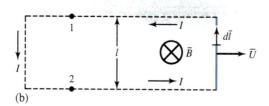

(b)

(a) U, B, and l being perpendicular to each other, as per the right-hand rule, the motional voltage is $e = BlU$. 1 is positive with respect to 2, since the resulting current (when the switch is closed) produces a flux opposing the original B, thereby satisfying Lenz's law.

(b) $I = BlU/R$; $P = I^2 R = (BlU)^2/R$.

(c) The magnitude of the induced magnetic-field force exerted on the moving bar is BlI, and it opposes the direction of motion. The mechanical force is equal and opposite to the induced field force. Hence the mechanical power required to move the bar is

$$(BlI)U = Bl\frac{BlU}{R}U = \frac{(BlU)^2}{R}$$

which is the same as the electric power dissipated in the resistor. Energy (from the mechanical source) that is put in to move the conductor bar is expended (or transferred) as heat in the resistor, thereby satisfying the principle of energy conservation.

(d) In order to move the conductor bar at a constant velocity, it is necessary to impress a mechanical force equal and opposite to the induced field force. Hence the net force on the bar is equal to zero.

EXAMPLE 12.1.3

A loudspeaker is a common electrochemical transducer in which vibration is caused by changes in the input current to a coil which, in turn, is coupled to a magnetic structure that can produce time-varying forces on the loudspeaker diaphragm. Figure E12.1.3(a) shows the schematic diagram of a loudspeaker, Figure E12.1.3(b) is a simplified model, Figure E12.1.3(c) is a free-body diagram of the forces acting on the loudspeaker diaphragm, and Figure E12.1.3(d) is the electrical model. The force exerted on the coil is also exerted on the mass of the loudspeaker diaphragm.

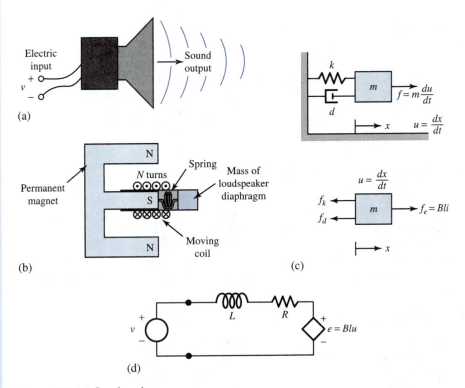

Figure E12.1.3 Loudspeaker.

Develop the equation of motion for the electrical and mechanical sides of the device and determine the frequency response $U(j\omega)/V(j\omega)$ of the loudspeaker using phase analysis, and neglecting the coil inductance.

Solution

The electrical side is described by

$$v = L\frac{di}{dt} + Ri + e = L\frac{di}{dt} + Ri + Blu$$

where B is the flux density and l is the length of the coil given by $2\pi Nr$, in which N is the number of coil turns and r the coil radius. Note that e is the emf generated by the motion of the coil in the magnetic field.

The mechanical side is described by

$$f = m\frac{du}{dt} = f_e - f_d - f_k = f_e - du - kx = Bli - du - kx$$

where d represents the damping coefficient, k represents the spring constant, and f_e is the magnetic force due to current flow in the coil. Using phasor techniques, we have

$$V(j\omega) = j\omega LI(j\omega) + RI(j\omega) + BlU(j\omega)$$

and

$$(j\omega m + d)U(j\omega) + \frac{k}{j\omega}U(j\omega) = BlI(j\omega)$$

Neglecting the coil inductance L, we get

$$I(j\omega) = \frac{V(j\omega) - BlU(j\omega)}{R}$$

The frequency response of the loudspeaker is then given by

$$\frac{U(j\omega)}{V(j\omega)} = \frac{Bl}{Rm}\frac{j\omega}{(j\omega)^2 + j\omega(d/m + B^2l^2/Rm) + k/m}$$

12.2 emf PRODUCED BY WINDINGS

The time variation of emf for a single conductor corresponds to the spatial variation of air-gap flux density. By suitable winding design, the harmonics can be reduced appreciably, and the waveform of the generated emf can be made to approach a pure sine shape.

Figure 12.2.1 shows an elementary single-phase, two-pole synchronous machine. In almost all cases, the armature winding of a synchronous machine is on the stator and the field winding is on the rotor, because it is constructionally advantageous to have the low-power field winding on the rotating member. The field winding is excited by direct current, which is supplied by a dc source connected to carbon brushes bearing on slip rings (or collector rings). The armature windings, though distributed in the slots around the inner periphery of the stator in an actual machine, are shown in Figure 12.2.1(a) for simplicity as consisting of a single coil of N turns, indicated in cross section by the two sides a and $-a$ placed in diametrically opposite narrow slots. The conductors forming these coil sides are placed in slots parallel to the machine shaft and connected in series by means of the end connections. The coil in Figure 12.2.1(a) spans 180° (or a complete *pole pitch*, which is the peripheral distance from the centerline of a north pole to the centerline of an adjacent south pole) and is hence known as a *full-pitch* coil. For simplicity and convenience, Figure 12.2.1(a) shows only a two-pole synchronous machine with salient-pole construction; the flux paths are shown by dashed lines. Figure 12.2.1(b) illustrates a nonsalient-pole or cylindrical-rotor construction. The stator winding details are not shown and the flux paths are indicated by dashed lines.

The space distribution of the radial air-gap flux density around the air-gap periphery can be made to approximate a sinusoidal distribution by properly shaping the rotor pole faces facing the air gap,

$$B = B_m \cos \beta \tag{12.2.1}$$

where B_m is the peak value at the rotor pole center and β is measured in electrical radians from the rotor pole axis (or the magnetic axis of the rotor), as shown in Figure 12.2.1. The air-gap flux per pole is the integral of the flux density over the pole area. For a two-pole machine,

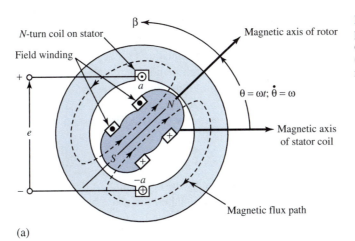

N-turn coil on stator

Field winding

$+$

a

e

$\theta = \omega t; \ \dot{\theta} = \omega$

$-a$

Magnetic axis of rotor

Magnetic axis
of stator coil

Magnetic flux path

(a)

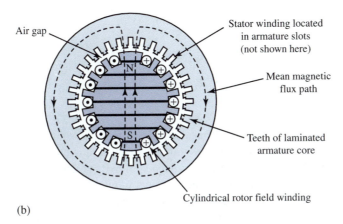

Air gap

Stator winding located
in armature slots
(not shown here)

Mean magnetic
flux path

Teeth of laminated
armature core

Cylindrical rotor field winding

(b)

Figure 12.2.1 Elementary single-phase two-pole synchronous machine. **(a)** Salient-pole machine. **(b)** Nonsalient-pole or cylindrical-rotor machine.

$$\phi = \int_{-\pi/2}^{+\pi/2} B_m \cos \beta \ lr \ d\beta = 2B_m lr \qquad (12.2.2)$$

and for a P-pole machine,

$$\phi = \frac{2}{P} 2B_m lr \qquad (12.2.3)$$

where l is the axial length of the stator and r is the average radius at the air gap. For a P-pole machine, the pole area is $2/P$ times that of a two-pole machine of the same length and diameter.

The flux linkage with the stator coil is $N\phi$ when the rotor poles are in line with the magnetic axis of the stator coil. If the rotor is turned at a constant speed ω by a source of mechanical power connected to its shaft, the flux linkage with the stator coil varies as the cosine of the angle between the magnetic axes of the stator coil and the rotor,

$$\lambda = N\phi \cos \omega t \qquad (12.2.4)$$

where time t is arbitrarily taken as zero when the peak of the flux-density wave coincides with the magnetic axis of the stator coil. By Faraday's law, the voltage induced in the stator coil is given by

$$e = -\frac{d\lambda}{dt} = \omega N\phi \ \sin \ \omega t - N\frac{d\phi}{dt} \ \cos \ \omega t \qquad (12.2.5)$$

The minus sign associated with Faraday's law in Equation (12.2.5) implies generator reference directions, as explained earlier. Considering the right-hand side of Equation (12.2.5), the first term is a speed voltage caused by the relative motion of the field and the stator coil. The second term is a transformer voltage, which is negligible in most rotating machines under normal steady-state operation because the amplitude of the air-gap flux wave is fairly constant. The induced voltage is then given by the speed voltage itself,

$$e = \omega N\phi \ \sin \ \omega t \qquad (12.2.6)$$

Equation (12.2.6) may alternatively be obtained by the application of the cutting-of-flux concept given by Equation (12.1.3), from which the motional emf is given by the product of B_{coil} times the total active length of the conductors l_{eff} in the two coil sides times the linear velocity of the conductor relative to the field, provided that these three are mutually perpendicular. For the case under consideration, then,

$$e = B_{\text{coil}}l_{\text{eff}}v = (B_m \ \sin \ \omega t)(2lN)(r\omega_m)$$

or

$$e = (B_m \ \sin \ \omega t)(2lN)\left(\frac{r2\omega}{P}\right) = \omega N\frac{2}{P}2B_mlr \ \sin \ \omega t \qquad (12.2.7)$$

which is the same as Equation (12.2.6) when the expression for ϕ, given by Equation (12.2.3), is substituted.

The resulting coil voltage is thus a time function having the same sinusoidal waveform as the spatial distribution B. The coil voltage passes through a complete cycle for each revolution of the two-pole machine of Figure 12.2.1. So its frequency in hertz is the same as the speed of the rotor in revolutions per second (r/s); that is, the electrical frequency is synchronized with the mechanical speed of rotation. Thus, a two-pole synchronous machine, under normal steady-state conditions of operation, revolves at 60 r/s, or 3600 r/min, in order to produce 60-Hz voltage. For a P-pole machine in general, however, the coil voltage passes through a complete cycle every time a pair of poles sweeps, i.e., $P/2$ times in each revolution. The frequency of the voltage wave is then given by

$$f = \frac{P}{2} \cdot \frac{n}{60} \ \text{Hz} \qquad (12.2.8)$$

where n is the mechanical speed of rotation in r/min. The *synchronous speed* in terms of the frequency and the number of poles is given by

$$n = \frac{120f}{P} \ \text{r/min} \qquad (12.2.9)$$

The radian frequency ω of the voltage wave in terms of ω_m, the mechanical speed in radians per second (rad/s), is given by

$$\omega = \frac{P}{2} \ \omega_m \qquad (12.2.10)$$

Figure 12.2.2 shows an elementary single-phase synchronous machine with four salient poles; the flux paths are shown by dashed lines. Two complete wavelengths (or cycles) exist in the flux distribution around the periphery, since the field coils are connected so as to form poles of alternate north and south polarities. The armature winding now consists of two coils ($a_1, -a_1$)

and $(a_2, -a_2)$, connected in series by their end connections. The span of each coil is one-half wavelength of flux, or 180 electrical degrees for the full-pitch coil. Since the generated voltage now goes through two complete cycles per revolution of the rotor, the frequency is then twice the speed in revolutions per second, consistent with Equation (12.2.8).

The field winding may be concentrated around the salient poles, as shown in Figures 12.2.1(a) and 12.2.2, or distributed in slots around the cylindrical rotor, as in Figure 12.2.1(b). By properly shaping the pole faces in the former case, and by appropriately distributing the field winding in the latter, an approximately sinusoidal field is produced in the air gap.

A salient-pole rotor construction is best suited mechanically for hydroelectric generators because hydroelectric turbines operate at relatively low speeds, and a relatively large number of poles is required in order to produce the desired frequency (60 Hz in the United States), in accordance with Equation (12.2.9). Salient-pole construction is also employed for most synchronous motors.

The nonsalient-pole (smooth or cylindrical) rotor construction is preferred for high-speed turbine-driven alternators (known also as turbo alternators or turbine generators), which are usually of two or four poles driven by steam turbines or gas turbines. The rotors for such machines may be made either from a single steel forging or from several forgings shrunk together on the shaft.

Going back to Equation (12.2.6), the maximum value of the induced voltage is

$$E_{max} = \omega N \phi = 2\pi f \, N\phi \tag{12.2.11}$$

and the rms value is

$$E_{rms} = \frac{2\pi}{\sqrt{2}} f \, N\phi = 4.44 \, f \, N\phi \tag{12.2.12}$$

which are identical in form to the corresponding emf equations for a transformer. The effect of a time-varying flux in association with stationary transformer windings is the same as that of the relative motion of a coil and a constant-amplitude spatial flux-density wave in a rotating machine. The space distribution of flux density is transformed into a time variation of voltage because of the time element introduced by mechanical rotation. The induced voltage is a single-phase voltage for single-phase synchronous machines of the nature discussed so far. As pointed out earlier, to avoid the pulsating torque, the designer could employ polyphase windings and polyphase sources to develop constant power under balanced conditions of operation.

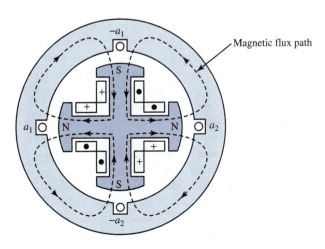

Figure 12.2.2 Elementary single-phase, four-pole synchronous machine.

In fact, with very few exceptions, three-phase synchronous machines are most commonly used for power generation. In general, three-phase ac power systems, including power generation, transmission, and usage, have grown most popular because of their economic advantages. An elementary three-phase, two-pole synchronous machine with one coil per phase (chosen for simplicity) is shown in Figure 12.2.3(a). The coils are displaced by 120 electrical degrees from each other in space so that the three-phase voltages of positive phase sequence a–b–c, displaced by 120 electrical degrees from each other in time, could be produced. Figure 12.2.3(b) shows an elementary three-phase, four-pole synchronous machine with one slot per pole per phase. It has 12 coil sides or six coils in all. Two coils belong to each phase, which may be connected in series in either wye or delta, as shown in Figures 12.2.3(c) and (d). Equation (12.2.12) can be applied to give the rms voltage per phase when N is treated as the total series turns per phase. The coils may also be connected in parallel to increase the current rating of the machine. In actual ac machine windings, instead of concentrated full-pitch windings, distributed fractional-pitch armature windings are commonly used to make better use of iron and copper and to make waveforms of the generated voltage (in time) and the armature mmf (in space) as nearly sinusoidal as possible. That is to say, the armature coils of each phase are distributed in a number of slots, and the coil span may be shorter than a full pitch. In such cases, Equation (12.2.12) is modified to be

$$E_{\text{rms}} = 4.44 k_W \ f \ N_{\text{ph}} \phi \ \text{V/phase} \tag{12.2.13}$$

where k_W is a winding factor (less than unity, usually about 0.85 to 0.95), and N_{ph} is the number of series turns per phase.

Special mechanical arrangements must be provided when making electrical connections to the rotating member. Such connections are usually made through carbon brushes bearing on either a *slip ring* or a *commutator,* mounted on—but insulated from—the rotor shaft and rotating with the rotor. A slip ring is a continuous ring, usually made of brass, to which only one electrical connection is made. For example, two slip rings are used to supply direct current to the field winding on the rotor of a synchronous machine. A commutator, on the other hand, is a mechanical switch consisting of a cylinder formed of hard-drawn copper segments separated and insulated from each other by mica.

In the conventional *dc machine* (with a closed continuous commutator winding on its armature), for example, full-wave rectification of the alternating voltage induced in individual armature coils is achieved by means of a commutator, which makes a unidirectional voltage available to the external circuit through the stationary carbon brushes held against the commutator surface. The armature windings of dc machines are located on the rotor because of this necessity for commutation and are of the closed continuous type, known as *lap* and *wave* windings. The simplex lap winding has as many parallel paths as there are poles, whereas the simplex wave winding always has two parallel paths. The winding connected to the commutator, called the *commutator winding,* can be viewed as a pseudostationary winding because it produces a stationary flux when carrying a direct current, as a stationary winding would. The direction of the flux axis is determined by the position of the brushes. In a conventional dc machine, in fact, the flux axis corresponds to the brush axis (the line joining the two brushes). The brushes are located so that commutation (i.e., reversal of current in the commutated coil) occurs when the coil sides are in the neutral zone, midway between the field poles. The axis of the armature mmf is then in the quadrature axis, whereas the stator mmf acts in the field (or direct) axis. Figure 12.2.4 shows schematic representations of a dc machine. The commutator is thus a device for changing the connections between a rotating closed winding and an external circuit at the instants when the individual coil-generated voltages reverse. In a dc machine, then, this arrangement enables a constant and unidirectional output voltage. The armature mmf axis is fixed in space because of the switching

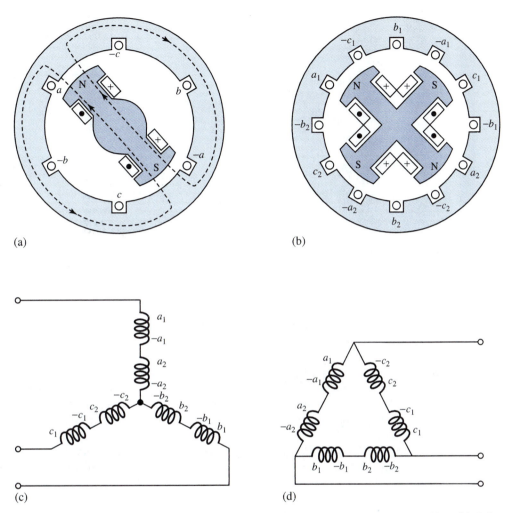

Figure 12.2.3 Elementary three-phase synchronous machines. **(a)** Salient two-pole machine. **(b)** Salient four-pole machine. **(c)** Phase windings connected in wye. **(d)** Phase windings connected in delta.

action of the commutator (even though the closed armature winding on the rotor is rotating), so the commutator winding becomes pseudostationary.

The action of slip rings and that of a commutator differ in only one way. The conducting coil connected to the slip ring is always connected to the brush, regardless of the mechanical speed ω_m of the rotor and the rotor position, but with the commutator, the conducting coil conducts current only when it is physically under the commutator brush, i.e., when it is stationary with respect to the commutator brush. This difference is illustrated in Figure 12.2.5.

A dc machine then operates with direct current applied to its field winding (generally located on the salient-pole stator of the machine) and to a commutator (via the brushes) connected to the armature winding situated inside slots on the cylindrical rotor, as shown schematically in Figure 12.2.4. In a dc machine the stator mmf axis is fixed in space, and the rotor mmf axis is also fixed in space, even when the rotor winding is physically rotating, because of the commutator action, which was briefly discussed earlier. Thus, the dc machine will operate under steady-state conditions,

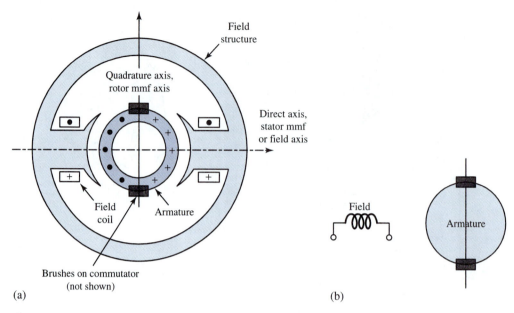

Field
structure

Quadrature axis,
rotor mmf axis

Direct axis,
stator mmf
or field axis

Field
coil   Armature

Field

Armature

Brushes on commutator
(not shown)

(a)   (b)

Figure 12.2.4 Schematic representations of a dc machine. **(a)** Schematic arrangement. **(b)** Circuit representation.

whatever the rotor speed ω_m. The armature current in the armature winding is alternating. The action of the commutator is to change the armature current from a frequency governed by the mechanical speed of rotation to zero frequency at the commutator brushes connected to the external circuit.

For the case of a dc machine with a flux per pole of ϕ, the total flux cut by one conductor in one revolution is given by ϕP, where P is the number of poles of the machine. If the speed of rotation is n r/min, the emf generated in a single conductor is given by

$$e = \frac{\phi P n}{60} \qquad (12.2.14)$$

For an armature with Z conductors and α parallel paths, the total generated armature emf E_a is given by

$$E_a = \frac{P \phi n Z}{60 \alpha} \qquad (12.2.15)$$

Since the angular velocity ω_m is given by $2\pi n/60$, Equation (12.2.15) becomes

$$E_a = \frac{P Z}{2\pi \alpha} \phi \omega_m = K_a \phi \omega_m \qquad (12.2.16)$$

where K_a is the design constant given by $PZ/2\pi\alpha$. The value obtained is the speed voltage appearing across the brush terminals in the quadrature axis (see Figure 12.2.4) due to the field excitation producing ϕ in the direct axis. For this reason, in the schematic circuit representation of a dc machine, the field axis and the brush axis are shown in quadrature, i.e., perpendicular to each other. The generated voltage, as observed from the brushes, is the sum of the rectified voltages of all the coils in series between brushes. If the number of coils is sufficiently large, the ripple in the waveform of the armature voltage (as a function of time) becomes very small, thereby making the voltage direct or constant in magnitude.

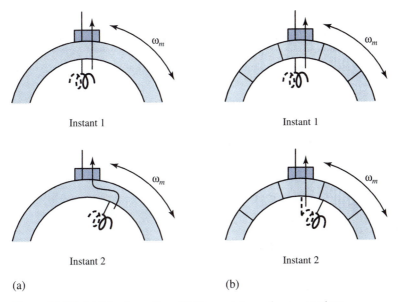

Instant 1   Instant 1

Instant 2   Instant 2

(a)   (b)

Figure 12.2.5 (a) Slip-ring action. **(b)** Commutator action connections.

The instantaneous electric power associated with the speed voltage should be equal to the instantaneous mechanical power associated with the electromagnetic torque T_e, the direction of power flow being determined by whether the machine is operating as a motor or a generator,

$$T_e \omega_m = E_a I_a \qquad (12.2.17)$$

where I_a is the armature current (dc). With the aid of Equation (12.2.15), it follows that

$$T_e = K_a \phi I_a \qquad (12.2.18)$$

which is created by the interaction of the magnetic fields of stator and rotor. If the machine is acting as a generator, this torque opposes rotation; if the machine is acting as a motor, the electromagnetic torque acts in the direction of the rotation.

In a conventional dc machine, the brush axis is fixed relative to the stator. If, however, there is continuous relative motion between poles and brushes, the voltage at the brushes will in fact be alternating. This principle is made use of in ac commutator machines.

EXAMPLE 12.2.1

A two-pole, three-phase, 60-Hz, wye-connected, round-rotor synchronous generator has $N_a = 12$ turns per phase in each armature phase winding and flux per pole of 0.8 Wb. Find the rms induced voltage in each phase and the terminal line-to-line rms voltage.

Solution

$$E_{\max} = 2\pi f N_a \phi = 2\pi \times 60 \times 12 \times 0.8 = 3619 \text{ V}$$
$$E_{\mathrm{rms}} = 3619/\sqrt{2} = 2559.5 \text{ V}$$
$$V_T = 2559.5\sqrt{3} = 4433 \text{ V line-to-line rms}$$

EXAMPLE 12.2.2

The armature of a four-pole dc machine has a simplex lap wound commutator winding (which has the number of parallel paths equal to the number of poles) with 120 two-turn coils. If the flux per pole is 0.02 Wb, calculate the dc voltage appearing across the brushes located on the quadrature axis when the machine is running at 1800 r/min.

Solution

This example can be solved by the direct application of Equation (12.2.15),

$$E_a = \frac{P\phi n Z}{\alpha \times 60}$$

For our example, $P = 4$, $\phi = 0.02$, $n = 1800$, the number of conductors $Z = 120 \times 2 \times 2 = 480$ and the number of parallel paths $\alpha = 4$ for the simplex lap winding (the same as the number of poles). Therefore,

$$E_a = \frac{4 \times 0.02 \times 1800 \times 480}{4 \times 60} = 288 \text{ V dc}$$

12.3 ROTATING MAGNETIC FIELDS

When a machine has more than two poles, only a single pair of poles needs to be considered because the electric, magnetic, and mechanical conditions associated with every other pole pair are repetitions of those for the pole pair under consideration. The angle subtended by one pair of poles in a P-pole machine (or one cycle of flux distribution) is defined to be 360 *electrical degrees*, or 2π *electrical radians*. So the relationship between the mechanical angle Θ_m and the angle Θ in electrical units is given by

$$\theta = \frac{P}{2} \theta_m \tag{12.3.1}$$

because one complete revolution has $P/2$ complete wavelengths (or cycles). In view of this relationship, for a two-pole machine, electrical degrees (or radians) will be the same as mechanical degrees (or radians).

In this section we set out to show that a rotating field of constant amplitude and sinusoidal space distribution of mmf around a periphery of the stator is produced by a three-phase winding located on the stator and excited by balanced three-phase currents when the respective phase windings are wound $2\pi/3$ electrical radians (or 120 electrical degrees) apart in space. Let us consider the two-pole, three-phase winding arrangement on the stator shown in Figure 12.3.1. The windings of the individual phases are displaced by 120 electrical degrees from each other in space around the air-gap periphery. The reference directions are given for positive phase currents. The concentrated full-pitch coils, shown here for simplicity and convenience, do in fact represent the actual distributed windings producing sinusoidal mmf waves centered on the magnetic axes of the respective phases. Thus, these three sinusoidal mmf waves are displaced by 120 electrical degrees from each other in space. Let a balanced three-phase excitation be applied with phase sequence a–b–c,

$$i_a = I \cos \omega_s t; \quad i_b = I \cos(\omega_s t - 120°); \quad i_c = I \cos(\omega_s t - 240°) \quad (12.3.2)$$

where I is the maximum value of the current, and the time $t = 0$ is chosen arbitrarily when the a-phase current is a positive maximum. Each phase current is an ac wave varying in magnitude sinusoidally with time. Hence, the corresponding component mmf waves vary sinusoidally with time. The sum of these components yields the resultant mmf.

Analytically, the resultant mmf at any point at an angle θ from the axis of phase a is given by

$$F(\theta) = F_a \cos \theta + F_b \cos(\theta - 120°) + F_c \cos(\theta - 240°) \tag{12.3.3}$$

But the mmf amplitudes vary with time according to the current variations,

$$F_a = F_m \cos \omega_s t; \qquad F_b = F_m \cos(\omega_s t - 120°); \qquad F_c = F_m \cos(\omega_s t - 240°) \tag{12.3.4}$$

Then, on substitution, it follows that

$$F(\theta, t) = F_m \cos \theta \cos \omega_s t + F_m \cos(\theta - 120°) \cos(\omega_s t - 120°)$$
$$+ F_m \cos(\theta - 240°) \cos(\omega_s t - 240°) \tag{12.3.5}$$

By the use of the trigonometric identity

$$\cos \alpha \cos \beta = \frac{1}{2} \cos(\alpha - \beta) + \frac{1}{2} \cos(\alpha + \beta)$$

and noting that the sum of three equal sinusoids displaced in phase by 120° is equal to zero, Equation (12.3.5) can be simplified as

$$F(\theta, t) = \frac{3}{2} F_m \cos(\theta - \omega_s t) \tag{12.3.6}$$

which is the expression for the resultant mmf wave. It has a constant amplitude $3/2 \, F_m$, is a sinusoidal function of the angle θ, and rotates in synchronism with the supply frequency; hence it is called a *rotating field*. The constant amplitude is $3/2$ times the maximum contribution F_m of any one phase. The angular velocity of the wave is $\omega_s = 2\pi f_s$ electrical radians per second, where f_s is the frequency of the electric supply in hertz. For a P-pole machine, the rotational speed is given by

$$\omega_m = \frac{2}{P} \omega_s \text{ rad/s} \qquad \text{or} \qquad n = \frac{120 f_s}{P} \text{r/min} \tag{12.3.7}$$

which is the synchronous speed.

The same result may be obtained graphically, as shown in Figure 12.3.2, which shows the spatial distribution of the mmf of each phase and that of the resultant mmf (given by the algebraic sum of the three components at any given instant of time). Figure 12.3.2(a) applies

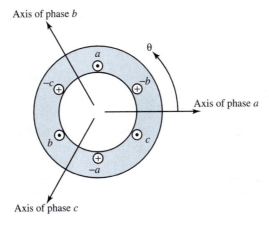

Axis of phase b

Axis of phase a

Axis of phase c

Figure 12.3.1 Simple two-pole, three-phase winding arrangement on a stator.

for that instant when the *a*-phase current is a positive maximum; Figure 12.3.2(*b*) refers to that instant when the *b*-phase current is a positive maximum; the intervening time corresponds to 120 electrical degrees. It can be seen from Figure 12.3.2 that during this time interval, the resultant sinusoidal mmf waveform has traveled (or rotated through) 120 electrical degrees of the periphery of the stator structure carrying the three-phase winding. That is to say, the resultant mmf is rotating in synchronism with time variations in current, with its peak amplitude remaining constant at $^3/_2$ times that of the maximum phase value. Note that the peak value of the resultant stator mmf wave coincides with the axis of a particular phase winding when that phase winding carries its peak current. The graphical process can be continued for different instants of time to show that the resultant mmf is in fact rotating in synchronism with the supply frequency.

Although the analysis here is carried out only for a three-phase case, it holds good for any *q*-phase ($q > 1$; i.e., polyphase) winding excited by balanced *q*-phase currents when the respective phases are wound $2\pi/q$ electrical radians apart in space. However, in a balanced two-phase case, note that the two phase windings are displaced 90 electrical degrees in space, and the phase currents in the two windings are phase-displaced by 90 electrical degrees in time. The constant amplitude of the resultant rotating mmf can be shown to be $q/2$ times the maximum contribution of any one phase. Neglecting the reluctance of the magnetic circuit, the corresponding flux density in the air gap of the machine is then given by

$$B_g = \frac{\mu_0 F}{g} \qquad (12.3.8)$$

where *g* is the length of the air gap.

Production of Rotating Fields from Single-Phase Windings

In this subsection we show that a single-phase winding carrying alternating current produces a *stationary pulsating flux* that can be represented by two counterrotating fluxes of constant and equal magnitude.

Let us consider a single-phase winding, as shown in Figure 12.3.3(a), carrying alternating current $i = I \cos \omega t$. This winding will produce a flux-density distribution whose axis is fixed along the axis of the winding and that pulsates sinusoidally in magnitude. The flux density along the coil axis is proportional to the current and is given by $B_m \cos \omega t$, where B_m is the peak flux density along the coil axis.

Let the winding be on the stator of a rotating machine with uniform air gap, and let the flux density be distributed sinusoidally around the air gap. Then the instantaneous flux density at any position θ from the coil axis can be expressed as

$$B(\theta) = (B_m \cos \omega t) \cos \theta \qquad (12.3.9)$$

which may be rewritten by the use of the trigonometric identity following Equation (12.3.5) as

$$B(\theta) = \frac{B_m}{2} \cos(\theta - \omega t) + \frac{B_m}{2} \cos(\theta + \omega t) \qquad (12.3.10)$$

The sinusoidal flux-density distribution given by Equation (12.3.9) can be represented by a vector $B_m \cos \omega t$ of pulsating magnitude on the axis of the coil, as shown in Figure 12.3.3(a). Alternatively, as suggested by Equation (12.3.10), this stationary pulsating flux-density vector can be represented by two counterrotating vectors of constant magnitude $B_m/2$, as shown in Figure 12.3.3(b). While Equation (12.3.9) represents a standing space wave varying sinusoidally with time, Equation (12.3.10) represents the two rotating components of constant and equal magnitude,

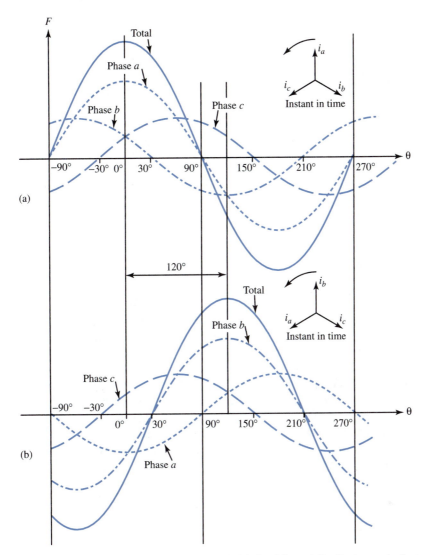

Figure 12.3.2 Generation of a rotating mmf. **(a)** Spatial mmf distribution at the instant in time when the a-phase current is a maximum. **(b)** Spatial mmf distribution at the instant in time when the b-phase current is a maximum.

rotating in opposite directions at the same angular velocity given by $d\theta/dt = \omega$. The vertical components of the two rotating vectors in Figure 12.3.3(b) always cancel, and the horizontal components always yield a sum equal to $B_m \cos \omega t$, the instantaneous value of the pulsating vector.

This principle is often used in the analysis of single-phase machines. The two rotating fluxes are considered separately, as if each represented the rotating flux of a polyphase machine, and the effects are then superimposed. If the system is linear, the principle of superposition holds and yields correct results. In a nonlinear system with saturation, however, one must be careful in reaching conclusions since the results are not as obvious as in a linear system.

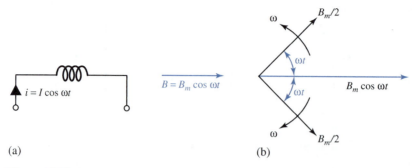

(a)   (b)

Figure 12.3.3 Single-phase winding carrying alternating current, producing a stationary pulsating flux or equivalent rotating flux components.

12.4 FORCES AND TORQUES IN MAGNETIC-FIELD SYSTEMS

We mentioned earlier that the greater ease of storing energy in magnetic fields largely accounts for the common use of electromagnetic devices for electromechanical energy conversion. In a magnetic circuit containing an air-gap region, the energy stored in the air-gap space is several times greater than that stored in the iron portion, even though the volume of the air gap is only a small fraction of that of the iron.

The energy-conversion process involves an interchange between electric and mechanical energy via the stored energy in the magnetic field. This stored energy, which can be determined for any configuration of the system, is a *state function* defined solely by functional relationships between variables and the final values of these variables. Thus, the *energy method* is a powerful tool for determining the coupling forces of electromechanics.

In a *singly excited system,* a change in flux density from a value of zero initial flux density to B requires an energy input to the field occupying a given volume,

$$W_m = \text{vol} \int_0^{B_1} H \, dB \tag{12.4.1}$$

which can also be expressed by

$$W_m = \int_0^{\lambda_1} i(\lambda) \, d\lambda = \int_0^{\phi_1} \mathscr{F}(\phi) \, d\phi \tag{12.4.2}$$

Note that the current i is a function of the flux linkages λ and that the mmf $\mathscr{F}$ is a function of the flux ϕ; their relations depend on the geometry of the coil, the magnetic circuit, and the magnetic

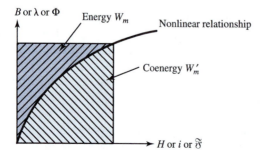

Figure 12.4.1 Graphical interpretation of energy and coenergy in a singly excited nonlinear system.

properties of the core material. Equations (12.4.1) and (12.4.2) may be interpreted graphically as the area labeled *energy* in Figure 12.4.1. The other area, labeled *coenergy* in the figure, can be expressed as

$$W'_m = \text{vol} \int_0^{H_1} B \, dH = \int_0^{i_1} \lambda(i) \, di = \int_0^{\mathcal{F}_1} \phi(\mathcal{F}) \, d\mathcal{F} \tag{12.4.3}$$

For a linear system in which B and H, λ and i, or ϕ and $\mathcal{F}$ are proportional, it is easy to see that the energy and the coenergy are numerically equal. For a nonlinear system, on the other hand, the energy and the coenergy differ, as shown in Figure 12.4.1, but the sum of the energy and the coenergy for a singly excited system is given by

$$W_m + W'_m = \text{vol} \cdot B_1 H_1 = \lambda_1 i_1 = \phi_1 \mathcal{F}_1 \tag{12.4.4}$$

The energy stored in a singly excited system can be expressed in terms of self-inductance, and that stored in a *doubly excited system* in terms of self and mutual inductances, for the circuit-analysis approach, as we pointed out earlier.

Let us now consider a model of an ideal (lossless) electromechanical energy converter that is doubly excited, as shown in Figure 12.4.2, with two sets of electrical terminal pairs and one mechanical terminal, schematically representing a motor. Note that all types of losses have been excluded to form a *conservative* energy-conversion device that can be described by state functions to yield the electromechanical coupling terms in electromechanics. A property of a conservative system is that its energy is a function of its state only, and is described by the same independent variables that describe the state. State functions at a given instant of time depend solely on the state of the system at that instant and not on past history; they are independent of how the system is brought to that particular state.

We shall now obtain an expression for the electromagnetic torque T_e from the principle of conservation of energy, which, for the case of a *sink* of electric energy (such as an electric motor), may be expressed as

$$W_e = W + W_m$$

or in differential form as

$$dW_e = dW + dW_m \tag{12.4.5}$$

where W_e stands for electric energy input from electrical sources, W_m represents the energy stored in the magnetic field of the two coils associated with the two electrical inputs, and W denotes the mechanical energy output. W_e and W may further be written in their differential forms,

$$dW_e = v_1 i_1 \, dt + v_2 i_2 \, dt \tag{12.4.6}$$

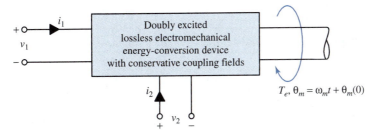

Figure 12.4.2 Model of an ideal, doubly excited electromechanical energy converter.

$$dW = T_e \, d\theta_m \qquad (12.4.7)$$

where θ_m is expressed in electrical radians. Neglecting the winding resistances, the terminal voltages are equal to the induced voltages given by

$$v_1 = e_1 = \frac{d\lambda_1}{dt} \qquad (12.4.8)$$

$$v_2 = e_2 = \frac{d\lambda_2}{dt} \qquad (12.4.9)$$

These are the volt-ampere equations, or the equations of motion, for the electrical side. Two volt-ampere equations result because of the two sets of electrical terminals. Substituting these into Equation (12.4.6), one gets

$$dW_e = i_1 \, d\lambda_1 + i_2 \, d\lambda_2 \qquad (12.4.10)$$

Substituting Equations (12.4.7) and (12.4.10) into the differential form of Equation (12.4.5), we have

$$dW_m = dW_e - dW = i_1 \, d\lambda_1 + i_2 \, d\lambda_2 - T_e \, d\theta_m \qquad (12.4.11)$$

By specifying one independent variable for each of the terminal pairs, i.e., two electrical variables (flux linkages or currents) and one mechanical variable θ_m for rotary motion, we shall attempt to express T_e in terms of the energy or the coenergy of the system.

Based on Equation (12.4.4) and the concepts of energy and coenergy, one has

$$W_m + W'_m = \lambda_1 i_1 + \lambda_2 i_2 \qquad (12.4.12)$$

Expressing Equation (12.4.12) in differential form, the expression for the differential coenergy function dW'_m is obtained as

$$dW'_m = d(\lambda_1 i_1) + d(\lambda_2 i_2) - dW_m$$
$$= \lambda_1 \, di_1 + i_1 \, d\lambda_1 + \lambda_2 \, di_2 + i_2 \, d\lambda_2 - dW_m \qquad (12.4.13)$$

Substituting Equation (12.4.11) into Equation (12.4.13) and simplifying, one gets

$$dW'_m = \lambda_1 \, di_1 + \lambda_2 \, di_2 + T_e \, d\theta_m \qquad (12.4.14)$$

Expressing the coenergy as a function of the independent variables i_1, i_2, and θ_m,

$$W'_m = W'_m(i_1, i_2, \theta_m) \qquad (12.4.15)$$

the total differential of the coenergy function can be written as

$$dW'_m = \frac{\partial W'_m}{\partial i_1} di_1 + \frac{\partial W'_m}{\partial i_2} di_2 + \frac{\partial W'_m}{\partial \theta_m} d\theta_m \qquad (12.4.16)$$

On comparing Equations (12.4.14) and (12.4.16) term by term, the expression for the electromagnetic torque T_e is obtained as

$$T_e = \frac{\partial W'_m(i_1, i_2, \theta_m)}{\partial \theta_m} \qquad (12.4.17)$$

On the other hand, choosing the independent variables λ_1, λ_2, and θ_m, it can be shown that

$$T_e = -\frac{\partial W_m(\lambda_1, \lambda_2, \theta_m)}{\partial \theta_m} \qquad (12.4.18)$$

Depending on the convenience in a given situation, either Equation (12.4.17) or (12.4.18) can be used. For the case of a translational electromechanical system consisting of only one-dimensional motion, say, in the direction of the coordinate x, the torque T_e and the angular displacement $d\theta_m$ are to be replaced by the force F_e and the linear displacement dx, respectively. Thus,

$$F_e = \frac{\partial W'_m(i_1, i_2, x)}{\partial x} \tag{12.4.19}$$

and

$$F_e = -\frac{\partial W_m(\lambda_1, \lambda_2, x)}{\partial x} \tag{12.4.20}$$

For a linear magnetic system, however, the magnetic energy and the coenergy are always equal in magnitude. Thus,

$$W_m = W'_m = \frac{1}{2}\lambda_1 i_1 + \frac{1}{2}\lambda_2 i_2 \tag{12.4.21}$$

In linear electromagnetic systems, the relationships between flux linkage and currents (in a doubly excited system) are given by

$$\lambda_1 = L_{11}i_1 + L_{12}i_2 \tag{12.4.22}$$

$$\lambda_2 = L_{21}i_1 + L_{22}i_2 \tag{12.4.23}$$

where L_{11} is the self-inductance of winding 1, L_{22} is the self-inductance of winding 2, and $L_{12} = L_{21} = M$ is the mutual inductance between windings 1 and 2. All of these inductances are generally functions of the angle θ_m (mechanical or spatial variable) between the magnetic axes of windings 1 and 2. Neglecting the iron-circuit reluctances, the electromagnetic torque can be found from either the energy or the coenergy stored in the magnetic field of the air-gap region by applying either Equation (12.4.17) or (12.4.18). For a linear system, the energy or the coenergy stored in a pair of mutually coupled inductors is given by

$$W'_m(i_1, i_2, \theta_m) = \frac{1}{2}L_{11}i_1^2 + L_{12}i_1 i_2 + \frac{1}{2}L_{22}i_2^2 \tag{12.4.24}$$

The instantaneous electromagnetic torque is then given by

$$T_e = \frac{i_1^2}{2}\frac{dL_{11}}{d\theta_m} + i_1 i_2 \frac{dL_{12}}{d\theta_m} + \frac{i_2^2}{2}\frac{dL_{22}}{d\theta_m} \tag{12.4.25}$$

The first and third terms on the right-hand side of Equation (12.4.25), involving the angular rate of change of self-inductance, are the *reluctance-torque* terms; the middle term, involving the angular rate of change of mutual inductance, is the *excitation torque* caused by the interaction of fields produced by the stator and rotor currents in an electric machine. It is this mutual inductance torque (or excitation torque) that is most commonly exploited in practical rotating machines. Multiply excited systems with more than two sets of electrical terminals can be handled in a manner similar to that for two pairs by assigning additional independent variables to the terminals.

If none of the inductances is a function of the mechanical variable θ_m, no electromagnetic torque is developed. If, on the other hand, the self-inductances L_{11} and L_{22} are independent of the angle θ_m, the reluctance torque is zero and the torque is produced only by the mutual term $L_{12}(\theta_m)$, as seen from Equation (12.4.25). Let us consider such a case in the following example on the basis of the *coupled-circuit viewpoint* (or *coupled-coils approach*).

EXAMPLE 12.4.1

Consider an elementary two-pole rotating machine with a uniform (or smooth) air gap, as shown in Figure E12.4.1, in which the cylindrical rotor is mounted within the stator consisting of a hollow cylinder coaxial with the rotor. The stator and rotor windings are distributed over a number of slots so that their mmf can be approximated by space sinusoids. As a consequence of a construction of this type, we can fairly assume that the self-inductances L_{ss} and L_{rr} are constant, but the mutual inductance L_{sr} is given by

$$L_{sr} = L \cos \theta$$

where θ is the angle between the magnetic axes of the stator and rotor windings. Let the currents in the two windings be given by

$$i_s = I_s \cos \omega_s t \quad \text{and} \quad i_r = I_r \cos(\omega_r t + \alpha)$$

and let the rotor rotate at an angular velocity

$$\omega_m = \dot{\theta} \text{ rad/s}$$

such that the position of the rotor at any instant is given by

$$\theta = \omega_m t + \theta_0$$

Assume that the reluctances of the stator and rotor-iron circuits are negligible, and that the stator and rotor are concentric cylinders neglecting the effect of slot openings.

(a) Derive an expression for the instantaneous electromagnetic torque developed by the machine.

(b) Find the condition necessary for the development of an average torque in the machine.

(c) Obtain the expression for the average torque corresponding to the following cases, where ω_s and ω_r are different angular frequencies:

(1) $\omega_s = \omega_r = \omega_m = 0; \alpha = 0$

(2) $\omega_s = \omega_r; \omega_m = 0$

(3) $\omega_r = 0; \omega_s = \omega_m; \alpha = 0$

(4) $\omega_m = \omega_s - \omega_r$

Solution

(a) Equations (12.4.22) through (12.4.25) apply. With constant L_{ss} and L_{rr}, and the variation of L_{sr} as a function of θ substituted, Equation (12.4.25) simplifies to

$$T_e = i_s i_r \frac{dL_{sr}}{d\theta} = -i_s i_r L \sin \theta$$

Note: For a P-pole machine this expression would be modified as $-(P/2)i_s i_r L \sin [(P/2)\theta_m]$.

For the given current variations, the instantaneous electromagnetic torque developed by the machine is given by

$$T_e = -L I_s I_r \cos \omega_s t \cos(\omega_r t + \alpha) \sin(\omega_m t + \theta_0)$$

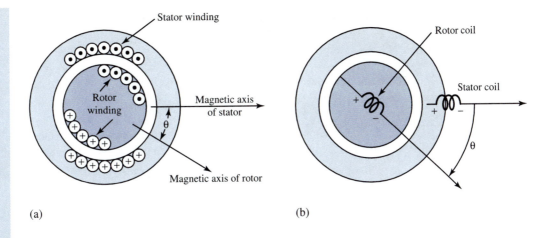

(a)   (b)

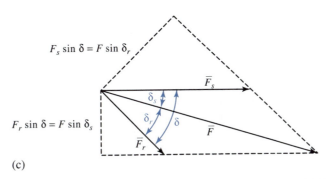

(c)

Figure E12.4.1 Elementary two-pole rotating machine with uniform air gap. (a) Winding distribution. (b) Schematic representation. (c) Vector diagram of mmf waves.

Using trigonometric identities, the product of the three trigonometric terms in this equation may be expressed to yield

$$T_e = \frac{-LI_sI_r}{4} \; [\sin\{[\omega_m + (\omega_s + \omega_r)]\,t + \alpha + \theta_0\}$$
$$+ \sin\{[\omega_m - (\omega_s + \omega_r)]\,t - \alpha + \theta_0\}$$
$$+ \sin\{[\omega_m + (\omega_s - \omega_r)]\,t - \alpha + \theta_0\}$$
$$+ \sin\{[\omega_m - (\omega_s - \omega_r)]\,t + \alpha + \theta_0\}]$$

(b) The average value of each of the sinusoidal terms in the previous equation is zero, unless the coefficient of t is zero in that term. That is, the average torque $(T_e)_{av}$ developed by the machine is zero unless

$$\omega_m = \pm(\omega_s \pm \omega_r)$$

which may also be expressed as

$$|\omega_m| = |\omega_s \pm \omega_r|$$

(c) (1) The excitations are direct currents I_s and I_r. For the given conditions of $\omega_s = \omega_r = \omega_m = 0$ and $\alpha = 0$,

$$T_e = -LI_sI_r \sin \theta_0$$

which is a constant. Hence,

$$(T_e)_{av} = -LI_sI_r \sin \theta_0$$

The machine operates as a *dc rotary actuator*, developing a constant torque against any displacement θ_0 produced by an external torque applied to the rotor shaft.

(2) With $\omega_s = \omega_r$, both excitations are alternating currents of the same frequency. For the conditions $\omega_s = \omega_r$ and $\omega_m = 0$,

$$T_e = -\frac{LI_sI_r}{4} [\sin(2\omega_st + \alpha + \theta_0) + \sin(-2\omega_st - \alpha + \theta_0) + \sin(-\alpha + \theta_0)$$
$$+ \sin(\alpha + \theta_0)]$$

The machine operates as an *ac rotary actuator*, and the developed torque is fluctuating. The average value of the torque is

$$(T_e)_{av} = -\frac{LI_sI_r}{2} \sin \theta_0 \cos \alpha$$

Note that α becomes zero if the two windings are connected in series, in which case $\cos \alpha$ becomes unity.

(3) With $\omega_r = 0$, the rotor excitation is a direct current I_r. For the conditions $\omega_r = 0$, $\omega_s = \omega_m$, and $\alpha = 0$,

$$T_e = -\frac{LI_sI_r}{4} [\sin(2\omega_st + \theta_0) + \sin \theta_0 + \sin(2\omega_st + \theta_0) + \sin \theta_0]$$

or

$$T_e = -\frac{LI_sI_r}{2} [\sin(2\omega_st + \theta_0) + \sin \theta_0)]$$

The device operates as an idealized *single-phase synchronous machine*, and the instantaneous torque is pulsating. The average value of the torque is

$$(T_e)_{av} = -\frac{LI_sI_r}{2} \sin \theta_0$$

since the average value of the double-frequency sine term is zero. If the machine is brought up to *synchronous* speed ($\omega_m = \omega_s$), an average unidirectional torque is established. Continuous energy conversion takes place at synchronous speed. Note that the machine is not self-starting, since an average unidirectional torque is not developed at $\omega_m = 0$ with the specified electrical excitations.

(4) With $\omega_m = \omega_s - \omega_r$, the instantaneous torque is given by

$$T_e = -\frac{LI_sI_r}{4} [\sin(2\omega_st + \alpha + \theta_0) + \sin(-2\omega_rt - \alpha + \theta_0)$$
$$+ \sin(2\omega_st - 2\omega_rt - \alpha + \theta_0) + \sin(\alpha + \theta_0)]$$

The machine operates as a *single-phase induction machine,* and the instantaneous torque is pulsating. The average value of the torque is

$$(T_e)_{av} = -\frac{LI_sI_r}{4} \sin(\alpha + \theta_0)$$

If the machine is brought up to a speed of $\omega_m = \omega_s - \omega_r$, an average unidirectional torque is established, and continuous energy conversion takes place at the *asynchronous* speed of ω_m. Again, note that the machine is not self-starting, since an average unidirectional torque is not developed at $\omega_m = 0$ with the specified electrical excitations.

The pulsating torque, which may be acceptable in small machines, is, in general, an undesirable feature in a rotating machine working as either a generator or a motor, since it may result in speed fluctuation, vibration, noise, and a waste of energy. In magnetic-field systems excited by single-phase alternating sources, the torque pulsates while the speed is relatively constant. Consequently, pulsating power becomes a feature. This calls for improvement; in fact, by employing polyphase windings and polyphase sources, constant power is developed in a balanced system.

EXAMPLE 12.4.2

Consider an electromagnet, as shown in Figure E12.4.2, which is used to support a solid piece of steel and is excited by a coil of $N = 1000$ turns carrying a current $i = 1.5$ A. The cross-sectional area of the fixed magnetic core is $A = 0.01$ m^2. Assume magnetic linearity, infinite permeability of the magnetic structure, and negligible fringing in the air gap.

(a) Develop a general expression for the force f acting to pull the bar toward the fixed magnetic core, in terms of the stored energy, from the basic principle of conservation of energy.

(b) Determine the force that is required to support the weight from falling for $x = 1.5$ mm.

Figure E12.4.2 Electromagnet.

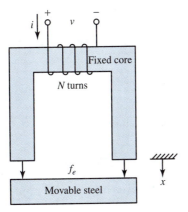

(a) A change in the energy stored in the electromagnetic field dW_m is equal to the sum of the incremental work done by the electric circuit and the incremental work done by the mechanical system. Thus,

$$dW_m = ei\, dt - f_e\, dx = \frac{d\lambda}{dt} i\, dt - f_e\, dx = i\, d\lambda - f_e\, dx$$

or

$$f_e\, dx = i\, d\lambda - dW_m$$

where e is the electromotive force across the coil and the negative sign is due to the sign convention shown in Figure E.12.4.2. Noting that the flux in the magnetic structure depends on two independent variables, namely, the current i through the coil and the displacement x of the bar, one can rewrite the equation,

$$f_e\, dx = i \left(\frac{\partial \lambda}{\partial i} di + \frac{\partial \lambda}{\partial x} dx \right) - \left(\frac{\partial W_m}{\partial i} di + \frac{\partial W_m}{\partial x} dx \right)$$

where W_m is a function of i and x. Since i and x are independent variables, one gets

$$f_e = i \frac{\partial \lambda}{\partial x} - \frac{\partial W_m}{\partial x} \qquad \text{and} \qquad 0 = i \frac{\partial \lambda}{\partial i} - \frac{\partial W_m}{\partial i}$$

We can then see that

$$f_e = \frac{\partial}{\partial x} (i\lambda - W_m) = \frac{\partial}{\partial x} W_m'$$

where W_m' is the coenergy, which is equal to the energy W_m in structures that are magnetically linear.

The force f acting to pull the bar toward the fixed magnet core is given by

$$f = -f_e = -\frac{\partial W_m}{\partial x}$$

The stored energy in a linear magnetic structure is given by

$$W_m = \frac{\phi \mathcal{F}}{2} = \frac{\phi^2 \mathcal{R}(x)}{2}$$

where ϕ is the flux, $\mathcal{F}$ is the mmf, and $\mathcal{R}(x)$ is the reluctance, which is a function of displacement. Finally, we get

$$f = -\frac{\partial W_m}{\partial x} = -\frac{\phi^2}{2} \frac{d\mathcal{R}(x)}{dx}$$

(b) The reluctance of the air gaps in the magnetic structure is given by

$$\mathcal{R}(x) = \frac{2x}{\mu_0 A} = \frac{2x}{4\pi \times 10^{-7} \times 0.01} = \frac{x}{0.6285 \times 10^{-8}}$$

The magnitude of the force in the air gap is then given by

$$|f| = \frac{\phi^2}{2}\frac{d\mathcal{R}(x)}{dx} = \frac{1}{2}\left(\frac{Ni}{\mathcal{R}}\right)^2\frac{d\mathcal{R}}{dx}$$

$$= \frac{i^2}{2}\frac{N^2}{\mathcal{R}^2}\frac{d\mathcal{R}}{dx} = \frac{1.5^2}{2}1000^2\frac{0.6285\times10^{-8}}{x^2}$$

For $x = 1.5$ mm $= 1.5\times10^{-3}$ m,

$$|f| = \frac{1.5^2}{2}1000^2\frac{0.6285\times10^{-8}}{(1.5\times10^{-3})^2} = 3142.5 \text{ N}$$

The student should recognize the practical importance of force-generating capabilities of electromechanical transducers.

EXAMPLE 12.4.3

Solenoids find application in a variety of electrically controlled valves. The magnetic structure shown in Figure E12.4.3 is a simplified representation of a solenoid in which the flux in the air gap activates the motion of the iron plunger.

(a) Develop a general expression for the force exerted on the iron plunger and comment on its dependence on position x.

(b) Determine the current through the coil of $N = 100$ turns to pull the plunger to $x = a$, given that $a = 1$ cm, $l_g = 1$ mm, and the spring constant is $k = 1$ N/m. Assume the permeability of the magnetic structure to be infinite and neglect fringing.

Fixed magnetic core structure ($\mu \to \infty$)

Figure E12.4.3 Simplified representation of a solenoid.

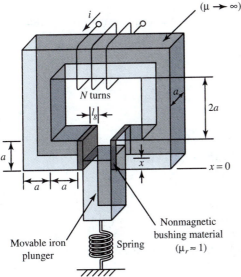

N turns

$2a$

$x = 0$

Nonmagnetic bushing material ($\mu_r \approx 1$)

Movable iron plunger   Spring

Solution

(a) The force in the air gap can be expressed by

$$f = -\frac{1}{2}\phi^2 \frac{d\mathcal{R}}{dx}$$

(See Example 12.4.2.) The gap reluctance, which in this case is due to the nonmagnetic bushing, is given by

$$\mathcal{R} = 2\mathcal{R}_g = \frac{2l_g}{\mu_0 ax}$$

where the area ax is variable, depending on the position of the plunger, and l_g is the thickness of the bushing on either side of the plunger. The air-gap magnetic flux is given by

$$\phi = \frac{\mathcal{F}}{\mathcal{R}} = \frac{Ni}{\mathcal{R}} = \frac{\mu_0 Niax}{2l_g}$$

Now,

$$\frac{d\mathcal{R}}{dx} = \frac{d}{dx}\left(\frac{2l_g}{\mu_0 ax}\right) = \frac{-2l_g}{\mu_0 ax^2}$$

The force is then given by

$$f = -\frac{1}{2}\phi^2 \frac{d\mathcal{R}}{dx} = -\frac{1}{2}\left(\frac{\mu_0 Niax}{2l_g}\right)^2 \left(\frac{-2l_g}{\mu_0 ax^2}\right)$$

or

$$f = \frac{1}{4}\frac{\mu_0 a}{l_g}(Ni)^2$$

which is independent of the position x, and is a constant for a given exciting mmf and geometry.

(b) For $x = a = 1$ cm $= 0.01$ m,

$$f = kx = 1 \times 0.01 = \frac{1}{4}\frac{4\pi \times 10^{-7} \times 0.01}{0.001}(100\ i)^2$$

or

$$i^2 = \frac{4 \times 0.001 \times 0.01}{4\pi \times 10^{-7} \times 0.01}\frac{1}{(100)^2} = 0.3182$$

or

$$i = 0.564\ \text{A}$$

The student should recognize the practical importance of determining the approximate mmf or current requirements for electromechanical transducers.

EXAMPLE 12.4.4

A relay is essentially an electromechanical switch that opens and closes electrical contacts. A simplified relay is represented in Figure E12.4.4. It is required to keep the fenomagnetic plate at a distance of 0.25 cm from the electromagnet excited by a coil of $N = 5000$ turns, when the torque is 10 N · m at a radius $r = 10$ cm. Estimate the current required, assuming infinitely permeable magnetic material and negligible fringing as well as leakage.

(a) Express the stored magnetic energy W_m as a function of the flux linkage λ and the position x.

(b) Express the coenergy W_m' as a function of the current i and the position x.

Figure E12.4.4 Simplified representation of a relay.

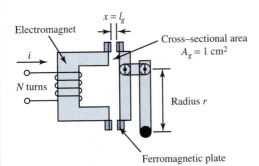

Solution

For the stated torque at a radius of 10 cm, the force on the plate is

$$f = \frac{10}{0.1} = 100 \text{ N}$$

The magnitude of the force developed by the electromagnet must balance this force. The reluctance due to the two air gaps is given by

$$Q = \frac{2x}{\mu_0 A_g} = \frac{2x}{4\pi \times 10^{-7} \times 1 \times 10^{-4}} = \frac{x}{2\pi \times 10^{-11}}$$

The inductance is then given by

$$L = \frac{N^2}{R} = \frac{5000^2 \times 2\pi \times 10^{-11}}{x} = \frac{5\pi \times 10^{-4}}{x} = \frac{1.57 \times 10^{-3}}{x}$$

In a magnetically linear circuit, the stored magnetic energy W_m is equal to the coenergy W_m',

$$W_m = W_m' = \frac{1}{2} L i^2 = \frac{1}{2} \frac{\lambda^2}{L}$$

(a) $W_m(\lambda, x) = \frac{1}{2} \lambda^2 \frac{x}{5\pi \times 10^{-4}} = \frac{\lambda^2 x}{\pi \times 10^{-3}}$

Since $f = -\partial W_m(\lambda, x)/\partial x$, the magnitude of the developed force is given by

$$|f| = \frac{\lambda^2}{\pi \times 10^{-3}} = \frac{(Li)^2}{\pi \times 10^{-3}} = \frac{(5\pi \times 10^{-4})^2 i^2}{\pi \times 10^{-3} \times x^2} = 2.5 \,\pi \times 10^{-4} \frac{i^2}{x^2}$$

For $x = 0.25$ cm and $|f| = 100$ N,

$$i^2 = \frac{100(0.25 \times 10^{-2})^2}{2.5\pi \times 10^{-4}} = \frac{2.5}{\pi} = 0.795$$

or

$$i = 0.89 \text{ A}$$

(b) $W'_m(i, x) = \frac{1}{2}Li^2 = \frac{1}{2}\frac{5\pi \times 10^{-4}}{x}i^2$

Since $f = \partial W'_m(i, x)/\partial x$, the magnitude of the developed force is given by

$$|f| = \frac{1}{2}\frac{5\pi \times 10^{-4}}{x^2}i^2$$

For $x = 0.25$ cm and $|f| = 100$ N,

$$i^2 = \frac{100 \times 2 \times (0.25 \times 10^{-2})^2}{5\pi \times 10^{-4}} = \frac{2.5}{\pi} = 0.795$$

or

$$i = 0.89 \text{ A}$$

Such relays find common application in industrial practice to remotely switch large industrial loads. The student should recognize that a relatively low-level current can be used to operate the relay, which in turn controls the opening and closing of a circuit that carries large currents.

Starting with the flux linkages given by Equations (12.4.22) and (12.4.23), one can develop the volt–ampere equations for the stator and rotor circuits. While the voltage and torque equations for the idealized elementary machine of Example 12.4.1 with a uniform air gap are now obtained from the coupled-circuit viewpoint, these can also be formulated from the *magnetic-field viewpoint* based on the interaction of the magnetic fields of the stator and rotor windings in the air gap.

Since the mmf waves of the stator and rotor are considered spatial sine waves, they can be represented by the space vectors $\bar{F}_s$ and $\bar{F}_r$, drawn along the magnetic axes of the stator and rotor mmf waves, as in Figure E12.4.1, with the phase angle δ (in electrical units) between their magnetic axes. The resultant mmf $\bar{F}$ acting across the air gap is also a sine wave, given by the vector sum of $\bar{F}_s$ and $\bar{F}_r$, so that

$$F^2 = F_s^2 + F_r^2 + 2F_s F_r \cos \delta \tag{12.4.26}$$

where F's are the peak values of the mmf waves. Assuming the air-gap field to be entirely radial, the resultant $\bar{H}$-field is a sinusoidal space wave whose peak is given by

$$H_{\text{peak}} = \frac{F}{g} \tag{12.4.27}$$

where g is the radial length of the air gap. Because of linearity, the coenergy is equal to the energy. The average coenergy density obtained by averaging over the volume of the air-gap region is

$$(w')_{\text{av}} = \frac{\mu_0}{2}(\text{average value of } H^2) = \frac{\mu_0}{2}\frac{H^2_{\text{peak}}}{2} = \frac{\mu_0}{4}\frac{F^2}{g^2} \tag{12.4.28}$$

since the average value of the square of a sine wave is one-half of the square of its peak value. The total coenergy for the air-gap region is then given by

$$W' = (w')_{\text{av}}(\text{volume of air-gap region}) = \frac{\mu_0}{4}\frac{F^2}{g^2}\pi Dlg \qquad (12.4.29)$$

where D is the average diameter at the air gap and l is the axial length of the machine. Equation (12.4.29) may be rewritten as follows by using Equation (12.4.26):

$$W' = \frac{\pi Dl\mu_0}{4g}(F_s^2 + F_r^2 + 2F_sF_r \cos \delta) \qquad (12.4.30)$$

The torque in terms of the interacting magnetic fields is obtained by taking the partial derivative of the field coenergy with respect to the angle δ. For a two-pole machine, such a torque is given by

$$T_e = \frac{\partial W'}{\partial \delta} = -\frac{\pi Dl\mu_0}{2g} F_sF_r \sin \delta = -KF_sF_r \sin \delta \qquad (12.4.31)$$

in which K is a constant determined by the dimensions of the machine. The torque for a P-pole cylindrical machine with a uniform air gap is then

$$T_e = -\frac{P}{2}KF_sF_r \sin \delta \qquad (12.4.32)$$

Equations (12.4.31) and (12.4.32) have shown that the torque is proportional to the peak values of the interacting stator and rotor mmfs and also to the sine of the space-phase angle δ between them (expressed in electrical units). The interpretation of the negative sign is the same as before, in that the fields tend to align themselves by decreasing the displacement angle δ between the fields.

Equation (12.4.32) shows that it is possible to obtain a constant torque, varying neither with time nor with rotor position, provided that the two mmf waves are of constant amplitude and have constant angular displacement from each other. While it is easy to conceive of the two mmf waves having constant amplitudes, the question would then be how to maintain a constant angle between the stator and rotor mmf axes if one winding is stationary and the other is rotating. Three possible answers arise:

1. If the stator mmf axis is fixed in space, the rotor mmf must also be fixed in space, even when the rotor winding is physically rotating, as is the case with a dc machine.

2. If the rotor mmf axis is fixed relative to the rotor, the stator mmf axis must rotate at the rotor speed relative to the stationary stator windings, as is the case with a polyphase synchronous machine.

3. The two mmf axes must rotate at such speeds relative to their windings that they remain stationary with respect to each other, as is the case with a polyphase induction machine (which we will explain later).

12.5   BASIC ASPECTS OF ELECTROMECHANICAL ENERGY CONVERTERS

Whereas detailed differences and particularly challenging problems emerge among various machine types, this section briefly touches on the interrelated problems that are common to all machine types, such as losses and efficiency, ventilation and cooling, machine ratings, magnetic saturation, leakage and harmonic fluxes, and machine applications. Various *standards*, developed by IEEE (Institute of Electrical and Electronics Engineers), NEMA (National Electrical Manufacturers Association), ANSI (American National Standards Institute), and IEC (International Electrotechnical Commission), deal with machine ratings, insulation and allowable temperature rise, testing methods, losses, and efficiency determination.

Besides the stator and rotor iron-core losses, *friction and windage* losses (which are generally functions of machine speed, and are usually assumed to be practically constant for small speed variations) are included in *no-load rotational losses*, which are effectively constant. Besides copper losses (stator and rotor winding I^2R losses), *stray-load losses* (which arise from various causes that are not usually accounted for, and are usually taken to be about 0.5 to 5% of the machine output) are included for the determination of efficiency (= output/input).

Much of the considerable progress made over the years in electric machinery is due to the improvements in the quality and characteristics of steel and insulating materials, as well as to innovative cooling methods. Modern large turbo alternators have direct water cooling (cooling water circulated through hollow passages in their conductors, being in direct contact with the copper conductors) in the stator (and the rotor in a few cases) and hydrogen cooling (with hydrogen under 1 to 5 atmospheres of pressure) in the rotor. With hydrogen under pressure, sealing the bearings appropriately needs particular attention for turbogenerators. For hydroelectric generators, on the other hand, designing the thrust bearings for vertical mounting becomes a prominent issue.

In general, every machine has a nameplate attached to the frame inscribed with relevant information regarding voltage, current, power, power factor, speed, frequency, phases, and allowable temperature rise. The nameplate rating is the *continuous rating*, unless otherwise specified, such as *short-time rating*. Motors are rated in hp (horsepower); dc generators in kW; and alternators and transformers in terms of kVA rather than kW (because their losses and heating are approximately determined by the voltage and current, regardless of the power factor). The physical size and cost of ac power-system apparatus are roughly proportional to the kVA rating.

In order to fully utilize the magnetic properties of the iron and optimize the machine design, the machine iron is worked at fairly saturated levels of flux density, such that the normal operating point on the open circuit is near the knee of the *open-circuit characteristic* (or the *no-load saturation curve,* which is similar to the magnetization *B–H* characteristic). Magnetic saturation does influence the machine performance to a considerable degree. Leakage and harmonic fluxes, which exist in addition to the mutual flux (generally assumed to be sinusoidally distributed) and which may develop *parasitic torques* causing vibration and noise, also have to be considered. Accounting for, and including, these effects becomes too involved to be discussed here.

For motors the major consideration is the *torque–speed characteristics*. The requirements of motor loads generally vary from one application to another. Some may need constant speed or horsepower, while some others may require adjustable varying speeds with different torque capabilities. For any motor application, the starting torque, maximum torque, and running characteristics (along with current requirements) should be looked into.

For generators it is the volt–ampere, or voltage–load, characteristics. For machines in general, it is also vital to know the limits between which characteristics can be varied and how to obtain such variations. Relevant economic features, such as efficiency, power factor, relative costs, and the effect of losses on heating and machine rating, need to be investigated.

Finally, since a generator or a motor may only be one component of a complicated modern power system, system-related dynamic applications and behavior (both steady-state and transient-state) and proper models to study such behavior become very important when designing electric machines.

12.6   LEARNING OBJECTIVES

The *learning objectives* of this chapter are summarized here, so that the student can check whether he or she has accomplished each of the following.

- Basic principles of electromechanical energy conversion, including induction, interaction, alignment, and torque production.
- Simplified analysis of electromechanical transducers, such as electromagnets, position and velocity sensors, relays, solenoids, and loudspeakers, which convert electric signals to mechanical forces, or mechanical motion to electric signals.
- Calculating the emf produced in ac and dc machines.
- Generation of rotating magnetic fields.
- Evaluating forces and torques in magnetic-field systems, from the coupled-circuit viewpoint and the magnetic-field viewpoint.
- Basic aspects of motors and generators.

12.7 PRACTICAL APPLICATION: A CASE STUDY

Sensors or Transducers

In almost all engineering applications there arises a need to measure some physical quantities, such as positions, displacements, speeds, forces, torques, temperatures, pressures, or flows. Devices known as *sensors* or *transducers* convert a physical quantity to a more readily manipulated electrical quantity (such as voltage or current), such that changes in physical quantity usually produce proportional changes in electrical quantity. The direct output of the sensor may often need additional manipulation, known as *signal conditioning*, for the output to be in a useful form free from noise and intereference. The conditioned sensor signal may then be *sampled* and converted to *digital* form and stored in a *computer* for additional manipulation or display. A typical computer-based *measurement system* using a sensor is represented in block diagram form in Figure 12.7.1. Table 12.7.1 lists various sensors and their uses.

TABLE 12.7.1 Sensors classified according to variable sensed

Variable Sensed	Sensor or Transducer
Variations in dimensions due to motion	Strain gauge Resistive potentiometer Variable reluctance sensor Moving-coil transducer Seismic sensor
Temperature	Thermocouple Resistance temperature detector Radiation detector
Force, torque, and pressure	Capacitive sensor Piezoelectric transducer
Flow	Magnetic flow meter Turbine flow meter Hot-wire anemometer Ultrasonic sensor
Light intensity	Photoelectric sensor
Humidity	Semiconductor transducer
Chemical composition	Solid-state gas sensor
Liquid level	Differential-pressure transducer

Data gathering with various sensors and data reduction have become much easier, automatic, and quite sophisticated due to the advent of the computer, since the computer can monitor many inputs, process and record data, furnish displays, and produce control outputs.

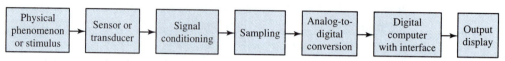

Figure 12.7.1 Block diagram of measurement system using a sensor.

PROBLEMS

12.1.1 (a) Show by applying Ampere's circuital law that the magnetic field associated with a long straight, current-carrying wire is given by $B_\phi = \mu_0 I/(2\pi r)$, where the subscript ϕ denotes the ϕ-component in the circular cylindrical coordinate system, μ_0 is the free-space permeability, I is the current carried by the wire, and r is the radius from the current-carrying wire. What is the net force on the wire due to the interaction of the B-field (produced by the current I) and the current I?

(b) A magnetic force exists between two adjacent, parallel, current-carrying wires. Let I_1 and I_2 be the currents carried by the wires and r the separation between them. Use the result of part (a) to find the force between the wires. Discuss the nature of the force when the wires carry currents in the same direction, and in opposite directions.

*12.1.2 A rectangular loop is placed in the field of an infinitely long straight conductor carrying a current of I amperes, as shown in Figure P12.1.2. Find

an expression for the total flux linking the loop, assuming a medium of permeability μ.

12.1.3 Consider a conducting loop of length l and width w, as shown in Figure P12.1.3, rotated about its axis (shown by the broken line) at a speed of ω_m rad/s under the influence of a magnetic field B.

(a) Obtain an expression for the induced emf, assuming the angle θ_m is zero at $t = 0$ and $\theta_m = \omega_m t$.

(b) If a resistor R is connected across terminals a–a', explain what happens and how the principle of energy conservation is satisfied.

12.1.4 A coil is formed by connecting 15 conducting loops, or turns, in series. Each loop has length $l = 2.5$ m and width $w = 10$ cm. The 15-turn coil is rotated at a constant speed of 30 r/s (or 1800 r/min) in a magnetic field of density $B = 2$ T. The configuration of Figure P12.1.3 applies.

(a) Find the induced emf across the coil.

(b) Determine the average power delivered to the resistor $R = 500$ Ω, which is connected between the terminals of the coil.

Figure P12.1.2

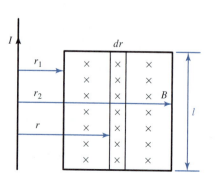

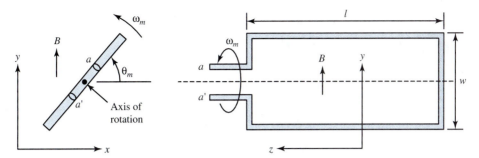

Figure P12.1.3 Rotating conducting loop.

(c) Calculate the average mechanical torque needed to turn the coil and generate power for the resistor. Identify the action of the device as that of a motor or a generator.

12.1.5 The machine of Problem 12.1.4 can be used as a motor. Let the terminals of the coil be connected to a voltage source of 1 kV rms. If the motor runs at 1800 r/min and draws a current of 2 A, find the torque supplied to the mechanical load.

12.1.6 The 50-turn coil in the configuration of Figure P12.1.3 is rotated at a constant speed of 300 r/min. The axis of rotation is perpendicular to a uniform magnetic flux density of 0.1 T. The loop has width $w = 10$ cm and length $l = 1$ m. Compute:

(a) The maximum flux passing through the coil.

(b) The flux linkage as a function of time.

(c) The maximum instantaneous voltage induced in the coil.

(d) The time-average value of the induced voltage.

(e) The induced voltage when the plane of the coil is 30° from the vertical.

*12.1.7 A 100-turn coil in the configuration of Figure P12.1.3 is rotated at a constant speed of 1200 r/min in a magnetic field. The rms induced voltage across the coil is 1 kV, and each turn has a length $l = 112.5$ cm and a width $w = 10$ cm. Determine the required value of the flux density.

12.1.8 A sectional view of a cylindrical iron-clad plunger magnet is shown in Figure P12.1.8. The small air gap between the sides of the plunger and the iron shell is uniform and 0.25 mm long. Neglect leakage and fringing, and consider the iron to be infinitely permeable. The coil has 1000 turns and carries a direct current of 3 A. Compute the pull on the plunger for $g = 1.25$ cm.

12.1.9 Figure E12.1.1 can also be considered as a simple model of a magnetically operated relay that is commonly used for the automatic control and protection of electric equipment. Consider the core and armature (shown as "Bar" in Figure E12.1.1) of the relay to be constructed out of infinitely permeable magnetic material. The core has a circular cross section and is 1.25 cm in diameter, while the armature has a rectangular cross section. The armature is so supported that the two air gaps are always equal and of uniform length over their areas. A spring (not shown in the figure), whose force opposes the magnetic pull, restrains the motion of the armature. If the operating coil has 1800 turns carrying a current of 1 A, and the gaps are set at 0.125 cm each, find the pull that must be exerted by the spring. You may neglect leakage and fringing.

12.1.10 The coil is placed so that its axis of revolution is perpendicular to a uniform field, as shown in Figure P12.1.10. If the flux per pole is 0.02 Wb, and the coil, consisting of 2 turns, is revolving at 1800 r/min, compute the maximum value of the voltage induced in the coil.

12.2.1 An elementary two-pole, single-phase, synchronous machine, as illustrated in Figure P12.2.1, has a field winding on its rotor and an armature winding on its stator, with $N_f = 400$ turns and $N_a = 50$ turns, respectively. The uniform air gap is of length 1 mm, while the armature diameter is 0.5 m, and the axial length of the machine is 1.5 m. The field winding carries a current of 1 A (dc) and the rotor is driven at 3600 r/min. Determine:

(a) The frequency of the stator-induced voltage.

(b) The rms-induced voltage in the stator winding.

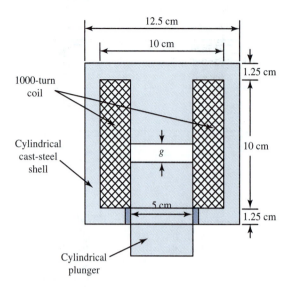

Figure P12.1.8 Plunger magnet.

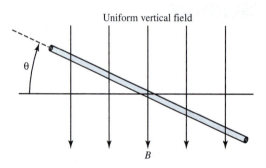

Figure P12.1.10

Uniform vertical field

B

12.2.2 The flux-density distribution produced in a two-pole synchronous generator by an ac-excited field winding is

$$B(\theta, t) = B_m \sin \omega_1 t \cos \theta$$

Find the nature of the armature voltage induced in an N-turn coil if the rotor (or field) rotates at ω_2 rad/s. Comment on the special case when $\omega_1 = \omega_2 = \omega$.

12.2.3 The flux-density distribution in the air gap of a 60-Hz, two-pole, salient-pole machine is sinusoidal, having an amplitude of 0.6 T. Calculate the instantaneous and rms values of the voltage induced in a 150-turn coil on the armature, if the axial length of the armature and its inner diameter are both 100 mm.

*12.2.4 Consider an elementary three-phase, four-pole alternator with a wye-connected armature winding, consisting of full-pitch concentrated coils, as shown in Figures 12.2.3(b) and (c). Each phase

coil has three turns, and all the turns in any one phase are connected in series. The flux per pole, distributed sinusoidally in space, is 0.1 Wb. The rotor is driven at 1800 r/min.

(a) Calculate the rms voltage generated in each phase.

(b) If a voltmeter were connected across the two line terminals, what would it read?

(c) For the a–b–c phase sequence, take $t = 0$ at the instant when the flux linkages with the a-phase are maximum.

   (i) Express the three phase voltages as functions of time.

   (ii) Draw a corresponding phasor diagram of these voltages with the a-phase voltage as reference.

   (iii) Represent the line-to-line voltages on the phasor diagram in part (ii).

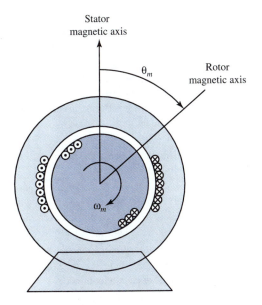

Stator
magnetic axis

θ_m

Rotor
magnetic axis

ω_m

Figure P12.2.1 Rotating machine with uniform air gap.

(iv) Obtain the time functions of the line-to-line voltages.

12.2.5 A wye-connected, three-phase, 50-Hz, six-pole synchronous alternator develops a voltage of 1000 V rms between the lines when the rotor dc field current is 3 A. If this alternator is to generate 60-Hz voltages, find the new synchronous speed, and calculate the new terminal voltage for the same field current.

12.2.6 Determine the synchronous speed in revolutions per minute and the useful torque in newton-meters of a 200-hp, 60-Hz, six-pole synchronous motor operating at its rated full load (1 hp ≅ 746 W).

12.2.7 From a three-phase, 60-Hz system, through a motor–generator set consisting of two directly coupled synchronous machines, electric power is supplied to a three-phase, 50-Hz system.

(a) Determine the minimum number of poles for the motor.

(b) Determine the minimum number of poles for the generator.

(c) With the number of poles decided, find the speed in r/min at which the motor–generator set will operate.

*12.2.8 Two coupled synchronous machines are used as a motor–generator set to link a 25-Hz system to a 60-Hz system. Find the three highest speeds at which this linkage would be possible.

12.2.9 A 10-turn square coil of side 200 mm is mounted on a cylinder 200 mm in diameter. If the cylinder rotates at 1800 r/min in a uniform 1.2-T field, determine the maximum value of the voltage induced in the coil.

12.3.10 A four-pole dc machine with 728 active conductors and 30 mWb flux per pole runs at 1800 r/min.

(a) If the armature winding is lap wound, find the voltage induced in the armature, given that the number of parallel paths is equal to the number of poles for lap windings.

(b) If the armature is wave wound, calculate the voltage induced in the armature, given that the number of parallel paths is equal to 2 for wave windings.

(c) If the lap-wound armature is designed to carry a maximum line current of 100 A, compute the maximum electromagnetic power and torque developed by the machine.

(d) If the armature were to be reconnected as wave wound, while limiting per-path current to the same maximum as in part (c), will either the maximum developed power or the torque be changed from that obtained in part (c)?

12.2.11 A four-pole, lap-wound armature has 144 slots with two coil sides per slot, each coil having two turns. If the flux per pole is 20 mWb and the armature rotates at 720 r/min, calculate the induced voltage, given that the number of parallel paths is equal to the number of poles for lap windings.

12.2.12 A four-pole dc generator is lap wound with 326 armature conductors. It runs at 650 r/min on full load, with an induced voltage of 252 V. If the bore of the machine is 42 cm in diameter, its axial length is 28 cm, and each pole subtends an angle of 60°, determine the air-gap flux density. (Note that the number of parallel paths is equal to the number of poles for lap windings.)

12.2.13 A six-pole, double-layer dc armature winding in 28 slots has five turns per coil. If the field flux is 0.025 Wb per pole and the speed of the rotor is 1200 r/min, find the value of the induced emf when the winding is (a) lap connected, and (b) wave-connected. (Note: The number of parallel paths is equal to the number of poles for lap windings, while it is equal to 2 for wave windings.)

*12.2.14 A four-pole, dc series motor has a lap-connected, two-layer armature winding with a total of 400 conductors. Calculate the gross torque developed for a flux per pole of 0.02 Wb and an armature current of 50 A. (Note: The number of parallel paths is equal to the number of poles for lap windings.)

*12.3.1 For a balanced two-phase stator supplied by balanced two-phase currents, carry out the steps leading up to an equation such as Equation (12.3.6) for the rotating mmf wave.

12.3.2 The N-coil windings of a three-phase, two-pole machine are supplied with currents i_a, i_b, and i_c, which produce mmfs given by $F_a = Ni_a \cos \theta_m$; $F_b = Ni_b \cos(\theta_m - 120°)$; and $F_c = Ni_c \cos(\theta_m - 240°)$, respectively.

(a) If the three-phase windings are connected in series and supplied by a single-phase voltage source, find the resultant mmf due to all the three windings as a function of θ_m.

(b) If the three-phase windings are connected to a balanced three-phase voltage supply of f Hz (with positive sequence), determine the resultant mmf.

(c) Letting $\theta_m = \omega_m t + \alpha$, obtain the relationship between ω_m and $\omega(= 2\pi f)$ that results in maximum mmf.

12.3.3 A two-pole, three-phase synchronous generator has a balanced three-phase winding with 15 turns per phase. If the three-phase currents are given by $i_a = 100 \cos 377t$, $i_b = 100 \cos(377t - 120°)$, and $i_c = 100 \cos(377t - 240°)$, determine:

(a) The peak fundamental component of the mmf of each winding.

(b) The resultant mmf.

12.3.4 Consider the balanced three-phase alternating currents, shown in Figure P12.3.4(a), to be flowing in phases a, b, and c, respectively, of the two-pole stator structure shown in Figure P12.3.4(b) with balanced three-phase windings. For instants $t = t_1$, t_3, and t_5 of Figure P12.3.4(a), sketch the individual phase flux contributions and their resultants in vectorial form.

12.4.1 Consider the electromagnetic plunger shown in Figure P12.4.1. The λ–i relationship for the normal working range is experimentally found to be $\lambda = Ki^{2/3}/(x+t)$, where K is a constant. Determine the electromagnetic force on the plunger by the application of Equations (12.4.19) and (12.4.20). Interpret the significance of the sign that you obtain in the force expression.

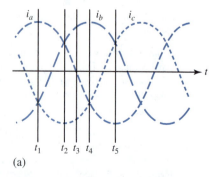

(a)

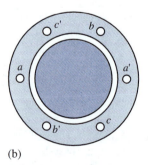

(b)

Figure P12.3.4

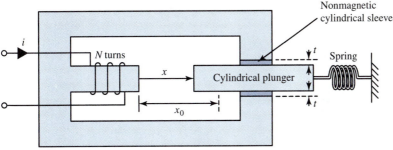

x_0: position of plunger when spring is relaxed

Figure P12.4.1

12.4.2 In Problem 12.4.1 neglect the saturation of the core, leakage, and fringing. Neglect also the reluctance of the ferromagnetic circuit. Assuming that the cross-sectional area of the center leg is twice the area of the outer legs, obtain expressions for the inductance of the coil and the electromagnetic force on the plunger.

*12.4.3 For the electromagnet shown in Figure P12.4.3, the λ–i relationship for the normal working range is given by $i = a\lambda^2 + b\lambda(x - d)^2$, where a and b are constants. Determine the force applied to the plunger by the electric system.

12.4.4 A solenoid of cylindrical geometry is shown in Figure P12.4.4.

(a) If the exciting coil carries a steady direct current I, derive an expression for the force on the plunger.

(b) For the numerical values $I = 10$ A, $N = 500$ turns, $g = 5$ mm, $a = 20$ mm, $b = 2$ mm,

and $l = 40$ mm, find the magnitude of the force. Assume infinite permeability of the core and neglect leakage.

12.4.5 Let the solenoid of Problem 12.4.4 carry an alternating current of 10 A (rms) at 60 Hz instead of the direct current.

(a) Find an expression for the instantaneous force.

(b) For the numerical values of N, g, a, b, and l given in Problem 12.4.4(b) compute the average force. Compare it to that of Problem 12.4.4(b).

12.4.6 Consider the solenoid with a core of square cross section shown in Figure P12.4.6.

(a) For a coil current I (dc), derive an expression for the force on the plunger.

(b) Given $I = 10$ A, $N = 500$ turns, $g = 5$ mm, $a = 20$ mm, and $b = 2$ mm, calculate the magnitude of the force.

Cylindrical magnetic shell   **Figure P12.4.3**

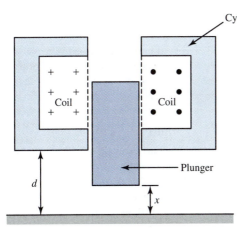

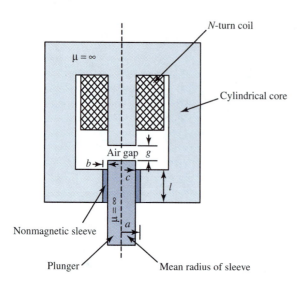

N-turn coil

Figure P12.4.4

$\mu = \infty$

Cylindrical core

Air gap g

b

c

l

$\mu = \infty$

a

Nonmagnetic sleeve

Plunger

Mean radius of sleeve

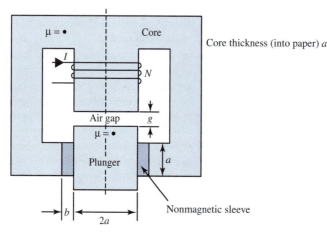

Figure P12.4.6

$\mu = \bullet$

Core

Core thickness (into paper) a

I

N

Air gap g

$\mu = \bullet$

Plunger

a

b

$2a$

Nonmagnetic sleeve

12.4.7 Let the coil of the solenoid of Problem 12.4.6 have a resistance R and be excited by a voltage $v = V_m \sin \omega t$. Consider a plunger displacement of $g = g_0$.

(a) Obtain the expression for the steady-state coil current.

(b) Obtain the expression for the steady-state electric force.

*12.4.8 A two-pole rotating machine with a singly excited magnetic field system as its stator and a rotor (that carries no coil) has a stator-coil inductance that can be approximated by $L(\theta) = (0.02 - 0.04 \cos 2\theta - 0.03 \cos 4\theta)$ H, where θ is the angle between the stator-pole axis and the rotor axis. A current of 5 A (rms) at 60 Hz is passed through the coil, and the rotor is driven at

a speed, which can be controlled, of ω_m rad/s. See Figure P12.4.8 for the machine configuration.

(a) Find the values of ω_m at which the machine can develop average torque.

(b) At each of the speeds obtained in part (a), determine the maximum value of the average torque and the maximum mechanical power output.

Note: $T_e = \partial W'_m(i, \theta)/\partial\theta = 1/2 \; i^2 \; \partial L(\theta)/\partial\theta$. Let $\theta = \omega_m t - \delta$.

12.4.9 Consider Example 12.4.1. With the assumed current-source excitations of part (c), determine the voltages induced in the stator and rotor windings at the corresponding angular velocity ω_m at which an average torque results.

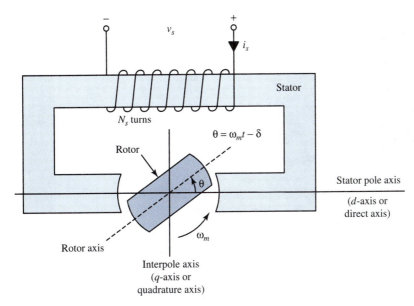

Figure P12.4.8

12.4.10 A two-winding system has its inductances given by

$$L_{11} = \frac{k_1}{x} = L_{22}; \quad L_{12} = L_{21} = \frac{k_2}{x}$$

where k_1 and k_2 are constants. Neglecting the winding resistances, derive an expression for the electric force when both windings are connected to the same voltage source $v = V_m \sin \omega t$. Comment on its dependence on x.

12.4.11 Two mutually coupled coils are shown in Figure P12.4.11. The inductances of the coils are $L_{11} = A$, $L_{22} = B$, and $L_{12} = L_{21} = C \cos \theta$. Find the electric torque for:

(a) $i_1 = I_0, i_2 = 0$.

(b) $i_1 = i_2 = I_0$.

(c) $i_1 = I_m \sin \omega t, i_2 = I_0$.

(d) $i_1 = i_2 = I_m \sin \omega t$.

(e) Coil 1 short-circuited and $i_2 = I_0$.

*12.4.12 Consider an elementary cylindrical-rotor two-phase synchronous machine with uniform air gap, as illustrated in the schematic diagram in Figure P12.4.12. It is similar to that of Figure E12.4.1, except that Figure P12.4.12 has two identical stator windings in quadrature instead of one. The self-inductance of the rotor or field winding is a constant given by L_{ff} H; the self-inductance of each stator winding is a constant

given by $L_{aa} = L_{bb}$. The mutual inductance between the stator windings is zero since they are in space quadrature; the mutual inductance between a stator winding and the rotor winding depends on the angular position of the rotor,

$$L_{af} = L \cos \theta; \quad L_{bf} = L \sin \theta$$

where L is the maximum value of the mutual inductance, and θ is the angle between the magnetic axes of the stator a-phase winding and the rotor field winding.

(a) Let the instantaneous currents be i_a, i_b, and i_f in the respective windings. Obtain a general expression for the electromagnetic torque T_e in terms of these currents, angle θ, and L.

(b) Let the stator windings carry balanced two-phase currents given by $i_a = I_a \cos \omega t$, and $i_b = I_a \sin \omega t$, and let the rotor winding be excited by a constant direct current I_f. Let the rotor revolve at synchronous speed so that its instantaneous angular position θ is given by $\theta = \omega t + \delta$. Derive the torque expression under these conditions and describe its nature.

(c) For conditions of part (b), neglect the resistance of the stator windings. Obtain the volt–ampere equations at the terminals of stator phases a and b, and identify the speed–voltage terms.

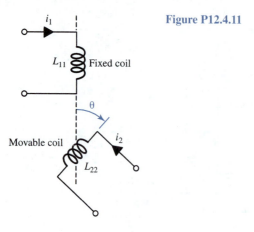

Figure P12.4.11

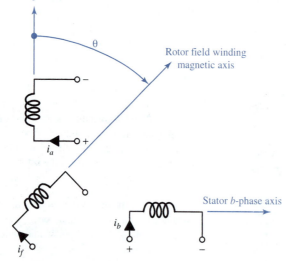

Figure P12.4.12

12.4.13 Consider the machine configuration of Problem 12.4.12. Let the rotor be stationary and constant direct currents I_a, I_b, and I_f be supplied to the windings. Further, let I_a and I_b be equal. If the rotor is now allowed to move, will it rotate continuously or will it tend to come to rest? If the latter, find the value of θ for stable equilibrium.

12.4.14 Let us consider an elementary salient-pole, two-phase, synchronous machine with nonuniform air gap. The schematic representation is the same as for Problem 12.4.12. Structurally the stator is similar to that of Figure E12.4.1 except that this machine has two identical stator windings in quadrature instead of one. The salient-pole rotor with two poles carries the field winding connected to slip rings. The inductances are given as

$$L_{aa} = L_0 + L_2 \cos 2\theta; \qquad L_{af} = L \cos \theta$$
$$L_{bb} = L_0 - L_2 \cos 2\theta; \qquad L_{bf} = L \sin \theta;$$
$$L_{ab} = L_2 \sin 2\theta$$

L_{ff} is a constant, independent of θ; L_0, L_2, and L are positive constants; and θ is the angle between the magnetic axes of the stator a-phase winding and the rotor field winding.

(a) Let the stator windings carry balanced two-phase currents given by $i_a = I_a \cos \omega t$ and $i_b = I_a \sin \omega t$, and let the rotor winding be excited by a constant direct current I_f. Let the rotor revolve at synchronous speed so that its instantaneous angular position θ is given by $\theta = \omega t + \delta$. Derive an expression for the torque under these conditions and describe its nature.

(b) Compare the torque with that of Problem 12.4.12(b).

(c) Can the machine be operated as a motor? As a generator? Explain.

(d) Suppose that the field current I_f is brought to zero. Will the machine continue to run?

12.4.15 Consider the analysis leading up to Equation (12.4.32) for the torque of an elementary cylindrical machine with uniform air gap.

(a) Express the torque in terms of F, F_s, and δ_s, where δ_s is the angle between $\bar{F}$ and $\bar{F}_s$.

(b) Express the torque in terms of F, F_r, and δ_r, where δ_r is the angle between $\bar{F}$ and $\bar{F}_r$.

(c) Neglecting magnetic saturation, obtain the torque in terms of B, F_r, and δ_r, where B is the peak value of the resultant flux-density wave.

(d) Let ϕ be the resultant flux per pole given by the product of the average value of the flux density over a pole and the pole area. Express the torque in terms of ϕ, F_r, and δ_r, where ϕ is the resultant flux produced by the combined effect of the stator and rotor mmfs.

12.4.16 An electromagnetic structure is characterized by the inductances $L_{11} = L_{22} = 4 + 2 \cos 2\theta$ and $L_{12} = L_{21} = 2 + \cos \theta$. Neglecting the resistances of the windings, find the torque as a function of θ when both windings are connected to the same ac voltage source such that

$$v_1 = v_2 = 220\sqrt{2} \sin 314t$$

12.4.17 By using the concept of interaction between magnetic fields, show that the electromagnetic torque cannot be obtained by using a four-pole rotor in a two-pole stator.

12.4.18 For each of the following devices, is a reluctance torque produced when their coils carry direct current?

(a) Salient-pole stator carrying a coil, and a salient-pole rotor.

(b) Salient-pole rotor carrying a coil, and a salient-pole stator.

(c) Salient-pole stator carrying a coil, and a cylindrical rotor.

(d) Salient-pole rotor carrying a coil, and a cylindrical stator.

(e) Cylindrical stator carrying a coil, and a cylindrical rotor.

(f) Cylindrical rotor carrying a coil, and a cylindrical stator.

(g) Cylindrical stator carrying a coil, and a salient-pole rotor.

(h) Cylindrical rotor carrying a coil, and a salient-pole stator.

*12.4.19 An elementary two-pole rotating machine with uniform air gap, as shown in Figure E12.4.1, has a stator-winding self-inductance L_{ss} of 50 mH, a rotor-winding self-inductance L_{rr} of 50 mH, and a maximum mutual inductance L of 45 mH. If the stator were excited from a 60-Hz source, and the rotor were excited from a 25-Hz source, at what speed or speeds would the machine be capable of converting energy?

12.4.20 A rotating electric machine with uniform air gap has a cylindrical rotor winding with inductance $L_2 = 1$ H and a stator winding with inductance $L_1 = 3$ H. The mutual inductance varies sinusoidally with the angle θ between the winding axes, with a maximum of 2 H. Resistances of the windings are negligible. Compute the mean torque if the stator current is 10 A (rms), the rotor is short-circuited, and the angle between the winding axes is $45°$.

12.4.21 The self and mutual inductances of a machine with two windings are given by $L_{11} = (1 + \sin \theta)$, $L_{22} = 2(1 + \sin \theta)$, and $L_{12} = L_{21} = M = (1 - \sin \theta)$. Assuming $\theta = 45°$, and letting coils 1 and 2 be supplied by constant currents $I_1 = 15$ A and $I_2 = -4$ A, respectively, find the following:

(a) Magnitude and direction of the developed torque.

(b) Amount of energy supplied by each source.

(c) Rms value of the current in coil 2, if the current of coil 1 is changed to a sinusoidal current of 10 A (rms) at 60 Hz and coil 2 is short-circuited.

(d) Instantaneous torque produced in part (c).

(e) Average torque in part (c).

12.4.22 Consider the elementary two-pole rotating machine with uniform air gap shown in Figure E12.4.1. Let N_s be the number of turns on the stator, N_r the number of turns on the rotor, l the axial length of the machine, r the radius of the rotor, and g the length of the air gap.

(a) Obtain expressions for self and mutual inductances, assuming infinite permeability for the magnetic cores.

(b) Find the expression for the electromagnetic torque in terms of currents i_s and i_r and angle θ.

(c) Show that the expression of part (b) is equivalent to

$$T_e = -\frac{\pi \mu_0 r l}{g} F_s F_r \sin \delta$$

12.5.1 A certain 10-hp, 230-V motor has a rotational loss of 600 W, a stator copper loss of 350 W, a rotor copper loss of 350 W, and a stray load loss of 50 W. It is not known whether the motor is an induction, synchronous, or dc machine.

(a) Calculate the full-load efficiency.

(b) Compute the efficiency at one-half load, assuming that the stray-loss and rotational losses do not change with load.

12.5.2 A synchronous motor operates continuously on the following duty cycle: 50 hp for 8 min, 100 hp for 8 min, 150 hp for 10 min, 120 hp for 20 min, and no load for 14 min. Specify the required continuous-rated hp of the motor. (*Note:* A motor rating is normally chosen on the basis of the rms value given by

$$\text{rms hp} = \sqrt{\frac{\sum(\text{hp})^2(\text{time})}{\text{running time} + \text{standstill time}/k}}$$

where k is a constant accounting for reduced ventilation at standstill.)

13 Rotating Machines

The most widely used electromechanical device is a rotating machine, which utilizes the magnetic field to store energy. The main purpose of most rotating machines is to convert electromechanical energy, i.e., to convert energy between electrical and mechanical systems, either for electric power generation (as in generators or sources) or for the production of mechanical power to perform useful tasks (as in motors or sinks). Rotating machines range in size and capacity from small motors that consume only a fraction of a watt to large generators that produce several hundred megawatts. In spite of the wide variety of types, sizes, and methods of construction, all such machines operate on the same principle, namely, the tendency of two magnets to align themselves.

Most space is devoted to induction, synchronous, and direct-current machines. In spite of the distinguishing features peculiar to each class of machines, there are several striking similarities among the main kinds of machines.

13.1 ELEMENTARY CONCEPTS OF ROTATING MACHINES

Some of the basic features of conventional ac (particularly synchronous) and dc machines have been introduced in Chapter 12. More will be presented in this section. Three modes of operation of a rotating electric machine—*motoring, generating,* and *braking*—are illustrated in Figure 13.1.1.

1. The *motoring mode* has electric power input and mechanical power output. The electromagnetic torque T_e drives the machine against the load torque T. The input voltage v drives the current into the winding against the generated emf e.

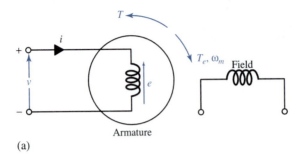

(a)

Figure 13.1.1 Three modes of operation of a rotating electric machine. (a) Motoring mode. (b) Generating mode. (c) Braking mode.

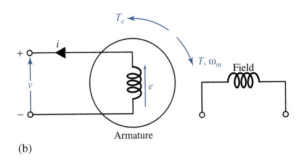

(b)

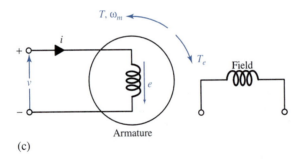

(c)

2. The *generating mode* has mechanical power input and electric power output. The torque T applied externally to the shaft drives the machine against the electrically developed torque T_e. The generated emf e drives current out of the winding against the terminal voltage v.

3. The *braking mode* has both mechanical and electric energy input. The total input is dissipated as heat. The machine is driven by the externally applied torque T, while the electromagnetic torque T_e is opposing T, thereby braking the machine. The electric braking of motor drives is achieved by causing the motor to act as a generator, receiving mechanical energy from the moving parts and converting it to electric energy, which is dissipated in a resistor or pumped back into the power line. Note that the applied voltage v and the generated emf e do not oppose each other.

As mentioned in Section 12.3, polyphase windings can be arranged to yield sinusoidally distributed current sheets and rotating mmfs. A number of possible doubly excited combinations, satisfying $\omega_s = \omega_m \pm \omega_r$, which relates electrical and mechanical angular speeds, exist, giving rise to the names by which the machines are generally known, synchronous, induction, or dc machines.

Elementary Synchronous Machines

The polyphase synchronous machine operates with direct current supplied to the field winding (assumed to be on the rotor, which is usually the case) through two slip rings and with polyphase alternating current supplied to the armature (assumed to be on the stator). The rotor mmf, which is obtained from a dc source, is stationary with respect to the rotor structure. When carrying balanced polyphase currents, as shown in Section 12.3, the armature winding produces a magnetic field in the air gap rotating at synchronous speed [Equation (12.2.9) or (12.2.10)], as determined by the system frequency and the number of poles in the machine. But the field produced by the dc rotor winding revolves with the rotor. To produce a steady unidirectional torque, the rotating fields of stator and rotor must be traveling at the same speed. Therefore, the rotor must turn precisely at the *synchronous speed*. Such conditions satisfy the second possibility for the development of a constant torque suggested in the discussion following Equation (12.4.32). The resulting physical system is shown in Figure 13.1.2.

The *polyphase synchronous machine* is one in which the rotor rotates in synchronism with and in the same direction as the rotating mmf wave produced by the stator. Thus for synchronous machines with polyphase stator winding, the following equation relating mechanical and electrical angular speeds is satisfied:

$$\omega_m = \omega_s = \frac{2\pi f}{P/2} \tag{13.1.1}$$

or alternatively, the steady-state speed at which the machine operates is known as the *synchronous speed*, given by

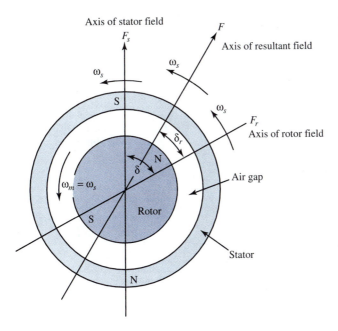

Figure 13.1.2 Simplified two-pole synchronous machine with non-salient poles.

$$\text{Synchronous speed} = \frac{120\,f}{P} \text{ r/min} \tag{13.1.2}$$

where f is the frequency of the system of which the machine is a part and P is the number of poles of the machine.

The electromagnetic torque, produced by the nonsalient-pole (or cylindrical-rotor) machine, can be expressed in terms of the resultant flux ϕ per pole produced by the combined effect of the stator and rotor mmfs (see Problem 12.4.15),

$$T_e = K\phi\,F_r\,\sin\,\delta_r \tag{13.1.3}$$

where K is a constant, F_r is the rotor mmf, and δ_r is the angle between the rotor mmf and the resultant flux or mmf axis. When the armature terminals are connected to a balanced polyphase *infinite bus* (which is a high-capacity, constant-voltage, constant-frequency system), the resultant air-gap flux ϕ is approximately constant, independent of the shaft load. Under normal operating conditions the resultant air-gap flux ϕ, which is given by Equation (12.2.13) as

$$\phi = \frac{\text{terminal phase voltage}}{4.44\,k_W\,f\,N_{\text{ph}}} \tag{13.1.4}$$

is essentially constant. The rotor mmf F_r determined by the direct field current is also a constant under normal operating conditions. So, as seen from Equation (13.1.3), any variation in the torque requirements of the load has to be accounted for entirely by variation of the angle δ_r, which is why δ_r is known as the *torque angle* (or *load angle*) of a synchronous machine. The effect of salient poles on the torque-angle characteristic is discussed in Section 13.3.

The torque-angle characteristic curve of a cylindrical-rotor synchronous machine is shown in Figure 13.1.3 as a function of the angle δ_r. For $\delta_r < 0$, $T_e > 0$, the developed torque is positive and acts in the direction of the rotation; the machine operates as a motor. If, on the other hand, the machine is driven by a prime mover so that δ_r becomes positive, the torque is then negative, and the machine operates as a generator. Note that at standstill, i.e., when $\omega_m = 0$, no average unidirectional torque is developed by the synchronous machine. Such a synchronous motor is not capable of self-starting because it has no starting torque. The designer must provide a method for bringing the machine up to synchronous speed.

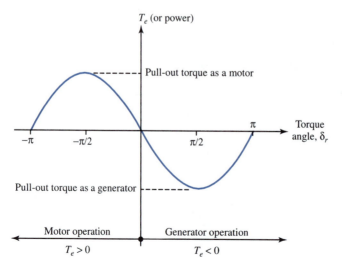

Figure 13.1.3 Torque-angle characteristic curve of a cylindrical-rotor synchronous machine (for a given field current and a fixed terminal voltage, i.e., for a constant resultant air-gap flux).

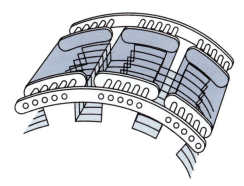

Figure 13.1.4 Sketch of damper bars located on salient-pole shoes of a synchronous machine.

Because a synchronous machine operates only at synchronous speed under steady-state conditions, the machine cannot operate at synchronous speed during the load transition when the load on the synchronous machine is to be changed; the readjustment process is in fact dynamic. In the case of a generator connected to an infinite bus, an increase in the electric output power of the generator is brought about by increasing the mechanical input power supplied by the prime mover. The speed of the rotor increases momentarily during the process, and the axis of the rotor-field mmf advances relative to the axes of both the armature mmf and the resultant air-gap mmf. The increase in torque angle results in an increase in electric power output. The machine then locks itself into synchronism and continues to rotate at synchronous speed until the load changes further. The same general argument can be made for the operation of a synchronous motor, except that an increase in mechanical load decreases the speed of the rotor during the transition period, so that the axis of the rotor-field mmf falls behind that of the stator mmf or the resultant air-gap mmf by the required value of the load angle.

When the load on a synchronous machine changes, the load angle changes from one steady value to another. During this transition, oscillations in the load angle and consequent associated mechanical oscillations (known as *hunting*) occur. To damp out these oscillations, it is common practice to provide an additional short-circuited winding on the field structure, made out of copper or brass bars located in pole-face slots on the pole shoes of a salient-pole machine and connected together at the ends of the machine. This additional winding (known as *damper* or *amortisseur winding*) is shown in Figure 13.1.4. The winding is also useful in getting a synchronous motor started, as explained in the subsection on elementary induction machines. When the machine is operating under steady-state conditions at synchronous speed, however, this winding has no effect. Because there is no rate of change of flux linkage, no voltage is induced in it.

As seen in Figure 13.1.3, when δ_r is $\pm\pi/2$ or $90°$, the maximum torque or power (called *pull-out torque* or *pull-out power*) is reached for a fixed terminal voltage and a given field current. If the load requirements exceed this value, the motor slows down because of the excess shaft torque. Synchronous-motor action is lost because the rotor and stator fields are no longer stationary with respect to each other. Any load requiring a torque greater than the maximum torque results in unstable operation of the machine; the machine pulls out of synchronism, known as *pulling out of step* or losing synchronism. The motor is usually disconnected from the electric supply by automatic circuit breakers, and the machine comes to a standstill. Note that the pull-out torque can be increased by increasing either the field current or the terminal voltage. In the case of a generator connected to an infinite bus, synchronism will be lost if the torque applied by the prime mover exceeds the maximum generator pull-out torque. The speed will then increase rapidly unless the quick-response governor action comes into play on the prime mover to control the speed.

Elementary Induction Machines

In the discussion that followed Equation (12.4.32), the third possible method of producing constant torque was to cause the mmf axes of stator and rotor to rotate at such speeds relative to their windings that they remain stationary with respect to each other. If the stator and rotor windings are polyphase and carry polyphase alternating current, then both the stator mmf and the rotor mmf axes may be caused to rotate relative to their windings. Such a machine will have polyphase stator ac excitation at ω_s, polyphase rotor ac excitation at ω_r, and the rotor speed ω_m satisfying

$$\omega_m = \omega_s - \omega_r \tag{13.1.5}$$

Let us consider the rotor speed given by Equation (13.1.5) and the same phase sequence of sources. A rotating magnetic field of constant amplitude, rotating at ω_s rad/s relative to the stator, is produced because of polyphase stator excitation. A rotating magnetic field of constant amplitude, rotating at ω_s rad/s relative to the rotor, is also produced because of polyphase rotor excitation. The speed of rotation of the rotor magnetic field relative to the stator is $\omega_m + \omega_r$, or ω_s, if the rotor is rotating with a positive speed of rotation ω_m in the direction of the rotating fields. If so, the condition for energy conversion at constant torque is satisfied. Such a situation is illustrated diagrammatically in Figure 13.1.5. Under these conditions the machine is operating as a double-fed polyphase machine. Normally, in an induction machine with polyphase stator and rotor windings, only a source to excite the stator is employed, and the rotor excitation at the appropriate frequency is induced from the stator winding. The device is thus known as an *induction machine*.

In an induction machine, the stator winding is essentially the same as in a synchronous machine. Equation (12.2.13) and the considerations leading to it hold here as in the case of a synchronous machine. When excited from a balanced polyphase source, the polyphase windings produce a magnetic field in the air gap rotating at a synchronous speed determined by the number of poles and the applied stator frequency given by Equation (12.2.9). On the rotor, the winding is electrically closed on itself (i.e., short-circuited) and often has no external terminals.

The induction machine rotor may be one of two types: the *wound rotor* or the *squirrel-cage rotor*. The wound rotor has a polyphase winding similar to and wound for the same number of poles as the stator winding. The terminals of the rotor winding (wye- or delta-connected, in the case of three-phase machines) are brought to insulated slip rings mounted on the shaft. Carbon brushes bearing on these slip rings make the rotor terminals available to the circuitry external to

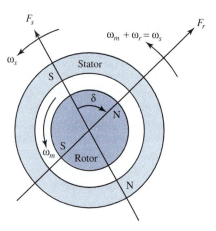

Figure 13.1.5 Mmf axes of an induction machine.

the motor. The rotor winding is usually short-circuited through external resistances that can be varied. A squirrel-cage rotor has a winding consisting of conducting bars of copper or aluminum embedded in slots cut in the rotor iron and short-circuited at each end by conducting end rings. The squirrel-cage induction machine is the electromagnetic machine most widely used as a motor because of its extreme simplicity and ruggedness. Although the induction machine in the motor mode is the most common of all motors, the induction machine has very rarely been used as a generator because its performance characteristics as a generator are not satisfactory for most applications; however, it has recently been used as a wind-power generator. The induction machine with a wound rotor is also used as a *frequency changer.*

The polyphase induction motor operates with polyphase alternating current applied to the primary winding, usually located on the stator of the polyphase machines. Three-phase motors are most used commercially in practice, whereas two-phase motors are used in control systems. The induction machine has emf (and consequently current) induced in the short-circuited secondary (or rotor) winding by virtue of the primary rotating mmf. Such a machine is then singly excited. The induction machine may be regarded as a generalized transformer in which energy conversion takes place, and electric power is transformed between the stator and the rotor along with a change of frequency and a flow of mechanical power.

Let us assume that the rotor is turning at a steady speed of n r/min in the same direction as the rotating stator field. Let the synchronous speed of the stator field be n_1 r/min, as given by Equation (12.2.9), corresponding to the applied stator frequency f_s Hz, or ω_s rad/s. It is convenient to introduce the concept of *per-unit slip* $\mathbf{S}$ given by

$$\mathbf{S} = \frac{\text{synchronous speed} - \text{actual rotor speed}}{\text{synchronous speed}} = \frac{n_1 - n}{n_1} = \frac{\omega_s - \omega_m}{\omega_s} \qquad (13.1.6)$$

The rotor is then traveling at a speed of $n_1 - n$ or $n_1\mathbf{S}$ r/min in the backward direction with respect to the stator field. The relative motion of the flux and rotor conductors induces voltages of frequency $\mathbf{S}f_s$, known as the *slip frequency,* in the rotor winding. Thus, the induction machine is similar to a transformer in its electrical behavior but with an additional feature of frequency change. The frequency f_r Hz of the secondary (or rotor) currents is then given by

$$f_r = \frac{\omega_r}{2\pi} = \mathbf{S}f_s \qquad (13.1.7)$$

At standstill, $\omega_m = 0$, so that the slip $\mathbf{S} = 1$ and $f_r = f_s$; that is, the machine then acts as a simple transformer with an air gap and a short-circuited secondary winding. A steady starting torque is produced because the condition for energy conversion at constant torque is satisfied; hence the polyphase induction motor is self-starting. At synchronous speed, however, $\omega_m = \omega_s$, so that the slip $\mathbf{S} = 0$ and $f_r = 0$; no induction takes place because there is no relative motion between flux and rotor conductors. Thus, at synchronous speed, the value of the secondary mmf is zero, and no torque is produced; that is, the induction motor cannot run at synchronous speed. The no-load speed of the induction motor is usually on the order of 99.5% of synchronous speed so that the no-load per-unit slip is about 0.005, and the full-load per-unit slip is on the order of 0.05. Thus, the polyphase induction motor is effectively a constant-speed machine.

An induction machine, connected to a polyphase exciting source on its stator side, can be made to generate (i.e., with the power flow reversed compared to that of a motor) if its rotor is driven mechanically by an external means at above synchronous speed, so that $\omega_m > \omega_s$ and the slip becomes negative. If the machine is driven mechanically in the direction opposite to its primary rotating mmf, then the slip is greater than unity and the machine acts as a brake. For example, let the machine be operating normally as a loaded motor; if two of the three phase supply lines to the stator are reversed, the direction of the stator rotating mmf will reverse. The rotor will then be rotating in the direction opposite that of the rotating mmf, so the machine will act as a brake and the speed will rapidly come to zero, at which time the electric supply can be removed from the machine. Such a reversal of two supply lines of the three-phase system, a useful method of stopping the motor rapidly, is generally referred to as *plugging* or *plug-braking*. If the electric supply is not removed at zero speed, however, the machine will reverse its direction of rotation because of the change of phase sequence of the supply resulting from the interchange of the two stator leads. The general form of the torque–speed curve (or torque–slip characteristic) for the polyphase induction machine between rotor speed limits of $-\omega_s \leq \omega_m \leq 2\omega_s$, corresponding to a range of slips $-1 \leq \mathbf{S} \leq 2$, is shown in Figure 13.1.6.

The torque that exists at any mechanical speed other than synchronous speed is known as an *asynchronous torque*. The induction machine is also known as an *asynchronous machine,* since no torque is produced at synchronous speed and the machine runs at a speed other than synchronous speed. In fact, as a motor, the machine runs only at a speed that is less than synchronous speed with positive slip. The factors influencing the general shape of the torque-speed characteristic (shown in Figure 13.1.6) can be appreciated in terms of the torque equation, Equation (13.1.3). Noting that the resultant air-gap flux ϕ is nearly constant when the stator-applied voltage and frequency are constant, as seen by Equation (13.1.4), and that the rotor mmf F_r is proportional to the rotor current i_r, the torque may be expressed as

$$T_e = K_1 i_r \sin \delta_r \qquad (13.1.8)$$

where K_1 is a constant and δ_r is the angle between the rotor mmf axis and the resultant flux or mmf axis. The rotor current i_r is determined by the rotor-induced voltage (proportional to slip) and the rotor impedance. Since the slip is small under normal running conditions, as already mentioned, the rotor frequency $f_r = \mathbf{S}f_s$ is very low (on the order of 3 Hz in 60-Hz motors with a per-unit slip of 0.05). Hence, in this range the rotor impedance is largely resistive, and the rotor current is very nearly proportional to and in phase with the rotor voltage; that is, the rotor current is very nearly proportional to slip. An approximately linear torque–speed relationship can be observed in the range of low values of slip in Figure 13.1.6. Further, with the rotor-leakage reactance being very small compared with the rotor resistance, the rotor mmf wave lags approximately 90 electrical degrees behind the resultant flux wave, and therefore $\sin \delta_r$ is approximately equal to unity.

As slip increases, the rotor impedance increases because of the increasing effect of rotor-leakage inductance; the rotor current is then somewhat less than proportional to slip. The rotor current lags further behind the induced voltage, and the rotor mmf wave lags further behind the resultant flux wave, so that $\sin \delta_r$ decreases. The torque increases with increasing values of slip up to a point and then decreases, as shown in Figure 13.1.6 for the motor region. The maximum torque that the machine can produce is sometimes referred to as the *breakdown torque,* because it limits the short-time overload capability of the motor. Higher starting torque can be obtained by inserting external resistances in the rotor circuit, as is usually done in the case of the wound-rotor induction motor. These resistances can be cut out for the normal running conditions in order to operate the machine with a higher efficiency. Recall that a synchronous motor has no starting torque. It is usually provided with a damper or amortisseur winding located in the rotor pole faces. Such a winding acts like a squirrel-cage winding to make the synchronous motor start as an induction motor and come up almost to synchronous speed, with the dc field winding unexcited. If the load and inertia are not too large, the motor will pull into synchronism and act as a synchronous motor when the field winding is energized from a dc source.

So far the discussion of induction machines applies only to machines operating from a polyphase supply. Of particular interest is the single-phase induction machine, which is widely used as a fractional-horsepower ac motor supplying the motive power for all kinds of equipment

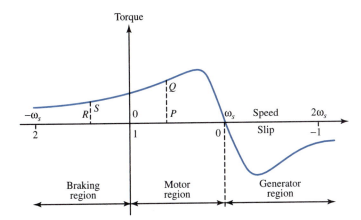

Figure 13.1.6 General form of torque–speed curve (or torque–slip characteristic) for a polyphase induction machine.

in the home, office, and factory. For the sake of simplicity, let us consider a single-phase induction motor with a squirrel-cage rotor and a stator carrying a single-phase winding, connected to a single-phase ac supply. The primary mmf cannot be rotating, but, in fact, it is pulsating in phase with the variations in the single-phase primary current. It can be shown, however, that any pulsating mmf can be resolved in terms of two rotating mmfs of equal magnitude, rotating in synchronism with the supply frequency but in opposite directions (see Section 12.3). The wave rotating in the same direction as the rotor is known as the *forward-rotating wave,* whereas the one rotating in the opposite direction is the *backward-rotating wave.* Then the slip S of the machine with respect to the forward-rotating wave is given by

$$S_f = \frac{\omega_s - \omega_m}{\omega_s} \tag{13.1.9}$$

which is the same as Equation (13.1.6). The slip S_b of the machine with respect to the backward-rotating wave, however, is given by

$$S_b = \frac{-\omega_s - \omega_m}{\omega_s} = 2 - S_f \tag{13.1.10}$$

S_f and S_b are known as forward (or positive-sequence) slip and backward (or negative-sequence) slip, respectively. Assuming that the two component mmfs exist separately, the frequency and magnitude of the component emfs induced in the rotor by their presence will, in general, be different because S_f is not equal to S_b. Then the machine can be thought of as producing a steady total torque as the algebraic sum of the component torques. The equivalent circuit based on the *revolving-field theory* is pursued in Section 13.2. At standstill, however, $\omega_m = 0$ and the component torques are equal and opposite; no starting torque is produced. Thus, it is clear that a single-phase induction motor is not capable of self-starting, but it will continue to rotate once started in any direction. In practice, additional means are provided to get the machine started (usually as an asymmetrical two-phase motor) from a single-phase source, and the machine is then run as a single-phase motor. An approximate shape of the torque–speed curve for the single-phase motor can readily be obtained from that of the three-phase machine shown in Figure 13.1.6. Corresponding to a positive slip of $S_f = OP$ in Figure 13.1.6, the positive-sequence torque is PQ. Then, corresponding to $S_b = 2 - S_f = OR$, the negative-sequence torque is RS. The resultant torque is given by $PQ - RS$. This procedure can be repeated for a range of slips $1 \leq S \leq 0$ to give the general form of the torque–speed characteristic of a single-phase induction motor, as shown in Figure 13.1.7.

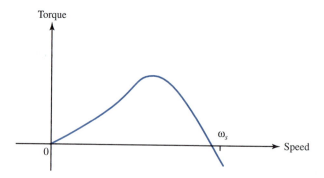

Figure 13.1.7 Approximate shape of torque–speed curve for a single-phase induction motor.

Elementary Direct-Current Machines

A preliminary discussion of dc machines, including Equations (12.2.14) through (12.2.18), was presented in Section 12.2. The location of the brushes on the commutator arrangement connected to the armature winding ensures that the rotor and stator mmf axes are at all times at right angles to one another, as shown in Figure 12.2.4. The expression for torque given by Equation (12.4.32) is applicable to the dc machine, provided the constant K is adjusted for the nonsinusoidal mmf distributions and $\sin \delta$ is made equal to unity. Not only are the conditions for constant torque fulfilled, but also the condition for maximum torque and hence for maximum energy conversion is satisfied, since $\delta = \pm\pi/2$, depending upon whether generator or motor action occurs for a given direction of rotation.

Since both armature and field circuits carry direct current in the case of a dc machine, they can be connected either in series or in parallel. When the armature and field circuits are connected in parallel, the machine is known as a *shunt machine*. In the shunt machine, the field coils are wound with a large number of turns carrying a relatively small current. When the circuits are connected in series, the machine is known as a *series machine*. The field winding in the series machine carries the full armature current and is wound with a smaller number of turns. A dc machine provided with both a series-field and a shunt-field winding is known as a *compound machine*. In the compound machine, the series field may be connected either *cumulatively,* so that its mmf adds to that of the shunt field, or *differentially,* so that the mmf opposes. The differential connection is used very rarely. The voltage of both shunt and compound generators (or the speed, for motors) is controlled over reasonable limits by means of a field rheostat in the shunt field. The machines are said to be *self-excited* when the machine supplies its own excitation of the field windings, as in the cases just defined. The field windings may be *excited separately* from an external dc source, however. A small amount of power in the field circuit can control a large amount of power in the armature circuit. The dc generator may then be viewed as a power amplifier. Some possible field-circuit connections of dc machines are shown in Figure 13.1.8.

For self-excited generators, residual magnetism must be present in the ferromagnetic circuit of the machine in order to start the self-excitation process. For a dc *generator,* the relationship between steady-state generated emf E_a and the terminal voltage V_t is given by

$$V_t = E_a - I_a R_a \tag{13.1.11}$$

where I_a is the armature current *output,* and R_a is the armature circuit resistance. For a dc *motor,* the relationship is given by

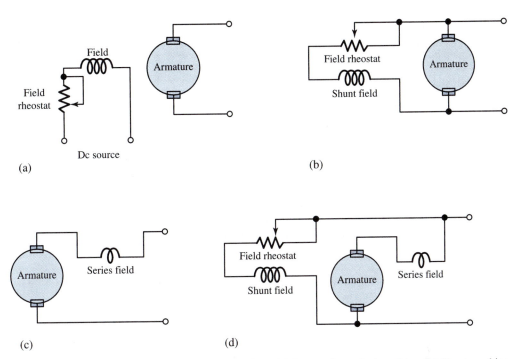

Figure 13.1.8 Field-circuit connections of dc machines. (**a**) Separately excited machine. (**b**) Shunt machine. (**c**) Series machine. (**d**) Compound machine.

$$V_t = E_a + I_a R_a \tag{13.1.12}$$

where I_a is now the armature current *input*. Under steady-state conditions, *volt–ampere* characteristic curves are of interest for dc generators, and *speed–torque* characteristics are of interest for dc motors. Depending on the method of excitation of the field windings, a wide variety of operating characteristics can be obtained. These possibilities make the dc machine both versatile and adaptable for control. More about dc machines is presented in Section 13.4.

13.2 INDUCTION MACHINES

We shall first consider polyphase induction motors and then single-phase induction motors. The polyphase induction motors used in industrial applications are almost without exception three-phase. The stator winding is connected to the ac source, and the rotor winding is either short-circuited, as in squirrel-cage machines, or closed through external resistances, as in wound-rotor machines. The cage machines are also known as *brushless* machines; the wound-rotor machines are also called *slip-ring* machines.

A review of the subsection on elementary induction machines in Section 13.1 can be helpful at this stage to recall the operating principle of polyphase induction machines.

Equivalent Circuit of a Polyphase Induction Machine

The induction machine may be regarded as a generalized transformer in which energy is converted and electric power is transferred between stator and rotor, along with a change of frequency and a flow of mechanical power. At standstill, however, the machine acts as a simple transformer with an air gap and a short-circuited secondary winding. The frequency of the rotor-induced emf is the same as the stator frequency at standstill. At any value of the slip under balanced steady-state operation, the rotor current reacts on the stator winding at the stator frequency because the rotating magnetic fields caused by the stator and rotor are stationary with respect to each other.

The induction machine may thus be viewed as a transformer with an air gap and variable resistance in the secondary; the stator of the induction machine corresponds to the transformer primary, and the rotor corresponds to the secondary. For analysis of the balanced steady state, it is sufficient to proceed on a *per-phase basis* with some phasor concepts; so we will now develop an equivalent circuit on a per-phase basis. Only machines with symmetrical polyphase windings excited by balanced polyphase voltages are considered. As in other discussions of polyphase devices, let us think of three-phase machines as wye-connected, so that currents are always line values and voltages are always line-to-neutral values (on a per-phase basis).

The resultant air-gap flux is produced by the combined mmfs of the stator and rotor currents. For the sake of conceptual and analytical convenience, the total flux is divided into a mutual flux (linking both the stator and the rotor) and leakage fluxes, represented by appropriate reactances.

Considering the conditions in the stator, the synchronously rotating air-gap wave generates balanced polyphase counter emfs in the phases of the stator. The volt–ampere equation for the phase under consideration *in phasor notation* is given by

$$\bar{V}_1 = \bar{E}_1 + \bar{I}_1 (R_1 + jX_{l1}) \tag{13.2.1}$$

where $\bar{V}_1$ is the stator terminal voltage, $\bar{E}_1$ is the counter emf generated by the resultant air-gap flux, $\bar{I}_1$ is the stator current, R_1 is the stator effective resistance, and X_{l1} is the stator-leakage reactance.

As in a transformer, the stator (primary) current can be resolved into two components: a load component $\bar{I}_2'$ (which produces an mmf that exactly counteracts the mmf of the rotor current) and an excitation component $\bar{I}_0$ (required to create the resultant air-gap flux). This excitation component itself can be resolved into a core-loss component $\bar{I}_c$ in phase with $\bar{E}_1$

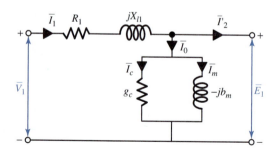

Figure 13.2.1 Equivalent circuit for the stator phase of a polyphase induction motor.

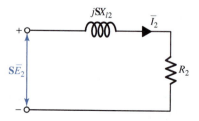

Figure 13.2.2 Slip-frequency equivalent circuit for the rotor phase of a polyphase induction motor.

and a magnetizing component $\bar{I}_m$ lagging $\bar{E}_1$ by 90°. A shunt branch formed by the core-loss conductance g_c and magnetizing susceptance b_m in parallel, connected across $\bar{E}_1$, will account for the exciting current in the equivalent circuit, as shown in Figure 13.2.1, along with the positive directions in a motor.

Thus far the equivalent circuit representing the stator phenomenon is exactly like that of the transformer primary. Because of the air gap, however, the value of the magnetizing reactance tends to be relatively low compared to that of a transformer, and the leakage reactance is larger in proportion to the magnetizing reactance than it is in transformers. To complete the equivalent circuit, the effects of the rotor must be incorporated, which we do by referring the rotor quantities to the stator.

Because the frequency of the rotor voltages and currents is the slip frequency, the magnitude of the voltage induced in the rotor circuit is proportional to the slip. Also, in terms of the standstill per-phase rotor-leakage reactance X_{l2}, the leakage reactance at a slip S is given by SX_{l2}. With R_2 as the per-phase resistance of the rotor, the slip-frequency equivalent circuit for a rotor phase is shown in Figure 13.2.2, in which E_2 is the per-phase voltage induced in the rotor at standstill. The rotor current I_2 is given by

$$I_2 = \frac{SE_2}{\sqrt{R_2^2 + (SX_{l2})^2}} \tag{13.2.2}$$

which may be rewritten as

$$I_2 = \frac{E_2}{\sqrt{(R_2/S)^2 + X_{l2}^2}} \tag{13.2.3}$$

resulting in the alternate form of the per-phase rotor equivalent circuit shown in Figure 13.2.3.

All rotor electrical phenomena, when viewed from the stator, become stator-frequency phenomena because the stator winding sees the mmf and flux waves traveling at synchronous

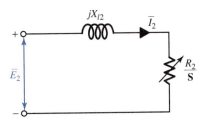

Figure 13.2.3 Alternate form of a per-phase rotor equivalent circuit.

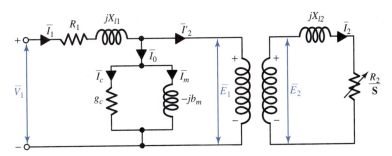

Figure 13.2.4 Per-phase equivalent coupled circuit of a polyphase induction motor.

speed. Returning to the analogy of a transformer, and considering that the rotor is coupled to the stator in the same way the secondary of a transformer is coupled to its primary, we may draw the equivalent circuit as shown in Figure 13.2.4.

Referring the rotor quantities to the stator, we can now draw the per-phase equivalent circuit of the polyphase induction motor, as shown in Figure 13.2.5(a). The combined effect of the shaft load and the rotor resistance appears as a reflected resistance R_2'/S, which is a function of slip and therefore of the mechanical load. The quantity R_2'/S may conveniently be split into two parts:

$$\frac{R_2'}{S} = R_2' + \frac{R_2'(1-S)}{S} \tag{13.2.4}$$

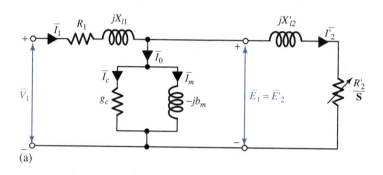

(a)

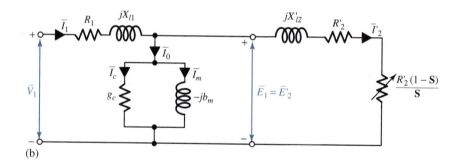

(b)

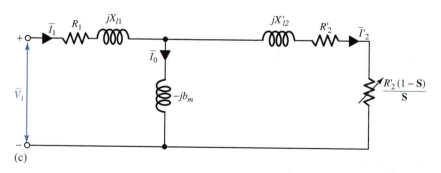

(c)

Figure 13.2.5 Per-phase equivalent circuits of a polyphase induction motor, referred to the stator.

and the equivalent circuit may be redrawn as in Figure 13.2.5(b). R_2' is the per-phase standstill rotor resistance referred to the stator, and $R_2'[(1 - S)/S]$ is a dynamic resistance that depends on the rotor speed and corresponds to the load on the motor. [See the discussion following Equation (13.2.6).]

When power aspects need to be emphasized, the equivalent circuit is frequently redrawn as in Figure 13.2.5(c), in which the shunt conductance g_c is omitted. The core losses can be included in efficiency calculations along with the friction, windage, and stray-load losses.

Recall that, in static transformer theory, analysis of the equivalent circuit is often simplified either by neglecting the exciting shunt branch entirely, or by adopting the approximation of moving it out directly to the terminals. For the induction machine, however, such approximations might not be permissible under normal running conditions because the air gap leads to a much higher exciting current (30 to 50% of full-load current) and relatively higher leakage reactances.

The parameters of the equivalent circuit of an induction machine can be obtained from the *no-load* (in which the motor is allowed to run on no load) and *blocked-rotor* (in which the rotor of the induction motor is blocked so that the slip is equal to unity) tests. These tests correspond to the no-load and short-circuit tests on the transformer, and are very similar in detail.

Polyphase Induction Machine Performance

Some of the important steady-state performance characteristics of a polyphase induction motor include the variation of current, speed, and losses as the load–torque requirements change, and the starting and maximum torque. Performance calculations can be made from the equivalent circuit. All calculations can be made on a *per-phase basis,* assuming balanced operation of the machine. Total quantities can be obtained by using an appropriate multiplying factor.

The equivalent circuit of Figure 13.2.5(c), redrawn for convenience in Figure 13.2.6, is usually employed for the analysis. The core losses, most of which occur in the stator, as well as friction, windage, and stray-load losses, are included in the efficiency calculations. The power-flow diagram for an induction motor is given in Figure 13.2.7, in which m_1 is the number of stator phases, ϕ_1 is the power factor angle between $\bar{V}_1$ and $\bar{I}_1$, ϕ_2 is the power factor angle between $\bar{E}_1$ and $\bar{I}_2'$, T is the internal electromagnetic torque developed, ω_s is the synchronous angular velocity in mechanical radians per second, and ω_m is the actual mechanical rotor speed given by $\omega_s(1 - S)$.

The total power P_g transferred across the air gap from the stator is the difference between the electric power input P_i and the stator copper loss. P_g is thus the total rotor input power, which is dissipated in the resistance R_2'/S of each phase so that

$$P_g = m_1(I_2')^2 \frac{R_2'}{S} = T\omega_s \qquad (13.2.5)$$

Subtracting the total rotor copper loss, which is $m_1(I_2')^2 R_2'$ or SP_g, from Equation (13.2.5) for P_g, we get the internal mechanical power developed,

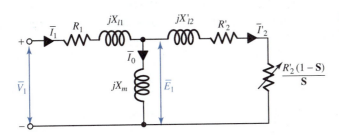

Figure 13.2.6 Per-phase equivalent circuit of a polyphase induction motor used for performance calculations.

$$P_m = P_g(1 - S) = T\omega_m = m_1(I_2')^2 \frac{R_2'(1 - S)}{S} \qquad (13.2.6)$$

This much power is absorbed by a resistance of $R_2'(1 - S)/S$, which corresponds to the load. For this reason, the resistance term R_2'/S has been split into two terms, as in Equation (13.2.4) and shown in the equivalent circuit of Figure 13.2.6. From Equation (13.2.6) we can see that of the total power delivered to the rotor, the fraction $1 - S$ is converted to mechanical power and the fraction S is dissipated as rotor copper loss. We can then conclude that an induction motor operating at high slip values will be inefficient.

The total rotational losses, including the core losses, can be subtracted from P_m to obtain the mechanical power output P_o that is available in mechanical form at the shaft for useful work,

$$P_o = P_m - P_{rot} = T_o\omega_m \qquad (13.2.7)$$

The per-unit efficiency of the induction motor is then given by

$$\eta = P_o/P_i \qquad (13.2.8)$$

Let us now illustrate this procedure and the analysis of the equivalent circuit in the following example.

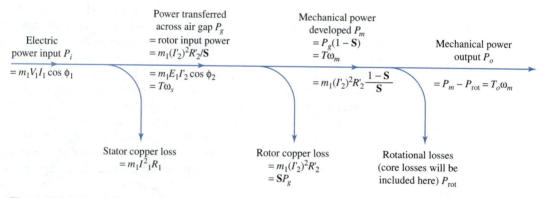

Figure 13.2.7 Power flow in an induction motor.

EXAMPLE 13.2.1

The parameters of the equivalent circuit shown in Figure 13.2.6 for a three-phase, wye-connected, 220-V, 10-hp, 60-Hz, six-pole induction motor are given in ohms per phase referred to the stator: $R_1 = 0.3$, $R_2' = 0.15$, $X_{l1} = 0.5$, $X_{l2}' = 0.2$, and $X_m = 15$. The total friction, windage, and core losses can be assumed to be constant at 400 W, independent of load. For a per-unit slip of 0.02, when the motor is operated at rated voltage and frequency, calculate the stator input current, the power factor at the stator terminals, the rotor speed, output power, output torque, and efficiency.

Solution

From the equivalent circuit of Figure 13.2.6, the total impedance per phase, as viewed from the stator input terminals, is given by

$$Z_t = R_1 + jX_{l1} + \frac{jX_m\left(\dfrac{R_2'}{S} + jX_{l2}'\right)}{\dfrac{R_2'}{S} + j(X_M + X_{l2}')} = 0.3 + j0.5 + \frac{j15(7.5 + j0.2)}{7.5 + j(15 + 0.2)}$$

$$= (0.3 + j0.5) + (5.87 + j3.10) = 6.17 + j3.60$$

$$= 7.14\angle 30.26° \ \Omega$$

Phase voltage $= 220/\sqrt{3} = 127$ V
Stator input current $= 127/7.14 = 17.79$ A
Power factor $= \cos 30.26° = 0.864$
Synchronous speed $= 120 \times 60/6 = 1200$ r/min
Rotor speed $= (1 - 0.02)1200 = 1176$ r/min
Total input power $= \sqrt{3} \times 220 \times 17.79 \times 0.864 = 5856.8$ W
Stator copper loss $= 3 \times 17.79^2 \times 0.3 = 284.8$ W
Power transferred across air gap $P_g = 5856.8 - 284.8 = 5572$ W

P_g can also be obtained as follows:

$$P_g = m_1(I_2')^2 R_2'/S = m_1 I_1^2 R_f$$

where R_f is the real part of the parallel combination of jX_m and $R_2'/S + jX_{l2}'$. Thus,

$$P_g = 3 \times 17.79^2 \times 5.87 = 5573 \text{ W}$$

Internal mechanical power developed $= P_g(1 - S) = 0.98 \times 5572 = 5460$ W
Total mechanical power output $= 5460 - 400 = 5060$ W, or $5060/745.7 = 6.8$ hp
Total output torque $= \dfrac{\text{output power}}{\omega_m} = \dfrac{\text{output power}}{(1-S)\omega_s}$

Since,

$$\omega_s = \frac{4\pi f}{\text{poles}} = \frac{4\pi \times 60}{6} = 40\pi = 125.7 \text{ mechanical rad/s}$$

it follows that

Total output torque $= \dfrac{5060}{0.98 \times 125.7} = 41.08$ N · m

Efficiency $= \dfrac{5060}{5856.8} = 0.864$, or 86.4%

The efficiency may alternatively be calculated from the losses:

Total stator copper loss $= 284.8$ W
Rotor copper loss $= m_1(I_2')^2 R_2' = SP_g = 0.02 \times 5572 = 111.4$ W
Friction, windage, and core losses $= 400$ W
Total losses $= 284.8 + 111.4 + 400 = 796.2$ W
Output $= 5060$ W
Input $= 5060 + 796.2 = 5856.2$ W
Efficiency $= 1 - \dfrac{\text{losses}}{\text{input}} = 1 - \dfrac{796.2}{5856.2} = 1 - 0.136 = 0.864$, or 86.4%

Torque-Speed Characteristics of 3-Phase Induction Motors

Because the torque–slip characteristic is one of the most important aspects of the induction motor, we will now develop an expression for torque as a function of slip and other equivalent circuit parameters. Recalling Equation (13.2.8) and the equivalent circuit of Figure 13.2.6, we will obtain an expression for I_2'. To that end, let us redraw the equivalent circuit in Figure 13.2.8. By applying Thévenin's theorem, we have the following from Figures 13.2.6 and 13.2.8:

$$\bar{V}_{1a} = \bar{V}_1 - \bar{I}_0(R_1 - jX_{l1}) = \bar{V}_1 \frac{jX_m}{R_1 + j(X_{l1} + X_m)} \tag{13.2.9}$$

$$R_1'' + jX_1'' = \frac{(R_1 + jX_{l1})jX_m}{R_1 + j(X_{l1} + X_m)} \tag{13.2.10}$$

$$I_2' = \frac{V_{1a}}{\sqrt{[R_1'' + (R_2'/S)]^2 + (X_1'' + X_{l2}')^2}} \tag{13.2.11}$$

$$T = \frac{1}{\omega_s} \frac{m_1 V_{1a}^2 (R_2'/S)}{[R_1'' + (R_2'/S)]^2 + (X_1'' + X_{l2}')^2} \tag{13.2.12}$$

Neglecting the stator resistance in Equation (13.2.9) results in negligible error for most induction motors. If X_m of the equivalent circuit shown in Figure 13.2.6 is sufficiently large that the shunt branch need not be considered, calculations become much simpler; R_1'' and X_1'' are then equal to R_1 and X_{l1}, respectively; also, V_{1a} is then equal to V_1.

The general shape of the torque–speed or torque–slip characteristic is shown in Figure 13.1.6, in which the motor region ($0 < \mathbf{S} \leq 1$), the generator region ($\mathbf{S} < 0$), and the breaking region ($\mathbf{S} > 1$) are included for completeness. The performance of an induction motor can be characterized by such factors as efficiency, power factor, starting torque, starting current, pull-out (maximum) torque, and maximum internal power developed. Starting conditions are those corresponding to $\mathbf{S} = 1$.

The maximum internal (or breakdown) torque $T_{\max}$ occurs when the power delivered to $R_2'/\mathbf{S}$ in Figure 13.2.8 is a maximum. Applying the familiar impedance-matching principle of circuit theory, this power will be a maximum when the impedance $R_2'/\mathbf{S}$ equals the magnitude of the impedance between that and the constant voltage V_{1a}. That is to say, the maximum occurs at a value of slip $\mathbf{S}_{\max T}$ for which the following condition is satisfied:

$$\frac{R_2'}{\mathbf{S}_{\max T}} = \sqrt{(R_1'')^2 + (X_1'' + X_{l2}')^2} \tag{13.2.13}$$

The same result can also be obtained by differentiating Equation (13.2.12) with respect to $\mathbf{S}$, or, more conveniently, with respect to $R_2'/\mathbf{S}$, and setting the result equal to zero. This calculation has been left to the enterprising student. The slip corresponding to maximum torque $\mathbf{S}_{\max T}$ is thus given by

$$\mathbf{S}_{\max T} = \frac{R_2'}{\sqrt{(R_1'')^2 + (X_1'' + X_{l2}')^2}} \tag{13.2.14}$$

and the corresponding maximum torque from Equation (13.2.12) results in

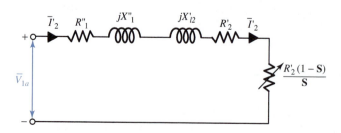

Figure 13.2.8 Another form of per-phase equivalent circuit for the polyphase induction motor of Figure 13.2.6.

$$T_{max} = \frac{1}{\omega_s} \frac{0.5 m_1 V_{1a}^2}{R_1'' + \sqrt{(R_1'')^2 + (X_1'' + X_{l2}')^2}} \qquad (13.2.15)$$

which can be verified by the reader. Equation (13.2.15) shows that the maximum torque is independent of the rotor resistance. The slip corresponding to the maximum torque is directly proportional to the rotor resistance R_2', however, as seen from Equation (13.2.14). Thus, when the rotor resistance is increased by inserting external resistance in the rotor of a wound-rotor induction motor, the maximum internal torque is unaffected, but the speed or slip at which it occurs is increased, as shown in Figure 13.2.9. Also note that maximum torque and maximum power do not occur at the same speed. The student is encouraged to work out the reason.

A conventional induction motor with a squirrel-cage rotor has about 5% drop in speed from no load to full load, and is thus essentially a constant-speed motor. Employing a wound-rotor motor and inserting external resistance in the rotor circuit achieves speed variation but results in poor efficiency. Variations in the starting torque (at $\mathbf{S} = 1$) with rotor-circuit resistance can also be seen from Figure 13.2.9. As stated in Section 13.1, we can obtain a higher starting torque by inserting external resistances in the rotor circuit and then cutting them out eventually for the normal running conditions in order to operate the machine at a higher efficiency. Creating a sufficiently large rotor-circuit resistance might make it possible to achieve an almost linear torque–slip relationship for the slip range of 0 to 1. For instance, two-phase servo motors (used as output actuators in feedback control systems) are usually designed with very high rotor resistance to ensure a negative slope for the torque–speed characteristic over the entire operating range.

Since the stator resistance is quite low and has only a negligible influence, let us set $R_1 = R_1'' = 0$, in which case, from Equations (13.2.12) and (13.2.15), we can show that

$$\frac{T}{T_{max}} = \frac{2}{(\mathbf{S}/\mathbf{S}_{max\ T}) + (\mathbf{S}_{max\ T}/\mathbf{S})} \qquad (13.2.16)$$

where $\mathbf{S}$ and $\mathbf{S}_{max\ T}$ are the slips corresponding to T and T_{max}, respectively.

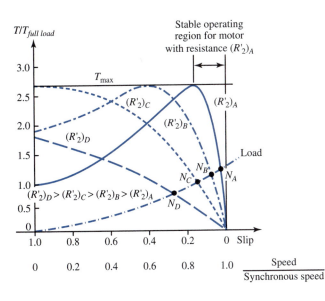

Figure 13.2.9 Effect of changing rotor-circuit resistance on the torque–slip characteristic of a polyphase induction motor.

EXAMPLE 13.2.2

For the motor specified in Example 13.2.1, compute the following:

(a) The load component I_2' of the stator current, the internal torque T, and the internal power P_m for a slip of 0.02.

(b) The maximum internal torque, and the corresponding slip and speed.

(c) The internal starting torque and the corresponding stator load current I_2'.

Solution

Let us first reduce the equivalent circuit of Figure 13.2.6 to its Thévenin-equivalent form shown in Figure 13.2.8. With the aid of Equations (13.2.9) and (13.2.10), we obtain

$$\bar{V}_{1a} = \bar{V}_1 \frac{jX_m}{R_1 + j(X_{l1} + X_m)} \cong \bar{V}_1 \frac{X_m}{X_{l1} + X_m} = \frac{220}{\sqrt{3}} \frac{15}{0.5 + 15} \angle 0° = 122.9 \angle 0° \text{ V}$$

$$R_1'' + jX_1'' = \frac{(0.3 + j0.5)j15}{0.3 + j(0.5 + 15)} = 0.281 + j0.489$$

(a) Corresponding to a slip of 0.02, $R_2'/S = 0.15/0.02 = 7.5$. From Equation (13.2.11), we get

$$I_2' = \frac{122.9}{\sqrt{7.8^2 + 0.689^2}} = \frac{122.9}{7.83} = 15.7 \text{ A}$$

The internal torque T can be calculated from either Equation (13.2.5) or (13.2.12),

$$T = \frac{1}{125.7} 3 \times 15.7^2 \times 7.5 = \frac{5546}{125.7} = 44.12 \text{ N} \cdot \text{m}$$

From Equation (13.2.6), the internal mechanical power is calculated as

$$P_m = 3 \times 15.7^2 \times 7.5 \times 0.98 = 5435 \text{ W}$$

which is also the same as $T\omega_m = T\omega_s(1-S) = 44.12 \times 125.7 \times 0.98$. In Example 13.2.1, this value is calculated as 5460 W; the small discrepancy is due to the approximations.

(b) From Equation (13.2.14) it follows that

$$S_{\text{max } T} = \frac{0.15}{\sqrt{0.281^2 + 0.689^2}} = \frac{0.15}{0.744} = 0.202$$

or the corresponding speed at T_{max} is $(1 - 0.202)1200 = 958$ r/min. From Equation (13.2.15), the maximum torque can be calculated as

$$T_{\text{max}} = \frac{1}{125.7} \frac{0.5 \times 3 \times 122.9^2}{0.281 + \sqrt{0.281^2 + 0.689^2}} = \frac{1}{125.7} \frac{0.5 \times 3 \times 122.9^2}{0.281 + 0.744} = 175.8 \text{ N} \cdot \text{m}$$

(c) Assuming the rotor-circuit resistance to be constant, with $S = 1$ at starting, from Equations (13.2.11) and (13.2.12) we get

$$I'_{2 \text{ start}} = \frac{122.9}{\sqrt{(0.281 + 0.15)^2 + (0.489 + 0.2)^2}} = \frac{122.9}{0.813} = 151.2 \text{ A}$$

$$T_{\text{start}} = \frac{1}{125.7} 3 \times 151.2^2 \times 0.15 = 81.8 \text{ N} \cdot \text{m}$$

For such applications as fans and blowers, a motor needs to develop only a moderate starting torque. Some loads, like conveyors, however, require a high starting torque to overcome high static torque and load inertia. The motor designer sometimes makes the starting torque equal to the maximum torque by choosing the rotor-circuit resistance at startup to be

$$R'_{2 \text{ start}} = \sqrt{(R''_1)^2 + (X''_1 + X'_{l2})^2} \qquad (13.2.17)$$

which can easily be obtained from Equation (13.2.13) with $S_{\max T} = 1$.

Speed and Torque Control of Polyphase Induction Motors

The induction motor is valuable in so many applications because it combines simplicity and ruggedness. Although a good number of industrial drives run at substantially constant speed, quite a few applications need variable speed. Speed-control capability is essential in such applications as conveyors, hoists, and elevators. Because the induction motor is essentially a constant-speed machine, designers have sought creative ways to easily and efficiently vary its speed continuously over a wide range of operating conditions. We only indicate the methods of speed control here.

The appropriate equation to be examined, based on Equation (13.1.8), is

$$n = (1 - S)n_1 = (1 - S)120 f_s/P \qquad (13.2.18)$$

where n is the actual speed of the machine in revolutions per minute, S is the per-unit slip, f_s is the supply frequency in hertz, P is the number of poles, and n_1 is the synchronous speed in revolutions per minute. Equation (13.2.18) suggests that the speed of the induction motor can be varied by varying either the slip or the synchronous speed, which in turn can be varied by changing either the number of poles or the supply frequency. Any method of speed control that depends on the variation of slip is inherently inefficient because the efficiency of the induction motor is approximately equal to $1 - S$. On the other hand, if the supply frequency is constant, varying the number of poles results only in discrete and stepped variation in motor speed. Indeed, all methods of speed control require some degree of sacrifice in performance, cost, and simplicity. These disadvantages must be weighed carefully against the advantages of speed variability.

The following are methods available for speed and torque control of induction motors.

- Pole-changing method
- Variable-frequency method
- Variable-line-voltage method
- Variable-rotor-resistance method
- Rotor-slip frequency control
- Rotor-slip energy recovery method
- Control by auxiliary devices (Kramer control, Scherbius control, Schrage motor)

• Solid-state control (variable-terminal-voltage control, variable-frequency control, rotor-resistance control for wound-rotor motors, injecting voltage into rotor circuit of wound-rotor motors) usually discussed under power-semiconductor controlled drives (see Section 16.1).

Figure 13.2.10 lists various ac motors as well as techniques for their speed and torque control.

Starting Methods for Polyphase Induction Motors

When high starting torques are required, a wound-rotor induction motor, with external resistances inserted in its rotor circuits, can be used. The starting current can be reduced and high values of starting torque per ampere of starting current can be obtained with *rotor-resistance starting*. The external resistances are generally cut out in steps as the machine runs up to speed.

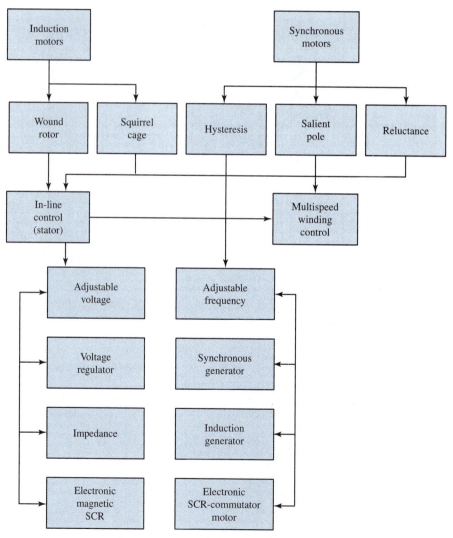

Figure 13.2.10 Ac motors and techniques for their speed and torque control.

For squirrel-cage-rotor machines, the problem is to keep down the starting current while maintaining adequate starting torque. The input current, for example, can be no more than 6 times the full-load current, while the starting torque may be about 1.5 times the full-load torque. Depending on the capacity of the available supply system, *direct-on-line starting* may be suitable only for relatively small machines, up to 10-hp rating. Other starting methods include *reduced-voltage starting* by means of *wye–delta starting, autotransformer starting,* or *stator-impedance starting.*

For employing the wye–delta starting method, a machine designed for delta operation is connected in wye during the starting period. Because the impedance between line terminals for wye connection is three times that for delta connection for the same line voltage, the line current at standstill for wye connection is reduced to one-third of the value for delta connection. Since the phase voltage is reduced by a factor of $\sqrt{3}$ during starting, it follows that the starting torque will be one-third of normal. For autotransformer starting, the setting of the autotransformer can be predetermined to limit the starting current to any desired value. An autotransformer, which reduces the voltage applied to the motor to x times the normal voltage, will reduce the starting current in the supply system as well as the starting torque of the motor to x^2 times the normal values.

Stator-impedance starting may be employed if the starting-torque requirement is not severe. Series resistances (or impedances) are inserted in the lines to limit the starting current. These resistances are short-circuited out when the motor gains speed. This method has the obvious disadvantage of inefficiency caused by the extra losses in the external resistances.

EXAMPLE 13.2.3

An induction motor has a starting current that is 6 times the full-load current and a per-unit full-load slip of 0.04. The machine is to be provided with an autotransformer starter. If the minimum starting torque must be 0.3 times the full-load torque, determine the required tapping on the transformer and the per-unit supply-source line current at starting.

Solution

From Equation (13.2.5), with slip at starting equal to 1, we have

$$\frac{T_s}{T_{fl}} = \left(\frac{I_s}{I_{fl}}\right)^2 S_{fl}$$

where subscripts s and fl correspond to starting and full-load conditions.

Substituting values, $0.3 = (I_s/I_{fl})^2 \times 0.04$, from which

$$I_s/I_{fl} = \sqrt{0.03/0.04} = 2.74$$

or $I_s = 2.74$ per unit, with I_{fl} taken as 1 per unit.

The applied voltage to the motor must reduce the current from 6 per unit to 2.74 per unit, i.e., the tapping must be sufficient to reduce the voltage to $2.74/6 = 0.456$ of the full value. The supply-source line current will then be $2.74 \times 0.456 = 1.25$ per unit, where 0.456 is the secondary-to-primary turns ratio.

Single-Phase Induction Motors

For reasons of simplicity and cost, a single-phase power supply is universally preferred for fractional-horsepower motors. It is also used widely for motors up to about 5 hp. Single-phase induction motors are usually two-pole or four-pole, rated at 2 hp or less, while slower and larger motors can be manufactured for special purposes. Single-phase induction motors are widely used in domestic appliances and for a very large number of low-power drives in industry. The single-phase induction machine resembles a small, three-phase, squirrel-cage motor, except that at full speed only a single winding on the stator is usually excited.

The single-phase stator winding is distributed in slots so as to produce an approximately sinusoidal space distribution of mmf. As discussed in Section 13.1, such a motor inherently has no starting torque, and as we saw in Section 12.3, it must be started by an auxiliary winding, by being displaced in phase position from the main winding, or by some similar device. Once started by auxiliary means, the motor will continue to run. Thus, nearly all single-phase induction motors are actually two-phase motors, with the main winding in the direct axis adapted to carry most or all of the current in operation, and an auxiliary winding in the quadrature axis with a different number of turns adapted to provide the necessary starting torque.

Since the power input in a single-phase circuit pulsates at twice the line frequency, all single-phase motors have a double-frequency torque component, which causes slight oscillations in rotor speed and imparts vibration to the motor supports. The design must provide a means to prevent this vibration from causing objectionable noise.

The viewpoint adopted in explaining the operation of the single-phase motor, based on the conditions already established for polyphase motors, is known as the *revolving-field theory*. (The other viewpoint, *cross-field theory*, is not presented here.) For computational purposes, the revolving-field point of view, already introduced in Section 13.1, is followed hereafter to parallel the analysis we applied to the polyphase induction motor. As stated in Section 13.1 and shown in Section 12.3, a stationary pulsating field can be represented by two counterrotating fields of constant magnitude. The equivalent circuit of a single-phase induction motor, then, consists of the series connection of a forward rotating field equivalent circuit and a backward rotating one. Each circuit is similar to that of a three-phase machine, but in the backward rotating field circuit, the parameter S is replaced by $2 - S$, as shown in Figure 13.2.11(a). The forward and backward torques are calculated from the two parts of the equivalent circuit, and the total torque is given by the albegraic sum of the two. As shown in Figure 13.2.11(b), the torque–speed characteristic of a single-phase induction motor is thus obtained as the sum of the two curves, one corresponding to the forward rotating field and the other to the backward rotating field.

The slip S_f of the rotor with respect to the forward rotating field is given by

$$S_f = S = \frac{n_s - n}{n_s} = 1 - \frac{n}{n_s} \tag{13.2.19}$$

where n_s is the synchronous speed and n is the actual rotor speed. The slip S_b of the rotor with respect to the backward rotating field is given by

$$S_b = \frac{n_s - (-n)}{n_s} = 1 + \frac{n}{n_s} = 2 - S \tag{13.2.20}$$

Since the amplitude of the rotating fields is one-half of the alternating flux, as seen from Equation (12.3.10), the total magnetizing and leakage reactances of the motor can be divided equally so as to correspond to the forward and backward rotating fields. In the equivalent circuit shown in Figure 13.2.11(a), then, R_1 and X_{l1} are, respectively, the resistance and the leakage reactance of the main winding, X_m is the magnetizing reactance, and R_2' and X_{l2}' are the standstill values of the rotor resistance and the leakage reactance referred to the main stator winding by the use of the appropriate turns ratio. The core loss, which is omitted here, can be accounted for later as if it were a rotational loss. The resultant torque of a single-phase induction motor can thus be expressed as

$$T_e = \frac{I_f^2(1-\mathbf{S})}{\omega_m \mathbf{S}} R_2' - \frac{I_b^2(1-\mathbf{S})}{\omega_m(2-\mathbf{S})} R_2' , \qquad (13.2.21)$$

The following example illustrates the usefulness of the equivalent circuit in evaluating the motor performance.

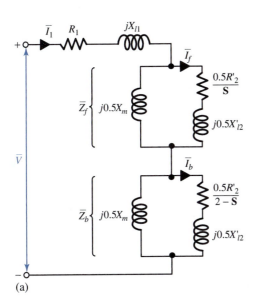

(a)

Figure 13.2.11 Single-phase induction motor based on the revolving-field theory. (a) Equivalent circuit. (b) Torque–speed characteristics.

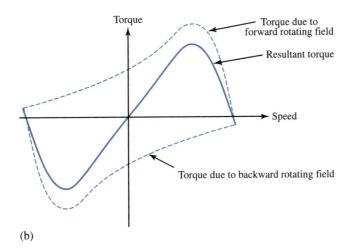

(b)

EXAMPLE 13.2.4

A $1/4$-hp, 230-V, 60-Hz, four-pole, single-phase induction motor has the following parameters and losses: $R_1 = 10\ \Omega$, $X_{l1} = X_{l2}' = 12.5\ \Omega$, $R_2' = 11.5\ \Omega$, and $X_m = 250\ \Omega$. The core loss at 230 V is 35 W and the friction and windage loss is 10 W. For a slip of 0.05, determine the stator current, power factor, developed power, shaft output power, speed, torque, and efficiency when the motor is running as a single-phase motor at rated voltage and frequency with its starting winding open.

Solution

From the given data applied to the equivalent circuit of Figure 13.2.11, we see that

$$\frac{0.5 R'_2}{S} = \frac{11.5}{2 \times 0.05} = 115 \ \Omega$$

$$\frac{0.5 R'_2}{2 - S} = \frac{11.5}{2(2 - 0.05)} = 2.95 \ \Omega$$

$$j0.5 X_m = j125 \ \Omega$$

$$j0.5 X'_{l2} = j6.25 \ \Omega$$

For the forward-field circuit, the impedance is

$$Z_f = \frac{(115 + j6.25) j125}{115 + j131.25} = 59 + j57.65 = R_f + jX_f$$

and for the backward-field circuit, the impedance is

$$Z_b = \frac{(2.95 + j6.25) j125}{2.95 + j131.25} = 2.67 + j6.01 = R_b + jX_b$$

The total series impedance Z_e is then given by

$$Z_e = Z_1 + Z_f + Z_b = (10 + j12.5) + (59 + j57.65) + (2.67 + j6.01)$$

$$= 71.67 + j76.16 = 104.6 \angle -46.74°$$

The input stator current is

$$\bar{I}_1 = \frac{230}{104.6 \angle 46.74°} = 2.2 \angle -46.74° \ \text{A}$$

also,

Power factor $= \cos 46.74° = 0.685$ lagging,

Developed power $P_d = I_1^2 R_f (1 - S) + I_1^2 R_b [1 - (2 - S)] = I_1^2 (R_f - R_b)(1 - S) = 2.2^2 (59 - 2.67)(1 - 0.05) = 259$ W

Shaft-output power $P_o = P_d - P_{rot} - P_{core} = 259 - 10 - 35 = 214$ W, or 0.287 hp

Speed $= (1 - S)$ synchronous speed $= 0.95 \times 120 \times 60/4 = 1710$ r/min, or 179 rad/s

Torque $= 214/179 = 1.2$ N · m

Efficiency $= \dfrac{\text{output}}{\text{input}} = \dfrac{214}{230 \times 2.2 \times 0.685} = \dfrac{214}{346.6} = 0.6174$, or 61.74%

Starting Methods for Single-Phase Induction Motors

The various forms of a single-phase induction motor are grouped into three principal types, depending on how they are started.

1. *Split-phase* or *resistance-split-phase motors:* Split-phase motors have two stator windings (a main winding and an auxiliary winding) with their axes displaced 90 electrical degrees in space. As represented schematically in Figure 13.2.12(a), the auxiliary winding in this type of motor has a higher resistance-to-reactance ratio than the main winding. The two currents are thus out of phase, as indicated in the phasor diagram of Figure 13.2.12(b). The motor is equivalent to an unbalanced two-phase motor. The rotating stator field produced by the unbalanced two-phase

winding currents causes the motor to start. The auxiliary winding is disconnected by a centrifugal switch or relay when the motor comes up to about 75% of the synchronous speed. The torque–speed characteristic of the split-phase motor is of the form shown in Figure 13.2.12(c). A split-phase motor can develop a higher starting torque if a series resistance is inserted in the starting auxiliary winding. A similar effect can be obtained by inserting a series inductive reactance in the main winding; this additional reactance is short-circuited when the motor builds up speed.

2. *Capacitor motors:* Capacitor motors have a capacitor in series with the auxiliary winding and come in three varieties: capacitor start, two-value capacitor, and permanent-split capacitor. As their names imply, the first two use a centrifugal switch or relay to open the circuit or reduce the size of the starting capacitor when the motor comes up to speed. A two-value-capacitor motor, with one value for starting and one for running, can be designed for optimum starting and running performance; the starting capacitor is disconnected after the motor starts. The relevant schematic diagrams and torque–speed characteristics are shown in Figures 13.2.13, 13.2.14, and 13.2.15. Motors in which the auxiliary winding and the capacitor are not cut out during the normal running conditions operate, in effect, as unbalanced two-phase induction motors.

3. *Shaded-pole motors:* The least expensive of the fractional-horsepower motors, generally rated up to $\frac{1}{20}$ hp, they have salient stator poles, with one-coil-per-pole main windings. The auxiliary winding consists of one (or rarely two) short-circuited copper straps wound on a portion of the pole and displaced from the center of each pole, as shown in Figure 13.2.16(a). The shaded-

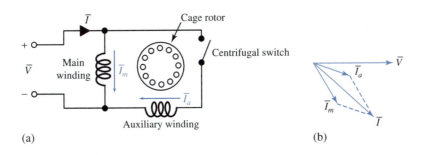

(a)

(b)

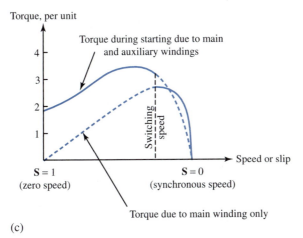

(c)

Figure 13.2.12 Split-phase motor. (**a**) Schematic diagram. (**b**) Phasor diagram at starting. (**c**) Typical torque–speed (or slip) characteristic.

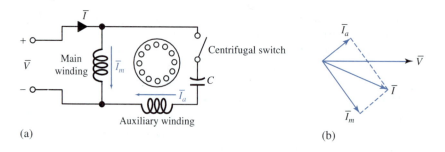

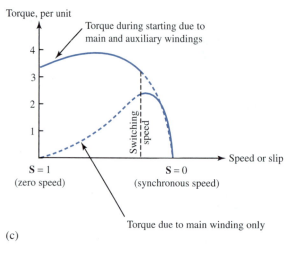

Figure 13.2.13 Capacitor-start motor. **(a)** Schematic diagram. **(b)** Phasor diagram at starting. **(c)** Typical torque–speed characteristic.

pole motor got its name from these shading bands. Induced currents in the shading coil cause the flux in the shaded portion of the pole to lag the flux in the other portion in time. The result is then like a rotating field moving in the direction from the unshaded to the shaded portion of the pole. A low starting torque is produced. A typical torque-speed characteristic is shown in Figure 13.2.16(b). Shaded-pole motors have a rather low efficiency.

Applications for Induction Motors

Before specifying a particular motor for a given application, the designer must know the load characteristics, such as horsepower requirement, starting torque, acceleration capability, speed variation, duty cycle, and the environment in which the motor is to operate. Typical speed–torque curves for squirrel-cage induction motors with NEMA design classification A, B, C, and D are shown in Figure 13.2.17. Having selected the appropriate motor for a given application, the next step is to specify a controller for the motor to furnish proper starting, stopping, and reversing without damaging the motor, other connected loads, or the power system. The ranges of standard power ratings given by NEMA for single-phase motors are listed in Table 13.2.1. Two-phase induction motors with high rotor resistance are employed as servomotors for control-system applications that usually require positive damping over the full speed range.

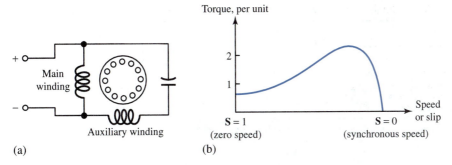

(a)

(b)

Figure 13.2.14 Permanent-split-capacitor motor. **(a)** Schematic diagram. **(b)** Typical torque–speed characteristic.

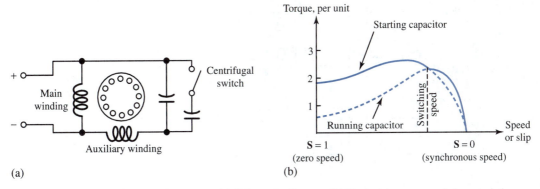

(a)

(b)

Figure 13.2.15 Two-value-capacitor motor. **(a)** Schematic diagram. **(b)** Typical torque–speed characteristic.

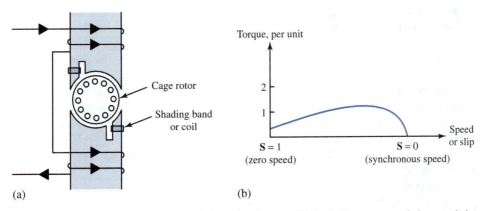

(a)

(b)

Figure 13.2.16 Shaded-pole motor. **(a)** Schematic diagram. **(b)** Typical torque–speed characteristic.

TABLE 13.2.1 Ranges of Standard Power Ratings for Single-Phase Induction Motors

Motor	Power Range
Capacitor start	1 mhp to 10 hp
Resistance start	1 mhp to 10 hp
Two-value capacitor	1 mhp to 10 hp
Permanent-split capacitor	1 mhp to 1.5 hp
Shaded pole	1 mhp to 1.5 hp

Source: NEMA Standards Publication MG 1, *Motors and Generators,* New York, 1987.

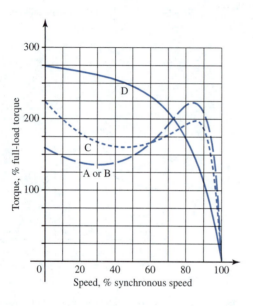

Figure 13.2.17 Typical speed–torque curves for squirrel-cage induction motors with NEMA design classifications A, B, C, and D. (Adapted from NEMA Standards Publication MG 10, *Energy Management Guide for Selection and Use of Polyphase Motors,* New York, 1988.)

13.3 SYNCHRONOUS MACHINES

Large ac power networks operating at a constant frequency of 60 Hz in the United States (50 Hz in Europe) rely almost exclusively on synchronous generators to generate electric energy. They can also have synchronous compensators or condensers at key points for reactive power control. Generators are the largest single-unit electric machines in production, having power ratings in the range of 1500 MVA, and we can expect machines of several thousand MVA to come into use in future decades. Private, standby, and peak-load plants with diesel or gas-turbine prime movers also have alternators. Non-land-based synchronous plants can be found on oil rigs, on large aircraft with hydraulically driven alternators operating at 400 Hz, and on ships for variable frequency supply to synchronous propeller motors. Synchronous motors provide constant speed industrial drives with the possibility of power factor correction, although they are not often built in small ratings, for which the induction motor is cheaper.

This section develops analytical methods of examining the steady-state performance of synchronous machines. We first consider cylindrical-rotor machines and discuss the effects of salient poles subsequently.

Equivalent Circuit of a Synchronous Machine

A review of the material about elementary synchronous machines presented in Section 13.1 is very helpful at this stage to recall the principles of operation for synchronous machines.

To investigate the equivalent circuit of a synchronous machine, for the sake of simplicity, let us begin by considering an unsaturated cylindrical-rotor synchronous machine. Effects of saliency and magnetic saturation can be considered later. Since for the present we are only concerned with the steady-state behavior of the machine, circuit parameters of the field and damper windings need not be considered. The effect of the field winding, however, is taken care of by the flux produced by the dc field excitation and the ac voltage generated by the field flux in the armature circuit. Thus, let $\bar{E}_f$ be the ac voltage, known as *excitation voltage*, generated by the field flux. As for the armature winding, under balanced conditions of operation we will analyze it on a *per-phase* basis. The armature winding obviously has a *resistance* R_a and a *leakage reactance* X_l of the armature per phase.

The armature current $\bar{I}_a$ produces the armature reaction flux, the effect of which can be represented by an inductive reactance X_ϕ, known as *magnetizing reactance* or *armature reaction reactance*. Thus Figure 13.3.1 shows the equivalent circuits in phasor notation, with all per-phase quantities, for a cylindrical-rotor synchronous machine working as either a generator or a motor. The sum of the armature leakage reactance and the armature reaction reactance is known as the *synchronous reactance*,

$$X_s = X_\phi + X_l \tag{13.3.1}$$

and $R_a + jX_s$ is called the *synchronous impedance* Z_s. Thus, the equivalent circuit for an unsaturated cylindrical-rotor synchronous machine under balanced polyphase conditions reduces to that shown in Figure 13.3.1(b), in which the machine is represented on a per-phase basis by its excitation voltage $\bar{E}_f$ in series with the synchronous impedance. Note that $\bar{V}_t$ is the terminal per-phase rms voltage, usually taken as reference.

In all but small machines, the armature resistance is usually neglected except for its effect on losses (and hence the efficiency) and heating. With this simplification, Figure 13.3.2 shows the four possible cases of operation of a round-rotor synchronous machine, in which the following relation holds:

$$\bar{V}_t + j\,\bar{I}_a X_s = \bar{E}_f \tag{13.3.2}$$

Observe that the motor armature current is taken in the direction opposite to that of the generator. The machine is said to be *overexcited* when the magnitude of the excitation voltage exceeds that of the terminal voltage; otherwise it is said to be *underexcited*. The angle δ between the excitation voltage $\bar{E}_f$ and the terminal voltage $\bar{V}_t$ is known as the *torque angle* or *power angle* of

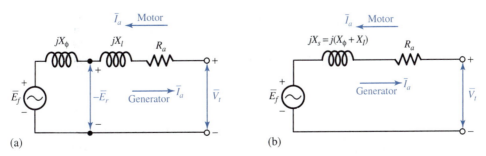

(a)   (b)

Figure 13.3.1 Per-phase equivalent circuits of a cylindrical-rotor synchronous machine.

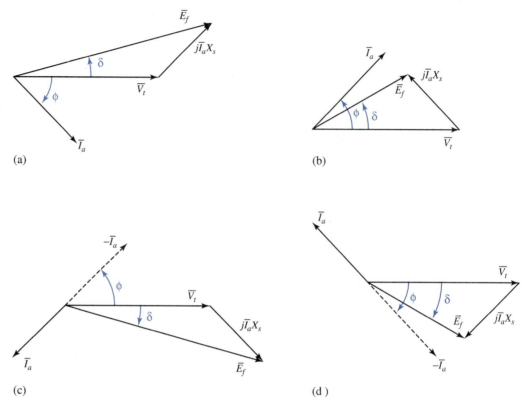

Figure 13.3.2 Four possible cases of operation of a round-rotor synchronous machine with negligible armature resistance. **(a)** Overexcited generator (power factor lagging), $P > 0, Q > 0, \delta > 0$. **(b)** Underexcited generator (power factor leading), $P > 0, Q < 0, \delta > 0$. **(c)** Overexcited motor (power factor leading), $P < 0, Q > 0, \delta < 0$. **(d)** Underexcited motor (power factor lagging), $P < 0, Q < 0, \delta < 0$.

the synchronous machine. The power-angle performance characteristics are discussed later in this section. The dc *excitation* can be provided by a self-excited dc generator, known as the *exciter*, mounted on the same shaft as the rotor of the synchronous machine.

The voltage regulation of a synchronous generator at a given load, power factor, and rated speed is defined as

$$\% \text{ voltage regulation} = \frac{E_f - V_t}{V_t} \times 100 \tag{13.3.3}$$

where V_t is the terminal voltage on the load, and E_f is the no-load terminal voltage at rated speed when the load is removed without changing the field current.

EXAMPLE 13.3.1

The per-phase synchronous reactance of a three-phase, wye-connected, 2.5-MVA, 6.6-kV, 60-Hz turboalternator is 10 Ω. Neglect the armature resistance and saturation. Calculate the voltage regulation when the generator is operating at full load with (a) 0.8 power factor lagging, and (b) 0.8 power factor leading.

Solution

$$\text{Per-phase terminal voltage } V_t = \frac{6.6 \times 1000}{\sqrt{3}} = 3811 \text{ V}$$

$$\text{Full-load per-phase armature current } I_a = \frac{2.5 \times 10^6}{\sqrt{3} \times 6.6 \times 1000} = 218.7 \text{ A}$$

(a) Referring to Figure 13.3.2(a) for an overexcited generator operating at 0.8 power factor lagging, and applying Equation (13.3.2), we have

$$\bar{E}_f = 3811 + j218.7(0.8 - j0.6)(10) = 5414\angle\tan^{-1}0.3415$$

$$\% \text{ voltage regulation } = \frac{5414 - 3811}{3811} \times 100 = 0.042 \times 100 = 4.2\%$$

(b) Referring to Figure 13.3.2(b) for an underexcited generator operating at 0.8 power factor leading, and applying Equation (13.3.2), we have

$$\bar{E}_f = 3811 + j218.7(0.8 + j0.6)(10) = 3050\angle\tan^{-1}0.7$$

$$\% \text{ voltage regulation } = \frac{3050 - 3811}{3811} \times 100 = -0.2 \times 100 = -20\%$$

As we have just seen, the voltage regulation for a synchronous generator can become negative.

Power Angle and Other Performance Characteristics

The real and reactive power delivered by a synchronous generator, or received by a synchronous motor, can be expressed in terms of the terminal voltage V_t, the generated voltage E_f, the synchronous impedance Z_s, and the power angle or torque angle δ. Referring to Figure 13.3.2, it is convenient to adopt a convention that makes positive the real power P and the reactive power Q delivered by an overexcited generator. Accordingly, the generator action corresponds to positive values of δ, whereas the motor action corresponds to negative values of δ. With the adopted notation it follows that $P > 0$ for generator operation, whereas $P < 0$ for motor operation. Further, positive Q means delivering inductive VARs for a generator action, or receiving inductive VARs for a motor action; negative Q means delivering capacitive VARs for a generator action, or receiving capacitive VARs for a motor action. It can be observed from Figure 13.3.2 that the power factor is lagging when P and Q have the same sign, and leading when P and Q have opposite signs.

The complex power output of the generator in volt-amperes per phase is given by

$$\bar{S} = P + jQ = \bar{V}_t \, \bar{I}_a^* \tag{13.3.4}$$

where $\bar{V}_t$ is the terminal voltage per phase, $\bar{I}_a$ is the armature current per phase, and * indicates a complex conjugate. Referring to Figure 13.3.2(a), in which the effect of armature resistance has been neglected, and taking the terminal voltage as reference, we have the terminal voltage,

$$\bar{V}_t = V_t + j\,0 \tag{13.3.5}$$

the excitation voltage or generated voltage,

$$\bar{E}_f = E_f(\cos \delta + j \sin \delta) \tag{13.3.6}$$

and the armature current,

$$\bar{I}_a = \frac{\bar{E}_f - \bar{V}_t}{jX_s} = \frac{E_f \cos \delta - V_t + jE_f \sin \delta}{jX_s} \tag{13.3.7}$$

where X_s is the synchronous reactance per phase. Since

$$\bar{I}_a^* = \frac{E_f \cos \delta - V_t - jE_f \sin \delta}{-jX_s} = \frac{E_f \sin \delta}{X_s} + j\frac{E_f \cos \delta - V_t}{X_s} \tag{13.3.8}$$

therefore,

$$P = \frac{V_t E_f \sin \delta}{X_s} \tag{13.3.9}$$

and

$$Q = \frac{V_t E_f \cos \delta - V_t^2}{X_s} \tag{13.3.10}$$

Equations (13.3.9) and (13.3.10) hold for a cylindrical-rotor synchronous generator with negligible armature resistance. To obtain the total power for a three-phase generator, Equations (13.3.9) and (13.3.10) should be multiplied by 3 when the voltages are line to neutral. If the line-to-line magnitudes are used for the voltages, however, these equations give the total three-phase power.

The power-angle or torque-angle characteristic of a cylindrical-rotor synchronous machine is shown in Figure 13.3.3, neglecting the effect of armature resistance. The maximum real power output per phase of the generator for a given terminal voltage and a given excitation voltage is

$$P_{\max} = \frac{V_t E_f}{X_s} \tag{13.3.11}$$

Any further increase in the prime-mover input to the generator causes the real power output to decrease. The excess power goes into accelerating the generator, thereby increasing its speed and causing it to pull out of synchronism. Hence, the *steady-state stability limit* is reached when $\delta = \pi/2$. For normal steady operating conditions, the power angle or torque angle is well below 90°. The maximum torque or *pull-out torque* per phase that a round-rotor synchronous motor can develop for a *gradually applied* load is

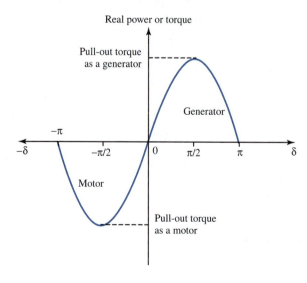

Figure 13.3.3 Steady-state power-angle or torque-angle characteristic of a cylindrical-rotor synchronous machine (with negligible armature resistance).

$$T_{max} = \frac{P_{max}}{\omega_m} = \frac{P_{max}}{2\pi \, n_s/60} \qquad\qquad (13.3.12)$$

where n_s is the synchronous speed in r/min.

In the steady-state theory of the synchronous machine, with known terminal bus voltage V_t and a given synchronous reactance X_s, the six operating variables are P, Q, δ, ϕ, I_a, and E_f. The synchronous machine is said to have two degrees of freedom, because the selection of any two, such as ϕ and I_a, P and Q, or δ and E_f, determines the operating point and establishes the other four quantities.

The principal steady-state operating characteristics are the interrelations among terminal voltage, field current, armature current, real power, reactive power, torque angle, power factor, and efficiency. These characteristics can be computed for application studies by means of phasor diagrams, such as those shown in Figure 13.3.2, corresponding to various conditions of operation.

The *efficiency* of a synchronous generator at a specified power output and power factor is determined by the ratio of the output to the input; the input power is given by adding the machine losses to the power output. The efficiency is conventionally computed in accordance with a set of rules agreed upon by ANSI. Six losses are included in the computation:

- *Armature winding copper loss* for all phases, calculated after correcting the dc resistance of each phase for an appropriate allowable temperature rise, depending on the class of insulation used.

- *Field copper loss,* based on the field current and measured field-winding dc resistance, corrected for temperature in the same way armature resistance is corrected. Note that the losses in the field rheostats that are used to adjust the generated voltage are not charged to the synchronous machine.

- *Core loss,* which is read from the open-circuit core-loss curve at a voltage equal to the internal voltage behind the resistance of the machine.

- *Friction and windage loss.*

- *Stray-load losses,* which account for the fact that the effective ac resistance of the armature is greater than the dc resistance because of the skin effect, and for the losses caused by the armature leakage flux.

- *Exciter loss,* but only if the exciter is an integral component of the alternator, i.e., shares a common shaft or is permanently coupled. Losses from a nonintegral exciter are not charged to the alternator.

EXAMPLE 13.3.2

A 1000-hp, 2300-V, wye-connected, three-phase, 60-Hz, 20-pole synchronous motor, for which cylindrical-rotor theory can be used and all losses can be neglected, has a synchronous reactance of 5.00 Ω/phase.

(a) The motor is operated from an infinite bus supplying rated voltage and rated frequency, and its field excitation is adjusted so that the power factor is unity when the shaft load is such as to require an input of 750 kW. Compute the maximum torque that the motor can deliver, given that the shaft load is increased slowly with the field excitation held constant.

(b) Instead of an infinite bus as in part (a), let the power to the motor be supplied by a 1000-kVA, 2300-V, wye-connected, three-phase, 60-Hz synchronous generator whose synchronous reactance is also 5.00 Ω/phase. The generator is driven at rated speed, and the field excitations of the generator and motor are adjusted so that the motor absorbs 750 kW at unity power factor and rated terminal voltage. If the field excitations of both machines are then held constant, and the mechanical load on the synchronous motor is gradually increased, compute the maximum motor torque under the conditions. Also determine the armature current, terminal voltage, and power factor at the terminals corresponding to this maximum load.

(c) Calculate the maximum motor torque if, instead of remaining constant as in part (b), the field currents of the generator and motor are gradually increased so as to always maintain rated terminal voltage and unity power factor while the shaft load is increased.

Solution

The solution neglects reluctance torque because cylindrical-rotor theory is applied.

(a) The equivalent circuit and the corresponding phasor diagram for the given conditions are shown in Figure E13.3.2 (a), with the subscript m attached to the motor quantities,

$$\text{Rated voltage per phase } V_t = \frac{2300}{\sqrt{3}} = 1328 \text{ V line-to-neutral}$$

$$\text{Current per phase } I_{am} = \frac{750 \times 10^3}{\sqrt{3} \times 2300 \times 1.0} = 188.3 \text{ A}$$

$$I_{am} X_{sm} = 188.3 \times 5 = 941.5 \text{ V}$$

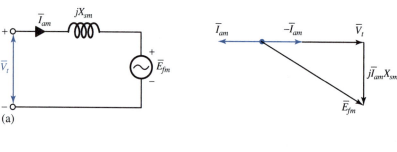

(a)

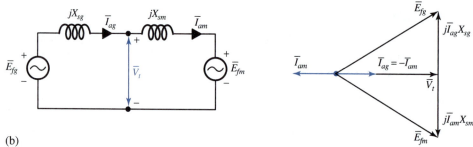

(b)

Figure E13.3.2

$$E_{fm} = \sqrt{1328^2 + 941.5^2} = 1628 \text{ V}$$

$$P_{max} = \frac{E_{fm} V_t}{X_{sm}} = \frac{1628 \times 1328}{5}$$

$$= 432.4 \text{ kW per phase, or } 1297.2 \text{ kW for three phases}$$

$$\text{Synchronous speed} = \frac{120 \times 60}{20} = 360 \text{ r/min, or } 6 \text{ r/s}$$

$$\omega_s = 2\pi \times 6 = 37.7 \text{ rad/s}$$

$$T_{max} = \frac{1297.2 \times 10^3}{37.7} = 34,408 \text{ N} \cdot \text{m}$$

(b) With the synchronous generator as the power source, the equivalent circuit and the corresponding phasor diagram for the given conditions are shown in Figure E13.3.2(b), with subscript g attached to the generator quantities,

$$E_{fg} = E_{fm} = 1628 \text{ V}$$

$$P_{max} = \frac{E_{fg} E_{fm}}{X_{sg} + X_{sm}} = \frac{1628 \times 1628}{10}$$

$$= 265 \text{ kW per phase, or } 795 \text{ kW for three phases}$$

$$T_{max} = \frac{795 \times 10^3}{37.7} = 21,088 \text{ N} \cdot \text{m}$$

If a load torque greater than this amount were applied to the motor shaft, synchronism would be lost; the motor would stall, the generator would tend to overspeed, and the circuit would be opened by circuit-breaker action.

Corresponding to the maximum load, the angle between $\bar{E}_{fg}$ and $\bar{E}_{fm}$ is 90°. From the phasor diagram it follows that

$$V_t = \frac{E_{fg}}{\sqrt{2}} = \frac{1628}{\sqrt{2}} = 1151.3 \text{ V line-to-neutral, or } 1994 \text{ V line-to-line}$$

$$I_{ag} X_{sg} = 1151.3 \quad \text{or} \quad I_{ag} = \frac{1151.3}{5} = 230 \text{ A}$$

The power factor is unity at the terminals.

(c) $V_t = 1328$ V, and the angle between $\bar{E}_{fg}$ and $\bar{E}_{fm}$ is 90°. Hence it follows that

$$E_{fg} = E_{fm} = 1328\sqrt{2} = 1878 \text{ V}$$

$$P_{max} = \frac{E_{fg} E_{fm}}{X_{sg} + X_{sm}} = \frac{1878 \times 1878}{10} = 352.7 \text{ kW per phase, or } 1058 \text{ kW for three phases}$$

$$T_{max} = \frac{1058 \times 10^3}{37.7} = 28,064 \text{ N} \cdot \text{m}$$

Effects of Saliency and Saturation

Because of saliency, the reactance measured at the terminals of a salient-pole synchronous machine as opposed to a cylindrical-rotor machine (with uniform air gap) varies as a function of the rotor position. The effects of saliency are taken into account by the *two-reactance theory*, in which the armature current $\bar{I}_a$ is resolved into two components: I_d in the direct or field axis, and I_q in the quadrature or interpolar axis. I_q will be in the time phase with the excitation speed voltage $\bar{E}_f$, whereas I_d will be in time quadrature with $\bar{E}_f$. Direct- and quadrature-axis reactances (X_d and X_q) are then introduced to model the machine in two axes. While this involved method of analysis is not pursued here any further, the steady-state power-angle characteristic of a salient-pole synchronous machine (with negligible armature resistance) is shown in Figure 13.3.4. The resulting power has two terms: one due to field excitation and the other due to saliency. The maximum torque that can be developed is somewhat greater because of the contribution due to saliency.

Saturation factors and saturated reactances can be developed to account approximately for saturation, or more involved field-plotting methods may be used if necessary. Such matters are obviously outside the scope of this text.

Parallel Operation of Interconnected Synchronous Generators

In order to assure continuity of the power supply within prescribed limits of frequency and voltage at all the load points scattered over the service area, it becomes necessary in any modern power system to operate several alternators in parallel, interconnected by various transmission lines, in a well-coordinated and optimized manner for the most economical operation. A generator can be paralleled with an infinite bus (or with another generator running at rated voltage and frequency supplying the load) by driving it at synchronous speed corresponding to the system frequency and adjusting its field excitation so that its terminal voltage equals that of the bus. If the frequency of the incoming machine is not exactly equal to that of the system, the phase relation between its voltage and the bus voltage will vary at a frequency equal to the difference between the frequencies of the machine and the bus voltages. In normal practice, this difference can usually be made quite small, to a fraction of a hertz; in polyphase systems, it is essential that the same phase sequence be

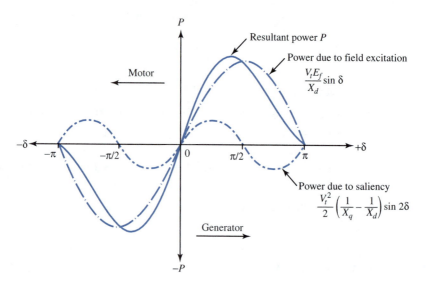

Figure 13.3.4 Steady-state power-angle characteristic of a salient-pole synchronous machine (with negligible armature resistance).

maintained on either side of the synchronizing switch. Thus, *synchronizing* requires the following conditions of the incoming machine:

- Correct phase sequence
- Phase voltages in phase with those of the system
- Frequency almost exactly equal to that of the system
- Machine terminal voltage approximately equal to the system voltage

A *synchroscope* is used for indicating the appropriate moment for synchronization. After the machine has been synchronized and is part of the system, it can be made to take its share of the active and reactive power by appropriate adjustments of its prime-mover throttle and field rheostat. The system frequency and the division of active power among the generators are controlled by means of prime-mover throttles regulated by governors and automatic frequency regulators, whereas the terminal voltage and the reactive volt-ampere division among the generators are controlled by voltage regulators acting on the generator-field circuits and by transformers with automatic tap-changing devices.

EXAMPLE 13.3.3

Two three-phase, 6.6-kV, wye-connected synchronous generators, operating in parallel, supply a load of 3000 kW at 0.8 power factor lagging. The synchronous impedance per phase of machine A is $0.5 + j10\ \Omega$, and of machine B it is $0.4 + j12\Omega$. The excitation of machine A is adjusted so that it delivers 150 A at a lagging power factor, and the governors are set such that the load is shared equally between the two machines. Determine the armature current, power factor, excitation voltage, and power angle of each machine.

Solution

One phase of each generator and one phase of the equivalent wye of the load are shown in Figure E13.3.3(a). The load current $\bar{I}_L$ is calculated as

$$\bar{I}_L = \frac{3,000}{\sqrt{3} \times 6.6 \times 0.8} \angle - \cos^{-1} 0.8 = 328(0.8 - j0.6) = 262.4 - j196.8\ \text{A}$$

For machine A,

$$\cos \phi_A = \frac{1500}{\sqrt{3} \times 6.6 \times 150} = 0.875 \text{ lagging}; \ \phi_A = 29°; \ \sin \phi_A = 0.485$$

$$\bar{I}_A = 150(0.874 - j0.485) = 131.1 - j72.75\ \text{A}$$

For machine B,

$$\bar{I}_B = \bar{I}_L - \bar{I}_A = 131.3 - j124 = 180.6\angle - \cos\left(\frac{131.3}{180.6}\right)$$

$$\cos \phi_B = \frac{131.3}{180.6} = 0.726 \text{ lagging}$$

With the terminal voltage $\bar{V}_r$ as reference, we have

$$\bar{E}_{fA} = \bar{V}_t + \bar{I}_A \bar{Z}_A = (6.6/\sqrt{3}) + (131.1 - j72.75)(0.5 + j10) \times 10^{-3}$$

$$= 4.6 + j1.27 \text{ kV per phase}$$

Power angle $\delta_A = \tan^{-1}(1.27/4.6) = 15.4°$

The line-to-line excitation voltage for machine A is $\sqrt{3}\sqrt{4.6^2 + 127^2} = 8.26$ kV.

$$\bar{E}_{fB} = \bar{V}_t + \bar{I}_B\bar{Z}_B = (6.6/\sqrt{3}) + (131.1 - j124)(0.4 + j12) \times 10^{-3}$$
$$= 5.35 + j1.52 \text{ kV per phase}$$

Power angle $\delta_B = \tan^{-1}(1.52/5.35) = 15.9°$

The line-to-line excitation voltage for machine B is $\sqrt{3}\sqrt{5.35^2 + 1.52^2} = 9.6$ kV. The corresponding phasor diagram is sketched in Figure E13.3.3(b).

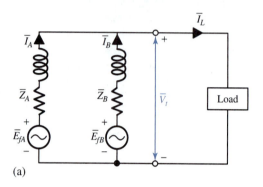

(a)

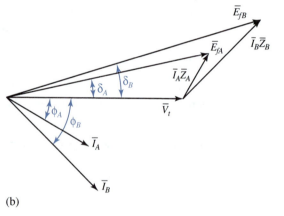

(b)

Figure E13.3.3

Steady-State Stability

The property of a power system that ensures that it will remain in equilibrium under both normal and abnormal conditions is known as *power-system stability. Steady-state stability* is concerned with slow and gradual changes, whereas *transient stability* is concerned with severe disturbances, such as sudden changes in load or fault conditions. The largest possible flow of power through a particular point, without loss of stability, is known as the *steady-state stability limit* when the power is increased gradually, and as the *transient-stability limit* when a sudden disturbance occurs.

For a generator connected to a system that is very large compared to its own size, the system in Figure 13.3.5 can be used. The power-angle equation (neglecting resistances) for the system under consideration becomes

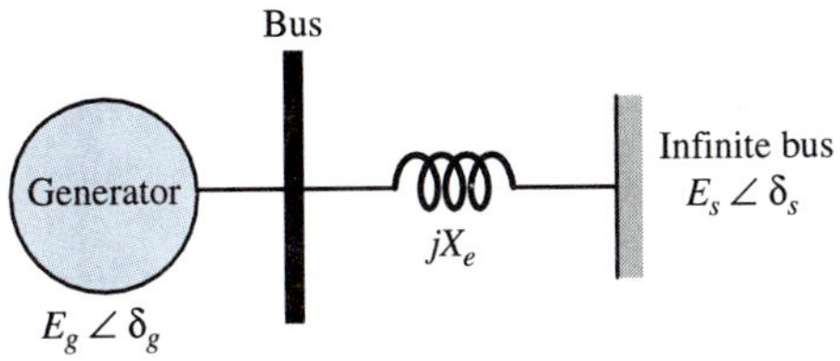

Bus

Infinite bus
$E_s \angle \delta_s$

jX_e

$E_g \angle \delta_g$

Figure 13.3.5 Generator connected to an infinite bus.

$$P = \frac{E_g E_s}{X} \sin \delta_{gs} \qquad (13.3.13)$$

where δ_{gs} is the angular difference between δ_g and δ_s, and $X = X_d + X_e$, with X_d being the direct-axis reactance of the synchronous generator. The power angle characteristic given by Equation (13.3.13) is plotted in Figure 13.3.6. The peak of the power-angle curve, given by P_{max}, is known as the *steady-state power limit* (shown by point b in Figure 13.3.6), representing the maximum power that can theoretically be transmitted in a stable manner. A machine is usually operated at less than the power limit (such as at point a in Figure 13.3.6), thereby leaving a steady-state margin, as otherwise even a slight increase in the angle δ_{gs} (such as at point c in Figure 13.3.6) would lead to instability. Installing parallel transmission lines [which effectively reduces X in Equation (13.3.13)] or adding series capacitors in lines raises the stability limit.

Applications for Synchronous Motors

With constant-speed operation, power factor control, and high operating efficiency, three-phase synchronous motors are employed in a wide range of applications. An overexcited synchronous motor, known as a synchronous condenser, is used to improve the system power factor. Synchronous motors are used as prime movers of dc generators and variable-frequency ac generators. Typical applications include pumps, compressors, mills, mixers, and crushers.

Single-phase reluctance motors find application in such devices as clocks, electric shavers, electric clippers, vibrators, sandpapering machines, and engraving tools. A hysteresis motor is employed for driving high-quality record players and tape recorders, electric clocks, and other timing devices. The horsepower range is up to 1 hp for hysteresis motors, and up to 100 hp for reluctance motors.

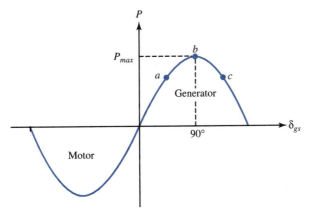

Figure 13.3.6 Power-angle curve of Equation (13.3.13).

13.4 DIRECT-CURRENT MACHINES

Generally speaking, conventional dc generators are becoming obsolete and increasingly often are being replaced by solid-state rectifiers in most applications for which an ac supply is available. The same is not true for dc motors. The torque–speed characteristics of dc motors are what makes them extremely valuable in many industrial applications. The significant features of the dc drives include adjustable motor speed over wide ranges, constant mechanical power output or torque, rapid acceleration or deceleration, and responsiveness to feedback signals.

The dc commutator machines are built in a wide range of sizes, from small control devices with a 1-W power rating up to the enormous motors of 10,000 hp or more used in rolling-mill applications. The dc machines today are principally applied as industrial drive motors, particularly when high degrees of flexibility in controlling speed and torque are required. Such motors are used in steel and aluminum rolling mills, traction motors, overhead cranes, forklift trucks, electric trains, and golf carts. Commutator machines are also used in portable tools supplied from batteries, in automobiles as starter motors, in blower motors, and in control applications as actuators and speed-sensing or position-sensing devices.

In this section, we examine the steady-state performance characteristics of dc machines to help us understand their applications and limitations, illustrating the versatility of the dc machine.

Constructional Features of DC Machines

A dc generator or motor may have as many as four field windings, depending on the type and size of the machine and the kind of service intended. These field windings consist of two normal exciting fields, the *shunt* and *series* windings, and two fields that act in a corrective capacity to combat the detrimental effects of armature reaction, called the *commutating* (*compole* or *interpole*) and *compensating* windings, which are connected in series with the armature. The type of machine, whether shunt, series, or compound, is determined solely by the normal exciting-field circuit connections, as shown in Figure 13.1.8. Figure 13.4.1 illustrates how various field windings are arranged with respect to one another in part of a cross section of a dc machine, whereas Figure 13.4.2 shows the schematic connection diagram of a dc machine. The commutating and compensating windings, their purpose, as well as the circuit connections are presented later in this section.

Equivalent Circuit of a DC Machine

A review of the material presented with regard to elementary direct-current machines in Section 13.1 can be helpful at this stage to recall the principles of operation for dc machines.

The circuit representations of a dc generator and a dc motor are shown in Figure 13.4.3. Under *steady-state conditions* the interrelationships between voltage and current are given by

$$V_f = I_f R_f \tag{13.4.1}$$

and

$$V_t = E_a \pm I_a R_a \tag{13.4.2}$$

where the plus sign signifies a motor and the minus sign a generator. V_f is the voltage applied to the field circuit, I_f is the field current, and R_f is the field-winding resistance. V_t is the terminal voltage, E_a is the generated emf, I_a is the armature current, and R_a is the armature resistance. The generated emf E_a is given by Equation (12.2.16) as

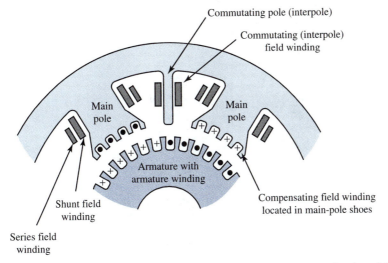

Figure 13.4.1 Section of a dc machine illustrating the arrangement of various field windings.

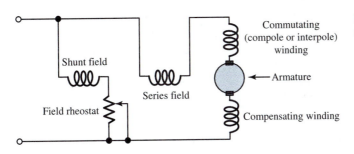

Figure 13.4.2 Schematic connection diagram of a dc machine.

$$E_a = K_a \phi \omega_m \qquad (13.4.3)$$

which is the speed (motional) voltage induced in the armature circuit due to the flux of the stator-field current. The electromagnetic torque T_e is given by Equation (12.2.18) as

$$T_e = K_a \phi I_a \qquad (13.4.4)$$

where K_a is the design constant. The product $E_a I_a$, known as the electromagnetic power being converted, is related to the electromagnetic torque by the relation

$$P_{em} = E_a I_a = T_e \omega_m \qquad (13.4.5)$$

For a motor, the terminal voltage is always greater than the generated emf, and the electromagnetic torque produces rotation against a load. For a generator, the terminal voltage is less than the generated emf, and the electromagnetic torque opposes that applied to the shaft by the prime mover. If the magnetic circuit of the machine is not saturated, note that the flux ϕ in Equations (13.4.3) and (13.4.4) is proportional to the field current I_f producing the flux.

Commutator Action

As a consequence of the arrangement of the commutator and brushes, the currents in all conductors under the north pole are in one direction and the currents in all conductors under the south pole

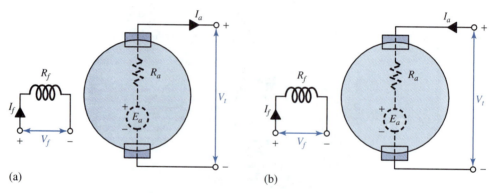

(a) (b)

Figure 13.4.3 (a) Circuit representation of a dc generator under steady-state conditions, $V_f = R_f I_f$ and $V_t = E_a - I a R_a$. (b) Circuit representation of a dc motor under steady-state conditions, $V_f = R_f I_f$ and $V_t = E_a + I_a R_a$.

are in the opposite direction. Thus, the magnetic field of the armature currents is stationary in space in spite of the rotation of the armature.

The process of reversal of currents in the coil is known as *commutation*. The current changes from $+I$ to $-I$ in time Δt. Ideally, the current in the coils being commutated should reverse linearly with time, as shown in Figure 13.4.4. Serious departure from *linear commutation* results in sparking at the brushes. Means for achieving sparkless commutation are touched upon later. As shown in Figure 13.4.4, with linear commutation, the waveform of the current in any coil as a function of time is trapezoidal.

Interpoles and Compensating Windings

The most generally used method for aiding commutation is by providing the machine with *interpoles*, also known as *commutating poles*, or simply as *compoles*. These are small, narrow auxiliary poles located between the main poles and centered on the interpolar gap. The commutating (interpole) winding is connected in series with the armature.

The demagnetizing effect of the armature mmf under pole faces can be compensated for, or neutralized, by providing a *compensating (pole-face) winding*, arranged in slots in the pole face in series with the armature, having a polarity opposite to that of the adjoining armature winding and having the same axis as that of the armature. Because they are costly, pole-face windings

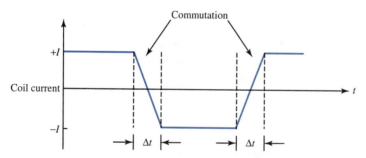

Figure 13.4.4 Waveform of current in an armature coil undergoing linear commutation.

are usually employed only in machines designed for heavy overload or rapidly changing loads, such as steel-mill motors subjected to reverse duty cycles or in motors intended to operate over wide speed ranges by shunt-field control. The schematic diagram of Figure 13.4.2 shows the relative positions of various windings, indicating that the commutating and compensating fields act along the armature axis (i.e., the quadrature axis), and the shunt as well as series fields act along the axis of the main poles (i.e., the direct axis). It is thus possible to achieve rather complete control of the air-gap flux around the entire armature periphery, along with smooth sparkless commutation.

DC Generator Characteristics

Figure 13.1.8 shows schematic diagrams of field-circuit connections for dc machines without including commutating pole or compensating windings. Shunt generators can be either *separately excited* or *self-excited,* as shown in Figures 13.1.8 (a) and (b), respectively. Compound machines can be connected either *long shunt,* as in Figure 13.1.8(d), or *short shunt,* in which the shunt-field circuit is connected directly across the armature, without including the series field.

In general, three characteristics specify the *steady-state* performance of a dc generator:

1. The *open-circuit characteristic* (abbreviated as OCC, and also known as *no-load magnetization curve*), which gives the relationship between generated emf and field current at constant speed.

2. The *external characteristic,* which gives the relationship between terminal voltage and load current at constant speed.

3. The *load characteristic,* which gives the relationship between terminal voltage and field current, with constant armature current and speed.

All other characteristics depend on the form of the open-circuit characteristic, the load, and the method of field connection. Under steady-state conditions, the currents being constant or, at most, varying slowly, voltage drops due to inductive effects are negligible.

As stated earlier and shown in Figure 13.4.3, the terminal voltage V_t of a dc generator is related to the armature current I_a and the generated emf E_a by

$$V_t = E_a - I_a R_a \tag{13.4.6}$$

where R_a is the total internal armature resistance, including the resistance of interpole and compensating windings as well as that of the brushes. The value of the generated emf E_a, by Equation (13.4.3), is governed by the direct-axis field flux (which is a function of the field current and armature reaction) and the angular velocity ω_m of the rotor.

The open-circuit and load characteristics of a separately excited dc generator, along with its schematic diagram of connections, are shown in Figures 13.4.5(a) and (b). It can be seen from the form of the external volt–ampere characteristic, shown in Figure 13.4.5(c), that the terminal voltage falls slightly as the load current increases. *Voltage regulation* is defined as the percentage change in terminal voltage when full load is removed, so that, from Figure 13.4.5(c) it follows that

$$\text{Voltage regulation} = \frac{E_a - V_t}{V_t} \times 100\% \tag{13.4.7}$$

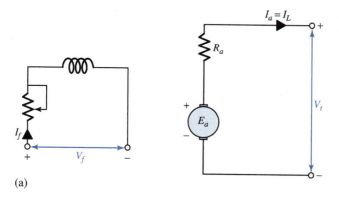

Figure 13.4.5 Open-circuit and load characteristics of a separately excited dc generator. **(a)** Schematic diagram of connections. **(b)** Terminal voltage versus field current relationships. **(c)** External volt-ampere characteristic.

(a)

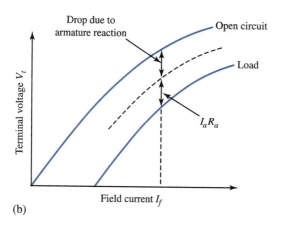

(b)

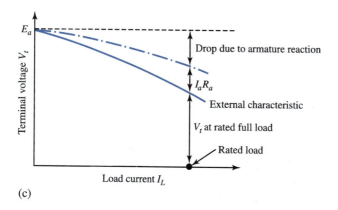

(c)

Because the separately excited generator requires a separate dc field supply, its use is limited to applications in which a wide range of controlled voltage is essential.

A shunt generator maintains approximately constant voltage on load. It finds wide application as an exciter for the field circuit of large ac generators. The shunt generator is also sometimes used as a tachogenerator when a signal proportional to the motor speed is required for control or display purposes.

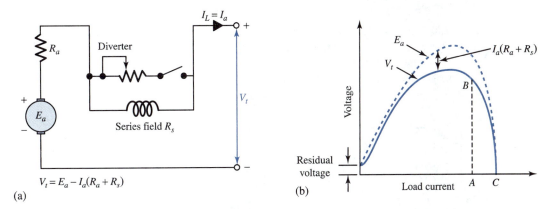

$$V_t = E_a - I_a(R_a + R_s)$$

(a)

(b)

Figure 13.4.6 Dc series generator. **(a)** Schematic diagram of connections. **(b)** Volt–ampere characteristic at constant speed.

The schematic diagram and the volt–ampere characteristic of a dc series generator at constant speed are shown in Figure 13.4.6. The resistance of the series-field winding must be low for efficiency as well as for low voltage drop. The series generator was used in early constant-current systems by operating in the range *B–C*, where the terminal voltage fell off very rapidly with increasing current.

The volt–ampere characteristics of dc compound generators at constant speeds are shown in Figure 13.4.7. *Cumulatively compound* generators, in which the series- and shunt-field winding mmfs are aiding, may be *overcompounded, flat-compounded,* or *undercompounded,* depending on the strength of the series field. Overcompounding can be used to counteract the effect of a decrease in the prime-mover speed with increasing load, or to compensate for the line drop when the load is at a considerable distance from the generator. *Differentially compounded* generators, in which the series-winding mmf opposes that of the shunt-field winding, are used in applications in which wide variations in load voltage can be tolerated, and when the generator might be exposed to load conditions approaching short circuit.

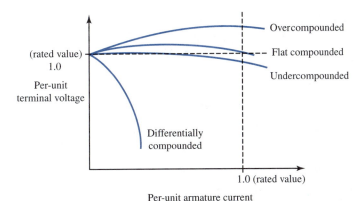

Figure 13.4.7 Volt–ampere characteristics of dc compound generators at constant speed.

EXAMPLE 13.4.1

A 250-V, 50-kW *short-shunt* compound dc generator, whose schematic diagram is shown in Figure E13.4.1(a), has the following data: armature resistance 0.05 Ω, series-field resistance 0.05 Ω, and shunt-field resistance 130 Ω. Determine the induced armature emf at rated load and terminal voltage, while taking 2 V as the total brush-contact drop.

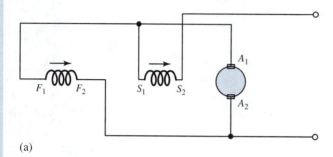

Figure E13.4.1 Short-shunt compound generator. **(a)** Schematic diagram. **(b)** Equivalent circuit.

(a)

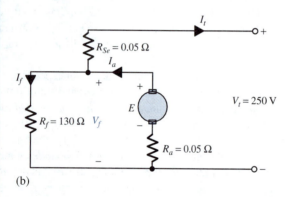

(b)

Solution

The equivalent circuit is shown in Figure E13.4.1(b).

$$I_t = \frac{50 \times 10^3}{250} = 200 \text{ A}$$

$$I_t R_{Se} = 200 \times 0.05 = 10 \text{ V}$$

$$V_f = 250 + 10 = 260 \text{ V}$$

$$I_f = \frac{260}{130} = 2.0 \text{ A}$$

$$I_a = 200 + 2 = 202 \text{ A}$$

$$I_a R_a = 202 \times 0.05 = 10.1 \text{ V}$$

$$E = 250 + 10.1 + 10 + 2 = 272.1 \text{ V}$$

DC Motor Characteristics

We gain an understanding of the speed–torque characteristics of a dc motor from Equations (13.4.2) through (13.4.4). In shunt motors, the field current can be simply controlled by the use of a variable resistance in series with the field winding; the load current influences the flux only through armature reaction, and its effect is therefore relatively small. In series motors, the flux is largely determined by the armature current, which is also the field current; it is somewhat difficult to control the armature and field currents independently. In the compound motor, the effect of the armature current on the flux depends on the degree of compounding. Most motors are designed to develop a given horsepower at a specified speed, and it follows from Equations (13.4.2) and (13.4.3) that the angular velocity ω_m can be expressed as

$$\omega_m = \frac{V_t - I_a R_a}{K_a \phi} \tag{13.4.8}$$

Thus, the speed of a dc motor depends on the values of the applied voltage V_t, the armature current I_a, the resistance R_a, and the field flux per pole ϕ.

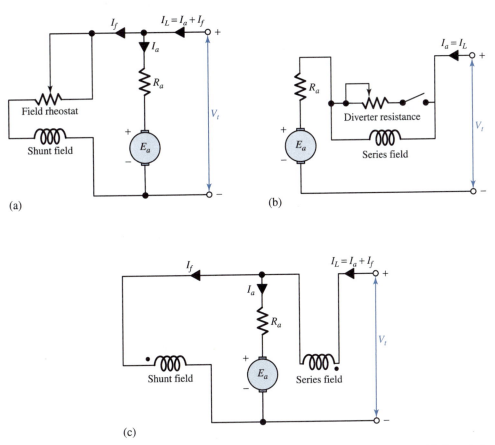

Figure 13.4.8 Schematic diagrams of dc motors. **(a)** Shunt motor. **(b)** Series motor. **(c)** Cumulatively compounded motor.

The schematic arrangement of a shunt motor is shown in Figure 13.4.8(a). For a given applied voltage and field current, Equations (13.4.4) and (13.4.3) can be rewritten as

$$T_e = K_a\phi I_a = K_m I_a \tag{13.4.9}$$

$$E_a = K_a\phi\omega_m = K_m\omega_m \tag{13.4.10}$$

Because $V_t = E_a + I_a R_a$, or $I_a = (V_t - E_a)/R_a$, it follows that

$$T_e = \frac{K_m V_t}{R_a} - \frac{K_m^2 \omega_m}{R_a} \tag{13.4.11}$$

$$P_{em} = E_a I_a = T_e \omega_m = \frac{V_t K_m \omega_m}{R_a} - \frac{K_m^2 \omega_m^2}{R_a} \tag{13.4.12}$$

The forms of the torque–armature current, speed–torque, and speed–power characteristics for a shunt-connected dc motor are illustrated in Figure 13.4.9.

The shunt motor is essentially a constant-speed machine with a low speed regulation. As seen from Equation (13.4.8), the speed is inversely proportional to the field flux, and thus it can be varied by controlling the field flux. When the motor operates at very low values of field flux, however, the speed will be high, and if the field becomes open-circuited, the speed will rise rapidly beyond the permissible limit governed by the mechanical structure. In order to limit the speed to a safe value, when a shunt motor is to be designed to operate with a low value of shunt-field flux, it is usually fitted with a small cumulative series winding, known as a *stabilizing winding*.

The schematic diagram of a series motor is shown in Figure 13.4.8(b). The field flux is directly determined by the armature current so that

$$T_e = K_a\phi I_a = K I_a^2 \tag{13.4.13}$$

and with negligible armature resistance,

$$V_t = E_a = K_a\phi\omega_m = K I_a\omega_m \tag{13.4.14}$$

$$T_e = \frac{V_t^2}{K\omega_m^2} \tag{13.4.15}$$

$$P_{em} = \omega_m T_e = \frac{V_t^2}{K\omega_m} \tag{13.4.16}$$

and the speed–power curve is a rectangular hyperbola. The forms of the torque–armature current, speed–torque, and speed–power characteristics for a series-connected dc motor are also illustrated

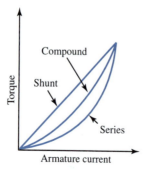
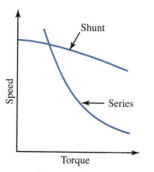
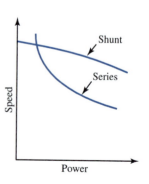

Figure 13.4.9 Characteristic curves for dc motors.

in Figure 13.4.9. Note that the no-load speed is very high; care must be taken to ensure that the machine always operates on load. In practice, however, the series machine normally has a small shunt-field winding to limit the no-load speed. The assumption that the flux is proportional to the armature current is valid only on light load in the linear region of magnetization. In general, performance characteristics of the series motor must be obtained by using the magnetization curve. The series motor is ideally suited to traction, when large torques are required at low speeds and relatively low torques are needed at high speeds.

A schematic diagram for a cumulatively compounded dc motor is shown in Figure 13.4.8(c). The operating characteristics of such a machine lie between those of the shunt and series motors, as shown in Figure 13.4.9. (The differentially compounded motor has little application, since it is inherently unstable, particularly at high loads.) Figure 13.4.10 compares the speed–torque characteristics of various types of electric motors; 1.0 per unit represents rated values.

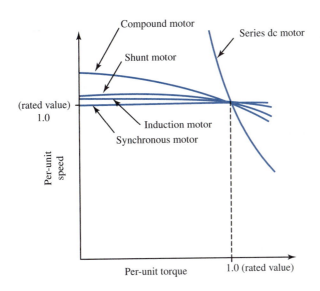

Figure 13.4.10 Typical speed–torque characteristics of various electric motors.

EXAMPLE 13.4.2

A 200-V dc shunt motor has a field resistance of 200 Ω and an armature resistance of 0.5 Ω. On no load, the machine operates with full field flux at a speed of 1000 r/min with an armature current of 4 A. Neglect magnetic saturation and armature reaction.

(a) If the motor drives a load requiring a torque of 100 N · m, find the armature current and speed of the motor.

(b) If the motor is required to develop 10 hp at 1200 r/min, compute the required value of external series resistance in the field circuit.

Solution

(a) Full field current $I_f = 200/200 = 1$ A. On no load, $E_a = V_t - I_a R_a = 200 - (4 \times 0.5) = 198$ V. Since $E_a = k_1 I_f \omega_m$, where k_1 is a constant,

$$k_1 = \frac{198}{1[(2\pi/60) \times 1000]} = 1.89$$

On load, $T_e = k_1 I_f I_a$, or $100 = 1.89 \times 1.0 \times I_a$. Therefore, the armature current $I_a = 100/1.89 = 52.9$ A. Now, $V_t = E_a + I_a R_a$, or $E_a = 200 - (52.9 \times 0.5) = 173.55$ V. Since $E_a = k_1 I_f \omega_m$, it follows that

$$\omega_m = \frac{173.55}{1.89 \times 1.0} = 91.8 \text{ rad/s}$$

That is, the load speed is $91.8 \times 60/2\pi = 876$ r/min.

(b) For 10 hp at 1200 r/min,

$$T_e = \frac{10}{(2\pi/60) \times 1200} = 59.34 \text{ N} \cdot \text{m}$$

Then, $59.34 = 1.89 I_f I_a$, or $I_f I_a = 31.4$. Since $V_t = E_a + I_a R_a$, it follows that

$$200 = 1.89 \left(\frac{2\pi}{60} \times 1200 \right) I_f + 0.5 I_a$$

$$= 237.6 I_f + 0.5 I_a = 237.6 I_f + \frac{0.5 \times 31.4}{I_f}$$

Hence, $I_f = 0.754$ A or 0.088 A; and $I_a = 31.4/I_f = 41.6$ A or 356.8 A. Since the value of $I_f = 0.088$ A will produce very high armature currents, it will not be considered. Thus, with $I_f = 0.754$ A,

$$R_f = 200/0.754 = 265.25 \ \Omega$$

The external resistance required is $265.25 - 200 = 65.25 \ \Omega$.

Speed Control of DC Motors

Equation (13.4.8) showed that the speed of a dc motor can be varied by control of the field flux, the armature resistance, and the armature applied voltage. The three most common speed-control methods are *shunt-field rheostat control, armature circuit-resistance control,* and *armature terminal-voltage control.* The *base speed* of the machine is defined as the speed with rated armature voltage and normal armature resistance and field flux. Speed control above the base value can be obtained by varying the field flux. By inserting a series resistance in the shunt-field circuit of a dc shunt motor (or a compound motor), we can achieve speed control over a wide range above the base speed. It is important to note, however, that a reduction in the field flux causes a corresponding increase in speed, so that the generated emf does not change appreciably while the speed is increased, but the machine torque is reduced as the field flux is reduced. The dc motor with shunt-field rheostat speed control is accordingly referred to as a *constant-horsepower drive.* This method of speed control is suited to applications in which the load torque falls as the speed increases. For a machine with a series field, speed control above the base value can be achieved by placing a diverter resistance in parallel with the series winding, so that the field current is less than the armature current.

When speed control below the base speed is required, the effective armature resistance can be increased by inserting external resistance in series with the armature. This method can be applied to shunt, series, or compound motors. It has the disadvantage, however, that the series resistance, carrying full armature current, will cause significant power loss with an associated reduction in overall efficiency. The speed of the machine is governed by the value of the voltage drop in the series resistor and is therefore a function of the load on the machine. The application of this method of control is thus limited. Because of its low initial cost, however, the series-resistance method, or a variation of it, is often attractive economically for short-time or intermittent slowdowns.

Unlike shunt-field control, armature-resistance control offers a *constant-torque drive* because both flux and, to a first approximation, allowable armature current remain constant as speed varies. The *shunted-armature method* is a variation of this control scheme. This is illustrated in Figure 13.4.11(a) as applied to a series motor, and in Figure 13.4.11(b) as applied to a shunt motor. Resistors R_1 and R_2 act as voltage dividers applying a reduced voltage to the armature. They offer greater flexibility in their adjustments to provide the desired performance.

The overall output limitations are as shown in Figure 13.4.12. With base speed defined as the full-field speed of the motor at the normal armature voltage, speeds above base speed are obtained by motor-field control at approximately constant horsepower, and speeds below base speed are obtained by armature-voltage control at approximately constant torque. The development of *solid-state controlled rectifiers* capable of handling many kilowatts has opened up a whole new field of *solid-state dc motor drives* with precise control of motor speed. The control resistors (in which energy is wasted) are eliminated through the development of power semiconductor devices and

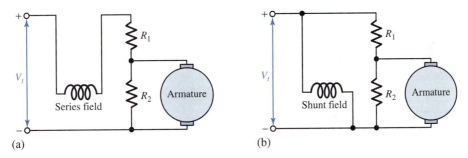

(a) (b)

Figure 13.4.11 Shunted-armature method of speed control. **(a)** As applied to a series motor. **(b)** As applied to a shunt motor.

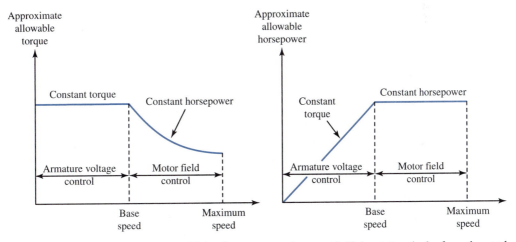

Figure 13.4.12 Output limitations combining the armature voltage and field rheostat methods of speed control.

the evolution of flexible and efficient converters (see Section 16.1). Thus, the inherently good controllability of a dc machine has been increased significantly.

EXAMPLE 13.4.3

Figure E13.4.3 shows a simplified Ward–Leonard system for controlling the speed of a dc motor. Discuss the effects of varying R_{fg} and R_{fm} on the motor speed. Subscripts g and m correspond to generator and motor, respectively.

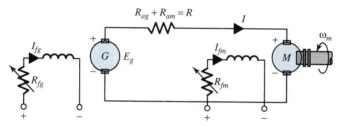

Figure E13.4.3 Simplified Ward–Leonard system.

Solution

Increasing R_{fg} decreases I_{fg} and hence E_g. Thus, the motor speed will decrease. The opposite will be true if R_{fg} is decreased.

Increasing R_{fm} will increase the speed of the motor. Decreasing R_{fm} will result in a decrease of the speed.

DC Motor Starting

When voltage is applied to the armature of a dc motor with the rotor stationary, no emf is generated and the armature current is limited only by the internal armature resistance of the machine. So, to limit the starting current to the value that the motor can commutate successfully, all except very small dc motors are started with variable external resistance in series with their armatures. This starting resistance is cut out manually or automatically as the motor comes up to speed.

EXAMPLE 13.4.4

A 10-hp, 230-V, 500-r/min shunt motor, having a full-load armature current of 37 A, is started with a four-point starter. The resistance of the armature circuit, including the interpole winding, is 0.39 Ω, and the resistances of the steps in the starting resistor are 1.56, 0.78, and 0.39 Ω, in the order in which they are successively cut out. When the armature current has dropped to its rated value, the starting box is switched to the next point, thus eliminating a step at a time in the starting resistance. Neglecting field-current changes, armature reaction, and armature inductance, find the initial and final values of the armature current and speed corresponding to each step.

Solution

Step 1: At this point, the entire resistance of the starting resistor is in series with the armature circuit. Thus,

$$R_{T1} = 1.56 + 0.78 + 0.39 + 0.39 = 3.12 \ \Omega$$

At starting, the counter emf is zero; the armature starting current is then

$$I_{st} = \frac{V_t}{R_{T1}} = \frac{230}{3.12} = 73.7 \ A$$

By the time the armature current drops to its rated value of 37 A, the counter emf is

$$E = 230 - 3 \times 12 \times 37 = 230 - 115.44 = 215.57 \ V$$

The counter emf, when the motor is delivering rated load at a rated speed of 500 r/min with the series starting resistor completely cut out, is

$$230 - 0.39 \times 37 = 230 - 14.43 = 215.57 \ V$$

The speed corresponding to the counter emf of 114.56 V is then given by

$$N = \frac{114.56}{215.57} \times 500 = 265.7 \ r/min$$

Step 2: The 1.56-Ω step is cut out, leaving a total resistance of $3.12 - 1.56 = 1.56 \ \Omega$ in the armature circuit. The initial motor speed at this step is 265.7 r/min, which means that the counter emf is still 114.56 V, if the effect of armature reaction is neglected. Accordingly, the resistance drop in the armature circuit is still 115.44 V, so that

$$I_a R_{T2} = 115.44 \quad \text{or} \quad I_a = \frac{115.44}{1.56} = 74A$$

That is, if the inductance of the armature is neglected, the initial current is 74 A. With the final current at 37 A, the counter emf is

$$E = 230 - 37 \times 1.56 = 230 - 57.72 = 172.28 \ V$$

corresponding to which the speed is

$$N = \frac{172.28}{215.57} \times 500 = 399.6 \ rpm$$

Step 3: The total resistance included in the armature circuit is 0.78 Ω at the beginning of this step. The initial motor speed is 399.6 r/min, and the counter emf is 172.28 V. The initial armature current is then $(230 - 172.28)/0.78 = 57.72/0.78 = 74$ A. With the final current at 37 A, the counter emf is

$$E = 230 - 37 \times 0.78 = 230 - 28.86 = 201.14 \ V$$

corresponding to which the speed is

$$N = \frac{201.14}{215.57} \times 500 = 466.5 \ r/min$$

Thus, we have the results shown in Table E13.4.4.

TABLE E13.4.4

Step Number	Current (A)		Speed (r/min)	
	Initial	Final	Initial	Final
1	74	37	0	266
2	74	37	266	400
3	74	37	400	467
4	74	37	467	500

Efficiency

As is true for any other machine, the efficiency of a dc machine can be expressed as

$$\text{Efficiency} = \frac{\text{output}}{\text{input}} = \frac{\text{input} - \text{losses}}{\text{input}} = 1 - \frac{\text{losses}}{\text{input}} \qquad (13.4.17)$$

The losses are made up of rotational losses (3 to 15%), armature-circuit copper losses (3 to 6%), and shunt-field copper losses (1 to 5%). Figure 13.4.13 shows the schematic diagram of a dc machine, along with the power division in a generator and a motor. The resistance voltage drop, also known as *arc drop*, between brushes and commutator is generally assumed constant at 2 V, and the brush-contact loss is therefore calculated as $2I_a$. In such a case, the resistance of the armature circuit should not include the resistance between brushes and commutator.

EXAMPLE 13.4.5

The following data apply to a 100-kW, 250-V, six-pole, 1000-r/min long-shunt compound generator: no-load rotational losses 4000 W, armature resistance at 75°C = 0.015 Ω, series-field resistance at 75°C = 0.005 Ω, interpole field resistance at 75°C = 0.005 Ω, and shunt-field current 2.5 A. Assuming a stray-load loss of 1% of the output and a brush-contact resistance drop of 2 V, compute the rated-load efficiency.

Solution

The total armature-circuit resistance (not including that of brushes) is

$$R_a = 0.015 + 0.005 + 0.005 = 0.025 \ \Omega$$

$$I_a = I_L + I_f = \frac{100,000}{250} + 2.5 = 402.5 \ \text{A}$$

The losses are then computed as follows:

```
No-load rotational loss                               4000 W
Armature-circuit copper loss = 402.5² × 0.025         4050 W
Brush-contact loss = 2Iₐ = 2 × 402.5                   805 W
Shunt-field circuit copper loss = 250 × 2.5            625 W
Stray-load loss = 0.01 × 100,000                      1000 W
Total losses                                        10,480 W
```

The efficiency at rated load is then given by

$$\eta = 1 - \frac{10,480}{100,000 + 10,480} = 1 - 0.095 = 0.905, \text{ or } 90.5\%$$

It is usual to determine the efficiency by some method based on the measurement of losses, according to test codes and standards.

Applications for DC Machines

Dc motors find wide applications in which control of speed, voltage, or current is essential. Shunt motors with constant speed are used for centrifugal pumps, fans, blowers, and conveyors, whereas shunt motors with adjustable speed are employed in rolling mills and paper mills. Compound motors find their use for plunger pumps, crushers, punch presses, and hoists. Series

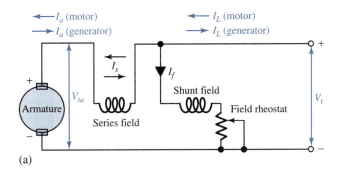

(a)

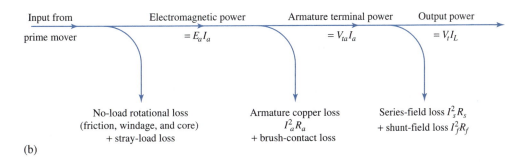

(b)

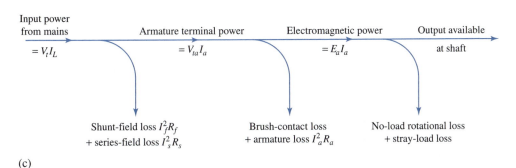

(c)

Figure 13.4.13 **(a)** Schematic diagram of a dc machine. **(b)** Power division in a dc generator. **(c)** Power division in a dc motor.

motors (known as traction motors) are utilized for electric locomotives, cranes, and car dumpers. Universal motors, operating with either dc or ac excitation, are employed in vacuum cleaners, food processors, hand tools, and several other household applications. They are available in sizes of fractional horsepower up to, and well beyond, 1 hp, in speeds ranging between 2000 and 12,000 r/min.

The dc shunt generators are often used as exciters to provide dc supply. The series generator is employed as a voltage booster and also as a constant-current source in welding machines. In applications for which a constant dc voltage is essential, the cumulative-compound generator finds its use. The differential-compound generator is used in applications such as arc welding, where a large voltage drop is desirable when the current increases.

13.5   LEARNING OBJECTIVES

The *learning objectives* of this chapter are summarized here, so that the student can check whether he or she has accomplished each of the following.

- Principles of operation of elementary synchronous, induction, and dc machines.
- Modeling a polyphase induction machine and evaluating its steady-state performance.
- Torque–speed characteristics of three-phase induction motors.
- Basic notions of speed and torque control of polyphase induction motors.
- Starting methods for polyphase induction motors.
- Analysis of single-phase induction motors by revolving-field theory.
- Starting methods for single-phase induction motors.
- Basic ideas about applications for induction motors.
- Modeling a polyphase synchronous machine and evaluating its steady-state performance.
- Parallel operation of interconnected synchronous generators.
- Basic notions about applications for synchronous motors.
- Modeling a dc machine and evaluating its steady-state performance as a motor and generator.
- Basic ideas about speed control of dc motors, dc motor starting, and applications for dc machines.

13.6 PRACTICAL APPLICATION: A CASE STUDY

Wind-Energy-Conversion Systems

It has been well recognized that renewable energy sources would have to play a key role in solving the world energy problem. Wind energy with an estimated potential of 130 million MW far exceeds the world's hydraulic supply of about 3 million MW. Consequently, researchers have been looking into the economic utilization of wind energy on a large scale by developing cost-competitive and reliable wind-energy-conversion systems (WECSs) for various applications such as electricity generation, agriculture, heating, and cooling.

The power coefficient C_p of wind turbines varies with the tip-speed ratio λ, as shown in Figure 13.6.1. Maximum power transfer is achieved by ensuing operation of λ_{opt}, where the turbine is most efficient. The mechanical power P_m available at the shaft of the wind turbine may be expressed as a function of the wind speed v and the shaft speed w,

$$P_m(v, w) = C_1 v w^2 + C_2 v^2 w + C_3 v^3 \tag{13.6.1}$$

where $C_1, C_2,$ and C_3 are constants to be determined by curve-fitting techniques. Figure 13.6.2 depicts typical wind-turbine characteristics of mechanical power versus wind speed for different values of shaft speed. The two nonzero roots of the curves represent a lower limit for the cut-in wind speed v_{ci} and an upper limit for the cut-out wind speed v_{co}. If the variable-speed operation is opted and the resulting system is able to follow the wind-speed variations, the operation at optimum tip-speed ratio can be ensured. In such a case, the power versus wind-speed characteristic will be as shown by the dashed line in Figure 13.6.2.

WECSs developed for the generation of electricity are generally classified as:

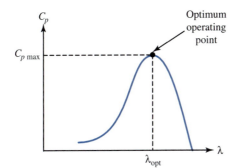

Figure 13.6.1 Typical power coefficient versus tip-speed ratio characteristic.

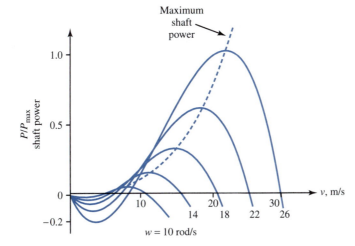

Figure 13.6.2 Typical mechanical power versus wind speed characteristics of a wind turbine for various values of shaft speed.

- Constant-speed, constant-frequency (CSCF) systems
- Variable-speed, constant-frequency (VSCF) systems
- Variable-speed, variable-frequency (VSVF) systems

The generating units in these WECSs are commonly the induction and synchronous generators.

Constant shaft-speed operation requires more complex and expensive mechanical and/or hydraulic control systems for accurate control of the shaft speed. This is usually accomplished by controlling the turbine blades. Since the turbine operates with a low efficiency for wind speeds other than the rated speed, only a small portion of the available wind energy is extracted. Hence variable-shaft-speed systems have been developed.

In variable-shaft-speed operation, the turbine is allowed to rotate at different speeds with the varying wind speed. Optimum power transfer is possible, while the actual speed of rotation is determined by the torque–speed characteristics of both the turbine and the generator. In such an operating mode, major control means are inevitably placed on the electrical side, since the control of electric systems is easy to implement, more reliable, and less costly than the control of mechanical systems.

Figure 13.6.3 illustrates a typical double-output induction generator (DOIG) scheme, in which the DOIG is equipped with two controlled converters to allow power flow in both directions. Power

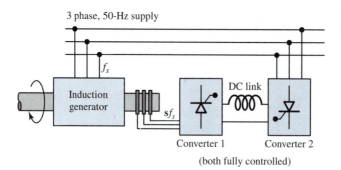

3 phase, 50-Hz supply

f_s

Induction
generator

sf_s

DC link

Converter 1   Converter 2

(both fully controlled)

Figure 13.6.3 Double-output induction-generator scheme.

generation over a wide range of shaft speeds, in the slip range of −1 to +1, is possible. The net output power is maximized by varying the firing angles of both converters.

Induction generates are also employed nowadays for power generation in conjunction with helical water turbines developed for operating in water streams.

PROBLEMS

13.1.1 A three-phase, 50-Hz induction motor has a full-load speed of 700 r/min and a no-load speed of 740 r/min.

(a) How many poles does the machine have?

(b) Find the slip and the rotor frequency at full load.

(c) What is the speed of the rotor field at full load (i) with respect to the rotor? and (ii) with respect to the stator?

13.1.2 A three-phase, 60-Hz induction motor runs at almost 1800 r/min at no load, and at 1710 r/min at full load.

(a) How many poles does the motor have?

(b) What is the per-unit slip at full load?

(c) What is the frequency of rotor voltages at full load?

(d) At full load, find the speed of (i) the rotor field with respect to the rotor, (ii) the rotor field with respect to the stator, and (iii) the rotor field with respect to the stator field.

*13.1.3 A four-pole, three-phase induction motor is energized from a 60-Hz supply. It is running at a load condition for which the slip is 0.03. Determine:

(a) The rotor speed in r/min.

(b) The rotor current frequency in Hz.

(c) The speed of the rotor rotating magnetic field with respect to the stator frame in r/min.

(d) The speed of the rotor rotating magnetic field with respect to the stator rotating magnetic field in r/min.

13.1.4 Consider a three-phase induction motor with a normal torque–speed characteristic. Neglecting the effects of stator resistance and leakage reactance, discuss the approximate effect on the characteristic, if:

(a) The applied voltage and frequency are halved.

(b) Only the applied voltage is halved, but the frequency is at its normal value.

13.1.5 Induction motors are often braked rapidly by a technique known as *plugging*, which is the reversal of the phase sequence of the voltage supplying the motor. Assume that a motor with four poles is operating at 1750 r/min from an infinite bus (a load-independent voltage supply) at 60 Hz. Two of the stator supply leads are suddenly interchanged.

(a) Find the new slip.

(b) Calculate the new rotor current frequency.

13.1.6 A four-pole, three-phase, wound-rotor induction machine is to be used as a variable-frequency supply. The frequency of the supply connected to the stator is 60 Hz.

(a) Let the rotor be driven at 3600 r/min in either direction by an auxiliary synchronous motor. If the slip-ring voltage is 20 V, what frequencies and voltages can be available at the slip rings when the rotor is at standstill?

(b) If the slip-ring voltage is 400 V, when the rotor frequency is 120 Hz, at what speed must the rotor be driven in order to give 150 Hz at the slip-ring terminals? What will the slip-ring voltage be in this case?

*13.1.7 A three-phase, wound-rotor induction machine, with its shaft rigidly coupled to the shaft of a three-phase synchronous motor, is used to change balanced 60-Hz voltages to other frequencies at the wound-rotor terminals brought out through slip rings. Both machines are electrically connected to the same balanced three-phase, 60-Hz source. Let the synchronous motor, which has four poles, drive the interconnecting shaft in the clockwise direction, and let the eight-pole balanced three-phase stator winding of the induction machine produce a counterclockwise rotating field, i.e., opposite that of the synchronous motor. Determine the frequency of the rotor voltages of the induction machine.

13.1.8 A synchronous generator has a rotor with six poles and operates at 60 Hz.

(a) Determine the speed of prime mover of the generator.

(b) Repeat part (a) if the generator has 12 poles.

(c) Repeat part (a) if the generator has 2 poles.

13.1.9 Repeat Problem 13.1.8 if the generator operates at 50 Hz.

13.1.10 A three-phase ac motor, used to drive a draft fan, is connected to a 60-Hz voltage supply. At no-load, the speed is 1188 r/min; at full load, the speed drops to 1128 r/min.

(a) Determine the number of poles of this ac motor.

(b) Comment on whether this motor is an induction motor or a synchronous motor.

13.1.11 A dc shunt machine has an armature winding resistance of 0.12 Ω and a shunt-field winding resistance of 50 Ω. The machine may be run on 250-V mains as either a generator or a motor. Find the ratio of the speed of the generator to the

speed of the motor when the total line current is 80 A in both cases.

13.1.12 A 100-kW, dc shunt generator, connected to a 220-V main, is belt-driven at 300 r/min, when the belt suddenly breaks and the machine continues to run as a motor, taking 10 kW from the mains. The armature winding resistance is 0.025 Ω, and the shunt field winding resistance is 60 Ω. Determine the speed at which the machine runs as a motor.

*13.1.13 A dc shunt motor runs off a constant 200-V supply. The armature winding resistance is 0.4 Ω, and the field winding resistance is 100 Ω. When the motor develops rated torque, it draws a total line current of 17.0 A.

(a) Determine the electromagnetic power developed by the armature under these conditions, in watts and in horsepower.

(b) Find the total line current drawn by the motor when the developed torque is one-half the rated value.

13.1.14 A dc machine, operating as a generator, develops 400 V at its armature terminals, corresponding to a field current of 4 A, when the rotor is driven at 1200 r/min and the armature current is zero.

(a) If the machine produces 20 kW of electromagnetic power, find the corresponding armature current and the electromagnetic torque produced.

(b) If the same machine is operated as a motor supplied from a 400-V dc supply, and if the motor is delivering 30 hp to the mechanical load, determine: (i) the current taken from the supply, (ii) the speed of the motor, and (iii) the electromagnetic torque, if the field current is maintained at 4 A. Assume that the armature circuit resistance and the mechanical losses are negligible.

(c) If this machine is operated from a 440-V dc supply, determine the speed at which this motor will run and the current taken from the supply, assuming no mechanical load, no friction and windage losses, and that the field current is maintained at 4 A.

13.1.15 Consider the operation of a dc shunt motor that is affected by the following changes in its operating conditions. Explain the corresponding approximate changes in the armature current and speed of the machine for each change in operating conditions.

(a) The field current is doubled, with the armature terminal voltage and the load torque remaining the same.

(b) The armature terminal voltage is halved, with the field current and load torque remaining the same.

(c) The field current and the armature terminal voltage are halved, with the horsepower output remaining the same.

(d) The armature terminal voltage is halved, with the field current and horsepower output remaining the same.

(e) The armature terminal voltage is halved and the load torque varies as the square of the speed, with the field current remaining the same.

13.2.1 A balanced three-phase, 60-Hz voltage is applied to a three-phase, two-pole induction motor. Corresponding to a per-unit slip of 0.05, determine the following:

(a) The speed of the rotating-stator magnetic field relative to the stator winding.

(b) The speed of the rotor field relative to the rotor winding.

(c) The speed of the rotor field relative to the stator winding.

(d) The speed of the rotor field relative to the stator field.

(e) The frequency of the rotor currents.

(f) Neglecting stator resistance, leakage reactance, and all losses, if the stator-to-rotor turns ratio is 2:1 and the applied voltage is 100 V, find the rotor-induced emf at standstill and at 0.05 slip.

13.2.2 No-load and blocked-rotor tests are conducted on a three-phase, wye-connected induction motor with the following results. The line-to-line voltage, line current, and total input power for the no-load test are 220 V, 20 A, and 1000 W; and for the blocked-rotor test they are 30 V, 50 A, and 1500 W. The stator resistance, as measured on a dc test, is 0.1 Ω per phase.

(a) Determine the parameters of the equivalent circuit shown in Figure 13.2.5(c).

(b) Compute the no-load rotational losses.

*13.2.3 A three-phase, 5-hp, 220-V, six-pole, 60-Hz induction motor runs at a slip of 0.025 at full load.

Rotational and stray-load losses at full load are 5% of the output power. Calculate the power transferred across the air gap, the rotor copper loss at full load, and the electromagnetic torque at full load in newton-meters.

13.2.4 The power transferred across the air gap of a two-pole induction motor is 24 kW. If the electromagnetic power developed is 22 kW, find the slip. Calculate the output torque if the rotational loss at this slip is 400 W.

13.2.5 The stator and rotor of a three-phase, 440-V, 15-hp, 60-Hz, eight-pole, wound-rotor induction motor are both connected in wye and have the following parameters per phase: $R_1 = 0.5$ Ω, $R_2 = 0.1$ Ω, $X_{l1} = 1.25$ Ω, and $X_{l2} = 0.2$ Ω. The magnetizing impedance is 40 Ω and the core-loss impedance is 360 Ω, both referred to the stator. The ratio of effective stator turns to effective rotor turns is 2.5. The friction and windage losses total 200 W, and the stray-load loss is estimated as 100 W. Using the equivalent circuit of Figure 13.2.5(a), calculate the following values for a slip of 0.05 when the motor is operated at rated voltage and frequency applied to the stator, with the rotor slip rings short-circuited: stator input current, power factor at the stator terminals, current in the rotor winding, output power, output torque, and efficiency.

13.2.6 Considering only the rotor equivalent circuit shown in Figure 13.2.2 or 13.2.3, find:

(a) The R_2 for which the developed torque would be a maximum.

(b) The slip corresponding to the maximum torque.

(c) The maximum torque.

(d) R_2 for the maximum starting torque.

*13.2.7 A three-phase induction motor, operating at its rated voltage and frequency, develops a starting torque of 1.6 times the full-load torque and a maximum torque of 2 times the full-load torque. Neglecting stator resistance and rotational losses, and assuming constant rotor resistance, determine the slip at maximum torque and the slip at full load.

13.2.8 A three-phase, wye-connected, 400-V, four-pole, 60-Hz induction motor has primary leakage impedance of $1 + j2$ Ω and secondary leakage impedance referred to the primary at standstill of $1 + j2$ Ω. The magnetizing impedance is $j40$ Ω and the core-loss impedance is 400 Ω. Using the T-equivalent circuit of Figure 13.2.5(a):

(a) Calculate the input current and power (i) on the no-load test ($\mathbf{S} \cong 0$) at rated voltage, and (ii) on a blocked-rotor test ($\mathbf{S} = 1$) at rated voltage.

(b) Corresponding to a slip of 0.05, compute the input current, torque, output power, and efficiency.

(c) Determine the starting torque and current; the maximum torque and the corresponding slip; and the maximum output power and the corresponding slip.

For the following parts, use the approximate equivalent circuit obtained by transferring the shunt core loss/magnetizing branch to the input terminals.

(d) Find the same values requested in part (b).

(e) When the machine is driven as an induction generator with a slip of -0.05, calculate the primary current, torque, mechanical power input, and electric power output.

(f) Compute the primary current and the braking torque at the instant of plugging (i.e., reversal of the phase sequence) if the slip immediately before plugging is 0.05.

13.2.9 A 500-hp, wye-connected, wound-rotor induction motor, when operated at rated voltage and frequency, develops its rated full-load output at a slip of 0.02; maximum torque of 2 times the full-load torque at a slip of 0.06, with a referred rotor current of 3 times that at full load; and 1.2 times the full-load torque at a slip of 0.2, with a referred rotor current of 4 times that at full load. Neglect rotational and stray-load losses. If the rotor-circuit resistance in all phases is increased to 5 times the original resistance, determine the following:

(a) The slip at which the motor will develop the same full-load torque.

(b) The total rotor-circuit copper loss at full-load torque.

(c) The horsepower output at full-load torque.

(d) The slip at maximum torque.

(e) The rotor current at maximum torque.

(f) The starting torque.

(g) The rotor current at starting.

13.2.10 The per-phase equivalent circuit shown in Figure 13.2.6 of a three-phase, 600-V, 60-Hz, four-pole, wye-connected, wound-rotor induction motor has the following parameters: $R_1 = 0.75\ \Omega$, $R_2' = 0.80\ \Omega$, $X_{l1} = X_{l2}' = 2.0\ \Omega$, and $X_m = 50\ \Omega$. Neglect the core losses.

(a) Find the slip at which the maximum developed torque occurs.

(b) Calculate the value of the maximum torque developed.

(c) What is the range of speed for stable operation of the motor?

(d) Determine the starting torque.

(e) Compute the per-phase referred value of the additional resistance that must be inserted in the rotor circuit in order to obtain the maximum torque at starting.

13.2.11 A three-phase, wye-connected, 220-V, 10-hp, 60-Hz, six-pole induction motor (using Figure 13.2.6 for notation) has the following parameters in ohms per phase referred to the stator: $R_1 = 0.294$, $R_2' = 0.144$, $X_{l1} = 0.503$, $X_{l2}' = 0.209$, and $X_m = 13.25$. The total friction, windage, and core losses can be assumed to be constant at 403 W, independent of load. For a slip of 2.00%, compute the speed, output torque and power, stator current, power factor, and efficiency when the motor is operated at rated voltage and frequency. Neglect the impedance of the source.

*13.2.12 A squirrel-cage induction motor operates at a slip of 0.05 at full load. The rotor current at starting is five times the rotor current at full load. Neglecting stator resistance and rotational and stray-load losses, and assuming constant rotor resistance, calculate the starting torque and the maximum torque in per-unit of full-load torque, as well as the slip at which the maximum torque occurs.

13.2.13 Using the approximate equivalent circuit in which the shunt branch is moved to the stator input terminals, show that the rotor current, torque, and electromagnetic power of a polyphase induction motor vary almost directly as the slip, for small values of slip.

13.2.14 A three-phase, 50-hp, 440-V, 60-Hz, four-pole, wound-rotor induction motor operates at a slip of 0.03 at full load, with its slip rings short-circuited. The motor is capable of developing a maximum torque of two times the full-load torque at rated voltage and frequency. The rotor resistance per phase referred to the stator is 0.1 Ω. Neglect the stator resistance and rotational and stray-load losses. Find the rotor copper loss at full load and the speed at maximum torque. Compute the value of the per-phase rotor resistance (referred to the stator) that must be added in series to produce a starting torque equal to the maximum torque.

13.2.15 A three-phase, 220-V, 60-Hz, four-pole, wye-connected induction motor has a per-phase stator resistance of 0.5 Ω. The following no-load and blocked rotor test data on the motor are given:

  • No-load test: line-to-line voltage 220 V, total input power 600 W, of which 200 W is the friction and windage loss, and line current 3 A
  • Blocked-rotor test: line-to-line voltage 35 V, total input power 720 W, and line current 15 A

(a) Calculate the parameters of the equivalent circuit shown in Figure 13.2.5(c).

(b) Compute the output power, output torque, and efficiency if the machine runs as a motor with a slip of 0.05.

(c) Determine the slip at which maximum torque is developed, and obtain the value of the maximum torque.

Note: It may help the student to solve Problems 13.2.15 through 13.2.17 if the background given in the solutions manual as part of the solution to Problem 13.2.15 is provided.

13.2.16 The synchronous speed of a wound-rotor induction motor is 900 r/min. Under a blocked-rotor condition, the input power to the motor is 45 kW at 193.6 A. The stator resistance per phase is 0.2 Ω, and the ratio of effective stator turns to effective rotor turns is 2. The stator and rotor are both wye-connected. Neglect the effect of the core-loss and magnetizing impedances. Calculate:

(a) The value in ohms of the rotor resistance per phase.

(b) The motor starting torque.

13.2.17 The no-load and blocked-rotor tests on a three-phase, wye-connected induction motor yield the following results:

  • No-load test: line-to-line voltage 400 V, input power 1770 W, input current 18.5 A, and friction and windage loss 600 W

  • Blocked-rotor test: line-to-line voltage 45 V, input power 2700 W, and input current 63 A

Determine the parameters of the equivalent circuit of Figure 13.2.5(a), assuming $R_1 = R'_2$ and $X_{l1} = X'_{l2}$.

*13.2.18 A three-phase induction motor has the per-phase circuit parameters shown in Figure P13.2.18. At what slip is the maximum power developed?

13.2.19 A large induction motor is usually started by applying a reduced voltage across the motor; such a voltage may be obtained from an autotransformer. A motor is to be started on 50% of full-load torque, and the full-voltage starting current is 5 times the full-load current. The full-load slip is 4%. Determine the percentage reduction in the applied voltage (i.e., the percentage tap on the autotransformer).

13.2.20 A three-phase, 400-V, wye-connected induction motor takes the full-load current at 45 V with the rotor blocked. The full-load slip is 4%. Calculate the tappings k on a three-phase autotransformer to limit the starting current to 4 times the full-load current. For such a limitation, determine the ratio of starting torque to full-load torque.

13.2.21 A three-phase, 2200-V, 60-Hz, delta-connected, squirrel-cage induction motor, when started at full rated voltage, takes a starting current of 693 A from the line and develops a starting torque of 6250 N · m.

(a) Neglect the impedance and the exciting current of the compensator. Calculate the ratio of a starting compensator (i.e., an auto-transformer starter) such that the current supplied by the 2200-V line is 300 A. Compute the starting torque with the starting compensator.

Figure P13.2.18

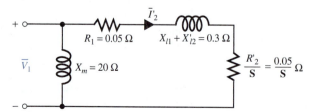

(b) If a wye–delta starting method is employed, find the starting current and the starting torque.

*13.2.22 A three-phase, four-pole, 220-V, 60-Hz induction machine with a per-phase resistance of 0.5 Ω is operating at rated voltage as a generator at a slip of -0.04, delivering 12 A of line current and a total output of 4000 W. The constant losses from a no-load run as a motor are given to be 220 W, of which 70 W represents friction and windage losses. Calculate the efficiency of the induction generator.

13.2.23 A 2200-V, 1000-hp, three-phase, 60-Hz, 16-pole, wye-connected, wound-rotor induction motor is connected to a 2200-V, three-phase, 60-Hz bus that is supplied by synchronous generators. The per-phase equivalent circuit of Figure 13.2.6 has the following parameters: $R_1 = 0.1\ \Omega = R_2'$, $X_{l1} = 0.625\ \Omega = X_{l2}'$, and $X_m = 20\ \Omega$. If the machine is driven at a speed of 459 r/min to act as a generator of real power, find the rotor current referred to the stator and the real and reactive power outputs of the induction machine.

13.2.24 A three-phase, 440-V, 60-Hz, four-pole induction motor operates at a slip of 0.025 at full load, with its rotor circuit short-circuited. This motor is to be operated on a 50-Hz supply so that the air-gap flux wave has the same amplitude at the same torque as on a 60-Hz supply. Determine the 50-Hz applied voltage and the slip at which the motor will develop a torque equal to its 60-Hz full-load value.

13.2.25 The rotor of a wound-rotor induction motor is rewound with twice the number of its original turns, with a cross-sectional area of the conductor in each turn of one-half the original value. Determine the ratio of the following in the rewound motor to the corresponding original quantities:

(a) Full-load current.

(b) Actual rotor resistance.

(c) Rotor resistance referred to the stator. Repeat the problem, given that the original rotor is rewound with the same number of turns as originally, but with one-half the original cross-section of the conductor. Neglect the changes in the leakage flux.

13.2.26 A wound-rotor induction machine, driven by a dc motor whose speed can be controlled, is operated as a frequency changer. The three-phase

stator winding of the induction machine is excited from a 60-Hz supply, while the variable-frequency three-phase power is taken out of the slip rings. The output frequency range is to be 120 to 420 Hz; the maximum speed is not to exceed 3000 r/min; and the maximum power output at 420 Hz is to be 70 kW at 0.8 power factor. Assuming that the maximum-speed condition determines the machine size, and neglecting exciting current, losses, and voltage drops in the induction machine, calculate:

(a) The minimum number of poles for the induction machine.

(b) The corresponding minimum and maximum speeds.

(c) The kVA rating of the induction-machine stator winding.

(d) The horsepower rating of the dc machine.

13.2.27 A $1/4$-hp, 110-V, 60-Hz, four-pole, capacitor-start, single-phase induction motor has the following parameters and losses: $R_1 = 2\ \Omega$, $X_{l1} = 2.8\ \Omega$, $X_{l2}' = 2\ \Omega$, $R_2' = 4\ \Omega$, $X_m = 70\ \Omega$. The core loss at 110 V is 25 W, and friction and windage is 12 W. For a slip of 0.05, compute the output current, power factor, power output, speed, torque, and efficiency when the motor is running at rated voltage and rated frequency with its starting winding open.

13.2.28 The no-load and blocked-rotor tests conducted on a 110-V, single-phase induction motor yield the following data:

• No-load test: input voltage 110 V, input current 3.7 A, and input power 50 W

• Blocked-rotor test: input voltage 50 V and input current 5.6 A

Taking the stator resistance to be 2.0 Ω, friction and windage loss to be 7 W, and assuming $X_{l1} = X_{l2}'$, determine the parameters of the double-revolving-field equivalent circuit.

*13.2.29 The impedance of the main and auxiliary windings of a $1/3$-hp, 120-V, 60-Hz, capacitor-start motor are given as $\bar{Z}_m = 4.6 + j3.8\ \Omega$ and $\bar{Z}_a = 9.6 + j3.6\ \Omega$. Determine the value of the starting capacitance that will cause the main and auxiliary winding currents to be in quadrature at starting.

*13.3.1 A three-phase, wye-connected, cylindrical-rotor synchronous generator rated at 10 kVA and 230

V has a synchronous reactance of 1.5 Ω per phase and an armature resistance of 0.5 Ω per phase.

(a) Determine the voltage regulation at full load with: (i) 0.8 lagging power factor, and (ii) 0.8 leading power factor.

(b) Calculate the power factor for which the voltage regulation becomes zero on full load.

13.3.2 A three-phase, wye-connected, 2300-V, four-pole, 1000-kVA, 60-Hz synchronous machine has a synchronous reactance $X_s = 5$ Ω, a field resistance $R_f = 10$ Ω, and an approximately linear magnetization characteristic (E_f versus I_f) with a slope $K_{ag} = 200$ Ω. The machine is connected to a balanced three-phase ac system and is used as a generator. Determine the following:

(a) The rated stator current.

(b) The exciter setting $V_{ex} = I_f R_f$ for operating the machine at rated conditions, at a power factor of (i) 0.866 lagging and (ii) 0.866 leading.

(c) V_{ex} in part (b) for unity power factor and the same real power output as in part (b).

(d) The complex power delivered by the generator to the system for parts (b) and (c).

13.3.3 The loss data for the synchronous generator of Problem 13.3.2 are:

```
Open-circuit core
  loss at 13.8 kV          70 kW
Short-circuit load
  loss at 418 A, 75°C      50 kW
Friction and windage
  loss                     80 kw
Field-winding resis-
  tance at 75°C            0.3 Ω
Stray-load loss at
  full load                20 kW
```

Determine the efficiency of the generator at rated load, rated voltage, and 0.8 power factor lagging.

13.3.4 A three-phase, six-pole, wye-connected synchronous generator is rated at 550 V and has a synchronous reactance $X_s = 2$ Ω. When the generator supplies 50 kVA at rated voltage and a power factor of 0.95 lagging, find the armature current I_a and the excitation voltage E_f. Sketch the phasor diagram of $\bar{V}_t$, $\bar{I}_a$, and $\bar{E}_f$. Also, determine the regulation corresponding to the operating conditions.

13.3.5 The synchronous machine of Problem 13.3.2 is to be used as a motor. Determine the following:

(a) The exciter setting V_{ex} for operation at rated conditions and a power factor of (i) 0.866 lagging, and (ii) 0.866 leading.

(b) V_{ex} in part (a) for unity power factor and the same real power input as in part (a).

(c) The complex power absorbed by the machine in parts (a) and (b).

13.3.6 For a 45-kVA, three-phase, wye-connected, 220-V synchronous machine at rated armature current, the short-circuit load loss (total for three phases) is 1.80 kW at a temperature of 25°C. The dc resistance of the armature at this temperature is 0.0335 Ω per phase. Compute the effective armature ac resistance in per unit and in ohms per phase at 25°C.

*13.3.7 A 4000-V, 5000-hp, 60-Hz, 12-pole synchronous motor, with a synchronous reactance of 4 Ω per phase (based on cylindrical-rotor theory), is excited to produce unity power factor at rated load. Neglect all losses.

(a) Find the rated and maximum torques.

(b) What is the armature current corresponding to the maximum torque?

13.3.8 A three-phase, wye-connected, four-pole, 400-V, 60-Hz, 15-hp synchronous motor has a synchronous reactance of 3 Ω per phase and negligible armature resistance. The data for its no-load magnetization curve follow:

Field current, A:							
	2	3.5	4.4	6	8	10	12
Line-to-neutral voltage, V:							
	100	175	200	232	260	280	295

(a) When the motor operates at full load at 0.8 leading power factor, determine the power angle and the field current. Neglect all losses.

(b) Compute the minimum line current for the motor operating at full load and the corresponding field current.

(c) When the motor runs with an excitation of 10 A while taking an armature current of 25 A, calculate the power developed and the power factor.

(d) If the excitation is adjusted such that the magnitudes of the excitation voltage and the terminal voltage are equal, and if the motor is taking 20 A, find the torque developed.

13.3.9 A three-phase, wye-connected, 2300-V, 60-Hz, round-rotor synchronous motor has a synchronous reactance of 2 Ω per phase and negligible armature resistance.

(a) If the motor takes a line current of 350 A operating at 0.8 power factor leading, calculate the excitation voltage and the power angle.

(b) If the motor is operating on load with a power angle of $-20°$, and the excitation is adjusted so that the excitation voltage is equal in magnitude to the terminal voltage, determine the armature current and the power factor of the motor.

13.3.10 A 2300-V, three-phase, wye-connected, round-rotor synchronous motor has a synchronous reactance of 3 Ω per phase and an armature resistance of 0.25 Ω per phase. The motor operates on load with a power angle of $-15°$, and the excitation is adjusted so that the internally induced voltage is equal in magnitude to the terminal voltage. Determine:

(a) The armature current.

(b) The power factor of the motor.
Neglect the effect of armature resistance.

13.3.11 An induction motor takes 350 kW at 0.8 power factor lagging while driving a load. When an overexcited synchronous motor taking 150 kW is connected in parallel with the induction motor, the overall power factor is improved to 0.95 lagging. Determine the kVA rating of the synchronous motor.

*13.3.12 An industrial plant consumes 500 kW at a lagging power factor of 0.6.

(a) Find the required kVA rating of a synchronous capacitor to improve the power factor to 0.9.

(b) If a 500-hp, 90% efficient synchronous motor, operating at full load and 0.8 leading power factor, is added instead of the capacitor in part (a), calculate the resulting power factor.

13.3.13 A three-phase, wye-connected, cylindrical-rotor, synchronous motor, with negligible armature resistance and a synchronous reactance of 1.27 Ω per phase, is connected in parallel with a three-phase, wye-connected load taking 50 A at 0.707 lagging power factor from a three-phase, 220-V, 60-Hz source. If the power developed by the motor is 33 kW at a power angle of 30°, determine:

(a) The overall power factor of the motor and the load.

(b) The reactive power of the motor.

13.3.14 Two identical three-phase, 33-kV, wye-connected, synchronous generators operating in parallel share equally a total load of 12 MW at 0.8 lagging power factor. The synchronous reactance of each machine is 8 Ω per phase, and the armature resistance is negligible.

(a) If one of the machines has its field excitation adjusted such that it delivers 125 A lagging current, determine the armature current, power factor, excitation voltage, and power angle of each machine.

(b) If the power factor of one of the machines is 0.9 lagging, find the power factor and current of the other machine.

13.3.15 A three-phase, wye-connected, round-rotor, 220-V, 60-Hz, synchronous motor, having a synchronous reactance of 1.27 Ω per phase and negligible armature resistance, is connected in parallel with a three-phase, wye-connected load that takes a current of 50 A at 0.707 lagging power factor and 220 V line-to-line. At a power angle of 30°, the power developed by the motor is 33 kW. Determine the reactive kVA of the motor, and the overall power factor of the motor and the load.

13.3.16 A three-phase, wye-connected, 2500-kVA, 6600-V, 60-Hz turboalternator has a per-phase synchronous reactance and an armature resistance of 10.4 and 0.071 Ω, respectively. Compute the power factor for zero voltage regulation on full load.

13.4.1 A 100-kW, 250-V shunt generator has an armature-circuit resistance of 0.05 Ω and a field-circuit resistance of 60 Ω. With the generator operating at rated voltage, determine the induced voltage at (a) full load, and (b) one-half full load. Neglect brush-contact drop.

13.4.2 A 100-kW, 230-V shunt generator has $R_a = 0.05$ Ω and $R_f = 57.5$ Ω. If the generator operates at rated voltage, calculate the induced voltage at (a) full load, and (b) one-half full load. Neglect brush-contact drop.

*13.4.3 A 10-hp, 250-V shunt motor has an armature-circuit resistance of 0.5 Ω and a field resistance of 200 Ω. At no load, rated voltage, and 1200 r/min, the armature current is 3 A. At full load and rated voltage, the line current is 40 A, and the flux is 5% less than its no-load value because of armature reaction. Compute the full-load speed.

13.4.4 When delivering rated load a 10-kW, 230-V self-excited shunt generator has an armature-circuit voltage drop that is 6% of the terminal voltage and a shunt-field current equal to 4% of the rated load current. Calculate the resistance of the armature circuit and the field circuit.

13.4.5 A 20-hp, 250-V shunt motor has a total armature-circuit resistance of 0.25 Ω and a field-circuit resistance of 200 Ω. At no load and rated voltage, the speed is 1200 r/min, and the line current is 4.5 A. At full load and rated voltage, the line current is 65 A. Assume the field flux to be reduced by 6% from its value at no load, due to the demagnetizing effect of armature reaction. Compute the full-load speed.

13.4.6 A dc series motor is connected to a load. The torque varies as the square of the speed. With the diverter-circuit open, the motor takes 20 A and runs at 500 r/min. Determine the motor current and speed when the diverter-circuit resistance is made equal to the series-field resistance. Neglect saturation and the voltage drops across the series-field resistance as well as the armature resistance.

13.4.7 A 50-kW, 230-V compound generator has the following data: armature-circuit resistance 0.05 Ω, series-field circuit resistance 0.05 Ω, and shunt-field circuit resistance 125 Ω. Assuming the total brush-contact drop to be 2 V, find the induced armature voltage at rated load and rated terminal voltage for: (a) short-shunt, and (b) long-shunt compound connection.

*13.4.8 A 50-kW, 250-V, short-shunt compound generator has the following data: $R_a = 0.06\ \Omega$, $R_S = 0.04\ \Omega$, and $R_f = 125\ \Omega$. Calculate the induced armature voltage at rated load and terminal voltage. Take 2 V as the total brush-contact drop.

13.4.9 Repeat the calculations of Problem 13.4.8 for a machine that is a long-shunt compound generator.

13.4.10 A 10-kW, 230-V shunt generator, with an armature-circuit resistance of 0.1 Ω and a field-circuit resistance of 230 Ω, delivers full load at rated voltage and 1000 r/min. If the machine is run as a motor while absorbing 10 kW from 230-V mains, find the speed of the motor. Neglect the brush-contact drop.

13.4.11 The magnetization curve taken at 1000 r/min on a 200-V dc series motor has the following data:

```
Field current, A:     5   10    15    20    25    30
Voltage, A:          80  160   202   222   236   244
```

The armature-circuit resistance is 0.25 Ω and the series-field resistance is 0.25 Ω. Calculate the speed of the motor (a) when the armature current is 25 A, and (b) when the electromagnetic torque is 36 N · m. Neglect the armature reaction.

*13.4.12 A dc series motor operates at 750 r/min with a line current of 100 A from the 250-V mains. Its armature-circuit resistance is 0.15 Ω and its series-field resistance is 0.1 Ω. Assuming that the flux corresponding to a current of 25 A is 40% of that corresponding to a current of 100 A, determine the motor speed at a line current of 25 A at 250 V.

13.4.13 A 7.5-hp, 250-V, 1800-r/min shunt motor, having a full-load line current of 26 A, is started with a four-point starter. The resistance of the armature circuit, including the interpole winding, is 0.48 Ω; and the resistance of the shunt-field circuit, including the field rheostat, is 350 Ω. The resistances of the steps in the starting resistor are 2.24, 1.47, 0.95, 0.62, 0.40, and 0.26 Ω, in the order in which they are successively cut out. When the armature current is dropped to its rated value, the starting box is switched to the next point, thus eliminating a step in the starting resistance. Neglecting field-current changes, armature reaction, and armature inductance, find the initial and final values of the armature current and speed corresponding to each step.

13.4.14 Two shunt generators operate in parallel to supply a total load current of 3000 A. Each machine has an armature resistance of 0.05 Ω and a field resistance of 100 Ω. If the generated emfs are 200 and 210 V, respectively, determine the load voltage and the armature current of each machine.

13.4.15 Three dc generators are operating in parallel with excitations such that their external characteristics are almost straight lines over the working range with the following pairs of data points:

Load Current (A)	Terminal Voltage (V)		
	Generator I	Generator II	Generator III
0	492.5	510	525
2000	482.5	470	475

Compute the terminal voltage and current of each generator.

(a) When the total load current is 4350 A.

(b) When the load is completely removed without change of excitation currents.

13.4.16 The external-characteristics data of two shunt generators in parallel are given as follows:

Load Current, A:	0	5	10	15	20	25	30
Terminal voltage I, V:	270	263	254	240	222	200	175
Terminal voltage II, V:	280	277	270	263	253	243	228

Calculate the load current and terminal voltage of each machine.

(a) When the generators supply a load resistance of 6 Ω.

(b) When the generators supply a battery of emf 300 V and resistance of 1.5 Ω.

*13.4.17 A separately excited dc generator with an armature-circuit resistance R_a is operating at a terminal voltage V_t, while delivering an armature current I_a, and has a constant loss P_c. Find the value of I_a for which the generator efficiency is a maximum.

13.4.18 A 100-kW, 230-V, dc shunt generator, with $R_a = 0.05$ Ω, and $R_f = 57.5$ Ω has no-load rotational loss (friction, windage, and core loss) of 1.8 kW. Compute:

(a) The generator efficiency at full load.

(b) The horsepower output from the prime mover to drive the generator at this load.

13.4.19 A dc series motor, with a design constant $K_a = 40$ and flux per pole of 46.15 mWb, operates at 200 V while taking a current of 325 A. The total series-field and armature-circuit resistances are 25 and 50 mΩ, respectively. The core loss is 220 W; friction and windage loss is 40 W. Determine:

(a) The electromagnetic torque developed.

(b) The motor speed.

(c) The mechanical power output.

(d) The motor efficiency.

13.4.20 A 230-V dc shunt motor delivers 30 hp at the shaft at 1120 r/min. If the motor has an efficiency of 87% at this load, find:

(a) The total input power.

(b) The line current.

(c) If the torque lost due to friction and windage is 7% of the shaft torque, calculate the developed torque.

13.4.21 A 10-kW, 250-V dc shunt generator, having an armature resistance of 0.1 Ω and a field resistance of 250 Ω, delivers full load at rated voltage and 800 r/min. The machine is now run as a motor while taking 10 kW at 250 V. Neglect the brush-contact drop. Determine the speed of the motor.

13.4.22 A 10-hp, 230-V dc shunt motor takes a full-load line current of 40 A. The armature and field resistances are 0.25 and 230 Ω, respectively. The total brush-contact drop is 2 V, and the core and rotational losses are 380 W. Assume that stray-load loss is 1% of output. Compute the efficiency of the motor.

PART FOUR

INFORMATION SYSTEMS

14 Signal Processing

The essential feature of communication, control, computation, and instrumentation systems is the processing of *information*. Because of the relative ease and flexibility of processing and transmitting electrical quantities, usually the information obtained from a nonelectrical source is converted into electrical form. An electric *signal* is a voltage or current waveform whose time or frequency variations correspond to the desired information. The information-bearing signals are processed either for purposes of measurement in an instrumentation system, or for transmitting over long distance in a communication system. All such systems, regardless of their particular details, share certain basic concepts and common problems.

Continuous signals (shown in Figure 6.0.1) are described by time functions which are defined for all values of t (a continuous variable). Commercial broadcast systems, analog computers, and various control and instrumentation systems process continuous signals. The information processed in *analog systems* is contained in the time function which defines the signal. Analog systems are often thought of as performing signal processing in the *frequency domain*.

Discrete signals (shown in Figure 6.0.2), on the other hand, exist only at specific instances of time, and as such, their functional description is valid only for discrete-time intervals. Discrete signals are invariably a sequence of pulses in which the information is contained in the pulse characteristics and the relation amidst the pulses in the sequence during a specified time interval. Digital computers, pulsed-communication systems (modern telephone and radar), and microprocessor-based control systems utilize discrete signals. *Digital systems* process digits, i.e., *pulse trains,* in which the information is carried in the pulse sequence rather than the amplitude–time characterization of the pulses. Digital systems are often thought of as performing signal processing in the *time domain.* Because of the advantages of economy in time, low power consumption, accuracy, and reliability, digital communication systems are increasingly used for transmitting information.

Foremost among signal concepts is *spectral analysis* (representation of signals in terms of their frequency components), a concept that serves as a unifying thread in signal processing and communication systems. Signals and spectral analysis are considered first in Section 14.1. Then in Section 14.2, processing techniques such as equalization, filtering, sampling, modulation, and multiplexing are presented, while topics on interference and noise are exposed in Section 14.3. The circuit functions required for time-domain processing parallel those needed for frequency-domain processing.

14.1 SIGNALS AND SPECTRAL ANALYSIS

Figure 14.1.1 shows the functional block diagram of a signal-processing system. The information source may be a speech (voice), an image (picture), or plain text in some language. The output of a source that generates information may be described in probabilistic terms by a *random variable*, when the random or *stochastic* signal is defined by a probability density function. The output of a source may not be *deterministic*, given by a real or complex number at any instant of time. However, in view of the scope of this text, random signals and random processes are not discussed here.

A transducer is usually required to convert the output of a source into an electrical signal that is suitable for transmission. Typical examples include a microphone converting an acoustic speech or a video camera converting an image into electric signals. A similar transducer is needed at the destination to convert the received electric signals into a form (such as voice, image, etc.) that is suitable for the user.

The heart of any communication system consists of three basic elements: *transmitter, transmission medium* or *channel,* and *receiver.* The transmitter (input processor) converts the electric signal into a form that is suitable for transmission through the physical channel or transmission medium. For example, in radio and TV broadcasts, since the FCC (Federal Communications Commission) specifies the frequency range for each transmitting station, the transmitter must translate the information signal to be transmitted into the appropriate frequency range that matches the frequency allocation assigned to the transmitter. This process is called *modulation*, which usually involves the use of the information signal to vary systematically the amplitude, frequency, or phase of a sinusoidal carrier. Thus, in general, carrier modulation such as *amplitude modulation* (AM), *frequency modulation* (FM), or *phase modulation*

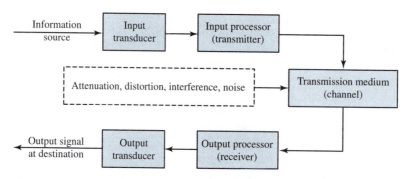

Figure 14.1.1 Functional block diagram of a signal-processing system.

(PM) is performed primarily at the transmitter. For example, for a radio station found at a setting of AM820, the carrier wave transmitted by the radio station is at the frequency of 820 kHz.

The function of the receiver is to recover the message signal contained in the received signal. If the message signal is transmitted by carrier modulation, the receiver performs carrier *demodulation* to extract the message from the sinusoidal carrier.

The communication channel (transmission medium) is the physical medium that is utilized to send the signal from the transmitter to the receiver. In wireless transmission, such as microwave radio, the transmission medium is usually the atmosphere or free space. Telephone channels, on the other hand, employ a variety of physical media such as wire lines and optical fiber cables. Irrespective of the type of physical medium for signal transmission, the essential feature is that the transmitted signal is corrupted in a random manner by a variety of possible mechanisms. For simplicity, the effects of these phenomena (*attenuation, distortion, interference, noise,* etc.) are shown at the center of Figure 14.1.1, since the transmission medium is often the most vulnerable part of a communication system, particularly over long distances.

Attenuation, caused by losses within the system, reduces the size or strength of the signal, whereas distortion is any alteration of the waveshape itself due to energy storage and/or non-linearities. Contamination by extraneous signals causes interference, whereas noise emanates from sources both internal and external to the system. To eliminate any one of these may pose a challenge to the design engineer.

Successful information recovery, while handling the aforementioned problems, invariably calls for *signal processing* at the input and output. Common signal-processing operations include the following:

- *Amplification* to compensate for attenuation
- *Filtering* to reduce interference and noise, and/or to obtain selected facets of information
- *Equalization* to correct some types of distortion
- *Frequency translation* or *sampling* to get a signal that better suits the system characteristics
- *Multiplexing* to permit one transmission system to handle two or more information-bearing signals simultaneously

In addition, to enhance the quality of information recovery, several specialized techniques, such as linearizing, averaging, compressing, peak detecting, thresholding, counting, and timing, are used.

Analog signals in an *analog communication system* can be transmitted directly via carrier modulation over the communication channel and demodulated accordingly at the receiver. Alternatively, an analog source output may be converted into a digital form and the message can be transmitted via digital modulation and demodulated as a digital signal at the receiver. Potential advantages in transmitting an analog signal by means of digital modulation are the following:

- Signal fidelity is better controlled through digital transmission than through analog transmission; effects of noise can be reduced significantly.
- Since the analog message signal may be highly redundant, with digital processing, redundancy may be removed prior to modulation.
- Digital communication systems are often more economical to implement.

Figure 14.1.2 illustrates the basic elements of a *digital communication system*. For each function in the transmitting station, there is an inverse operation in the receiver. The analog input

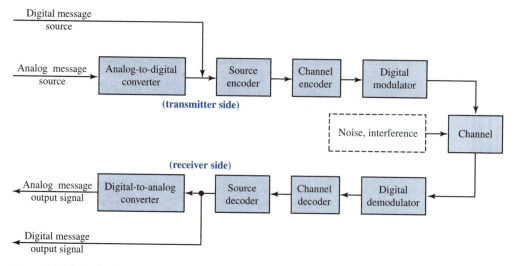

Figure 14.1.2 Basic elements of a digital communication system.

signal (such as an audio or video signal) must first be converted to a digital signal by an analog-to-digital (A/D) converter. If no analog message is involved, a digital signal (such as the output of a teletype machine, which is discrete in time and has a finite number of output characters) can be directly input.

Encoding is a critical function in all digital systems. The messages produced by the source are usually converted into a sequence of binary digits. The process of efficiently converting the output of either an analog or a digital source into a sequence of binary digits is called *source encoding* or *data compression*.

The sequence of binary digits from the source encoder, known as the *information sequence*, is passed on to the *channel encoder*. The purpose of the channel encoder is to introduce some redundancy in a controlled manner in the binary information sequence, so that the redundancy can be used at the receiver to overcome the effects of noise and interference encountered in the transmission of the signal through the channel. Thus, redundancy in the information sequence helps the receiver in decoding the desired information sequence, thereby increasing the reliability of the received data and improving the fidelity of the received signal.

The binary sequence at the output of the channel encoder is passed on to the *digital modulator*, which functions as the interface to the communication channel. The primary purpose of the digital modulator is to map the binary information sequence into signal waveforms, since nearly all the communication channels used in practice are capable of transmitting electric signals (waveforms). Because the message has only two amplitudes in a binary system, the modulation process is known as *keying*. In *amplitude-shift keying* (ASK), a carrier's amplitude is shifted or keyed between two levels. *Phase-shift keying* (PSK) involves keying between two phase angles of the carrier, whereas *frequency-shift keying* (FSK) consists of shifting a carrier's frequency between two values. Many other forms of modulation are also possible.

The functions of the receiver in Figure 14.1.2 are the inverse of those in the transmitter. At the receiving end of a digital communication system, the *digital demodulator* processes the channel-corrupted transmitted waveform and reduces each waveform to a single number, which represents an estimate of the transmitted data symbol. For example, when binary modulation is used, the demodulator may process the received waveform and decide on whether the transmitted bit is a 0 or 1. The *source decoder* accepts the output sequence from the *channel decoder*, and from the knowledge of the source encoding method used, attempts to reconstruct the original signal from the source. Errors due to noise, interference, and practical system imperfections do occur. The digital-to-analog (D/A) converter reconstructs an analog message that is a close approximation to the original message. The difference, or some function of the difference, between the original signal and the reconstructed signal is a measure of the *distortion* introduced by the digital communication system.

The remainder of this chapter deals with basic methods for analyzing and processing analog signals. A large number of building blocks in a communication system can be modeled by linear time-invariant (LTI) systems. LTI systems provide good and accurate models for a large class of communication channels. Some basic components of transmitters and receivers (such as filters, amplifiers, and equalizers) are LTI systems.

Periodic Signals and Fourier Series

In the study of analog systems, predicting the response of circuits to a general time-varying voltage or current waveform $x(t)$ is a difficult task. However, if $x(t)$ can be expressed as a *sum of sinusoids,* then the principle of superposition can be invoked on linear systems and the frequency response of the circuit can be utilized to expedite calculations. Expressing a signal in terms of sinusoidal components is known as *spectral analysis*. Let us begin here by considering the Fourier-series expansion of periodic signals, which has been introduced in Section 3.1.

A *periodic signal* has the property that it repeats itself in time, and hence, it is sufficient to specify the signal in the basic time interval called the period. A periodic signal $x(t)$ satisfies the property

$$x(t + kT) = x(t) \tag{14.1.1}$$

for all t, all integers k, and some positive real number T, called the period of the signal. For discrete-time periodic signals, it follows that

$$x(n + kN) = x(n) \tag{14.1.2}$$

for all integers n, all integers k, and a positive integer N, called the period. A signal that does not satisfy the condition of periodicity is known as nonperiodic.

EXAMPLE 14.1.1

Consider the following signals, sketch each one of them and comment on the periodic nature:

(a) $x(t) = A \cos(2\pi f_0 t + \theta)$, where A, f_0, and θ are the amplitude, frequency, and phase of the signal.

(b) $x(t) = e^{j(2\pi f_0 t + \theta)}$, $A > 0$.

(c) Unit step signal $u_{-1}(t)$ defined by $u_{-1}(t) = \begin{cases} 1, & t > 0 \\ \frac{1}{2}, & t = 0 \\ 0, & t < 0 \end{cases}$.

(d) Discrete-time signal $x[n] = A \cos(2\pi f_0 n + \theta)$, where n is an integer.

Solution

(a) This is a continuous-time signal (called a sinusoidal signal) that is real and periodic with period $T = 1/f_0$, as sketched in Figure E14.1.1(a).

(b) This is a complex periodic exponential signal. Its real part is

$$x_r(t) = A \, \cos(2\pi f_0 t + \theta)$$

and its imaginary part is

$$x_i(t) = A \, \sin(2\pi f_0 t + \theta)$$

This signal could also be described in terms of its modulus and phase. The absolute value of $x(t)$ is

$$|x(t)| = \sqrt{x_r^2(t) + x_i^2(t)} = A$$

and its phase is

$$\angle x(t) = 2\pi f_0 t + \theta$$

Sketches of these functions are shown in Figure E14.1.1(b). In addition, the following relations apply:

$$x_r(t) = |x(t)| \, \cos[\angle x(t)]$$

$$x_i(t) = |x(t)| \, \sin[\angle x(t)]$$

$$\angle x(t) = \arctan \frac{x_i(t)}{x_r(t)}$$

(c) The unit step signal is a nonperiodic signal, sketched in Figure E14.1.1(c).

(d) A sketch of this discrete-time signal is shown in Figure E14.1.1(d).

   This is not periodic for all values of f_0. The condition for it to be periodic is

$$2\pi f_0(n + kN) + \theta = 2\pi f_0 n + \theta + 2m\pi$$

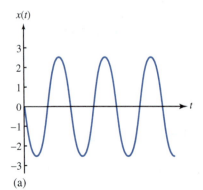

(a)

Figure E14.1.1 **(a)** Sinusoidal signal. **(b)** Real–imaginary and magnitude–phase graphs of the complex exponential signal. **(c)** Unit step signal. **(d)** Discrete sinusoidal signal.

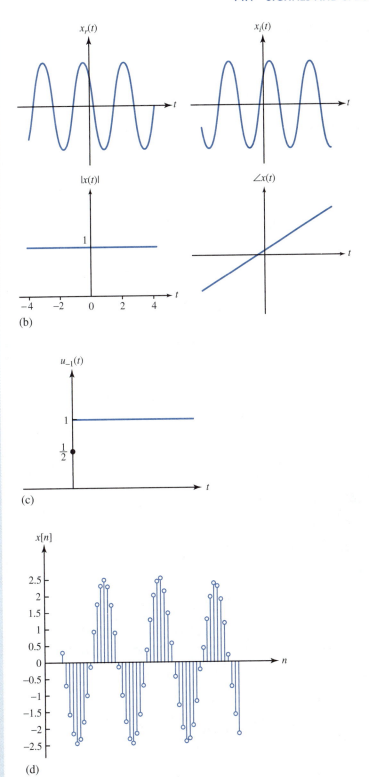

(b)

(c)

(d)

Figure E14.1.1 Continued

for all integers n and k, some positive integer N, and some integer m. From this it follows that

$$2\pi f_0 k N = 2m\pi \quad \text{or} \quad f_0 = m/(kN)$$

that is, the discrete sinusoidal signal is periodic only for rational values of f_0.

Evenness and oddness are expressions of various types of symmetry present in signals. A signal $x(t)$ is *even* if it has a mirror symmetry with respect to the vertical axis. A signal is *odd* if it is symmetric with respect to the origin. The signal $x(t)$ is even if and only if, for all t, it satisfies

$$x(-t) = x(t) \tag{14.1.3}$$

and is odd if and only if, for all t,

$$x(-t) = -x(t) \tag{14.1.4}$$

Any signal $x(t)$, in general, can be expressed as the sum of its even and odd parts,

$$x(t) = x_e(t) + x_0(t) \tag{14.1.5}$$

$$x_e(t) = \frac{x(t) + x(-t)}{2} \tag{14.1.6}$$

$$x_o(t) = \frac{x(t) - x(-t)}{2} \tag{14.1.7}$$

The *half-wave symmetry* is expressed by

$$x\left(t \pm \frac{T}{2}\right) = -x(t) \tag{14.1.8}$$

EXAMPLE 14.1.2

Discuss the nature of evenness and oddness of:

(a) The sinusoidal signal $x(t) = A\ \cos(2\pi f_0 t + \theta)$.

(b) The complex exponential signal $x(t) = e^{j 2\pi f_0 t}$.

Solution

(a) The signal is, in general, neither even nor odd. However, for the special case of $\theta = 0$, it is even; for the special case of $\theta = \pm\pi/2$, it is odd. In general,

$$x(t) = A\ \cos\theta\ \cos 2\pi f_0 t - A\ \sin\theta\ \sin 2\pi f_0 t$$

Since $\cos 2\pi f_0 t$ is even and $\sin 2\pi f_0 t$ is odd, it follows that

$$x_e(t) = A\ \cos\theta\ \cos 2\pi f_0 t$$

and

$$x_o(t) = -A\ \sin\theta\ \sin 2\pi f_0 t$$

(b) From the sketches of Figure E14.1.1(b), for $\theta = 0$, $x(t) = Ae^{j2\pi f_0 t}$, the real part and the magnitude are even; the imaginary part and the phase are odd. Noting that a complex signal $x(t)$ is called *hermitian* if its real part is even and its imaginary part is odd, the signal and symmetry are then said to be hermitian.

A signal $x(t)$ is said to be *causal* if, for all $t < 0$, $x(t) = 0$; otherwise, the signal is noncausal. An anticausal signal is identically equal to zero for $t > 0$. A discrete-time signal is a causal signal if it is identically equal to zero for $n < 0$. Note that the unit step multiplied by any signal produces a causal version of the signal.

Signals can also be classified as *energy-type* and *power-type* signals based on the finiteness of their energy content and power content, respectively. A signal $x(t)$ is an energy-type signal if and only if the energy E_x of the signal,

$$E_x = \int_{-\infty}^{\infty} |x(t)|^2 \, dt = \lim_{T \to \infty} \int_{-T/2}^{T/2} |x(t)|^2 \, dt \qquad (14.1.9)$$

is well defined and finite. A signal is a power-type signal if and only if the power P_x of the signal,

$$P_x = \lim_{T \to \infty} \frac{1}{T} \int_{-T/2}^{T/2} |x(t)|^2 \, dt \qquad (14.1.10)$$

is well defined and $0 \le P_x < \infty$. For real signals, note that $|x(t)|^2$ can be replaced by $x^2(t)$.

EXAMPLE 14.1.3

(a) Evaluate whether the sinusoidal signal $x(t) = A \cos(2\pi f_0 t + \theta)$ is an energy-type or a power-type signal.

(b) Show that any periodic signal is not typically energy type, and the power content of any periodic signal is equal to the average power in one period.

Solution

(a) $E_x = \lim_{T \to \infty} \int_{-T/2}^{T/2} A^2 \cos^2(2\pi f_0 t + \theta) \, dt = \infty$

Therefore, the sinusoidal signal is not an energy-type signal. However, the power of this signal is

$$P_x = \lim_{T \to \infty} \frac{1}{T} \int_{-T/2}^{T/2} A^2 \cos^2(2\pi f_0 t + \theta) \, dt$$

$$= \lim_{T \to \infty} \frac{1}{T} \int_{-T/2}^{T/2} \frac{A^2}{2} [1 + \cos(4\pi f_0 t + 2\theta)] \, dt$$

$$= \lim_{T \to \infty} \left\{ \frac{A^2 T}{2T} + \left[\frac{A^2}{8\pi f_0 T} \sin(4\pi f_0 t + 2\theta) \right]_{-T/2}^{+T/2} \right\}$$

$$= \frac{A^2}{2} < \infty$$

Hence, the given $x(t)$ is a power-type signal with its power given by $A^2/2$.

(b) For any periodic signal with period T_0, the energy is given by

$$E_x = \lim_{T \to \infty} \int_{-T/2}^{+T/2} |x(t)|^2\, dt = \lim_{n \to \infty} \int_{-nT_0/2}^{+nT_0/2} |x(t)|^2\, dt$$

$$= \lim_{n \to \infty} n \int_{-T_0/2}^{+T_0/2} |x(t)|^2\, dt = \infty$$

Therefore, periodic signals are not typically energy type. The power content of any periodic signal is

$$P_x = \lim_{T \to \infty} \frac{1}{T} \int_{-T/2}^{T/2} |x(t)|^2\, dt = \lim_{n \to \infty} \frac{1}{nT_0} \int_{-nT_0/2}^{nT_0/2} |x(t)|^2\, dt$$

$$= \lim_{n \to \infty} \frac{n}{nT_0} \int_{-T_0/2}^{T_0/2} |x(t)|^2\, dt = \frac{1}{T_0} \int_{-T_0/2}^{T_0/2} |x(t)|^2\, dt$$

which shows that the power content of a periodic signal is equal to the average power in one period.

The *Fourier-series representation* states that almost any periodic signal can be decomposed into an infinite series of the form

$$x(t) = a_0 + \sum_{n=1}^{\infty} (a_n \cos n\omega t + b_n \sin n\omega t) \tag{14.1.11}$$

where $a_0, a_1, b_1, \ldots$ are the Fourier coefficients, and ω is the fundamental angular frequency related to the period T by $\omega = 2\pi/T = 2\pi f$. The integer multiples of ω are known as harmonics: 2ω being the second harmonic that is even, 3ω being the third harmonic that is odd, and so forth. The dc component is given by

$$a_0 = \frac{1}{T} \int_0^T x(t)\,dt \tag{14.1.12}$$

which is seen to be the average value of $x(t)$. The remaining coefficients can be computed from the following integrals:

$$a_n = \frac{2}{T} \int_0^T x(t) \cos n\omega t\, dt, \qquad \text{for } n = 1, 2, \ldots \tag{14.1.13}$$

$$b_n = \frac{2}{T} \int_0^T x(t) \sin n\omega t\, dt, \qquad \text{for } n = 1, 2, \ldots \tag{14.1.14}$$

It can be seen that $b_n = 0$ for $n = 1, 2, 3, \ldots$ for even symmetry. Similarly, for odd symmetry, $a_n = 0$ for $n = 0, 1, 2, 3, \ldots$. For half-wave symmetry, $a_n = b_n = 0$ for $n = 2, 4, 6, \ldots$ so that the series contains only the odd-harmonic components. For relatively smooth signals, the higher harmonic components tend to be smaller than the lower ones. Discontinuous signals have more significant high-frequency content than continuous signals.

EXAMPLE 14.1.4

Consider the following periodic waveforms shown in Figure E14.1.4:

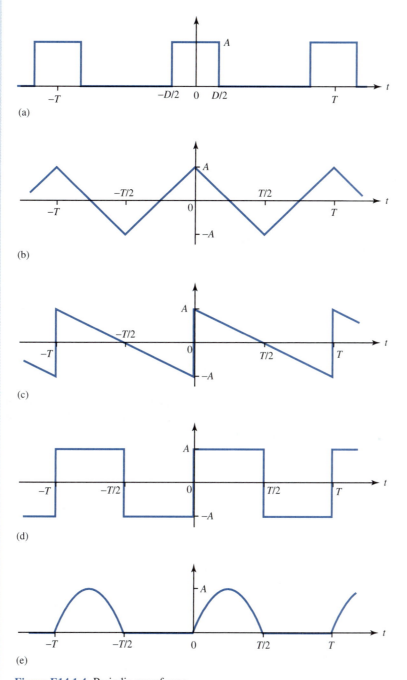

(a)

(b)

(c)

(d)

(e)

Figure E14.1.4 Periodic waveforms.

(a) Rectangular pulse train

(b) Triangular wave

(c) Sawtooth wave

(d) Square wave

(e) Half-rectified sine wave

Identify the waveform symmetry and find expressions for the Fourier coefficients.

Solution

Waveform	Symmetry	a_0	a_n or b_n	
(a) Rectangular pulse train	Even	$\dfrac{DA}{T}$	$a_n = \dfrac{2A}{\pi n} \sin \dfrac{\pi Dn}{T}$	$n = 1, 2, 3, \ldots$
(b) Triangular wave	Even and half-wave	0	$a_n = \dfrac{8A}{\pi^2 n^2}$	$n = 1, 3, 5, \ldots$
(c) Sawtooth wave	Odd	0	$b_n = \dfrac{2A}{\pi n}$	$n = 1, 2, 3, \ldots$
(d) Square wave	Odd and half-wave	0	$b_n = \dfrac{4A}{\pi n}$	$n = 1, 3, 5, \ldots$
(e) Half-rectified sine wave	None	$\dfrac{A}{\pi}$	$b_1 = \dfrac{A}{2}$	
			$a_n = -\dfrac{2A}{\pi(n^2 - 1)}$	$n = 2, 4, 6, \ldots$

Spectral Analysis and Signal Bandwidth

Spectral analysis is based on the fact that a sinusoidal waveform is completely characterized by three quantities: amplitude, phase, and frequency. By plotting amplitude and phase as a function of frequency $f(= \omega/2\pi = 1/T)$, the frequency-domain picture conveys all the information, including the signal's bandwidth and other significant properties about the signal that consists entirely of sinusoids. The two plots together (amplitude versus frequency and phase versus frequency) constitute the *line spectrum* of the signal $x(t)$.

If $x(t)$ happens to be a periodic signal whose Fourier coefficients are known, Equation (14.1.11) can be rewritten as

$$x(t) = a_0 + \sum_{n=1}^{\infty} A_n \cos(n\omega t + \phi_n) \tag{14.1.15}$$

where

$$A_n = \sqrt{a_n^2 + b_n^2} \quad \text{and} \quad \phi_n = -\arctan\left(\frac{b_n}{a_n}\right)$$

The phase angles are referenced to the cosine function, in agreement with our phasor notation. The corresponding phasor diagram of Fourier coefficients is shown in Figure 14.1.3.

Equation (14.1.15) reveals that the spectrum of a periodic signal contains lines at frequencies of $0, f, 2f$, and all higher harmonics of f, although some harmonics may be missing in particular cases. The zero-frequency or dc component represents the average value a_0, the components corresponding to the first few harmonics represent relatively slow time variations, and the higher harmonics represent more rapid time variations.

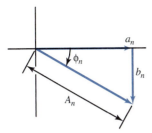

Figure 14.1.3 Phasor diagram of Fourier coefficients.

EXAMPLE 14.1.5

The rectangular pulse train of Figure E14.1.4(a) consists of pulses of height A and duration D. Such pulse trains are employed for timing purposes and to represent digital information. For a particular pulse train, $A = 3$ and the duty cycle $D/T = 1/3$.

(a) Find the Fourier-series expansion of the pulse train.

(b) Sketch the line spectrum of the rectangular pulse train when $T = 1$ ms.

Solution

(a) Using the solution of Example 14.1.4, we have for the particular pulse train:

$$a_0 = \frac{DA}{T} = 1; \qquad a_n = \frac{2A}{\pi n} \sin \frac{2Dn}{T} = \frac{6}{\pi n} \sin \frac{\pi n}{3}; \qquad b_n = 0$$

Thus, the Fourier-series expansion results,

$$x(t) = 1 + 1.65 \cos \omega t + 0.83 \cos 2\omega t - 0.41 \cos 4\,\omega t - 0.33 \cos 5\,\omega t + \dots$$

Note that the terms corresponding to 3ω, 6ω, . . . are missing because $a_n = 0$ for $n = 3$, 6, ...

(b) $T = 1$ ms $= 10^{-3}$ s and $f = 10^3$ Hz $= 1$ kHz. The line spectrum of the particular rectangular pulse train is shown in Figure E14.1.5. As seen from the spectrum, most of the time variation comes from the large-amplitude components below 6 kHz. The higher frequency components have much smaller amplitudes, which account for the stepwise jumps in the pulse train.

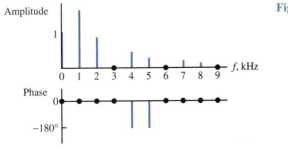

Figure E14.1.5

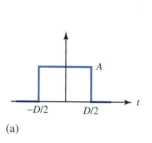

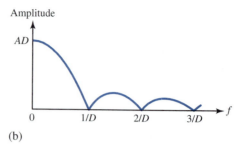

(a)   (b)

Figure 14.1.4 Nonperiodic signal and its spectrum. **(a)** Single rectangular pulse. **(b)** Continuous amplitude spectrum.

By letting $T \rightarrow \infty$ for the rectangular pulse train of Figure E14.1.4(a), so that all pulses vanish except the one centered at $t = 0$, we would obtain a single rectangular pulse, as shown in Figure 14.1.4(a). Because $f = 1/T \rightarrow 0$ when $T \rightarrow \infty$, intuitively, the amplitude lines will merge to form a continuous plot, as shown in Figure 14.1.4(b). Spectral analysis of nonperiodic signals involves *Fourier transform theory*, which goes beyond the scope of this text. Smooth curves, such as one in Figure 14.1.4(b), suggest that the signal energy is spread over a continuous frequency range, rather than being concentrated in discrete sinusoidal components. Note that the amplitude at $f = 0$ is equal to the net area DA of the nonperiodic signal.

When spectral peaks occur at or near $f = 0$, and their amplitude spectra become progressively smaller as frequency increases, such waveforms are known as *low-pass signals,* for which there exists a *signal bandwidth W* such that all significant frequency content falls within the range of $0 \leq f \leq W$. The concept of signal bandwidth plays a significant role in signal-processing and communication systems. Table 14.1.1 lists the nominal bandwidths of a few selected signals.

The approximate reciprocal relationship $W \cong 1/D$ conveys the salient point that long pulses have small bandwidths while short pulses have large bandwidths. This agrees qualitatively with the rectangular pulse spectrum of Figure 14.1.4(b), although it ignores the components above $f = 1/D$. In order to preserve the square corners of the rectangular pulse shape, we would have to take $W \gg 1/D$.

Filtering, Distortion, and Equalization

Frequency response and *filters* were discussed in Section 3.4. Any undesired waveform alteration produced by a frequency-selective network is known as *linear distortion.* It is so designated to distinguish it from the distortion caused by nonlinear elements. Let us now investigate filtering and linear distortion from the point of view of spectral analysis.

TABLE 14.1.1 Lowpass Signal Bandwidths of Selected Signals

Signal Type	Bandwidth
Telephone-quality voice	3 kHz
Moderate-quality audio	5 kHz
High-fidelity audio	20 kHz
Television video	4 MHz

Figure 14.1.5 shows a block diagram in which an arbitrary linear network characterized by its ac transfer function $H(j\omega)$ has an input signal $x(t)$ yielding an output signal $y(t)$. Note that $H(j\omega)$ is represented here in terms of the *amplitude ratio* and *phase shift* as a function of frequency f given by $|H(f)| = |H(j\omega)|$ and $\theta(f) = \angle H(j\omega)$, respectively, where ω is related to f through the relation $\omega = 2\pi f$.

If $x(t)$ contains a sinusoidal component of magnitude A_1 and phase ϕ_1 at frequency f_1, the corresponding output component of the linear network will have amplitude $|H(f_1)| A_1$ and phase $\phi_1 + \theta(f_1)$. If, on the other hand, the input should consist of several sinusoids given by

$$x(t) = \sum_n A_n \cos(2\pi f_n t + \theta_n) \tag{14.1.16}$$

By superposition, the steady-state response at the output will be

$$y(t) = \sum_n |H(f_n)| A_n \cos[2\pi f_n t + \phi_n + \theta(f_n)] \tag{14.1.17}$$

By letting $f_n = nf_1$ with $n = 0, 1, 2, \ldots$, *periodic steady-state response* due to a periodic signal can be obtained.

Whether periodic or nonperiodic, the output waveform signal is said to be *undistorted* if the output is of the form

$$y(t) = Kx(t - t_d) \tag{14.1.18}$$

That is to say, the output has the same shape as the input *scaled* by a factor K and *delayed* in time by t_d. For *distortionless transmission* through a network it follows then that

$$|H(f)| = K \qquad \text{and} \qquad \theta(f) = -360°(t_d f) \tag{14.1.19}$$

which must hold for all frequencies in the input signal $x(t)$. Thus, a distortionless network will have a constant amplitude ratio and a negative linear phase shift over the frequency range in question.

When a *low-pass signal* having a bandwidth W is applied to a *low-pass filter* (see Section 3.4) with bandwidth B, essentially distortionless output is obtained when $B \geq W$. Figure 14.1.6 illustrates the frequency-domain interpretation of distortionless transmission.

The preceding observation is of practical interest because many information-bearing waveforms are low-pass signals, and transmission cables often behave like low-pass filters. Also notice that unwanted components at $f > W$ contained in a low-pass signal can be eliminated by low-pass filtering without distorting the filtered output waveform.

If $|H(f)| \neq K$, one of the conditions given by Equation (14.1.19) for distortionless transmission is not satisfied. Then the output suffers from *amplitude or frequency distortion*, i.e., the amplitudes of different frequency components are selectively increased or decreased. If $\theta(f)$ does not satisfy the condition given in Equation (14.1.19), then the output suffers from *phase or delay distortion*, i.e., different frequency components are delayed by different amounts of time. Both types of *linear distortion* usually occur together.

The linear distortion occurring in signal transmission can often be corrected or reduced by using an *equalizer* network. The concept is illustrated in Figure 14.1.7, in which an equalizer is connected at the output of the transmission medium, such that

$$|H(f)| \, |H_{eq}(f)| = K \qquad \text{and} \qquad \theta(f) + \theta_{eq}(f) = -360°(t_d f) \tag{14.1.20}$$

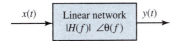

Figure 14.1.5 Linear network with input and output signals.

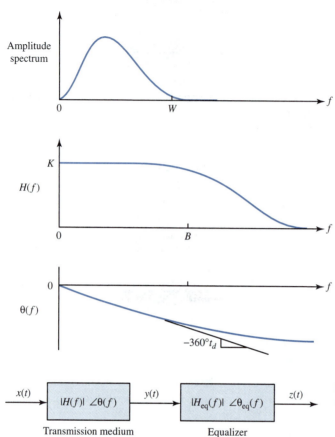

Figure 14.1.6 Frequency-domain interpretation of distortionless transmission when $B \geq W$.

Figure 14.1.7 Transmission system with equalizer.

so that the equalized output signal is then $z(t) = Kx(t - t_d)$, i.e., undistorted, regardless of the distortion in $y(t)$. For example, for correcting electrical and acoustical frequency distortion, audio equalizers in high-fidelity systems are used to adjust the amplitude ratio over several frequency bands. Sometimes, as in audio systems, phase equalization is not that critical since the human ear is not that sensitive to delay distortion. However, human vision is quite sensitive to delay distortion. Equalizers are also applied whenever energy storage in a transducer, or some other part of a signal-processing system, causes linear distortion.

14.2   MODULATION, SAMPLING, AND MULTIPLEXING

Modulation is the process whereby the amplitude (or another characteristic) of a wave is varied as a function of the instantaneous value of another wave. The first wave, which is usually a single-frequency wave, is called the *carrier wave*; the second is called the *modulating wave*. *Demodulation* or *detection* is the process whereby a wave resulting from modulation is so operated upon that a wave is obtained having substantially the characteristics of the original modulating wave. Modulation and demodulation are then reverse processes.

The information from a signal $x(t)$ is impressed on a carrier waveform whose characteristics suit a particular application. If the carrier is a sinusoid, we will see that a phenomenon known as *frequency translation* occurs. If, on the other hand, the carrier is a pulse train, the modulating signal needs to be *sampled* as part of the modulation process. Frequency translation and sampling have extensive use in communication systems. Both of these lend to *multiplexing*, which permits a transmission system to handle two or more information-bearing signals simultaneously.

Frequency Translation and Product Modulation

The basic operation needed to build modulators is the multiplication of two signals. Whenever sinusoids are multiplied, frequency translation takes place. Figure 14.2.1(a) shows a *product modulator*, which multiplies the signal $x(t)$ and a sinusoidal carrier wave at frequency f_c to yield

$$x_c(t) = x(t) \cos 2\pi f_c t \qquad (14.2.1)$$

Choosing $x(t)$ to be a low-pass signal with bandwidth $W << f_c$, Figure 14.2.1(b) depicts the relationship between $x_c(t)$ and $x(t)$. The modulated wave $x_c(t)$ can now be seen to have a *bandpass* spectrum resulting from frequency translation, which will be explained later.

If $x(t)$ contains a sinusoidal component $A_m \cos 2\pi f_m t$, multiplication by a sinusoidal carrier wave $\cos 2\pi f_c t$ with $f_c >> f_m$ yields

$$(A_m \cos 2\pi f_m t) \times (\cos 2\pi f_c t) = \frac{A_m}{2} \cos 2\pi (f_c - f_m)t + \frac{A_m}{2} \cos 2\pi (f_c + f_m)t \qquad (14.2.2)$$

Waveforms of the signal, the carrier wave, and the product, as well as their respective line spectra, are shown in Figure 14.2.2. Notice that the low frequency f_m has been translated to the higher frequencies $f_c \pm f_m$.

Next, let us consider an arbitrary low-pass signal $x(t)$ with the typical amplitude spectrum of Figure 14.2.3(a). The amplitude spectrum of the modulated wave $x_c(t)$ will now have *two sidebands* (lower and upper sidebands), each of width W on either side of f_c, as illustrated in Figure 14.2.3(b). Thus, we have a signal that can be transmitted over a *bandpass* system with a minimum bandwidth of

$$B = 2W \qquad (14.2.3)$$

which is twice the bandwidth of the modulating signal. This process is then known as *double-sideband modulation* (DSB). Either the lower or the upper sideband may be removed by filtering so as to obtain single-sideband modulation (SSB) with $B = W$, if the bandwidth needs to be conserved. By choosing the carrier frequency f_c at a value where the system has favorable characteristics, the frequency translation by product modulation helps in minimizing the distortion and other problems in system design.

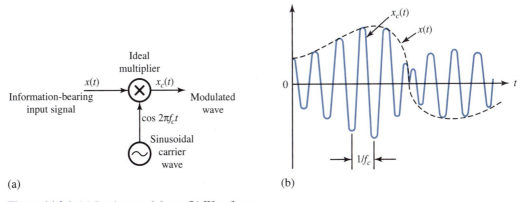

(a)

(b)

Figure 14.2.1 (a) Product modulator. (b) Waveforms.

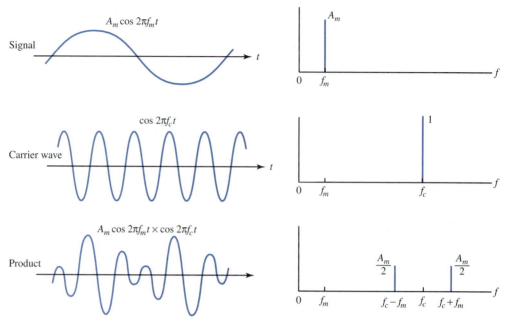

Figure 14.2.2 Frequency translation waveforms and line spectra.

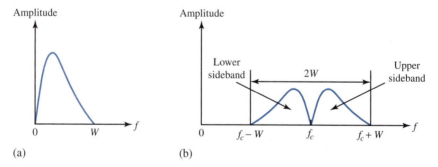

(a)　　　　　　　　　　(b)

Figure 14.2.3 Amplitude spectra in double-sideband modulation (DSB). **(a)** Amplitude spectrum of low-pass modulation signal. **(b)** Amplitude spectrum of bandpass modulated signal.

Now, in order to recover $x(t)$ from $x_c(t)$, the *product demodulator* shown in Figure 14.2.4(a), which has a local oscillator synchronized in frequency and phase with the carrier wave, can be used. The input $y(t)$ to the low-pass filter is given by

$$x(t) \cos 2\pi f_c t = x(t) \cos^2 2\pi f_c t$$

$$= \frac{1}{2}x(t) + \frac{1}{2}x(t) \cos 2\pi (2f_c)t \qquad (14.2.4)$$

indicating that the multiplication has produced both upward and downward frequency translation. In Equation (14.2.4), the first term is proportional to $x(t)$, while the second looks like DSB at carrier frequency $2f_c$. Then, if the low-pass filter in Figure 14.2.4(a) rejects the high-frequency components and passes $f \leq W$, the filtered output $z(t)$ will have the desired form $z(t) = Kx(t)$.

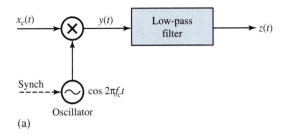

Figure 14.2.4 (a) Product demodulator. (b) Spectrum prior to low-pass filtering.

(a)

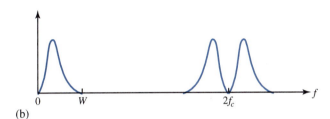

(b)

Sampling and Pulse Modulation

In most analog circuits, signals are processed in their entirety. However, in many modern electric systems, especially those that convert waveforms for processing by digital circuits, such as digital computers, only *sample values* of signals are utilized for processing. Sampling makes it possible to convert an analog signal to discrete form, thereby permitting the use of discrete processing methods. Also, it is possible to sample an electric signal, transmit only the sample values, and use them to interpolate or reconstruct the entire waveform at the destination. Sampling of signals and signal reconstruction from samples have widespread applications in communications and signal processing.

One of the most important results in the analysis of signals is the *sampling theorem*, which is formally presented later. Many modern signal-processing techniques and the whole family of digital communication methods are based on the validity of this theorem and the insight it provides. The idea leading to the sampling theorem is rather simple and quite intuitive. Let us consider a relatively smooth signal $x_1(t)$, which varies slowly and has its main frequency content at low frequencies, as well as a rapidly changing signal $x_2(t)$ due to the presence of high-frequency components. Suppose we are to approximate these signals with samples taken at regular intervals, so that linear interpolation of the sampled values can be used to obtain an approximation of the original signals. It is obvious that the sampling interval for the signal $x_1(t)$ can be much larger than the sampling interval necessary to reconstruct signal $x_2(t)$ with comparable distortion. This is simply a direct consequence of the smoothness of the signal $x_1(t)$ compared to $x_2(t)$. Therefore, the sampling interval for the signals of smaller bandwidths can be made larger, or the sampling frequency can be made smaller. The sampling theorem is, in fact, a statement of this intuitive reasoning.

Let us now look from another point of view by considering a simple switching sampler and waveforms shown in Figure 14.2.5(a). Let the switch alternate between the two contacts at the *sampling frequency* $f_s = 1/T_s$. The lower contact of the switch is grounded.

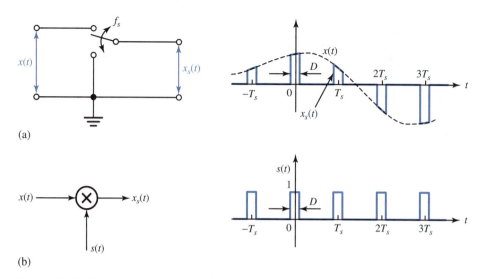

(a)

(b)

Figure 14.2.5 (a) Switching sampler. (b) Model using switching function $s(t)$.

While the switch is in touch with the upper contact for a short interval of time $D \ll T_s$, and obtains a sample piece of the input signal $x(t)$ every T_s seconds, the output sampled waveform $x_s(t)$ will look like a train of pulses with their tops carrying the sample values of $x(t)$, as shown in the waveform in Figure 14.2.5(a). This process can be modeled by using a *switching function* $s(t)$, shown in the waveform of Figure 14.2.5(b), and a multiplier in the form

$$x_s(t) = x(t)s(t) \tag{14.2.5}$$

shown in Figure 14.2.5(b). The periodic switching function $s(t)$ is simply a rectangular pulse train of unit height, whose Fourier expansion is given by

$$s(t) = a_0 + a_1 \cos 2\pi f_s t + a_2 \cos 2\pi (2 f_s) t + \ldots \tag{14.2.6}$$

with $a_0 = D/T_s$ and $a_n = (2/\pi n) \sin(\pi Dn/T_s)$ for $n = 1, 2, \ldots$ [see Figure 14.1.4(a)]. Using Equation (14.2.6) in Equation (14.2.5), we get

$$x_s(t) = a_0 x(t) + a_1 x(t) \cos 2\pi f_s t + a_2 x(t) \cos 2\pi (2 f_s) t + \ldots \tag{14.2.7}$$

By employing the frequency-domain methods, one can gain insight for signal analysis and easily interpret the results. Supposing that $x(t)$ has a low-pass amplitude spectrum, as shown in Figure 14.2.6(a), the corresponding spectrum of the sampled signal $x_s(t)$ is depicted in Figure 14.2.6(b). Taking Equation (14.2.7) term by term the first term will have the same spectrum as $x(t)$ scaled by the factor a_0; the second term corresponds to product modulation with a scale factor a_1 and carrier frequency f_s, so that it will have a DSB spectrum over the range $f_s - W \le f \le f_s + W$; the third and all other terms will have the same DSB interpretation with progressively higher carrier frequencies $2f_s, 3f_s, \ldots$.

Note that provided the sampling frequency satisfies the condition

Amplitude

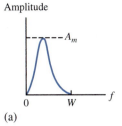

(a)

Amplitude

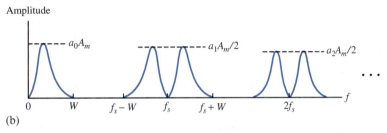

(b)

Figure 14.2.6 (a) Spectrum of low-pass signal. (b) Spectrum of sampled signal.

$$f_s \geq 2W \tag{14.2.8}$$

none of the translated components falls into the signal range of $0 \leq f \leq W$, as seen from Figure 14.2.6(b). Therefore, if the sampled signal $x_s(t)$ is passed through a low-pass filter, all components at $f \geq f_s - W$ will be removed so that the resulting output signal is of the same shape as $a_0 x(t)$, where a_0 is given by D/T_s. These observations are summarized in the following *uniform sampling theorem:*

> A signal that has no frequency components at $f \geq W$ is completely described by uniformly spaced sample values taken at the rate $f_s \geq 2W$. The entire signal waveform can be reconstructed from the sampled signal put through a low-pass filter that rejects $f \geq f_s - W$.

The importance of the sampling theorem lies in the fact that it provides a method of reconstruction of the original signal from the sampled values and also gives a precise upper bound on the sampling interval (or equivalently, a lower bound on the sampling frequency) needed for distortionless reconstruction. The minimum sampling frequency $f_s = 2W$ is known as *Nyquist rate.*

When Equation (14.2.8) is not satisfied, spectral overlap occurs, thereby causing unwanted spurious components in the filtered output. In particular, if any component of $x(t)$ originally at $f' > f_s/2$ appears in the output at the lower frequency $|f_s - f'| < W$, it is known as *aliasing.* In order to prevent aliasing, one can process the signal $x(t)$ through a low-pass filter with bandwidth $B_p \leq f_s/2$ *prior to* sampling.

The elements of a typical *pulse modulation system* are shown in Figure 14.2.7(a). The pulse generator produces a pulse train with the sampled values carried by the pulse amplitude, duration, or relative position, as illustrated in Figure 14.2.7(b). These are then known

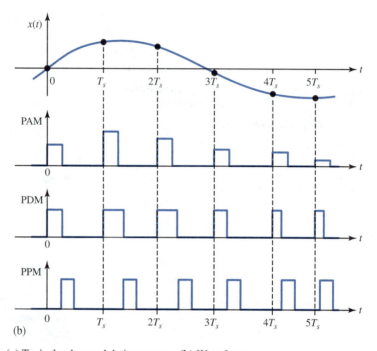

Figure 14.2.7 **(a)** Typical pulse modulation system. **(b)** Waveforms.

as *pulse amplitude modulation* (PAM), *pulse duration modulation* or *pulse width modula-tion* (PDM or PWM), and *pulse position modulation* (PPM), respectively. At the output end, the modulated pulses are converted back to sample values for reconstruction by low-pass filtering.

EXAMPLE 14.2.1

In order to demonstrate aliasing, make a plot of the signal

$$x(t) = 3 \cos 2\pi 10t - \cos 2\pi 30t$$

which approximates a square wave with $W = 30$ Hz. If the sample points are taken at

$$t = 0, \ \frac{1}{60}, \ \frac{2}{60}, \ \ldots, \ \frac{6}{60}$$

corresponding to $T_s = 1/(2W)$, you can see that $x(t)$ could be recovered from those samples. However, if the sample points are taken at

$$t = 0, \ \frac{1}{40}, \ \frac{2}{40}, \ \cdots$$

corresponding to $T_s = 1/(2W)$ and $f_s < 2W$, a smooth curve drawn through these points will show the effect of aliasing.

Solution

The waveforms are sketched in Figure E14.2.1.

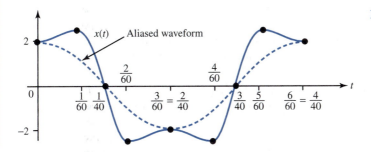

Figure E14.2.1

Multiplexing Systems

A multiplexing system is one in which two or more signals are transmitted jointly over the same transmission channel. There are two commonly used methods for signal multiplexing. In *frequency-division multiplexing* (FDM), various signals are translated to nonoverlapping frequency bands. The signals are demultiplexed for individual recovery by bandpass filtering at the destination. FDM may be used with either analog or discrete signal transmission. *Time-division multiplexing* (TDM), on the other hand, makes use of the fact that a sampled signal is off most of the time and the intervals between samples are available for the insertion of samples from other signals. TDM is usually employed in the transmission of discrete information. Let us now describe basic FDM and TDM systems.

Figure 14.2.8(a) shows a simple FDM system which is used in telephone communication systems. Each input is passed through a low-pass filter (LPF) so that all frequency components above 3 kHz are eliminated. It is then modulated onto individual *subcarriers* with 4-kHz spacing. While all subcarriers are synthesized from a master oscillator, the modulation is achieved with single sideband (SSB). The multiplexed signal, with a typical spectrum as shown in Figure 14.2.8(b), is formed by summing the SSB signals and a 60-kHz *pilot carrier*. The bandpass filters (BPFs) at the destination separate each SSB signal for product demodulation. Synchronization is achieved by obtaining the local oscillator waveforms from the pilot carrier. Telephone signals are often multiplexed in this fashion.

A basic TDM system is illustrated in Figure 14.2.9(a). Let us assume for simplicity that all three input signals have equal bandwidths W. A *commutator* or an electronic switch subsequently obtains a sample from each input every T_s seconds, thereby producing a multiplexed waveform with interleaved samples, as shown in Figure 14.2.9(b). Another synchronized commutator at the destination isolates and distributes the samples to a bank of low-pass filters (LPFs) for individual signal reconstruction. More sophisticated TDM systems are available in which the sampled values are converted to pulse modulation prior to multiplexing and carrier modulation is included after

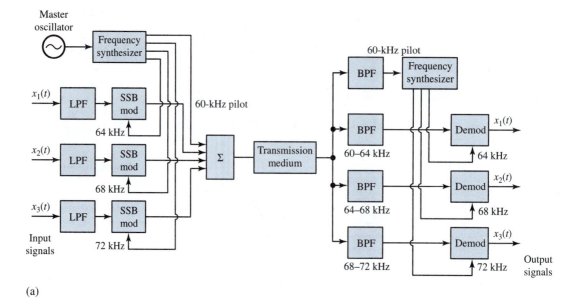

(a)

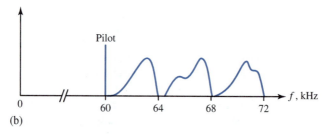

(b)

Figure 14.2.8 (a) Simple FDM system. (b) Typical spectrum of multiplexed signal with pilot.

multiplexing. Integrated switching circuits have made the TDM implementation much simpler than FDM.

EXAMPLE 14.2.2

Find the transmission bandwidth required of a data telemetry system that is to handle three different signals with bandwidths $W_1 = 1$ kHz, $W_2 = 2$ kHz, and $W_3 = 3$ kHz, by employing:

(a) FDM with DSB subcarrier modulation.

(b) TDM with pulse duration $D = T_s/6$.

Solution

(a) $B \geq 2W_1 + 2W_2 + 2W_3 = 12 kHz$.

(b) $f_s \geq 2W_3 = 6$ kHz, so $B \geq 1/D = 6f_s \geq 36$ kHz.

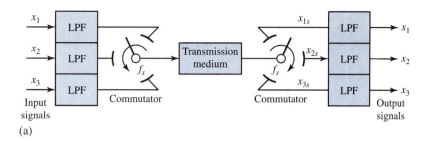

(a)

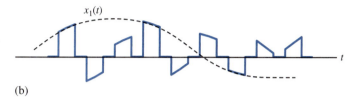

(b)

Figure 14.2.9   (a) Basic TDM system. **(b)** Multiplexed waveform.

14.3   INTERFERENCE AND NOISE

An information-bearing signal often becomes contaminated by externally generated interference and noise and/or by internally generated noise. The demodulated message signal is generally degraded to some extent by the presence of these distortions (attenuation, interference, and noise) in the received signal. The fidelity of the received message signal is then a function of the type of modulation, the strength of the additive noise, the type and strength of any other additive interference, and the type of any nonadditive interference or noise. This section only introduces some of the major causes of interference and noise, and touches upon some methods of dealing with their effects in order to minimize the interference and suppress the noise.

Interference may take several forms: ac hum, higher frequency pulses and "whistles," or erratic waveforms commonly known as static. Interfering signals can be seen to enter the system primarily through the following mechanisms:

- *Capacitive coupling,* because of the stray capacitance between the system and an external voltage
- *Magnetic coupling,* because of the mutual inductance between the system and an external current
- *Radiative coupling,* because of electromagnetic radiation impinging on the system, particularly in the channel
- *Ground-loop coupling,* because of the currents flowing between different ground points

To minimize coupling from the inevitable sources, all exposed elements are usually enclosed within conducting *shields,* which offer low-resistance paths to ground. When held at a common potential, these shields are quite effective in reducing most types of interference. However, low-frequency magnetic coupling can induce unwanted current flow through the shields themselves. Then the shield connection has to be interrupted to avoid a closed-loop current path. An additional layer of special magnetic shielding material may become necessary sometimes in extreme cases of magnetic-coupling interference. The grounding terminals, the equipment cases, and the shields are generally tied together at a single system ground point so as to prevent ground-loop current-coupling interference.

The transducer in some cases may have a local ground that cannot be disconnected. In such a case, a separate *ground strap* (braided-wire straps used because of their low inductance) is used to connect the local ground and the system ground point. The shield is also disconnected from the amplifier so as to prevent ground-loop current through the shield. Because the ground strap has nonzero resistance, any stray current through the strap will cause an interference voltage v_{cm} known as *common-mode voltage* since it appears at both the transducer and the shield terminals. v_{cm} is generally quite small; however, when the information-bearing signal voltage itself is rather small, the common-mode voltage may pose a problem, which can be eliminated by using the *differential amplifier*, as shown in Figure 14.3.1. The analysis with the virtual-short model of the op-amp reveals that

$$v_{out} = K(v_2 - v_1) \tag{14.3.1}$$

amplifying the difference voltage $v_2 - v_1$. With reasonable assumptions that $R_s << R_1$ and $R_t <<$ R_2, we have $v_1 \cong v_{cm}$ and $v_2 \cong x + v_{cm}$, so that

$$v_{out} = K[(x + v_{cm}) - v_{cm}] = Kx \tag{14.3.2}$$

in which v_{cm} has been eliminated as desired. Such an op-amp circuit is also known as an *instrumentation* or *transducer amplifier*.

Any interference at frequencies outside the signal band can be eliminated by appropriate filtering. However, in order to combat interference within the signal band (after proper shielding and grounding), a *notch filter* is sometimes used to avoid the bothersome interference at a single frequency f_0. Figure 14.3.2 illustrates the point: Part (a) shows the composite amplitude spectrum including the single-frequency interference; part (b) depicts the amplitude ratio of the notch filter. The notch-filtering technique does, of course, introduce some inevitable signal distortion, and such filtering should precede amplification to prevent possible saturation of the amplifier due to the interference.

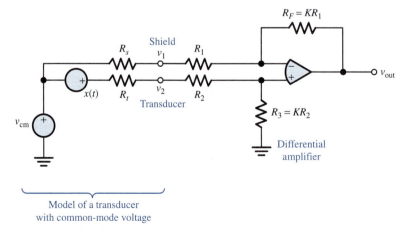

Figure 14.3.1 Differential amplifier to eliminate the common-mode voltage.

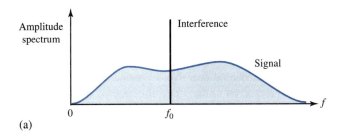

Interference

Amplitude
spectrum

Signal

f

0

f_0

(a)

Figure 14.3.2 (a) Composite amplitude spectrum including single-frequency interference. **(b)** Amplitude ratio of notch filter.

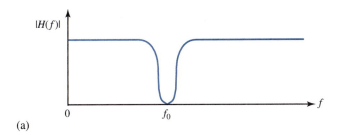

$|H(f)|$

f

0

f_0

(a)

Noise

In any communication system there are usually two dominant factors that limit the performance of the system:

1. *Additive noise,* generated by electronic devices that are used to filter and amplify the communication signal.
2. *Signal attenuation,* as the signal travels through a lossy channel.

A simple mathematical model of a channel with attenuation and additive noise is shown in Figure 14.3.3. If the transmitted signal is $s(t)$, the received signal is given by

$$r(t) = \alpha s(t) + n(t) \tag{14.3.3}$$

where $\alpha < 1$, and $n(t)$ represents an additive random noise process corrupting the transmitted signal. Physically, the additive noise may arise from electronic components and amplifiers at the

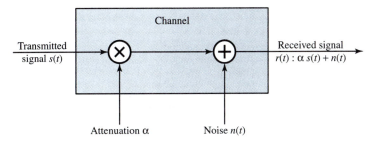

Channel

Transmitted
signal $s(t)$

$\times$

$+$

Received signal
$r(t) : \alpha\, s(t) + n(t)$

Attenuation α

Noise $n(t)$

Figure 14.3.3 Mathematical model of channel with attenuation and additive noise.

receiver of the communication system, or from interference encountered in transmission as in the case of radio-signal transmission. The effect of signal attenuation is to reduce the amplitude of the desired signal $s(t)$ and, thus, to render the information-bearing signal more vulnerable to additive noise.

Signal attenuation in many channels can be offset by using amplifiers to boost the signal level during transmission. However, the amplifier also introduces additive noise in the process of amplification, thereby corrupting the signal. The additional noise must also be taken into consideration in the design of the communication system.

Any conductive two-terminal device is characterized generally as lossy, with some resistance R. A resistor, which is at a temperature τ above absolute zero, contains free electrons that exhibit random motion and, thus, result in a noise voltage across the terminals of the resistor. Such a noise voltage is known as *thermal noise*. If the noise is introduced primarily by electronic components and amplifiers at the receiver, it may be characterized as thermal noise.

In general, any physical resistor or lossy device can be modeled by a noise source in series with a noiseless resistor, as shown in Figure 14.3.4. The noise source is usually characterized as a sample function of a random process. Since random processes involving probability and random variables are outside the scope of this text, we will resort to simpler explanations. Figure 14.3.5

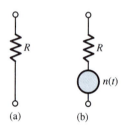

Figure 14.3.4 Physical resistor or lossy device. **(a)** Noiseless resistor. **(b)** Noiseless resistor in series with a noise source.

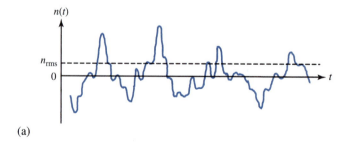

(a)

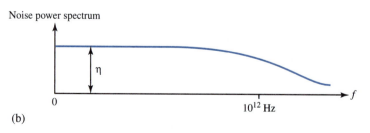

(b)

Figure 14.3.5 Thermal or white noise. **(a)** Typical waveform. **(b)** Typical power spectrum.

(a) illustrates a typical thermal noise waveform $n(t)$. In view of the unpredictable behavior, since the average value of $n(t)$ may be equal to zero, a more useful quantity is the rms value n_{rms} so that the *average noise power* is given by

$$N = n_{rms}^2/R, \qquad \text{if } n_{rms} \text{ is noise voltage} \qquad (14.3.4)$$

or

$$N = n_{rms}^2 R, \qquad \text{if } n_{rms} \text{ is noise current} \qquad (14.3.5)$$

The spectrum of thermal noise power is uniformly spread over frequency up to the infrared region around 10^{12} Hz, as shown in Figure 14.3.5(b). Such a distribution indicates that $n(t)$ contains all electrical frequencies in equal proportion, and an equal number of electrons is vibrating at every frequency. By analogy to white light, which contains all visible frequencies in equal proportion, thermal noise is also referred to as *white noise*.

The constant η in Figure 14.3.5(b) stands for the *noise power spectral density*, expressed in terms of power per unit frequency (W/Hz). Statistical theory shows that

$$\eta = kT \qquad (14.3.6)$$

where k is the Boltzmann constant given by 1.381×10^{-23} J/K and T is the source temperature in kelvins. Equation (14.3.6) suggests that a hot resistance is noisier than a cool one, which is compatible with our notion of thermally agitated electrons. At room temperature $T_0 \cong 290$ K ($17°$ C), η_0 works out as 4×10^{-21} W/Hz.

When we employ amplifiers in communication systems to boost the level of a signal, we are also amplifying the noise corrupting the signal. Because any amplifier has some finite passband, we may model an amplifier as a filter with frequency response characteristic $H(f)$. Let us evaluate the effect of the amplifier on an input thermal noise source.

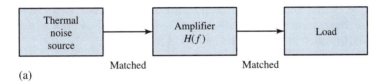

(a)

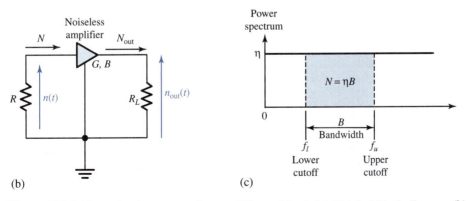

(b)   (c)

Figure 14.3.6 Thermal noise converted to amplifier and load. **(a)** Matched block diagram. **(b)** Circuit representing thermal noise at amplifier input. **(c)** Power spectrum.

Figure 14.3.6(a), in block diagram form, illustrates a thermal noise source connected to a matched two-port network having frequency response $H(f)$ and the output of the network connected to a matched load. Figure 14.3.6(b) shows a thermal noise source represented by a resistance R connected to an amplifier with a matched input resistance. Presuming the amplifier to be noiseless, with power gain G and bandwidth B, the output noise power is

$$N_{\text{out}} = GN = G\eta B \qquad (14.3.7)$$

where $N = \eta B$ represents the source noise power [the area under the power-spectrum curve falling within the passband, as shown in Figure 14.3.6(c)] accepted by the amplifier as input. The rms noise voltage for a thermal source connected to a matched resistance is then given by

$$\eta_{\text{rms}} = \sqrt{RkTB} \qquad (14.3.8)$$

The open-circuit voltage would be twice this value.

Amplifier noise arises from both thermal sources (resistances) and nonthermal sources (semiconductor devices). Although nonthermal noise is not related to physical temperature and does not necessarily have a uniform spectrum like that of thermal noise, one still refers to an amplifier's *noise temperature* T_a, for convenience, as a measure of noisiness referred to the input. The model of a noisy amplifier is shown in Figure 14.3.7(a) with input noise $N = \eta B = kTB$ from a source at temperature T, and the output power given by

$$N_{\text{out}} = GN + N_a = GN + GkT_a B = Gk(T + T_a)B \qquad (14.3.9)$$

where $N_a = GkT_a B$ is the output noise power caused only by the amplifier, G is the power gain of the amplifier, and B is the bandwidth of the amplifier. Note that the amplifier noise N_a is added to the amplified source noise to yield the output power in Equation (14.3.9). If $T = T_0$ (i.e., room temperature), then $N_{\text{out}} \cong N_a$, and the amplifier noise dominates the source noise, which is a common occurrence. When $T_a >> T_0$, the amplifier is very noisy, although not physically hot.

Figure 14.3.7(b) depicts the variation of noise temperature with frequency for a nonthermal source. Several phenomena lumped together under the term *one-over-f* ($1/f$) *noise* lead to the pronounced low-frequency rise in Figure 14.3.7(b). Such $1/f$ noise is produced by transistors and certain transducers, such as photodiodes and optical sensors.

Signals in Noise

Let us now consider a weak information signal that is to be amplified by a noisy amplifier. The *signal-to-noise ratio* (SNR), usually expressed in decibels, becomes an important system performance measure. It is given by

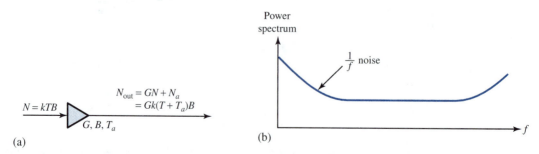

Figure 14.3.7 (a) Model of a noisy amplifier. (b) Power spectrum of nonthermal noise.

$$\text{SNR} = \frac{P_{\text{out}}}{N_{\text{out}}} = \frac{G P_{\text{in}}}{Gk(T + T_a)B} = \frac{P_{\text{in}}}{k(T + T_a)B} \qquad (14.3.10)$$

where the amplified signal power is $P_{\text{out}} = GP_{\text{in}}$ in the numerator, which includes source and amplifier noise given by Equation (14.3.9); P_{in} is the average power of the input signal; and the denominator N_{out} is the total output noise power given by Equation (14.3.9). Notice that the amplifier's power gain G does not appear in the final result of the SNR. A large SNR indicates that the signal is strong enough to mask the noise and possibly make the noise inconsequential. For example, with SNR $\geq$ 20 dB, intelligible voice communication results; otherwise, "static" in the voice signal; with SNR $\geq$ 50 dB, noisefree television image results; otherwise, "snowy" TV picture.

For a good system performance, Equation (14.3.10) suggests a large value of P_{in} and/or small values for $T + T_a$ and B. However, one should be reminded here that the amplifier's bandwidth B should not be less than the signal bandwidth W. That simply means that with large-bandwidth signals, one would expect noise to be more troublesome.

Frequency translation can be used effectively to reduce the effect of $1/f$ noise by putting the signal in a less noisy frequency band. Figure 14.3.8(a) shows the schematic implementation with product modulation and demodulation, whereas Figure 14.3.8(b) illustrates the noise reduction in terms of the areas under the noise power curve. The two multipliers in Figure 14.3.8(a) are normally implemented by using a pair of synchronized switches. It turns out that the product modulation requires bandwidth $B = 2W$, and the synchronized product demodulation doubles the final SNR.

Another way of improving the SNR is by *preemphasis* and *deemphasis filtering*. Generally, for low-frequency components of the message signal FM (frequency modulation) performs better, and for high-frequency components PM (phase modulation) is a better choice. Hence, if one can design a system that performs FM for low-frequency components of the message signal, and

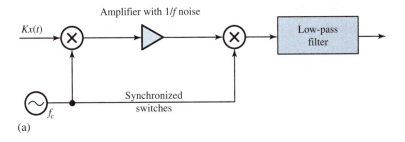

(a)

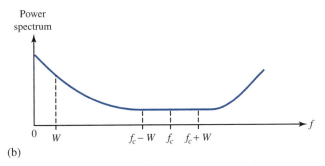

(b)

Figure 14.3.8 **(a)** Schematic arrangement with frequency translation to reduce the effect of $1/f$ noise. **(b)** Noise power spectrum.

works as a phase modulator for high-frequency components, a better overall system performance results compared to each system (FM or PM) alone. This is the idea behind preemphasis and deemphasis filtering techniques.

Figure 14.3.9(a) shows a typical noise power spectrum at the output of the demodulator in the frequency interval $|f| < W$ for PM, whereas Figure 14.3.9(b) shows that for FM. The preemphasis and deemphasis filter characteristics (i.e., frequency responses) are shown in Figure 14.3.10.

Due to the high level of noise at high-frequency components of the message in FM, it is desirable to attenuate the high-frequency components of the demodulated signal. This results in a reduction in the noise level, but causes the higher frequency components of the message signal to be attenuated also. In order to compensate for the attenuation of the higher components of the message signal, one can amplify these components at the transmitter before modulation. Thus, at the transmitter we need a high-pass filter, and at the receiver we must use a low-pass filter. The net effect of these filters is to have a flat frequency response. The receiver filter should therefore be the inverse of the transmitter filter. The modulator filter, which emphasizes high frequencies, is called the preemphasis filter, and the demodulator filter, which is the inverse of the modulator filter, is called the deemphasis filter.

If the signal in question is a constant whose value we seek, as is the case sometimes in simple measurement systems, the measurement accuracy will be enhanced by a low-pass filter with the smallest bandwidth B. Low-pass filtering, in a sense, carries out the operation of averaging, since a constant corresponds to the average value (or dc component) and since noise usually has zero average value. However, some noise will get through the filter and cause the processed signal $z(t)$ to fluctuate about the true value x, as shown in Figure 14.3.11. Allowing any sample to fall somewhere between $x - \varepsilon$ and $x + \varepsilon$, the *rms error* G is defined by

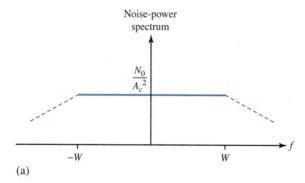

(a)

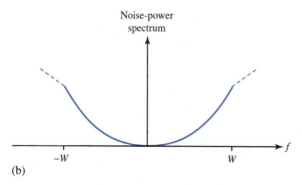

(b)

Figure 14.3.9 Noise power spectrum at demodulator output. **(a)** In PM. **(b)** In FM.

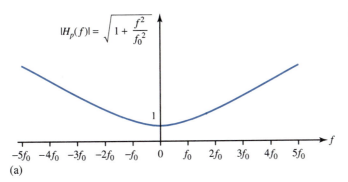

$$|H_p(f)| = \sqrt{1 + \frac{f^2}{f_0^2}}$$

(a)

Figure 14.3.10 **(a)** Preemphasis filter characteristic. **(b)** Deemphasis filter characteristic.

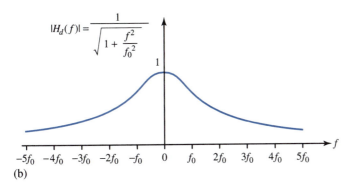

$$|H_d(f)| = \frac{1}{\sqrt{1 + \frac{f^2}{f_0^2}}}$$

(b)

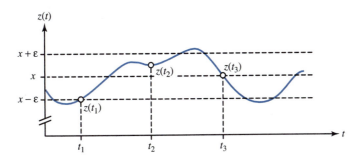

z(t)

Figure 14.3.11 Constant signal x with noise fluctuations.

$$G = \frac{x}{\sqrt{P_{out}/N_{out}}} \qquad (14.3.11)$$

By taking M different samples of $z(t)$, the arithmetic average can be seen to be

$$z_{av} = \frac{1}{M}(z_1 + z_2 + \ldots + z_M) \qquad (14.3.12)$$

If the samples are spaced in time by at least $1/B$ seconds, then the noise-induced errors tend to cancel out and the rms error of z_{av} becomes

$$\varepsilon_M = \varepsilon/\sqrt{M} \qquad (14.3.13)$$

This averaging method amounts to reducing the bandwidth to B/M.

When the signal in question is a sinusoid whose amplitude is to be measured, averaging techniques can also be used by utilizing a narrow bandpass filter, or a special processor known as a *lock-in amplifier*. For extracting information from signals deeply buried in noise, more sophisticated methods based on digital processing are available.

EXAMPLE 14.3.1

A low-noise transducer is connected to an instrumentation system by a cable that generates thermal noise at room temperature. The information-bearing signal has a bandwidth of 6 kHz. The signal power delivered is $P_{\text{in}} = 120$ pW. Evaluate the condition on the amplifier noise temperature T_a such that the signal-to-noise ratio (SNR) is greater than or equal to 50 dB.

Solution

Applying Equation (14.3.10) with $T = T_0$,

$$\text{SNR} = \frac{P_{\text{out}}}{N_{\text{out}}} = \frac{P_{\text{in}}}{k(T + T_a)B} = \frac{P_{\text{in}}}{kT_0\left(1 + \frac{T_a}{T_0}\right)B} = \frac{120 \times 10^{-12}}{4 \times 10^{-21}\left(1 + \frac{T_a}{T_0}\right)6 \times 10^3} \geq 10^5$$

Hence,

$$T_a \leq 49T_0$$

This condition can easily be satisfied in the case of a well-designed amplifier.

14.4 LEARNING OBJECTIVES

The *learning objectives* of this chapter are summarized here so that the student can check whether he or she has accomplished each of the following.

- Basic ideas of analog and digital communication systems.
- Constructing the line spectrum of a periodic signal from its Fourier-series expansion.
- Conditions for distortionless transmission.
- Sketching spectra at various points in a system using product modulation and filtering.
- Conditions under which a signal can be sampled and then reconstructed from a pulse-modulated waveform.
- Basic notions of multiplexing systems.
- Causes of interference and noise, and techniques for minimizing their effects.

14.5 PRACTICAL APPLICATION: A CASE STUDY

Antinoise Systems—Noise Cancellation

Traditionally sound-absorbing materials have been used quite effectively to reduce noise levels in aircraft, amphitheaters, and other locations. An alternate way is to develop an electronic system that cancels the noise. Ear doctors and engineers have successfully developed ear devices that will nearly eliminate the bothersome and irritating noise (so-called tinnitus) experienced by patients suffering from Ménière's disease. For passengers in airplanes, helicopters, and other flying equipment, a proper headgear is being developed in order to eliminate the annoying noise.

Applications could conceivably extend to people residing near airports and bothered by airplane takeoffs and landings. For industrial workers who are likely to develop long-term ill effects due to various noises they may be subjected to in their workplace, and even for persons who are irritated by the pedestrian noise levels in certain locations, antinoise systems that nearly eliminate or nullify noise become very desirable.

Figure 14.5.1 illustrates in block-diagram form the principle of noise cancellation as applied to an aircraft carrying passengers. The electric signal resulting after sampling the noise at the noise sources is passed through a filter whose transfer function is continuously adjusted by a special-purpose computer to match the transfer function of the sound path. An inverted version of the signal is finally applied to loudspeakers, which project the sound waves out of phase with those from the noise sources, nearly canceling the noise. Microphones on the headrests monitor the sound experienced by the airline passengers so that the computer can determine the proper filter adjustments.

Signal processing, which is concerned with manipulating signals to extract information and to use that information to generate other useful electric signals, is indeed an important and far-reaching subject.

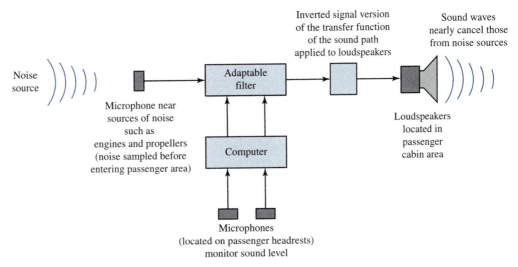

Figure 14.5.1 Block diagram of antinoise system to suppress the noise in an aircraft.

PROBLEMS

14.1.1 (a) A rectangular pulse is denoted by $\Pi(t)$ and defined as

$$\Pi(t) = \begin{cases} 1, & -\dfrac{1}{2} < t < \dfrac{1}{2} \\ \frac{1}{2}, & t = \pm\frac{1}{2} \\ 0, & \text{otherwise} \end{cases}$$

Sketch the signal. Also express it in terms of unit-step signals.

(b) The sinc signal is given by

$$\text{sinc}(t) = \begin{cases} \frac{\sin \pi t}{\pi t}, & t \neq 0 \\ 1, & t = 0 \end{cases}$$

Sketch the waveform and comment on its salient features.

(c) The sign or signum signal is represented by

$$\text{sgn}(t) = \begin{cases} 1, & t > 0 \\ -1 & t < 0 \\ 0, & t = 0 \end{cases}$$

which can be expressed as the limit of the signal $x_n(t)$ defined by

$$x_n(t) = \begin{cases} e^{-1/n}, & t > 0 \\ -e^{1/n}, & t < 0 \\ 0, & t = 0 \end{cases}$$

as $n \rightarrow \infty$. Sketch the waveform as the limit of $x_n(t)$.

*14.1.2 A large number of building blocks in a communication system can be modeled by LTI (linear time-invariant) systems, for which the impulse response completely characterizes the system. Consider the system described by

$$y(t) = \int_{-\infty}^{t} x(\tau) \, d\tau$$

which is called an integrator. Investigate whether the system is LTI by finding its response to $x(t - t_0)$.

14.1.3 For a real periodic signal $x(t)$ with period T_0, three alternative ways to represent the Fourier series expansion are:

$$x(t) = \sum_{-\infty}^{+\infty} x_n e^{j2\pi \frac{n}{T_0} t}$$

$$= \frac{a_0}{2} + \sum_{n=1}^{\infty} \left[a_n \cos\left(2\pi \frac{n}{T_0} t\right) \right.$$

$$\left. + b_n \sin\left(2\pi \frac{n}{T_0} t\right) \right]$$

$$= x_0 + 2\sum_{n=1}^{\infty} |x_n| \cos\left(2\pi \frac{n}{T_0} t + \angle x_n\right)$$

where the corresponding coefficients are obtained from

$$x_n = \frac{1}{T_0} \int_{\alpha}^{\alpha+T_0} x(t) e^{-j2\pi \frac{n}{T_0} t} \, dt = \frac{a_n}{2} - j\frac{b_n}{2}$$

$$a_n = \frac{2}{T_0} \int_{\alpha}^{\alpha+T_0} x(t) \cos\left(2\pi \frac{n}{T_0} t\right) dt$$

$$b_n = \frac{2}{T_0} \int_{\alpha}^{\alpha+T_0} x(t) \sin\left(2\pi \frac{n}{T_0} t\right) dt$$

$$|x_n| = \frac{1}{2}\sqrt{a_n^2 + b_n^2}$$

$$\angle x_n = -\arctan\left(\frac{b_n}{a_n}\right)$$

in which the parameter α in the limits of the integral is arbitrarily chosen as $\alpha = 0$ or $\alpha = -T_0/2$, for convenience.

(a) Show that the Fourier-series representation of an impulse train is given by

$$x(t) = \sum_{n=-\infty}^{+\infty} \delta(t - nT_0) = \frac{1}{T_0} \sum_{n=-\infty}^{+\infty} e^{j2\pi \frac{n}{T_0} t}$$

Also sketch the impulse train.

(b) Obtain the Fourier-series expansion for the signal $x(t)$ sketched in Figure P14.1.3 with $T_0 = 2$, by choosing $\alpha = -1/2$.

14.1.4 (a) Show that the sum of two discrete periodic signals is periodic.

(b) Show that the sum of two continuous periodic signals is not necessarily periodic; find the condition under which the sum of two continuous periodic signals is periodic.

14.1.5 Classify the following signals into even and odd signals:

(a)
$$x_1(t) = \begin{cases} e^{-t}, & t > 0 \\ -e^{-t}, & t < 0 \\ 0, & t = 0 \end{cases}$$

(b) $x_2(t) = e^{-|t|}$

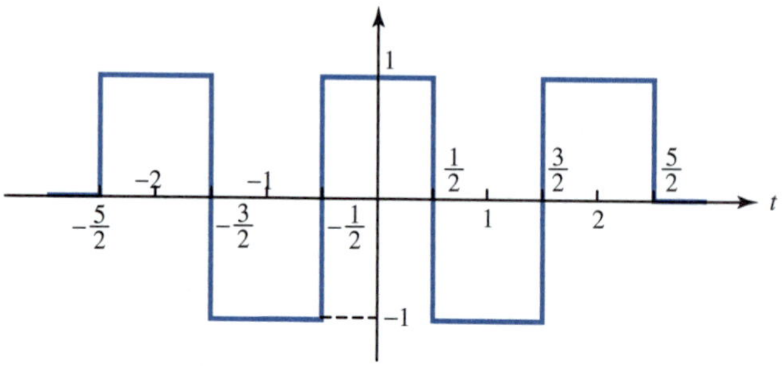

Figure P14.1.3

(c)

$$x_3(t) = \begin{cases} \frac{t}{|t|}, & t \neq 0 \\ 0, & t = 0 \end{cases}$$

*14.1.6 For the following neither even nor odd signals, find the even and odd parts of the signals.

(a)

$$x_4(t) = \begin{cases} t, & t \geq 0 \\ 0, & t < 0 \end{cases}$$

(b) $x_5(t) = \sin t + \cos t$

14.1.7 Classify the following signals into energy-type or power-type signals, and determine the energy or power content of the signal.

(a) $x_1(t) = e^{-t} \cos t \, u_{-1}(t)$

(b) $x_2(t) = e^{-t} \cos t$

(c)

$$x_3(t) = \text{sgn}(t) = \begin{cases} 1, & t > 0 \\ -1, & t < 0 \\ 0, & t = 0 \end{cases}$$

Note: $\int e^{ax} \cos^2 x \, dx = \frac{1}{4+a^2}[(a \cos^2 x + \sin 2x) + \frac{2}{a}]e^{ax}$.

14.1.8 (a) Based on Example 14.1.3, comment on whether $x(t) = A \cos 2\pi f_1 t + B \cos 2\pi f_2 t$ is an energy- or a power-type signal.

(b) Find its energy or power content for $f_1 = f_2$ and $f_1 \neq f_2$.

14.1.9 For the power-type signals given, find the power content in each case.

(a) $x(t) = Ae^{j(2\pi f_0 t + \theta)}$.

(b) $x(t) = u_{-1}(t)$, the unit-step signal.

14.1.10 Show that the product of two even or two odd signals is even, whereas the product of an even and an odd signal is odd.

14.1.11 The triangular signal is given by

$$\Lambda(t) = \begin{cases} t + 1, & -1 \leq t \leq 0 \\ -t + 1, & 0 \leq t \leq 1 \\ 0, & \text{otherwise} \end{cases}$$

(a) Sketch the triangular pulse.

(b) Sketch $x(t) = \sum_{n=-\infty}^{+\infty} \Lambda(t - 2n)$.

(c) Sketch $x(t) = \sum_{n=-\infty}^{+\infty} (-1)^n \Lambda(t - n)$.

14.1.12 For real, even, and periodic functions with period T_0, the Fourier-series expansion can be expressed as

$$x(t) = \frac{a_0}{2} + \sum_{n=1}^{\infty} a_n \cos\left(2\pi \frac{n}{T_0} t\right)$$

where

$$a_n = \frac{2}{T_0} \int_{\alpha}^{\alpha+T_0} x(t) \cos\left(2\pi \frac{n}{T_0} t\right) dt$$

Determine a_n for the following signals:

(a) $x(t) = |\cos 2\pi f_0 t|$, full wave rectifier output.

(b) $x(t) = \cos 2\pi f_0 t + |\cos 2\pi f_0 t|$, half-wave rectifier output.

*14.1.13 For real $x(t)$ given by Equation (14.1.11), identify the even and odd parts of $x(t)$.

14.1.14 Three alternative ways of representing a real periodic signal $x(t)$ in terms of Fourier-series expansion are given in Problem 14.1.3. Determine the expansion coefficients x_n of each of the periodic signals shown in Figure P14.1.14, and for each signal also determine the trigonometric Fourier-series coefficients a_n and b_n.

14.1.15 Certain waveforms can be viewed as a combination of some other waveforms for which the Fourier coefficients are already known. Example 14.1.4 shows some periodic waveforms for which the coefficients a_0, a_n, and b_n are given by Equations (14.1.12) through (14.1.14), respectively. Use those to find the nonzero Fourier-series coefficients for the waveforms given in Figure P14.1.15.

14.1.16 Determine the bandwidth W from two criteria, (i) $A_n < (A_n)_{max}/10$, for $nf_1 > W$, and (ii) $A_n < (A_n)_{max}/20$, for $nf_1 > W$, for the following cases. (Note that A stands for amplitude.)

(a) Waveform of Figure E14.1.4(a), with $A = \pi$, $D = 0.25 \, \mu s$, and $T = 0.5 \, \mu s$.

(b) Waveform of Figure E14.1.4(b), with $A = \pi^2$ and $T = 2.5 \, \mu s$.

(c) Waveform of Figure E14.1.4(d), with $A = \pi$ and $T = 10$ ms.

(d) Waveform of Figure E14.1.4(e), with $A = \pi$ and $T = 800 \, \mu s$.

14.1.17 Consider the rectangular pulse train $x(t)$ of Figure E14.1.4(a), with $A = 2$ and $D = T/2$. Let $v(t) = x(t) - 1$.

(a) Sketch $x(t)$ and $v(t)$.

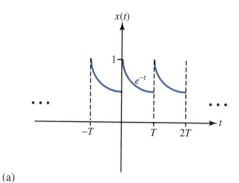

(a)

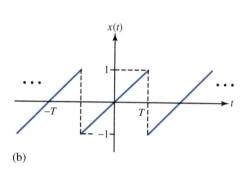

(b)

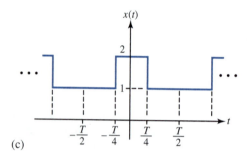

(c)

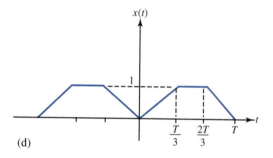

(d)

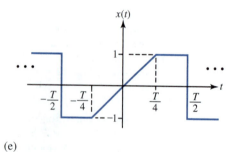

(e)

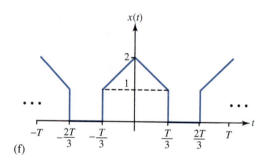

(f)

Figure P14.1.14

(b) Using the result of Figure E14.1.4(a), find the Fourier coefficients of $v(t)$.

*14.1.18 The waveforms of Figure E14.1.4(b) and (c) are given to have $A = \pi$ and $T = 0.2$ ms.

(a) For $0 \leq f \leq 30$ kHz, sketch and label the amplitude spectra.

(b) For $A_n < A_1/5$ for all $nf_1 < W$ (where A stands for amplitude), determine the value of W in each case.

14.1.19 Consider Figure 14.1.5, with $x(t) = 3 \cos 2\pi t$

$+ \cos(2\pi 3t + 180°)$, $|H(f)| = 1$, and constant phase shift $\theta(f) = -90°$. Sketch $x(t)$ and $y(t)$.

14.1.20 The frequency response of a transmission system is given by

$$|H(f)| = \frac{1}{\sqrt{1 + (f/f_{co})^2}};$$

$$\theta(f) = -\tan^{-1} \frac{f}{f_{co}}$$

where $f_{co} = \omega_{co}/2\pi = 5$ kHz. In order to satisfy Equation (14.1.20), over the range of $0 \leq f \leq$

Figure P14.1.15

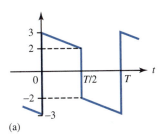

(a)

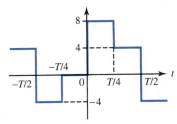

(b)

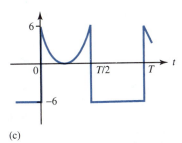

(c)

10 kHz, with $K = 1$, find and sketch the required equalizer characteristics.

14.1.21 The frequency response of a high-pass transmission system is given by

$$|H(f)| = \frac{f/f_{co}}{1 + (f/f_{co})^2};$$

$$\theta(f) = 90° - \tan^{-1}(f/f_{co})$$

with $f_{co} = \omega_{co}/2\pi = 100$ Hz. If $x(t)$ is a triangular wave of Figure E14.1.4(b), with $A = \pi^2/8$ and $T = 25$ ms, obtain an approximate expression for the periodic steady-state response $y(t)$. See Figure 14.1.5.

14.2.1 (a) Let $x(t) = 12 \cos 2\pi 100t + 8 \cos 2\pi 150t$, and $x_c(t) = x(t) \cos 2\pi f_c t$, where $f_c = 600$ Hz. Sketch the amplitude spectrum.

(b) List all the frequencies in the product $x_c(t) \cos 2\pi 500t$, where $x_c(t)$ is given in part (a).

14.2.2 The input to the product modulator of Figure 14.2.1 is given by $x(t) = 8 \cos 2\pi 3000t + 4 \cos 2\pi 7000t$. The frequency of the carrier wave is 6 kHz. Sketch the amplitude line spectrum of the modulated wave $x_c(t)$.

***14.2.3** Consider the following system with $x(t) = 12 \cos 2\pi 100t + 4 \cos 2\pi 300t$ and two ideal filters as shown in Figure P14.2.3. Find $x_a(t)$ and $x_b(t)$.

14.2.4 Consider the product modulator of Figure 14.2.4(a), where the oscillator generates $\cos[2\pi(f_c + \Delta f)t + \Delta\phi]$ in which Δf and $\Delta\phi$ are synchronization errors. Find $z(t)$ produced by the following inputs, when $f_m = 1$ kHz, $\Delta f = 200$ Hz, and $\Delta\phi = 0$:

(a) DSB input $x_c(t) = 4 \cos 2\pi f_m t \cos 2\pi f_c t$.

(b) Upper-sideband SSB input $x_c(t) = 2 \cos 2\pi(f_c + f_m)t$.

(c) Lower-sideband SSB input $x_c(t) = 2 \cos 2\pi(f_c - f_m)t$.

14.2.5 Repeat Problem 14.2.4 when $f_m = 1$ kHz, $\Delta f = 0$, and $\Delta\phi = 90°$.

14.2.6 (a) Consider Figure 14.2.5, in which the signal to be sampled is $x(t) = 18 \cos 2\pi 20t + 12 \cos 2\pi 60t$. With $f_s = 100$ and $D = T_s/2$, for $0 \leq f \leq 2f_s$, sketch the amplitude line spectrum of $x_s(t)$.

(b) Then find the signal $y(t)$ that would be reconstructed by an ideal low-pass filter that rejects all $f > f_s/2$.

14.2.7 (a) The continuous amplitude spectrum of the input to a switching sampler (Figure 14.2.5) is shown in Figure P14.2.7. For $0 \leq f \leq 100$, with $f_s = 70$ and $D = T_s/4$, sketch the resulting spectrum of $x_s(t)$.

(b) Suggest how $x(t)$ can be reconstructed from $x_s(t)$.

14.2.8 PDM and PPM (see Figure 14.2.7) have the advantage of being immune to nonlinear distortion, because the pulse is either on or off. However, in exchange, the transmission bandwidth must be $B \geq 1/D \gg W$, which is needed to accommodate pulses with duration $D \ll T_s \leq 1/(2W)$. Let D be the pulse duration of the PPM waveform in Figure 14.2.7. Let the maximum position shift be $\pm\Delta$. For $\Delta = 2D$ and $f_s = 20$ kHz, find the maximum allowed value of D and the minimum required transmission bandwidth.

*14.2.9 The TDM system of Figure 14.2.9 has a transmission bandwidth $B = 250$ kHz.

(a) Find the maximum number of voice signals with $W = 4$ kHz that can be multiplexed.

(b) Repeat part (a) with the additional constraint that the TDM waveform be off for at least 50% of the time.

14.3.1 An amplifier has a gain of 50 dB, a bandwidth of 9 MHz, and a noise temperature $T_a = 25T_0$, where T_0 is the room temperature, 290 K. Find the output noise power and rms voltage across a 100-Ω load resistor when the source temperature $T = T_0$.

*14.3.2 If the signal in Example 14.3.1 has a bandwidth of 600 kHz, determine P_{out}/N_{out} assuming a noiseless amplifier. Then check to see whether it is possible to obtain $P_{out}/N_{out} \geq 10^5$.

14.3.3 A simple RC filter yields $|H(\omega)|^2 = 1/[1 + (\omega/\omega_{co})^2]$, whereas a more sophisticated and relatively more expensive Butterworth filter gives $|H(\omega)|^2 = 1/[1 + (\omega/\omega_{co})^4]$. Either of these can be used to reduce the hum amplitude. Let an information signal with significant frequency content for $f \leq 30$ Hz be contaminated by ac hum at 120 Hz. The contaminated signal is then applied to a low-pass filter to reduce the hum amplitude by a factor of α. For (i) $\alpha = 0.25$, and (ii) $\alpha = 0.1$, determine the required cutoff frequency f_{co} for (a) the RC filter, and (b) the Butterworth filter, and suggest the filter to be used in each case.

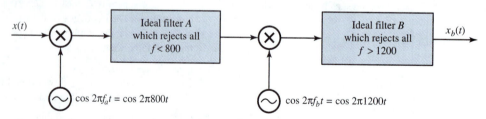

Figure P14.2.3

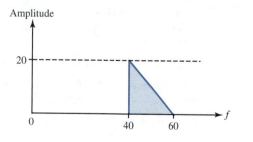

Figure P14.2.7

14.3.4 (a) The transfer function of a notch filter is given by $H(j\omega) = (100 + j\omega - \omega^2)/(10 + j\omega)^2$. Sketch H_{dB} versus ω.

(b) A notch filter centered at $f_0 = \omega_0/2\pi$ can be formed by using a resonant circuit arranged such that

$$H(j\omega) = \frac{jQ(\omega/\omega_0 - \omega_0/\omega)}{1 + jQ(\omega/\omega_0 - \omega_0/\omega)}$$

Investigate $|H(f)|$ at $f = 0$ and $f = \infty$; also at $f_l = f_0(1 - 1/(2Q))$, and $f_u = f_0(1 + 1/(2Q))$, for $Q >> 1$.

(c) With $\omega_0 = 1/\sqrt{LC}$ and $Q = (1/R)\sqrt{L/C}$, show that a series RLC circuit can perform as a notch filter when the output voltage is taken across L and C.

(d) For the purpose of rejecting 1-kHz interference, in order to get $f_l = 980$ Hz and $f_u = 1020$ Hz, find the values of L and C of the series RLC circuit with $R = 50\ \Omega$.

(e) Now consider a tuned circuit in which R is connected in series with a parallel combination of L and C, and the output voltage is taken across R. With $\omega_0 = 1/\sqrt{LC}$ and $Q = R\sqrt{C/L}$, show that the circuit can perform as a notch filter.

(f) For the purpose of rejecting 60-Hz interference, in order to get $f_l = 58$ Hz and $f_u = 62$ Hz, find the values of L and C of the circuit in part (e) with $R = 1$ kΩ.

14.3.5 A noisy amplifier has $N_{out} = 600\ \mu$W when $T = T_0$; but N_{out} drops to 480 μW when the source is immersed in liquid nitrogen at $T = 80$ K. Find the amplifier's noise temperature T_a.

14.3.6 Two noisy amplifiers (with noise temperatures T_{a1} and T_{a2}) having the same bandwidth are connected in cascade so that the overall gain is $G = G_1G_2$. If the input noise to the first amplifier is $N = kTB$, determine the total output noise power and the effective noise temperature T_a of the cascade, so that $N_{out} = Gk(T + T_a)B$.

*14.3.7 In order to measure the amplifier noise temperature T_a, a thermal source at temperature T_0 is connected and the corresponding N_{out} is observed; then the source temperature is increased to T_R when the corresponding N_{out} is doubled. Find T_a in terms of T_R and T_0.

14.3.8 A signal (with $P = 1\ \mu$W and $B = 250$ kHz) contaminated by white noise at noise temperature $T = 2T_0$ is applied to an amplifier. Obtain the condition on T_a such that $P_{out}/N_{out} \geq 80.8$ dB.

14.3.9 The noise figure of an amplifier is given by $F = 1 + T_a/T_0$. Express the output signal-to-noise ratio in terms of F, input noise power N, and input noise temperature T. See how the result is simplified when $T = T_0$.

14.3.10 A system for measuring the constant signal value x has $P_{out}/N_{out} = 40$ dB and $B = 8$ Hz. In order to obtain an accuracy of $\pm 0.2\%$, how long must the output be observed?

14.3.11 An amplifier has a noise equivalent bandwidth $B = 25$ kHz and a maximum available power gain of $G = 30$ dB. If its output noise power is $10^8 kT_0$, determine the effective noise temperature and the noise figure (given by $F = 1 + T_a/T_0$). Assume the input noise source temperature to be T_0.

*14.3.12 The overall noise figure of a cascade of K amplifiers with gains G_k and corresponding noise figures F_k, $1 \leq k \leq K$, is

$$F = F_1 + \frac{F_2 - 1}{G_1} + \frac{F_3 - 1}{G_1G_2} + \cdots$$
$$+ \frac{F_K - 1}{G_1G_2 \ldots G_{K-1}}$$

If an amplifier is designed having three identical states, each of which has a gain of $G_i = 5$ and a noise figure of $F_i = 6$, $i = 1, 2, 3$, determine the overall noise figure of the cascade of the three stages. Looking at the result, justify the statement that the front end of a receiver should have a low noise figure and a high gain. (Note that the noise figure of an amplifier is $F = 1 + T_a/T_0$.)

14.3.13 A radio antenna with a noise temperature of 60 K is pointed in the direction of the sky. The antenna feeds the received signal to the preamplifier, which has a gain of 30 dB over a bandwidth of 10 MHz and a noise figure ($F = 1 + T_a/T_0$) of 2 dB.

(a) Determine the effective noise temperature at the input to the preamplifier.

(b) Determine the noise power at the output of the preamplifier.

15 Communication Systems

Even though most modern communication systems have only been invented and developed during the eighteenth and nineteenth centuries, it is difficult to imagine a world without telephones, radio, and television. After the invention of the electric battery by Alessandro Volta in 1799, Samuel Morse developed the electric telegraph and demonstrated it in 1837. Morse devised the variable-length binary code, in which letters of the English alphabet are represented by a sequence of dots and dashes (code words). In this code, more frequently occurring letters are represented by short code words, whereas letters occurring less frequently are represented by larger code words. *Morse code* (variable-length binary code), developed in 1837, became the precursor to variable-length source coding methods. *Telegraphy*, the earliest form of electrical communication, was a binary digital communication system in which the letters of the English alphabet were efficiently encoded into corresponding variable-length code words having binary elements. In the *baudout code*, developed in 1875, the binary code elements were of equal length, and each letter was encoded into fixed-length binary code words of length 5.

With the invention of the telephone, *telephony* came into being and the Bell Telephone Company was established in 1877. Transcontinental telephone transmission became operational in 1915. Automatic switching was another important advance, and a digital switch was placed in service in 1960. Numerous significant advances have taken place in telephone communications over the past four decades. For example, fiber-optic cables have replaced copper wire, and electronic switches have come into use in place of electromechanical devices.

Marconi is credited with the development of *wireless telegraphy* in the late 1890s, after Maxwell's theory of electromagnetic radiation was verified experimentally by Hertz in 1887. The invention of the vacuum tube was particularly instrumental in the development of radio communication systems. Amplitude modulation (AM) broadcast was initiated in 1920, while frequency modulation (FM) broadcast was developed commercially by the end of World War II. Commercial television broadcasting began in 1936 by the BBC (British Broadcasting Corporation), and the FCC (Federal Communications Commission) authorized television broadcasting five years later in the United States.

The growth in communications over the past 60 years has been phenomenal. The invention of the transistor in 1947 and the integrated circuit and laser in 1958 have paved the way to *satellite communication systems*. Most of the wire-line communication systems are being replaced by fiber-optic cables (providing extremely high bandwidth), which makes the transmission of a wide variety of information sources (voice, data, and video) possible. High-speed communication networks linking computers and the greater need for personal communication services are just the beginning of the modern telecommunications era.

Today *digital communication systems* are in common use, carrying the bulk of our daily information transmission through a variety of communication media, such as wire-line telephone channels, microwave radio, fiber-optic channels, and satellite channels. Even the current analog AM and FM radio and television broadcasts will be replaced in the near future by digital transmission systems. High-speed integrated circuits (ICs), programmable digital signal processing chips, microelectronic IC fabrication, and sophisticated digital modulation techniques have certainly helped digital communications as a means of transmitting information.

In spite of the general trend toward digital transmission of analog signals, a significant amount of analog signal transmission still takes place, especially in audio and video broadcasting. Historically, *analog communication systems* were placed first, and then came digital communication systems.

In any communication system, the communication channel provides the connection between the transmitter and the receiver. The physical channel (medium) may be any of the following:

- A pair of wires, which carry the electric signal

- Optical fiber, which carries the information on a modulated light beam

- An underwater ocean channel, in which the information is transmitted acoustically

- Free space, over which the information-bearing signal is radiated by using an antenna

- Data storage media, such as magnetic tape, magnetic disks, and optical disks.

The available channel bandwidth, as well as the noise and interference, limit the amount of data that can be transmitted reliably over any communication channel.

Figure 15.0.1 illustrates the various frequency bands of the electromagnetic spectrum (radio and optical portions) along with types of transmission media and typical applications.

Wire-line channels are used extensively by the telephone network for voice, data, and video transmission. Twisted-pair wire lines (with a bandwidth of several hundred kHz) and coaxial cable (with a usable bandwidth of several MHz) are basically guided electromagnetic channels.

Fiber-optic channels offer a channel bandwidth that is several orders of magnitude larger than coaxial cable channels. The transmitter or modulator in a fiber-optic communication system is a light source, such as a light-emitting diode (LED) or a laser, whose intensity is varied (modulated) with the message signal. The light propagates through the fiber as a light wave and is amplified periodically along the transmission path to compensate for signal attenuation. At the receiver end, the light intensity is detected by a photodiode, whose output is an electric signal that varies

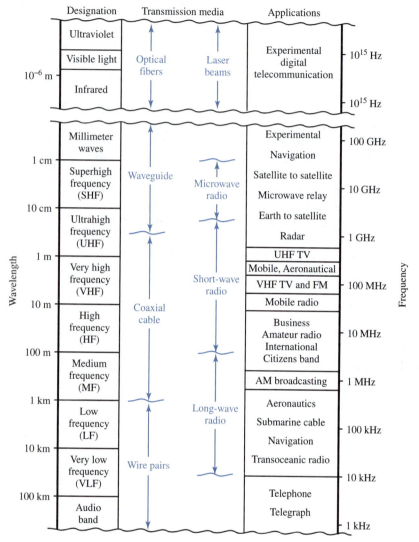

Figure 15.0.1 Frequency bands of the electromagnetic spectrum. (*Source:* A. Carlson, *Communication Systems*, 3rd ed., McGraw-Hill, New York, 1986.)

in direct proportion to the power of light striking on the photodiode. Optical fiber channels are replacing nearly all wire-line channels in the telephone network.

Wireless electromagnetic channels are used in wireless communication systems, in which the electromagnetic energy is coupled to the propagation medium through an antenna that serves as a radiator. The physical size and configuration of the antenna depend mainly on the frequency of operation. For example, a radio station transmitting AM frequency band of 1 MHz (with a corresponding wavelength of $\lambda = c/f_c = 300$ m) requires an antenna of at least 30 m (approximately one-tenth of the wavelength).

The mode of propagation of electromagnetic waves in free space and atmosphere may be subdivided into three categories:

- Ground-wave propagation
- Sky-wave propagation
- Line-of-sight (LOS) propagation.

In the frequency bands that are primarily used to provide navigational aids from shore to ships around the world (VLF to LF to MF), the available channel bandwidths are relatively small, and hence the information that is transmitted through these channels is relatively slow speed and generally confined to digital transmission. Noise at these frequencies is caused by thunderstorm activity around the globe, whereas interference is caused by the many users.

For frequencies of 0.3 to 3 MHz, in the MF band, ground-wave (or surface-wave) propagation, illustrated in Figure 15.0.2, is the dominant mode used for AM broadcasting and maritime radio broadcasting. Dominant disturbances include atmospheric noise, human-made noise, and thermal noise from electronic components. The range is limited to about 100 miles for even the more powerful radio stations.

In the ionosphere, the rarefied air becomes ionized, mainly due to ultraviolet sunlight. The D-region, usually falling between 50 and 90 km in altitude, will reflect waves below 300 kHz or so, and attenuate higher frequency waves (300 kHz $< f <$ 30 MHz), especially in the daytime. The D-region mostly disappears at night. The E-region (about 110 km in altitude) reflects high frequencies (3 MHz $< f <$ 30 MHz) during the daytime, and medium frequencies (300 kHz $< f <$ 3 MHz) at night. The F_1-region (about 175 to 250 km in altitude) is distinct from the F_2-region (250 to 400 km in altitude) only during the day; at night they merge. Waves that penetrate the E-region usually go through the F_1-region as well, with some attenuation being the primary effect. The F_2-region provides the main means of long-distance, high-frequency (3 MHz $< f <$ 30 MHz) communication by wave reflection. Sky-wave propagation is illustrated in Figure 15.0.3.

Signal multipath occurs with electromagnetic wave propagation via sky wave in the HF range. When the transmitted signal reaches the receiver through multiple propagation paths with different delays, *signal fading* may result. Both atmospheric noise and thermal noise become the additive noise at high frequencies. It is possible to have ionospheric scatter propagation in the frequency range of 30 to 60 MHz, and tropospheric scattering in the range of 40 to 300 MHz; but relatively large antennas are needed with a large amount of transmitted power, because of large signal propagation losses.

Frequencies above 30 MHz, propagating through the ionosphere with relatively little loss, make satellite and extraterrestrial communications possible. In the VHF band and higher, the dominant mode is line-of-sight (LOS) propagation, in which the transmitter and receiver antennas must be in direct LOS with relatively little or no obstruction. That is why television stations

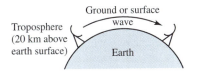

Figure 15.0.2 Illustration of ground-wave propagation in MF band.

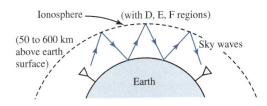

Figure 15.0.3 Illustration of sky-wave propagation.

transmitting in the VHF and UHF bands have their antennas mounted on high towers to achieve a broad coverage area. A television antenna mounted on a tower of 1200 feet in height ($=h$) provides a coverage of about $d = \sqrt{2h} \cong 50$ miles. Microwave radio relay systems (for telephone and video transmission at about 1 GHz) also have antennas mounted on tall towers.

In the VHF and UHF bands, thermal noise and cosmic noise, picked up by the antenna, become predominant. Above 10 GHz in the SHF band, atmospheric conditions (such as precipitation and heavy rains) play a major role in signal propagation. In the infrared and visible light regions of the electromagnetic spectrum, LOS optical communication in free space is being experimented with for satellite-to-satellite links.

A good understanding of a communication system can be achieved by studying electro-magnetic wave propagation (via transmission lines and antennas), and modulation as well as demodulation involved in analog and digital communication systems. Toward that end, this chapter is divided into three sections. Since the wave concepts that apply to transmission lines are easily understood, the first section deals with waves, transmission lines, and antenna fundamentals. Then we go on to discuss analog and digital communication systems in Sections 15.2 and 15.3, respectively.

15.1 WAVES, TRANSMISSION LINES, WAVEGUIDES, AND ANTENNA FUNDAMENTALS

In basic circuit theory we neglect the effects of the finite time of transit of changes in current and voltage and the finite distances over which these changes occur. We assume that changes occur simultaneously at all points in the circuits. But there are situations in which we must consider the finite time it takes for an electrical or magnetic wave to travel and the distance it will travel. It is in these situations that one must employ *traveling-wave theory*. Traveling-wave concepts must be used whenever the distance is so great or the frequency so high that it takes an appreciable portion of a cycle for the wave to travel the distance.

For sinusoidal signals, a *wavelength* λ is defined as the distance that a wave travels in one cycle or period. Since electric waves in free space travel at the velocity of light $c(\cong 3 \times 10^8$ m/s), the free-space wavelength is given by c/f. Table 15.1.1 shows some free-space wavelengths at selected frequencies. If the traveling-wave technique is to be employed for distances greater than 1/10 wavelength, a distance of 3 mm at 10 GHz would require the use of this technique, whereas the same distance at 100 MHz would not. On the other hand, a distance of 1 km is insignificant at power-line frequencies, but not in the broadcast band.

The connection of the high-power output of a transmitter located on a building to the transmitting antenna on a tower is often made by special conductors called *transmission lines,* which guide the waves and usually consist of two or more parallel conductors, which are separated by insulating (dielectric) materials. While transmission lines are available in many forms, Figure 15.1.1 illustrates cross sections of some common types. The two-wire line of Figure 15.1.1(a) is

TABLE 15.1.1 Free-Space Wavelengths at Selected Frequencies

Application	Frequency	Wavelength
Power transmission	60 Hz	5000 km
Voice	1 kHz	300 km
Broadcast band	1 MHz	300 m
FM, television	100 MHz	3 m
X-band radar	10 GHz	3 cm

used to connect some television antennas. The coaxial cable of Figure 15.1.1(b) is the most widely used of the many possible cable-type transmission lines. For printed-circuit and integrated-circuit applications, transmission lines sketched in Figures 15.1.1(c) through (f) are commonly employed.

At higher frequencies, when power levels are large and attenuation in transmission lines is significant, connections between system components are often made through *waveguides*, which are usually hollow, closed, rigid conductor configurations (much like water pipes) through which waves propagate. The most common waveguides are either *rectangular* or *circular* in cross section, as depicted in Figure 15.1.2, but other shapes and flexible varieties are also possible.

Coaxial transmission lines commonly operate in what is called the *transverse èlectric magnetic* (TEM) *mode*, in which both the electric and the magnetic fields are perpendicular

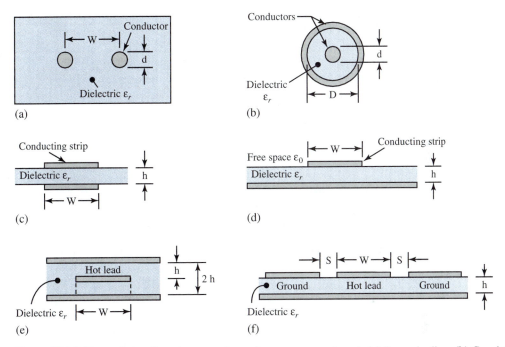

Figure 15.1.1 Transmission lines (cross sections of some common types). (**a**) Two-wire line. (**b**) Coaxial line (cable). (**c**) Parallel strip line. (**d**) Microstrip line. (**e**) Strip line. (**f**) Coplanar waveguide.

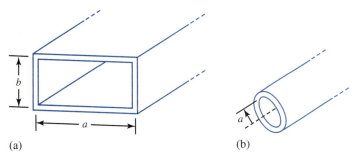

Figure 15.1.2 Waveguides. (**a**) Rectangular. (**b**) Circular.

(transverse) to the direction of propagation, which is along the axial line. That is to say, no electromagnetic field component exists in the axial direction. In the case of single-conductor hollow (pipelike) waveguides, either the TE (*transverse electric*) or the TM (*transverse magnetic*) *mode* can be energized. In the TE configuration, the electric field is transverse to the direction of propagation (which is along the axial line of the waveguide); that is to say, no electric field exists in the direction of propagation, while an axial component of the magnetic field is present. On the other hand, in the TM configuration, the magnetic field is transverse to the direction of propagation: i.e., no magnetic field exists in the axial direction, whereas an axial component of the electric field is present. Within either grouping, a number of configurations or *modes* can exist, either separately or simultaneously. However, we are generally concerned with the so-called *dominant mode*, which is defined as the lowest frequency mode that can exist in the waveguide. When operating at a frequency above f_c, known as the *cutoff frequency,* a wave propagates down the waveguide and the mode is called *propagation mode*. When operating below the cutoff frequency, the field decays exponentially and there is no wave propagation.

Figure 15.1.3 represents a transmission line of length l connecting a signal source to a distant load. The line may be a two-wire line, a coaxial cable, or a hollow waveguide. The voltage $v_1(t)$ between the source-side (input) terminals of the transmission line gives rise to an electric field, while the current $i_1(t)$ produces a magnetic field. The *characteristic impedance* $\bar{Z}_0$ of the line is given by

$$\bar{Z}_0 = \frac{\bar{V}_1}{\bar{I}_1} \tag{15.1.1}$$

which relates the voltage and current of the wave traveling along the line. Although $\bar{Z}_0$ in general could be complex, for distortionless transmission, $\bar{Z}_0\,(=R_0)$ must be constant and resistive over the frequency range of the signal. The signal-source voltage $v_S(t)$ is related to $v_1(t)$ by

$$v_S(t) = v_1(t) + i_1(t)R_S \tag{15.1.2}$$

where R_S is the signal-source internal resistance. From Equations (15.1.1) and (15.1.2), we get

$$v_1(t) = \frac{R_0}{R_S + R_0} v_S(t) \tag{15.1.3}$$

$$i_1(t) = \frac{1}{R_S + R_0} v_S(t) \tag{15.1.4}$$

As the electromagnetic fields associated with $v_1(t)$ and $i_1(t)$ propagate down the line, they carry along the associated voltages and currents that are no different from $v_1(t)$ and $i_1(t)$ except for a delay in time, because the charges (and therefore current) move down the line at finite velocity v_g. Thus, at output terminals c and d in Figure 15.1.3, which are a distance l apart, the voltage

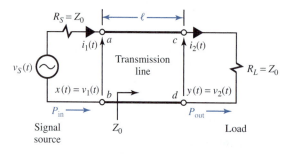

Figure 15.1.3 Transmission line (with matched impedances).

and current are given by

$$v_2(t) = v_1(t - t_l) \tag{15.1.5}$$

$$i_2(t) = i_1(t - t_l) \tag{15.1.6}$$

where the delay time t_l is

$$t_l = l/v_g \tag{15.1.7}$$

On substituting Equations (15.1.3) and (15.1.4), we obtain

$$v_2(t) = \frac{R_0}{R_S + R_0} v_S \left(t - \frac{l}{v_g} \right) \tag{15.1.8}$$

$$i_2(t) = \frac{1}{R_S + R_0} v_S \left(t - \frac{l}{v_g} \right) \tag{15.1.9}$$

Because of the line's behavior, the impedance at terminals c and d, which is $\bar{V}_2/\bar{I}_2$, is still $\bar{Z}_0(= R_0)$. As a consequence, the line can be terminated by an impedance $\bar{Z}_0(= R_0)$ at the output terminals with no effect on voltages and currents anywhere else on the line. This result is very significant, because if a wave is launched on a real, finite-length transmission line, the wave will dissipate itself in the terminating load impedance if that impedance is equal to the line's characteristic impedance. The load is then said to be *matched* to the line. In Figure 15.1.3, the source and load resistances are shown matched so that $R_S = R_0 = R_L$. Otherwise, a mismatch at the load ($R_L \neq R_0$) reflects some of the signal energy back toward the source where any mismatch ($R_S \neq R_0$) further reflects energy in the forward direction. Impedance matching on both ends eliminates these undesired multiple reflections. Note that the value of R_0 will be different for different types of transmission lines.

EXAMPLE 15.1.1

An RG-213/U (radio guide 213/universal coaxial cable) is a small-sized, flexible, double-braided cable with silvered-copper conductors, and a characteristic impedance of 50 Ω. The characteristic impedance $\bar{Z}_0(= R_0)$ is related to the cable's geometrical parameters by

$$\bar{Z}_0 = R_0 = \frac{60}{\sqrt{\varepsilon_r}} \ln \frac{b}{a}$$

where ε_r is the relative permittivity (dielectric constant) of the dielectric, and b and a are the radii of the outer and inner conductors, respectively. The velocity of wave propagation in a coaxial line is $v_g = c/\sqrt{\varepsilon_r}$, where $c = 3 \times 10^8$ m/s is the velocity of light. The cutoff frequency, given by

$$f_c = \frac{c}{\pi \sqrt{\varepsilon_r}(a + b)} \text{ Hz}$$

puts an upper bound for wave propagation. The attenuation due to conductor losses is approximately given by

$$\text{Attenuation}\Big|_c = \frac{(1.373 \times 10^{-3})\sqrt{\rho f}}{Z_0} \left(\frac{1}{a} + \frac{1}{b} \right) \text{ dB/m}$$

where ρ is the resistivity of the conductors. Attenuation due to dielectric losses is given by

$$\text{Attenuation}\Big|_d = (9.096 \times 10^{-8})\sqrt{\varepsilon_r} \, f \tan \delta \text{ dB/m}$$

where $\tan \delta$ is known as the loss tangent of the dielectric.

With $a = 0.445$ mm, $b = 1.473$ mm, $\varepsilon_r = 2.26$ for polyethylene dielectric, $\tan \delta = 2 \times 10^{-4}$, and $\rho = 1.63 \times 10^{-8}$ $\Omega \cdot$m, calculate the total line attenuation at 100 MHz, and check the value of $\bar{Z}_0(= R_0)$ in the specification. If the cable connects an antenna to a receiver 30 m away, determine the time delay of the cable, the velocity of wave propagation, and the cutoff frequency.

Solution

$$\text{Attenuation}\Big|_c = \frac{(1.373 \times 10^{-3})\sqrt{1.63 \times 10^{-8} \times 10^8}}{50} \left(\frac{1}{0.445 \times 10^{-3}} + \frac{1}{1.473 \times 10^{-3}} \right)$$

$$\cong 0.103 \text{ dB/m}$$

$$\text{Attenuation}\Big|_d = (9.096 \times 10^{-8})\sqrt{2.26} \times 10^8 \times 2 \times 10^{-4}$$

$$\cong 0.00273 \text{ dB/m}$$

Losses due to conductors obviously dominate.

$$\text{Total attenuation} = 0.103 + 0.00273 \cong 0.106 \text{ dB/m}$$

$$\bar{Z}_0 = R_0 = \frac{60}{\sqrt{2.26}} \ln \frac{1.473}{0.445} \cong 50 \text{ } \Omega$$

$$\text{Time delay} = \tau = \frac{l}{v_g} = \frac{30\sqrt{2.26}}{3 \times 10^8} \cong 0.15 \text{ } \mu s$$

The velocity of wave propagation $v_g = c/\sqrt{\varepsilon_r} = c/\sqrt{2.26} = 0.665$ times the speed of light, or

$$v_g \cong 2 \times 10^8 \text{ m/s}$$

$$\text{Cutoff frequency } f_c = \frac{3 \times 10^8}{\pi \sqrt{2.26}(0.445 + 1.473)10^{-3}}$$

$$= \frac{3 \times 10^{11}}{\pi \sqrt{2.26} \times 1.918} \cong 33 \text{ GHz}$$

In practice, $f < 0.95 f_c$ is usually maintained.

EXAMPLE 15.1.2

Unlike transmission lines, which operate at any frequency up to a cutoff value, waveguides have both upper and lower cutoff frequencies. For rectangular air-filled waveguides [see Figure 15.1.2(a)], the lower cutoff frequency (for propagation by the dominant mode) is given by $f_c = c/2a$, where c is the speed of light. Since the upper limit cannot be larger than $2f_c$, practical waveguides are designed with $b \cong a/2$ with a suggested frequency of $1.25 f_c \leq f \leq 1.9 f_c$. For a circular air-filled waveguide [see Figure 15.1.2(b)] with inside radius a, the lower cutoff frequency (for propagation by the dominant mode) is $f_c = 0.293c/a$. The operating band is usually $f_c < f < 1.307 f_c$. The characteristic impedance $\bar{Z}_0(= R_0)$ in waveguides is not constant with frequency, as it is in transmission lines. For rectangular or circular air-filled waveguides, the expression for $\bar{Z}_0(= R_0)$ is given by

$$\bar{Z}_0 = R_0 = \frac{377}{\sqrt{1 - (f_c/f)^2}}$$

(a) For a rectangular air-filled waveguide with $a = 4.8$ cm and $b = 2.4$ cm, compute the cutoff frequency. If the operating frequency is 4 GHz, find the waveguide's characteristic impedance.

(b) Calculate the diameter of an air-filled circular waveguide that will have a lower cutoff frequency of 10 GHz.

Solution

(a) $f_c = \dfrac{c}{2a} = \dfrac{3 \times 10^8}{2 \times 4.8 \times 10^{-2}} = 3.125$ GHz

$\bar{Z}_0 = R_0 = \dfrac{377}{\sqrt{1 - (3.125/4)^2}} = \dfrac{377}{\sqrt{1 - 0.8839}} = \dfrac{377}{0.3407} = 1106.5 \ \Omega$

(b) $a = \dfrac{0.293c}{f_c} = \dfrac{0.293 \times 3 \times 10^8}{10 \times 10^9} = 0.88$ cm

Diameter $= 2a = 1.76$ cm

Now referring to Figure 15.1.3, assuming matched and distortionless conditions, with signal voltage $x(t)$ across the line input, the resulting output voltage $y(t)$ is given by

$$y(t) = Kx(t - t_d) \tag{15.1.10}$$

where K is the attenuation factor (less than unity) due to ohmic heating in the line dissipating part of the input signal energy and t_d is the delay time. Working with the average signal powers P_{in} and P_{out}, the *transmission loss L* is defined as the power ratio

$$L = P_{in}/P_{out} \tag{15.1.11}$$

where $L = 1/K^2 > 1$ and $P_{out} < P_{in}$. Regardless of the type of transmission line, the transmission loss increases exponentially with distance l such that

$$L = 10^{\alpha l/10)} \tag{15.1.12}$$

where α is the *attenuation coefficient* of the line in decibels per unit length. Expressing L in dB, similar to power gain, we obtain

$$L_{dB} = 10 \log \dfrac{P_{in}}{P_{out}} = \alpha l \tag{15.1.13}$$

Typical values of α range from 0.05 to 100 dB/km, depending on the type of transmission line and the frequency. Rewriting Equation (15.1.13) as

$$\dfrac{P_{out}}{P_{in}} = 10^{-L_{dB}/10} = 10^{-\alpha l/10} \tag{15.1.14}$$

Equation (15.1.14) reveals that P_{out} will be one-tenth of P_{in} when $l = 10/\alpha$, showing thereby how rapidly the output power falls off as the distance increases.

The transmission loss can be overcome with the aid of one or more amplifiers connected in cascade with the transmission line. Figure 15.1.4 shows such a system with a preamplifier (transmitting amplifier) at the source, a receiving amplifier at the destination, and a repeater

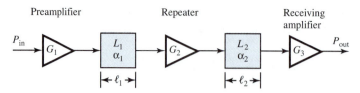

Figure 15.1.4 Transmission system with preamplifier, repeater, and receiving amplifier.

(an additional amplifier) at some intermediate point in the line. All amplifiers will of course be impedance-matched for maximum power transfer. The final output power P_{out} in Figure 15.1.4 is then given by

$$\frac{P_{out}}{P_{in}} = \frac{G_1 G_2 G_3}{L_1 L_2} \tag{15.1.15}$$

where the Gs are the power gains of the amplifiers, and the Ls are the transmission losses of the two parts of the line. Equation (15.1.15) reveals that we can compensate for the line loss and get $P_{out} \geq P_{in}$ if $G_1 G_2 G_3 \geq L_1 L_2$. Noise considerations often call for a preamplifier to boost the signal level before the noise becomes significant. As in the case of transcontinental telephone links, several repeaters are generally required for long-distance transmission.

EXAMPLE 15.1.3

From a source with $P_{in} = 2.4$ mW, we want to get $P_{out} = 60$ mW at a distance $l = 20$ km from the source. α for the transmission line is given to be 2.3 dB/km. The available amplifiers have adjustable power gain, but are subject to two limitations: (i) the input signal power must be at least 1 μW to overcome internal noise, and (ii) the output signal power must not be greater than 1 W to avoid nonlinear distortion. Design an appropriate system.

Solution

$$\alpha l = 2.3 \times 20 = 46 \text{ dB}$$

$$L_{dB} = 46 \qquad \text{or} \qquad L = 10^{46/10} \cong 40,000$$

Hence, we need a total gain of

$$G_{total} = L(P_{out}/P_{in}) = 40,000 \times (60/2.4) = 10^6, \text{ or } 60 \text{ dB}$$

We cannot put all the amplification at the destination, because the signal power at the output of the line would be $P_{in}/L = 2.4 \times 10^{-3}/(40 \times 10^3) = 0.06 \mu$ W, which falls below the amplifier noise level. Nor can we put all amplification at the source, because the amplified source power $G P_{in} = 10^6 \times 2.4 \times 10^{-3} = 2.4$ kW would exceed the amplifier power rating. But we could try a preamplifier with $G_1 = 400$, so as to get $G_1 P_{in} = 400 \times 2.4 \times 10^{-3} = 0.96$ W at the input of the line, and $G_1 P_{in}/L = 24 \mu$ W at the output. The output amplifier should then have $G_2 = P_{out}/(24 \mu W) = 60 \times 10^{-3}/(24 \times 10^{-6}) = 2500$, and a repeater is not needed.

Antenna Fundamentals

We shall discuss here only the fundamental concepts needed to understand the role of an antenna as a power-coupling element of a system. Figure 15.1.5 illustrates the elements of a communication system that involves antennas (with no conductors in the propagation medium) and the following nomenclature:

P_t Power generated by the transmitter

L_t, L_r Transmitting-path loss (representing the power reduction caused by the transmission line or waveguide that connects to the transmitting antenna) and receiving-path loss

L_{ta}, L_{ra} Transmitting antenna loss and receiving antenna loss

R Distance of separation between the antennas

S Signal power available at the lossless antenna output

S_r Signal power available at the receiver input

     Radio transmission consists of antennas at the source and at the destination. It requires the signal to be modulated on a high-frequency carrier, which usually is a sinusoid. Driven by an appropriate carrier, the transmitting antenna launches an electromagnetic wave that propagates through space without the help of a transmission line. A portion of the radiated power is collected at the receiving antenna.

     The *wavelength* λ of the radio wave in air is related to the carrier frequency f_c by

$$f_c \lambda = c = 3 \times 10^8 \text{ m/s} \qquad (15.1.16)$$

The radio-transmission loss differs from that of a transmission line in two ways: (i) it increases as the square of the separating distance instead of exponentially, and (ii) it can be partly compensated by the antenna gains.

     Antenna gain depends on both shape and size. *Dipole antennas,* commonly used at lower radio frequencies, are made up of a rod or wire of length $\lambda/10$ to $\lambda/2$, and have an antenna-gain range of 1.5 to 1.64 (1.8 to 2.1 dB). *Horn antennas* and *parabolic dishes* (so named after their shapes) have much more gain at higher frequencies. A useful, although approximate, expression for gain is given by

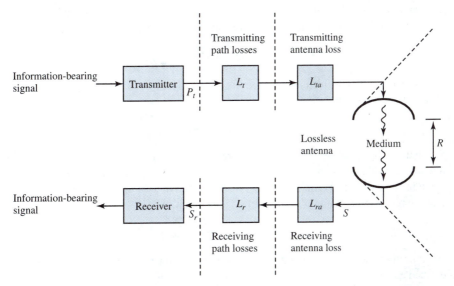

Figure 15.1.5 Communication system elements involving antennas.

$$G = \frac{4\pi A_e}{\lambda^2} \tag{15.1.17}$$

where λ is the wavelength being transmitted (or received), and A_e is the *effective aperture* or *effective area* of the antenna. Power amplifiers are needed to overcome the radio-transmission loss just as in a transmission-line system. Similar considerations hold for *optical* radiation, in which the electromagnetic wave takes the form of a coherent light beam.

Radio transmission is inherently a *bandpass* process with a limited bandwidth B nominally centered at the carrier frequency f_c. The *fractional bandwidth* B/f_c is a key design factor with a general range of $1/100 \le B/f_c \le 1/10$. It is obvious then that large signal bandwidths require high carrier frequencies to satisfy $f_c \ge 10B$. You can reason why television signals are transmitted at f_c of 100 MHz, whereas AM radio signals are transmitted at f_c of 1 MHz. Since optical communication systems offer tremendous bandwidth potential on the order of 10^{12} Hz, and a corresponding high information rate, they have become topics of current research interest.

Antennas do not radiate power equally in all directions in space. The *radiation intensity pattern* describes the power intensity (which is power per unit solid angle, expressed in units of watts per steradian) in any spatial direction. Conceptually it is convenient to define an *isotropic antenna* as a lossless antenna that radiates its power uniformly in all directions. Although an isotropic antenna cannot be realized in practice, it serves as a reference for comparison with real antennas. The radiation intensity for such an antenna, with input power P, is a constant in any direction, given by $P/4\pi$. The *power gain G* of a realistic antenna is a measure of the maximum radiation intensity of the antenna as compared with the intensity that would result from an isotropic antenna, with the same power input. G is then expressed as

$$G = \frac{\text{maximum radiation intensity}}{\text{radiation intensity of isotropic source (with the same power input)}}$$

$$= \frac{4\pi(\text{maximum radiation intensity})}{P} \tag{15.1.18}$$

Referring to Figure 15.1.5, when a power P_t/L_t is applied to the transmitting antenna, let us find the signal power S_r available to the receiver from the receiving antenna. An isotropic transmitting antenna would cause a radiation power density (power per unit area of a sphere) of

$$\text{Power density} = \frac{P_t}{4\pi R^2 L_t} \tag{15.1.19}$$

For a practical antenna that has power gain G_t and loss L_{ta} relative to an isotropic antenna, Equation (15.1.19) would be modified as

$$\text{Power density} = \frac{P_t G_t}{4\pi R^2 L_t L_{ta}} \tag{15.1.20}$$

We shall assume that the transmitting and receiving antennas (*reciprocal elements*) point directly toward each other, so that their gains are maximum. Letting L_{ch} denote any losses incurred by the wave in the channel (medium), and A_{re} be the effective area of the receiving antenna, the power that the receiving antenna is able to produce is given by

$$S = \frac{P_t G_t A_{re}}{4\pi R^2 L_t L_{ta} L_{ch}} \tag{15.1.21}$$

Accounting for the receiving antenna loss and receiving-path losses, a total system loss L can be defined as

$$L = L_t L_{ta} L_{ch} L_{ra} L_r \tag{15.1.22}$$

S_r would then be given by

$$S_r = \frac{P_t G_t A_{re}}{4\pi R^2 L} \tag{15.1.23}$$

Using Equation (15.1.17) with G_r representing the receiving antenna gain, Equation (15.1.23) becomes

$$S_r = \frac{P_t G_t G_r \lambda^2}{(4\pi)^2 R^2 L} \tag{15.1.24}$$

Quite often, for simplicity, in the case of LOS radio transmission illustrated in Figure 15.1.6, the transmission loss for a path of length R is given by

$$L_{TR} = \frac{P_{in}}{P_{out}} = \frac{1}{G_t G_r} \left(\frac{4\pi R}{\lambda} \right)^2 \tag{15.1.25}$$

which justifies the statements following Equation (15.1.16).

EXAMPLE 15.1.4

Let a LOS radio system and a transmission-line system both have $L = 60$ dB when the distance R between transmitter and receiver is 15 km. Compute the loss of each when the distance is increased to 30 km.

Solution

R is doubled. L_{dB} for the transmission line is proportional to the distance, as per Equation (15.1.13). Hence, L_{dB} for the new transmission line = $2 \times 60 = 120$ dB.

L_{TR} for the LOS radio system is proportional to R^2, as per Equation (15.1.25). Noting that $L_{dB} = 10 \log L$ and $L = KR^2$, where K is a constant, it follows that

$$L_{dB} = 20 \log KR$$

When R is doubled,

$$L_{dB\ new} = 20 \log (K2R) = 20(\log KR + \log 2)$$
$$= 20 \log KR + 20 \log 2 = L_{dB\ old} + 6$$

Hence, L_{dB} for the new LOS radio system = $60 + 6 = 66$ dB.

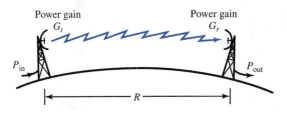

Power gain G_t Power gain G_r **Figure 15.1.6** Illustration of LOS radio transmission.

P_{in} P_{out}

R

EXAMPLE 15.1.5

(a) Some antennas have a physical aperture area A that can be identified and is related to the effective area A_e by $A_e = \rho_a A$, where ρ_a is known as the *aperture efficiency*. For a circular-aperture antenna with a diameter of 2 m and an aperture efficiency of 0.5 at 4 GHz, calculate the power gain.

(b) Referring to Figure 15.1.5, let two such antennas be used for transmitting and receiving, while the two stations are separated by 50 km. Let the total loss over the link be 9 dB, while the transmitter generates 0.5 W. Find the available received power.

Solution

(a) $\lambda = \dfrac{c}{f} = \dfrac{3 \times 10^8}{4 \times 10^9} = 0.075$ m. From Equation (15.1.17),

$$G = \frac{4\pi A_e}{\lambda^2} = \frac{4\pi \rho_a A}{\lambda^2} = \frac{4\pi \times 0.5 \times \pi}{0.075^2} = 3509$$

(b) $G = G_t = G_r = 3509$; $L_{dB} = 10 \log L$, or $L = 7.94$. From Equation (15.1.24),

$$S_r = \frac{0.5 \times 3509^2 \times 0.075^2}{(4\pi)^2 \times 50^2 \times 10^6 \times 7.94} \cong 1.1 \times 10^{-8} \text{ W}$$

The variety and number of antennas are almost endless. However, for our introductory purposes, they may be divided into the following types:

- *Wire antennas,* such as half-wavelength dipole, folded half-wave dipole, and helical antenna
- *Array antennas,* such as YAGI-UDA array
- *Aperture antennas,* such as pyramidal horn, conical horn, paraboloidal antenna, and Cassegrain antenna
- *Lens-type antennas* in radar and other applications.

Some of their geometries are illustrated in Figure 15.1.7.

The *radiation-intensity pattern* describing the power intensity in any spatial direction is an important antenna characteristic, since the antenna does not radiate power equally in all directions in space. Such patterns are three-dimensional in nature. One normally chooses spherical coordinates centered on the antenna at A, and represents the power-intensity function $P(\theta, \phi)$ at any distant point R as a magnitude P from A, which appears as a surface with a large *main lobe* and several *side lobes* (minor lobes), as shown in Figure 15.1.8.

Generally speaking, in most of the communication systems the transmitting and receiving antennas (reciprocal elements) face each other directly such that their large main lobes point toward each other, and the received-power output will be maximum. When scaled such that the maximum intensity is unity, the radiation-intensity pattern is commonly called the *radiation pattern*. In many problems in practice, the radiation pattern occurs with one dominant main lobe, and as such, instead of considering the full three-dimensional picture, the behavior may adequately and conveniently be described in two orthogonal planes containing the maximum of the main lobe. These are known as *principal-plane patterns* in terms of angles θ and ϕ. Figure 15.1.9 illustrates one such pattern in polar and linear angle plots as a function of θ. The angular separation between points on the radiation pattern that are 3 dB down from the maximum is called

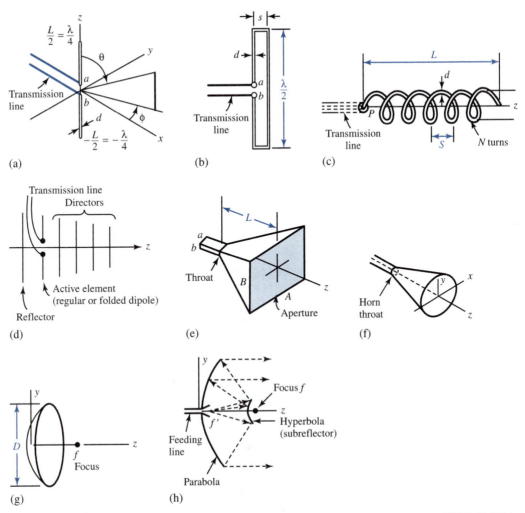

Figure 15.1.7 **(a)** Half-wave dipole. **(b)** Folded half-wave dipole. **(c)** Helical-beam antenna. **(d)** YAGI-UDA array antenna. **(e)** Pyramidal horn. **(f)** Conical horn. **(g)** Paraboloidal antenna. **(h)** Cassegrain antenna (section).

the *beamwidth,* which is designated as θ_B or ϕ_B in the principal-plane patterns. Another useful, although approximate, expression for gain in terms of the beamwidths θ_B and ϕ_B of the pattern's main lobe is given by

$$G = \frac{4\pi}{\theta_B \phi_B} \tag{15.1.26}$$

when beamwidths are expressed in radians, or

$$G = \frac{41.3 \times 10^3}{\theta_B \phi_B} \tag{15.1.27}$$

when beamwidths are expressed in degrees.

The *antenna impedance,* looking into the feed-point terminals of an antenna, will in general have both resistive and reactive components. The *antenna resistance* is the combination of the radiation resistance (accounting for the power radiated by the antenna) and the *loss resistance*

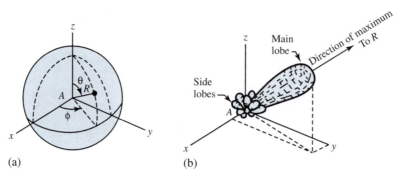

Figure 15.1.8 (a) Spherical coordinates centered on the antenna at A. (b) Typical radiation-intensity pattern with main and side lobes (with all possible values of θ and ϕ considered).

(accounting for the power dissipated in the antenna itself as losses). Ideally, the antenna resistance should be equal (matched) to the characteristic impedance of the feeding line or guide to prevent reflected power, and the antenna reactance should be zero.

Most antennas transmit only one *polarization* of electromagnetic wave. That is to say, the electric field of the propagating wave is oriented with respect to the antenna in only one direction. The main lobe in most antennas is directed normal to the plane of the aperture. For phased-array antennas, however, the main lobe is electronically steered to other angles away from the so-called broadside. Figure 15.1.10 illustrates *vertical, horizontal,* and *arbitrarily linear polarizations* of the electric field. In Figure 15.1.10(a), with the electric field lying in the vertical plane, the radiation is said to be vertically polarized. With the electric field being in the horizontal plane, as in Figure 15.1.10(b), the radiation is said to be horizontally polarized. Since both vertical and horizontal polarizations are simply special cases of linear polarization, the electric field, having both horizontal and vertical components (that are in time phase), can still be in a plane, as shown in Figure 15.1.10(c).

Some systems transmit simultaneously two linear orthogonal polarizations that are not in time phase. *Elliptical polarization* results when the two linear components have arbitrary relative amplitudes and arbitrary time phase. *Circular polarization* (probably the most useful type) is a special case in which the horizontal and vertical electric fields are 90° out of time phase and have equal magnitude. *Left-hand* circular polarization results when the horizontal radiation component lags the vertical one by 90° and the resultant field appears to rotate counterclockwise in the *xy*-plane with time, as one located at the antenna views the wave leaving the antenna. If the horizontal component leads the vertical one by 90°, *right-hand* circular polarization is said to take place.

Referring to Figure 15.1.7, the half-wave dipole is a relatively narrow-band antenna with its radiated wave linearly polarized. The dipole can be driven by a transmission line of 75-Ω characteristic impedance. The folded half-wave dipole (a variation of the half-wave dipole) is used in television, broadcast FM, and other applications. This antenna is well suited for use with 300-Ω television cable. The helical antenna of Figure 15.1.7(c) yields a *pencil-beam pattern* in the axis of the helix with circular wave polarization.

The YAGI-UDA array of Figure 15.1.7(d) is commonly used for television reception. It is usually seen with 3 to 12 elements, although even 40 elements are sometimes employed. Design frequencies from 100 to 1000 MHz are typical. A half-wave or folded half-wave dipole is the *active element*. The array consists of parallel dipoles, all lying in the same plane. The *reflector*, which reflects waves back toward the active element, enhances radiation in the axis of the array

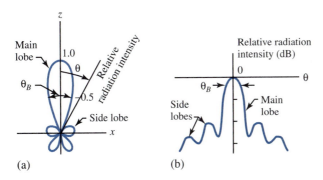

Figure 15.1.9 Typical radiation pattern in one principal plane as a function of θ. **(a)** Polar-coordinate plot. **(b)** Linear-coordinate plot.

($+z$-direction). The radiation pattern, with linear polarization, exhibits a principal lobe in the $+z$-direction. The other elements, called *directors*, are designed to enhance radiation in the $+z$ direction. Gain increases with the number of elements and is often in the range of 10 to 20 dB. The bandwidth is usually small.

Both conical-horn and pyramidal-horn (aperture-type) antennas are mainly used as *illuminators* for large-aperture paraboloidal antennas, which are capable of generating very narrow beamwidth patterns (with even less than 1° in angle-tracking radars). When a parabola is rotated about the z-axis [see Figure 15.1.17(g)], a surface of revolution known as a paraboloid results. With the source at the focus called the *feed*, the radiation pattern is mainly a dominant lobe in the z-direction with smaller side lobes. Paraboloidal antennas have found wide use as antennas for radar and communications. The Cassegrain antenna [shown in section in Figure 15.1.17(h)] is a variation of the paraboloid that gives improved system performance. The feed in this case is moved to the rear of the antenna, and it illuminates a conducting surface (*subreflector* in the shape of a hyperboloid) placed near the focus.

Noise in receiving systems exists in two broad categories: (i) that originated *external to the system* (i.e., the one generated by the antenna in response to random waves from cosmic sources and atmospheric effects), and (ii) *internally generated noise* (i.e., the one generated within all circuits making up the receiver, including transmission lines and amplifiers). It is common to model internal noise as having been generated by an external source.

Figure 15.1.11(a) shows a typical receiving system with noise. The antenna is a source of noise with effective noise temperature T_a, known as the *antenna temperature*. Whatever receiving path components (such as transmission lines, waveguides, and filters) are present prior to the receiver's amplifier, their noise effect is represented by a noisy loss $L_r\ (\geq 1)$ between points A and B, while the loss is assumed to have a physical temperature T_L. The noisy receiver is supposed to operate at a nominal center frequency f_0, and have available power gain $G_a(f)$ as a function of the frequency, with $G_a(f_0)$ as center-frequency power gain. N_{ao} represents the total available output noise power in Figure 15.1.11(a).

Figure 15.1.11(b) shows the noise-free model in a small frequency band df. Here k is the Boltzmann constant (see Section 14.3), and $T_R(f)$ is the *effective input noise temperature* (to the noise-free receiver).

Figure 15.1.11(c) gives the noise-free model with *noise bandwidth* B_N (a rectangular passband of width B_N in hertz centered on f_0). The actual receiver is replaced by an idealized one with the same nominal power gain $G_a(f_0)$ and a *constant* (average) effective input noise temperature $\bar{T}_R$, which is related to the *average standard noise figure* F_0;

$$\bar{T}_R = 290(F_0 - 1) \tag{15.1.28}$$

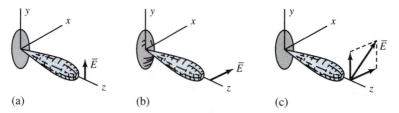

(a)   (b)   (c)

Figure 15.1.10 Polarizations of the electric field. **(a)** Vertical polarization. **(b)** Horizontal polarization. **(c)** General linear polarization.

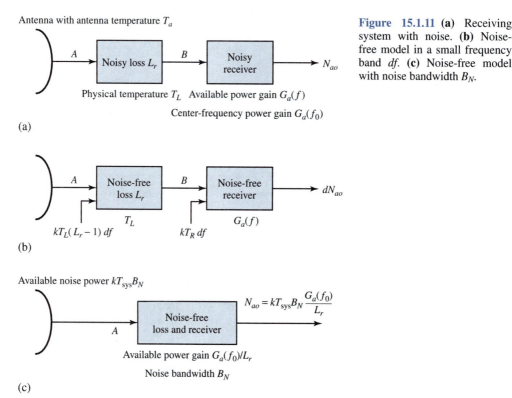

Antenna with antenna temperature T_a

Figure 15.1.11 (a) Receiving system with noise. **(b)** Noise-free model in a small frequency band df. **(c)** Noise-free model with noise bandwidth B_N.

(a)

(b)

(c)

Note that both F_0 and $\bar{T}_R$ are measures of the noisiness of an amplifier. The noise figure F_0 is usually available, whereas the temperature $\bar{T}_R$ may not be given. In an ideal noise-free unit, $F_0 = 1$ and $\bar{T}_R = 0$.

The *system noise temperature* T_{sys}, which is the equivalent noise temperature of the antenna, can now be introduced such that

$$T_{\text{sys}} = T_a + T_L(L_r - 1) + \bar{T}_R L_r \qquad (15.1.29)$$

and all available output noise power is emanating from the antenna, as illustrated in Figure 15.1.11(c). Now the available system noise power becomes $kT_{\text{sys}}B_N$ at point A; the noise-free receiver has available power gain $G_a(f_0)/L_r$ with noise bandwidth B_N; and the output noise power is given by

$$N_{ao} = \frac{kT_{sys}B_N G_a(f_0)}{L_r} \tag{15.1.30}$$

The signal-to-noise power ratio at the system output is an excellent measure of performance for many communication systems. The available signal power at point A in Figure 15.1.11(a) is given by

$$S_A = \frac{P_t G_t G_r \lambda^2}{(4\pi)^2 R^2 L_t L_{ta} L_{ch} L_{ra}} \tag{15.1.31}$$

based on Equation (15.1.24). The available noise power at point A is given by

$$N_{aA} = kT_{sys}B_N \tag{15.1.32}$$

Thus, the system performance with noise is measured by the signal-to-noise power ratio,

$$\left(\frac{S}{N}\right)_A = \frac{P_t G_t G_r \lambda^2}{(4\pi)^2 R^2 L_t L_{ta} L_{ch} L_{ra} kT_{sys}B_N} \tag{15.1.33}$$

15.2 ANALOG COMMUNICATION SYSTEMS

An *analog message* is a continuum of possible amplitudes at any given time, and analog signals are continuous in time and in amplitude, such as audio and video signals. When the message to be sent over a communication system is analog, we refer to the system as analog. The transmitted waveform must be some function of the message so that the receiver could decipher the message. Usually the transmitted waveform is the result of varying either the amplitude, phase, or frequency of a basic signal called a *carrier*. Combinations of amplitude, phase, and frequency variations are also possible.

The carrier usually is sinusoidal of the form $A_c \cos(\omega_c t + \phi_c)$, where A_c, ϕ_c, and $f_c = \omega_c/2\pi$ are the carrier's amplitude, phase, and frequency, respectively. When A is varied as a linear function of the message, *amplitude modulation* (AM) occurs. In *phase modulation* (PM) a phase term that is a linear function of the message is added to the carrier. When the added phase is a linear function of the integral of the message, the result is known as *frequency modulation* (FM). Note that the carrier's frequency in FM is a linear function of the message, because instantaneous angular frequency is the time derivative of instantaneous phase. Thus, FM and PM are closely related.

Every communication system has a *modulator* at the transmitting station to structure the transmitted waveform, and a *demodulator (detector)* at the receiving end to recover the message from the received signal. Radio (AM and FM) and television broadcasting are the most familiar forms of communication through analog signal transmission. The FCC in the United States regulates the carrier-frequency assignments in order to minimize the interference between nearby stations. Commercial AM radio broadcasting utilizes the frequency band of 535 to 1605 kHz for the transmission of voice and music. The carrier-frequency allocations range from 540 kHz to 1600 kHz, with 10-kHz spacing. Each station can occupy a channel bandwidth of only 10 kHz centered on its carrier. Even though the baseband message signal is limited to a bandwidth of about 5 kHz, the AM broadcasting system adequately meets the need for low-cost mass communication and general audio entertainment, in spite of lacking high-fidelity behavior.

Commercial FM radio broadcasting utilizes the frequency band of 88 to 108 MHz for the transmission of music and voice signals. The carrier frequencies are separated by 200 kHz, and the peak frequency deviation is fixed at 75 kHz. Each station (out of the possible 100) broadcasts in a channel bandwidth of 200 kHz centered on the carrier. The FM system is capable of presenting higher quality audio to the user than AM because of the larger audio band allowed from 50 Hz to

15 kHz. Distortion at most stations is below 1%. Besides broadcast applications, FM is also used in satellite links, aircraft altimetry, radars, amateur radio, and various two-way radio applications. Because of the larger bandwidth (up to 20 times more than in AM), an FM system has more freedom from interference and better performance in noise than any AM station.

Commercial television broadcasting is allocated frequencies that fall in the VHF and UHF bands. Table 15.2.1 lists the television channel allocations in the United States, with the channel bandwidth of 6 MHz. In contrast to radio broadcasting, television signal-transmission standards vary from country to country. The U.S. standard is set by the National Television Systems Committee (NTSC). Commercial television broadcasting began as black-and-white (monochrome) picture transmission in London in 1936 by the British Broadcasting Corporation (BBC). Although color television was demonstrated a few years later, due to the high cost of color television receivers, color television signal transmission was slow in its development. With the advent of the transistor and microelectronic components, the cost of color television receivers decreased significantly, and by the middle 1960s, color television broadcasting was widely used by the industry. The NTSC color system is compatible with monochrome receivers, so that the older monochrome receivers still function receiving black-and-white images out of the transmitted color signal.

Amplitude Modulation (AM)

In AM the message signal is impressed on the amplitude of the carrier signal. There are several different ways of amplitude modulating the carrier signal by the message signal, each of which results in different spectral characteristics for the transmitted signal. Four methods are:

- Conventional (standard) double-sideband AM
- Double-sideband suppressed-carrier AM
- Single-sideband AM
- Vestigial-sideband AM.

AM, as its name implies, carries the modulating signal $x(t)$ in a *time-varying amplitude* of the form

$$a(t) = A_c[1 + m_A x(t)] \tag{15.2.1}$$

where the constant A_c stands for the unmodulated carrier amplitude, and m_A is the *modulation index*. The resulting modulated wave is

$$
\begin{aligned}
x_c(t) &= a(t)\cos(2\pi f_c t + \phi_c) \\
&= A_c \cos(2\pi f_c t + \phi_c) + A_c m_A x(t) \cos(2\pi f_c t + \phi_c)
\end{aligned} \tag{15.2.2}
$$

TABLE 15.2.1 VHF and UHF Allocations for Commercial TV in the United States

Channel	Frequency Band (6-MHz bandwidth/station)
VHF 2–4	54–72 MHz
VHF 5–6	76–88 MHz
VHF 7–13	174–216 MHz
UHF 14–69	470–896 MHz

whereas the sinusoidal carrier waveform is given by

$$c(t) = A_c \cos(\omega_c t + \phi_c) = A_c \cos(2\pi f_c t + \phi_c)$$

Sometimes the standard AM signal is expressed in the form of

$$x_c(t) = s_{AM}(t) = [A_c + f(t)] \cos(\omega_c t + \phi_c) \qquad (15.2.3)$$

in which the total amplitude of the carrier $[A_c + f(t)]$ is a linear function of the message. Figure 15.2.1 illustrates the AM waveforms.

If the maximum amplitude of $f(t)$ in Equation (15.2.3) exceeds A_c, *overmodulation* is said to occur. In order to prevent excessive distortion in the receiver, overmodulation is avoided in standard AM by requiring

$$|f(t)|_{max} \leq A_c \qquad (15.2.4)$$

such that the envelope of $s_{AM}(t)$ or $x_c(t)$ will never go negative and may at most become zero. The spectral behavior is depicted in Figure 15.2.2, assuming $\phi_0 = 0$ for simplicity. The spectral impulses are due to the carrier and will always be present, even if the message were to disappear. The effect of AM is to shift half-amplitude replicas of the signal spectrum out to angular frequencies ω_c and $-\omega_c$. The band of frequencies above ω_c or below $-\omega_c$ is called the *upper sideband* (USB). The one on the opposing side is known as the *lower sideband* (LSB). Since AM transmits both USB and LSB, it is known as double-sideband modulation. If the maximum circular (radian) frequency extent of $f(t)$ is W_f rad/s, the frequency extent of the standard AM waveform is $2W_f$. Thus, the low-pass signal is translated in frequency to the passband of the channel so that the spectrum of the transmitted bandpass signal will match the passband characteristics of the channel.

There are several different methods for generating amplitude-modulated signals. Since the process of modulation involves the generation of new frequency components, modulators are generally characterized as nonlinear and/or time-variant systems, because a linear or a time-invariant system cannot create new frequencies other than those contained in its input signal. Figure 15.2.3 shows a block diagram of *power-law modulation* that is nonlinear. Let the voltage input to such a device be the sum of the message signal and the carrier, as illustrated in Figure 15.2.3. The nonlinear device (that has an input–output characteristic of the form of a square law) will generate a product of the message $x(t)$ with the carrier, plus additional terms. The desired modulated signal can be filtered out by passing the output of the nonlinear device through a bandpass filter. The signal generated by this method is a conventional DSB AM signal.

$x(t)$

(a)

(b)

Figure 15.2.1 AM waveforms. **(a)** Information (message or modulating) signal. **(b)** Modulated wave (corresponding standard AM signal).

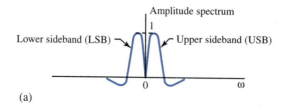

Amplitude spectrum

Lower sideband (LSB) ⟶ ⟵ Upper sideband (USB)

(a)

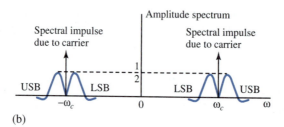

Amplitude spectrum

Spectral impulse due to carrier

Spectral impulse due to carrier

USB   LSB   LSB   USB

$-\omega_c$ 0 ω_c ω

(b)

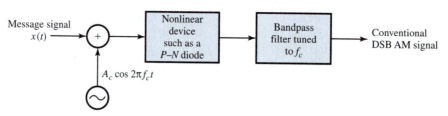

Message signal x(t) ⟶ (+) ⟶ Nonlinear device such as a P–N diode ⟶ Bandpass filter tuned to f_c ⟶ Conventional DSB AM signal

$A_c \cos 2\pi f_c t$

Figure 15.2.3 Block diagram of power-law modulation.

Let $s_{AM}(t)$ of Equation (15.2.3) be a voltage representing the wave that excites the transmitting station's antenna, which is assumed to represent a resistive impedance R_0 to the transmitter that feeds it. The power in $s_{AM}(t)$ is found by

$$P_{AM} = \frac{\overline{s_{AM}^2(t)}}{R_0} = P_c + P_f = \frac{1}{2R_0}\left[A_c^2 + \overline{f^2(t)}\right] \qquad (15.2.5)$$

in which the overbar represents the time average, and $f(t)$ is assumed to have no dc component, as is the usual case. $P_c = A_c^2/2R_0$ is the power in the carrier, and $P_f = \overline{f^2(t)}/2R_0$ is the added power caused by modulation. P_f is called the useful power, since only this power caused by the message contributes toward message quality. P_c, the carrier power, is not useful power in the sense that it carries no information. However, it is important to the receiver's ability to recover the message with low-cost circuitry. *Modulation efficiency* η_{AM} of the transmitted signal is defined as the ratio of the useful power to the total power,

$$\eta_{AM} = \frac{P_f}{P_c + P_f} = \frac{\overline{f^2(t)}}{A_c^2 + \overline{f^2(t)}} \qquad (15.2.6)$$

When $f(t)$ is a square wave of peak amplitude A_c, the largest possible value of η_{AM} equal to 0.5, or 50%, occurs. When $f(t)$ is a sinusoid, $\eta_{AM} \leq 1/3$, as shown in Example 15.2.1. For practical audio (voice and music) messages, efficiency could be less than $1/3$.

EXAMPLE 15.2.1

Let $f(t)$ be a sinusoid given by $f(t) = A_m \cos \omega_m t$ with period $T_m = 2\pi/\omega_m$. Apply Equation (15.2.6) and obtain the value of η_{AM}.

Solution

$$\overline{f^2(t)} = \frac{1}{T_m} \int_{-T_m/2}^{T_m/2} A_m^2 \cos^2(\omega_m t)\, dt$$

$$= \frac{A_m^2}{8\pi} \int_{-2\pi}^{2\pi} (1 + \cos x)\, dx = \frac{A_m^2}{2}$$

$$\eta_{AM} = \frac{A_m^2}{2A_c^2 + A_m^2} = \frac{(A_m/A_c)^2}{2 + (A_m/A_c)^2}$$

For no overmodulation, $A_m \le A_c$, so that $\eta_{AM} \le \frac{1}{3}$.

Suppressed-carrier AM, also known as double-sideband suppressed-carrier (DSB SC AM), results when the carrier term A_c in Equation (15.2.3) is eliminated. The DSB waveform is then given by

$$s_{DSB}(t) = f(t) \cos(\omega_c t + \phi_c) \tag{15.2.7}$$

Figure 15.2.4(a) depicts the waveforms of the message signal and its amplitude spectrum; Figure 15.2.4(b) shows the DSB waveform $s_{DSB}(t)$, and Figure 15.2.4(c) displays the amplitude spectrum of $s_{DSB}(t)$. From Equation (15.2.6) with $A_c = 0$, the efficiency of DSB SC AM comes out as 1.0, or 100%. While the power efficiency is increased, there will be added complexity, especially in the demodulator.

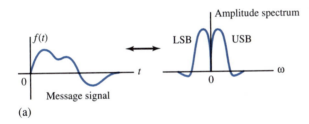

(a)

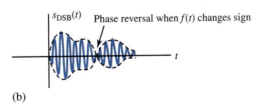

(b)

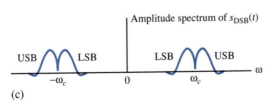

(c)

Figure 15.2.4 (a) Message signal and its amplitude spectrum. (b) DSB SC signal $s_{DSB}(t)$ corresponding to the message. (c) Corresponding SC AM signal spectrum.

A relatively simple method to generate a DSB SC AM signal is by employing two (identical) conventional AM modulators, such as two square-law AM modulators, arranged in the configuration of Figure 15.2.5, which is known as a *balanced modulator*. $\phi_c = 0$ is assumed here for simplicity.

Single-sideband (SSB) AM is obtained by filtering one sideband of the DSB signal. As pointed out earlier, a DSB SC AM signal required a channel bandwidth of $2W$ for transmission, where W is the bandwidth of the baseband signal. However, the two sidebands are redundant. The transmission of either sideband is sufficient to reconstruct the message signal at the receiver. Thus, the bandwidth of the transmitted signal is reduced to that of the baseband signal. Figure 15.2.6 illustrates the generation of an SSB AM signal by filtering one of the sidebands (either USB or LSB) of a DSB SC AM signal. SSB is popular with amateur radio operators because of its high efficiency ($\eta_{SSB} = 1$) and bandwidth savings.

Vestigial-sideband (VSB) AM is a variation of SSB where a small portion (or vestige) of the filtered sideband is allowed to remain. The stringent frequency-response requirements on the sideband filter in an SSB AM system can be relaxed by allowing a part, known as a vestige, of the unwanted sideband to appear at the output of the modulator. Thus, the design of the sideband filter is simplified at the cost of a modest increase in the channel bandwidth required to transmit the signal. Figure 15.2.7 illustrates the generation of a VSB AM signal. VSB is mainly used in the

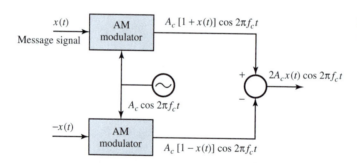

Figure 15.2.5 Block diagram of a balanced modulator.

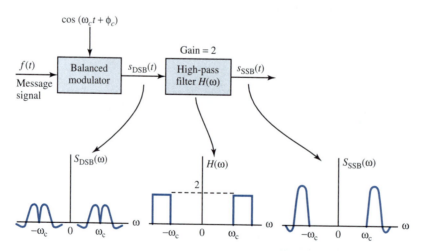

Figure 15.2.6 Generation of a SSB AM signal by filtering one of the sidebands of a DSB SC AM signal.

television broadcast system. The generation of VSB is similar to the generation of SSB, except that the sideband-removal filter has a slightly different transfer function.

Message Demodulation

The major advantage of the conventional (standard) DSB AM signal transmission is the ease with which the signal can be demodulated. The message signal is received by passing the rectified signal through a low-pass filter whose bandwidth matches that of the message signal. The combination of the rectifier and the low-pass filter is known as an *envelope detector*, which is almost universally used. A standard AM envelope detector consisting of a diode and an RC circuit (which is basically a simple low-pass filter) and its response are shown in Figure 15.2.8.

During the positive half-cycle of the input signal, the diode is conducting and the capacitor charges up to the peak value of the input signal. When the input falls below the voltage on the capacitor, the diode becomes reverse-biased and the input becomes disconnected from the output. During this period, the capacitor discharges slowly through the load resistor R. On the next cycle of the carrier, the diode conducts again when the input signal exceeds the voltage across the capacitor. The capacitor charges up again to the peak value of the input signal, and the process is continued. The time constant RC must be selected to follow the variations in the envelope of the carrier-modulated signal. Generally, it is so chosen that

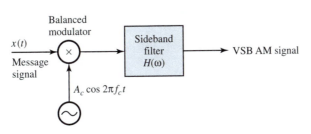

Figure 15.2.7 Generation of a VSB AM signal.

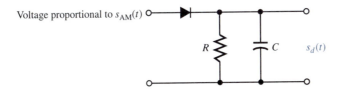

(a)

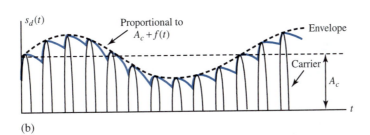

(b)

Figure 15.2.8 Envelope detection of conventional (standard) AM signal. **(a)** Standard AM envelope detector. **(b)** Its response.

$$\frac{1}{f_c} << RC << \frac{1}{W} \tag{15.2.8}$$

The simplicity of the demodulator has made conventional DSB AM a practical choice for AM radio broadcasting. A relatively inexpensive demodulator is very important in view of the several billions of radio receivers. With only a few broadcast transmitters and numerous receivers, the power inefficiency of conventional AM is justified. Thus, it is cost-effective to construct powerful transmitters and sacrifice power efficiency in order to have simpler signal demodulation at the receivers.

Demodulation of DSB SC AM signals requires a *synchronous demodulator,* which is also known as *coherent* or *synchronous detector.* That is, the demodulator must use a coherent phase reference. This is usually generated by means of a *phase-locked loop* (PLL), which forces the phase of the voltage-controlled oscillator to follow the phase of the reference signal, to demodulate the received signal. The need for such a device is a chief disadvantage of the DSB SC scheme.

Figure 15.2.9 shows the general configuration. A PLL is utilized to generate a phase-coherent carrier signal that is mixed with the received signal in a balanced modulator. The output of the balanced modulator is fed into a low-pass filter of bandwidth W, which passes the desired signal and rejects all signal (and noise) components above W Hz.

Demodulation of SSB signals also requires the use of a phase-coherent reference. Figure 15.2.10 shows the general configuration to demodulate the SSB signal. A small carrier component, which is transmitted along with the message, is inserted. A balanced modulator is used for frequency conversion of the bandpass signal to low pass or baseband.

Demodulation of VSB signals generally uses a synchronous detector in the configuration of Figure 15.2.10. In VSB, a carrier component is generally transmitted along with the message sidebands. The existence of the carrier component makes it possible to extract a phase-coherent reference for demodulation in a balanced modulator. In some applications, such as television broadcasting, however, a large carrier component is transmitted along with the message in the VSB signal, and in such a case it is possible to recover the message by passing the received signal through an envelope detector.

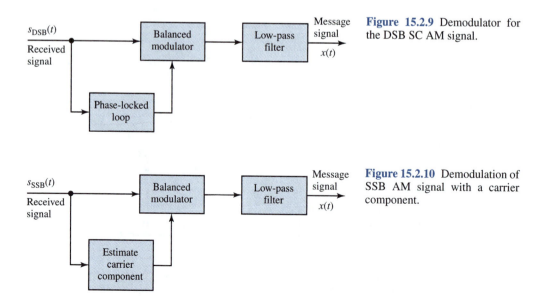

Figure 15.2.9 Demodulator for the DSB SC AM signal.

Figure 15.2.10 Demodulation of SSB AM signal with a carrier component.

From the *noise performance* point of view, all the AM systems yield the same performance for the same messages and transmitted powers. Note, however, that the bandwidth of the DSB system is twice that of the SSB or VSB systems, so its input noise power is twice as large.

Frequency Modulation

So far we have considered AM of the carrier as a means for transmitting the message signal. AM methods are also known as *linear modulation methods*, although conventional AM is not linear in the strict sense. Other classes of modulation methods are frequency modulation (FM) and phase modulation (PM). In FM systems the frequency of the carrier f_c is changed by the message signal, and in PM systems the phase of the carrier is changed according to the variations of the message signal. FM and PM, which are quite *nonlinear*, are referred to as *angle-modulation methods*. Angle modulation is more complex to implement and much more difficult to analyze because of its inherent nonlinearity. FM and PM systems generally expand the bandwidth such that the effective bandwidth of the modulated signal is usually many times the bandwidth of the message signal. The major benefit of these systems is their high degree of noise immunity. Trading off bandwidth for high-noise immunity, the FM systems are widely used in high-fidelity music broadcasting and point-to-point communication systems where the transmitter power is rather limited.

A sinusoid is said to be frequency-modulated if its instantaneous angular frequency $\omega_{FM}(t)$ is a linear function of the message,

$$\omega_{FM}(t) = \omega_c + k_{FM} f(t) \tag{15.2.9}$$

where k_{FM} is a constant with units of radians per second per volt when $f(t)$ is a message-signal voltage, and ω_c is the carrier's nominal angular frequency. Instantaneous phase being the integral of instantaneous angular frequency, the FM signal can be expressed as

$$S_{FM}(t) = A_c \cos\left[\omega_c t + \phi_c + k_{FM} \int f(t)\, dt\right] \tag{15.2.10}$$

where A_c is a constant amplitude and ϕ_c is an arbitrary constant phase angle. The maximum amount of deviation that $\omega_{FM}(t)$ of Equation (15.2.9) can have from its nominal value is known as *peak frequency deviation*, given by

$$\Delta\omega = k_{FM} |f(t)|_{max} \tag{15.2.11}$$

Because FM involves more than just direct frequency translation, spectral analysis and bandwidth calculations are difficult in general, except for a few message forms. However, practical experience indicates that the following relations hold for the FM transmission bandwidth:

$$W_{FM} \cong 2(\Delta\omega + W_f) \tag{15.2.12}$$

known as *Carson's rule* for *narrow-band* FM with $\Delta\omega < W_f$ or

$$W_{FM} \cong 2(\Delta\omega + 2W_f) \tag{15.2.13}$$

for *wide-band* FM with $\Delta\omega \gg W_f$, where W_f is the spectral extent of $f(t)$; i.e., the message signal has a low-pass bandwidth W_f. For example, commercial FM broadcasting utilizes $W_f = 2\pi(15 \times 10^3)$ rad/s (corresponding to 15 kHz) and $\Delta\omega = 5W_f$ such that $\omega_{FM} = 14W_f$. Because the performance of narrow-band FM with noise is roughly equivalent to that of AM systems, only wide-band FM that exhibits a marked improvement will be considered here.

Figure 15.2.11 illustrates the close relationship between FM and PM. Phase modulating the integral of a message is equivalent to the frequency modulation of the original message,

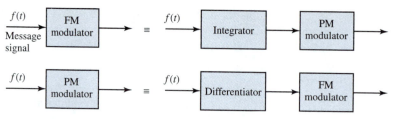

Figure 15.2.11 Close relationship between FM and PM.

and frequency modulating the derivative of a message is equivalent to phase modulation of the message itself.

Generation of wide-band FM can be done by various means. However, only the most common and conceptually the simplest one, known as the *direct method*, is considered here. It employs a *voltage-controlled oscillator* (VCO) as a modulator. A VCO is an oscillator whose oscillation frequency is equal to the resonant frequency of a tuned circuit, as shown in Figure 15.2.12. The frequency can be varied if either the inductance or the capacitance is made voltage-sensitive to the message signal $f(t)$.

One approach to obtaining a voltage-variable reactance is through a *varactor diode*, whose capacitance changes with the applied voltage, such as the junction capacitance of a reverse-biased diode, which depends on the amount of bias. Figure 15.2.13 illustrates the varactor-diode implementation of an angle modulator. The frequency of the tuned circuit and the oscillator will change in accordance with the message signal $f(t)$. Varactor-controlled VCOs can have a nearly linear frequency–voltage characteristic, but often yield only small frequency deviations, i.e., small $\Delta\omega$. Since any VCO is inherently unstable as its frequency is varied to produce FM, it becomes necessary in many applications to stabilize the carrier's frequency.

Demodulators for FM

Figure 15.2.14 shows a block diagram of a general FM demodulator, which is implemented by generating an AM signal whose amplitude is proportional to the instantaneous frequency of the FM signal, and then using an AM demodulator to recover the message signal. Transforming

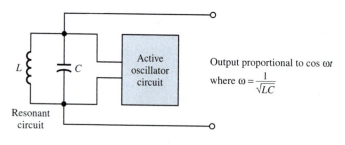

Figure 15.2.12 Oscillator for use in a VCO.

Output proportional to cos ωt

where $\omega = \dfrac{1}{\sqrt{LC}}$

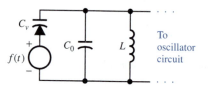

Figure 15.2.13 Varactor-diode implementation of an angle modulator.

the FM signal into an AM signal can be achieved by passing the FM signal through an LTI system whose frequency response is nearly a straight line in the frequency band of the FM signal.

FM demodulators mainly fall into two categories: *frequency discriminators* and *locked-loop demodulators*. While both give the same performance for relatively high signal levels, locked-loop demodulators provide better performance than the discriminator when signal levels are low and noise is a problem. The discriminator produces an output voltage proportional to the frequency variations that occur in the FM signal at its input. Locked-loop demodulators are more cost-efficient when implemented in IC form.

A *balanced discriminator* with the corresponding frequency characteristics is depicted in Figure 15.2.15. The rising half of the frequency characteristic of a tuned circuit, shown in Figure 15.2.15(b), may not have a wide enough linear region. In order to obtain a linear characteristic over a wider range of frequencies [see Figure 15.2.15(d)], usually two circuits tuned at two frequencies f_1 and f_2 [with the frequency response shown in Figure 15.2.15(c)] are used in a balanced discriminator [Figure 15.2.15(a)].

Figure 15.2.16 shows a block diagram of an FM demodulator with feedback (FMFB), in which the FM discrimination is placed in a feedback system that uses a VCO path for feedback. The bandwidth of the discriminator and the subsequent low-pass filter is matched with that of the message signal, which is the output of the low-pass filter.

An alternative to the FMFB demodulator is the use of a PLL, as shown in Figure 15.2.17, in which the phase of the VCO's output signal is forced to follow (or lock to) the phase of the input FM waveform with small error. Since the VCO acts as an integrator, and phase is the integral of frequency, the amplified error voltage appearing at the VCO input will be proportional to the message signal $f(t)$. The filter is selected with a closed-loop bandwidth that is wide enough to yield demodulation with minor distortion of $f(t)$, and narrow enough to reject noise.

The signal and noise components, particularly at low SNRs, are so intermingled that one may not be able to distinguish the signal from the noise. In such a case, a *mutilation* or *threshold effect* is said to be present. There exists a specific SNR at the input of the demodulator known as the *threshold SNR*, beyond which signal mutilation occurs. The threshold effect then places an upper limit on the tradeoff between bandwidth and power in an FM system. Since the thresholds for locked loops are lower than for the discriminator, loop-type receivers, operating at smaller signal-power levels, find wide application in space communications where transmitter power is at a premium.

At the output of the discriminator in an FM receiver, higher frequency components of output *noise* power are accentuated. A low-pass filter, known as a *deemphasis filter*, is added so that the large-amplitude noise can be greatly reduced and the output SNR increased. Since the filter also acts on the message, causing distortion, the message at the transmitter is passed through a compensating filter, called a *preemphasis filter*, before modulation occurs. It accentuates the higher frequencies in the message so as to exactly compensate for the effect of the deemphasis filter, so that there is no overall effect on the message (see also Section 14.3). The scheme is illustrated in Figure 15.2.18.

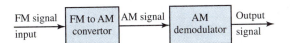

Figure 15.2.14 Block diagram of a general FM demodulator.

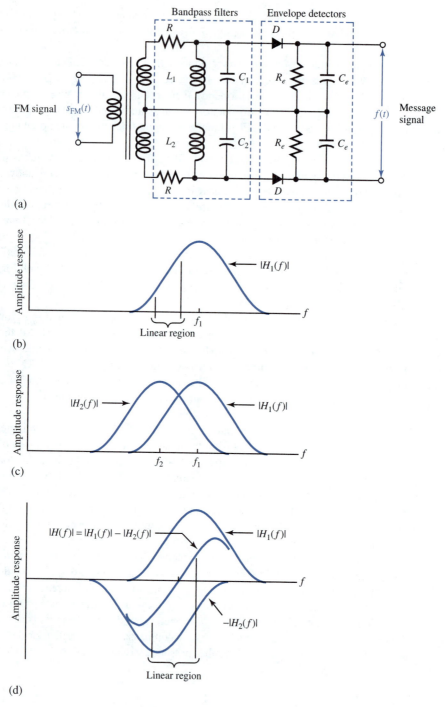

Figure 15.2.15 Balanced discriminator with the corresponding frequency response.

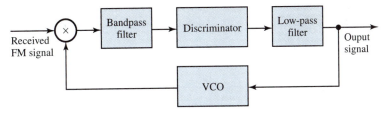

Figure 15.2.16 Block diagram of an FMFB demodulator.

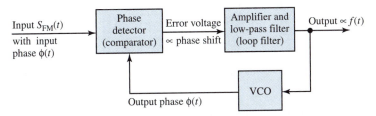

Figure 15.2.17 Block diagram of a PLL FM demodulator.

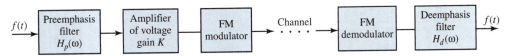

Figure 15.2.18 Schematic block diagram of an FM system with emphasis filters.

EXAMPLE 15.2.2

A common deemphasis filter used in FM broadcast has a transfer function

$$H_d(\omega) = \frac{1}{1 + j(\omega/W_1)}$$

where $W_1/2\pi = 2.12$ kHz. For perfect message recovery, the preemphasis filter must have a transfer function

$$H_p(\omega) = \frac{1}{H_d(\omega)} = 1 + j\left(\frac{\omega}{W_1}\right)$$

over all important frequencies (out to about 15–20 kHz). The system performance improvement with these filters for voice-type messages is given by

$$R_{FM} = \frac{\text{SNR with emphasis}}{\text{SNR with no emphasis}} = \frac{(W_f/W_1)^3}{3\left[(W_f/W_1) - \tan^{-1}(W_f/W_1)\right]}$$

where W_f is the spectral extent of $f(t)$. However, for broader band messages, such as music audio, since the channel bandwidth in FM broadcast is limited to 200 kHz, a reduction in the improvement factor occurs, when it can be shown that

$$\frac{R_{FM} \text{ with bandwidth limitation}}{R_{FM} \text{ with no bandwidth limitation}} = \frac{1}{1 + (W_{\text{rms}}/W_f)^2(W_f/W_1)^2}$$

where W_{rms} is the rms bandwidth of $f(t)$. Given $W_1/2\pi = 2.12$ kHz, $W_f/2\pi = 15$ kHz, and $W_{rms}/2\pi = 4.25$ kHz, determine R_{FM} with and without bandwidth constraints.

Solution

$$\underset{\text{(with no bandwidth constraint)}}{R_{FM}} = \frac{(15/2.12)^3}{3\left[(15/2.12) - \tan^{-1}(15/2.12)\right]} \cong 20.92, \text{ or } 13.2 \text{ dB}$$

$$\underset{\text{(with bandwidth limitation)}}{R_{FM}} = \frac{20.92}{1 + (4.25/15)^2(15/2.12)^2} \cong 4.17, \text{ or } 6.2 \text{ dB}$$

Thus, the bandwidth constraint has resulted in a loss in emphasis improvement of $13.2 - 6.2 = 7$ dB.

FM Stereo

Figure 15.2.19 shows the block diagram of an FM stereo transmitter and an FM stereo receiver. The following notation is used:

- $f_L(t), f_R(t)$: Left and right messages that undergo preemphasis and are then added to yield $f_1(t)$ and differenced to give $f_d(t)$.
- $f_2(t)$: Given by signal $f_d(t)$ when it *DSB*-modulates (with carrier suppressed) a 38-kHz subcarrier.
- $f_3(t)$: A low-level pilot carrier at 19 kHz that is included to aid in the receiver's demodulation process.
- $f_s(t)$: Final composite message when $f_1(t), f_2(t), f_3(t)$, and SCA (*subsidiary communications authorization*) are all added up.
- SCA: A narrow-band FM waveform on a 67-kHz subcarrier with a total bandwidth of 16 kHz. It is a special signal available to fee-paying customers who may desire to have background music free of commercials or nonaudio purposes such as paging.
- NBPF: Narrow-band band-pass filter.
- LPF, BPF, and NBPF: Appropriate filters that select the spectrum portions corresponding to $f_1(t), f_2(t)$, and $f_3(t)$, respectively.

The output signal-to-noise power ratio is smaller in FM stereo than in a monaural system with the same transmitted power, messages, and other parameters. With a loss as high as 22 dB, many FM stations can tolerate the loss because of the high power being transmitted.

Comparison of Analog-Modulation Systems

The comparison of analog modulation systems (conventional AM, DSB SC, SSB SC, VSB, FM, and PM) can be based on three practical criteria:

1. Bandwidth efficiency of the system.
2. Power efficiency of the system, as reflected in its performance in the presence of noise. The output SNR at a given received signal power is used as a criterion for comparing the power efficiencies of various systems.
3. Ease of implementation of the system (transmitter and receiver).

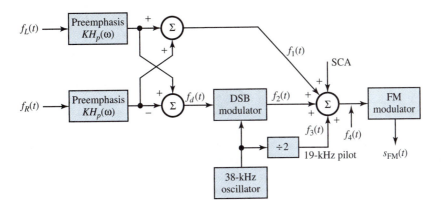

(a)

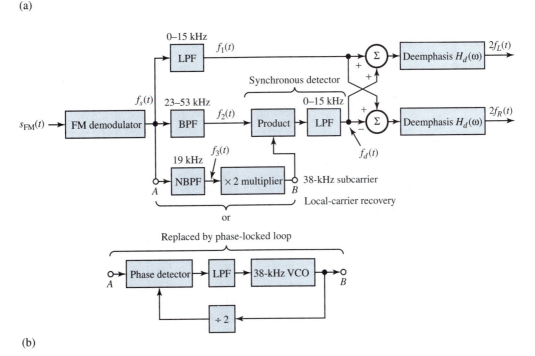

(b)

Figure 15.2.19 (a) FM stereo transmitter. (b) FM stereo receiver.

The most bandwidth-efficient analog communication system is the SSB SC system with a transmission bandwidth equal to the signal bandwidth. In bandwidth-critical applications, such as voice transmission over microwave and satellite links and some point-to-point communication systems in congested areas, this system is used widely. When transmission signals have a significant dc component, such as image signals, SSB SC cannot be used because it cannot effectively transmit direct current. A good compromise is the VSB system (with its bandwidth slightly larger than SSB and a capability for transmitting dc values), which is widely used in television broadcasting and some data-communication systems. When bandwidth is the major concern, PM and particularly FM systems are least favorable. Only in terms of their high level of noise immunity, their usage may sometimes be justified.

FM, with its high level of noise immunity, and hence power efficiency, is widely used on high-fidelity radio broadcasting and power-critical communication links such as point-to-point communication systems. It is also employed for satellite links and voice transmission on microwave LOS systems. When the transmitted power is a major concern, conventional AM and VSB (being the least power-efficient systems) are not used, unless their development can be justified by the simplicity of the receiver structure.

From the viewpoint of ease of implementation, the simplest receiver structure is that of conventional AM. Standard AM, VSB, and FM are widely used for AM, television, and high-fidelity FM broadcasting. The relative power inefficiency of the AM transmitter is compensated for by the extremely simple structure of several billions of receivers. The receiver structure is much more complicated for DSB SC and SSB SC systems, since they require synchronous demodulation. These systems, therefore, are never used for broadcasting purposes. Note that DSB SC also suffers from its relative bandwidth inefficiency.

Radio and Television Broadcasting

Radio (AM and FM) and television broadcasting are the most familiar forms of communication via analog transmission systems. The receiver most commonly used in AM radio broadcasting is the *superheterodyne receiver,* shown in Figure 15.2.20, which consists of a radio-frequency (RF) tuned amplifier, mixer, local oscillator, intermediate frequency (IF) amplifier, envelope detector, audio-frequency amplifier, and a loudspeaker. Tuning at the desired radio frequency f_c is achieved by a variable capacitor, which simultaneously tunes the RF amplifier and the frequency f_{LO} of the local oscillator. Every AM radio signal, in a superheterodyne receiver, is converted to a common IF frequency of $f_{IF} = |f_c - f_{LO}| = 455$ kHz, which allows the use of a single tuned IF amplifier for signals from any radio station in the frequency band. Matching the bandwidth of the transmitted signal, the IF amplifier is set to have a bandwidth of 10 kHz.

The frequency conversion to IF is done by the combination of the RF amplifier and the mixer. The tuning range of the local oscillator is 955–2,055 kHz. f_{LO} could be either higher or lower than f_c. If f_{LO} is higher than f_c, then $f_{LO} = f_c + f_{IF}$. By tuning the RF amplifier to the frequency f_c and mixing its output with the local oscillator frequency f_{LO}, we obtain two signal components: one centered at the difference frequency f_{IF}; and the other centered at the sum frequency $2f_c + f_{IF}$, known as the *image frequency.* Only the first component is passed on by the IF amplifier. The

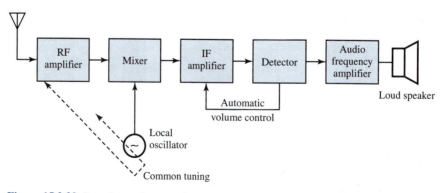

Figure 15.2.20 Superheterodyne receiver.

image response is suppressed by using the antenna and other RF-tuned circuits as a filter. By limiting the bandwidth of the RF amplifier to the range $B_c < B_{RF} < 2f_{IF}$, where B_c is the bandwidth of the AM radio signal (10 kHz), the radio signal transmitted at the image frequency ($f_c' = f_{LO} + f_{IF}$) is rejected. A similar behavior occurs when f_{LO} is lower than f_c. Figure 15.2.21 illustrates the AM station and image frequencies for a high-side and a low-side local oscillator, whereas Figure 15.2.22 depicts the frequency response characteristics of IF and RF amplifiers for the case $f_{LO} > f_c$.

Amplifiers in the IF circuits provide most of the gain needed to raise the small antenna signal to a level sufficient to drive the envelope detector. The output of the detector contains a dc component proportional to A_c and a component proportional to the audio message $f(t)$, the amplified signal of which is used to drive the loudspeaker. The dc component is utilized in an *automatic volume control* (AVC), otherwise known as *automatic gain control* (AGC), loop to control the gain of RF and IF amplifiers by controlling their operating bias points. The loop action is to maintain nearly a constant IF level at the detector's input, even for large variations in antenna voltage.

The IF amplifier, with its narrow bandwidth, provides signal rejection from adjacent channels, and the RF amplifier provides signal rejection from image channels.

An *FM radio superheterodyne receiver* is shown in block diagram form in Figure 15.2.23. The part consisting of the antenna, RF amplifier, mixer, and local oscillator functions in a manner similar to that of an AM receiver, except that the frequencies involved are different. $f_{IF} = 10.7$ MHz in FM, so that the image is 21.4 MHz from the carrier frequency f_c. The RF amplifier must eliminate the image-frequency band $2f_{IF}$ away from the station to which the receiver is tuned.

The IF amplifier is generally divided into two parts. The higher level stage is set to limit at a proper level to drive the demodulator. More expensive FM receivers may have AGC added to

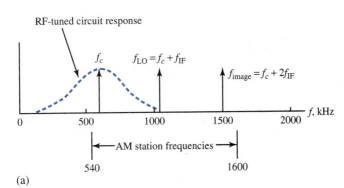

(a)

(b)

Figure 15.2.21 AM radio station and image frequencies. (a) High-side local oscillator, $f_{LO} > f_c$. (b) Low-side local oscillator, $f_{LO} < f_c$.

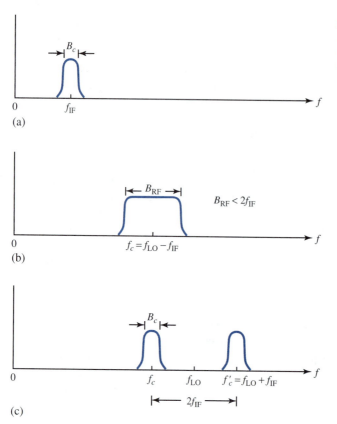

Figure 15.2.22 Frequency response characteristics of IF and RF amplifiers. **(a)** IF amplifier. **(b)** RF amplifier. **(c)** Desired signal and image signal to be rejected.

reduce the gains of the RF and IF amplifier stages. A heavily filtered output from the demodulator is often used to provide an automatic frequency control (AFC) loop through the local oscillator that can be implemented to have electronic tuning by using a varactor. After manual tuning, the AFC loop locks the receiver to the selected station. Finally, the response of the demodulator is fed into the stereo demodulator, which is implemented as shown in Figure 15.2.19(b).

Television signals in television signal transmission are the electric signals generated by converting visual images through *raster (TV image area) scanning*. The two-dimensional image or picture is converted into a one-dimensional electric signal by sequentially scanning the image and producing an electrical signal that is proportional to the brightness level of the image. A television camera, which optically focuses the image on a photo cathode tube that consists of a photosensitive surface, is used for scanning. An electron beam produces an output current or voltage that is proportional to the brightness of the image, known as a *video signal*. The scanning of the electron beam is controlled by two voltages, as shown in Figure 15.2.24, applied across the horizontal and vertical deflection plates. In the raster scanning in an NTSC TV system, the image is divided into 525 lines which define a *frame*, as illustrated in Figure 15.2.25. The resulting signal is transmitted in $1/30$ second. The number of lines determines the picture resolution and, along with the rate of transmission, sets the channel bandwidth needed for image transmission. However, the time interval of $1/30$ second to transmit a complete image is not generally fast enough to avoid flickering, which is annoying to the average viewer. Therefore, to overcome the flickering, the scanning of the image is performed in an *interlaced pattern*, as shown in Figure

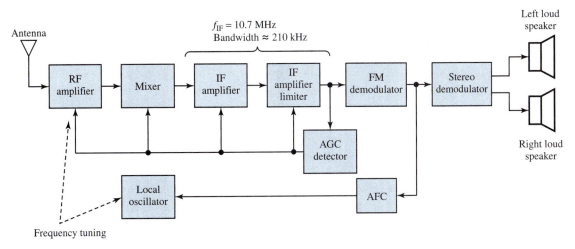

Figure 15.2.23 Block diagram of a superheterodyne FM radio receiver.

15.2.26, consisting of two fields, each of 262.5 lines. Each field is transmitted in $1/60$ second. The first field begins at point a and terminates at point b, whereas the second field begins at point c and terminates at point d.

The image is scanned left to right and top to bottom in a system of closely spaced parallel lines. When 242.5 lines are completed at the rate of 15,734.264 lines per second (63.556 μs per line), the raster's visual area is scanned once; this scan is called a *field*. While the next 20 lines are not used for visual information, during that time of 1.27 ms special signals (testing, closed captions, etc.) are inserted and the beam is retraced vertically to begin a new field (shown as the second field in Figure 15.2.26). The raster (TV image area) has a standardized *aspect ratio* of four units of width for each three units of height. Good performance is achieved when the raster is scanned with 525 lines at a rate of 29.97 frames per second.

The television waveform representing one scan is illustrated in Figure 15.2.27. A blanking pulse with a duration of 0.18 of the horizontal-sweep period T_h is added to the visual voltage generated by the camera. While the blanking pulse turns off the electron beam in the receiver's picture tube during the horizontal retrace time, an added sync (synchronization) pulse helps the receiver to synchronize its horizontal scanning rate with that of the transmitter. Also, a burst of at least 8 cycles of 3.579545 MHz, called the *color burst*, is added to the "back porch" of the blanking pulse for synchronizing the receiver's color circuits. The visual information fluctuates according to the image between the "black level" and the "white level" set at 70.3% and 12.5%, respectively, of the peak amplitude. An array of various sync pulses are added on top for both horizontal and vertical synchronization purposes.

If a filter is added to the television camera optics, so that only the red color passes through, the camera's voltage becomes proportional to the intensity of the amount of red in the image. Three such cameras, all synchronized and viewing the same image, are employed in color television to decompose the image into its primary color components of red R, green G, and blue B. The color receiver utilizes a picture tube with three electron beams and a phosphor having R, G, and B components. While each beam excites one color of phosphor, at any spot in the image the three colors separately glow with proper intensities in response to the three transmitted color signals. The viewer's eye effectively adds the three colors together to reproduce the original scene in color.

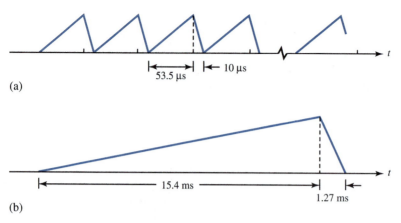

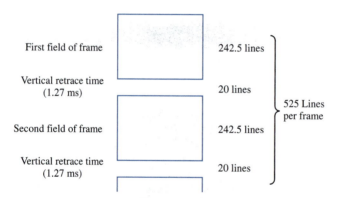

Figure 15.2.24 Voltage waveforms. **(a)** Applied to horizontal deflection plate. **(b)** Applied to vertical deflection plate.

First field of frame — 242.5 lines

Vertical retrace time (1.27 ms) — 20 lines

Second field of frame — 242.5 lines

Vertical retrace time (1.27 ms) — 20 lines

525 Lines per frame

Figure 15.2.25 Raster scanning in an NTSC television system.

Now that the fundamentals needed in color television have been explained, it remains to be seen how the signals are processed by the transmitter and the receiver.

A *color television transmitter* in a transmitting station is shown in Figure 15.2.28 in a block diagram indicating the most important functions. A mixture of three primary-color signals (having the visual signal bandwidth of about 4.2 MHz) are transmitted in the standard color television system in terms of the following three linearly independent combinations generated by the matrix circuit:

$$m_Y(t) = 0.30 m_R(t) + 0.59 m_G(t) + 0.11 m_B(t) \qquad (15.2.14)$$

$$m_I(t) = 0.60 m_R(t) - 0.28 m_G(t) - 0.32 m_B(t) \qquad (15.2.15)$$

$$m_Q(t) = 0.21 m_R(t) - 0.52 m_G(t) + 0.31 m_B(t) \qquad (15.2.16)$$

The following notation is being used:

- $m_Y(t)$: Luminance signal, to which monochrome receivers respond, and which defines the brightness (white or gray level) of the image.

- $m_I(t)$, $m_Q(t)$: Chrominance signals, which relate only to the color content of the image and have bandwidths of about 1.6 and 0.6 MHz, respectively.

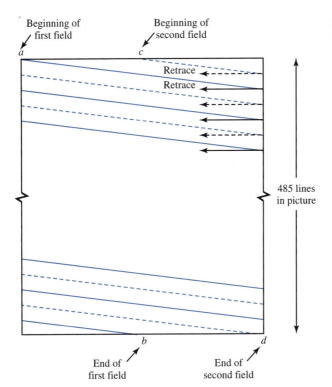

Beginning of
first field

Beginning of
second field

a

c

Retrace

Retrace

485 lines
in picture

b

d

End of
first field

End of
second field

Figure 15.2.26 Interlaced scanning pattern to avoid flickering.

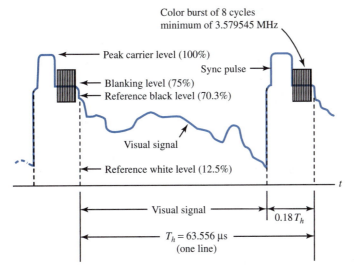

Color burst of 8 cycles
minimum of 3.579545 MHz

Peak carrier level (100%)

Sync pulse

Blanking level (75%)
Reference black level (70.3%)

Visual signal

Reference white level (12.5%)

t

Visual signal

$0.18\,T_h$

$T_h = 63.556\ \mu s$
(one line)

Figure 15.2.27 TV waveform representing one horizontal line of a raster scan.

- $[m_I^2(t) + m_Q^2(t)]^{1/2}$: Saturation or color intensity. A very deep red is saturated while red diluted with white to give a light pink is nearly unsaturated.

- $\tan^{-1}[m_Q(t)/m_I(t)]$: Hue or tint.

- $S_I(t)$, $S_Q(t)$: Filtered chrominance signals of $m_I(t)$, $m_Q(t)$ by low-pass filters. $S_I(t)$ modulates a color subcarrier at a frequency of 3.579545 MHz $\pm$ 10 Hz via DSB.

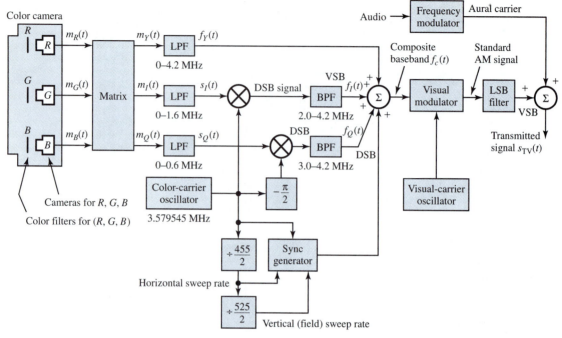

Figure 15.2.28 Color TV transmitter in a TV-transmitting station.

- $f_I(t)$: Nearly a VSB signal, when the DSB signal is filtered by the BPF of passband 2–4.2 MHz to remove part of the USB in the DSB.

- $f_Q(t)$: DSB signal that is produced when the other chrominance signal modulates a quadrature-phase version of the color subcarrier. This DSB signal passes directly through the BPF with passband 3–4.2 MHz without any change.

- $f_Y(t)$: Filtered luminance signal of $m_Y(t)$ by an LPF.

- $f_c(t)$: Composite baseband waveform by adding $f_Y(t)$, $f_I(t)$, $f_Q(t)$, and sync pulses. This has a bandwidth of about 4.2 MHz and modulates a visual carrier by standard AM.

The standard AM signal is then filtered to remove part of the lower sideband. The resulting VSB signal and the audio-modulated aural carrier are added to form the final transmitted signal $S_{TV}(t)$. Figure 15.2.29 illustrates the spectrum of a color television signal.

A *color television receiver* is shown in Figure 15.2.30 in block diagram form, indicating only the basic functions. The early part forms a straightforward superheterodyne receiver, except for the following changes:

- The frequency-tuning local oscillator is typically a push-button-controlled frequency synthesizer.

- IF circuitry in television is tuned to give a filter characteristic required in VSB modulation.

The filter shapes the IF signal spectrum so that envelope detection is possible. The output of the envelope detector contains the composite visual signal $f_c(t)$ and the frequency-modulated aural carrier at 4.5 MHz. The latter is processing in a frequency demodulator to recover the audio information for the loudspeaker. The former is sent through appropriate filters to separate

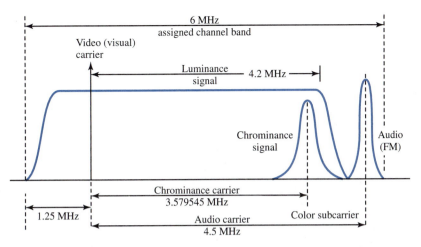

Figure 15.2.29 Spectrum of a color television signal.

out signals $f_Y(t)$ and $[f_I(t) + f_Q(t)]$, which is further processed by two synchronous detectors in quadrature to recover $S_I(t)$ and $S_Q(t)$. An appropriate matrix combines $f_Y(t)$, $S_I(t)$, and $S_Q(t)$ to yield close approximations of the originally transmitted $m_R(t)$, $m_G(t)$, and $m_B(t)$. These three signals control the three electron beams in the picture tube.

The output of the envelope detector is also applied to circuits that separate the sync signals needed to lock in the horizontal and vertical sweep circuits of the receiver. The bursts of color carriers are isolated such that a PLL can lock to the phase of the color carrier, and thereby provide the reference signals for the chrominance synchronous detectors.

Mobil Radio Systems (Cellular Telephone Systems)

Today radio-based systems make it possible for mobile people to communicate via cellular telephone systems while traveling on airplanes and motor vehicles. For radio telephone service, the FCC in the United States has assigned parts of the UHF band in the range of 806–890 MHz. For efficient use of the available frequency spectrum, especially in highly populated metropolitan areas with the greatest demand for mobile telephone services, the cellular radio concept has been adopted, in which a geographic area is subdivided into *cells*, each of which contains a *base station*, as shown in Figure 15.2.31. Each base station is connected by telephone lines to a *mobile telephone switching office* (MTSO), which in turn is connected through telephone lines to a telephone central office of the terrestrial telephone network.

When a mobile user (identified by the telephone number and telephone serial number assigned by the manufacturer) communicates via radio with the base station within the cell, the base station routes the call through the MTSO to another base station if the called party is located in another cell, or to the central office if the called party is not mobile. Once the desired telephone number is keyed and the "send" button is pressed, the MTSO checks the authentication of the mobile user and assigns (via a supervisory control channel) an available frequency channel for radio transmission of the voice signal from the mobile telephone to the base station. A second frequency is assigned for radio transmission from the base station to the mobile user. Simultaneous transmission between two parties is known as *full-duplex operation*. In order to complete the connection to the called party, the MTSO interfaces with the central office of the telephone network by means of wide-

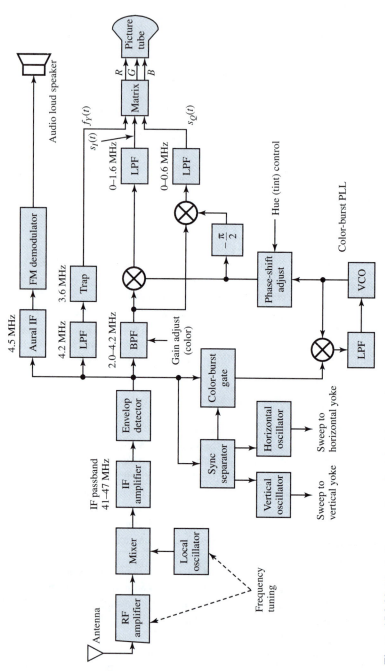

Figure 15.2.30 Block diagram of a color television receiver.

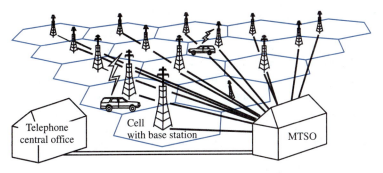

Figure 15.2.31 Cellular telephone concept in mobile radio system.

band trunk lines, which carry speech signals from many users. When the two parties hang up upon completion of the telephone call, the radio channel then becomes available for another user.

During the telephone conversation, if the signal strength drops below a preset threshold, the MTSO monitors and finds a neighboring cell that receives a stronger signal and automatically switches (in a fraction of a second) the mobile user to the base station of the adjacent cell. If a mobile user is outside of the assigned service area, the mobile telephone may be placed in a roam mode, which allows the user to initiate and receive calls.

In analog transmission between the base station and the mobile user, the 3-kHz wide audio signal is transmitted via FM using a channel bandwidth of 30 kHz. Such a large bandwidth expansion (by a factor of 10) is needed to obtain a sufficiently large SNR at the output of the FM demodulator. Since the use of FM is indeed wasteful of the radio frequency spectrum, cellular telephone systems based on digital transmission of digitized compressed speech are later developed. With the same available channel bandwidth, the system then accommodates a four- to tenfold increase in the number of simultaneous users.

Cellular systems employed cells with a radius in the range of 5–18 km. The base station usually transmitted at a power level of 35 W or less, and the mobile users transmitted at a power level of about 3 W, so that signals did not propagate beyond immediately adjacent cells. By making the cells smaller and reducing the radiated power, frequency reuse, bandwidth efficiency, and the number of mobile users have been increased. With the advent of small and powerful integrated circuits (which consume very little power and are relatively inexpensive), the cellular radio concept has been extended to various types of personal communication services using low-power hand-held sets (radio transmitter and receivers).

With *analog cellular*, or AMPS (Advanced Mobile Phone System), calls are transmitted in sound waves at 800 MHz to 900 MHz. This was the first mobile phone technology available in early 1980s. *Digital cellular*, or D-AMPS (Digital AMPS), transmits calls in bits at the same frequency as analog cellular, with improved sound quality and security. To send numerous calls at once, D-AMPS phones use either CDMA (Code Division Multiple Access) or TDMA (Time Division Multiple Access) technology; but CDMA phones won't work in TDMA areas, and vice versa. A *dual-mode* unit can switch to analog transmission outside of the more limited digital network.

PCS (Personal Communications Service) phones transmit at 1800 MHz to 1900 MHz and are smaller and more energy efficient. To get around the limited coverage, a *dual-band* digital phone (which switches to the lower digital frequency) and a *trimode* phone (which works in

AMPS, D-AMPS, or PCS areas) has been developed. The GSM (Global System for Messaging communications) is the most widely accepted transmission method for PCS phones.

The latest wireless communications technology is the *personal satellite phone.* The coverage is planetary and one can reach anywhere on earth. Examples include the Iridium satellite handset developed by Motorola and others from the Teledesic constellation. Special requirements, such as sending and receiving data or faxes, can be handled by the new handheld *computer-and-mobile-phone hybrids* such as Nokia's 9000i and the Ericsson DI27. Mobile phones with a Web browser capability are also available. We have yet to see the more exciting new developments in the telecommunications industry with computers, networking, and wireless technology.

15.3 DIGITAL COMMUNICATION SYSTEMS

A digital signal can be defined as having any one of a finite number of discrete amplitudes at any given time. The signal could be a voltage or current, or just a number such as 0 or 1. A signal for which only two amplitudes are possible is known as a *binary digital signal,* the type of which is commonly used in computers and most digital communication systems. A communication system that is designed to process only digital signals (or messages) to convey information is said to be digital. The recent trend is to make as much of the system digital as possible, because:

- Discrete data are efficiently processed.
- Analog messages can also be converted to digital form.
- Digital systems interface well with computers.
- Digital systems offer great reliability and yield high performance at low cost.
- Being flexible, digital systems can accommodate a variety of messages with ease.
- Security techniques are available to offer message privacy to users.
- Advanced signal-processing techniques can be added on.

However, the most serious disadvantages are the increased complexity needed for system synchronization and the need for larger bandwidths than in an equivalent analog system. A digital system can directly interface with a source having only discrete messages, because of the inherent characteristic of the digital system. With suitable conversion methods, however, systems currently exist that can simultaneously transmit audio, television, and digital data over the same channel. Figure 14.1.2 illustrates the basic elements of a digital communication system, which was introduced in Section 14.1.

Before we begin discussing digital systems, it is helpful to talk about the methods by which analog messages are converted into digital form. *Sampling, quantization,* and *coding* are the three operations needed for the transmission of an analog signal over a digital system.

Sampling

This method was introduced in Section 14.2. Sampling of an analog signal makes it discrete in time. A bandlimited signal can be recovered exactly from its samples, taken periodically in time at a rate at least equal to twice the signal's bandwidth. If a message $f(t)$ has a spectral extent of W_f rad/s, the sampling rate f_s (samples per second) must satisfy

$$f_s \geq \frac{W_f}{\pi} \tag{15.3.1}$$

from the sampling theorem. The minimum rate W_f/π (samples per second) is known as the *Nyquist rate.* If the exact message samples could be transmitted through the digital system, the original $f(t)$ could be exactly reconstructed at all times, with no error by the receiver. However, since the exact samples cannot be conveyed, they must be converted to discrete samples in a process known as *quantization*.

Quantization

Let Figure 15.3.1(a) illustrate a message $f(t)$ with values between 0 and 7 V. A sequence of exact samples taken at uniform intervals of time is shown: 1.60, 3.70, 4.75, 3.90, 3.45, and 5.85 V. Quantization consists of rounding exact sample values to the nearest of a set of discrete amplitudes called *quantum levels*. Assuming the quantizer to have eight quantum levels (0, 1, 2, . . . , 7 V), a sequence of *quantized samples* (2, 4, 5, 4, 3, and 6 V) is shown in Figure 15.3.1(a). Obviously, the scheme is not limited to messages with only nonnegative voltages. The quantizer is said to be *uniform* when the *step size* between any two adjacent quantum levels is a constant, denoted by δv volts. Quantizers with nonuniform step size are also designed for improved system performance. An L-level quantizer can have even or odd L. A quantizer is said to be saturated or overloaded when

$$|f(t)| > \left(\frac{L-2}{2}\right)\delta v + \delta v = \frac{L}{2}\delta v \qquad (15.3.2)$$

Figure 15.3.2 shows the output quantum levels versus input voltage characteristic (stairstep in shape) of an L-level quantizer, when the message signal has both positive and negative amplitudes of the same maximum magnitude. Case (a) corresponds to L being an even integer, when the *midriser* can be observed; and case (b) corresponds to L being an odd integer, when the *midtread* can be seen.

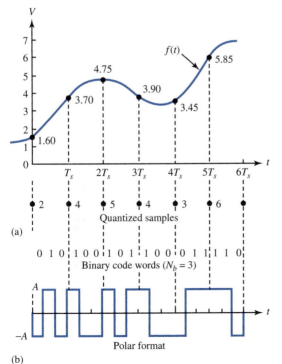

Figure 15.3.1 **(a)** Analog message $f(t)$ with exact and quantized samples. **(b)** Coding and waveform formatting of quantized samples.

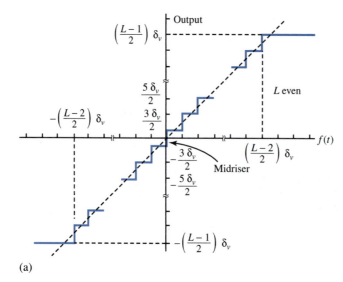

(a)

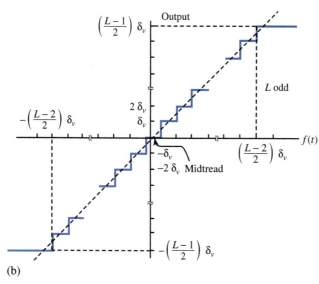

(b)

Figure 15.3.2 Uniform quantizer characteristics. (a) Midriser with even L. (b) Midtread with odd L.

EXAMPLE 15.3.1

A uniform quantizer is said to have 16 levels, and hence is called a midriser. The saturation levels are to correspond to extreme values of the message of $1 \text{ V} \le f(t) \le 17 \text{ V}$. Find the quantum levels.

Solution

The tread width is $\delta v = (17 - 1)/16 = 1$ V. The first quantum level is then given by $1 + (\delta v/2) = 1.5$ V. Other quantum levels, denoted by l_i, are given by

$$l_i = 1.5 + (i - 1)\delta v = 1.5 + (i - 1)1, \qquad i = 1, 2, \ldots, 16$$

Quantization Error

Sampling followed by quantization is equivalent to quantization followed by sampling. Figure 15.3.3 illustrates a message signal $f(t)$ and its quantized version denoted by $f_q(t)$. The difference between $f_q(t)$ and $f(t)$ is known as the *quantization error* $\varepsilon_q(t)$,

$$\varepsilon_q(t) = f_q(t) - f(t) \tag{15.3.3}$$

Theoretically, $f_q(t)$ can be recovered in the receiver without error. The recovery of $f_q(t)$ can be viewed as the recovery of $f(t)$ with an error (or noise) $\varepsilon_q(t)$ present. For a small δv with a large number of levels, it can be shown that the mean-squared value of $\varepsilon_q(t)$ is given by

$$\overline{\varepsilon_q^2(t)} = \frac{(\delta v)^2}{12} \tag{15.3.4}$$

When a digital communication system transmits an analog signal processed by a uniform quantizer, the best SNR that can be attained is given by

$$\left(\frac{S_o}{N_q}\right) = \frac{\overline{f^2(t)}}{\overline{\varepsilon_q^2(t)}} = \frac{12\,\overline{f^2(t)}}{(\delta v)^2} \tag{15.3.5}$$

where S_0 and N_q represent the average powers in $f(t)$ and $\varepsilon_q(t)$, respectively. When $f(t)$ fluctuates symmetrically between equal-magnitude extremes, i.e., $-|f(t)|_{\max} \leq f(t) \leq |f(t)|_{\max}$, choosing a sufficiently large number of levels L, the step size δv comes out as

$$\delta v = \frac{2\,|f(t)|_{\max}}{L} \tag{15.3.6}$$

and the SNR works out as

$$\left(\frac{S_0}{N_q}\right) = \frac{3L^2\,\overline{f^2(t)}}{|f(t)|_{\max}^2} \tag{15.3.7}$$

By defining the message *crest factor* K_{CR} as the ratio of peak amplitude to rms value,

$$K_{CR}^2 = \frac{|f(t)|_{\max}^2}{\overline{f^2(t)}} \tag{15.3.8}$$

Equation (15.3.7) can be rewritten as

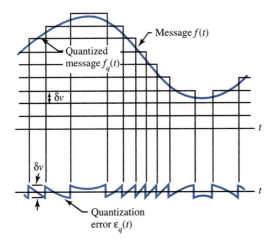

Figure 15.3.3 Message signal $f(t)$, its quantized version $f_q(t)$, and quantization error $\varepsilon_q(t)$.

Message $f(t)$

Quantized message $f_q(t)$

δv

t

δv

t

Quantization error $\varepsilon_q(t)$

$$\left(\frac{S_o}{N_q}\right) = \frac{3L^2}{K_{CR}^2} \tag{15.3.9}$$

It can be seen that messages with large crest factors will lead to poor performance.

Companding

In order to lower the crest factor of a waveform, so as to produce better performance, a process known as *companding* is used. It works like a *compressor* which progressively compresses the larger amplitudes of a message when it is passed through a nonlinear network. The inverse operation in the receiver is known as an *expandor,* when it restores the original message. Figure 15.3.4 illustrates a typical set of input–output characteristics for a form of compandor. One can see that the action of the compressor is to increase the rms–signal value for a given peak magnitude.

Source Encoding

After the quantization of message samples, the digital system will then code each quantized sample into a sequence of binary digits (bits) 0 and 1. Using the *natural binary code* is a simple approach. For a code with N_b bits, integers N (from 0 to $2^{N_b} - 1$) are represented by a sequence of digits, $b_{N_b}, b_{N_b-1}, \ldots, b_2, b_1$, such that

$$N = b_{N_b}(2^{N_b-1}) + \ldots + b_2(2^1) + b_1(2^0) \tag{15.3.10}$$

Note that b_1 is known as the least significant bit (LSB), and b_{N_b} as the most significant bit (MSB). Since a natural binary code of N_b bits can encode $L_b = 2^{N_b}$ levels, it follows that

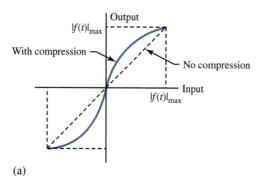

(a)

Figure 15.3.4 Compandor input–output characteristics. **(a)** Compressor. **(b)** Expandor.

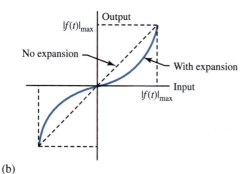

(b)

$$L \le L_b = 2^{N_b} \tag{15.3.11}$$

if L levels span the message variations. For example, in Figure 15.3.1(b), $N_b = 3$ and $L = 8$; the binary code word 010 represents $0(2^{3-1}) + 1(2^1) + 0(2^0) = 2$ V, and 110 represents $1(2^{3-1}) + 1(2^1) + 0(2^0) = 6$ V. Thus, binary code words with $N_b = 3$ are shown in Figure 15.3.1(b).

EXAMPLE 15.3.2

A symmetrical fluctuating message, with $|f(t)|_{max} = 6.3$ V and $K_{CR} = 3$, is to be encoded by using an encoder that employs an 8-bit natural binary code to encode 256 voltage levels from -7.65 V to $+7.65$ V in steps of $\delta v = 0.06$ V. Find L, $\overline{f^2(t)}$, and S_0/N_q.

Solution

From Equation (15.3.6),

$$L = \frac{2|f(t)|_{max}}{\delta v} = \frac{2 \times 6.3}{0.06} = 210$$

From Equation (15.3.8),

$$\overline{f^2(t)} = \frac{|f(t)|_{max}^2}{K_{CR}^2} = \frac{6.3^2}{9} = 4.41 \text{ V}^2$$

From Equation (15.3.9),

$$\left(\frac{S_0}{N_q}\right) = \frac{3L^2}{K_{CR}^2} = \frac{3 \times 210^2}{9} = 14{,}700 \qquad \text{or} \qquad 41.67 \text{ dB}$$

Digital Signal Formatting

After quantization and coding the samples of the message, a suitable waveform has to be chosen to represent the bits. This waveform can then be transmitted directly over the channel (if no carrier modulation is involved), or used for carrier modulation. The waveform selection process is known as formatting the digital sequence. Three kinds of waveforms are available:

1. *Unipolar waveform,* which assigns a pulse to code 1 and no pulse to code 0. The duration of a pulse is usually chosen to be equal to T_b, if binary digits occur each T_b seconds (the bit interval's duration).

2. *Polar waveform,* which consists of a pulse of duration T_b for a binary 1 and a negative pulse of the same magnitude and duration for a 0. This yields better system performance in noise than the unipolar format, because of the wider distinction between the two values.

3. *Manchester waveform,* which transmits a pulse of duration $T_b/2$ followed by an equal magnitude, but negative pulse of duration $T_b/2$ for each binary 1, and the negative of this two-pulse sequence for a binary 0. Even when a long string of 0s or 1s may occur in the digital sequence, the advantage of this format is that it never contains a dc component.

Figure 15.3.5 illustrates these formats for a sequence of binary digits. Figure 15.3.1 (b) shows the polar format corresponding to the coding in that case. Since it is important that a digital system not lose track of polarity while processing polar or Manchester waveforms, a technique called *differential encoding* (see Problem 15.3.9) is employed so as to remove the need to maintain polarity.

Because the digits in a typical digital sequence fluctuate randomly between 0s and 1s with time, the formatted waveform is then a randomly fluctuating set of pulses corresponding to the selected format. With such random waveforms, one uses the *power spectral density* (with units of V^2/Hz) to define the spectral content. On comparing the three waveform formats, the unipolar and polar formats both have the same bandwidth and relative side-lobe level, whereas the Manchester waveform has no spectral component at direct current, but requires twice the bandwidth of the other two signals.

Pulse-Code Modulation (PCM)

PCM is the simplest and oldest waveform coding scheme for processing an analog signal by sampling, quantizing, and binary encoding. Figure 15.3.6 shows a functional block diagram of a PCM system transmitter. In order to guarantee that the message is band-limited to the spectral extent for which the system is designed, a low-pass filter is introduced. The compressor is rather optional for better performance. Let us assume that the PCM signal is transmitted directly over the baseband channel. Corrupted by the noise generated within the receiver, the PCM signal is shown as the input to the PCM reconstruction function in Figure 15.3.7, which depicts a block diagram of functions (including an optional expandor) needed to receive PCM. The operations of the receiver are basically the inverse of those in the transmitter. The first and most critical receiver operation is to reconstruct the originally transmitted PCM signal as nearly as possible from the noise-contaminated received waveform. The effect of noise is to be minimized through a careful selection of circuit implementation.

The only knowledge required of the receiver to reconstruct the original PCM signal is whether the various transmitted bits are 0s and 1s, depending on the voltage levels transmitted, assuming that the receiver is synchronized with the transmitter. The two levels associated with unipolar

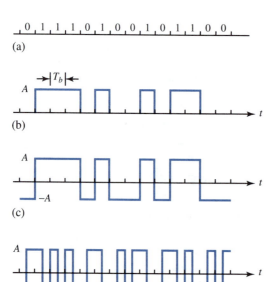

(a)

(b)

(c)

(d)

Figure 15.3.5 Waveform formats for a binary digital sequence. (**a**) Binary digital sequence $\{b_k\}$. (**b**) Its unipolar format. (**c**) Its polar format. (**d**) Its Manchester format.

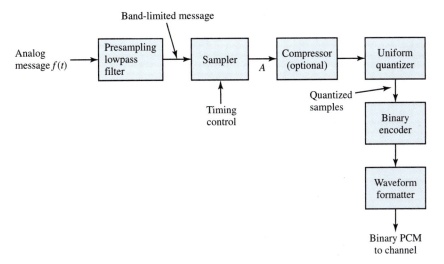

Figure 15.3.6 Block diagram of a PCM system transmitter.

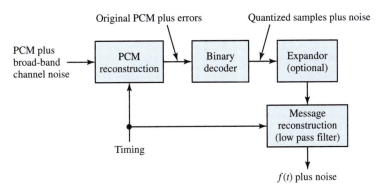

Figure 15.3.7 Block diagram of a PCM system receiver.

pulses of amplitude A are 0 and A, whereas those associated with polar pulses (of amplitudes $\pm A$) are A and $-A$. It is, of course, better for the receiver if the ratio of the pulse-caused voltage to the noise rms voltage is the largest possible at the time of measurement. Figure 15.3.8 shows PCM reconstruction circuits for unipolar, polar, and Manchester waveforms. The following notation is used:

- V_T: Preset threshold, which is zero for polar and Manchester PCM. In the unipolar system, it is equal to half the signal component of the integrator's output level ($A^2 T_b/2$) at the sampling time when the input has a binary 1. (After the sample is taken, the integrator is discharged to 0 V in preparation for integration over the next bit interval.)

- D: The difference between the integrator's output and V_T at the end of each bit interval of duration T_b. If $D \geq 0$, binary 1 is declared; if $D < 0$, a 0 is declared.

- Square-wave clock: Generates a voltage A for $0 < t < T_b/2$ and $-A$ for $T_b/2 < t < T_b$, with its fundamental frequency $1/T_b$ Hz. The product of the clock and the incoming

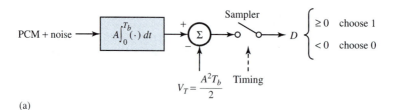

(a)

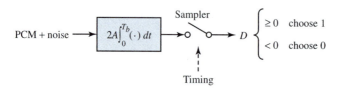

(b)

(c)

Figure 15.3.8 PCM reconstruction circuits. **(a)** For unipolar waveform. **(b)** For polar waveform. **(c)** For Manchester waveform.

Manchester PCM waveform becomes a polar PCM signal; the product with the clock inverted is the negative of a polar signal.

In Figure 15.3.8(c), after differencing in the summing junction, the response is then a double-amplitude polar PCM signal. The rest of the circuit is similar to that of a polar PCM, as in Figure 15.3.8(b).

The very presence of noise suggests that the PCM reconstruction circuits may occasionally make a mistake in deciding what input pulse was received in a given bit interval. How often an error is made is determined by the *bit-error probability*. Bit errors in a unipolar system occur much more frequently than in a polar system having the same signal-to-noise ratio.

With negligible receiver noise, only quantization error is present when Equation (15.3.3) applies. With not so negligible receiver noise, the recovered signal $f_q^R(t)$ can be expressed as

$$f_q^R(t) = f_q(t) + \varepsilon_n(t) = f(t) + \varepsilon_q(t) + \varepsilon_n(t) \tag{15.3.12}$$

in which an error $\varepsilon_n(t)$ due to noise is introduced in the reconstructed message $f_q^R(t)$. The ratio of desired output signal power to total output noise power is given by

$$\left(\frac{S_0}{N_0}\right)_{\text{PCM}} = \frac{S_0/N_q}{1 + \left[\overline{\varepsilon_n^2(t)/\varepsilon_q^2(t)}\right]} \tag{15.3.13}$$

where the numerator S_0/N_q is given by Equation (15.3.5), or by Equation (15.3.9) when Equation (15.3.6) applies. If $\overline{\varepsilon_n^2(t)} \ll \overline{\varepsilon_q^2(t)}$, noise has no effect on performance; otherwise it has a significant effect. Thus, $\varepsilon_n(t)$ gives rise to a *threshold effect* in PCM.

Signal Multiplexing

Frequency-division multiplexing (FDM) and time-division multiplexing (TDM) systems were introduced in Section 14.2. When data from many sources in time are interlaced, the interlacing of data is called *time multiplexing*, in which case a single link can handle all sources. Figure 15.3.9(a) illustrates time multiplexing soon after sampling for N similar messages. With proper interleaving of sampling pulses [see Figures 15.3.9(b) and (c) for individual message signal waveforms], the train of samples can be added for the signal at point A in Figure 15.3.9(a), as shown in Figure 15.3.9(d). If we consider N similar messages of spectral extent W_f rad/s, the sampling interval T_s must satisfy

$$T_s \leq \frac{\pi}{W_f} \tag{15.3.14}$$

based on the sampling theorem (see Section 14.2). A *time slot* is the time per sampling interval that is allowed per message. It is equal to the sum of the sampling-pulse duration τ and separation τ_g, called the *guard time*. Thus, we have

$$\tau + \tau_g = \frac{T_s}{N} \leq \frac{\pi}{NW_f} \tag{15.3.15}$$

The time that is required to gather at least one sample of each message is known as a *frame*, which is T_s, as shown in Figure 15.3.9(d). Now, with a single composite source of the waveform shown in Figure 15.3.9(d) at point A of Figure 15.3.9(a), the time multiplexer of Figure 15.3.9(a) operates beyond A. For N_b-bit encoding, each time slot in the output PCM signal will have N_b bits of duration,

$$T_b = \frac{T_s}{NN_b} \tag{15.3.16}$$

It is assumed that all sample trains are derived from the same timing source, called a *clock*, and hence have the same frequency. Instead of using up all frame time for messages, some time is usually allocated for synchronization so that the receiver will know the start times of frames. The American Telephone and Telegraph Company (AT & T) employs a device known as a *D3 channel bank*, which is a synchronous multiplexer, whose characteristics are as follows:

- It multiplexes 24 telephone messages, each having an 8-kHz sampling rate so that $T_s = 1/(8 \times 10^3) = 125 \ \mu s$. The digital structure of each input message is determined by a single master clock.

- Each sample uses 8-bit encoding, so that there are $8 \times 24 = 192$ message bits; one extra bit, known as a *framing bit*, of less than one time slot is allowed for frame synchronization. Thus, a total of 193 bits per frame, with a total bit rate of $193 \times 8000 = 1.544$ megabits per second (Mbit/s), is available.

- The frame structure is illustrated in Figure 15.3.10.

- *Bit robbing* or *bit stealing*, which is the occasional borrowing of a message bit for purposes other than message information, is done for signaling, which refers to conveying information concerning telephone number dialed, dial tone, busy signal, ringing, and so on.

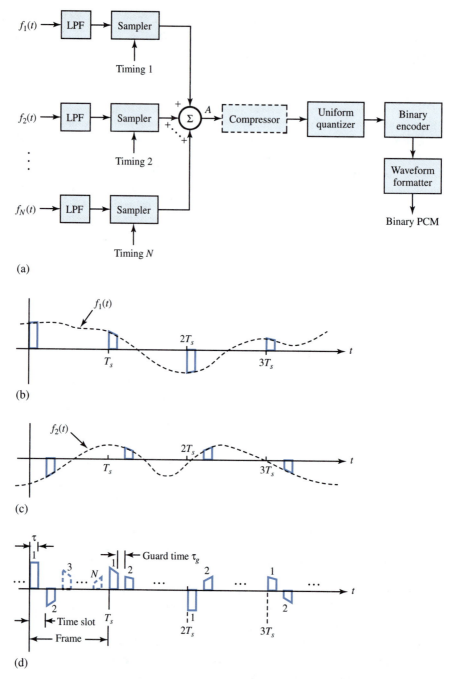

Figure 15.3.9 (a) Time multiplexing for N similar analog signals. (b) Waveform at point A for message signal 1. (c) Waveform at point A for message signal 2. (d) Full waveform at point A.

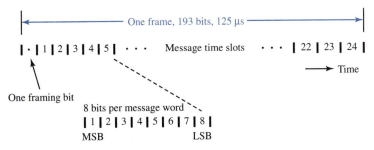

Figure 15.3.10   Frame structure for a D3 channel bank.

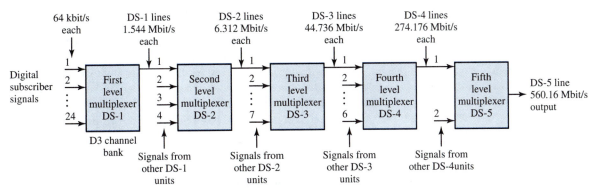

Figure 15.3.11   Digital TDM hierarchy for North American telephone communication system.

In digital speech transmission over telephone lines via PCM, a standard TDM hierarchy has been established for accommodating multiple subscribers. Figure 15.3.11 illustrates the TDM hierarchy for the North American telephone system. The output from the channel bank is a digital signal (DS) on a line said to carry level 1 multiplexing. In the first level of the TDM hierarchy, 24 digital subscriber signals are time-division multiplexed into a single high-speed data stream of 1.544 Mbit/s (nominal bit rate). The resulting combined signal is usually called a DS-1 channel. In the second level of TDM, four DS-1 channels are multiplexed into a DS-2 channel, having the nominal bit rate of 6.312 Mbit/s. In a third level of hierarchy, seven DS-2 channels are combined via TDM to produce a DS-3 channel, which has a nominal bit rate of 44.736 Mbit/s. Beyond DS-3, there are two more levels, as shown in Figure 15.3.11. All multiplexers except the channel bank are asynchronous.

In a mobile cellular radio system (see Section 15.2) for the transmission of speech signals, since the available channel bandwidth per user is small and cannot support the high bit rates needed by waveform encoding methods such as PCM, the analysis–synthesis method based on *linear predictive coding* (LPC) is used to estimate the set of model parameters from short segments of the speech signal. The speech-model parameters are then transmitted over the channel. With LPC, a bit rate of 4800–9600 bit/s is achieved.

In mobile cellular communication systems, LPC speech compression is only needed for the radio transmission between the mobile transcriber and the base station in any cell. At the base station interface, the LPC-encoded speech is converted to analog form and resampled and digitized (by using PCM) for transmission over the terrestrial telephone system. Thus, a speech signal transmitted from a mobile subscriber to a fixed subscriber will undergo two different types of analog-to-digital (A/D) encoding, whereas speech-signal communication between two mobiles serviced by different base stations will undergo four translations between the analog and digital domains.

In the conversion of analog audio signals to digital form, with the development of the compact disc (CD) player and the digital audio tape recorder, the most dramatic changes and benefits have been experienced by the entertainment industry. The CD system, from a systems point of view, embodies most of the elements of a modern digital communication system: A/D and D/A conversion, modulation/demodulation, and channel coding/decoding. Figure 15.3.12 shows a general block diagram of the elements of a CD digital audio system. The sampling rate in a CD system is chosen to be 44.1 kHz, which is compatible with the video recording equipment commonly used for digital recording of audio signals on magnetic tape. The samples of both the L and R signals are quantized using PCM with 16 bits per sample. While the D/A conversion of the two 16-bit signals at the 44.1-kHz sampling rate is relatively simple, the practical implementation of a 16-bit D/A converter is very expensive. Because inexpensive D/A converters with 12-bit (or less) precision are readily available, a method is to be devised for D/A conversion that employs a low precision (and hence a low-cost D/A converter), while maintaining the 16-bit precision of the digital audio signal. Without going into details, the practical solution to this problem is to expand the bandwidth of the digital audio signal by oversampling through interpolation and digital filtering prior to analog conversion.

Time-division multiple access (TDMA) is an important means by which each station on earth timeshares the communication satellite in the sky, and broadcasts to all other stations during its assigned time. Figure 15.3.13 shows the communication links of several (N) earth stations that communicate with each other through satellite. All stations use the same up-link frequency, and all receive a single down-link frequency from the satellite.

Carrier Modulation by Digital Signals

Digitally modulated signals with low-pass spectral characteristics can be transmitted directly through *baseband* channels (having low-pass frequency-response characteristics) without the need for frequency translation of the signal. However, there are many communication *bandpass* channels (telephone, radio, and satellite channels) that pass signals within a band of frequencies (that is far removed from dc). Digital information may be transmitted through such channels by using a sinusoidal carrier that is modulated by the information sequence in either amplitude, phase, or frequency, or some other combination of amplitude and phase. The effect of impressing the information signal on one or more of the sinusoidal parameters is to shift the frequency content of the transmitted signal to the appropriate frequency band that is passed by the channel. Thus, the signal is transmitted by carrier modulation. There are several carrier-modulation methods. However, we shall limit our discussion to the following, assuming only binary modulation in all cases:

- Amplitude-shift keying (ASK)
- Phase-shift keying (PSK)
- Differential phase-shift keying (DPSK)
- Frequency-shift keying (FSK).

AMPLITUDE-SHIFT KEYING (ASK)

A carrier's amplitude is keyed between two levels (binary 1 and 0) in binary ASK. Figure 15.3.14 shows the functions of a coherent ASK communication system. Let us consider a bit interval from

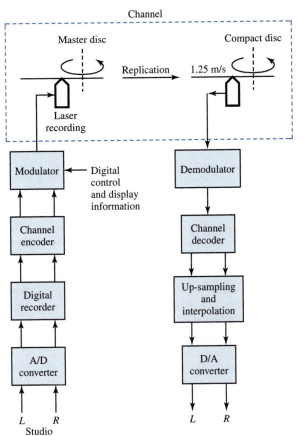

Figure 15.3.12 General block diagram of a CD digital audio system.

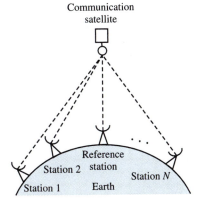

Figure 15.3.13 Communication links of several earth stations communicating through a satellite.

$t = 0$ to $t = T_b$, since the operation of any other interval will be similar. The desired ASK signal is given by

$$s_{\text{ASK}}(t) = \begin{cases} A_c \cos(\omega_c t + \phi_c), & 0 < t < T_b, \text{ for } 1 \\ 0, & 0 < t < T_b, \text{ for } 0 \end{cases} \quad (15.3.17)$$

where A_c, ω_c, and ϕ_c are the peak amplitude, angular frequency, and phase angle of the modulated carrier, respectively. Equation (15.3.17) can be viewed as a carrier $A_c \cos(\omega_c t + \phi_c)$ modulated by a digital signal $d(t)$ that is 0 or 1 in a given bit interval. The digital signal and product device shown in Figure 15.3.14 are then equivalent to the waveform formatter and modulator. Assuming that the received signal $s_R(t)$ differs only in amplitude from $s_{ASK}(t)$, one can write

$$s_R(t) = \begin{cases} A_c \cos(\omega_c t + \phi_c), & 0 < t < T_b, \text{ for } 1 \\ 0, & 0 < t < T_b, \text{ for } 0 \end{cases} \tag{15.3.18}$$

The product device in the receiver's demodulator acts like a synchronous detector that removes the input carrier. The major disadvantage of the coherent ASK is that the required local carrier must be phase-connected with the input signal. The input to the integrator is a unipolar PCM signal, and the remainder of Figure 15.3.14 is a PCM receiver [see Figure 15.3.8(a)].

The noncoherent ASK system eliminates the need for a coherent local oscillator. Figure 15.3.15 shows the demodulator and code generator for a noncoherent ASK system. A matched filter (that has its impulse response matched to have the same form as the carrier pulse at its input) and envelope detector take the place of the synchronous detector and integrator of Figure 15.3.14.

While ASK systems are not as widely used as other systems for various reasons, the ASK concept remains significant, particularly as applied to modern optical communication systems which use intensity modulation of a light source.

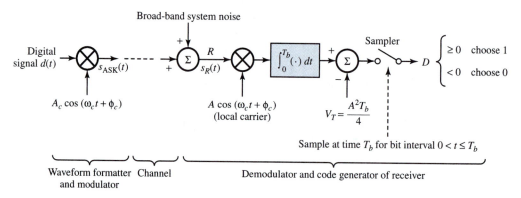

Figure 15.3.14 Functions of a coherent ASK communication system.

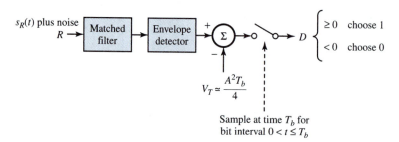

Figure 15.3.15 Demodulator and code generator for a noncoherent ASK system.

PHASE-SHIFT KEYING (PSK)

In PSK, the phase angle of a carrier is keyed between two values. When the values are separated by π radians, it is known as *phase-reversal keying* (PRK). The PSK waveform in the bit interval is given by

$$s_{PSK}(t) = \begin{cases} A_c \cos(\omega_c t + \phi_c), & 0 < t < T_b, \text{ for } 1 \\ -A_c \cos(\omega_c t + \phi_c), & 0 < t < T_b, \text{ for } 0 \end{cases} \qquad (15.3.19)$$

Equation (15.3.18) can be viewed as a carrier $A_c \cos(\omega_c t + \phi_c)$ modulated in amplitude by a digital signal $d(t)$ with amplitudes of 1 when the binary digit in the bit interval is 1 and 0, respectively. Figure 15.3.16 illustrates the functions of a coherent PSK communication system, which is similar to the ASK system of Figure 15.3.14 with the following differences:

- The received signal is now given by

$$s_R(t) = \begin{cases} A_c \cos(\omega_c t + \phi_c), & 0 < t < T_b, \text{ for } 1 \\ -A_c \cos(\omega_c t + \phi_c), & 0 < t < T_b, \text{ for } 0 \end{cases} \qquad (15.3.20)$$

- A gain of 2 is assigned to the integrator.
- A threshold is absent when binary digits 0 and 1 occur with equal probability.

No truly noncoherent PSK version is possible because the PSK signal carries its information in the carrier's phase, whereas a noncoherent system purposely disregards phase and operates only on signal amplitude.

DIFFERENTIAL PHASE-SHIFT KEYING (DPSK)

In order to eliminate the need of a local carrier, DPSK has been developed in which the receiver uses the received signal to act as its own carrier. Figure 15.3.17 shows the functions of a DPSK system in which the leftmost product operation along with 1-bit delay is the differential encoder. The digital signal $d(t)$ is a polar waveform of levels ± 1, corresponding to binary digits 1 and 0, respectively. The output signal $a(t)$ from the differential encoder PSK-modulates a carrier to produce the DPSK signal $s_{DPSK}(t)$. The product device followed by the wide-band low-pass filter acts as the coherent detector. The two inputs to the product device are obtained from the output of a filter matched to the input pulse in a single bit interval. Note that the 1-bit delayed input to the product device serves the purpose of the local oscillator for the coherent detector; that is to

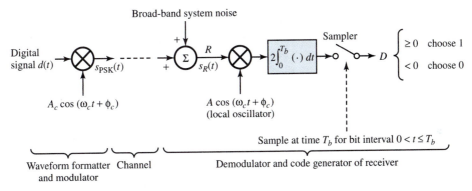

Figure 15.3.16 Functions of a coherent PSK communication system.

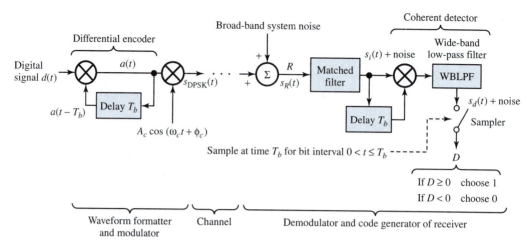

Figure 15.3.17 Functions of a DPSK system.

say, the DPSK waveform in a given bit interval serves as its own local-oscillator signal in the following bit interval.

In Figure 15.3.17, $s_i(t)$ is the signal component of the matched filter output, and $s_d(t)$ is the signal component of the detector output. If the phases of both $s_i(t)$ and $s_i(t - T_b)$ are the same, $s_d(t)$ is then a positive voltage; if their phases differ by π radians, $s_d(t)$ will then be a negative voltage. These voltages will have maximum amplitudes at the sample time at the end of the bit interval. Because the sign of the voltage at the sampler depends upon the phase relationship between $s_i(t)$ and its delayed replica, and the sign of $s_d(t)$ is of the same form as $d(t)$, the original digital bit sequence can be determined by sampling to decide the sign of the detector output.

Figure 15.3.18 illustrates an example sequence of message binary digits, modulator waveforms in DPSK, and phase and polarity relationships as applied to DPSK message recovery.

FREQUENCY-SHIFT KEYING (FSK)

Digital transmission by FSK is a nonlinear modulation method that is appropriate for channels that lack the phase stability needed to perform carrier-phase estimation. Figure 15.3.19 shows the functions of a coherent FSK system in which the transmitted signal $s_{FSK}(t)$ is generated by frequency modulation of a voltage-controlled oscillator (VCO). The digital signal $d(t)$ has a polar format with amplitudes ± 1, corresponding to binary digits 1 and 0, respectively. Modulation keys the VCO's angular frequency between two values, such that

$$\omega_2 = \omega_c + \Delta\omega \qquad \text{when } d(t) = 1 \tag{15.3.21}$$

$$\omega_1 = \omega_c - \Delta\omega \qquad \text{when } d(t) = -1 \tag{15.3.22}$$

where $\Delta\omega$ is the frequency deviation from a nominal or carrier angular frequency ω_c. In order to conserve the bandwidth in the signal $s_{FSK}(t)$, $\Delta\omega$ is usually selected not much larger than the minimum allowable value given by

$$\Delta\omega = \frac{\pi}{T_b} = \frac{\omega_b}{2} \tag{15.3.23}$$

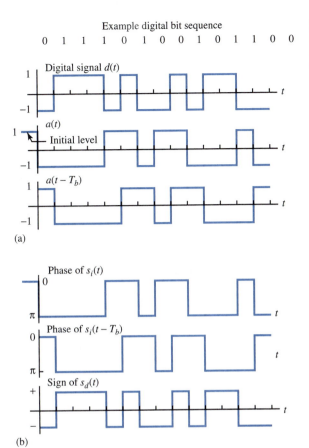

Example digital bit sequence

0 1 1 1 0 1 0 0 1 0 1 1 0 0

(a)

Figure 15.3.18 (a) Example sequence of message binary digits and modulator waveforms in DPSK. **(b)** Phase and polarity relationships as applied to DPSK message recovery.

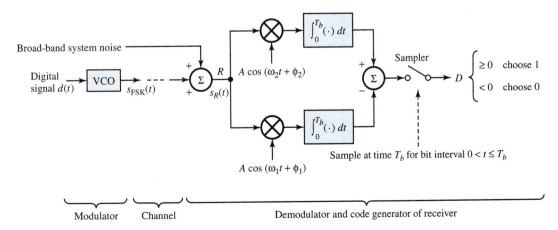

Figure 15.3.19 Functions of a coherent FSK system.

When Equation (15.3.23) is satisfied, the bandwidth of the channel needed to support $s_{FSK}(t)$ is about $2\omega_b$ rad/s. The transmitted signal in Figure 15.3.19 can be seen to be two unipolar ASK signals in parallel, one at carrier frequency ω_2 and the other at ω_1. The receiver then becomes two coherent ASK receivers in parallel (see Figure 15.3.14).

The noncoherent FSK version results when each product device and the following integrator in the receiver are replaced by a corresponding matched filter followed by an envelope detector.

Comparison of Digital Communication Systems

The noise performances of different digital communication systems are usually compared on the basis of their bit-error probabilities P_e. Keeping the ratio E_b/N_0 as an independent parameter, where E_b represents the average signal energy per bit and N_0 stands for the noise level, the following expressions are given here with no derivation:

$$
P_e = \begin{cases}
\frac{1}{2}\left[1 + \sqrt{\frac{1}{2\pi(E_b/N_0)}}\right]\exp\left(\frac{-E_b}{2N_0}\right) & \text{noncoherent ASK} \\[2mm]
\frac{1}{2}\exp\left(\frac{-E_b}{2N_0}\right) & \text{noncoherent FSK} \\[2mm]
\frac{1}{2}\operatorname{erfc}\left[\sqrt{\frac{E_b}{2N_o}}\right] & \left.\begin{array}{l}\text{coherent ASK}\\\text{coherent FSK}\\\text{unipolar PCM}\end{array}\right\} \\[2mm]
\frac{1}{2}\exp\left(\frac{-E_b}{N_o}\right) & \text{DPSK} \\[2mm]
\frac{1}{2}\operatorname{erfc}\left[\sqrt{\frac{E_b}{N_o}}\right] & \left.\begin{array}{l}\text{PSK}\\\text{polar PCM}\\\text{Manchester PCM}\end{array}\right\}
\end{cases}
\tag{15.3.24}
$$

where

$$
\operatorname{erfc}(x) = 1 - \frac{2}{\sqrt{\pi}} \int_0^x e^{-\xi^2}\, d\xi
\tag{15.3.25}
$$

is known as the *complementary error function*, which is equal to 1 at $x = 0$ and decreases to 0 as $x \to \infty$; its behavior may be approximated by

$$
\operatorname{erfc}(x) \cong \frac{e^{-x^2}}{x\sqrt{\pi}} \qquad \text{when } x > 2
\tag{15.3.26}
$$

and

$$
\operatorname{erfc}(-x) = 2 - \operatorname{erfc}(x)
\tag{15.3.27}
$$

Figure 15.3.20 gives the plots of Equation (15.3.24). Of the PCM (noncarrier) systems, the polar and Manchester-formatted systems are superior to the unipolar system, since no energy is transmitted during half the bits (on average) in the unipolar system. Of the carrier-modulated systems, coherent systems perform slightly better (by about 1 dB) than their corresponding noncoherent versions. However, in practice, to avoid having to generate a local carrier in the receiver, it may be worth the expense of 1 dB more in transmitted power. The PSK system can be seen to be superior by about 3 dB for a given P_e to both FSK and ASK (which exhibit nearly the same performance).

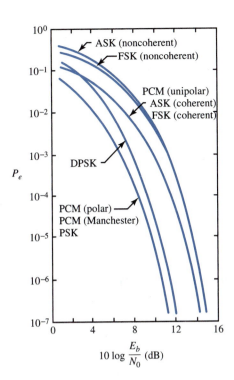

EXAMPLE 15.3.3

Let both coherent ASK and coherent PSK systems transmit the same average energy per bit interval and operate on the same channel such that $E_b/N_0 = 18$. Determine the bit-error probability P_e for the two systems.

Solution

From Equation (15.3.24) it follows for coherent ASK,

$$P_e = \frac{1}{2} \operatorname{erfc} \sqrt{\frac{E_b}{2N_0}}$$

$$= \frac{1}{2} \operatorname{erfc} \sqrt{\frac{18}{2}} = \frac{1}{2} \operatorname{erfc}(3)$$

$$\cong \frac{1}{2} \frac{e^{-9}}{3\sqrt{\pi}} \qquad \text{[from Equation (15.3.26)]}$$

$$\cong 1.16 \times 10^{-5}$$

and for coherent PSK,

$$P_e = \frac{1}{2} \operatorname{erfc} \sqrt{\frac{E_b}{N_o}}$$

$$= \frac{1}{2} \operatorname{erfc} \sqrt{18} = \frac{1}{2} \operatorname{erfc}(4.243)$$

$$\cong \frac{1}{2} \frac{e^{-18}}{4.243 \sqrt{\pi}}$$

$$\cong 1.01 \times 10^{-9}$$

The PSK system is clearly superior to the ASK system, based on equal values of E_b/N_0.

Digital Transmission on Fading Multipath Channels

We have discussed thus far digital modulation and demodulation methods for the transmission of information over two types of channels, namely, an additive noise channel and a linear filter channel. While these channel models are appropriate for a large variety of physical channels, they become inadequate in characterizing signal transmission over radio channels whose transmission characteristics vary with time. Time-varying behavior of the channel is exhibited by the following:

- *Signal transmission via ionospheric propagation in the HF band* (see Figure 15.0.3), in which signal fading is a result of multipath signal propagation that makes the signal arrive at the receiver via different propagation paths with different delays.

- *Mobile cellular transmission*, between a base station and a telephone-equipped automobile, in which the signal transmitted by the base station to the automobile is reflected from surrounding buildings, hills, and other obstructions.

- *Line-of-sight microwave radio transmission*, in which signals may be reflected from the ground to the receiving antenna due to tall obstructions or hilly terrain in the path of propagation.

- *Airplane-to-airplane radio communications*, in which secondary signal components may be received from ground reflections.

Such channels may be treated as linear systems with time-varying impulse response by adopting a statistical characterization. Models for time-variant multipath channels will not be considered here.

15.4 LEARNING OBJECTIVES

The *learning objectives* of this chapter are summarized here so that the student can check whether he or she has accomplished each of the following.

- Basic ideas about waveguides.
- Transmission line with matched impedances.
- Calculating the signal power at the output of a radio transmission system.
- Fundamentals of antennas.
- Signal-to-noise power ratio at the system output, as a measure of performance of the communication system.
- Amplitude modulation and message demodulation in analog communication systems.
- Frequency modulation and demodulation in analog communication systems.
- Block diagrams of superheterodyne radio receivers.
- Basic notions of television broadcasting.
- Cellular-telephone-system concepts.
- Sampling, quantization, and coding in digital communication systems.

- Pulse-code modulation and demodulation for processing analog signals.
- Time-division multiplexing.
- Digital modulation and demodulation methods for the transmission of information and the comparison of digital communication systems.

15.5 PRACTICAL APPLICATION: A CASE STUDY

Global Positioning Systems

Modern communication systems abound in practice: cellular phones, computer networks, television satellites, and optical links for telephone service. Communication with instruments has been made possible between our planet, the earth, and Mars.

A global positioning system (GPS) is a modern and sophisticated system in which signals are broadcast from a network of 24 satellites. A receiver, which contains a special-purpose computer to process the received signals and compare their phases, can establish its position quite accurately. Such receivers are used by flyers, boaters, and bikers. Various radio systems that utilize phase relationships among signals received from several radio transmitters have been employed in navigation, surveying, and accurate time determination. An early system of this type, known as LORAN, was developed so that the receivers could determine their latitudes and longitudes.

Figure 15.5.1 illustrates in a simple way the working principle of LORAN, which consists of three transmitters (a master and two slaves) that periodically broadcast 10-cycle pulses of 100-kHz sine waves in a precise phase relationship. The signal received from each transmitter is phase-shifted in proportion to the distance from that transmitter to the receiver. A phase reference is established at the receiver by the signal from the master transmitter; then the receiver determines the differential time delay between the master and each of the two slaves. The difference in time delay between the master and a given slave yields a line of position (LOP), as shown in Figure 15.5.1.

If the time delays of the signals from the master and a given slave are equal (i.e., no differential delay), then the line of position is the perpendicular bisector of the line joining the master and that particular slave. On the other hand, if the time delay from the master is smaller by a certain amount, the line of position happens to be a hyperbola situated toward and near the master, as illustrated by LOP for slave 1 in Figure 15.5.1, and LOP for slave 2, where the time delay from slave 2 is smaller. The intersection of the LOPs for the two slaves determines the location of the receiver.

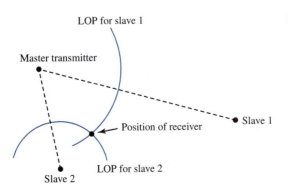

Figure 15.5.1 Working principle of LORAN.

The simple concept of relative phase relationships is exploited by engineers and scientists in several systems that are beneficial to the general public. For example, remote measurements of the height of Greenland's ice cap (by using high-quality GPS receivers) are being made to assess the possible effects of global warming.

PROBLEMS

15.1.1 A coaxial cable with polyethylene dielectric ($\varepsilon_r = 2.26$) connects an antenna to a receiver 30 m away. Determine the velocity of wave propagation in the cable and the delay of the cable. (*Hint:* See Example 15.1.1.)

15.1.2 If b and a are the radii of the outer and inner conductors, respectively, of a coaxial cable using a polyethylene dielectric ($\varepsilon_r = 2.26$), what ratio b/a is needed for the cable to have a characteristic impedance $\bar{Z}_0 = R_0 = (60/\sqrt{\varepsilon_r}) \ln (b/a)$ of 50 Ω?

*15.1.3 A rigid 50-Ω coaxial transmission line has air as dielectric. If the radius of the outer conductor is 1 cm, find the cutoff frequency

$$f_c = \frac{c}{\pi \sqrt{\varepsilon_r}(a + b)}$$

Note that $\bar{Z}_0 = R_0 = (60/\sqrt{\varepsilon_r}) \ln (b/a)$.

15.1.4 If the line of Problem 15.1.3 is made of copper whose resistivity $\rho = 1.72 \times 10^{-8} \ \Omega \cdot m$, determine the maximum length that can be used if losses are not to exceed 3 dB when $f = 3$ GHz. For the expression of attenuation, see Example 15.1.1.

15.1.5 For an RG-290/U aluminum rectangular waveguide, $a = 58.42$ cm and $b = 29.21$ cm. Compute the theoretical and practical frequency ranges of operation for the guide. (See Example 15.1.2.)

15.1.6 A rectangular air-filled RG-52/U is made of brass ($\rho = 3.9 \times 10^{-8} \ \Omega \cdot m$) and has dimensions $a = 22.86$ mm and $b = 10.16$ mm.

(a) Determine $\bar{Z}_0(= R_0)$ at the limits of the practical operating frequency range of the guide. (See Example 15.1.2.)

(b) Compute the attenuation given by

$$\frac{(0.458 \times 10^{-4})\sqrt{f\rho}\left[1 + (2b/a)(f_c/f)^2\right]}{b\sqrt{1 - (f_c/f)^2}} \ \text{dB}$$

per unit length corresponding to those limiting frequencies.

15.1.7 By using the expression for attenuation given in Problem 15.1.6(b), find the attenuation of the air-

filled waveguide of Problem 15.1.5 if $\rho_{\text{aluminum}} = 2.83 \times 10^{-8} \ \Omega \cdot m$, at frequencies of $1.25f_c$ and $1.9f_c$.

*15.1.8 An RG-139/U rectangular waveguide is given to have dimensions $a = 0.8636$ mm and $b = 0.4318$ mm. Compute the theoretical and practical frequency ranges of operation for the guide. (See Example 15.1.2.)

15.1.9 Find the diameter of a circular waveguide that will have a lower cutoff frequency of 10 GHz and also specify its largest usable frequency. (See Example 15.1.2.)

15.1.10 It is desired to cut a $\lambda/4$ length of RG58A/U cable ($\varepsilon_r = 2.3$) at 150 MHz. What is the physical length of this cable?

15.1.11 A transmission line with air dielectric is 25 m long. Find, in wavelengths, how long the line is at frequencies of 1 kHz, 10 MHz, and 100 MHz.

15.1.12 A transmission line with a dielectric ($\varepsilon_r = 3.5$) is 100 m long. At a frequency of 10 GHz, how many wavelengths long is the line?

*15.1.13 Comment briefly on the following:

(a) Why are waveguides not used at low frequencies?

(b) Why are open-wire lines not generally used as guiding structures at very high frequencies?

(c) What is the velocity of wave propagation in a Teflon ($\varepsilon_r = 2.1$) coaxial transmission line?

15.1.14 (a) What is the difference between a TEM mode and a TE mode?

(b) Explain the terms "cutoff wavelength" and "dominant mode" as applied to waveguides. Find the cutoff wavelength for an air-filled rectangular waveguide for the propagation by the dominant mode.

15.1.15 The cutoff frequency of a dominant mode in an air-filled rectangular waveguide is 3 GHz. What would the cutoff frequency be if the same waveguide were filled with a lossless dielectric having an $\varepsilon_r = 3.24$?

15.1.16 A transmitter is connected to an antenna by a transmission line for which $\bar{Z}_0 = R_0 = 50\ \Omega$. The transmitter source impedance is matched to the line, but the antenna is known to be unmatched and has a *reflection coefficient*

$$\Gamma_L = \frac{R_L - R_0}{R_L + R_0} = 0.3$$

where $R_L\ (= \bar{Z}_L)$ is the load impedance. The transmitter produces a power of 15 kW in the incident wave to the antenna and will be destroyed due to overheat if the reflected wave's power exceeds 2 kW. Determine the antenna's radiated power in this case and comment on whether the transmitter will survive.

15.1.17 A source of impedance $\bar{Z}_S = R_S = 100\ \Omega$ has an open-circuit voltage $v_S(t) = 12.5 \cos \omega_o t$ and drives a 75-Ω transmission line terminated with a 75-Ω load. Find the current and voltage at the input terminals of the line.

15.1.18 The model of an elemental length of a lossy transmission line is shown in Figure P15.1.18(a),

along with its parameters, where R is series resistance per unit length, L is series inductance per unit length, G is shunt conductance per unit length, and C is shunt capacitance per unit length. The *characteristic impedance* $\bar{Z}_0$ of the line is given by

$$\bar{Z}_0 = \sqrt{\frac{\bar{Z}}{\bar{Y}}} = \sqrt{\frac{R + j\omega L}{G + j\omega C}}$$

The *propagation constant* V is given by

$$\bar{V} = \sqrt{\bar{Z}\,\bar{Y}} = \sqrt{(R + j\omega L)(G + j\omega C)}$$
$$= \alpha + j\beta$$

where α is the attenuation constant (nepers per unit length) and β is the phase constant (radians per unit length). The ac steady-state solution for the uniform line reveals the voltage on a matched line ($\bar{Z}_L = \bar{Z}_0$) to be $\bar{E}(x) = \bar{E}_S e^{-\bar{\gamma}x} = \bar{E}_S e^{-\alpha x} e^{-j\beta x}$, where $\bar{E}_S$ is the sending-end voltage and x is distance along the line from the sending end.

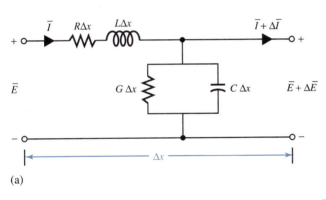

(a)

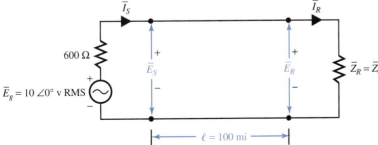

(b)

Figure P15.1.18

(a) Now consider a typical open-wire transmission line with parameters of $R = 14\ \Omega/\text{mi}$, $L = 4.6\ \text{mH/mi}$, $C = 0.01\ \mu\text{F/mi}$, and $G = 0.3 \times 10^{-6}\ \text{S/mi}$. If the line operates at 1 kHz, find the characteristic impedance $\bar{Z}_0$ and the propagation constant $\bar{\gamma}$.

(b) Then consider a 100-mi open-wire flat telephone line with the same parameters as those given in part (a). The matched transmission line is shown in Figure P15.1.18(b). If the frequency of the generator is 1 kHz, determine the following:

   (i) Sending-end current $\bar{I}_S$.

   (ii) Sending-end voltage $\bar{E}_S$.

   (iii) Sending-end power P_S.

   (iv) Receiving-end current $\bar{I}_R$.

   (v) Receiving-end voltage $\bar{E}_R$.

   (vi) Receiving-end power P_R.

   (vii) Power loss in dB, which is given by 10 $\log (P_S/P_R)$.

(c) Find the wavelength ($\lambda = 2\pi/\beta$) of the signal on the line in part (b) and the length of the line in terms of wavelengths. Also, find the transmission loss in nepers and decibels of the transmission line in part (b). (1 neper = 8.686 dB.)

(d) For a *lossless* line (with $R = G = 0$) in which the velocity of energy propagation is given by $v_p = \omega/\beta = 1/\sqrt{LC}$, find expressions for α and β. Consider the line in part (b) to be lossless, and calculate the corresponding α, β, and v_p.

*15.1.19 Describe the following phasor equations represented in the time domain:

$$\text{(a) } \bar{E} = K_1 e^{-\bar{\gamma}z} \qquad \text{(b) } \bar{E} = K_2 e^{\bar{\gamma}z}$$

where z is the space coordinate, K_1 and K_2 are constants, and $\bar{\gamma} = \alpha + j\beta$.

15.1.20 Consider a transmission system as shown in Figure P15.1.20. Determine the individual gains and $G = G_1 G_2 G_3$.

15.1.21 Consider a transmission system as shown in Figure P15.1.21. Taking G_1 as large as possible, find the needed gains G_1 and G_2.

15.1.22 In the transmission system of Figure 15.1.4, let $G_1 = 23$ dB; $\alpha_1 = \alpha_2 = 2.5$ dB/km, $l_1 + l_2 = 30$ km, $P_{in} = 1$ mW; and $P_{out} = 50$ mW. Determine l_1, l_2, G_2, and G_3 such that the signal power equals 20 μW at the input to G_2 and G_3.

15.1.23 A signal with bandwidth of 100 MHz is to be transmitted 40 km by LOS radio transmission. Taking $B/f_c = 1/30$, and using a circular-aperture parabolic dish with 50-cm radius at each end, compute the transmission loss.

*15.1.24 Consider the LOS radio system of Figure 15.1.6, with dipole antennas. Let $P_{in} = 10$ W, $R = 20$ km, $G_t = G_r = 2$ dB, and $f_c = 500$ MHz. Compute P_{out}.

15.1.25 A satellite radio transmitter has $P_{in} = 3$ W and $G_t = 30$ dB. The receiving antenna has a circular aperture with radius r at the ground station 30,000 km away. Find r in meters if $P_{out} = 30$ pW.

15.1.26 A microwave relay system uses two identical horn antennas mounted on towers spaced 40 km apart. If $f_c = 6$ GHz, and each relay hop has $L = 60$ dB, calculate the antenna aperture area A_e in square meters.

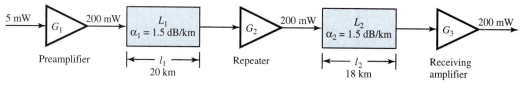

Figure P15.1.20

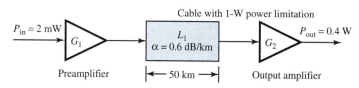

Figure P15.1.21

15.1.27 An antenna has an aperture area of 10 m², an aperture efficiency of 0.6, and negligible losses. If it is used at 5 GHz, find:

(a) Its power gain.

(b) The maximum power density that the antenna can generate at a distance of 20 km with an input power of 2 kW.

15.1.28 The power gain of an antenna is 10,000. If its input power is 1 kW, calculate the maximum radiation intensity that it can generate.

*15.1.29 An antenna has beam widths of 3° and 10° in orthogonal planes and has a radiation efficiency factor of 0.6. Find the maximum radiation intensity if 1 kW is applied to the antenna.

15.1.30 The radiation pattern of a half-wave dipole antenna [see Figure 15.1.7(a)] is given by

$$P(\theta, \phi) = \frac{\cos^2\left[(\pi/2)\cos\theta\right]}{\sin^2\theta}$$

(a) Sketch the radiation pattern in the principal plane of xz containing angle θ.

(b) Determine the beam width between the -10-dB points of its radiation pattern as well as the half-power beamwidth.

15.1.31 The effective area of a dipole is given by $A_e = 0.13\lambda^2$. Find the effective area of a half-wave dipole at 3 GHz.

15.1.32 For a helical antenna [see Figure 15.1.7(c)], the half-power beamwidth and directive gain are given by

$$\theta_B \cong \frac{52\lambda^{3/2}}{C\sqrt{NS}}$$

where $C = \pi D$, $N = L/S$, and $S = C\tan\alpha$, in which α is called the pitch angle, and

$$G_D \cong \frac{12NC^2S}{\lambda^3}$$

The input impedance seen by the transmission line at point P [Figure 15.1.7(c)] is almost purely resistive, given by $Z_a \cong 140C/\lambda$. Calculate the

antenna parameters of a 10-turn helix at $f = 500$ MHz by assuming that $C = \lambda$ and $\alpha = 14\pi/180$.

15.1.33 For a pyramidal-horn antenna [Figure 15.1.7(e)], the maximum directive gain is given by

$$G_D \cong \frac{2.05\pi AB}{\lambda^2}$$

occurring when the aperture dimensions are $A \cong \sqrt{3\lambda L}$ and $B \cong 0.81A$. The principal-plane beamwidths for the optimum horn with maximum gain are given by $\theta_B \cong 54\lambda/B$ in degrees in the yz-plane, and by $\phi_B \cong 78\lambda/A$ in degrees in the xz-plane. For a pyramidal horn, with $A = 6\lambda$ and $B = 4.86\lambda$, at 6 GHz, find G_D, θ_B, and ϕ_B. Comment on whether this horn is optimum.

15.1.34 A conical horn [Figure 15.1.7(f)] has a side view as shown in Figure P15.1.34. The maximum value of the directive gain is given by $G_D \cong 5.13(D/\lambda)^2$, where $D \cong \sqrt{3.33\lambda L_1}$ and $L_1 \cong L/(1-d/D)$, in which d is the inside diameter of the waveguide. Beamwidths for the main beam directed along the z-axis are given by $\theta_B \cong 60\lambda/D$ in degrees in the yz-plane, and by $\phi_B \cong 70\lambda/D$ in degrees in the xz-plane. Let the circular waveguide, with a 2.5-cm inside diameter, be expanded by adding a conical flare to have an aperture with an inside diameter of 5.771λ at 8 GHz. For $L_1 = 10\lambda$, find:

(a) The horn length L.

(b) The directive gain.

(c) The principal-plane beamwidths.

*15.1.35 A paraboloidal antenna [Figure 15.1.7(g)] has an aperture efficiency of 0.6 and a diameter $D = 100\lambda$ at 6 GHz. Illumination by the feed is such that the beamwidths of the principal-plane secondary patterns are equal. Determine the antenna's power gain and beamwidth. (Note that the radiation pattern of the feed is called the primary pattern, whereas that of the overall antenna is the secondary pattern.)

15.1.36 Two stations, located on mountain tops 40 km apart, communicate with each other using two

Figure P15.1.34

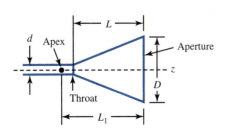

identical paraboloidal antennas with pencil beam-width of 1°, aperture efficiency of 0.8, and radiation efficiency of 0.85. The transmitting station has $P_t = 60$ W and $L_t = 2$ at 8 GHz. With $L_r = 1.5$ and $L_{ch} = 2.5$, find the diameter of the antennas and the signal power available at the receiver input.

15.1.37 Two stations (using identical antennas, with diameters of 50λ and aperture efficiencies of 0.6 at 35 GHz) are separated by 30 km. With negligible antenna losses, antenna connection-path losses are $L_t = L_r = 1.33$ at a physical temperature of 285 K, while the antenna temperature is 85 K. The receiver at either station has $G_a(f_0) = 10^7$, $\bar{T}_R = 250$ K, and $B_N = 12$ MHz. The rain attenuation and clear-air attenuation are given to be 3.9 dB/km and 0.072 dB/km, respectively. Determine:

(a) The required output power to guarantee a system signal-to-noise power ratio of 45 dB when heavy rain falls over a distance of 6 km.

(b) The available noise power occurring at the receiver output.

(c) The output signal power occurring with and without rain, when the transmitted power found in part (a) is used.

15.1.38 If an antenna has an available noise power of 1.6×10^{-15} W in a 1-MHz bandwidth, find the antenna temperature.

15.1.39 Determine the effective input noise temperature of a long piece of waveguide (that connects an antenna to a receiver) with a loss of 3.4 dB at 12 GHz and a physical temperature of 280 K.

*15.1.40 An antenna with an effective noise temperature of 130 K couples through a waveguide that has a loss of 0.8 dB to a receiver. Find the effective noise temperature presented by the waveguide output to the receiver if the waveguide's physical temperature is 280 K.

15.1.41 If the antenna and waveguide of Problem 15.1.40 feed a receiver for which $B_N = 10$ MHz, $G_a(f_0) =$

10^{12}, and $\bar{T}_R = 300$ K, determine the system noise temperature and N_{ao} at the receiver output.

15.1.42 (a) An amplifier with $F_0 = 3$ or 4.77 dB, $f_0 = 4$ GHz, and $B_N = 14$ MHz is used with an antenna for which $T_a = 200$ K. The connecting path loss is 1.45, or 1.61 dB at a physical temperature of 250 K. Find the available system noise power.

(b) If the antennas of Example 15.1.5 are used with the receiver of part (a), compute the signal-to-noise ratio.

*15.2.1 Find the number of possible station frequencies in the AM broadcast system in the United States.

15.2.2 Let $a(t) = [1 + m_A x(t)]$ and $x(t) = \cos 2\pi f_m t$, $f_m << f_c$, and $x_c(t) = A(t) \cos 2\pi f_c t$.

(a) With $m_A = 1$, sketch one full period of the AM wave and draw the envelope by connecting the positive peaks of $x_c(t)$.

(b) Repeat part (a) with $m_A = 2$, and notice that the carrier is overmodulated and the envelope does not have the same shape as $x(t)$.

15.2.3 Figure P15.2.3 illustrates a way to generate an AM wave of the form of Equation (15.2.2), using a nonlinear device and an appropriate bandpass filter. Comment on the nature of the BPF to be used.

15.2.4 If the nonlinear device in Problem 15.2.3 produces $z = ay^2$, determine how the system must be augmented to obtain an AM wave in the form of Equation (15.2.2).

15.2.5 If the message has a spectrum

$$F(\omega) = \begin{cases} K \cos \frac{\pi \omega}{2W_f}, & -W_f \leq \omega \leq W_f \\ 0, & \text{elsewhere} \end{cases}$$

where K and W_f are positive constants, sketch the spectrum of a standard AM signal that uses the message. Comment on the physical significance of K and W_f in the modulation process.

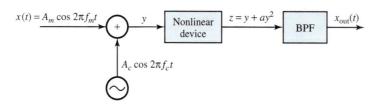

Figure P15.2.3

*15.2.6 In a standard AM system, represented by Equation (15.2.3), $A_c = 200$ V, $\left|\overline{f^2(t)}\right|^{1/2} = 50$ V, and $\bar{Z}_0 = R_0 = 50$ Ω. Calculate η_{AM}, P_c, P_f, and P_{AM}.

15.2.7 If a standard AM waveform [Equation (15.2.3)] of 1-kW average power is transmitted by an antenna with an input resistance of 75 Ω, evaluate $\overline{f^2(t)}$, carrier power, and sideband power, given the efficiency to be 0.1.

15.2.8 In the envelope detector [Figure 15.2.8(a)] of a standard AM receiver, the product RC is chosen to satisfy $\pi/\omega_{IF} \ll RC < 1/\omega_{m,\max}$, where $\omega_{m,\max}/2\pi = 5$ kHz in standard AM and $\omega_{IF} = 455$ kHz. If $R = 5000$ Ω, find the value of its capacitor (by applying a factor of 10 to represent $\ll$).

15.2.9 In an AM transmitter that transmits a total power of 1 kW, the unmodulated carrier power is 850 W. Compute the required value of $(S_i/N_i)_{AM}$ at a receiver, if $(S_0/N_0)_{AM} = 2\eta_{AM}(S_i/N_i)_{AM}$ must be 10^3 for good performance.

15.2.10 At the transmitter in a standard AM system, $P_f = 50$ W. In the receiver $(S_0/N_0)_{AM} = 250$ when $(S_i/N_i)_{AM} = 3000$. Find the transmitter's unmodulated carrier power and the total transmitted power if $(S_0/N_0)_{AM} = 2\eta_{AM}(S_i/N_i)_{AM}$.

*15.2.11 Determine the image frequency if an AM radio receiver is tuned to a station at 1030 kHz and has a high-side local oscillator.

15.2.12 Although the low side is not as good a choice as the high-side local oscillator frequency, let an AM broadcast receiver be designed with a low-side local oscillator. As f_c is varied from 540 to 1600 kHz, find the values f_{LO} and f_{image}.

15.2.13 For a DSB system for which $f(t) = A_m \cos \omega_m t$,

$\omega_c = 5\omega_m$, and $\phi_c = 0$, sketch the transmitted signal.

15.2.14 One of the many types of product device to produce suppressed-carrier AM is shown in Figure P15.2.14. Explain briefly how the device operates.

15.2.15 If the output signal from an AM modulator is given by $u(t) = 5 \cos 1800 \pi t + 20 \cos 2000 \pi t + 5 \cos 2200 \pi t$, determine:

(a) The modulating signal $m(t)$ and carrier $c(t)$.

(b) The modulation index.

(c) The ratio of the power in sidebands to the power in the carrier.

15.2.16 The signal $m(t)$, whose frequency spectrum $M(f)$ is shown in Figure P15.2.16, is to be transmitted from one station to another. Let the signal be normalized, i.e., $-1 \le m(t) \le 1$. Find the bandwidth of the modulated signal:

(a) If USSB (upper single sideband) is employed.

(b) If DSB (double sideband) is employed.

(c) If an AM modulation scheme with a modulation index of 0.8 is used.

(d) If an FM signal with the frequency deviation constant $k_f = 60$ kHz is used.

*15.2.17 An FM signal, for which $f(t) = 2 \cos \omega_m t$, is given by

$$s_{FM}(t) = 50 \cos \left\{ 185\pi(10^6)t + \frac{\pi}{3} + 6 \sin \left[\pi(10^4)t \right] \right\}$$

Determine: A_c, ω_c, ϕ_c, ω_m, k_{FM}, and the modulation index $\beta_{FM} = \Delta\omega/\omega_m$.

15.2.18 A commercial FM station broadcasts a signal with 180-kHz bandwidth when $|f(t)|_{\max} = 2$ V. Find k_{FM} for the modulator if the spectral extent of $f(t)$ is 30 kHz, by using Carson's rule.

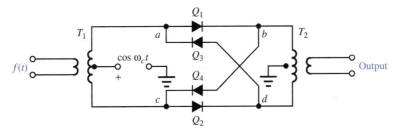

Figure P15.2.14

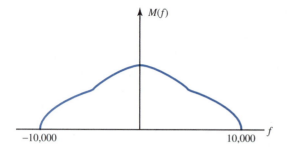

$M(f)$

−10,000   10,000

f

15.2.19 An FM station's modulator has a sensitivity $k_{FM} = 5\pi \times 10^4$ rad/s·V. A receiver uses a discriminator that has a gain constant of $10^{-5}/\pi$ V·s/rad. Neglecting noise, determine the signal at the receiver output.

15.2.20 A voice message with $W_f = 2\pi(3 \times 10^3)$ rad/s and $W_{rms} = 2\pi(1 \times 10^3)$ rad/s is transmitted over an FM broadcast system with standard emphasis. See Example 15.2.2, and compare the improvements due to emphasis.

15.2.21 Show that the image frequency for an FM station does not fall in the range of 88.1–107.9 MHz, regardless of the choice of high- or low-side local oscillator.

***15.2.22** Consider the signals given by Equations (15.2.12) through (15.2.14). Let them be normalized such that the red, green, and blue signals have a maximum amplitude of unity. If $m_I(t)$ and $m_Q(t)$ are the real and imaginary components of a color vector, defined by a magnitude $[m_I^2(t) + m_Q^2(t)]^{1/2}$ and phase $\tan^{-1}[-m_Q(t)/m_I(t)]$, sketch points corresponding to fully saturated R, G, and B colors.

15.2.23 If a television station operates on UHF channel 20 (band 506–512 MHz), determine the station's visual-carrier frequency.

15.2.24 For DSB and conventional AM, obtain expressions for the in-phase and quadrature components $x_d(t)$ and $x_q(t)$, and envelope and phase $v(t)$ and $\phi(t)$.

15.2.25 The normalized signal $x_n(t)$ has a bandwidth of 10 kHz and its power content is 0.5 W, while the carrier $A_c \cos 2\pi f_c t$ has a power content of 200 W. Find the bandwidth and the power content of the modulated signal, if the modulation scheme is:

(a) Conventional AM with a modulation index

of 0.6 and a transmitted signal of $A_c[1 + m_A x(t)] \cos 2\pi f_c t$.

(b) DSB SC with a transmitted signal of $A_c x(t) \cos 2\pi f_c t$.

(c) SSB with a transmitted signal of $A_c x(t) \cos 2\pi f_c t \mp A_c \bar{x}(t) \sin 2\pi f_c t$, where $\bar{x}(t)$ is a signal with a 90° phase shift in all frequency components of $x(t)$, and the upper − sign corresponds to the USSB signal, whereas the lower + sign corresponds to the LSSB signal.

(d) FM with $k_f = 50$ kHz.

15.2.26 Let the modulating signal $m(t)$ be a sinusoid of the form $m(t) = \cos 2\pi f_m t$, $f_m \ll f_c$, and let the carrier signal be $\cos(2\pi f_c t + \phi_c)$.

(a) Determine the conventional AM signal, its upper and lower sidebands, and its spectrum if the modulation index is α.

(b) Determine the DSB SC AM signal, its upper and lower sidebands, and its spectrum for $m(t) = \alpha \cos 2\pi f_m t$, $f_m \ll f_c$.

(c) Determine the two possible SSB AM signals and their spectra.

***15.2.27** Let the message signal $m(t) = \alpha \cos(2\pi f_m t)$ be used to either frequency-modulate or phase-modulate the carrier $A_c \cos(2\pi f_c t)$. Find the modulated signal in each case.

15.2.28 If an FM signal is given by $s_{FM}(t) = 100 \cos[2\pi f_c t + 100 \int_{-\infty}^{t} m(\tau) \, d\tau]$ and $m(t)$ is given in Figure P15.2.28, sketch the instantaneous frequency as a function of time and determine the peak frequency deviation.

15.2.29 If $m(t)$ of Figure P15.2.29 is fed into an FM modulator with peak frequency deviation $k_f = 25$ Hz/V, plot the frequency deviation in hertz and the phase deviation in radians.

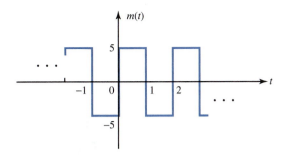

Figure P15.2.28

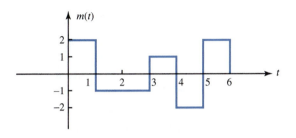

Figure P15.2.29

15.2.30 An angle-modulated signal has the form $u(t) = 100 \cos[2\pi f_c t + 4 \sin 2\pi f_m t]$, where $f_c = 10$ MHz and $f_m = 1$ kHz. Determine the modulation index β_f or β_p and the transmitted signal bandwidth, with $B_{FM} = 2(\beta_f + 1) f_m$ or $B_{PM} = 2(\beta_p + 1) f_m$,

(a) If this is an FM signal.

(b) If this is a PM signal.

15.2.31 Let $m_1(t)$ and $m_2(t)$ be two message signals, and let $u_1(t)$ and $u_2(t)$ be the corresponding modulated versions.

(a) When the combined message signal $m_1(t) + m_2(t)$ DSB modulates a carrier $A_c \cos 2\pi f_c t$, show that the result is the sum of the two DSB AM signals $u_1(t) + u_2(t)$. That is to say, AM satisfies the superposition principle.

(b) If $m_1(t) + m_2(t)$ frequency modulates a carrier, show that the modulated signal is not equal to $u_1(t) + u_2(t)$. That is to say, FM does not satisfy the principle of superposition, and angle modulation is not a linear modulation method.

15.2.32 Figure P15.2.32 shows an FM discriminator. Assume the envelope detector to be ideal, with an infinite input impedance. Choose the values for L and C if the discriminator is to be employed to demodulate an FM signal with a carrier frequency $f_c = 80$ MHz and a peak frequency deviation of 6 MHz.

*15.2.33 Let a message signal $m(t)$ have a bandwidth of 10 kHz and a peak magnitude of 1 V. Estimate the bandwidth, by using Carson's rule, of the signal $u(t)$ obtained when $m(t)$ modulates a carrier with a peak frequency deviation of:

(a) $f_d = 10$ Hz/V.

(b) 100 Hz/V.

(c) 1000 Hz/V.

15.2.34 The operating frequency range of a superheterodyne FM receiver is 88–108 MHz. The IF and LO frequencies are so chosen that $f_{IF} < f_{LO}$. If the image frequency f_c' must fall outside of the 88–108-MHz range, determine the minimum needed f_{IF} and the corresponding range of f_{LO}.

15.2.35 The television audio signal of Figure 15.2.29 is frequency modulated with $\Delta f = 25$ kHz and has $W_f \cong 10$ kHz. By using $W_{FM} \cong 2(\Delta \omega + 2W_f)$ for $\Delta \omega \gg W_f$ (for wide-band FM), find the percentage of the channel bandwidth occupied by the audio signal.

15.2.36 Suppose that a video signal, having $W = 5$ MHz, is transmitted via FM with $\Delta f = 20$ MHz. For $1/100 \le B/f_c \le 1/10$, determine the bounds on the carrier frequency. Use $W_{FM} \cong 2(\Delta \omega + 2W_f)$ for $\Delta \omega \gg W_f$ (for wide-band FM).

15.2.37 A TDM signal, like that of Figure 14.2.9(b), is formed by sampling 5 voice signals at $f_s = 8$ kHz. The signal is then transmitted via FM on a radio channel with 400-kHz bandwidth. Estimate the maximum allowable frequency deviation Δf [use

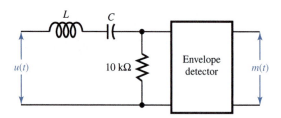

Equation (15.2.13) for wide-band FM if $\Delta f \geq 2W$).

*15.2.38 A radar system uses pulses of duration D to modulate the amplitude of a radio carrier wave. The system can distinguish between targets spaced apart by a distance of $d \geq cD$, where c is the velocity of light. In view of $1/100 \leq B/f_c \leq 1/10$, find the minimum practical value of f_c so that $d_{\min} = 30$ m.

15.2.39 A TDM signal of the type shown in Figure 14.2.9(b) is formed by sampling M voice signals at $f_s = 8$ kHz. If the TDM signal then modulates the amplitude of a 4-MHz carrier for radio transmission, determine the upper limit on M that satisfies $1/100 \leq B/f_c \leq 1/10$.

*15.3.1 Find the minimum rate of sampling that must be used to convert the message into digital form:

   (a) If an audio message has a spectral extent of 3 kHz.

   (b) If a television signal has a spectral extent of 4.5 MHz.

15.3.2 If an analog message that has a spectral extent of 15 kHz is sampled at three times the Nyquist rate, determine the sampling rate.

15.3.3 The quantum levels of a quantizer are separated by the step size $\delta v = 0.2$ V, with the lowest and highest levels of -3.3 V and $+3.3$ V, respectively. A sequence of message samples is given by -2.15 V, -0.53 V, 0.95 V, 0.17 V, and -0.76 V.

   (a) List the quantum levels.

   (b) Is the quantizer midriser or midtread?

   (c) Find the corresponding sequence of the quantized samples.

   (d) Determine the maximum amplitude that a message can have if the quantizer is not to be saturated.

15.3.4 When the quantum step size δv and the step size of $f(t)$ are the same as in Figure 15.3.2, the quantizer is said to have a gain of unity. If, on the other hand,

the quantizer has a gain of $K_q > 1$, what does that imply?

15.3.5 A symmetrically fluctuating message that has a crest factor $K_{\mathrm{CR}} = 3$ and $\overline{f^2(t)} = 2.25$ V^2 is to be quantized such that the signal-to-noise ratio $S_0/N_q = 2700$. If the quantizer is to use the smallest number of levels that is a power of 2, find the following:

   (a) The step size v.

   (b) The binary number of levels L_b.

   (c) The extreme quantum level voltages.

*15.3.6 A quantizer has 130 quantum levels that exactly span the extremes of a symmetrically fluctuating message with step size $\delta v = 0.04$ V. Determine the following:

   (a) $|f(t)|_{\max}$.

   (b) The largest crest factor the message can have if S_0/N_q must be at least 5500.

15.3.7 If a compressor in a system can change a message crest factor from 3.2 to 2, while maintaining its peak amplitude constant, find the decibels of improvement in signal-to-noise ratio that can be expected in the system.

15.3.8 Determine the number of bits of a natural binary encoder that works with the quantizer:

   (a) If the extreme levels are ±3.1 V, with a step size of 0.2 V.

   (b) If there are 128 quantum levels with $\delta v = 0.04$ V.

15.3.9 A differential encoding scheme converts the original sequence of digits, denoted by $\{b_k\}$, to a new sequence of digits, denoted by $\{a_k\}$, by using the *differential encoder* shown in Figure P15.3.9. The output digit in the kth interval is given by $a_k = a_{k-1} \oplus b_k$, where $\oplus$ represents *modulo-2 addition* (i.e., $0 \oplus 0 = 0; 0 \oplus 1 = 1; 1 \oplus 0 = 1$; and $1 \oplus 1 = 0$). The new sequence is used in waveform formatting. For the input sequence $\{b_k\}$ given below, find the sequences $\{a_k\}$ and $\{a_{k-1}\}$ by letting the initial value of the output be 1. Also, show the schematic representation of the required decoder in the receiver.

Input $\{b_k\}$: 0 1 1 1 0 1 0 0 1 0

                   1 1 0 0

15.3.10 (a) In the so-called *folded binary code* of 4 bits, the leftmost digit represents the sign of an analog signal's quantized samples (with 0 for negative and 1 for positive), and the next three digits are the natural binary code words for the magnitude of the quantized samples. Obtain a table of binary code words.

(b) By letting the digit sequence in part (a) be $b_4b_3b_2b_1$, which also represents a 4-bit natural binary code for 16 levels labeled 0 through 15, obtain the digit sequence $g_4g_3g_2g_1$, where

$$g_k = \begin{cases} b_4, & \text{for } k = 4 \\ b_{k+1} \oplus b_k, & \text{for } k = 1, 2, 3 \end{cases}$$

which is known as the *Gray code*. A unique characteristic of the Gray code is that code words change in only one digit between adjacent levels. Check the same in your result. [See Problem 15.3.9 for modulo-2 addition, represented by $\oplus$.]

15.3.11 Consider the digit sequence

$$\{b_k\} = 1 \ 0 \ 0 \ 1 \ 1 \ 1 \ 0 \ 0 \ 1 \ 0$$

           1 1 0 1 1 0 0 0

(a) Sketch the polar and unipolar waveforms for the sequence. Estimate the probability that a binary 1 will occur in the next digit interval.

(b) Let $\{b_k\}$ be the input to the differential encoder of Figure P15.3.9, and

(i) Find $\{a_k\}$ with the initial value 1 of the output.

(ii) Find $\{a_k\}$ with the initial value 0 of the output.

(iii) Check whether the original sequence $\{b_k\}$ is recovered in both cases, when $\{a_k\}$ of (i) and (ii) is put through a differential decoder, as shown in Figure P15.3.11.

(c) For the first 10 digits of the sequence $\{b_k\}$, illustrate the Manchester-formatted waveform.

*15.3.12 An audio message is band-limited to 15 kHz, sampled at twice the Nyquist rate, and encoded by a 12-bit natural binary code that corresponds to $L_b = 2^{12} = 4096$ levels, all of which span the message's variations. Find the first-null bandwidth (given by $\omega_b/2\pi = 1/T_b$) required to support a polar waveform format. Also, determine the best possible signal-to-noise ratio S_0/N_q with a crest factor of 3.8.

15.3.13 In Equation (15.3.24), for polar and Manchester PCM, $E_b = A^2T_b$. Find the noise level $N_0/2$ at the input to the polar PCM receiver that corresponds to $P_e = 10^{-5}$ when $A = 6$ V and $T_b = 0.5$ μs.

15.3.14 For a unipolar PCM, the bit-error probability is given by

$$P_e = \frac{1}{2} \text{erfc} \left[\sqrt{\frac{1}{8} \left(\frac{\hat{S}_i}{N_i} \right)} \right]_{\text{PCM}}$$

whereas for polar and Manchester PCM,

$$P_e = \frac{1}{2} \text{erfc} \left[\sqrt{\frac{1}{2} \left(\frac{\hat{S}_i}{N_i} \right)} \right]_{\text{PCM}}$$

Figure P15.3.9

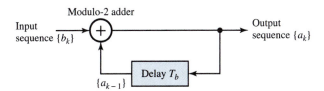

Modulo-2 adder

Input sequence $\{b_k\}$ Output sequence $\{a_k\}$

Delay T_b

$\{a_{k-1}\}$

Figure P15.3.11

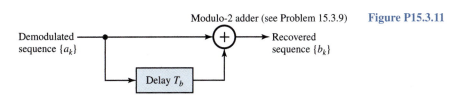

Modulo-2 adder (see Problem 15.3.9)

Demodulated sequence $\{a_k\}$ Recovered sequence $\{b_k\}$

Delay T_b

where $\hat{S}_i/N_i$ is the ratio of the power in the signal peak amplitude (at the integrator output at the sample time) to the average noise power at the same point. For $(\hat{S}_i/N_i)_{PCM} = 32$ at the sampler, find P_e for a unipolar and a polar system. In which system will the bit errors occur more frequently?

15.3.15 For a Manchester PCM system, expressions for E_b and P_e are given in Problems 15.3.13 and 15.3.14. Determine:

(a) The minimum value of $(\hat{S}_i/N_i)_{PCM}$ that is needed to realize $P_e = 10^{-4}$.

(b) The minimum input-pulse amplitude A required if $P_e = 10^{-4}$, $N_0 = 8 \times 10^{-7}$ V²/Hz, and $T_b = 0.4 \ \mu s$.

15.3.16 For polar or Manchester PCM, Equation (15.3.13) is sometimes expressed as

$$\left(\frac{S_0}{N_0}\right)_{PCM} = \frac{S_0/N_q}{1 + 2^{2N_b+1} \ \text{erfc}\left[\sqrt{\frac{1}{2}\left(\frac{\hat{S}_i}{N_i}\right)_{PCM}}\right]}$$

where $\hat{S}_i/N_i$ is the ratio of the power in the signal peak amplitude (at the integrator output at the

sample time) to the average noise power at the same point, and N_b is the number of bits in a natural binary code. In a polar PCM system, with $N_b = 8$ and $(\hat{S}_i/N_i)_{PCM} = 20$, find $(S_0/N_0)_{PCM}$ for the message if its crest factor is 3. Assume that all levels span the variations of the analog message. Comment on whether the system is operating above threshold.

*15.3.17 The expression given in Problem 15.3.16 for $(S_0/N_0)_{PCM}$ is plotted for a sinusoidal message in Figure P15.3.17 to show the performance curves for a PCM system using a polar waveform.

(a) Comment on the threshold effect.

(b) For the polar system of Problem 15.3.14, with $N_b = 8$ and $(\hat{S}_i/N_i)_{PCM} = 32$, determine whether the system noise has significant effect on performance.

15.3.18 Four voice messages, each with 3-kHz bandwidth, are to be sampled at Nyquist rate, and time-multiplexed with samples taken at twice the Nyquist rate from six analog monitoring signals, each with 0.5-kHz bandwidth. Determine the duration of the frame and suggest a sampling scheme indicating the timing of samples.

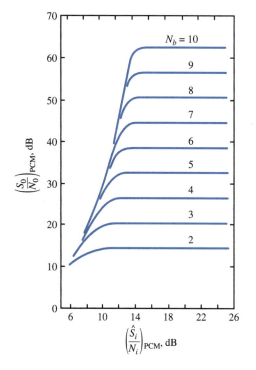

Figure P15.3.17

15.3.19 By a simple time multiplexing of natural samples over a single line, a large radar site transmits 85 analog signals, each with 200-Hz bandwidth. If the sampling is done at twice the Nyquist rate, one time slot is used for synchronization, and the guard time $\tau_g = 2\tau$, find the duration τ of the pulse.

15.3.20 Sketch an ASK signal for the binary sequence 1 0 1 1 0 0 1 0 0 1 if $\omega_c = 3\omega_b = 6\pi/T_b$.

15.3.21 In Equation (15.3.24), for coherent ASK, $E_b = A^2 T_b/4$, with $A = 2.2$ V, $T_b = 2.6$ μs, and $N_0 = 2 \times 10^{-7}$ V^2/Hz, find P_e for a coherent ASK system. Also, find E_b for a 1-Ω impedance.

*15.3.22 In Figure 15.3.14, when the pulse of a carrier of amplitude A during $0 < t \leq T_b$ arrives at point R, show that $D \cong A^2 T_b/4$ at time T_b if noise is neglected and $\omega_c T_b >> 1$.

15.3.23 If $E_b/N_0 = 20$ in a coherent ASK system, find the value of E_b/N_0 that is needed in a noncoherent ASK system to yield the same value of P_e as the coherent system.

15.3.24 To have the same value of P_e, show that E_b/N_0 in a coherent-ASK system has to be twice that in a coherent PSK system.

15.3.25 In a DPSK system, when the received pulses are 2 V in amplitude, $P_e = 3 \times 10^{-4}$. If the pulse amplitude increases such that $P_e = 2 \times 10^{-6}$, find the new amplitude.

15.3.26 Consider Figure 15.3.19, in which the receiver becomes two coherent ASK receivers in parallel. Justify why the difference, rather than the sum, occurs.

15.3.27 Consider Figure 15.3.19, in which the input FSK pulse at point R is given by $s_R(t) = A \cos(\omega_2 t + \phi_2)$, $0 < t < T_b$, and zero elsewhere in t, when the bit interval corresponds to a 1.

(a) Compute the outputs from the two integrators.

(b) If $s_R(t)$ changes in frequency and phase to ω_1 and ϕ_1, respectively, show that the two outputs of part (a) are reversed. (Assume $\omega_c T_b >> 1$, and make suitable approximations.)

*15.3.28 Apply Equation (15.3.24) for various digital communication systems with $E_b/N_0 = 12$, and using PSK as the reference, compare their performances.

15.3.29 A communication system for a voice-band (3 kHz) channel is designed for a received SNR E_b/N_0 at the detector of 30 dB when the transmitter power is $P_s = -3$ dBW. Find the value of P_s if it is desired to increase the bandwidth of the system to 10 kHz, while maintaining the same SNR at the detector.

PART

CONTROL SYSTEMS

FIVE

16 Basic Control Systems

Electric energy is widely used in practice, because of the ease with which the system and device performance can be reliably controlled. One of the major areas of electrical engineering of interest to all engineers is control and instrumentation. Instrumentation is integrated throughout the book in sections on electric circuits, electronic analog and digital systems, energy systems, and information systems. This final chapter serves to introduce a variety of methods by which the performance of physical systems is controlled. By focusing on control aspects, the integration of many of the concepts used in the preceding chapters is effected.

Systems is a term used in many fields of study: economics, ecology, social and physical sciences. The catchword is used to describe an assemblage of components, subsystems, and interfaces arranged or existing in such a fashion as to perform a function or functions in order to achieve a goal. *Control* refers to the function or purpose of the system we wish to discuss. Control is almost always realized not by a single component, such as a transistor, resistor, or motor, but by an entire system of components and interfaces.

Control systems influence our everyday lives just as much as some of the other areas of electrical engineering. Examples abound in practice: household appliances, manufacturing and processing plants, and navigational and guidance systems, in which concepts of the analysis and design of control systems are utilized. A control system, in general, can be viewed as an interconnection of components—electrical, mechanical, hydraulic, thermal, etc.—so as to obtain a desired function in an efficient and accurate manner. The control engineer is concerned with the control of industrial processes and systems. The concepts of control engineering are not limited to any particular branch of engineering. Hence, a basic understanding of control theory is essential to every engineer involved in the understanding of the dynamic behavior of various systems.

This chapter introduces different types of control systems, and some elementary methods for studying their behavior. Three classes of control systems are presented:

1. *Power semiconductor-controlled drives*, in which the electrical input to a motor is adjusted to control performance.

2. *Feedback control systems*, in which a measure of the actual performance has to be known in order to effect control.

3. *Digital control systems*, in which a digital processor becomes an essential element of the system, and the resulting processed output forms the basis for system control.

Many of the concepts and techniques used may be similar to those already developed earlier in the book. Indeed, any discussion of control methodology integrates much of the material on circuits, electronic devices, and electromechanical energy-conversion devices. Such control techniques are also employed in business, ecological, and social systems, as well as in problem areas related to inventory control, economic models, health-care delivery systems, and urban planning.

16.1   POWER SEMICONDUCTOR-CONTROLLED DRIVES

Power electronics deals with the applications of solid-state electronics for the control and conversion of electric power. Conversion techniques require switching power semiconductor devices on and off. The development of solid-state motor drive packages has progressed to the point at which they can be used to solve practically any power-control problem. This section describes fundamentals common to all electric drives: dc drives fed by controlled rectifiers and choppers; squirrel-cage induction motor drives controlled by ac voltage controllers, inverters, and cycloconverters; slip-power-controlled wound-rotor induction motor drives; and inverter-controlled and cycloconverter-controlled synchronous motor drives, including brushless dc and ac motor drives. Even though the detailed study of such power electronic circuits and components would require a book in itself, some familiarity becomes important to an understanding of modern motor applications. This section is only a very modest introduction.

The essential components of an electric drive controlled by a power semiconductor converter are shown in the block diagram of Figure 16.1.1. The converter regulates the flow of power from the source to the motor in such a way that the motor speed–torque and speed–current characteristics become compatible with the load requirements. The low-voltage control unit, which may consist of integrated transistorized circuits or a microprocessor, is electrically isolated from the converter-motor circuit and controls the converter. The sensing unit, required for closed-loop operation or protection, or both, is used to sense the power circuit's electrical parameters, such as converter current, voltage, and motor speed. The command signal forms an input to the control unit, adjusting the operating point of the drive. The complete electric drive system shown in Figure 16.1.1 must be treated as an integrated system.

A motor operates in two modes—motoring and braking. Supporting its motion, it converts electric energy to mechanical energy while motoring. In braking, while opposing the motion, it works as a generator converting mechanical energy to electric energy, which is consumed in some part of the circuit. The motor can provide motoring and braking operations in both forward and reverse directions. Figure 16.1.2 illustrates the four-quadrant operation of drives. The continuous as well as the transient torque and power limitations of a drive in the four quadrants of operation

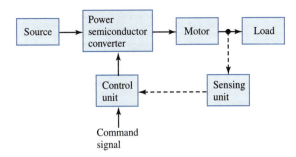

Figure 16.1.1 Essential components of an electric drive.

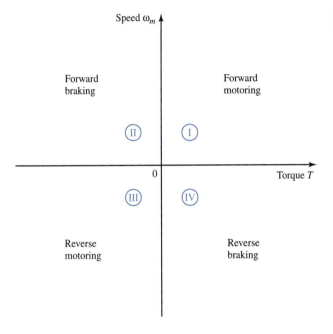

Figure 16.1.2 Four-quadrant operation of drives.

are shown in Figure 16.1.3 for speeds below and above *base speed* ω_{mb}, which is the highest drive speed at the rated flux.

Motors commonly used in variable-speed drives are induction motors, dc motors, and synchronous motors. For the control of the motors, various types of converters are needed, as exemplified in Table 16.1.1. A variable-speed drive can use a single converter or more than one. All converters have harmonics in their inputs and outputs. Some converters suffer from a poor power factor, particularly at low output voltages. The main advantages of converters are high efficiency, fast response, flexibility of control, easy maintenance, reliability, low weight and volume, less noise, and long life. The power semiconductor converters have virtually replaced the conventional power controllers such as mercury-arc rectifiers and magnetic amplifiers. Most of the drive specifications are governed by the load requirements, which in turn depend on normal running needs, transient operational needs, and needs related to location and environment. Other specifications are governed by the available source and its capacity, as well as other aspects like harmonics, power factor, reactive power, regenerated power, and peak current.

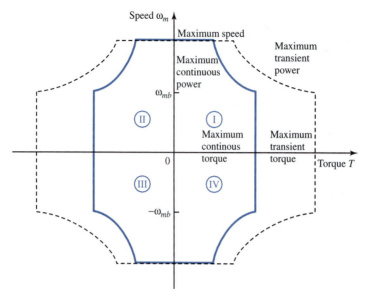

Figure 16.1.3 Continuous as well as transient torque and power limitations of a drive.

Power Semiconductor Devices

Since the advent of the first thyristor or silicon-controlled rectifier (SCR) in 1957, tremendous advances have been made in power semiconductor devices during the past four decades. The devices can be divided broadly into four types:

- Power diodes
- Thyristors
- Power bipolar junction transistors (BJTs)
- Power MOSFETs (metal-oxide semiconductor field-effect transistors).

Thyristors can be subdivided into seven categories:

- Forced-commutated thyristor
- Line-commutated thyristor
- Gate turn-off thyristor (GTO)
- Reverse-conducting thyristor (RCT)
- Static-induction thyristor (SITH)
- Gate-assisted turn-off thyristor (GATT)
- Light-activated silicon-controlled rectifier (LASCR).

Typical ratings of these devices are given in Table 16.1.2. Whereas this comprehensive table, as of 1983, is given here for illustration purposes only, several new developments have taken place since then: A single thyristor is currently available with the capability of blocking 6.5 kV and controlling 1 kA. The 200-V Schottky barrier diodes are commercially available with ratings of several hundred amperes, up to 1 kA, and with blocking voltages as high as 1–5 kV.

TABLE 16.1.1 Converters and Their Functions for the Control of Motors

Converter	Conversion Function	Applications
Controlled rectifiers	Ac to variable dc	Control of dc motors and synchronous motors
Choppers	Fixed voltage dc to variable voltage dc	Control of dc motors
Ac voltage controllers	Fixed voltage ac to variable-voltage ac at same frequency	Control of induction motors
Inverters (voltage source or current source)	Dc to fixed or variable voltage and frequency ac, voltage or current source	Control of induction motors and synchronous motors
Cycloconverters	Fixed voltage and frequency ac to variable voltage and frequency ac	Induction motors and synchronous motors

MOS-controlled thyristors (MCT) are rated for 3 kV, capable of interrupting around 300 A with a recovery time of 5 μs. Gate turn-off (GTO) thyristors can control 1 to 3 kA with a blocking voltage capability of 6 to 8 kV. Table 16.1.3 gives the symbols and the *v-i* characteristics of the commonly used power semiconductor devices.

Figure 16.1.4 illustrates the output voltages and control characteristics of some commonly used power switching devices. The switching devices can be classified as follows:

TABLE 16.1.2 Typical Ratings of Power Semiconductor Devices

Type		Voltage/Current Rating	Switching Time (μs)	On Voltage/Current[*]
Diodes	General purpose	3 kV/3.5 kA		1.6 V/10 kA
	High speed	3 kV/1 kA	2–5	3 V/3 kA
	Schottky	40 V/60 A	0.23	0.58 V/60 A
Forced-turned-off thyristors	Reverse blocking	3 kV/1 kA	400	2.5 V/10 kA
	High speed	1.2 kV/1.5 kA	20	2.1 V/4.5 kA
	Reverse blocking	2.5 kV/400 A	40	2.7 V/1.25 kA
	Reverse conducting	2.5 kV/1 kA/R400 A	40	2.1 V/1 kA
	GATT	1.2 kV/400 A	8	2.8 V/1.25 kA
	Light triggered	6 kV/1.5 kA	200–400	2.4 V/4.5 kA
TRIACs		1.2 kV/300 A		1.5 V/420 A
Self-turned-off thyristors	GTO	3.6 kV/600 A	25	2.5 V/1 kA
	SITH	4 kV/2.2 kA	6.5	2.3 V/400 A
Power transistors	Single	400 V/250 A	9	1 V/250 A
		400 V/40 A	6	1.5 V/49 A
		630 V/50 A	1.7	0.3 V/20 A
	Darlington	900 V/200 A	40	2 V
SITs		1.2 kV/10 A	0.55	1.2 Ω
Power MOSFETs		500 V/8.6 A	0.7	0.6 Ω
		1 kV/4.7 A	0.9	2 Ω
		500 V/10 A	0.6	0.4 Ω

Source: F. Harashima, "State of the Art on Power Electronics and Electrical Drives in Japan," in *Proc. 3rd IFAC Symposium on Control in Power Electronics and Electrical Drives* (Lausanne, Switzerland, 1983), Tutorial Session and Survey Papers, pp. 23–33.
[*] *Note:* On voltage is the on-state voltage drop of the device at a specified current.

TABLE 16.1.3 Symbols and v–i Characteristics of Some Power Semiconductor Devices

Device	Symbol	Characteristics
Diode	A, I_D, K; V_D	I_D vs V_D
Thyristor	I_D, $+G$, V_G, A, K	I_D vs V_D, Gate triggered
GTO	I_A, G, A, K	I_D vs V_D, Gate triggered
TRIAC	I_D, B, A, G	I_D vs V_D, Gate triggered
LASCR	A, K, I_D, G	I_D vs V_D, Gate triggered
NPN BJT	I_B, I_C, C, B, E, I_E	I_C vs V_{CE}; $I_{Bn} > I_{B1}$, I_{B1}
PNP BJT	I_B, C, I_C, B, E, I_E	I_C vs V_{EC}; $I_{Bn} > I_{B1}$, I_{B1}
n-channel MOSFET	D, I_D, G, S	I_D vs V_{DS}; V_{GS1}, $V_{GS1} > V_{GSn}$, V_{GSn}
p-channel MOSFET	D, I_D, G, S	I_D vs V_{DS}; V_{GS1}, $V_{GS1} < V_{GSn}$, V_{GSn}

- Uncontrolled turn-on and turn-off (diode)
- Controlled turn-on and uncontrolled turn off (SCR)
- Controlled turn-on and turn-off (BJT, GTO, MOSFET)
- Continuous gate signal requirement (BJT, MOSFET)
- Pulse gate requirement (GTO, SCR)
- Bipolar voltage capability (SCR)
- Unipolar voltage capability (BJT, GTO, MOSFET)
- Bidirectional current capability (RCT, TRIAC)
- Unidirectional current capability (BJT, diode, GTO, MOSFET, SCR).

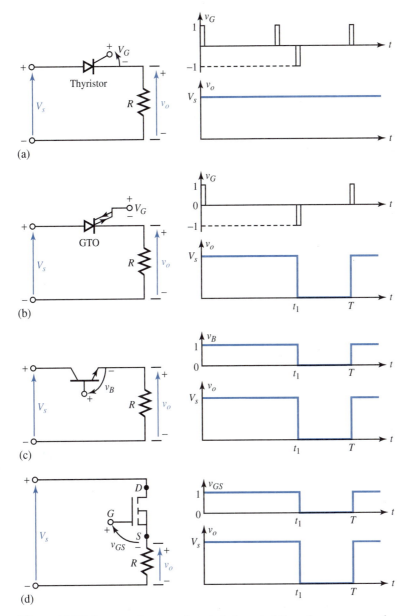

Figure 16.1.4 Output voltages and control characteristics of some commonly used power switching devices. **(a)** Thyristor switch. **(b)** GTO switch. **(c)** Transistor switch. **(d)** MOSFET switch.

Power Electronic Circuits

These circuits can be classified as follows:

- Diode rectifiers
- Ac–dc converters (controlled rectifiers)

- Ac–ac converters (ac voltage controllers)
- Dc–dc converters (dc choppers)
- Dc–ac converters (inverters)
- Static switches (contactors), supplied by either ac or dc.

Figure 16.1.5 shows a single-phase rectifier circuit converting ac voltage into a fixed dc voltage. It can be extended to three-phase supply. Figure 16.1.6 shows a single-phase ac–dc converter with two natural line-commutated thyristors. The average value of the output voltage is controlled by varying the conduction time of the thyristors. Three-phase input can also be converted.

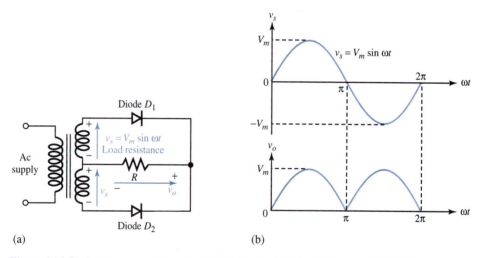

Figure 16.1.5 Single-phase rectifier circuit with diodes. **(a)** Circuit diagram. **(b)** Voltage waveforms.

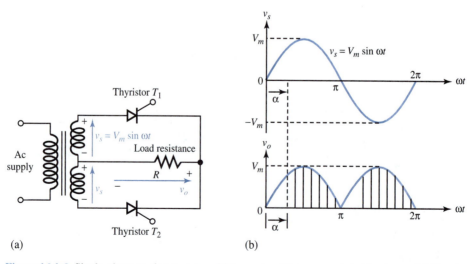

Figure 16.1.6 Single-phase ac–dc converter with two natural line-commutated thyristors. **(a)** Circuit diagram. **(b)** Voltage waveforms.

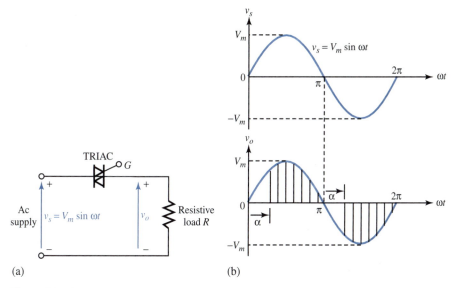

(a)   (b)

Figure 16.1.7 Single-phase ac–ac converter with TRIAC. **(a)** Circuit diagram. **(b)** Voltage waveforms.

Figure 16.1.7 shows a single-phase ac–ac converter with a TRIAC to obtain a variable ac output voltage from a fixed ac source. The output voltage is controlled by changing the conduction time of the TRIAC.

Figure 16.1.8 shows a dc–dc converter in which the average output voltage is controlled by changing the conduction time t_1 of the transistor. The chopping period is T, the duty cycle of the chopper is δ, and the conduction time t_1 is given by δT.

Figure 16.1.9 shows a single-phase dc–ac converter, known as an inverter, in which the output voltage is controlled by varying the conduction time of transistors. Note that the voltage is of alternating form when transistors Q_1 and Q_2 conduct for one-half period and Q_3 and Q_4 conduct for the other half.

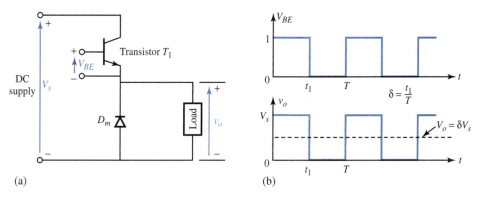

(a)   (b)

Figure 16.1.8 A dc–dc converter. **(a)** Circuit diagram. **(b)** Voltage waveforms.

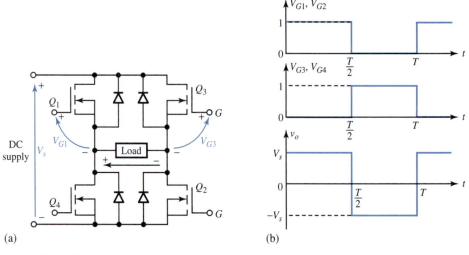

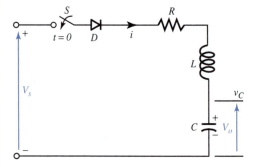

Figure 16.1.9 Single-phase dc–ac converter (inverter). (a) Circuit diagram. (b) Voltage waveforms.

EXAMPLE 16.1.1

Consider a diode circuit with an *RLC* load, as shown in Figure E16.1.1, and analyze it for $i(t)$ when the switch S is closed at $t = 0$. Treat the diode as ideal, so that the reverse recovery time and the forward voltage drop are negligible. Allow for general initial conditions at $t = 0$ to have nonzero current and a capacitor voltage $v_C = V_0$.

Figure E16.1.1 Diode circuit with *RLC* load.

Solution

The KVL equation for the load current is given by

$$L\frac{di}{dt} + Ri + \frac{1}{C} \int i\, dt + v_C \text{ (at } t = 0) = V_S$$

Differentiating and then dividing both sides by L, we obtain

$$\frac{d^2i}{dt^2} + \frac{R}{L}\frac{di}{dt} + \frac{i}{LC} = 0$$

Note that the capacitor will be charged to the source voltage V_S under steady-state conditions, when the current will be zero. While the forced component of the current is zero in the solution of the previous second-order homogeneous differential equation, we can solve for the natural component. The characteristic equation in the frequency domain is

$$s^2 + \frac{R}{L}s + \frac{1}{LC} = 0$$

whose roots are given by

$$s_{1,2} = -\frac{R}{2L} \pm \sqrt{\left(\frac{R}{2L}\right)^2 - \frac{1}{LC}}$$

For the second-order circuit, the damping factor α and the resonant frequency ω_0 are given by

$$\alpha = \frac{R}{2L}; \qquad \omega_0 = \frac{1}{\sqrt{LC}}$$

(*Note:* the ratio of α/ω_0 is known as *damping ratio* δ.) Substituting, we get

$$s_{1,2} = -\alpha \pm \sqrt{\alpha^2 - \omega_0^2}$$

Three possible cases arise for the solution of the current, which will depend on the values of α and ω_0:

Case 1—$\alpha = \omega_0$: The roots are then equal, $s_1 = s_2$. The current is said to be *critically damped*. The solution of the current is of the form

$$i(t) = (A_1 + A_2 t)e^{s_1 t}$$

Case 2—$\alpha > \omega_0$: The roots are unequal and real. The circuit is said to be *overdamped*. The solution is then

$$i(t) = A_1 e^{s_1 t} + A_2 e^{s_2 t}$$

Case 3—$\alpha < \omega_0$: the roots are complex. The circuit is *underdamped*. Let $s_{1,2} = -\alpha \pm j\omega_r$, where ω_r is known as the damped resonant frequency or *ringing frequency*, given by $\sqrt{\omega_0^2 - \alpha^2}$. The solution takes the form

$$i(t) = e^{-\alpha t}(A_1 \cos \omega_r t + A_2 \sin \omega_r t)$$

Observe that the current consists of a damped or decaying sinusoid. The constants A_1 and A_2 are determined from the initial conditions of the circuit.

The current waveform can be sketched, taking the conduction time of the diode into account.

Let us now consider a single-phase half-wave rectifier circuit as shown in Figure 16.1.10(a), with a purely resistive load R. We shall introduce and calculate the following quantities:

1. Efficiency
2. Form factor
3. Ripple factor
4. Transformer utilization factor
5. Peak inverse voltage (PIV) of diode
6. Displacement factor
7. Harmonic factor
8. Input power factor.

During the positive half-cycle of the input voltage, the diode D conducts and the input voltage appears across the load. During the negative half-cycle of the input voltage, the output voltage is

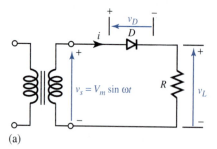

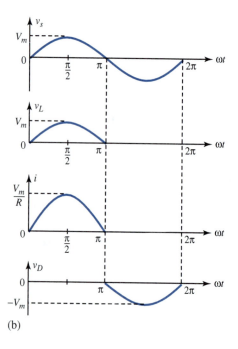

Figure 16.1.10 Single-phase half-wave rectifier. (a) Circuit. (b) Waveforms of voltages and current.

(a)

(b)

zero when the diode is said to be in a *blocking condition*. The waveforms of voltages v_S, v_L, v_D, and current i are shown in Figure 16.1.10(b).

The average voltage V_{dc} is given by

$$V_{dc} = \frac{1}{T} \int_0^T v_L(t) \, dt \tag{16.1.1}$$

because $v_L(t) = 0$ for $T/2 \le t \le T$. In our case,

$$V_{dc} = \frac{1}{T} \int_0^{T/2} v_m \sin \omega t \, dt = -\frac{V_m}{\omega T} \left(\cos \frac{\omega T}{2} - 1 \right) \tag{16.1.2}$$

Using the relationships $f = 1/T$ and $\omega = 2\pi f$, we obtain

$$V_{dc} = \frac{V_m}{\pi} = 0.318 V_m \tag{16.1.3}$$

$$I_{dc} = \frac{V_{dc}}{R} = \frac{0.318 V_m}{R} \tag{16.1.4}$$

The rms value of a periodic waveform is given by

$$V_{rms} = \left[\frac{1}{T} \int_0^T v_L^2(t) \, dt \right]^{1/2} \tag{16.1.5}$$

For a sinusoidal voltage $v_L(t) = V_m \sin \omega t$ for $0 \le t \le T/2$, the rms value of the output voltage is

$$V_{rms} = \left[\frac{1}{T} \int_0^{T/2} (V_m \sin \omega t)^2 \, dt \right]^{1/2} = \frac{V_m}{2} = 0.5 V_m \tag{16.1.6}$$

$$I_{rms} = \frac{V_{rms}}{R} = \frac{0.5 V_m}{R} \tag{16.1.7}$$

The output dc power P_{dc} is given by

$$P_{dc} = V_{dc} I_{dc} = \frac{(0.318 V_m)^2}{R} \tag{16.1.8}$$

The output average ac power P_{ac} is given by

$$P_{ac} = V_{rms} I_{rms} = \frac{(0.5 V_m)^2}{R} \tag{16.1.9}$$

The *efficiency* or the *rectification ratio* of the rectifier, which is a figure of merit used for comparison, is then

$$\text{Efficiency} = \eta = \frac{P_{dc}}{P_{ac}} = \frac{(0.318 V_m)^2}{(0.5 V_m)^2} = 0.404, \text{ or } 40.4\% \tag{16.1.10}$$

The *form factor* is FF $= V_{rms}/V_{dc}$, which gives a measure of the shape of the output voltage. In our case,

$$\text{FF} = \frac{V_{rms}}{V_{dc}} = \frac{0.5 V_m}{0.318 V_m} = 1.57, \text{ or } 157\% \tag{16.1.11}$$

The *ripple factor* is RF $= V_{ac}/V_{dc}$, which gives a measure of the ripple content. In our case,

$$\text{RF} = \frac{V_{ac}}{V_{dc}} = \sqrt{\left(\frac{V_{rms}}{V_{dc}} \right)^2 - 1} = \sqrt{\text{FF}^2 - 1} = \sqrt{1.57^2 - 1} = 1.21, \text{ or } 121\% \tag{16.1.12}$$

The *transformer utilization factor* (TUF) is given by

$$\text{TUF} = \frac{P_{dc}}{V_s I_s} \tag{16.1.13}$$

where V_s and I_s are the rms voltage and rms current, respectively, of the transformer secondary. In our case, the rms voltage of the transformer secondary is

$$V_s = \left[\frac{1}{T} \int_0^T (V_m \sin \omega t)^2 \, dt \right]^{1/2} = \frac{V_m}{\sqrt{2}} = 0.707 V_m \tag{16.1.14}$$

The rms value of the transformer secondary current is the same as that of the load,

$$I_s = \frac{0.5 V_m}{R} \tag{16.1.15}$$

Hence,

$$\text{TUF} = \frac{0.318^2}{0.707 \times 0.5} = 0.286 \tag{16.1.16}$$

The *peak inverse voltage* (PIV), which is the peak reverse blocking voltage, is given by

$$PIV = V_m \qquad (16.1.17)$$

The *displacement factor* DF is given by $\cos \phi$, where ϕ is the angle between the fundamental components of the input current and voltage.

The *harmonic factor* HF of the input current is given by

$$HF = \left(\frac{I_s^2 - I_1^2}{I_1^2}\right)^{1/2} = \left[\left(\frac{I_s}{I_1}\right)^2 - 1\right]^{1/2} \qquad (16.1.18)$$

where I_1 is the fundamental rms component of the input current.

The *input power factor* is given by

$$PF = \frac{I_1}{I_s} \cos \phi \qquad (16.1.19)$$

For an ideal rectifier,

$$\eta = 1.00; \qquad V_{ac} = 0; \qquad FF = 1.0; \qquad RF = 0;$$

$$TUF = 1.0; \qquad HF = 0; \qquad PF = 1.0 \qquad (16.1.20)$$

Solid-State Control of DC Motors

Dc motors, which are easily controllable, have historically dominated the adjustable-speed drive field. The torque–speed characteristics of a dc motor can be controlled by adjusting the armature voltage or the field current, or by inserting resistance into the armature circuit (see Section 13.4). Solid-state motor controls are designed to use each of these modes. The control resistors, in which much energy is wasted, are being eliminated through the development of power semiconductor devices and the evolution of flexible and efficient converters. Thus, the inherently good controllability of a dc machine has been significantly increased in recent years by *rectifier control, chopper control,* and *closed-loop control* of dc motors. When a dc source of suitable and constant voltage is already available, designers can employ dc-to-dc converters or choppers. When only an ac source is available, phase-controlled rectifiers are used. When the steady-state accuracy requirement cannot be satisfied in an open-loop configuration, the drive is operated as a closed-loop system. Closed-loop rectifier drives are more widely used than chopper drives. Only rectifier control of dc motors is considered here.

Controlled rectifier circuits are classified as fully controlled and half-controlled rectifiers, which are fed from either one-phase or three-phase supply. Figure 16.1.11 shows a fully controlled, rectifier-fed, separately excited dc motor drive and its characteristics. A transformer might be required if the motor voltage rating is not compatible with the ac source voltage. To reduce ripple in the motor current, a filter inductor can be connected in series between the rectifier and the motor armature. The field can be supplied from the same ac source supplying the armature, through a transformer and a diode bridge or a controlled rectifier. While single-phase controlled rectifiers are used up to a rating of 10 kW, and in special cases even up to 50 kW, three-phase controlled rectifiers are used for higher ratings.

V_a and I_a in Figure 16.1.11 denote the average values of the converter output voltage and current, respectively. Assuming continuous conduction when the armature currents flow continuously without becoming zero for a finite time interval, the variation of V_a with the firing angle is shown in Figure 16.1.11(b). Providing operation in the first and fourth quadrants of the V_a–I_a plane, as shown in Figure 16.1.11(c), the fully controlled rectifiers are two-quadrant

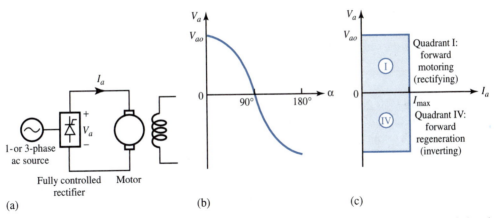

Figure 16.1.11 Fully controlled rectifier-fed separately excited dc motor drive and its characteristics. **(a)** Line diagram. **(b)** Output voltage versus firing angle curve. **(c)** Quadrants of operation.

converters. I_{max} is the rated rectifier current. In quadrant 4, the rectifier works like a line-commutated inverter with a negative output voltage, and the power flows from the load to the ac source.

Let us now consider the single-phase, fully controlled, rectifier-fed separately excited dc motor shown in Figure 16.1.12(a). Note that the armature has been replaced by its equivalent circuit, in which R_a and L_a, respectively, represent the armature-circuit resistance and inductance (including the effect of a filter, if connected), and E is the back emf.

Figure 16.1.12(b) shows the source voltage and thyristor firing pulses. The pair T_1 and T_3 receives firing pulses from α to π, and the pair T_2 and T_4 receives firing pulses from $(\pi + \alpha)$ to 2π.

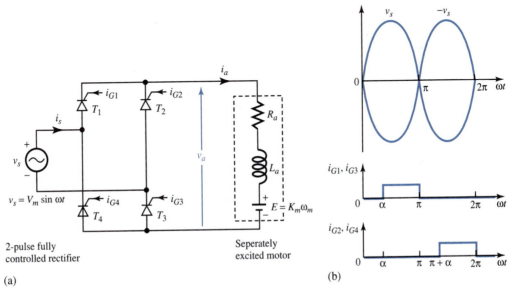

Figure 16.1.12 Single-phase two-pulse fully controlled rectifier-fed separately excited dc motor.

Only the *continuous conduction* mode of operation of the drive for motoring and regenerative braking will be considered here. The angle α can be greater or less than γ, which is the angle at which the source voltage v_s is equal to the back emf E,

$$\gamma = \sin^{-1}\left(\frac{E}{V_m}\right) \tag{16.1.21}$$

where V_m is the peak value of the supply voltage. For the case of $\alpha < \gamma$, waveforms are shown in Figure 16.1.13(a) for the motoring operation. It is possible to turn on thyristors T_1 and T_3 because $i_a > 0$, even though $v_s < E$. The same is true for thyristors T_2 and T_4.

When T_1 and T_3 conduct during the interval $\alpha \le \omega t \le (\pi + \alpha)$, the following volt–ampere equation holds:

$$v_s = E + i_a R_a + L_a \frac{di_a}{dt} \tag{16.1.22}$$

Multiplying both sides by $i_a \Delta t$, where Δt is a small time interval, we obtain

$$v_s i_a \Delta t = E i_a \Delta t + i_a^2 R_a \Delta t + L_a i_a \left(\frac{di_a}{dt}\right) \Delta t \tag{16.1.23}$$

in which the terms can be identified as the energy supplied or consumed by the respective elements. Figure 16.1.13(b) corresponds to the regenerative braking operation, in which

$$\gamma' = \pi - \gamma = \pi - \sin^{-1}\left(\frac{|E|}{V_m}\right) \tag{16.1.24}$$

and α can be greater or less than γ'.

When T_1 and T_3 conduct, Equation (16.1.22) describes the motor operation with $E = K_m \omega_m$. Thus,

$$v_a = L_a \frac{di_a}{dt} + R_a i_a + K_m \omega_m = V_m \sin \omega t \tag{16.1.25}$$

Similarly, when T_2 and T_4 conduct,

$$v_a = L_a \frac{di_a}{dt} + R_a i_a + K_m \omega_m = -V_m \sin \omega t \tag{16.1.26}$$

where ω is the supply frequency, ω_m is the motor speed, and K_m is the motor back emf constant.

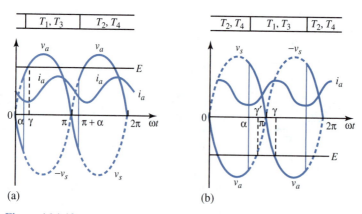

(a)　　　　　　　　　　　　　　(b)

Figure 16.1.13 Continuous conduction mode of operation. **(a)** Motoring (case shown for $\alpha < \gamma$). **(b)** regenerative braking (case shown for $\alpha < \gamma'$).

From α to $\pi + \alpha$ in the output voltage waveform of Figure 16.1.13(a), Equation (16.1.25) holds and its solution can be found as

$$i_a(\omega t) = \frac{V_m}{z} \sin(\omega t - \psi) - \frac{K_m \omega_m}{R_a} + K_1 e^{-t/\tau_a}, \qquad \text{for } \alpha \leq \omega t \leq \pi + \alpha \quad (16.1.27)$$

where

$$z = \left[R_a^2 + (\omega L_a)^2 \right]^{1/2} \qquad (16.1.28)$$

$$\tau_a = \frac{L_a}{R_a} \qquad (16.1.29)$$

$$\psi = \tan^{-1} \left(\frac{\omega L_a}{R_a} \right) \qquad (16.1.30)$$

and K_1 is a constant.

The first term on the right-hand side of Equation (16.1.27) is due to the ac source; the second term is due to the back emf; and the third represents the combined transient component of the ac source and back emf. In the steady state, however,

$$i_a(\alpha) = i_a(\pi + \alpha) \qquad (16.1.31)$$

Subject to the constraint in Equation (16.1.31), the steady-state expression of current can be obtained. Flux being a constant, recall that the average motor torque depends only on the average value or the dc component of the armature current, whereas the ac components produce only pulsating torques with zero-average value. Thus, the motor torque is given by

$$T_a = K_m I_a \qquad (16.1.32)$$

To obtain the average value of I_a under steady state, we can use the following equation:

Average motor voltage V_a = average voltage drop across R_a

$$+ \text{ average voltage drop across } L_a$$

$$+ \text{ back emf} \qquad (16.1.33)$$

in which

$$V_a = \frac{1}{\pi} \int_{\alpha}^{\pi+\alpha} V_m \sin(\omega t)\, d(\omega t) = \frac{2V_m}{\pi} \cos \alpha \qquad (16.1.34)$$

The rated motor voltage will be equal to the maximum average terminal voltage $2V_m/\pi$.

$$\text{Average drop across } R_a = \frac{1}{\pi} \int_{\alpha}^{\pi+\alpha} R_a i_a(\omega t)\, d(\omega t) = R_a i_a \qquad (16.1.35)$$

$$\text{Average drop across } L_a = \frac{1}{\pi} \int_{\alpha}^{\pi+\alpha} L_a \left(\frac{di_a}{dt} \right) d(\omega t) = \frac{\omega}{\pi} \int_{i_a(\alpha)}^{i_a(\pi+\alpha)} L_a\, di_a$$

$$= \frac{\omega L_a}{\pi} [i_a(\pi + \alpha) - i_a(\alpha)] = 0 \qquad (16.1.36)$$

Substituting, we obtain

$$V_a = I_a R_a + K_m \omega_m \qquad (16.1.37)$$

for the steady-state operation of a dc motor fed by any converter. From Equations (16.1.34) and (16.1.37), it follows that

$$I_a = \frac{(2V_m/\pi)\cos\alpha - K_m\omega_m}{R_a} \tag{16.1.38}$$

Substituting in Equation (16.1.32), we get the following equation after rearranging terms:

$$\omega_m = \frac{2V_m}{\pi K_m}\cos\alpha - \frac{R_a}{K_m^2}T_a \tag{16.1.39}$$

which is the relationship between speed and torque under steady state. The ideal no-load speed ω_{m0} is obtained when I_a is equal to zero,

$$\omega_{m0} = \frac{V_m}{K_m}, \qquad 0 \le \alpha \le \frac{\pi}{2} \tag{16.1.40}$$

$$\omega_{m0} = \frac{V_m \sin\alpha}{K_m}, \qquad \frac{\pi}{2} \le \alpha \le \pi \tag{16.1.41}$$

For torques less than the rated value, a low-power drive operates predominantly in the *discontinuous conduction* mode, for which a zero armature-current interval exists besides the duty interval. With continuous conduction, as seen from Equation (16.1.39), the speed–torque characteristics are parallel straight lines whose slope depends on R_a, the armature circuit resistance. A filter inductor is sometimes included to reduce the discontinuous conduction zone, although such an addition will lead to an increase in losses, armature circuit time constant, noise, cost, weight, and volume of the drive.

EXAMPLE 16.1.2

Consider a 3-hp, 220-V, 1800-r/min separately excited dc motor controlled by a single-phase fully controlled rectifier with an ac source voltage of 230 V at 60 Hz. Assume that the full-load efficiency of the motor is 88%, and enough filter inductance is added to ensure continuous conduction for any torque greater than 25% of rated torque. The armature circuit resistance is 1.5 Ω.

(a) Determine the value of the firing angle to obtain rated torque at 1200 r/min.
(b) Compute the firing angle for the rated braking torque at -1800 r/min.
(c) With an armature circuit inductance of 30 mH, calculate the motor torque for $\alpha = 60°$ and 500 r/min, assuming that the motor operates in the continuous conduction mode.
(d) Find the firing angle corresponding to a torque of 35 N·m and a speed of 480 r/min, assuming continuous conduction.

Solution

(a) $V_m = \sqrt{2} \times 230 = 325.27$ V

$$I_a = \frac{3 \times 746}{0.88 \times 220} = 11.56 \text{ A}$$

$E = 220 - (11.56 \times 1.5) = 220.66$ V at rated speed of 1800 r/min

$$\omega_m = \frac{1800 \times 2\pi}{60} = 188.57 \text{ rad/s}$$

$$K_m = \frac{E}{\omega_m} = \frac{202.66}{188.57} = 1.075$$

For continuous conduction, Equation (16.1.38) holds:

$$\frac{2V_m}{\pi} \cos \alpha = I_a R_a + K_m \omega_m = I_a R_a + E$$

At rated torque,

$$I_a = 11.56 \text{ A}$$

$$\text{Back emf at 1200 r/min} = E_1 = \frac{1200}{1800} \times 202.66 = 135.11$$

Substituting these values, one gets

$$\frac{2 \times 325.27}{\pi} \cos \alpha = 11.56 \times 1.5 + 135.11 = 152.45$$

or

$$\cos \alpha = \frac{152.45 \times \pi}{2 \times 325.27} = 0.7365, \text{ or } \alpha = 42.6°$$

(b) At -1800 r/min,

$$E = -202.66 \text{ V}$$

So it follows that

$$\frac{2 \times 325.27}{\pi} \cos \alpha = 11.56 \times 1.5 - 202.66 = -185.32$$

or

$$\cos \alpha = -\frac{185.32 \times \pi}{2 \times 325.27} = -0.8953, \text{ or } \alpha = 153.5°$$

(c) From part (a), $K_m = 1.075$. Corresponding to a speed of 500 r/min,

$$\omega_m = \frac{500 \times 2\pi}{60} = 52.38 \text{ rad/s}$$

From Equation (16.1.29), we have

$$52.38 = \frac{2 \times 325.27}{\pi \times 1.075} \cos 60° - \frac{1.5}{1.075^2} T_a, \text{ or } T_a = 33.82 \text{ N} \cdot \text{m}$$

(d) From part (a), $K_m = 1.075$. From Equation (16.1.39), we obtain

$$\frac{480 \times 2\pi}{60} = \frac{2\pi \times 325.27}{\pi \times 1.075} \cos \alpha - \frac{1.5}{1.075^2} \times 35$$

or,

$$\cos \alpha = 0.497 \quad \text{or} \quad \alpha = 60.2°$$

The most widely used dc drive is the three-phase, fully controlled, six-pulse, bridge-rectifier-fed, separately excited dc motor drive shown in Figure 16.1.14. With a phase difference of

60°, the firing of thyristors occurs in the same sequence as they are numbered. The line commutation of an even-numbered thyristor takes place with the turning on of the next even-numbered thyristor, and similarly for odd-numbered thyristors. Thus, each thyristor conducts for 120° and only two thyristors (one odd-numbered and one even-numbered) conduct at a time.

Considering the continuous conduction mode for the motoring operation, as shown in Figure 16.1.15, for the converter output voltage cycle from $\omega t = \alpha + \pi/3$ to $\omega t = \alpha + 2\pi/3$,

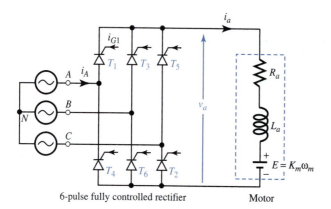

Figure 16.1.14 Three-phase fully controlled six-pulse bridge-rectifier-fed separately excited dc motor.

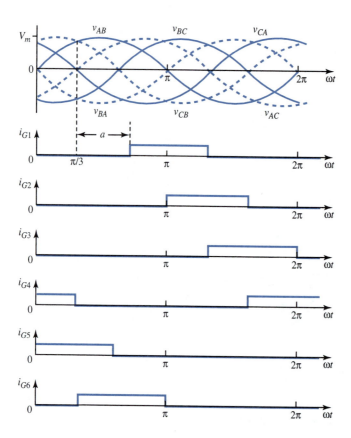

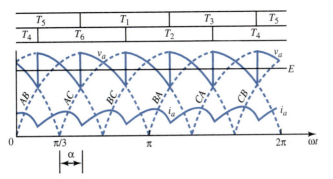

Figure 16.1.15 Continuous conduction mode for the motoring operation.

$$V_a = \frac{3}{\pi} \int_{\alpha+\pi/3}^{\alpha+2\pi/3} V_m \sin \omega t \, d(\omega t) = \frac{3}{\pi} V_m \cos \alpha = V_{ao} \cos \alpha \qquad (16.1.42)$$

where the line voltage $v_{AB} = V_m \sin \omega t$ is taken as the reference. From Equations (16.1.32), (16.1.37), and (16.1.42), we get

$$\omega_m = \frac{3V_m}{\pi K_m} \cos \alpha - \frac{R_a}{K_m^2} T_a \qquad (16.1.43)$$

For normalization, taking the base voltage V_B as the maximum average converter output voltage $V_{ao} = 3V_m/\pi$, and the base current as the average motor current (that will flow when $\omega_m = 0$ and $V_a = V_B$)$I_B = V_B/R_a = 3V_m/\pi R_a$, the normalized speed and torque are given by

$$\text{Speed } \omega_{mn} = \frac{E}{V_B} = \frac{\pi E}{3V_m} \qquad (16.1.44)$$

$$\text{Torque } T_{an} = I_{an} = \frac{\pi R_a}{3V_m} (I_a) \qquad (16.1.45)$$

EXAMPLE 16.1.3

Consider the 220-V, 1800-r/min dc motor of Example 16.1.2, controlled by a three-phase fully controlled rectifier from a 60-Hz ac source. The armature-circuit resistance and inductance are 1.5 Ω and 30 mH, respectively.

(a) When the motor is operating in continuous conduction, find the ac source voltage required to get rated voltage across the motor terminals.

(b) With the ac source voltage obtained in part (a), compute the motor speed corresponding to $\alpha = 60°$ and $T_a = 25$ N·m assuming continuous conduction.

(c) Let the motor drive a load whose torque is constant and independent of speed. The minimum value of the load torque is 1.2 N·m. Calculate the inductance that must be added to the armature circuit to get continuous conduction for all operating points, given that $\psi = \tan^{-1}(\omega L_a/R_a) = 1.5$ rad, $T_{an} > 0.006$, and all points on the $T_{an}-\omega_{mn}$ plane lie to the right of the boundary between continuous and discontinuous conductions.

Solution

(a) From Equation (16.1.42), with $\alpha = 0$, we get

$$220 = \frac{3\sqrt{2}\, V}{\pi} \quad \text{or} \quad V = 163 \text{ V line-to-line}; \quad V_m = 230.5 \text{ V}$$

(b) From Equation (16.1.43), we obtain

$$\omega_m = \frac{3 \times 230.5 \times \cos 60°}{\pi \times 1.075} - \frac{1.5}{1.075^2} 25 = 69.9 \text{ rad/s} = 667 \text{ r/min}$$

(c) From Equation (16.1.45), the normalized torque corresponding to 1.2 N·m is

$$T_{an} = \frac{\pi R_a}{3 V_m} \left(\frac{T_a}{K_m} \right) = \frac{\pi \times 1.5}{3 \times 230.5} \left(\frac{1.2}{1.075} \right) = 0.0076$$

The straight line $T_{an} = 0.0076$ is to the right of the boundary for

$$\psi = 1.5 \text{ rad} = \tan^{-1} \left(\frac{\omega L}{R_a} \right)$$

Therefore,

$$L_a = \frac{R_a}{\omega} \tan \psi = \frac{1.5}{2\pi \times 60} \tan 85.9° = 55.5 \text{ mH}$$

The external inductance needed is $55.5 - 30 = 25.5$ mH.

Solid-State Control of Induction Motors

For our next discussions you might find it helpful to review Section 13.2. The speed-control methods employed in power semiconductor-controlled induction motor drives are listed here:

- Variable terminal voltage control (for either squirrel-cage or wound-rotor motors)
- Variable frequency control (for either squirrel-cage or wound-rotor motors)
- Rotor resistance control (for wound-rotor motors only)
- Injecting voltage into rotor circuit (for wound-rotor motors only).

AC VOLTAGE CONTROLLERS

Common applications for these controllers are found in fan, pump, and crane drives. Figure 16.1.16 shows three-phase symmetrical ac voltage-controlled circuits for wye-connected and delta-connected stators, in which the thyristors are fired in the sequence that they are numbered, with a phase difference of 60°. The four-quadrant operation with plugging is obtained by the use of the typical circuit shown in Figure 16.1.17. Closed-loop speed-control systems have also been developed for single-quadrant and multiquadrant operation. Induction motor starters that realize energy savings are one of the ac voltage controller applications. However, one should look into the problems associated with harmonics.

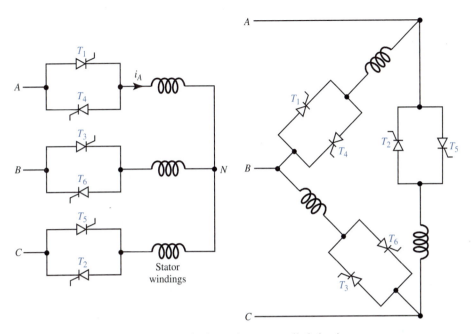

Figure 16.1.16 Three-phase symmetrical ac voltage-controlled circuits.

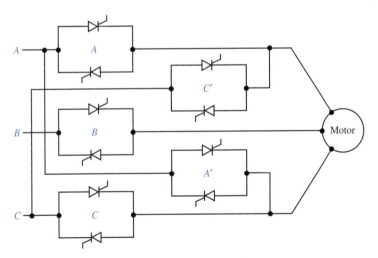

Figure 16.1.17 Four-quadrant ac voltage controller.

FREQUENCY-CONTROLLED INDUCTION-MOTOR DRIVES

The converters employed for variable-frequency drives can be classified as:

- Voltage-source inverter (which is the only one considered here)
- Current-source inverters
- Cycloconverters, which allow a variable-frequency ac supply with voltage-source or current-source characteristics obtained from a fixed-frequency voltage source.

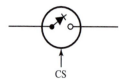

Figure 16.1.18 Symbol of a self-commutated semiconductor switch.

CS

A common symbol for the *self-commutated semiconductor switch* is shown in Figure 16.1.18. The control signal (either voltage or current) is denoted by CS, and the diode gives the direction in which the switch can conduct current. GTOs, power transistors, and MOSFETs are classified as self-commutated semiconductor devices because they can be turned off by their respective control signals: GTOs by a gate pulse, a power transistor by a base drive, and a MOSFET by a gate-to-source voltage. A thyristor, on the other hand, is a naturally commutated device that cannot be turned off by its gate signal. A thyristor combined with a forced commutation circuit behaves like a self-commutated semiconductor device, however. The self-commutation capability makes its turnoff independent of the polarity of the source voltage, the load voltage, or the nature of load. Self-commutated semiconductor switches are suitable for applications for converters fed from a dc source, such as inverters and choppers.

Figure 16.1.19 shows a three-phase voltage-source inverter circuit, along with the corresponding voltage and current waveforms. The motor connected to terminals *A, B,* and *C* can have wye or delta connection. Operating as a six-step inverter, the inverter generates a cycle of line or phase voltage in six steps. The following Fourier-series expressions describe the voltages v_{AB} and v_{AN}:

$$v_{AB} = \frac{2\sqrt{3}}{\pi} V_d \left[\sin\left(\omega t + \frac{\pi}{6}\right) + \frac{1}{5} \sin\left(5\omega t - \frac{\pi}{6}\right) + \frac{1}{7} \sin\left(7\omega t + \frac{\pi}{6}\right) + \cdots \right] \tag{16.1.46}$$

$$v_{AN} = \frac{2}{\pi} V_d \left[\sin \omega t + \frac{1}{5} \sin 5\omega t + \frac{1}{7} \sin 7\omega t + \cdots \right] \tag{16.1.47}$$

The rms value of the fundamental component of the phase voltage v_{AN} is given by

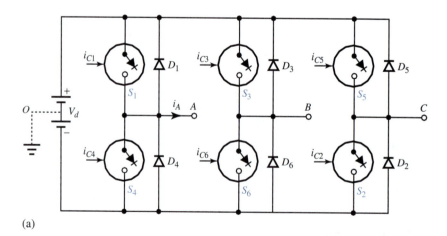

(a)

Figure 16.1.19 Three-phase voltage-source inverter circuit with corresponding voltage and current waveforms.

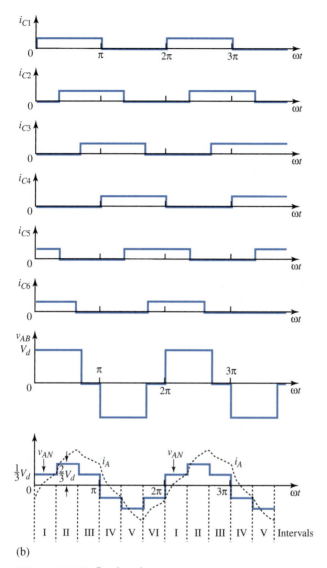

(b)

Figure 16.1.19 Continued

$$V_1 = \frac{\sqrt{2}}{\pi} V_d \qquad (16.1.48)$$

From Figure 16.1.18(b), the rms value of the phase voltage is given by

$$V = \left[\frac{1}{\pi} \left\{ \int_0^{\pi 3} \left(\frac{1}{3} V_d \right)^2 d(\omega t) + \int_{\pi/3}^{2\pi/3} \left(\frac{2}{3} V_d \right)^2 d(\omega t) \right. \right.$$

$$\left. \left. + \int_{2\pi/3}^{\pi} \left(\frac{1}{3} V_d \right)^2 d(\omega t) \right\} \right]^{1/2}$$

$$= \frac{\sqrt{2}}{3} V_d \qquad (16.1.49)$$

For a wye-connected stator, the waveform of the phase current is shown in Figure 16.1.19, which is also the output current i_A of the inverter. The output voltage of a six-step inverter can be controlled by controlling either the dc input voltage or the ac output voltage with multiple inverters.

EXAMPLE 16.1.4

A 440-V, 60-Hz, six-pole, wye-connected, squirrel-cage induction motor with a full-load speed of 1170 r/min has the following parameters per phase referred to the stator: $R_1 = 0.2\,\Omega$, $R'_2 = 0.1\,\Omega$, $X_{l1} = 0.75\,\Omega$, $X'_{l2} = 0.7\,\Omega$, and $X_m = 20\,\Omega$. (See Chapter 13, Figure 13.2.6, for the notation.) Consider the motor to be fed by a six-step inverter, which in turn is fed by a six-pulse, fully controlled rectifier.

(a) Let the rectifier be fed by an ac source of 440 V and 60 Hz. Find the rectifier firing angle that will obtain rated fundamental voltage across the motor.

(b) Calculate the inverter frequency at 570 r/min and rated torque when the motor is operated at a constant flux.

(c) Now let the drive be operated at a constant V/f ratio. Compute the inverter frequency at 570 r/min and half the rated torque. Neglect the derating due to harmonics and use the approximate equivalent circuit of Figure 13.2.6, with jX_m shifted to the supply terminals.

Solution

(a) From Equation (16.1.46), the fundamental rms line voltage of a six-step inverter is

$$V_L = \frac{\sqrt{6}}{\pi} V_d$$

For a six-pulse rectifier, from Equation (16.1.42), $V_d = (3/\pi)V_m \cos\alpha$, where V_m is the peak ac source line voltage. Thus,

$$V_L = \frac{3\sqrt{6}}{\pi^2} V_m \cos\alpha \qquad \text{or} \quad \cos\alpha = \frac{V_L}{V_m}\frac{\pi^2}{3\sqrt{6}}$$

With $V_L = 440$ V and $V_m = 440\sqrt{2}$ V,

$$\cos\alpha = \frac{\pi^2}{3\sqrt{6}\sqrt{2}} = 0.95 \qquad \text{or} \qquad \alpha = 18.26°$$

(b) For a given torque, the motor operates at a fixed slip speed for all frequencies as long as the flux is maintained constant. At rated torque, the slip speed $N_{Sl} = 1200 - 1170 = 30$ r/min. Hence, synchronous speed at 570 r/min is $N_S = N + N_{Sl} = 570 + 30 = 600$ r/min. Therefore, the inverter frequency is $(600/1200)60 = 30$ Hz.

(c) Based on the equivalent circuit, it can be shown that the torque for a constant V/f ratio is

$$T = \frac{3}{\omega_S}\left[\frac{V_{\text{rated}}^2 (R'_2/S)}{\left(R_1 + R'_2/S\right)^2 + \left(X_{l1} + X'_{l2}\right)^2}\right]$$

With $a = f/f_{\text{rated}}$,

$$T = \frac{3}{\omega_S}\left[\frac{V_{rated}^2(R_2'/aS)}{\left(\dfrac{R_1}{a} + \dfrac{R_2'}{aS}\right)^2 + (X_{l1} + X_{l2}')^2}\right], \qquad a < 1$$

Here,

$$\omega_S = \frac{120 \times 60}{6} \times \frac{2\pi}{60} = 40\pi = 125.66 \text{ rad/s}$$

$$V_{rated} = \frac{440}{\sqrt{3}} = 254 \text{ V}$$

$$S = \frac{30}{1200} = 0.025$$

With $a = 1$,

$$T_{rated} = \frac{3}{125.66}\left[\frac{254^2(0.1/0.025)}{\left(0.2 + \dfrac{0.1}{0.025}\right)^2 + (0.75 + 0.7)^2}\right] = 312.11 \text{ N} \cdot \text{m}$$

At half the rated torque,

$$0.5 \times 312.11 = \frac{3}{125.66}\left[\frac{254^2(0.1/aS)}{\left(\dfrac{0.2}{a} + \dfrac{0.1}{aS}\right)^2 + 1.45^2}\right]$$

or

$$\left(\frac{0.2}{a} + \frac{0.1}{aS}\right)^2 + 2.1 = \frac{0.987}{aS}$$

Also,

$$S = \frac{a\omega_S - \omega_m}{a\omega_S} \qquad \text{or} \qquad a = \frac{\omega_m}{(1 - S)\omega_S} = \frac{59.7}{(1 - S)125.66} = \frac{0.457}{1 - S}$$

(*Note*: $\omega_m = 570 \times 2\pi/60 = 59.7$ rad/s.)

Based on the last two equations, we can solve for a and S using an iterative procedure,

$$a = 0.4864; \quad S = 0.0235$$

Therefore, frequency $= 0.4864 \times 60 = 29.2$ Hz.

The inverter can also be operated as a pulse-width modulated (PWM) inverter. Figure 16.1.20 shows the schemes when the supply is dc and when it is ac.

SLIP-POWER CONTROLLED WOUND-ROTOR INDUCTION-MOTOR DRIVES

Slip power is that portion of the air-gap power that is not converted into mechanical power. The methods involving rotor-resistance control and voltage injection into the rotor circuit belong to this

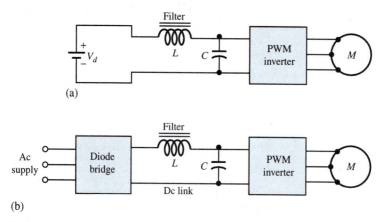

(a)

(b)

Figure 16.1.20 PWM inverter drives. **(a)** Supply is dc. **(b)** Supply is ac.

class of slip-power control. In order to implement these techniques using power semiconductor devices, some of the schemes are static rotor resistance control; the static Scherbius drive; and the static Kramer drive, of which only the first is considered here.

The rotor circuit resistance can be smoothly varied statically (instead of mechanically) by using the principles of a chopper. Figure 16.1.21 shows a scheme in which the slip-frequency ac rotor voltages are converted into direct current by a three-phase diode bridge and applied to an external resistance R. The self-commutated semiconductor switch S, operating periodically with a period T, remains on for an interval t_{on} in each period, with a duty ratio δ defined by t_{on}/T.

The filter inductor helps to reduce the ripple in current I_d and eliminate discontinuous conduction. Figure 16.1.22 shows input phase voltage waveforms of the diode bridge and a six-step rotor phase-current waveform, assuming ripple-free I_d. For a period T of the switch operation, the energy absorbed by resistance R is given by $I_d^2 R(T - t_{on})$, while the average power consumed by R is

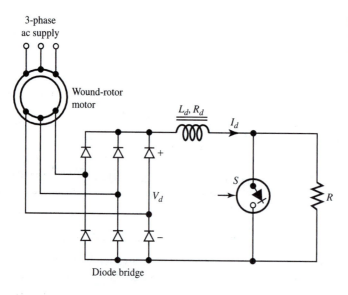

Figure 16.1.21 Static rotor resistance control of a wound-rotor induction motor.

Figure 16.1.22 Rotor voltage and current waveforms.

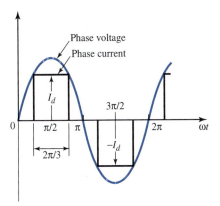

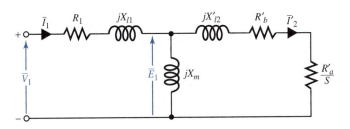

Figure 16.1.23 Per-phase fundamental equivalent circuit of the drive with static rotor resistance control.

$$\frac{1}{T}\left[I_d^2 R(T - t_{on})\right] \qquad \text{or} \qquad I_d^2 R(1 - \delta)$$

The effective value of resistance is then given by

$$R^* = (1 - \delta)R \tag{16.1.50}$$

The total resistance across the diode bridge is

$$R_t = R^* + R_d = R_d + (1 - \delta)R \tag{16.1.51}$$

The per-phase power absorbed by R_t is $\frac{1}{3} I_d^2 R_t$, where I_d is related to the rms value of the rotor phase current, $\sqrt{3/2}I_{rms}$. Thus, the effective per-phase resistance R_{eff} is given by

$$R_{eff} = 0.5R_t \tag{16.1.52}$$

From the Fourier analysis of the rotor phase current with quarter-wave symmetry, the fundamental rotor current is $(\sqrt{6}/\pi)I_d$, or $(3/\pi)I_{rms}$. It can be shown (as in Problem 16.1.14) that the per-phase fundamental equivalent circuit of the drive referred to the stator is given by Figure 16.1.23, where

$$R_a = R_2 + R_{eff} \tag{16.1.53}$$

and

$$R_b = \left(\frac{\pi^2}{9} - 1\right) R_a \tag{16.1.54}$$

As in Chapter 13, the primed notation denotes referral to the stator, and R_2 is the per-phase resistance of the rotor. The resistance R'_a/S accounts for the mechanical power developed and the fundamental rotor copper loss, whereas the resistance R'_b represents the effect of the rotor harmonic copper loss.

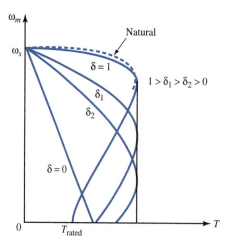

Figure 16.1.24 Effect of static rotor resistance control on speed–torque curves.

With jX_m shifted to the stator terminals in Figure 16.1.23 as an approximation, we get

$$\bar{I}'_2 = \frac{\bar{V}_1}{\left[R_1 + R'_b + (R'_a/S)\right] + j(X_{l1} + X'_{l2})} \tag{16.1.55}$$

and

$$T = \frac{3}{\omega_s}(I'_2)^2 \left(\frac{R'_a}{S}\right) \tag{16.1.56}$$

Figure 16.1.24 shows the nature of the speed–torque characteristics for different values of the duty ratio δ.

EXAMPLE 16.1.5

A 440-V, 60-Hz, six-pole, wye-connected, wound-rotor induction motor with a full-load speed of 1170 r/min has the following per-phase parameters referred to the stator: $R_1 = R'_2 = 0.5\,\Omega$, $X_{l1} = X'_{l2} = 2\,\Omega$, $X_m = 40\,\Omega$, and a stator-to-rotor turns ratio of 2.5.

The scheme of Figure 16.1.21 is employed for speed control with $R_d = 0.02\,\Omega$ and $R = 1\,\Omega$. For a speed of 1000 r/min at 1.5 times the rated torque, find the duty ratio δ, neglecting friction and windage, and using the equivalent circuit with jX_m moved adjacent to the stator terminals.

Solution

Full-load torque without rotor resistance control is

$$T = \frac{3}{\omega_s}\left[\frac{V_1^2(R'_2/S)}{(R_1 + R'_2/S)^2 + (X_{l1} + X'_{l2})^2}\right]$$

With $V_1 = 440/\sqrt{3} = 254$ V; $\omega_s = 125.7$ rad/s, and full-load slip $= (1200-1170)/1200 = 0.025$, then

$$T = \frac{3}{125.7}\left[\frac{254^2(0.5/0.025)}{(0.5 + 0.5/0.025)^2 + 4^2}\right] = 70.6\ \text{N}\cdot\text{m}$$

With rotor resistance control,

$$T = \frac{3}{\omega_S}\left[\frac{V_1^2(R_a'/S)}{\left(R_1 + R_b' + R_a'/S\right)^2 + (X_{l1} + X_{l2}')^2}\right]$$

With $R_b' = (\pi^2/9 - 1)R_a' = 0.0966R_a'$, and slip $= (1200 - 1000)/1200 = 0.167$, we obtain

$$1.5 \times 70.6 = \frac{3}{125.7}\left[\frac{254^2(R_a'/0.167)}{\left(0.5 + 0.0966R_a' + R_a'/0.167\right)^2 + 4^2}\right]$$

Thus, $R_a'^2 - 2.18R_a' + 0.44 = 0$, or $R_a' = 0.225$, or 1.955 Ω. The value of 0.225 being less than R_2' is not feasible; therefore $R_a' = 1.955$ Ω.

From Equation (16.1.53), $R_{\text{eff}} = (R_a' - R_2')/(\text{turns ratio})^2 = (1.955 - 0.5)/25^2 = 0.233$ Ω. From Equations (16.1.52) and (16.1.51), we calculate

$$(1 - \delta) = \frac{2R_{\text{eff}} - R_d}{R} = \frac{2 \times 0.233 - 0.22}{1} = 0.446, \text{ or } \delta = 0.554$$

Instead of wasting the slip power in the rotor circuit resistance, as was suggested by Scherbius, we can feed it back to the ac mains by using a scheme known as a *static Scherbius drive*. The slip power can be converted to mechanical power (with the aid of an auxiliary motor mounted on the induction-motor shaft), which supplements the main motor power, thereby delivering the same power to the load at different speeds, as in the Kramer drive.

Solid-State Control of Synchronous Motors

The speed of a synchronous motor can be controlled by changing its supply frequency. With variable-frequency control, two modes of operation are possible: true synchronous mode, employed with voltage source inverters, in which the supply frequency is controlled from an independent oscillator; and self-controlled mode, in which the armature supply frequency is changed proportionally so that the armature field always rotates at the same speed as the rotor. The true synchronous mode is used only in multiple synchronous-reluctance and permanent-magnet motor drives, in applications such as paper mills, textile mills, and fiber-spinning mills, because of the problems associated with hunting and stability. Variable-speed synchronous-motor drives, commonly operated in the self-controlled mode, are superior to or competitive with induction-motor or dc-motor variable-speed drives.

Drives fed from a load-commutated current-source inverter or a cycloconverter find applications in high-speed high-power drives such as compressors, conveyors, traction, steel mills, and ship propulsion. The drives fed from a line-commutated cycloconverter are used in low-speed gearless drives for mine hoists and ball mills in cement production. Self-controlled permanent-magnet synchronous-motor drives are replacing the dc-motor drives in servo applications.

Self-control can be applied to all variable-frequency converters, whether they are voltage-source inverters, current-source inverters, current-controlled PWM inverters, or cycloconverters. Rotor position sensors, i.e., rotor position encoders with optical or magnetic sensors, or armature terminal voltage sensors, are used for speed tracking. In the optical rotor position encoder shown in Figure 16.1.25 for a four-pole synchronous machine, the semiconductor switches are fired at a frequency proportional to the motor speed. A circular disk, with two slots S' and S'' on an inner radius and a large number of slots on the outer periphery, is mounted on the rotor shaft. Four stationary optical sensors P_1 to P_4 with the corresponding light-emitting diodes and photo transistors are placed as shown in Figure 16.1.25. Whenever the sensor faces a slot, an output results. Waveforms caused by the sensors are also shown in Figure 16.1.25. A detailed discussion of this kind of microprocessor control of current-fed synchronous-motor drives is outside the scope of this text.

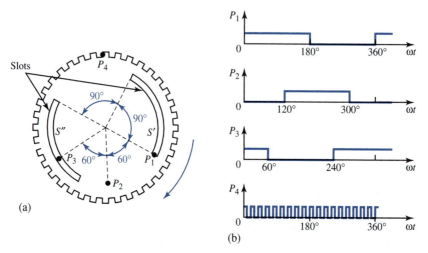

Figure 16.1.25 Optical rotor position encoder and its output waveforms.

EXAMPLE 16.1.6

A 1500-hp, 6600-V, six-pole, 60-Hz, three-phase, wye-connected, synchronous motor, with a synchronous reactance of 36 Ω, negligible armature resistance, and unity power factor at rated power, is controlled from a variable-frequency source with constant V/f ratio. Determine the armature current, torque angle, and power factor at full-load torque, one-half rated speed, and rated field current. (Neglect friction, windage, and core loss.)

Solution

The simplified per-phase equivalent circuits of the motor are shown in Figure E16.1.6. Phase voltage $V = 6600/\sqrt{3} = 3810.6$ V, and

$$\bar{I}'_m = \frac{V\angle 0°}{jX_s} = \frac{3810.6}{36\angle 90°} = 105.9\angle -90° \text{ A}$$

With unity power factor at rated power, we obtain

$$1500 \times 746 = \sqrt{3} \times 6600 \times I_s$$

Hence, the rated armature current $I_s = 97.9$ A. Then

$$\bar{I}'_f = \bar{I}'_m - \bar{I}'_s = 105.9\angle -90° - 97.9\angle 0° = -97.9 - j105.9 = 144.2\angle -132.8° \text{ A}$$

The synchronous speed is $(120 \times 60)/6 = 1200$ rpm, or $\omega_s = 125.7$ rad/s. The rated torque is then

$$T_{\text{rated}} = \frac{1500 \times 746}{125.7} = 8902 \text{ N} \cdot \text{m}$$

and

$$\sin \delta = \frac{1500 \times 746}{3 \times (6600/\sqrt{3}) \times 144.2} = 0.6788$$

or the torque angle $\delta = 42.75°$ electrical or $14.25°$ mechanical.

One-half of rated speed $= 600$ r/min. With a constant V/f ratio, I'_m is constant. Since torque $T = (3/\omega_s)VI'_f \sin \delta = (3X_s/\omega_s)I'_m I'_f \sin \delta$, δ is unchanged. Noting that $\bar{I}_s = \bar{I}'_m - \bar{I}'_f$, $I_s \sin \phi = -\bar{I}'_m + \bar{I}'_f \cos \delta$, and $I_s \cos \phi = \bar{I}'_f \sin \delta$, then, for constant values of δ and I'_m, both I_s and the power factor remain the same, which is true for any speed. Therefore,

$$I_s = 97.9 \text{ A}; \qquad \delta = 42.75° \text{ electrical}; \qquad \text{power factor} = 1.0$$

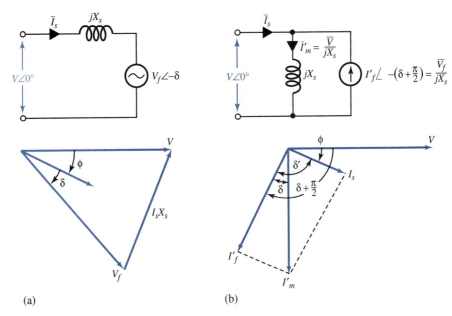

(a)   (b)

Figure E16.1.6

Applications for the Electronic Control of Motors

A wide range of variable-speed drive systems are now available, each having particular advantages and disadvantages. The potential user needs to evaluate each application on its own merit, with consideration given to both technical limitations of particular systems and long-term economics. A number of possible applications of the electronic control of motors are listed in Table 16.1.4.

16.2 FEEDBACK CONTROL SYSTEMS

An *open-loop* system is one in which the control action is independent of the output or desired result, whereas a *closed-loop (feedback)* system is one in which the control action is dependent upon the output. In the case of the open-loop system, the input command (actuating signal) is the sole factor responsible for providing the control action, whereas for a closed-loop system the control action is provided by the difference between the input command and the corresponding output.

TABLE 16.1.4 Applications for the Electronic Control of Motors

Converter	Motor	Typical Output Range (kW)	Practical Speed Range for a One- or Two-Quadrant Drive	Quadrants of Operation	Supply	Variables Governing Speed	Applications
Dc Drives							
Single-phase rectifier with center-tapped transformer	Separately excited or permanent magnet dc	0–5	1:50	Quad 1 driving Quad 2 dynamic or mechanical braking Reversal or armature connections or dual converter for four-quadrant operation	120-V single-phase ac	Armature terminal pd Field current	Small variable-speed drives in general
Single-phase fully controlled bridge rectifier	Separately excited or permanent magnet dc	0–10	1:50	Quad 1 driving Quad 2 dynamic braking Reversal of armature connections or dual converter for four-quadrant operation	240-V single-phase ac	Armature terminal pd Field current	Processing machinery, machine tools
Single-phase half-controlled bridge rectifier	Separately excited or permanent magnet dc	0–10	1:50	Quad 1 driving Quad 2 dynamic braking Reversal of armature connections for four-quadrant operation	240-V single-phase ac	Armature terminal pd Field current	Processing machinery, machine tools
Three-phase bridge rectifier	Separately excited or permanent magnet dc	10–200 (10–50 for PM motor)	1:50	Quad 1 driving Quad 2 dynamic braking Dual converter for regenerative braking with four-quadrant operation	240-V to 480-V three-phase ac	Armature terminal pd Field current	Hoists, machine tools, centrifuges, calender rollers
Three-phase bridge rectifier	Separately excited dc	200–1000	1:50	Quad 1 driving Quad 2 dynamic braking Dual converter for regenerative braking with four-quadrant operation	460-V to 600-V three-phase ac	Armature terminal pd Field current	Winders, rolling mills
Chopper (dc-to-dc converter)	Separately excited dc		1:50	Alternatives are quad 1 only; quads 1 and 2; quads 1 and 4; all four	600-V to 1000-V dc	Armature terminal pd Field current	Electric trains, rapid transit streetcars, trolleys, buses, cranes

Ac Drives

AC power controller	Class D squirrel-cage induction motor	0–25	1:2	Quad 1 driving / Quad 4 plugging with source phase-sequence reversal / Quad 3 driving / Quad 2 plugging	240-V three-phase ac	Stator terminal pd	Centrifugal pumps and fans
Slip-energy recovery system	Wound-rotor induction motor	Up to 20,000	1:2	Quad 1 driving, with source phase-sequence reversal / Quad 3 driving	Up to 5000-V three-phase ac	Rotor terminal pd	Centrifugal pumps and fans
Voltage-source inverter and additional converters	Class B or C squirrel-cage induction motor or synchronous motors	15–250	1:10	Quad 1 driving / Quad 2 regenerative braking, with appropriate additional converters / Quad 3 driving / Quad 4 regenerative breaking	240-V to 600-V three-phase ac	Stator terminal pd and frequency	Group drives in textile machinery and runout tables
Current-source inverter and additional converters	Class B or C squirrel cage induction motor	15–500	1:10	Quad 1 driving / Quad 2 regenerative braking, with inverter control signal sequence reversal / Quad 3 driving / Quad 4 regenerative braking	240-V to 600-V three-phase ac	Stator current and frequency	Single-motor dives, centrifuges, mixers, conveyors, etc.
Current-source inverter and additional converters	Three-phase synchronous motor	Up to 15,000	1:50	Quad 1 driving / Quad 2 regenerative breaking, with inverter control signal sequence reversal / Quad 3 driving / Quad 4 regenerative braking	Up to 5000-V three-phase ac	Stator current and frequency	Single-motor drives, processing machinery of all kinds

Adapted from S. A. Nasar, Editor, *Handbook of Electric Machines*, McGraw-Hill, New York, 1987.

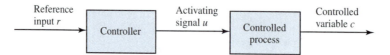

Figure 16.2.1 Elements of an open-loop control system.

The elements of an open-loop control system may be divided into two parts: the controller and the controlled process, as shown by the block diagram of Figure 16.2.1. An input signal or command r is applied to the controller, whose output acts as the actuating signal u. The actuating signal then controls the controlled process such that the controlled variable c will perform according to some prescribed standards. The controller may be an amplifier, mechanical linkage, or other basic control means in simple cases, whereas in more sophisticated electronics control, it can be an electronic computer such as a microprocessor.

Feedback control systems can be classified in a number of ways, depending upon the purpose of the classification.

1. According to the method of analysis and design.

   - *Linear versus nonlinear control systems.* Linear feedback control systems are idealized models that are conceived by the analyst for the sake of simplicity of analysis and design. For the design and analysis of linear systems there exist a wealth of analytical and graphical techniques. On the other hand, nonlinear systems are very difficult to treat mathematically, and there are no general methods that can be used for a broad class of nonlinear systems.

   - *Time-invariant versus time-varying systems.* When the parameters of a control system are stationary with respect to time during the operation of the system, the system is known as a time-invariant system. Even though a time-varying system without nonlinearity is still a linear system, the analysis and design of such a class of systems are generally much more complex than that of linear time-invariant systems.

2. According to the types of signal found in the system.

   - *Continuous-data and discrete-data systems.* A continuous-data system is one in which the signals at various parts of the system are all functions of the continuous-time variable t. When one refers to an *ac control system*, it usually means that the signals in the system are *modulated* by some kind of modulation scheme. On the other hand, a *dc control system* implies that the signals are *unmodulated*, but they are still ac signals according to the conventional definition. Typical components of a dc control system are potentiometers, dc amplifiers, dc motors, and dc tachometers; typical components of an ac control system are synchros, ac amplifiers, ac motors, gyroscopes, and accelerometers.

   - *Sampled-data and digital control systems.* These differ from the continuous-data systems in that the signals at one or more points of the system are in the form of either a pulse train or a digital code. Sampled-data systems usually refer to a more general class of systems whose signals are in the form of pulse data, whereas a digital control system refers to the use of a digital computer or controller in the system. The term *discrete-data control system* is used to describe both types of systems.

3. According to the type of system components.

   • *Electromechanical control systems.*

   • *Hydraulic control systems.*

   • *Pneumatic control systems.*

   • *Biological control systems.*

4. According to the main purpose of the system.

   • *Position control systems.* Here the output position, such as the shaft position on a motor, exactly follow the variations of the input position.

   • *Velocity control systems.*

   • *Regulators.* Their function consists in keeping the output or controlled variable constant in spite of load variations and parameter changes. Speed control of a motor forms a good example. Feedback control systems used as regulators are said to be *type 0 systems*, which have a steady-state position error with a constant position input.

   • *Servomechanisms.* Their inputs are time-varying and their function consists in providing a one-to-one correspondence between input and output. Position-control systems, including automobile power steering, form good examples. Servomechanisms are usually *type 1* or *higher order systems*. A type 1 system has no steady-state error with a constant position input, but has a position error with a constant velocity input (the two shafts running at the same velocity, but with an angular displacement between them).

Once the mathematical modeling of physical systems is done, while satisfying equations of electric networks and mechanical systems, as well as linearizing nonlinear systems whenever possible, feedback control systems can be analyzed by using various techniques and methods: *transfer function approach, root locus techniques, state-variable analysis, time-domain analysis,* and *frequency-domain analysis.* Although the primary purpose of the feedback is to reduce the error between the reference input and the system output, feedback also has effects on such system performance characteristics as stability, bandwidth, overall gain, impedance, transient response, frequency response, effect of noise, and sensitivity, as we shall see later.

Transfer Functions and Block Diagrams

The *transfer function* is a means by which the dynamic characteristics of devices or systems are described. The transfer function is a mathematical formulation that relates the output variable of a device to the input variable. For linear devices, the transfer function is independent of the input quantity and solely dependent on the parameters of the device together with any operations of time, such as differentiation and integration, that it may possess. To obtain the transfer function, one usually goes through three steps: (i) determining the governing equation for the device, expressed in terms of the output and input variables, (ii) Laplace transforming the governing equation, assuming all initial conditions to be zero, and (iii) rearranging the equation to yield the ratio of the output to input variable. The properties of transfer functions are summarized as follows.

   • A transfer function is defined only for a linear time-invariant system. It is meaningless for nonlinear systems.

   • The transfer function between an input variable and an output variable of a system is defined as the ratio of the Laplace transform of the output to the Laplace transform of the input, or as the Laplace transform of the impulse response. The impulse response of a linear system is defined as the output response of the system when the input is a unit impulse function.

- All initial conditions of the system are assumed to be zero.
- The transfer function is independent of the input.

The *block diagram* is a pictorial representation of the equations of the system. Each block represents a mathematical operation, and the blocks are interconnected to satisfy the governing equations of the system. The block diagram thus provides a chart of the procedure to be followed in combining the simultaneous equations, from which useful information can often be obtained without finding a complete analytical solution. The block-diagram technique has been highly developed in connection with studies of feedback control systems, often leading to programming a problem for solution on an analog computer.

The simple configuration shown in Figure 16.2.2 is actually the basic building block of a complex block diagram. The arrows on the diagram imply that the block diagram has a unilateral property; in other words, a signal can pass only in the direction of the arrows. A box is the symbol for multiplication; the input quantity is multiplied by the function in the box to obtain the output. With circles indicating summing points (in an algebraic sense), and with boxes or blocks denoting multiplication, any linear mathematical expression can be represented by block-diagram notation, as in Figure 16.2.3 for an *elementary feedback control system*. The expression for the output quantity with negative feedback is given by

$$C = GE = G(R - B) = G(R - HC) \qquad (16.2.1)$$

or

$$C = \frac{G}{1 + HG} R \qquad (16.2.2)$$

Figure 16.2.2 Basic building block of a block diagram.

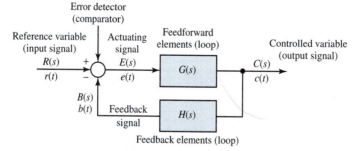

R(s)   reference variable (input signal)
C(s)   output signal (controlled variable)
B(s)   feedback signal, $= H(s)C(s)$
E(s)   actuating signal (error variable), $= R(s) - B(s)$
G(s)   forward path transfer function or open-loop transfer function, $= C(s)/E(s)$
M(s)   closed-loop transfer function, $= C(s)/R(s) = G(s)/1 + G(s)H(s)$
H(s)   feedback path transfer function
G(s)H(s)   loop gain

Figure 16.2.3 Block diagram of an elementary feedback control system.

The transfer function M of the closed-loop system is then given by

$$M = \frac{C}{R} = \frac{G}{1 + HG}$$

$$= \frac{\text{direct transfer function}}{1 + \text{loop transfer function}} \qquad (16.2.3)$$

The block diagrams of complex feedback control systems usually contain several feedback loops, which might have to be simplified in order to evaluate an overall transfer function for the system. A few of the block diagram reduction manipulations are given in Table 16.2.1; no attempt is made to cover all possibilities.

TABLE 16.2.1 Some Block Diagram Reduction Manipulations

Original Block Diagram	Manipulation	Modified Block Diagram
$R \rightarrow G_1 \rightarrow G_2 \rightarrow C$	Cascaded elements	$R \rightarrow G_1 G_2 \rightarrow C$
$R \rightarrow G_1,\ G_2 \rightarrow \pm \rightarrow C$	Addition or subtraction (eliminating auxiliary forward path)	$R \rightarrow G_1 \pm G_2 \rightarrow C$
$R \rightarrow G \rightarrow C$ (pick-off)	Shifting of pick-off point ahead of block	$R \rightarrow G \rightarrow C$; G in feedback
$R \rightarrow G \rightarrow C$	Shifting of pick-off point behind block	$R \rightarrow G \rightarrow C$; $1/G$
$R \rightarrow G \rightarrow (+,-) \rightarrow E$, C	Shifting of summing point ahead of block	$R \rightarrow (+,-) \rightarrow G \rightarrow E$; $1/G \leftarrow C$
$R \rightarrow (+,-) \rightarrow E \rightarrow G$, C	Shifting of summing point behind block	$R \rightarrow G \rightarrow (+,-) \rightarrow E$; $G \leftarrow C$
$R \rightarrow (+,-) \rightarrow G \rightarrow C$; H	Removing H from feedback path	$R \rightarrow 1/H \rightarrow (+,-) \rightarrow H \rightarrow G \rightarrow C$
$R \rightarrow (+,-) \rightarrow G \rightarrow C$; H	Eliminating feedback path	$R \rightarrow \dfrac{G}{1 + GH} \rightarrow C$

EFFECT OF FEEDBACK ON SENSITIVITY TO PARAMETER CHANGES

A good control system, in general, should be rather insensitive to parameter variations, while it is still able to follow the command quite responsively. It is apparent from Figure 16.2.1 that in an open-loop system the gain of the system will respond in a one-to-one fashion to the variation in G. Let us now investigate what effect feedback has on the sensitivity to parameter variations.

The sensitivity S of a closed-loop system with Equation (16.2.3) to a parameter change p in the direct transmission function G is defined as

$$S_p^M = \frac{\text{per-unit change in closed-loop transmission}}{\text{per-unit change in open-loop transmission}}$$

$$= \frac{\partial M/M}{\partial G_p/G_p} \tag{16.2.4}$$

where G_p denotes that the derivative is to be taken with respect to the parameter p of the function G. From Equation (16.2.3) it follows that

$$\partial M = \frac{-HG_p\, \partial G_p}{(1+HG_p)^2} + \frac{\partial G_p}{1+HG_p} = \frac{\partial G_p}{(1+HG_p)^2} \tag{16.2.5}$$

Substituting Equations (16.2.5) and (16.2.3) into Equation (16.2.4), one gets

$$S_p^M = \frac{\partial M/M}{\partial G_p/G_p} = \frac{\left[\partial G_p/(1+HG_p)^2\right]\left[(1+HG_p)/G_p\right]}{\partial G_p/G_p} = \frac{1}{1+HG_p} \tag{16.2.6}$$

or

$$\frac{\partial M}{M} = \left(\frac{1}{1+HG_p}\right)\frac{\partial G_p}{G_p} \tag{16.2.7}$$

Equation (16.2.6) shows that the sensitivity function can be made arbitrarily small by increasing HG_p, provided that the system remains stable.

EFFECT OF FEEDBACK ON STABILITY

Let us consider the direct transfer function of the unstable first-order system of Figure 16.2.4(a). The transfer function is given by

$$G = \frac{C}{E} = \frac{K}{1-p\tau} = \frac{-K/\tau}{p-1/\tau} \tag{16.2.8}$$

where $p = d/dt$ is the differential operator. If a unit-step function is applied as the input quantity E, the output becomes

$$C(p) = \frac{-K/\tau}{p-1/\tau}\, E(p) = \frac{-K/\tau}{p(p-1/\tau)} \tag{16.2.9}$$

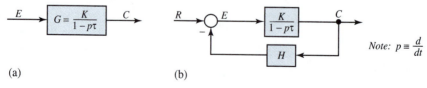

(a)　　　　　　　　　　　　(b)

Figure 16.2.4 (a) Block-diagram representation of an unstable first-order system. (b) System of part (a) modified with a feedback path.

for which the corresponding time solution is given by

$$c(t) = K(1 - e^{t/\tau}) \tag{16.2.10}$$

which is clearly unstable, since the response increases without limit as time passes. By placing a negative feedback path H around the direct transfer function, as shown in Figure 16.2.4(b), the closed-loop transfer function is then

$$M = \frac{C}{R} = \frac{\dfrac{K}{1 - p\tau}}{1 + \dfrac{KH}{1 - p\tau}} = \frac{K}{1 - p\tau + HK} \tag{16.2.11}$$

By choosing $H = ap$ and $aK > \tau$, it follows that

$$M = \frac{K}{1 - p\tau + paK} \tag{16.2.12}$$

and the corresponding time response of the closed-loop system, when subjected to a unit step, becomes

$$c_f(t) = K(1 - e^{-t/(aK-\tau)}) \tag{16.2.13}$$

where $c_f(t)$ denotes response with feedback. The system is now clearly stable with $aK > \tau$. Thus, the insertion of feedback causes the unstable direct transmission system of Figure 16.2.4(a) to become stable. Such a technique is often used to stabilize space rockets and vehicles, which are inherently unstable because of their large length-to-diameter ratios.

Note that the direct transfer function of Equation (16.2.8) has a pole located in the right half p-plane at $p = 1/\tau$, whereas the pole of the closed-loop transfer function of Equation (16.2.12) with $aK > \tau$ is located in the left half p-plane.

EFFECT OF FEEDBACK ON DYNAMIC RESPONSE AND BANDWIDTH

Let us consider the block-diagram representation of the open-loop system shown in Figure 16.2.5(a), whose direct transfer function is given by

$$G = \frac{C}{E} = \frac{K}{1 + p\tau} = \frac{K/\tau}{p + 1/\tau} \tag{16.2.14}$$

corresponding to which, the transient solution of the system is of the form given by

$$c(t) = Ae^{-t/\tau} \tag{16.2.15}$$

The transient in this system is seen to decay in accordance with a time constant of τ seconds. By placing a feedback path H around the direct transfer function, as shown in Figure 16.2.5(b), the closed-loop transfer function is then

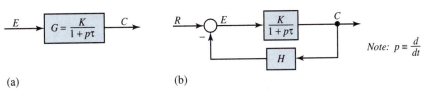

(a)   (b)

Figure 16.2.5 Block-diagram representation of system. **(a)** Without feedback. **(b)** With feedback.

$$M = \frac{C}{R} = \frac{\dfrac{K/\tau}{p + 1/\tau}}{1 + \dfrac{HK/\tau}{p + 1/\tau}} = \frac{K/\tau}{p + \dfrac{1 + HK}{\tau}} \qquad (16.2.16)$$

and the corresponding transient solution of the closed-loop system is of the form

$$c_f(t) = A_f e^{-[(1+HK)/\tau]t} \qquad (16.2.17)$$

where $c_f(t)$ represents the transient response of the output variable with feedback. Comparing Equations (16.2.15) and (16.2.17), it is clear that the time constant with feedback is smaller by the factor $1/(1 + HK)$, and hence the transient decays faster.

By treating the differential operator p as the sinusoidal frequency variable $j\omega$, i.e., $p = j\omega = j(1/\tau)$, it follows from Equation (16.2.14) that the bandwidth of the open-loop system spreads over a range from zero to a frequency of $1/\tau$ rad/s. On the other hand, for the system with feedback, Equation (16.2.16) reveals that the bandwidth spreads from zero to $(1 + HK)/\tau$ rad/s, showing that the bandwidth has been augmented by increasing the upper frequency limit by $1 + KH$.

Dynamic Response of Control Systems

The existence of transients (and associated oscillations) is a characteristic of systems that possess energy-storage elements and that are subjected to disturbances. Usually the complete solution of the differential equation provides maximum information about the system's dynamic performance. Consequently, whenever it is convenient, an attempt is made to establish this solution first. Unfortunately, however, this is not easily accomplished for high-order systems. Hence we are forced to seek out other easier and more direct methods, such as the frequency-response method of analysis.

Much of linear control theory is based on the frequency-response formulation of the system equations, and several quasi-graphical and algebraic techniques have been developed to analyze and design linear control systems based on frequency-response methods. Although frequency-response techniques are limited to relatively simple systems, and apply only to linear systems in the rigorous mathematical sense, they are still most useful in system design and the stability analysis of practical systems and can give a great deal of information about the relationships between system parameters (such as time constants and gains) and system response.

Once the transfer function of Equation (16.2.3) is developed in terms of the complex frequency variable s, by letting $s = j\omega$, the frequency-response characteristic and the loop gain $GH(j\omega)$ can be determined. The Bode diagram, displaying the frequency response and root-locus techniques, can be used to study the stability analysis of feedback control systems. The dc steady-state response, which becomes one component of the step response of the control system, can also be determined by allowing s to be zero in the transfer function. The step response, in turn, can be used as a measure of the speed of response of the control system. Thus, the transfer function obtained from the block diagram can be used to describe both the steady-state and the transient response of a feedback control system.

The matrix formulations associated with *state-variable* techniques have largely replaced the block-diagram formulations. Computer software for solving a great variety of state-equation formulations is available on most computer systems today. However, in the state-variable formulation, much of the physical reality of any system is lost, including the relationships between system response and system parameters.

With the development and widespread use of digital (discrete) control systems and the advent of relatively inexpensive digital computers, time-response methods have become more necessary and available. These may be divided into two broad methodologies:

1. *The actual simulation or modeling* of the system differential equations by either analog or digital computers.

2. *The state-variable formulation* of the system state equations and their solution by a digital computer. State-variable methods offer probably the most general approach to system analysis and are useful in the solution of both linear and nonlinear system equations.

The transient portion of the time response of a stable control system is that part which goes to zero as time increases and becomes sufficiently large. The transient behavior of a control system is usually characterized by the use of a unit-step input. Typical performance criteria that are used to characterize the transient response to a unit-step input include *overshoot, delay time, rise time,* and *settling time.* Figure 16.2.6 illustrates a typical unit-step response of a linear control system. The following four quantities give a direct measure of the transient characteristics of the step response:

- *Maximum* (or *peak*) *overshoot A* is the largest deviation of the output over the step input during the transient state. It is used as a measure of the relative stability of the system. The percentage maximum overshoot is given by the ratio of maximum overshoot to final value, expressed in percent.

- *Delay time* t_d is defined as the time required for the step response to reach 90% of its final value.

- *Rise time* t_r is defined as the time needed for the step response to rise from 10% to 90% of its final value.

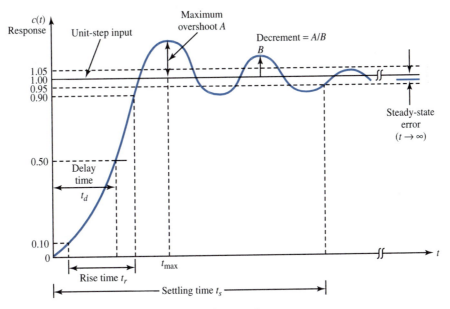

Figure 16.2.6 Typical unit-step response of a control system.

• *Settling time* t_s is the time required for the step response to decrease and stay within a specified percentage (usually 5%) of its final value.

While these quantities are relatively easy to measure once a step response is plotted, they cannot easily be determined analytically, except for simple cases.

Steady-State Error of Linear Systems

If the steady-state response of the output does not agree exactly with the steady state of the input, the system is said to have a *steady-state error*. Steady-state errors in practical control systems are almost unavoidable because of friction, other imperfections, and the nature of the system. The objective is then to keep the error to a minimum, or below a certain value. The steady-state error is a measure of system accuracy when a specific type of input is applied to a control system.

Referring to Figure 16.2.3, assuming the input and output signals are of the same dimension and are at the same level before subtraction, with a nonunity element $H(s)$ incorporated in the feedback path, the error of the feedback control system is defined as

$$e(t) = r(t) - b(t) \qquad \text{or} \qquad E(s) = R(s) - B(s) = R(s) - H(s)C(s)$$

or using Equation (16.2.2),

$$E(s) = \frac{R(s)}{1 + G(s)H(s)} \tag{16.2.18}$$

Applying the final-value theorem, the steady-state error of the system is

$$e_{SS} = \lim_{t \to \infty} e(t) = \lim_{s \to 0} sE(s) \tag{16.2.19}$$

in which $sE(s)$ is to have no poles that lie on the imaginary axis and in the right half of the s-plane. Substituting Equation (16.2.18) into Equation (16.2.19), we get

$$e_{SS} = \lim_{s \to 0} \frac{sR(s)}{1 + G(s)H(s)} \tag{16.2.20}$$

which apparently depends on the reference input $R(s)$ and the loop gain (loop transfer function) $G(s)H(s)$.

The *type* of feedback control system is decided by the order of the pole of $G(s)H(s)$ at $s = 0$. Thus, if the loop gain is expressed as

$$G(s)H(s) = \frac{KN(s)}{s^q D(s)} = \frac{K(1 + T_1 s)(1 + T_2 s) \cdots (1 + T_m s)}{s^q (1 + T_a s)(1 + T_b s) \cdots (1 + T_n s)} \tag{16.2.21}$$

where K and all of the T are constants, the exponent of s, i.e., q, in the denominator represents the number of integrations in the open loop, and the exponent q defines the system type. With $q = 0$, 1, or 2, the system is classified as *position, velocity,* or *acceleration* system, respectively. Table 16.2.2 summarizes the error response for different unit inputs and three system types. The development of these results is left as Problem 16.2.32 at the end of this chapter.

Classification of Feedback Control Systems by Control Action

A more common means of describing industrial and process controllers is by the way in which the error signal $E(s)$ is used in the forward loop of Figure 16.2.3. The basic control elements are

• A *proportional device,* such as an amplifier
• A *differentiating device,* such as an inductor

TABLE 16.2.2 Steady-State Error Response for Unit Inputs

System Type	Unit Step $\left(\dfrac{1}{s}\right)$	Unit Ramp $\left(\dfrac{1}{s^2}\right)$	Unit Acceleration $\left(\dfrac{1}{s^3}\right)$
0	Finite	Infinite	Infinite
1	0	Finite	Infinite
2	0	0	Finite

- An *integrating device,* such as an operational amplifier with feedback.

Many industrial controllers utilize two or more of the basic control elements. The classification by control action is summarized in Table 16.2.3.

Error-Rate Control, Output-Rate Control, and Integral-Error (Reset) Control

Let us consider a typical second-order servomechanism (containing two energy-storing elements) whose defining differential equation for obtaining the dynamic behavior of the system is given by

$$J\frac{d^2c}{dt^2} + F\frac{dc}{dt} + Kc + T_L = Kr \tag{16.2.22}$$

where J, F, and T_L represent, respectively, effective inertia, equivalent viscous friction, and resultant load torque appearing at the motor shaft; $K = K_p K_a K_m$, where K_p is the potentiometer transducer constant (V/rad), K_a is the amplifier gain factor (V/V), and K_m is the motor-developed torque constant (N·m/V) of the physical servomechanism shown in Figure 16.2.7; r is the input command and c is the output displacement, both in radians.

The block diagram of the servomechanism of Figure 16.2.7 is given in Figure 16.2.8, in which the transfer function of each component is shown. Since the actuating signal is given by $E = K_p(R - C)$, the block diagram may be simplified, as shown in Figure 16.2.9, in which T_L is assumed to be zero for simplicity. The direct transmission function (forward path transfer function) for the system can be seen to be

$$G(s) = \frac{K_p K_a K_m}{Js^2 + Fs} = \frac{K}{Js^2 + Fs} \tag{16.2.23}$$

The corresponding closed-loop transfer function is given by

$$M(s) = \frac{C(s)}{R(s)} = \frac{G(s)}{1 + HG(s)} = \frac{K/(Js^2 + Fs)}{1 + K/(Js^2 + Fs)} = \frac{K}{Js^2 + Fs + K} \tag{16.2.24}$$

TABLE 16.2.3 Classification by Control Action

Control Action	$G(s)$
Proportional	K_p
Differential	sK_d
Integral	K_i/s
Proportional and differential (PD)	$K_p + sK_d$
Proportional and integral (PI)	$K_p + K_i/s$
Proportional and integral and differential (PID)	$K_p + K_i/s + sK_d$

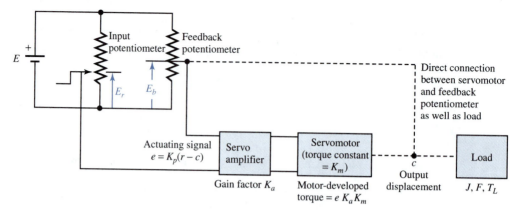

Figure 16.2.7 Second-order servomechanism.

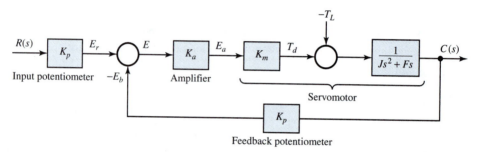

Figure 16.2.8 Block diagram of the servomechanism of Figure 16.2.7.

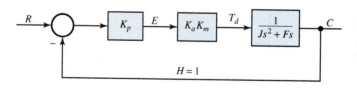

Figure 16.2.9 Simplified block diagram of the servomechanism of Figure 16.2.7 (with zero load torque).

or

$$M(s) = \frac{C(s)}{R(s)} = \frac{K/J}{s^2 + (F/J)s + K/J} \qquad (16.2.25)$$

or

$$C(s) = R(s) \frac{K/J}{s^2 + (F/J)s + K/J} = R(s) \frac{\omega_n^2}{s^2 + 2\xi\omega_n s + \omega_n^2} \qquad (16.2.26)$$

where $\omega_n \equiv \sqrt{K/J}$ is known as the *system natural frequency*, and

$$\xi \equiv \frac{F}{2\sqrt{KJ}} = \frac{\text{total damping}}{\text{critical damping}}$$

is known as the *damping ratio*. ξ and ω_n are the two figures of merit that describe the dynamic behavior of any linear second-order system. The damping ratio ξ provides information about the maximum overshoot (see Figure 16.2.6) in the system when it is excited by a step-forcing function, whereas the natural frequency ω_n of a system provides a measure of the settling time (see Figure 16.2.6). The settling time is no greater than $t_s = 3/\xi\omega_n$ for the response to a step input to reach 95% of its steady-state value. For a 1% tolerance band, the settling time is no greater than $t_s = 5/\xi\omega_n$. Figure 16.2.10 depicts the percent maximum overshoot as a function of the damping ratio ξ for a linear second-order system. For two different systems having the same damping ratio ξ, the one with the larger natural frequency will have the smaller settling time in responding to input commands or load disturbances.

The particular form of the forced solution of Equation (16.2.26) depends upon the type of forcing function used. If a step input of magnitude r_0 is applied, then $R(s) = r_0/s$, so that the complete transformed solution becomes

$$C(s) = \left(\frac{r_0}{s}\right) \frac{\omega_n^2}{s^2 + 2\xi\omega_n s + \omega_n^2} \tag{16.2.27}$$

Note that the steady-state solution is generated by the pole associated with $R(s)$, while the transient terms resulting from a partial-fraction expansion are associated with the poles of the denominator.

In order to meet the requirements for the steady-state performance, as well as dynamic performance of the feedback control system of Figure 16.2.7, it becomes necessary to provide independent control of both performances. Practical methods that have gained widespread acceptance are

- *Error-rate control*
- *Output-rate control*
- *Integral-error (or reset) control.*

ERROR-RATE CONTROL

A system is said to possess error-rate damping when the generation of the output in some way depends upon the rate of change of the actuating signal. For the system of Figure 16.2.7, if the

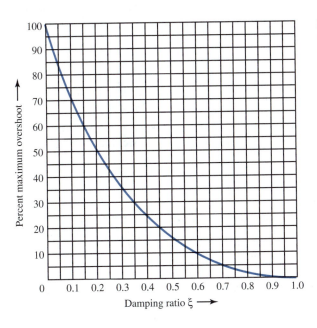

Figure 16.2.10 Plot of percent maximum overshoot as a function of the damping ratio for linear second-order system.

amplifier is so designed that it provides an output signal containing a term proportional to the derivative of the input, as well as one proportional to the input itself, error-rate damping will be introduced. For the system that includes error rate, the only modification needed is in the transfer function of the servoamplifier. Instead of the gain K_a, the new transfer function will be $K_a + sK_e$, where K_e denotes the error-rate gain factor of the amplifier. The complete block diagram is shown in Figure 16.2.11. In this case the direct transmission function becomes

$$G(s) = \frac{K_p K_m (K_a + sK_e)}{Js^2 + Fs} \tag{16.2.28}$$

and the closed-loop transfer function is given by

$$M(s) = \frac{C(s)}{R(s)} = \frac{G(s)}{1 + HG(s)} = \frac{K + sQ_e}{Js^2 + (F + Q_e)s + K} \tag{16.2.29}$$

where $K = K_p K_a K_m$ and $Q_e = K_p K_e K_m$ are known as the *loop proportional gain factor* and the *loop error-rate gain factor*, respectively.

Note that the steady-state solution for a step input r_0 is the same whether or not error-rate damping is present. The advantage of error-rate damping lies in the fact that it allows higher gains to be used without adversely affecting the damping ratio, and thereby makes it possible to satisfy the specifications for the damping ratio as well as for the steady-state performance. Also, the system's natural frequency is increased, which in turn implies smaller settling times.

OUTPUT-RATE CONTROL

A system is said to have output-rate damping when the generation of the output quantity in some way is made to depend upon the rate at which the controlled variable is varying. Output-rate control often involves the creation of an auxiliary loop, making the system multiloop. For the system of Figure 16.2.7, output-rate damping can be obtained by means of a tachometer generator, driven from the servomotor shaft. The complete block diagram of the servomechanism with output-rate damping is depicted in Figure 16.2.12, where K_o is the output-rate gain factor (V·s/rad) and the output-rate signal is given by $K_o(dc/dt)$. By applying the feedback relationship of Equation (16.2.3) to the minor (inner) loop, one gets

$$\frac{C(s)}{E(s)} = \frac{\dfrac{K_a K_m}{Js^2 + sF}}{1 + \dfrac{sK_o K_a K_m}{Js^2 + Fs}} = \frac{K_a K_m}{Js^2 + (F + Q_o)s} \tag{16.2.30}$$

where $Q_o = K_o K_a K_m$ is known as the loop output-rate gain factor. Figure 16.2.12 may then be simplified, as shown in Figure 16.2.13. The closed-loop transfer function for the complete system is then given by

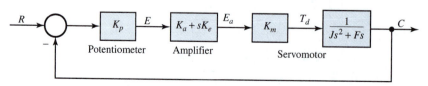

Figure 16.2.11 Block diagram of servomechanism with error-rate damping.

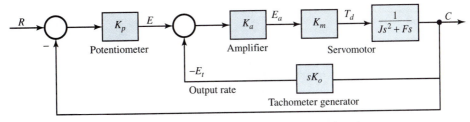

Figure 16.2.12 Block diagram of servomechanism with output-rate damping.

$$M(s) = \frac{C(s)}{R(s)} = \frac{\dfrac{K}{Js^2 + (F + Q_o)s}}{1 + \dfrac{K}{Js^2 + (F + Q_o)s}} = \frac{K}{Js^2 + (F + Q_o)s + K} \qquad (16.2.31)$$

where $K = K_p K_a K_m$ is known as the loop proportional gain factor, as stated earlier.

The manner in which output-rate damping effects in controlling the transient response are most easily demonstrated is by assuming a step input applied to the system. The output-rate signal appears in opposition to the proportional signal, thereby removing the tendency for an excessively oscillatory response. With error-rate damping and output-rate damping, the damping ratio can be seen to be

$$\xi = \frac{F + Q}{2\sqrt{KJ}} \qquad (16.2.32)$$

where $Q = Q_e$ or Q_o, in which a term is added to the numerator, while the natural frequency is unchanged (as given by $\omega_n = \sqrt{K/J}$). The damping ratio can then be adjusted independently through Q_e or Q_o, while K can be used to meet accuracy requirements.

INTEGRAL-ERROR (OR RESET) CONTROL

A control system is said to possess integral-error control when the generation of the output in some way depends upon the integral of the actuating signal. By designing the servoamplifier such that it makes available an output voltage that contains an integral term and a proportional term, integral-error control can be obtained. By modifying the transfer function of the servoamplifier in the block diagram, the complete block diagram of the servomechanism with integral-error control is shown in Figure 16.2.14, where K_i is the proportionality factor of the integral-error component. The direct transmission function is then given by

$$G(s) = \frac{K_p K_m (K_i + sK_a)}{Js^3 + Fs^2} \qquad (16.2.33)$$

and the corresponding closed-loop transfer function becomes

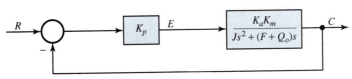

Figure 16.2.13 Simplified block diagram of Figure 16.2.12.

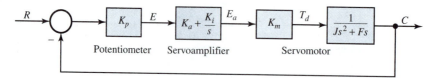

Figure 16.2.14 Block diagram of servomechanism with integral-error control.

$$M(s) = \frac{C(s)}{R(s)} = \frac{Q_i + sK}{Js^3 + Fs^2 + Ks + Q_i} \qquad (16.2.34)$$

where $Q_i = K_i K_p K_m$ and $K = K_p K_a K_m$, as defined earlier. Since the integral term stands alone without combining with the viscous-friction F term (as was the case with error-rate and output-rate controls), its influence on system performance differs basically from the previous compensation schemes. Also note that the order of the system is changed from second to third, because of the integral-error compensation. The inclusion of the integral term implies that a third independent energy-storing element is present. The position lag error, which exists with error-rate and output-rate control, disappears with integral-error control. This is a characteristic of integral control which greatly improves steady-state performance and system accuracy. However, it may make the dynamic behavior more difficult to cope with successfully.

Let us now present some illustrative examples.

EXAMPLE 16.2.1

Since dc motors of various types are used extensively in control systems, it is essential for analytical purposes that we establish a mathematical model for the dc motor. Let us consider the case of a separately excited dc motor with constant field excitation. The schematic representation of the model of a dc motor is shown in Figure E16.2.1(a). We will investigate how the speed of the motor responds to changes in the voltage applied to the armature terminals. The linear analysis involves electrical transients in the armature circuit and the dynamics of the mechanical load driven by the motor. At a constant motor field current I_f, the electromagnetic torque and the generated emf are given by

$$T_e = K_m i_a \qquad (1)$$

$$e_a = K_m \omega_m \qquad (2)$$

where $K_m = kI_f$ is a constant, which is also the ratio e_a/ω_m. In terms of the magnetization curve, e_a is the generated emf corresponding to the field current I_f at the speed ω_m. Let us now try to find the transfer function that relates $\Omega_m(s)$ to $V_t(s)$.

Solution

The differential equation for the motor armature current i_a is given by

$$v_t = e_a + L_a \frac{di_a}{dt} + R_a i_a \qquad (3)$$

or

$$R_a(1 + \tau_a p)i_a = v_t - e_a \qquad (4)$$

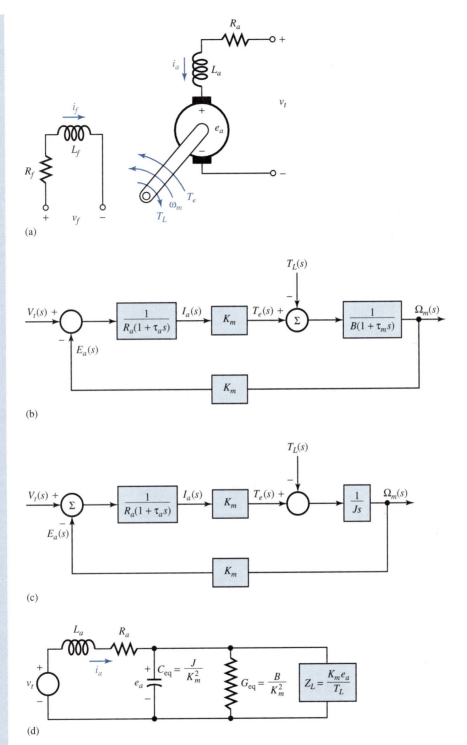

Figure E16.2.1 (a) Model of a separately excited dc motor. **(b)** Block diagram representing Equations (8) and (9). **(c)** Block diagram representing Equation (11). **(d)** Analog electric circuit for a separately excited dc motor.

where v_t is the terminal voltage applied to the motor, e_a is the back emf given by Equation (2), R_a and L_a include the series resistance and inductance of the armature circuit and electrical source put together, and $\tau_a = L_a/R_a$ is the *electrical time constant* of the armature circuit. Note that the operator p stands for (d/dt). The electromagnetic torque is given by Equation (1), and from the dynamic equation for the mechanical system given by

$$T_e = Jp\omega_m + B\omega_m + T_L \tag{5}$$

the acceleration is then given by

$$(B + Jp)\omega_m = T_e - T_L \tag{6}$$

or

$$B(1 + \tau_m p)\omega_m = T_e - T_L \tag{7}$$

where $\tau_m = J/B$ is the *mechanical time constant*. The load torque T_L, in general, is a function of speed, J is the combined polar moment of inertia of the load and the rotor of the motor, and B is the equivalent viscous friction constant of the load and the motor.

Laplace transforms of Equations (4) and (7) lead to the following:

$$I_a(s) = \frac{V_t(s) - E_a(s)}{R_a(1 + \tau_a s)} = \frac{V_t(s) - K_m \Omega_m(s)}{R_a(1 + \tau_a s)} \tag{8}$$

$$\Omega_m(s) = [T_e(s) - T_L(s)] \frac{1}{B} \frac{1}{(1 + \tau_m s)} \tag{9}$$

The corresponding block diagram representing these operations is given in Figure E16.2.1(b) in terms of the state variables $I_a(s)$ and $\Omega_m(s)$, with $V_t(s)$ as input.

The application of the closed-loop transfer function $M(s)$, shown in Figure 16.2.3, to the block diagram of Figure E16.2.1(b) yields the following transfer function relating $\Omega_m(s)$ and $V_t(s)$, with $T_L = 0$:

$$\frac{\Omega_m(s)}{V_t(s)} = \frac{K_m/[R_a(1 + \tau_m s)B(1 + \tau_m s)]}{1 + [K_m^2/R_a(1 + \tau_a s)B(1 + \tau_m s)]} \tag{10}$$

With mechanical damping B neglected, Equation (10) reduces to

$$\frac{\Omega_m(s)}{V_t(s)} = \frac{1}{K_m[\tau_i s(\tau_a s + 1) + 1]} \tag{11}$$

where $\tau_i = JR_a/K_m^2$ is the *inertial time constant*, and the corresponding block diagram is shown in Figure E16.2.1(c). The transfer function relating speed to load torque with $V_t = 0$ can be obtained from Figure E16.2.1(c) by eliminating the feedback path as follows:

$$\frac{\Omega_m(s)}{T_L(s)} = -\frac{1/Js}{1 + (1/Js)[K_m^2/R_a(1 + \tau_a s)]} = -\frac{\tau_a s + 1}{Js(1 + \tau_a s) + (K_m^2/R_a)} \tag{12}$$

Expressing the torque equation for the mechanical system as

$$T_e = K_m i_a = Jp\omega_m + B\omega_m + T_L \tag{13}$$

then dividing by K_m, and substituting $\omega_m = e_a/K_m$, we obtain

$$i_a = \frac{J}{K_m^2} \frac{de_a}{dt} + \frac{B}{K_m^2} e_a + \frac{T_L}{K_m} \tag{14}$$

Equation (14) can be identified to be the node equation for a parallel C_{eq}–G_{eq}–Z_L circuit with

$$C_{eq} = J/K_m^2; \qquad G_{eq} = B/K_m^2; \qquad Z_L = K_m e_a/T_L \tag{15}$$

and a common voltage e_a. An analog electric circuit can then be drawn as in Figure E16.2.1(d) for a separately excited dc motor, in which the inertia is represented by a capacitance, the damping by a shunt conductance, and the load–torque component of current is shown flowing through an equivalent impedance Z_L. The time constants, τ_i associated with inertia and τ_m associated with the damping-type load torque that is proportional to speed (in terms of the analog-circuit notations), are given by

$$\tau_i = R_a C_{eq} \qquad \text{and} \qquad \tau_m = C_{eq}/G_{eq} \qquad (16)$$

Note in the preceding analysis that K_m is proportional to the constant motor field current I_f.

The self-inductance of the armature can often be neglected except for a motor driving a load that has rapid torque pulsations of appreciable magnitude.

EXAMPLE 16.2.2

A voltage amplifier without feedback has a nominal gain of 500. The gain, however, varies in the range of 475 to 525 due to parameter variations. In order to reduce the per-unit change to 0.02, while maintaining the original gain of 500, if the feedback is introduced, find the new direct transfer gain G and the feedback factor H.

Solution

If G' is the gain without feedback, it follows that

$$\frac{\partial G'}{G'} = \frac{525 - 475}{500} = 0.1$$

From Equation (16.2.7),

$$\frac{\partial M}{M} = \frac{\partial G}{G}\left(\frac{1}{1 + HG}\right) = 0.02 \qquad (1)$$

where G stands for the new transmission gain with feedback. Also, from Equation (16.2.3),

$$M = \frac{G}{1 + HG} = 500 \qquad (2)$$

From Equation (1),

$$1 + HG = \frac{0.1}{0.02} = 5 \qquad (3)$$

Substituting Equation (3) into Equation (2), one obtains

$$G = 500(1 + HG) = 500(5) = 2500$$

and from Equation (3),

$$H = \frac{5 - 1}{2500} = 0.0016$$

EXAMPLE 16.2.3

The Ward–Leonard system, which is used in the control of large dc motors employed in rolling mills, is a highly flexible arrangement for effecting position and speed control of a separately excited dc motor. A simplified schematic diagram of such a system is shown in Figure E16.2.3(a), in which the tachometer, error detector, and amplifier are represented functionally rather than by electric circuit connections. Assume the motor field current is held constant, the load torque is ignored, and the generator speed is held constant under all conditions of operation.

Determine the transfer functions Ω_o/E_g and Ω_o/E_e, and develop a block diagram for the system.

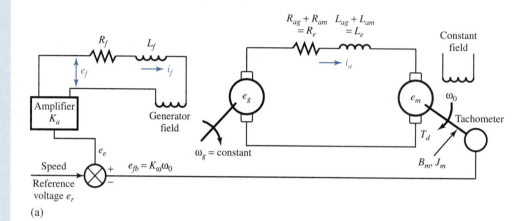

(a)

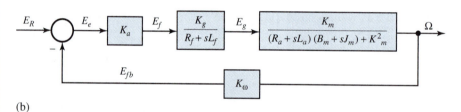

(b)

Figure E16.2.3 (a) Schematic representation of a simplified Ward–Leonard dc motor control. (b) Block diagram of (a).

Solution

For the dc generator with constant speed, we have

$$e_f = i_f R_f + L_f \frac{di_f}{dt}$$

$$e_g = K_g i_f$$

where K_g is a constant, and

$$\frac{E_g(s)}{E_f(s)} = \frac{K_g}{R_f + sL_f}$$

The differential equation in the armature circuits is given by

$$e_g = i_a R_e + L_e \frac{di_a}{dt} + e_m \tag{1}$$

where

$$e_m = K_m \omega_o \tag{2}$$

for the constant field current, with K_m a constant, and

$$T_d = K_m i_a = B_m \omega_o + J_m \frac{d\omega_o}{dt} \tag{3}$$

neglecting load torque. Transforming Equations (1), (2), and (3) and rearranging, we get

$$\frac{\Omega_o}{E_g} = \frac{K_m}{(R_a + sL_a)(B_m + sJ_m) + K_m^2}$$

Accounting for the amplifier gain K_a, we have

$$\frac{\Omega_o}{E_e} = \frac{K_a K_g K_m}{(R_f + sL_f)\left[(R_a + sL_a)(B_m + sJ_m) + K_m^2\right]}$$

The block diagram, including the feedback loop due to the tachometer, is shown in Figure E16.2.3(b).

EXAMPLE 16.2.4

Consider an elementary feedback control system, as shown in Figure 16.2.3, with $H = 1$. The output variable c and the input e to the direct transmission path are related by

$$\frac{d^2 c}{dt^2} + 8 \frac{dc}{dt} + 12c = 68e$$

(a) Assuming the system to be initially at rest, for $e = u(t)$, find the complete solution for $c(t)$ when the system is operated in open-loop fashion without any feedback.

(b) With the feedback loop connected and with a forcing function of a unit step, i.e., $R(s) = 1/s$, describe the nature of the dynamic response of the controlled variable by working with the differential equation for the closed-loop system, without obtaining a formal solution for $c(t)$.

(c) By working solely in terms of transfer functions, discuss the closed-loop behavior.

Solution

(a) Laplace transforming the given equation with zero initial conditions, we have

$$s^2 C(s) + 8s C(s) + 12 C(s) = \frac{68}{s} \tag{1}$$

or

$$C(s) = \frac{68}{s(s^2 + 8s + 12)} = \frac{68}{s(s+2)(s+6)} = \frac{K_0}{s} + \frac{K_1}{s+2} + \frac{K_2}{s+6} \tag{2}$$

Evaluating the coefficients of the partial-fraction expansion, we get

$$C(s) = \frac{17}{3}\left(\frac{1}{s}\right) - \frac{17}{2}\left(\frac{1}{s+2}\right) + \frac{17}{6}\left(\frac{1}{s+6}\right) \tag{3}$$

The corresponding time solution is then given by

$$c(t) = \frac{17}{3} - \frac{17}{2}e^{-2t} + \frac{17}{6}e^{-6t} \tag{4}$$

(b) The differential equation describing the closed-loop system operation is

$$\frac{d^2c}{dt^2} + 8\frac{dc}{dt} + 12c = 68(r - c) \tag{5}$$

The output and input variables of the closed-loop system are related by

$$\frac{d^2c}{dt^2} + 8\frac{dc}{dt} + 80c = 68r \tag{6}$$

The two figures of merit, damping ratio ξ and natural frequency ω_n, follow from the characteristic equation of the closed-loop system,

$$s^2 + 8s + 80 = 0 \tag{7}$$

(when the input is set equal to zero and the left side of Equation (6) is Laplace transformed with zero initial conditions).

Identifying Equation (7) with the general form that applies to all linear second order systems,

$$s^2 + 2\xi\omega_n s + \omega_n^2 = 0 \tag{8}$$

we obtain the following by comparing coefficients:

$$\omega_n = \sqrt{80} = 8.95 \text{ rad/s} \tag{9}$$

and

$$\xi = \frac{8}{2\omega_n} = \frac{4}{8.95} = 0.448 \tag{10}$$

Referring to Figure 16.2.10, the maximum overshoot can be seen to be 17%. With $\xi\omega_n = 4$, it follows that the controlled variable reaches within 1% of its steady-state value after the time elapse of

$$t_s = \frac{5}{\xi\omega_n} = \frac{5}{4} = 1.25 \text{ s}$$

(c) The transfer function of the direct transmission path is given by

$$G(s) = \frac{C(s)}{E(s)} = \frac{68}{s^2 + 8s + 12} \tag{11}$$

The closed-loop transfer function is

$$M(s) = \frac{C(s)}{R(s)} = \frac{G(s)}{1 + HG(s)} = \frac{\dfrac{68}{s^2 + 8s + 12}}{1 + \dfrac{68}{s^2 + 8s + 12}} \tag{12}$$

or

$$\frac{C(s)}{R(s)} = \frac{68}{s^2 + 8s + 80} \tag{13}$$

Noting that the denominator of Equation (13) is the same as the left side of Equation (7), it follows that the damping ratio and the natural frequency will have the same values as found in part (b). The oscillatory dynamic behavior can be described with the two figures of merit ξ and ω_n, along with percent maximum overshoot and settling time.

EXAMPLE 16.2.5

A feedback control system with the configuration of Figure 16.2.7 has the following parameters: $K_p = 0.5$ V/rad, $K_a = 100$ V/V, $K_m = 2.7 \times 10^{-4}$ N·m/V, $J = 1.5 \times 10^{-5}$ kg·m², and $F = 2 \times 10^{-4}$ kg·m²/s.

(a) With an applied step-input command, describe the dynamic response of the system, assuming the system to be initially at rest.

(b) Find the position lag error in radians, if the load disturbance of 1.5×10^{-3} N·m is present on the system when the step command is applied.

(c) In order that the position lag error of part (b) be no greater than 0.025 rad, determine the new amplifier gain.

(d) Evaluate the damping ratio for the gain of part (c) and the corresponding percent maximum overshoot.

(e) With the gain of part (c), in order to have the percent maximum overshoot not to exceed 25%, find the value of the output-rate gain factor.

Solution

(a) $K = K_p K_a K_m = 0.5 \times 100 \times 2.7 \times 10^{-4} = 0.0135$ N·m/rad

$$\xi = \frac{F}{2\sqrt{KJ}} = \frac{2 \times 10^{-4}}{2\sqrt{0.0135 \times 1.5 \times 10^{-5}}} = \frac{10^{-4}}{0.45 \times 10^{-3}} = 0.222$$

From Figure 16.2.10, the percent maximum overshoot is 48%,

$$\omega_n = \sqrt{\frac{K}{J}} = \sqrt{\frac{0.0135}{1.5 \times 10^{-5}}} = 30 \text{ rad/s}$$

The commanded value of the controlled variable reaches within 1% of its final value in

$$t_s = \frac{5}{\xi \omega_n} = \frac{5}{0.222 \times 30} = 0.75 \text{ s}$$

The roots of the characteristic equation

$$s^2 + \frac{F}{J}s + \frac{K}{J} = 0$$

are

$$s_{1,\,2} = -\frac{F}{2J} \pm \sqrt{\left(\frac{F}{2J}\right)^2 - \frac{K}{J}}$$

The system is:

(i) Overdamped if $\left(\dfrac{F}{2J}\right)^2 > \dfrac{K}{J}$

(ii) Underdamped if $\left(\dfrac{F}{2J}\right)^2 < \dfrac{K}{J}$

(iii) Critically damped if $\left(\dfrac{F}{2J}\right)^2 = \dfrac{K}{J}$

To make the servomechanism fast acting, the underdamped case is desirable. For the underdamped situation, the roots are given by

$$s_{1,\,2} = -\frac{F}{2J} \pm j\sqrt{\frac{K}{J} - \left(\frac{F}{2J}\right)^2}$$

which can be expressed as

$$s_{1,\,2} = -\xi\omega_n \pm j\omega_d$$

where $\omega_d = \omega_n\sqrt{1-\xi^2}$ is known as *damped frequency of oscillation*. The damped oscillations in our example occur at a frequency of

$$\omega_d = 30\sqrt{1 - 0.222^2} = 29.25 \text{ rad/s}$$

(b) Starting from Equation (16.2.22), with a step command of magnitude r_0 and a step change in load of magnitude T_L, assuming the system to be initially at rest, one obtains

$$C(s) \cdot (s^2 J + sF + K) = \frac{Kr_0}{s} - \frac{T_L}{s}$$

or

$$C(s) = \frac{(K/J)r_0}{s\left(s^2 + s\dfrac{F}{J} + \dfrac{K}{J}\right)} - \frac{T_L/J}{s\left(s^2 + s\dfrac{F}{J} + \dfrac{K}{J}\right)}$$

If $T_L = 0$, at steady state, $c_{SS} = r_0$. In the presence of a fixed load torque, $c_{SS} = r_0 - T_L/K$. The position lag error for our example is given by

$$\frac{T_L}{K} = \frac{1.5 \times 10^{-3}}{0.0135} = 0.111 \text{ rad}$$

(c) The loop gain must be increased by a factor of $0.111/0.025 = 4.44$. Hence the new value of the amplifier gain is

$$K'_a = 4.44\, K_a = 4.44 \times 100 = 444 \text{ V/V}$$

(d) Since $\xi = F/2\sqrt{KJ}$, it follows that

$$\xi' = \frac{\xi}{\sqrt{4.44}} = 0.4746\xi = 0.4746 \times 0.222 = 0.105$$

corresponding to which the percent maximum overshoot from Figure 16.2.10 is 72%.

(e) For a 25% overshoot, Figure 16.2.10 shows that $\xi = 0.4$. Then from Equation (16.2.32), $\xi = (F + Q_o)/2\sqrt{KJ}$, or

$$0.4 = \frac{2 \times 10^{-4} + Q_o}{2\sqrt{0.0135 \times 1.5 \times 10^{-5}}} = \frac{2 \times 10^{-4} + Q_o}{2 \times 0.45 \times 10^{-3}}$$

Hence,

$$Q_o = 0.36 \times 10^{-3} - 0.2 \times 10^{-3} = 0.16 \times 10^{-3}$$

Then

$$K_o = \frac{Q_o}{K'_a K_m} = \frac{0.16 \times 10^{-3}}{444 \times 2.7 \times 10^{-4}} = 0.0013, \text{ or } 1.3 \times 10^{-3} \text{ V·s/rad.}$$

16.3   DIGITAL CONTROL SYSTEMS

Significant progress has been made in recent years in discrete-data and digital control systems because of the advances made in digital computers and microcomputers, as well as the advantages found in working with digital signals. Discrete-data and digital control systems differ from the continuous-data or analog systems in that the signals in one or more parts of these systems are in the form of either a pulse train or a numerical (digital) code. The terms, sampled-data systems, discrete-data systems, discrete-time systems, and digital systems have been loosely used in the control literature. However, sampled-data systems usually refer to a general class of systems whose signals are in the form of pulse data; sampled data refers to signals that are pulse-amplitude modulated, i.e., trains of pulses with signal information carried by the amplitudes. Digital control systems refers to the use of a digital computer or controller in the system; digital data usually refers to signals that are generated by digital computers or digital transducers and are thus in some kind of coded form. A practical system such as an industrial process control is generally of such complexity that it contains analog and sampled as well as digital data. Hence the term discrete-data systems is used in a broad sense to describe all systems in which some form of digital or sampled signals occur. When a microprocessor receives and outputs digital data, the system then becomes a typical discrete-data or digital control system.

Figure 16.3.1(a) illustrates the basic elements of a typical closed-loop control system with sampled data; Figure 16.3.1(b) shows the continuous-data input $e(t)$ to the sampler, whereas Figure 16.3.1(c) depicts the discrete-data output $e^*(t)$ of the sampler. A continuous input signal $r(t)$ is applied to the system. The continuous error signal is sampled by a sampling device, the sampler, and the output of the sampler is a sequence of pulses. The pulse train may be periodic or aperiodic, with no information transmitted between two consecutive pulses. The sampler in the present case is assumed to have a uniform sampling rate, even though the rate may not be uniform in some other cases. The magnitudes of the pulses at the sampling instants represent the values of the input signal $e(t)$ at the corresponding instants. Sampling schemes, in general, may have many variations: periodic, cyclic-rate, multirate, random, and pulse-width modulated

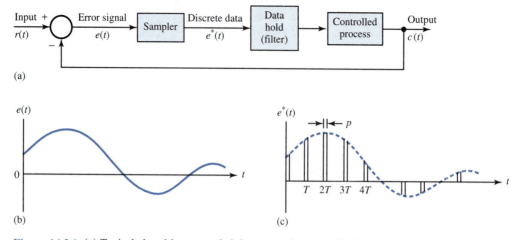

(a)

(b)   (c)

Figure 16.3.1 **(a)** Typical closed-loop sampled-data control system. **(b)** Continuous-data input to sampler of (a). **(c)** Discrete-data output of sampler of (a).

samplings. Incorporating sampling into a control system has several advantages, including that of time sharing of expensive equipment among various control channels.

The purpose of the filter located between the sampler and the controlled process is for smoothing, because most controlled processes are normally designed to receive analog signals.

Figure 16.3.2 illustrates a typical digital control system, in which the signal at one or more points of the system is expressed in a numerical code for digital-computer or digital-transducer processing in the system. Because of the digitally coded (such as binary-coded) signals in some parts of the system, it becomes necessary to employ digital-to-analog (D/A) and analog-to-digital (A/D) converters. In spite of the basic differences between the structures and components of a sampled-data and a digital control system, from an analytical standpoint both types of systems are treated by the same analytical tools.

Sampled data and digital control offer several advantages over analog systems:

- More compact and lightweight
- Improved sensitivity
- Better reliability, speed, and accuracy
- More flexibility and versatility (in programming)
- Lower cost
- More rugged in construction

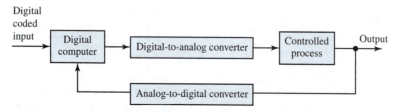

Figure 16.3.2 Typical digital control system.

- No drift
- Less effect due to noise and disturbance.

Many physical systems have inherent sampling, and hence their behavior can be described by sampled data or digital models. Even the dynamics of the social, economic, and biological systems can be modeled by sampled-data system models.

The block diagram of Figure 16.3.3 is a functional representation of a type of digital control system, in which G and H serve the same function as in any feedback system. Note that the error signal is sampled and a digital processor is used. The controller in this system is the digital processor whose output, reconverted to an analog signal, becomes the excitation for the block G. As usual, G is the subsystem that provides the output to be controlled. A central computer which controls several functions could be used as a digital processor; or a microprocessor (special-purpose computer) designed for the particular control function may also be used as a digital processor. Large, high-speed computers with their speed, memory, and computational ability, as well as programmability, are utilized for central control in large automated manufacturing facilities. The control process may be summarized as follows:

1. The computer is programmed to indicate the sequence of operations required. Data are fed continuously from various monitoring stations on the progress of the process.

2. By comparing the actual and desired performance, the computer generates a new set of instructions to correct for the deviations.

3. The new set of instructions are usually converted to an analog signal, which in turn forms the excitation applied to the machines that actually do the manufacturing.

Examples of Discrete-Data and Digital Control Systems

A few *simplified* examples are presented to illustrate some of the essential components of the control systems.

SINGLE-AXIS AUTOPILOT CONTROL SYSTEM OF AN AIRCRAFT OR MISSILE

Figure 16.3.4(a) shows the block diagram of a simplified single-axis (pitch, yaw, or roll) autopilot control system with digital data. The objective of the control is to make the position of the airframe follow the command signal. The rate loop included here helps to improve system stability. Figure 16.3.4(b) illustrates a digital autopilot control system in which the position and rate information are obtained by digital transducers. Sample-and-hold devices are shown on the block diagram. While a sampler essentially samples an analog signal at some uniform sampling period, the hold

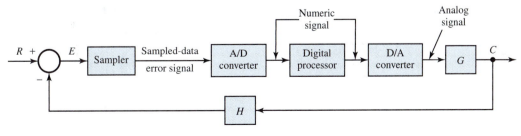

Figure 16.3.3 Block diagram of a type of digital control system.

device simply holds the value of the signal until the next sample comes along. Two samplers with two different sampling periods (T_1 and T_2) are shown in Figure 16.3.4(b), which is known as a *multirate sampled-data system*. If the rate of variation of the signal in one loop is relatively much less than that of the other loop, the sampling period of the sampler employed in the slower loop can be larger. With sampling and multirate sampling, some of the expensive components of the system can be utilized on a time-sharing basis.

DIGITAL CONTROLLER FOR A STEAM-TURBINE DRIVEN GENERATOR

Figure 16.3.5 illustrates the basic elements of a minicomputer system in terms of a block diagram for speed and voltage control (as well as data acquisition) of a turbine-generator unit. Typical output variables of the generator are speed, rotor angle, terminal voltage, field (excitation) current, armature current, and real and reactive power. Some output variables are measured by digital transducers, whose outputs are then digitally multiplexed and sent back to the minicomputer. Some other output variables may be measured by analog transducers, whose outputs are processed through an analog multiplexer which performs a time-division multiplexing operation. The output of the analog multiplexer is connected to a sample-and-hold device, which samples the output of the multiplexer at a fixed time interval and then holds the signal level at its output until the A/D converter performs the analog-to-digital conversion. Thus, following the multiplexer, time sharing is done amid a number of signal channels.

MACHINE-TOOL PROCESS TO DRILL OR PUNCH HOLES (DIGITALLY CONTROLLED DRILL PRESS)

Figure 16.3.6 depicts an elementary system, including the input, digital processor, drill-positioning mechanism, and sampled-data position-control system. The process is controlled as follows:

1. The processor receives the input data and determines the sequence of operations, while storing both the sequence and the hole positions in its memory.
2. The sampled-data system helps in positioning the drill at the first hole location.
3. If the actual drill position agrees with the desired location stored in the memory, the processor sends a signal, thereby causing the drilling process to begin.
4. At the end of the first drilling operation, the process is continued until all the required holes are drilled.

To the basic system illustrated in Figure 16.3.6, additional controls such as speed control (to accommodate different kinds of materials to be drilled) and timing control (to control the time duration of the drilling process) can be added rather easily.

STEP-MOTOR CONTROL SYSTEM FOR READ–WRITE HEAD POSITIONING ON A DISK DRIVE

Figure 16.3.7 illustrates a system in which the prime mover used in the disk-drive system is a step motor driven by pulse commands. In response to each pulse input, the step motor moves by some fixed displacement. This all-digital system does not need an A/D or D/A converter.

POSITION SERVO IN A RADAR SYSTEM

Figure 16.3.8 depicts the essential elements of a position servo system. In a radar system, an electromagnetic pulse is radiated from an antenna into space. An echo pulse is received back

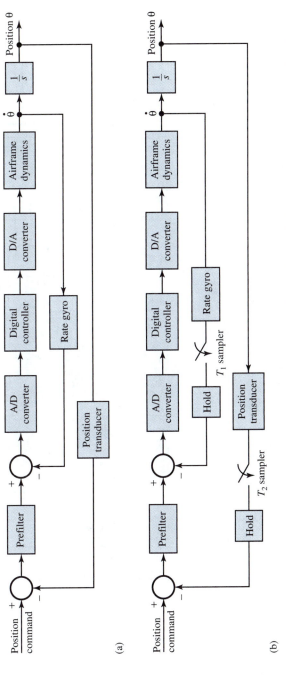

Figure 16.3.4 **(a)** Single-axis autopilot control system with digital data. **(b)** Single-axis digital autopilot control system with multirate sampling.

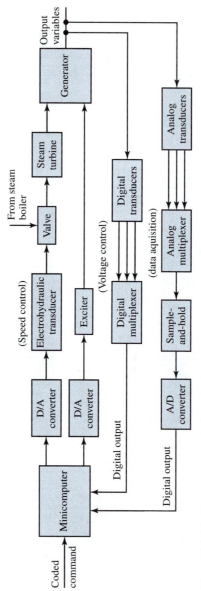

Figure 16.3.5 Computer control of a turbogenerator unit.

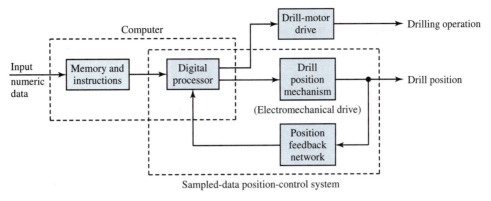

Figure 16.3.6 Simplified block diagram of a digitally controlled drill press.

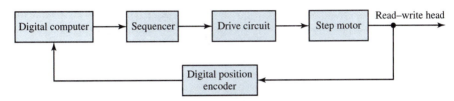

Figure 16.3.7 Step-motor control system for read–write head positioning on a disk drive.

when a conductive surface, such as an airplane, appears in the signal path. While the antenna is continuously rotated in search of a target, the antenna is stopped when the target is located. It is pointed toward the target by varying its angular direction until a maximum echo is heard. The position coordinate $\theta(t)$ of the antenna is usually controlled through a gear train by a dc servomotor. The motor torque is varied in both magnitude and direction by means of a control voltage obtained from the amplifier output. The control problem is then to command the motor such that the output $\theta(t)$ is nearly the same as the reference angular position $\theta_r(t)$. The closed loop can be controlled by a digital controller.

Microprocessor Control

The microprocessor, which has rapidly become a key component in digital control systems, and its associated circuits function as the digital processor. It is often used as a dedicated computer

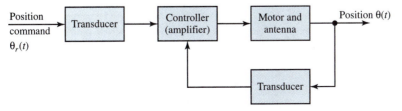

Figure 16.3.8 Plant model of a position servo in a radar system.

in which the chip itself is programmed to perform specific functions. Digital watches and auto fuel-injection systems are examples. With microprocessor control, not only the same hardware can be used to perform a variety of tasks by reprogramming, but also *distributed computation* (which results in less expensive and more reliable operation than centralized control) is possible with interconnection and intercommunication of several microprocessors.

Figure 16.3.9 shows a block diagram of a microprocessor-controlled dc motor system, in which the controlled process consists of a dc motor, the load, and the power amplifier.

The block diagram of Figure 16.3.10 illustrates a typical microprocessor system used to implement the digital PI controller. It would be simple to include the derivative operation to implement the PID controller. While an analog timer is shown in Figure 16.3.10 to determine the start of the next sampling period, a software timing loop can be used to keep track of when T seconds have elapsed. The output pulse, once every T seconds, is applied to the interrupt line of the microprocessor. This will cause the processor to execute the interrupt routine to output the next value of the control, $u[(k + 1)T]$, which is sent to the D/A converter, whose output in turn controls the power amplifier. The timing pulse from the timer is also sent to the "sample" command line, thereby triggering the sample-and-hold circuitry; the motor velocity $\omega(t)$ is sampled and held constant for one sampling period. The value of $\omega(kT)$ is then converted to an N-bit binary number by the A/D circuitry. The microprocessor is signaled via "data ready" line (which may be attached to the interrupt line of the microprocessor) that the sampled data have been converted. The second interrupt will cause the processor to read in the value of

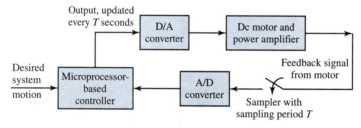

Figure 16.3.9 Block diagram of a microprocessor-controlled dc motor system.

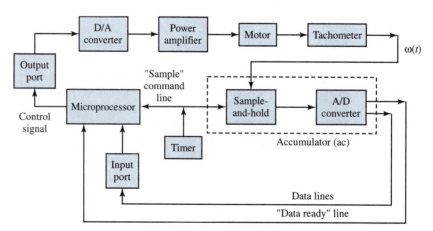

Figure 16.3.10 Block diagram of discretized proportional and integral (PI) controller.

$\omega(kT)$ and then compute the next value of control, $u[(k + 1)T]$. After calculation of the control, the microprocessor waits for another interrupt from the timer before it outputs the control at $t = (k + 1)T$. An *assembly-language* program can be developed for the implementation of the PI controller.

The control of a dc motor can be achieved with a PI controller discretized for microprocessor programming. The starting point is that the PI controller is described by a differential equation. The latter is discretized at the sampling instants by one of the numerical approximation methods, and then is programmed in the microprocessor machine language.

Practical limitations of microprocessor-based control systems stem from the following considerations:

- Finite-word-length characteristic, that is, an 8-bit word would only allow $2^8 = 256$ levels of resolution.

- Time delays encountered in executing the data handling, which may have a significant effect on the system response.

- Quantization effects, which affect accuracy and stability.

Adaptive Control

Another type of control system that makes use of the computer is known as *adaptive control*, which is functionally represented in Figure 16.3.11. While the block G provides the system output for the system input, the feedback is provided by H and is converted to a digital signal. This digital signal is processed by the digital processor to effect the necessary control. The characteristic feature of the adaptive control is that the characteristics of the block G are changed with time so that, for a given input, the desired output is obtained. Based on the prediction of the performance of the system, the digital processor determines how G is to be changed.

Figure 16.3.12 illustrates the adaptive control process as applied to a motor-speed control system, which is based on adjustments of the armature resistance. Using all of the data inputs, the computer computes the optimum value of R_a such that the motor speed is appropriate to the load. The signal corresponding to the computed value of R_a is in turn used to position the potentiometer arm of the variable resistance. The actual armature resistance is then controlled by the position of the potentiometer arm, which in turn may be controlled by a sampled-data version of the position-control system.

The speed and power of the digital computer (in predicting the response to a wide range of changes that affect the system) are the main reasons for the use of adaptive systems in industrial control. Large-scale systems may have multiple inputs applied simultaneously, and control is successfully achieved because the process of predicting the desired response, and comparing with the actual response, can be accomplished in a relatively short time compared to the response time of the system to be controlled.

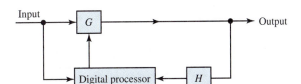

Figure 16.3.11 Functional representation of an adaptive control system.

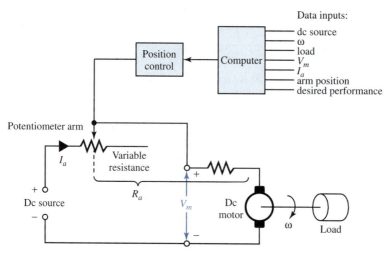

Figure 16.3.12 Adaptive control process as applied to a motor-speed control system.

Methods of Analysis

Just as differential equations are used to represent systems with analog signals, *difference equations* are used for systems with discrete or digital data. Difference equations are also used to approximate differential equations, since the former are more easily programmed on a digital computer, and are generally easier to solve.

One of the mathematical tools devised for the analysis and design of discrete-data systems is the *z-transform* with $z = e^{Ts}$. The role of the *z*-transform for digital systems is similar to that of the Laplace transform for continuous-data systems. While the Laplace transform can be used to solve linear ordinary differential equations, for linear difference equations and linear systems with discrete or digital data, the *z*-transform becomes more appropriate to use. Since it is not a simple matter to perform an inverse Laplace transform on transcendental functions which involve terms like e^{-kTs}, the need arises to convert transcendental functions in *s* into algebraic ones in *z*. The development of *z*-transform methods of analysis are considered to be outside the scope of this book.

Various techniques and methods mentioned earlier, such as state-variable analysis, time-domain analysis, frequency-domain analysis, root-locus techniques, and Bode diagrams, are applied to the analysis of digital control systems. Details of these are obviously outside the scope of this introductory text.

Finally, no one can be an expert in all areas discussed in this chapter, or indeed in the preceding chapters. Therefore, it is always good advice to *consult* with those who are. The basics you have been exposed to will help you to select such consultants, either in or out of house, who will provide the knowledge to solve the problem confronting you.

16.4 LEARNING OBJECTIVES

The *learning objectives* of this chapter are summarized here, so that the student can check whether he or she has accomplished each of the following.

- Basic notions of power semiconductor devices and analysis of power electronic circuits.
- Solid-state control of dc motors.
- Solid-state control of induction motors.
- Solid-state control of synchronous motors.
- Transfer functions and block diagrams of feedback control systems.
- Dynamic response of control systems.
- Steady-state error of linear systems.
- Error-rate control, output-rate control, and integral-error control.
- Elementary concepts of digital control systems and their applications.

16.5 PRACTICAL APPLICATION: A CASE STUDY

Digital Process Control

Figure 16.5.1 shows a block diagram for microcomputer-based control of a physical process, such as a chemical plant. A slight variation of the system can be used for automotive instrumentation in which sensors furnish various signals for speed, fuel reserve, battery voltage, oil pressure, engine temperature, and so on. The data are presented to the driver in one or more displays on the dashboard. In a physical process on the other hand, based on the display information, an operator can assess and direct the operation of the control process through a keyboard or other input devices to the microcomputer.

Various physical inputs, such as power and materials, are regulated by actuators, which are in turn controlled by the microcomputer. Electric signals related to the controlled-process parameters, such as pressure and temperature, are produced by various sensors, which in turn

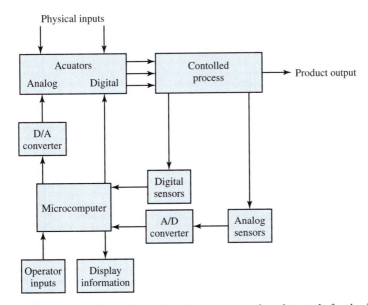

Figure 16.5.1 Block diagram for miocrocomputer-based control of a physical process.

feed the information to the microcomputer. Actuators and sensors may be either analog or digital. Digital-to-analog (D/A) converters are used to convert the digital signals to analog form so as to suit the analog actuators, where as analog-to-digital (A/D) converters are employed to convert the analog sensor signals to digital form so as to suit the microcomputer.

One can think of so many systems in daily practice controlled or monitored by microcomputers. Some examples include monitoring patients in intensive cardiac-care units of hospitals, nuclear-reactor controls, traffic signals, aircraft and automobile instrumentation, chemical plants, and various manufacturing processes.

PROBLEMS

16.1.1 (a) Consider a diode circuit with RC load as shown in Figure P16.1.1. With the switch closed at $t = 0$ and with the initial condition at $t = 0$ that $v_C = 0$, obtain the functional forms of $i(t)$ and $v_C(t)$, and plot them.

(b) Then consider a diode circuit with an RL load with the initial condition at $t = 0$ that $i = 0$. With the switch closed at $t = 0$, obtain the functional forms of $i(t)$ and $v_L(t)$.

(c) In part (b), if $t \gg L/R$, describe what happens. If an attempt is then made to open switch S, comment on what is likely to happen.

(d) Next consider a diode circuit with an LC load with the initial condition at $t = 0$ that $i = 0$ and $v_C = 0$. With the switch closed at $t = 0$, obtain the waveforms of $i(t)$ and $v_C(t)$.

16.1.2 (a) Figure P16.1.2(a) contains a *freewheeling diode* D_m, commonly connected across an inductive load to provide a path for the current in the inductive load when the switch S is opened after time t (during which the switch was closed). Consider the circuit operation in two modes, with mode 1 beginning when the switch is closed at $t = 0$, and mode 2 starting when the switch is opened after the current i

has reached its steady state in mode 1. Obtain the waveforms of the currents $i(t)$ and $i_f(t)$.

(b) Consider the energy-recovery diode circuit shown in Figure P16.1.2(b), along with a *feedback winding*. Assume the transformer to have a magnetizing inductance of L_m and an ideal turns ratio of $a = N_2/N_1$. Let mode 1 begin when the switch S is closed at $t = 0$, and mode 2 start at $t = t_1$, when the switch is opened. Let t_1 and t_2 be the durations of modes 1 and 2, respectively. Develop the equivalent circuits for the two modes of operation, and obtain the various waveforms for the currents and voltages.

16.1.3 (a) Consider a full-wave rectifier circuit with a center-tapped transformer, as shown in Figure P16.1.3, with a purely resistive load of R. Let $v_s = V_m \sin \omega t$. Determine: (i) efficiency, (ii) form factor, (iii) ripple factor, (iv) TUF, and (v) PIV of diode D_1. Compare the performance with that of a half-wave rectifier. (See Example 16.1.2.)

(b) Let the rectifier in part (a) have an RL load. Obtain expressions for output voltage $v_L(t)$ and load current $i_L(t)$ by using the Fourier series.

Figure P16.1.1

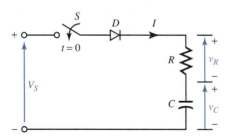

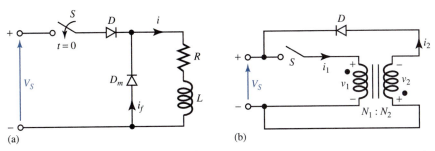

Figure P16.1.2 (a) Diode circuit with freewheeling diode. (b) Energy-recovery diode circuit with feedback winding.

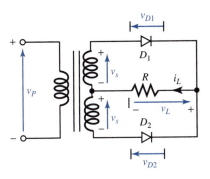

Figure P16.1.3

*16.1.4 Consider a full-wave single-phase bridge rectifier circuit with dc motor load, as shown in Figure P16.1.4(a). Let the transformer turns ratio be unity. Let the load be such that the motor draws a ripple-free armature current of I_a. Given the waveforms for the input current and input voltage of the rectifier, as in Figure P16.1.4(b), determine: (i) harmonic factor HF of the input current, and (ii) input power factor PF of the rectifier.

16.1.5 (a) Consider a three-phase star or wye half-wave rectifier with a purely resistive load R. Determine: (i) efficiency, (ii) form factor, (iii) ripple factor, (iv) TUF, and (v) PIV of each diode.

(b) Express the output voltage of the three-phase rectifier in Fourier series.

16.1.6 Consider a three-phase, full-wave bridge rectifier, as shown in Figure P16.1.6, with a purely resistive load R. For each diode, determine: (i) efficiency, (ii) form factor, (iii) ripple factor, (iv) TUF, and (v) PIV.

16.1.7 Consider Example 16.1.2 in the text.

(a) Calculate the firing angle corresponding to a torque of 35 N·m and a speed of −1350 r/min, assuming continuous conduction. What

is the quadrant of operation in the torque–speed relationship?

(b) Find the motor speed at the rated torque and $\alpha = 160°$ for the regenerative braking in the second quadrant.

16.1.8 For a three-phase, fully controlled, rectifier-fed, separately excited dc motor, corresponding to ideal no-load operation, find the expression for the no-load speeds. Comment on whether no-load speeds could be negative. Compared to the one-phase case, would you expect a considerable reduction in the zone of discontinuous conduction?

*16.1.9 Consider the motor of Example 16.1.3 in the text. Calculate the motor speed for: (a) $\alpha = 120°$, $T_a = 25$ N·m, and (b) $\alpha = 60°$, $T_a = 5$ N·m. Assume continuous conduction in both cases.

16.1.10 Consider Example 16.1.4 in the text. For part (c), compute the corresponding rms fundamental stator current.

16.1.11 A 60-Hz, six-pole, wye-connected, three-phase induction motor, with the parameters $R_1 = R'_2 = 0.025\ \Omega$ and $X_{l1} = X'_{l2} = 0.125\ \Omega$, is controlled by variable-frequency control with constant V/f ratio. Use the approximate equivalent circuit of Figure 13.2.6, with jX_m moved over to the supply

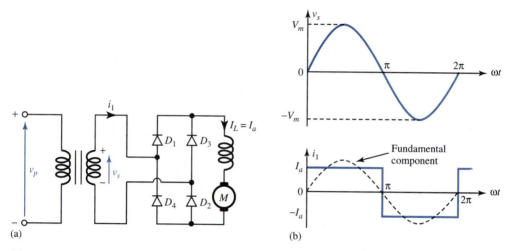

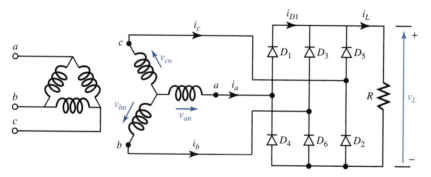

Figure P16.1.4 **(a)** Circuit diagram. **(b)** Waveforms.

Figure P16.1.6

terminals. For an operating frequency of 15 Hz, find:

(a) The maximum motoring torque as a ratio of its value at the rated frequency.

(b) The starting torque and rotor current in terms of their values at the rated frequency.

16.1.12 Given that the motor of Problem 16.1.11 has a full-load slip of 0.05, compute the motor speed corresponding to rated torque and a frequency of 30 Hz.

16.1.13 Consider the motor of Example 16.1.4 in the text. Let the motor be controlled by variable frequency at a constant flux of rated value. By using the equivalent circuit of Figure 13.2.6,

(a) Determine the motor speed and the stator current at one-half the rated torque and 30 Hz.

(b) Redo part (a), assuming the speed–torque curves to be straight lines for $S < S_{max}$.

16.1.14 For the static rotor resistance control considered in the text for an induction motor drive, develop the fundamental frequency equivalent circuit of Figure 16.1.23.

*16.1.15 Consider the induction motor drive of Example 16.1.5 in the text. Compute the motor speed corresponding to $\delta = 0.65$ and 1.5 times the rated full-load torque.

16.1.16 Let the simplified per-phase equivalent circuit of an underexcited, cylindrical-rotor, synchronous machine be given by a source $V_f \angle -\delta$ in series with an impedance jX_s. Let $V \angle 0°$ and $I_s \angle -\phi$ be the applied voltage at the terminals and the current drawn by the motor, respectively.

(a) With $\bar{V}$ as reference, draw a phasor diagram for the motor.

(b) Develop an equivalent circuit with a current source and draw the corresponding phasor diagram with $\bar{V}$ as reference.

(c) Obtain expressions for power and torque for parts (a) and (b).

(d) Now consider the variable-frequency operation of the synchronous motor, with the per-unit frequency a to be f/f_{rated}. Taking $\bar{V}_f$ as reference in the quadrature axis, draw the phasor diagram and comment on the effect of the variable frequency.

16.1.17 Consider the motor of Example 16.1.6 in the text.

(a) Determine the armature current and power factor at one-half the rated speed, one-half the rated torque, and rated field current.

(b) Find the torque and field current corresponding to rated armature current, 1.25 times the rated speed, and unity power factor.

16.2.1 Consider the electrical transients on a linear basis in a separately excited dc generator (whose model is shown in Figure P16.2.1), resulting from changes in excitation. Let the generator speed be a constant, so that the dynamics of the mechanical drive do not enter the problem. Obtain expressions for $E_a(s)/V_f(s)$ and $I_a(s)/V_f(s)$, and develop the corresponding block diagrams. Let R_L and L_L be the load resistance and load inductance, respectively.

16.2.2 A separately excited dc generator has the following parameters: $R_f = 100\ \Omega$, $L_f = 20\ \text{H}$, $R_a = 0.1\ \Omega$, $L_a = 0.1\ \text{H}$, and $K_g = 100\ \text{V}$ per field ampere at rated speed. The load connected to the generator has a resistance $R_L = 4.5\ \Omega$ and an inductance $L_L = 2.2\ \text{H}$. Assume that the prime mover is rotating at rated speed, the load switch is closed, and the generator is initially unexcited. Determine the armature current as a function of time when a 230-V dc supply is suddenly applied to the field winding, assuming the generator speed to be essentially constant.

16.2.3 A 5-hp, 220-V, separately excited dc motor has the following parameters: $R_a = 0.5\ \Omega$, $k = 2\ \text{H}$, $R_f = 220\ \Omega$, and $L_f = 110\ \text{H}$. The armature winding inductance is negligible. The torque required by the load is proportional to the speed, and the combined constants of the motor armature and the load are $J = 3\ \text{kg·m}^2$ and $B = 0.3\ \text{kg·m}^2/\text{s}$. Consider the armature-controlled dc motor, whose speed is made to respond to variations in the applied motor armature voltage v_t. Let the field current be maintained constant at 1 A.

(a) Develop a block diagram relating the motor speed and the motor applied voltage, and find the corresponding transfer function.

(b) Compute the steady-state speed corresponding to a step-applied armature voltage of 220 V.

(c) How long does the motor take to reach 0.95 of the steady-state speed of part (b)?

(d) Determine the value of the total effective viscous damping coefficient of the motor-load configuration.

Figure P16.2.1 Schematic representation of the model of a separately excited dc motor.

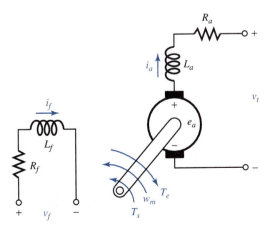

*16.2.4 A separately excited dc generator has the following parameters:

Field winding resistance $R_f = 60\ \Omega$
Field winding inductance $L_f = 60\ \mathrm{H}$
Armature resistance $R_a = 1\ \Omega$
Armature inductance $L_a = 0.4\ \mathrm{H}$
Generated emf constant $K_g = 120\ \mathrm{V}$ per field ampere at rated speed

The armature terminals of the generator are connected to a low-pass filter with a series inductance $L = 1.6\ \mathrm{H}$ and a shunt resistance $R = 1\ \Omega$. Determine the transfer function relating the output voltage $V_t(s)$ across the shunt resistance R, and the input voltage $V_f(s)$ applied to the field winding.

16.2.5 A separately excited dc generator, running at constant speed, supplies a load having a 1-Ω resistance in series with a 1-H inductance. The armature resistance is 0.1 Ω and its inductance is negligible. The field, having a resistance of 50 Ω and an inductance of 5 H, is suddenly connected to a 100-V source. Determine the armature current buildup as a function of time, if the generator voltage constant $K_g = 50\ \mathrm{V}$ per field ampere at rated speed.

16.2.6 The following test data are taken on a 20-hp, 250-V, 600-r/min dc shunt motor: $R_f = 150\ \Omega$, $\tau_f = 0.5\ \mathrm{s}$, $R_a = 0.15\ \Omega$, and $\tau_a = 0.05\ \mathrm{s}$. When the motor is driven at rated speed as a generator with no load, a field current of 2 A produces an armature emf of 250 V. Determine the following:

(a) L_f, the self-inductance of the field circuit.

(b) L_a, the self-inductance of the armature circuit.

(c) The coefficient K relating the speed voltage to the field current.

(d) The friction coefficient B_L of the load at rated load and rated speed, assuming that the torque required by the load T_L is proportional to the speed.

16.2.7 A separately excited dc generator can be treated as a power amplifier when driven at constant angular velocity ω_m. If the armature circuit is connected to a load having a resistance R_L, obtain an expression for the voltage gain $V_L(s)/V_f(s)$. With $R_a = 0.1\ \Omega$, $R_f = 10\ \Omega$, $R_L = 1\ \Omega$, and $K_g = 100$ V per field ampere at rated speed, determine the voltage gain and the power gain if the generator is operating at steady state with 25 V applied across the field.

16.2.8 Consider the motor of Problem 16.2.6 to be initially running at constant speed with an impressed armature voltage of 250 V, with the field separately excited by a constant field current of 2 A. Let the motor be driving a pure-inertia load with a combined polar moment of inertia of armature and load of 3 kg·m². The rotational losses of the motor can be neglected.

(a) Determine the speed.

(b) Neglecting the self-inductance of the armature, obtain the expressions for the armature current and the speed as functions of time, if the applied armature voltage is suddenly increased from 250 V to 260 V.

(c) Repeat part (b), including the effect of the armature self-inductance.

*16.2.9 A separately excited dc motor carries a load of $300\dot\omega_m + \omega_m$ N·m. The armature resistance is 1 Ω and its inductance is negligible. If 100 V is suddenly applied across the armature while the field current is constant, obtain an expression for the motor speed buildup as a function of time, given a motor torque constant $K_m = 10\ \mathrm{N\cdot m/A}$.

16.2.10 Consider the motor of Problem 16.2.3 to be operated as a separately excited dc generator at a constant speed of 900 r/min, with a constant field current of 1.5 A. Let the load current be initially zero. Determine the armature current and the armature terminal voltage as functions of time for a suddenly applied load impedance consisting of a resistance of 11.5 Ω and an inductance of 0.1 H.

16.2.11 Consider a motor supplied by a generator, each with a separate and constant field excitation. Assume the internal voltage E of the generator to be constant, and neglect the armature reaction of both machines. With the motor running with no external load, and the system being in steady state, let a load torque be suddenly increased from zero to T. The machine parameters are as follows:

Armature inductance of motor + generator, $L = 0.008\ \mathrm{H}$
Armature resistance of motor + generator, $R = 0.04\ \Omega$
Internal voltage of the generator $E = 400\ \mathrm{V}$
Moment of inertia of motor armature and load $J = 42\ \mathrm{kg\cdot m^2}$
Motor constant $K_m = 4.25\ \mathrm{N\cdot m/A}$
No-load armature current $i_0 = 35\ \mathrm{A}$
Suddenly applied torque $T = 2000\ \mathrm{N\cdot m}$
Determine the following:

(a) The undamped angular frequency of the transient speed oscillations, and the damping ratio of the system.

(b) The initial speed, final speed, and the speed drop in r/min.

16.2.12 Consider a separately excited dc motor having a constant field current and a constant applied armature voltage. It is accelerating a pure-inertia load from rest. Neglecting the armature inductance and the rotational losses, show that by the time the motor reaches its final speed, the energy dissipated in the armature resistance is equal to the energy stored in the rotating parts.

16.2.13 Determine the parameters of the analog capacitive circuit shown in Figure E16.2.1(d) for the motor in Problem 16.2.3 and its connected load. With the aid of the equivalent circuit, obtain the expression for the armature current, with a 3.5-Ω starting resistance included in series with the armature to limit the starting current.

*16.2.14 Neglecting the self-inductance of the armature circuit, show that the time constant of the equivalent capacitive circuit for a separately excited dc motor with no load is $R_a R_{eq} C_{eq}/(R_a + R_{eq})$, where $R_{eq} = 1/G_{eq}$, as shown in Figure E16.2.1(d).

16.2.15 Consider the dc motor of Problem 16.2.3 to be operating at rated voltage in steady state with a field current of 1 A, and with the starting resistance in series with the armature reduced to zero.

(a) Obtain the equivalent capacitive circuit neglecting the armature self-inductance and calculate the steady armature current.

(b) If the field current is suddenly reduced to 0.8 A while the armature applied voltage is constant at 220 V, compute the initial armature current $i_a(0)$ on the basis that the kinetic energy stored in the rotating parts cannot change instantaneously.

(c) Determine the final armature current $i_a(\infty)$ for the condition of part (b).

(d) Obtain the time constant τ'_{am} of the armature current for the condition of part (b), and ex-

press the armature current as a function of time on the basis that $i_a = i_a(\infty) + [i_a(0) - i_a(\infty)] e^{-t/\tau'_{am}}$.

16.2.16 A separately excited dc motor, having a constant field current, accelerates a pure inertia load from rest. If the system is represented by an electrical equivalent circuit, with symbols as shown in Figure P16.2.16, express R, L, and C in terms of the motor parameters.

16.2.17 Figure P16.2.17 represents the Ward–Leonard system for controlling the speed of the motor M. With the generator field voltage v_{fg} as the input and the motor speed ω_m as the output, obtain an expression for the transfer function for the system, assuming idealized machines. Let the load on the motor be given by $J\dot{\omega}_m + B\omega_m$. The generator runs at constant angular velocity ω_g.

16.2.18 A separately excited dc generator has the following parameters: $R_f = 100\,\Omega$, $R_a = 0.25\,\Omega$, $L_f = 25$ H, $L_a = 0.02$ H, and $K_g = 100$ V per field ampere at rated speed.

(a) The generator is driven at rated speed, and a field circuit voltage of $V_f = 200$ V is suddenly applied to the field winding.

   (i) Find the armature generated voltage as a function of time.

   (ii) Calculate the steady-state armature voltage.

   (iii) How much time is required for the armature voltage to rise to 90% of its steady-state value?

(b) The generator is driven at rated speed, and a load consisting of $R_L = 1\,\Omega$ and $L_L = 0.15$ H in series is connected to the armature terminals. A field circuit voltage $V_f = 200$ V is suddenly applied to the field winding. Determine the armature current as a function of time.

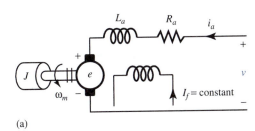

(a)

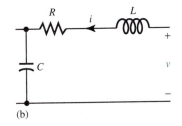

(b)

Figure P16.2.16

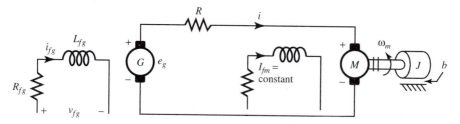

Figure P16.2.17

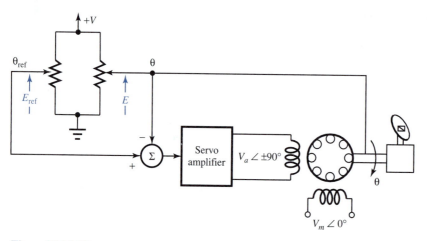

Figure P16.2.19

16.2.19 For the position control system shown in Figure P16.2.19, let the potentiometer transducers give a voltage of 1 V per radian of position. The transfer function of the servoamplifier is $G(s) = 10(1 + 0.01571s)/(7+s)$. Let the initial angular position of the radar be zero, and the transfer function between the motor control phase voltage V_a and the radar position θ be $M(s) = 2.733/s(1 + 0.01571s)$.

(a) Obtain the transfer function of the system.

(b) For a step change in the command angle of $180°$ ($= \pi$ radians), find the time response of the angular position of the antenna.

*__16.2.20__ A separately excited dc motor has the following parameters: $R_a = 0.5\ \Omega$, $L_a \cong 0$, and $B \cong 0$. The machine generates an open-circuit armature voltage of 220 V at 2000 r/min, and with a field current of 1.0 A. The motor drives a constant load torque $T_L = 25$ N·m. The combined inertia of motor and load is $J = 2.5$ kg·m². With field current $I_f = 1.0$ A, if the armature terminals are connected to a 220-V dc source:

(a) Obtain expressions for speed ω_m and armature current i_a as functions of time.

(b) Find the steady-state values of speed and armature current.

16.2.21 The process of plugging a motor involves reversing the polarity of the supply to the armature of the machine. *Plugging* corresponds to applying a step voltage of $-Vu(t)$ to the armature of the machine, where V is the rated terminal voltage. A separately excited 200-V dc motor operates at rated voltage with constant excitation on zero load. The torque constant of the motor is 2 N·m/A, its armature resistance is 0.5 Ω, and the total moment of inertia of the rotating parts is 4 kg·m². Neglect the rotational losses and the armature inductance. Obtain an expression for the speed of the machine after plugging as a function of time, and calculate the time taken for the machine to stop.

16.2.22 The schematic diagram of a Ward–Leonard system is shown in Figure P16.2.22, including a separately excited dc generator, the armature of which is connected directly to the armature of the separately excited dc motor driving a mechanical load. Let J be the combined polar moment of inertia of the load and motor, and B the combined viscous friction constant of the load and motor. Assuming that the mechanical angular speed of the generator ω_{mG} is a constant, develop the block diagram for the system and obtain an expression for the transfer function $\Omega_{mM}(s)/V_{fG}(s)$.

16.2.23 Consider the elementary motor-speed regulator scheme shown in Figure P16.2.23 for a separately excited dc motor, whose armature is supplied from a solid-state controlled rectifier. The motor speed is measured by means of a dc tachometer generator, and its voltage e_t is compared with a reference voltage E_R. The error voltage $E_R - e_t$ is amplified and made to control the output voltage of the power-conversion equipment, so as to maintain substantially constant speed at the value set by the reference voltage. Let the armature-circuit parameters be R_a and L_a, and the speed–voltage constant

of the motor be K_m, with units of V·s/rad. Assume that the combination of A and P is equivalent to a linear controlled voltage source $v_s = K_A$ (error voltage), with negligible time lag and gain K_A. Assume also that the load torque T_L is independent of the speed, with zero damping. Neglect no-load rotational losses.

(a) Develop the block diagram for the feedback speed-control system with E_R/K_t, the steady-state no-load speed setting, as input, and Ω_m as output. K_t is the tachometer speed–voltage constant in V/(r/min).

(b) With $T_L = 0$, evaluate the transfer function Ω_m/E_R.

(c) With $E_R = 0$, obtain the transfer function Ω_m/T_L.

(d) Find the expressions for the underdamped natural frequency ω_n, the damping factor α, and the damping ratio $\xi = \alpha/\omega_n$.

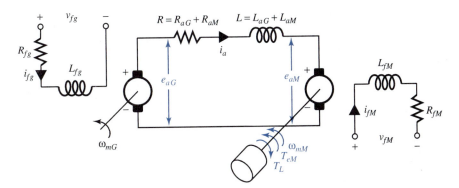

Figure P16.2.22

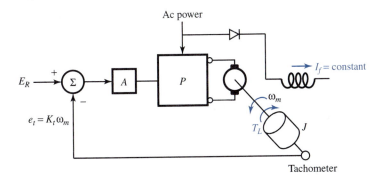

Figure P16.2.23

(e) For a step input ΔE_R, obtain the final steady-state response $\Delta\omega_m(\infty)$, i.e., evaluate $\Delta\omega_m(\infty)/\Delta E_R$.

(f) Evaluate $\Delta\omega_m(\infty)/\Delta T_L$ for the step input ΔT_L of a load torque.

16.2.24 Consider the motor of Problem 16.2.3 to be used as a field-controlled dc machine. Let the armature be energized from a constant current source of 15 A. Assume no saturation.

(a) Develop a block diagram relating the motor speed and the applied field voltage.

(b) Determine the steady-state speed for a step-applied field voltage of 220 V.

(c) How long does the motor take to reach 0.95 of the steady-state speed of part (b)?

*16.2.25 The output voltage of a 10-kW, 240-V dc generator is regulated by means of the closed-loop system shown in Figure P16.2.25. The generator parameters are $R_f = 150\ \Omega$, $R_a = 0.5\ \Omega$, $L_f = 75\ H$, and $K_E = 150$ V per field ampere at 1200 r/min. The self-inductance of the armature is negligible. The amplifier has an amplification factor $A = 10$, and the potentiometers are set such that a is unity and the reference voltage v_r is 250 V. The generator is driven by an induction motor, the speed of which is almost 1200 r/min when the generator output is zero, and 1140 r/min when the generator delivers an armature current of 42 A.

(a) Compute the steady-state armature terminal voltage at no load, and when the generator is delivering 42 A.

(b) Calculate the time constant for part (a).

(c) With the value of the field current as in part (a), for an armature current of 42 A, calculate the steady-state armature voltage.

16.2.26 (a) A common analog control element is the dc tachometer, which is basically a permanent

magnet generator, as illustrated schematically in Figure P16.2.26(a). Determine the transfer function of this device, with speed as the input and voltage as the output. See what happens to the transfer function if $R_L \rightarrow \infty$.

(b) A dc servomotor is another common analog element in control systems, shown schematically in Figure P16.2.26(b). Obtain an expression for $\Omega_m(s)/E_i(s)$, assuming constant-field configuration and linear elements.

(c) Add load torque to the load on the servomotor of Figure P16.2.26(b). Develop a block diagram with a voltage signal that will serve as a speed reference and a load torque as a second input (or load disturbance).

16.2.27 (a) Consider a single loop system of the configuration shown in Figure P16.2.27(a) with *positive feedback*. Although positive feedback generally results in instability, it is frequently employed at low magnitudes or in portions of a large control system, particularly in an inner loop. Obtain an expression for $C(s)/R(s)$.

(b) Figure P16.2.27(b) illustrates a multiple-loop feedback control system, where the argument s is omitted for simplicity. Reduce it to a single-loop configuration.

16.2.28 Loop topography that is relatively common in control systems is shown in Figure P16.2.28. Obtain its single-loop representation.

*16.2.29 A generalized two-input system is illustrated in Figure P16.2.29. Treating multiple inputs by means of the principle of superposition, which holds in linear control systems, find the system response C.

16.2.30 Simplify the loop topography shown in Figure P16.2.30 to that of a single-loop configuration.

16.2.31 Determine the steady-state error of a type 0 system with a unit-step reference input function for the Ward–Leonard system of Figure E16.2.3(a).

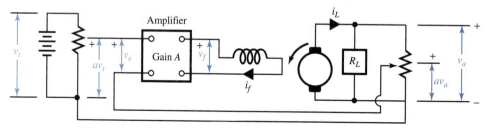

Figure P16.2.25

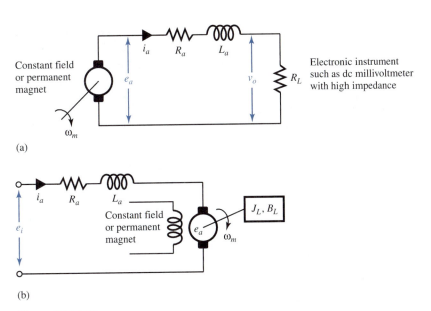

(a)

(b)

Figure P16.2.26

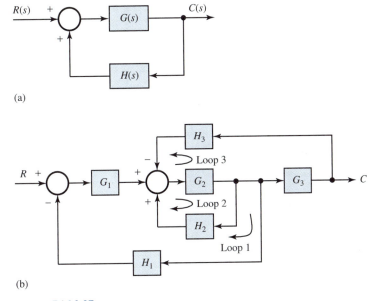

(a)

(b)

Figure P16.2.27

16.2.32 Verify the error-response conclusions listed in Table 16.2.2.

16.2.33 (a) Determine the system type, the steady-state error for a unit-step reference function, and the closed-loop time constant for the speed-control system of Figure P16.2.33(a), which uses proportional control.

(b) Determine the system type, the steady-state error for a unit-step reference function, and the damping ratio for the speed-control system of Figure P16.2.33(b), which uses integral control.

(c) Add proportional control to the system of part (b) by replacing K_i/s with $K_p + K_i/s$, and find the new damping ratio.

*16.2.34 Consider the speed-control system shown in

Figure P16.2.34, which uses proportional control. With $T_d(s) = 0$, if a unit-step function is applied to the reference input, find the expression for the time response $\omega_c(t)$.

16.2.35 Figure P16.2.35 shows a block diagram of a speed-control system that uses proportional as well as integral control action. Show that this will result in the elimination of the steady-state system error in response to a unit-step disturbance torque input.

16.2.36 Consider two type-1 systems, as shown in Figures P16.2.36(a) and (b), with the integrator located in two different locations. Check whether both

systems eliminate the steady-state error for unit-step function disturbance input $U(s)$.

16.2.37 It is desirable to have an amplifier system of several stages with an overall gain of 1000 ± 20. The gain of any one stage is given to drift from 10 to 20. Determine the required number of stages and the feedback function needed to meet the specifications. Consider the closed-loop system as one that has an initial value of 980 and a final value of 1020, while each stage has an initial value of 10 and a final value of 20.

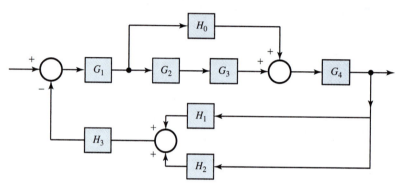

Figure P16.2.28

Figure P16.2.29

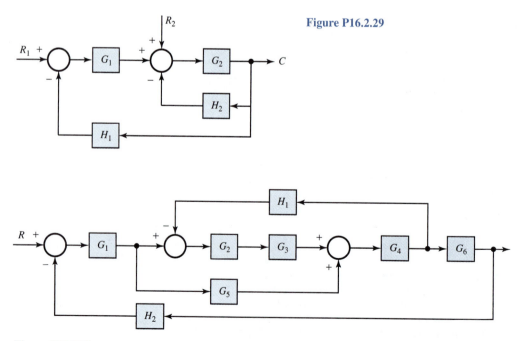

Figure P16.2.30

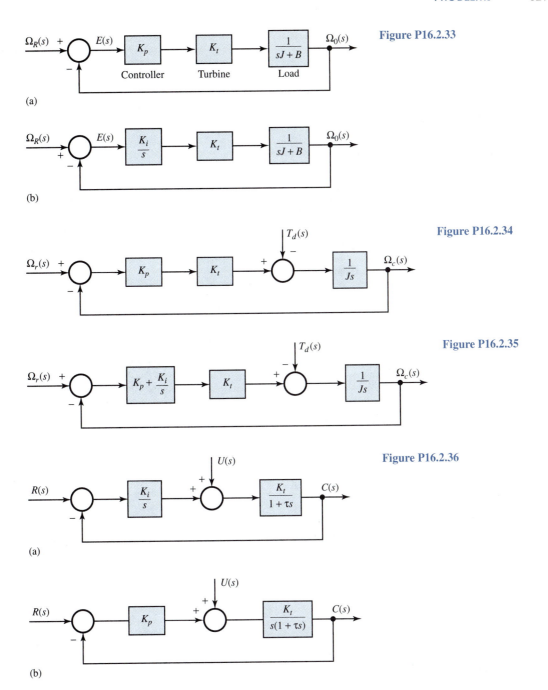

Figure P16.2.33

(a)

(b)

Figure P16.2.34

Figure P16.2.35

Figure P16.2.36

(a)

(b)

16.2.38 The parameters of the FET in a grounded-source amplifier are affected because aging causes a net per-unit decrease in gain of 25%. Each amplifier circuit is designed to yield a gain (between the input–output terminals) of 80, with the change at no time exceeding 0.1%.

(a) Find the minimum number of stages needed to meet the specification.

(b) Determine the corresponding feedback factor H.

16.2.39 (a) A system has a direct transmission function given by

$$G(s) = \frac{10}{s^2 + s - 2}$$

(i) Is the system stable?

(ii) If it is unstable, can it be made stable by employing a feedback path with a transfer function H around $G(s)$? If so, find H.

(b) Redo part (a) for the case where

$$G(s) = \frac{10}{s(s^2 + s - 2)}$$

16.2.40 The output response of a second-order servomechanism is given as

$$\frac{c(t)}{r_0} = 1 - 1.66e^{-8t} \sin(6t + 37°)$$

when the input is a step function of magnitude r_0.

(a) Determine the damped frequency of oscillation.

(b) Obtain the value of the damping ratio.

(c) Find the natural frequency of the system.

(d) Evaluate the loop gain if the inertia of the output member is 0.01 kg·m² and the viscous coefficient is 0.2 kg·m²/s.

(e) To what value should the loop gain be increased if the damping ratio is not to be less than 0.4?

(f) Obtain the closed-loop transfer function.

(g) Find the corresponding open-loop transfer function.

(h) When the system is operated in open-loop configuration, determine the complete output response for a unit-step input.

16.2.41 Figure P16.2.41 shows the block diagram of a control system. Determine:

(a) The closed-loop transfer function.

(b) The frequency of oscillation of the output variable in responding to a step command before reaching steady state.

(c) The percent maximum overshoot in part (b).

(d) The time required for the output to reach up to 99% of steady state in part (b).

***16.2.42** The feedback control system is characterized by

$$\frac{d^2c}{dt^2} + 6.4\frac{dc}{dt} = 160e$$

where c is the output variable and $e = r - 0.4c$. Determine the damping ratio ξ, the natural frequency ω_n, and percent maximum overshoot.

16.2.43 A second-order servomechanism with the configuration of Figure 16.2.7 has the following parameters:

Open-loop gain $K = 24 \times 10^{-4}$ N·m/rad
System inertia $J = 1.4 \times 10^{-5}$ kg·m²
System viscous-friction coefficient $F = 220 \times 10^{-6}$ kg·m²/s

(a) Find the damping ratio.

(b) If the loop gain has to be increased to 250×10^{-4} N·m/rad in order to meet the accuracy requirements during steady-state operation, determine the error-rate damping coefficient needed, while keeping the damping ratio unchanged.

16.2.44 The system of Problem 16.2.41 is modified, as shown in Figure P16.2.44, to include error-rate damping. Find the value of the error-rate factor K_e so that the damping ratio of the modified characteristic equation is 0.6.

16.2.45 For the system shown in Figure P16.2.45, determine the value of the output-rate factor that yields a response (to a step command) with a maximum overshoot of 10%.

***16.2.46** Redo Problem 16.2.45 for the system whose block diagram is depicted in Figure P16.2.46.

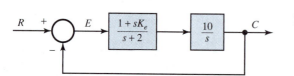

Figure P16.2.41

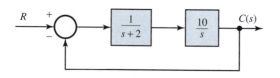

Figure P16.2.44

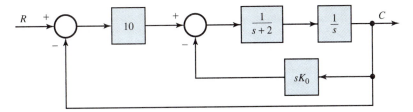

Figure P16.2.45

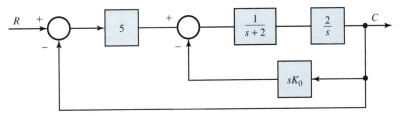

Figure P16.2.46

References

GENERAL ELECTRICAL ENGINEERING AND CIRCUITS

Bobrow, *Fundamentals of Electrical Engineering*, 2nd edition. Oxford University Press, New York, 1996

DeCarlo and Lin, *Linear Circuit Analysis*, 2nd edition. Oxford University Press, New York, 2001

Dorf and Svoboda, *Introduction to Electric Circuits*, Wiley, New York, 1999

Franco, *Electric Circuits Fundamentals*, Oxford University Press, New York, 1994

Irwin and Kerns, *Introduction to Electrical Engineering*, Prentice-Hall, Upper Saddle River, 1995

Nasar, *3000 Solved Problems in Electric Circuits*, McGraw-Hill, New York, 1992

Pratap, *Getting Started with MATLAB 5*, Oxford University Press, New York, 1999

Rizzoni, *Principles and Applications of Electrical Engineering*, 3rd edition. McGraw-Hill, New York, 2000

Roberts and Sedra, *SPICE*, 2nd edition. Oxford University Press, New York, 1999

Sadiku, *Elements of Electromagnetics*, 3rd edition. Oxford University Press, New York, 2001

Sarma, *Introduction to Electrical Engineering*, Oxford University Press, New York, 2001

Schwarz and Oldham, *Electrical Engineering: An Introduction*, 2nd edition. Oxford University Press, New York, 1993

Schwarz, *Electromagnetics for Engineers*, Oxford University Press, New York, 1990

Tuinenga, *SPICE: A Guide to Circuit Simulation and Analysis Using PSpice*, 3rd edition. Prentice-Hall, Upper Saddle River, 1995

INFORMATION AND COMMUNICATION SYSTEMS

Carlson, *Communication Systems: An Introduction to Signals and Noise in Electrical Communication*, 3rd edition. McGraw-Hill, NY, 1986

Chen, *Digital Signal Processing*, Oxford University Press, New York, 2001

Cooper and McGillem, *Probabilistic Methods of Signal and System Analysis*, 3rd edition. Oxford University Press, New York, 1999

Ingle and Proakis, *Digital Signal Processing Using MATLAB*, Version 4, 2nd edition. Brooks Cole, Monterey, 1999

Lathi, *Modern Digital and Analog Communications Systems*, 3rd edition. Oxford University Press, New York, 1998

Proakis, *Digital Communications*, 3rd edition. McGraw-Hill, New York, 1995

Proakis and Manolakis, *Digital Signal Processing: Principles, Algorithms and Applications*, 3rd edition. Prentice Hall, Upper Saddle River, 1995

Stallings, *Data and Computer Communications*, 6th edition, Prentice-Hall, Upper Saddle River, 2000

Stallings, *Local and Metropolitan Area Networks*, 6th edition, Prentice-Hall, Upper Saddle River, 2000

Yariv, *Optical Electronics in Modern Communications*, 5th edition, Oxford University Press, New York, 1997

ELECTRONIC SYSTEMS, DIGITAL AND ANALOG

Allen and Holberg, *CMOS Analog Circuit Design*, Oxford University Press, New York, 1987

Campbell, *The Science and Engineering of Microelectronic Fabrication*, 2nd edition. Oxford University Press, New York, 2001

Dimitrijev, *Understanding Semiconductor Devices*, Oxford University Press, New York, 2000

Franco, *Design with Operational Amplifiers and Analog Integrated Circuits*, 2nd edition. McGraw-Hill, New York, 1998

Gray and Meyer, *Analysis and Design of Analog Integrated Circuits*, 3rd edition. Wiley, New York, 1992

Hansalman and Littlefield, *Mastering MATLAB 5: A Comprehensive Tutorial and Reference*, Prentice-Hall, Upper Saddle River, 1998
Hodges and Jackson, *Analysis and Design of Digital Integrated Circuits,* 2nd edition. McGraw-Hill, New York, 1988
Johns and Martin, *Analog Integrated Circuit Design*, Wiley, New York, 2000
Martin, *Digital Integrated Circuit Design*, Oxford University Press, New York, 2000
Razavi, *Design of Analog CMOS Integrated Circuits*, McGraw-Hill, New York, 2001
Roulston, *An Introduction to the Physics of Semiconductor Devices*, Oxford University Press, New York, 1999
Roth, *Fundamentals of Logic Design*, 4th edition, Brooks Cole, Monterey, 1992
Schaumann and Van Valkenburg, *Design of Analog Filters*, Oxford University Press, New York, 2001
Sedra and Smith, *Microelectronic Circuits*, 4th edition, Oxford University Press, New York, 1998
Van Valkenburg, *Analog Filter Design*, Oxford University Press, New York, 1982
Wakerly, *Digital Design: Principles and Practices,* 3rd edition. Prentice-Hall, Upper Saddle River, 2000

CONTROL SYSTEMS

Bishop, *Modern Control Systems Analysis and Design Using MATLAB and Simulink*, Addison Wesley, Reading, 1997
Chen, *Linear System Theory and Design*, 3rd edition. Oxford University Press, New York, 1999
Dorf and Bishop, *Modern Control Systems,* 8th edition, Addison Wesley, Reading, 1997
Franklin et. al., *Digital Control of Dynamic Systems*, 3rd edition. Addison Wesley, Reading, 1998
Kuo, *Automatic Control Systems,* 7th edition, Prentice-Hall, Upper Saddle River, 1994
Kuo, *Digital Control Systems,* 3rd edition. Oxford University Press, New York, 1992
Ogata, *Modern Control Engineering*, 3rd edition. Prentice-Hall, Upper Saddle River, 1997
Stefani, Savant, Shahian, and Hostetter, *Design of Feedback Control Systems,* 4th edition, 2001

ELECTRIC POWER

Chapman, *Electric Machinery Fundamentals,* 3rd edition. McGraw-Hill, New York, 1998
Guru and Hiziroglu, *Electric Machinery & Transformers,* 3rd edition. Oxford University Press, New York, 2001
Glover and Sarma, *Power Systems Analysis and Design,* 2nd edition. Brooks-Cole, Monterey, 1994
Krein, *Elements of Power Electronics*, Oxford University Press, New York, 1998
Nasar, *Schaum's Outline of Theory and Problems of Electric Machines and Electromechanics*, 2nd edition. McGraw-Hill, New York, 1997
Sen, *Principles of Electric Machines and Power Electronics,* 2nd edition. Wiley, New York, 1996
Sarma, *Electric Machines: Steady-State Theory and Dynamic Performance,* 2nd edition. Brooks-Cole, Monterey, 1994

Brief Review of Fundamentals of Engineering (FE) Examination

Anyone who wishes to practice the engineering profession and offer professional services to the public must become registered as a professional engineer (PE). Boards in each of the 50 states regulate the profession of engineering. The process of registering or licensing oneself as a PE is a multistep process.

Engineering programs accredited by the Engineering Accreditation Commission (EAC) of the Accreditation Board for Engineering and Technology (ABET) are acceptable to all boards as qualifying education. After obtaining the necessary education and applying to the board, one will be allowed to take the Fundamentals of Engineering (FE) examination. The FE examination, formerly called the Engineer-in-Training (EIT) examination, is offered twice a year and is usually taken by engineering-college seniors or fresh graduates. After successfully passing the FE examination, one is known as an Engineer-Intern (EI), and is admitted to the preprofessional status as a newly trained engineer. The EI must then obtain a minimum of four years of acceptable experience before being qualified to take the Professional Engineering (PE) examination. This text should serve as an excellent reference to prepare for both of these examinations.

The FE examination consists of two parts, morning session and an afternoon session. Administered by the National Council of Examiners for Engineering and Surveying (NCEES), 140 multiple-choice problems with 5 choices for each problem out of which one correct answer is to be selected are given in the morning session; 70 problems are given in the afternoon session.

Candidates for the FE examination are provided with a booklet called the Reference Handbook, which contains relevant tables, formulas, and charts along with the question paper. Candidates are not supposed to bring their own reference books to the examination. The electric engineering topics covered on the FE exam will now be outlined in this appendix, and the location of these topics in this text will be identified to facilitate study and review.

FE EXAMINATION TOPICS

A combined topic and subtopic list for the morning and afternoon sessions of the FE examination is given here, along with the relevant sections of this text.

1. DC circuits
   Electrical quantities   Section 1.1
   Resistor combinations and Ohm's law   Section 1.2
   Maximum power transfer   Section 1.2
   Kirchhoff's laws   Section 1.3
   Node-voltage and mesh-current analyses   Section 2.2
   Thévenin and Norton equivalent circuits   Section 2.1
   Superposition and linearity   Section 2.3
2. Capacitance and inductance
   Series and parallel combinations   Section 1.2

For additional and/or up-dated information regarding the FE examination, one may contact directly the NCEES, P.O. Box 1686, Clemson, SC 29633–1686.

Technical Terms, Units, Constants, and Conversion Factors for the SI System

C.1 Physical Quantities

C.2 Prefixes

C.3 Physical Constants

C.4 Conversion Factors

TABLE C.1 Physical Quantities

Physical Quantity	SI Unit	Symbol
Length	meter	m
mass	kilogram	kg
time	second	s
current	ampere	A
admittance	siemen (A/V)	S
angle	radian	rad
angular acceleration	radian per second squared	rad/s^2
angular velocity	radian per second	rad/s
apparent power	voltampere (VA)	VA
area	square meter	m^2
capacitance	farad (C/V)	F
charge	coulomb (As)	C
conductance	siemen (A/V)	S
electric field intensity	volt/meter	V/m
electric flux	coulomb (As)	C
electric flux density	coulomb/square meter	C/m^2
energy	joule (Nm)	J
force	newton (kgm/s^2)	N
frequency	hertz (1/s)	Hz
impedance	ohm (V/A)	Ω
inductance	henry (Wb/A)	H
linear acceleration	meter per second squared	m/s^2
linear velocity	meter per second	m/s
magnetic field intensity	ampere per meter	A/m
magnetic flux	weber (Vs)	Wb
magnetic flux density	tesla (Wb/m^2)	T
magnetomotive force	ampere or ampere-turn	A or At
moment of inertia	kilogram-meter squared	kgm^2
power	watt (J/s)	W
pressure	pascal (N/m^2)	Pa
reactance	ohm (V/A)	Ω
reactive power	voltampere reactive	var
resistance	ohm (V/A)	$\Omega \cdot m$
resistivity	ohmmeter	m
susceptance	siemen (A/V)	S
torque	Newtonmeter	Nm
voltage	volt (W/A)	V
volume	cubic meter	m^3

TABLE C.2 Prefixes

Prefix	Symbol	Meaning
exa	E	10^{18}
peta	P	10^{15}
tera	T	10^{12}
giga	G	10^{9}
mega	M	10^{6}
kilo	k	10^{3}
hecto	h	10^{2}
deka	da	10^{1}
deci	d	10^{-1}
centi	c	10^{-2}
milli	m	10^{-3}
micro	μ	10^{-6}
nano	n	10^{-9}
pico	p	10^{-12}
femto	f	10^{-15}
alto	a	10^{-18}

TABLE C.3 Physical Constants

Quantity	Symbol	Value	Unit
permeability constant	μ_0	1.257×10^{-6}	H/m
permittivity constant	ε_0	8.854×10^{-12}	F/m
gravitational acceleration constant	g_0	9.807	m/s^2
speed of light in a vacuum	c	0.2998×10^{9}	m/s
charge of an electron	q	-1.602×10^{-19}	C
electron mass	m	9.108×10^{-31}	kg
Boltzmann constant	k	1.381×10^{-23}	J/K

TABLE C.4 Conversion Factors

Physical Quantity	SI Unit	Equivalents
length	1 meter (m)	3.281 feet (ft) 39.37 inches (in)
angle	1 radian (rad)	57.30 degrees
mass	1 kilogram (kg)	0.0685 slugs 2.205 pounds (lb) 35.27 ounces (oz)
force	1 newton (N)	0.2248 pounds (lbf) 7.233 poundals $0.1 \ 10^{6}$ dynes 102 grams
torque	1 newton-meter (Nm)	0.738 pound-feet (lbf-ft) 141.7 oz-in $10 \ 10^{6}$ dyne-centimeter $10.2 \ 10^{3}$ gram-centimeter^2
moment of inertia	1 kilogram-meter^2 (kgm^2)	0.738 slug-feet^2 23.7 pound-feet^2 (lb-ft^2) $54.6 \ 10^{3}$ ounce-inches^2 $10 \ 10^{6}$ gram-centimeter^2 (g· cm^2)

Continued

TABLE C.4 Continued

energy	1 joule (J)	1 watt-second
		0.7376 foot-pounds (ft-lb)
		$0.2778 \ 10^{-6}$ kilowatt-hours (kWh)
		0.2388 calorie (cal)
		$0.948 \ 10^{-3}$ British Thermal Units (BTU)
		$10 \ 10^{6}$ ergs
power	1 watt (W)	0.7376 foot-pounds/second
		$1.341 \ 10^{-3}$ horsepower (hp)
resistivity	1 ohm-meter ($\Omega \cdot$m)	$0.6015 \ 10^{9}$ ohm-circular mil/foot
		$0.1 \ 10^{9}$ micro-ohm-centimeter
magnetic flux	1 weber (Wb)	$0.1 \ 10^{9}$ maxwells or lines
		$0.1 \ 10^{6}$ kilolines
magnetic flux density	1 tesla (T)	$10 \ 10^{3}$ gauss
		64.52 kilolines/in^{2}
magnetomotive force	1 ampere (A) or ampere-turn (At)	1.257 gilberts
magnetic field intensity	1 ampere/meter (A/m)	$25.4 \ 10^{-3}$ ampere/in
		$12.57 \ 10^{-3}$ oersted
temperature	degree centigrade (°C)	$\frac{5}{9}$ (temperature, °F -32)
		(temperature, °K -273.18)

Mathematical Relations

D.1 TRIGONOMETRIC FUNCTIONS

$$\sin\left(x \pm \frac{\pi}{2}\right) = \pm\, \cos\,(x)$$

$$\cos\left(x \pm \frac{\pi}{2}\right) = \mp\, \sin\,(x)$$

$$\sin\,(x \pm y) = \sin\,(x)\,\cos\,(y) \pm \cos\,(x)\,\sin\,(y)$$

$$\cos\,(x \pm y) = \cos\,(x)\,\cos\,(y) \mp \sin\,(x)\,\sin\,(y)$$

$$\sin\,(2x) = 2\,\sin\,(x)\,\cos\,(x)$$

$$\cos\,(2x) = \cos^2\,(x) - \sin^2\,(x)$$

$$2j\,\sin\,(x) = e^{jx} - e^{-jx}$$

$$2\,\cos\,(x) = e^{jx} + e^{-jx}$$

$$2\,\sin\,(x)\,\sin\,(y) = \cos\,(x - y) - \cos\,(x + y)$$

$$2\,\sin\,(x)\,\cos\,(y) = \sin\,(x - y) + \sin\,(x + y)$$

$$2\,\cos\,(x)\,\cos\,(y) = \cos\,(x - y) + \cos\,(x + y)$$

$$\sin^2\,(x) + \cos^2\,(x) = 1$$

$$2\,\sin^2\,(x) = 1 - \cos\,(2x)$$

$$2\,\cos^2\,(x) = 1 + \cos\,(2x)$$

$$4\,\sin^3\,(x) = 3\,\sin\,(x) - \sin\,(3x)$$

$$4 \cos^3 (x) = 3 \, \cos \, (x) + \, \cos \, (3x)$$

$$8 \sin^4 (x) = 3 - 4 \, \cos \, (2x) + \, \cos \, (4x)$$

$$8 \cos^4 (x) = 3 + 4 \, \cos \, (2x) + \, \cos \, (4x)$$

$$A \, \cos \, (x) - B \, \sin \, (x) = R \, \cos \, (x + \theta)$$

$$\text{where,} \quad R = \sqrt{A^2 + B^2}$$

$$\theta = \tan^{-1} \frac{B}{A}$$

$$A = R \, \cos \, \theta$$

$$B = R \, \sin \, \theta$$

$$\text{If } x \ll 1, \quad \sin x \cong x; \, \cos x \cong 1 - \frac{x^2}{2}$$

D.2 EXPONENTIAL AND LOGARITHMIC FUNCTIONS

$$e^{\pm jx} = \, \cos \, x \pm j \, \sin \, x$$

$$e^x e^y = e^{(x+y)}$$

$$e^x / e^y = e^{(x-y)}$$

$$\log \, xy = \, \log \, x + \, \log \, y$$

$$\log \, \frac{x}{y} = \, \log \, x - \, \log \, y$$

$$\log \, x^n = n \, \log \, x$$

$$\log_a x = \log_b x / \log_b a = (\log_b x)(\log_a b)$$

Base of natural (Naperian or hyperbolic) logarithms: $e \cong 2.71828$

$$\ln \, x = \log_e \, x = (\log_e \, 10)(\log_{10} \, x) \cong 2.3026 \, \log_{10} \, x$$

$$\log_{10} \, x = (\log_{10} \, e)(\log_e \, x) = 0.4343 \, \log_e \, x = 0.4343 \, \ln x$$

$$a \ln e^b = ab$$

$$\text{If } x \ll 1, \quad e^x \cong 1 + x \, ; \ln(1 + x) \cong x$$

D.3 DERIVATIVES AND INTEGRALS

Derivatives:

$$\frac{d}{dx}(a) = 0, \text{ where } a \text{ is a fixed real number}$$

$$\frac{d}{dx}(x) = 1$$

$$\frac{d}{dx}(au) = a\frac{du}{dx}, \text{ where } u \text{ is a function of } x$$

$$\frac{d}{dx}(u \pm v) = \frac{du}{dx} \pm \frac{dv}{dx}, \text{ where } u \text{ and } v \text{ are functions of } x$$

$$\frac{d}{dx}(uv) = u\frac{dv}{dx} + v\frac{du}{dx}$$

$$\frac{d}{dx}\left(\frac{u}{v}\right) = \frac{v\frac{du}{dx} - u\frac{dv}{dx}}{v^2} = \frac{1}{v}\frac{du}{dx} - \frac{u}{v^2}\frac{dv}{dx}$$

$$\frac{d}{dx}(u^n) = nu^{n-1}\frac{du}{dx}$$

$$\frac{d}{dx}[f(u)] = \frac{d}{du}[f(u)] \cdot \frac{du}{dx}$$

$$\frac{d}{dx}(\ln u) = \frac{1}{u}\frac{du}{dx} \ ; \frac{d}{dx}(\ln x) = \frac{1}{x}$$

$$\frac{d}{dx}(\log_a u) = (\log_a e)\frac{1}{u}\frac{du}{dx}$$

$$\frac{d}{dx}(e^u) = e^u\frac{du}{dx} \ ; \frac{d}{dx}e^{ax} = ae^{ax}$$

$$\frac{d}{dx}(\sin u) = \frac{du}{dx}(\cos u) \ ; \frac{d}{dx}\sin ax = a\cos ax$$

$$\frac{d}{dx}(\cos u) = -\frac{du}{dx}(\sin u) \ ; \frac{d}{dx}\cos ax = -a\sin ax$$

Integrals:

$$\int (a + bx)^n dx = \frac{(a + bx)^{n+1}}{(n + 1)b}, \quad n \neq -1$$

$$\int \frac{dx}{a + bx} = \frac{1}{b}\ln|a + bx|$$

$$\int \frac{dx}{a^2 + b^2x^2} = \frac{1}{ab}\tan^{-1}\frac{bx}{a}$$

$$\int \frac{x\,dx}{a^2 + x^2} = \frac{1}{2}\ln(a^2 + x^2)$$

$$\int \frac{x^2\,dx}{a^2 + x^2} = x - a\tan^{-1}\frac{x}{a}$$

$$\int \frac{dx}{(a^2 + x^2)^2} = \frac{x}{2a^2(a^2 + x^2)} + \frac{1}{2a^3}\tan^{-1}\frac{x}{a}$$

$$\int \frac{x\,dx}{(a^2 + x^2)^2} = \frac{-1}{2(a^2 + x^2)}$$

$$\int \frac{x^2 \, dx}{(a^2 + x^2)^2} = \frac{-x}{2(a^2 + x^2)} + \frac{1}{2a} \tan^{-1} \frac{x}{a}$$

$$\int \ln x \, dx = x \ln x - x$$

$$\int e^{ax} \, dx = \frac{e^{ax}}{a}, \quad a \text{ real or complex}$$

$$\int x e^{ax} \, dx = e^{ax} (\frac{x}{a} - \frac{1}{a^2}), \quad a \text{ real or complex}$$

$$\int x^2 e^{ax} \, dx = e^{ax} (\frac{x^2}{a} - \frac{2x}{a^2} + \frac{2}{a^3}), \quad a \text{ real or complex}$$

$$\int x^3 e^{ax} \, dx = e^{ax} (\frac{x^3}{a} - \frac{3x^2}{a^2} + \frac{6x}{a^3} - \frac{6}{a^4}), \quad a \text{ real or complex}$$

$$\int e^{ax} \sin (x) \, dx = \frac{e^{ax}}{a^2 + 1} [a \sin (x) - \cos (x)]$$

$$\int e^{ax} \cos (x) \, dx = \frac{e^{ax}}{a^2 + 1} [a \cos (x) + \sin (x)]$$

$$\int \cos (x) \, dx = \sin (x); \int \cos ax \, dx = \frac{1}{a} \sin ax$$

$$\int x \cos (x) \, dx = \cos (x) + x \sin (x)$$

$$\int x^2 \cos (x) \, dx = 2x \cos (x) + (x^2 - 2) \sin (x)$$

$$\int \sin (x) \, dx = - \cos (x); \int \sin ax \, dx = -\frac{1}{a} \cos ax$$

$$\int x \sin (x) \, dx = \sin (x) - x \cos (x)$$

$$\int x^2 \sin (x) \, dx = 2x \sin (x) - (x^2 - 2) \cos (x)$$

$$\int_{-\infty}^{\infty} e^{-a^2 x^2 + bx} \, dx = \frac{\sqrt{\pi}}{a} e^{b^2/(4a^2)}, \quad a > 0$$

$$\int_0^{\infty} x^2 e^{-x^2} \, dx = \frac{\sqrt{\pi}}{4}$$

$$\int_0^{\infty} Sa(x) \, dx = \int_0^{\infty} \frac{\sin (x)}{x} \, dx = \frac{\pi}{2}$$

$$\int_0^{\infty} Sa^2(x) \, dx = \frac{\pi}{2}$$

D.4   SERIES EXPANSIONS AND FINITE SERIES

$$(1 \pm x)^n = 1 \pm nx + \frac{n(n-1)}{2!}x^2 \pm \frac{n(n-1)(n-2)}{3!}x^3 \pm \ldots \qquad (x^2 < 1) \; Binomial$$

$$(1 \pm x)^{-n} = 1 \mp nx + \frac{n(n-1)}{2!}x^2 \mp \frac{n(n+1)(n+2)}{3!}x^3 \mp \ldots \qquad (x^2 < 1) \; Binomial$$

$$e^x = 1 + x + \frac{1}{2!}x^2 + \ldots \qquad \text{(all real values of } x\text{)} \; Exponential$$

$$\sin x = x - \frac{1}{3!}x^3 + \frac{1}{5!}x^5 - \ldots \qquad \text{(all real values of } x\text{)} \; Trigonometric$$

$$\cos x = 1 - \frac{1}{2!}x^2 + \frac{1}{4!}x^4 - \ldots \qquad \text{(all real values of } x\text{)} \; Trigonometric$$

$$f(x) = f(a) + (x-a)f'(a) + \frac{(x-a)^2}{2!}f''(a) + \frac{(x-a)^3}{3!}f'''(a) + \ldots \; Taylor$$

$$f(x+h) = f(x) + hf'(x) + \frac{h^2}{2!}f''(x) + \frac{h^3}{3!}f'''(x) + \ldots$$

$$Taylor$$

$$= f(h) + xf'(h) + \frac{x^2}{2!}f''(h) + \frac{x^3}{3!}f'''(h) + \ldots$$

Finite Series:

$$\sum_{n=1}^{N} n = \frac{N(N+1)}{2}$$

$$\sum_{n=1}^{N} n^2 = \frac{N(N+1)(2N+1)}{6}$$

$$\sum_{n=1}^{N} n^3 = \frac{N^2(N+1)^2}{4}$$

$$\sum_{n=0}^{N} x^n = \frac{x^{N+1} - 1}{x - 1}$$

$$\sum_{n=0}^{N} e^{j(\theta + n\phi)} = \frac{\sin\left[(N+1)\phi/2\right]}{\sin(\phi/2)} e^{j[\theta + (N\phi/2)]}$$

APPENDIX E

Solution of Simultaneous Equations

CRAMER'S RULE

Cramer's rule provides an efficient organization for the work that is needed to solve a set of *simultaneous linear algebraic equations*. Here we develop formulae for the cases of two and three unknowns. When there are more than three unknowns, the arithmetic becomes quite tedious; in such cases, the solution is best carried out by a computer or calculator program.

Let us consider a pair of linear algebraic equations with two unknowns, x_1 and x_2, written in the form

$$a_{11}x_1 + a_{12}x_2 = b_1 \tag{1a}$$

$$a_{21}x_1 + a_{22}x_2 = b_2 \tag{1b}$$

Cramer's rule gives the solution for the unknowns as

$$x_1 = D_1/D \tag{2a}$$

$$x_2 = D_2/D \tag{2b}$$

where the Ds are the *determinants* given by

$$D = \begin{vmatrix} a_{11} & a_{12} \\ a_{21} & a_{22} \end{vmatrix} = a_{11}a_{22} - a_{12}a_{21} \tag{3a}$$

$$D_1 = \begin{vmatrix} b_1 & a_{12} \\ b_2 & a_{22} \end{vmatrix} = b_1a_{22} - a_{12}b_2 \tag{3b}$$

and

$$D_2 = \begin{vmatrix} a_{11} & b_1 \\ a_{21} & b_2 \end{vmatrix} = a_{11}b_2 - b_1a_{21} \tag{3c}$$

The extension of Cramer's rule to more than two equations is very similar to the results for two equations, but is slightly more involved in the evaluation of the resulting determinants. For example, let us consider a set of three linear simultaneous algebraic equations with three unknowns, x_1, x_2, and x_3:

$$a_{11}x_1 + a_{12}x_2 + a_{13}x_3 = b_1 \tag{4a}$$

$$a_{21}x_1 + a_{22}x_2 + a_{23}x_3 = b_2 \tag{4b}$$

$$a_{31}x_1 + a_{32}x_2 + a_{33}x_3 = b_3 \tag{4c}$$

Cramer's rule yields the solution for the three unknowns as

$$x_k = D_k/D , \qquad k = 1, 2, 3 \tag{5}$$

where

$$D = \begin{vmatrix} a_{11} & a_{12} & a_{13} \\ a_{21} & a_{22} & a_{23} \\ a_{31} & a_{32} & a_{33} \end{vmatrix} = a_{11}\begin{vmatrix} a_{22} & a_{23} \\ a_{32} & a_{33} \end{vmatrix} - a_{12}\begin{vmatrix} a_{21} & a_{23} \\ a_{31} & a_{33} \end{vmatrix} + a_{13}\begin{vmatrix} a_{21} & a_{22} \\ a_{31} & a_{32} \end{vmatrix}$$

$$= a_{11}(a_{22}a_{33} - a_{23}a_{32}) - a_{12}(a_{21}a_{33} - a_{23}a_{31}) + a_{13}(a_{21}a_{32} - a_{22}a_{31}) \tag{6a}$$

843

$$D_1 = \begin{vmatrix} b_1 & a_{12} & a_{13} \\ b_2 & a_{22} & a_{23} \\ b_3 & a_{32} & a_{33} \end{vmatrix} = b_1 \begin{vmatrix} a_{22} & a_{23} \\ a_{32} & a_{33} \end{vmatrix} - a_{12} \begin{vmatrix} b_2 & a_{23} \\ b_3 & a_{33} \end{vmatrix} + a_{13} \begin{vmatrix} b_2 & a_{22} \\ b_3 & a_{32} \end{vmatrix}$$

$$= b_1(a_{22}a_{33} - a_{23}a_{32}) - a_{12}(b_2a_{33} - a_{23}b_3) + a_{13}(b_2a_{32} - a_{22}b_3) \tag{6b}$$

$$D_2 = \begin{vmatrix} a_{11} & b_1 & a_{13} \\ a_{21} & b_2 & a_{23} \\ a_{31} & b_3 & a_{33} \end{vmatrix} = a_{11} \begin{vmatrix} b_2 & a_{23} \\ b_3 & a_{33} \end{vmatrix} - b_1 \begin{vmatrix} a_{21} & a_{23} \\ a_{31} & a_{33} \end{vmatrix} + a_{13} \begin{vmatrix} a_{21} & b_2 \\ a_{31} & b_3 \end{vmatrix}$$

$$= a_{11}(b_2a_{33} - a_{23}b_3) - b_1(a_{21}a_{33} - a_{23}a_{31}) + a_{13}(a_{21}b_3 - b_2a_{31}) \tag{6c}$$

$$D_3 = \begin{vmatrix} a_{11} & a_{21} & b_3 \\ a_{21} & a_{22} & b_2 \\ a_{31} & a_{32} & b_3 \end{vmatrix} = a_{11} \begin{vmatrix} a_{22} & a_{23} \\ a_{32} & a_{33} \end{vmatrix} - a_{12} \begin{vmatrix} a_{21} & b_2 \\ a_{31} & b_3 \end{vmatrix} + b_1 \begin{vmatrix} a_{21} & a_{22} \\ a_{31} & a_{32} \end{vmatrix}$$

$$= a_{11}(a_{22}b_3 - b_2a_{32}) - a_{12}(a_{21}b_3 - b_2a_{31}) + b_1(a_{21}a_{32} - a_{22}a_{31}) \tag{6d}$$

As an alternative approach to the solution of three simultaneous equations, typically given by Equations (4a), (4b), and (4c), one can always use one of the equations to express an unknown in terms of the other two; substitution into the remaining two equations reduces the problem to two equations with two unknowns. This method is most convenient when one does not need the unknown eliminated and one of the b-terms equals zero to simplify the elimination process.

GAUSS ELIMINATION

A simple technique, which is also used in digital computer methods, for solving linear simultaneous algebraic equations is the method of Gauss elimination. The basic idea is to reduce the equations through manipulations to an *equivalent* form which is *triangular*; for example, the equation set of Equation (4) would be simplified to its equivalent in a triangular form given typically by

$$c_{11}x_1 + c_{12}x_2 + c_{13}x_3 = d_1 \tag{7a}$$

$$c_{22}x_2 + c_{23}x_3 = d_2 \tag{7b}$$

$$c_{33}x_3 = d_3 \tag{7c}$$

The reduction of Equation (4) to that of Equation (7) is accomplished by the following tasks:

1. One can multiply any equation in the set by a nonzero number without changing the solution.

2. One can also add or subtract any two equations and replace one of the two with the result.

Once the triangular form of Equation (7) is achieved, one can solve for the unknowns by *back substitution*:

From Equation (7c),

$$x_3 = \frac{d_3}{c_{33}} \tag{8a}$$

Substituting Equation (8a) in Equation (7b), one gets

$$c_{22}x_2 + c_{23}\frac{d_3}{c_{33}} = d_2$$

$$\text{or} \quad x_2 = \frac{1}{c_{22}}\left[-c_{23}\frac{d_3}{c_{33}} + d_2\right] \tag{8b}$$

Substituting Equations (8a) and (8b) into Equation (7a), one can solve for x_1.

MATRIX METHOD

For those readers who have been introduced to matrix methods of analysis, the set of equations in Equation (4) can be expressed as

$$\mathbf{AX} = \mathbf{B} \tag{9}$$

where

$$\text{coefficient matrix, } \mathbf{A} = \begin{bmatrix} a_{11} & a_{12} & a_{13} \\ a_{21} & a_{22} & a_{23} \\ a_{31} & a_{32} & a_{33} \end{bmatrix} \tag{10a}$$

$$\text{column matrix of unknowns, } \mathbf{X} = \begin{bmatrix} x_1 \\ x_2 \\ x_3 \end{bmatrix} \tag{10b}$$

and

$$\text{column matrix, } \mathbf{B} = \begin{bmatrix} b_1 \\ b_2 \\ b_3 \end{bmatrix} \tag{10c}$$

The solution is accomplished by

$$\mathbf{X} = \mathbf{A}^{-1}\mathbf{B} \tag{11}$$

where $\mathbf{A}^{-1}$ is the inverse of the matrix $\mathbf{A}$.

One can find the inverse of a matrix by following the steps

1. Obtain the *determinant* of the matrix $\mathbf{A}$ as indicated in Equation (6a); it should be nonzero for the inverse to exist.
2. Replace each element of the matrix by its *cofactor*. For example, the cofactor of a_{11} in Equation (10a) is given by

$$(-1)^{1+1}\begin{vmatrix} a_{22} & a_{23} \\ a_{32} & a_{33} \end{vmatrix} = (a_{22}a_{33} - a_{23}a_{32}) ;$$

the cofactor of a_{12} in Equation (10a) is given by

$$(-1)^{1+2}\begin{vmatrix} a_{21} & a_{23} \\ a_{31} & a_{33} \end{vmatrix} = -(a_{21}a_{33} - a_{23}a_{31})$$

3. Find the *transpose* of the resultant matrix in STEP 2 by interchanging its rows and columns.
4. Divide the resultant matrix in STEP 3 by the determinant found in STEP 1.

As a check, one can verify that the product $\mathbf{AA}^{-1}$ results in a *unit matrix*.

Complex Numbers

$$j = \sqrt{-1} = 1\angle 90° \; ; j^2 = -1 = 1\angle 180° \; ; j^3 = -j = 1\angle 270° = 1\angle -90° \; ; j^4 = 1\angle 0°$$

$$e^{\pm j\theta} = \cos\theta \pm j \sin\theta = 1\angle \pm\theta$$

$$e^{j\theta} + e^{-j\theta} = 2 \cos\theta$$

$$e^{j\theta} - e^{-j\theta} = j2 \sin\theta$$

If $\bar{A} = A\angle\theta = Ae^{j\theta} = a + jb$, then

$$a = A \cos\theta = \mathrm{Re}\bar{A} \; ; b = A \sin\theta = \mathrm{Im}\bar{A}$$

$$A = \sqrt{a^2 + b^2} \; ; \theta = \tan^{-1}(b/a)$$

$$(\bar{A})^* = A\angle -\theta = Ae^{-j\theta} = a - jb$$

$$\bar{A} + (\bar{A})^* = a + jb + a - jb = 2a = 2\mathrm{Re}\bar{A}$$

$$\bar{A}(\bar{A})^* = Ae^{j\theta} \times Ae^{-j\theta} = A^2$$

If $\bar{A}_1 = a_1 + jb_1$ and $\bar{A}_2 = a_2 + jb_2$, it follows that

$$\bar{A}_1 \pm \bar{A}_2 = (a_1 + jb_1) \pm (a_2 + jb_2) = (a_1 \pm a_2) + j(b_1 \pm b_2)$$

$$\mathrm{Re}\{\bar{A}_1 \pm \bar{A}_2\} = \mathrm{Re}\bar{A}_1 \pm \mathrm{Re}\bar{A}_2$$

$$\mathrm{Im}\{\bar{A}_1 \pm \bar{A}_2\} = \mathrm{Im}\bar{A}_1 \pm \mathrm{Im}\bar{A}_2$$

$$\bar{A}_1\bar{A}_2 = (a_1 + jb_1)(a_2 + jb_2) = (a_1 a_2 - b_1 b_2) + j(a_1 b_2 + a_2 b_1)$$

$$\frac{\bar{A}_1}{\bar{A}_2} = \frac{a_1 + jb_1}{a_2 + jb_2} = \frac{a_1 + jb_1}{a_2 + jb_2} \times \frac{a_2 - jb_2}{a_2 - jb_2} = \frac{(a_1 a_2 + b_1 b_2) + j(a_2 b_1 - a_1 b_2)}{a_2^2 + b_2^2}$$

If $\bar{A}_1 = A_1 e^{j\theta_1} = A_1\angle\theta_1$ and $\bar{A}_2 = A_2 e^{j\theta_2} = A_2\angle\theta_2$, then

$$\bar{A}_1\bar{A}_2 = (A_1 e^{j\theta_1})(A_2 e^{j\theta_2}) = (A_1\angle\theta_1)(A_2\angle\theta_2) = A_1 A_2 e^{j(\theta_1+\theta_2)} = A_1 A_2\angle\theta_1 + \theta_2$$

$$\frac{\bar{A}_1}{\bar{A}_2} = \frac{A_1 e^{j\theta_1}}{A_2 e^{j\theta_2}} = \frac{A_1\angle\theta_1}{A_2\angle\theta_2} = \frac{A_1}{A_2} e^{j(\theta_1-\theta_2)} = \frac{A_1}{A_2}\angle\theta_1 - \theta_2$$

ASSOCIATIVE, COMMUTATIVE, AND DISTRIBUTIVE LAWS ARE SUMMARIZED

$$\bar{A} + (\bar{B} + \bar{C}) = (\bar{A} + \bar{B}) + \bar{C} \; ; (\bar{A}\,\bar{B})\bar{C} = \bar{A}(\bar{B}\,\bar{C})$$

$$\bar{A} + \bar{B} = \bar{B} + \bar{A} \; ; \bar{A}\bar{B} = \bar{B}\bar{A}$$

$$\bar{A}(\bar{B} + \bar{C}) = \bar{A}\bar{B} + \bar{A}\bar{C}$$

Fourier Series

A periodic waveform $f(t) = f(t + T)$, which is said to be periodic with a period T, fundamental frequency $f = 1/T$, and fundamental radian frequency $\omega = 2\pi/T$, can be expressed in terms of an infinite series (known as the *Fourier series*) of sinusoidal signals. Expressed mathematically, we have

$$f(t) = \frac{a_0}{2} + \sum_{n=1}^{\infty} (a_n \cos n\omega t + b_n \sin n\omega t) \tag{1}$$

where the coefficients a_n and b_n are given by

$$a_n = \frac{2}{T} \int_{-T/2}^{T/2} f(t) \cos n\omega t \, dt = \frac{2}{T} \int_{0}^{T} f(t) \cos n\omega t \, dt , \qquad n = 0, 1, 2, 3, \ldots \tag{2}$$

$$b_n = \frac{2}{T} \int_{-T/2}^{T/2} f(t) \sin n\omega t \, dt = \frac{2}{T} \int_{0}^{T} f(t) \sin n\omega t \, dt , \qquad n = 1, 2, 3, \ldots \tag{3}$$

The average value of $f(t)$, which is also known as the dc component, is given by

$$\frac{a_0}{2} = \frac{1}{T} \int_{0}^{T} f(t) \, dt$$

COMPLEX (EXPONENTIAL) FOURIER SERIES

By making use of the trigonometric identities

$$\cos t = \frac{1}{2}(e^{jt} + e^{-jt}) \tag{4}$$

$$\sin t = \frac{1}{2j}(e^{jt} - e^{-jt}) \tag{5}$$

where $j = \sqrt{-1}$, one can express Equation (1) as follows in terms of *complex Fourier series*:

$$f(t) = \sum_{n=-\infty}^{\infty} \bar{c}_n e^{jn\omega t} \tag{6}$$

where the complex coefficients $\bar{c}_n$ are given by

$$\bar{c}_n = \frac{1}{T} \int_{-T/2}^{T/2} f(t) e^{-jn\omega t} \, dt , \qquad n = 0, \pm 1, \pm 2, \ldots \tag{7}$$

The set of coefficients $\{\bar{c}_n\}$ is often referred to as the Fourier spectrum. These coefficients are related to the coefficients a_n and b_n of Equations (2) and (3) by

$$c_0 = \frac{a_0}{2} \tag{8}$$

$$\bar{c}_n = \frac{1}{2}(a_n - jb_n), \quad n = 1, 2, \dots \tag{9}$$

$$\bar{c}_{-n} = \frac{1}{2}(a_n + jb_n) = \bar{c}_n^*, \quad n = 1, 2, \dots \tag{10}$$

where the asterisk represents complex conjugation.

The periodic function $f(t)$ of Equation (1) can also be represented as

$$f(t) = \frac{a_0}{2} + \sum_{n=1}^{\infty} 2\bar{c}_n \cos(n\omega t + \phi_n) \tag{11}$$

where the coefficients are related by

$$a_n = 2\bar{c}_n \cos \phi_n \tag{12}$$

$$b_n = -2\bar{c}_n \sin \phi_n \tag{13}$$

$$\bar{c}_n = \frac{1}{2}\sqrt{a_n^2 + b_n^2} \tag{14}$$

$$\phi_n = \arctan(-b_n/a_n) \tag{15}$$

The periodic waveform of Equation (1) can also be expressed as

$$f(t) = \frac{a_0}{2} + \sum_{n=1}^{\infty} 2\bar{c}_n \sin(n\omega t + \phi_n) \tag{16}$$

where the coefficients are related by

$$a_n = 2\bar{c}_n \sin \phi_n \tag{17}$$

$$b_n = 2\bar{c}_n \cos \phi_n \tag{18}$$

$$\bar{c}_n = \frac{1}{2}\sqrt{a_n^2 + b_n^2} \tag{19}$$

$$\phi_n = \arctan(a_n/b_n) \tag{20}$$

note that the quadrant of ϕ_n is to be chosen so as to make the formulae for a_n, b_n, and $\bar{c}_n$ hold.

PROPERTIES OF FOURIER SERIES

Existence: If a bounded single-valued periodic function $f(t)$ of period T has at most a finite number of maxima, minima, and jump discontinuities in any one period, then $f(t)$ can be represented over a complete period by a Fourier series of simple harmonic functions, the frequencies of which are integral multiples of the fundamental frequency. This series will converge to $f(t)$ at all points where $f(t)$ is continuous and to the average of the right- and left-hand limits of $f(t)$ at each point where $f(t)$ is discontinuous.

Delay: If a periodic function $f(t)$ is delayed by any multiple of its period T, the waveform is unchanged. That is to say

$$f(t - nT) = f(t), \quad n = \pm 1, \pm 2, \pm 3, \dots \tag{21}$$

Symmetry: A periodic waveform $f(t)$ with even symmetry such that $f(-t) = f(t)$ will have a Fourier series with no sine terms; that is to say, all coefficients b_n go to zero. If, on the other hand,

$f(t)$ has odd symmetry such that $f(-t) = -f(t)$, its Fourier series will have no cosine terms; that is to say, all a_n coefficients become zero.

Decomposition: An arbitrary periodic waveform $f(t)$ can be expressed as

$$f(t) = f_e(t) + f_o(t) \tag{22}$$

where $f_e(t)$ represents a part with even symmetry, and $f_o(t)$ represents another part with odd symmetry. These parts may be evaluated from the original signal by

$$f_e(t) = \frac{1}{2}[f(t) + f(-t)] \tag{23}$$

$$f_o(t) = \frac{1}{2}[f(t) - f(-t)] \tag{24}$$

Integration: The integral of a periodic signal that has a valid Fourier series can be found by termwise integration of the Fourier series of the signal.

Differentiation: If a periodic function $f(t)$ is continuous everywhere and its derivative has a valid Fourier series, then wherever it exists, the derivative of $f(t)$ can be found by termwise differentiation of the Fourier series of $f(t)$.

SOME USEFUL AUXILIARY FORMULAE FOR FOURIER SERIES

$$\sin \frac{n\pi}{2} = \frac{(j)^{n+1}}{2}[(-1)^n - 1] \tag{25}$$

$$\cos \frac{n\pi}{2} = \frac{(j)^n}{2}[(-1)^n + 1] \tag{26}$$

The following table of trigonometric functions will be helpful for developing Fourier series:

Function	n Any integer	n Even	n Odd	$n/2$ Odd	$n/2$ Even
$\sin n\pi$	0	0	0	0	0
$\cos n\pi$	$(-1)^n$	1	-1	1	1
$\sin \dfrac{n\pi}{2}$		0	$(-1)^{(n-1)/2}$	0	0
$\cos \dfrac{n\pi}{2}$		$(-1)^{n/2}$	0	-1	1
$\sin \dfrac{n\pi}{4}$			$\dfrac{\sqrt{2}}{2}(-1)^{(n^2+4n+11)/8}$	$(-1)^{(n-2)/4}$	0

$$1 = \frac{4}{\pi}[\sin \frac{\pi t}{k} + \frac{1}{3} \sin \frac{3\pi t}{k} + \frac{1}{5} \sin \frac{5\pi t}{k} + \ldots] \qquad (0 < t < k) \tag{27}$$

$$t = \frac{2k}{\pi}[\sin \frac{\pi t}{k} - \frac{1}{2} \sin \frac{2\pi t}{k} + \frac{1}{3} \sin \frac{3\pi t}{k} + \ldots] \qquad (-k < t < k) \tag{28}$$

$$t = \frac{k}{2} - \frac{4k}{\pi^2}[\cos \frac{\pi t}{k} + \frac{1}{3^2} \cos \frac{3\pi t}{k} + \frac{1}{5^2} \cos \frac{5\pi t}{k} + \ldots] \qquad (0 < t < k) \tag{29}$$

$$t^2 = \frac{2k^2}{\pi^3}\left[(\frac{\pi^2}{1} - \frac{4}{1}) \sin \frac{\pi t}{k} - \frac{\pi^2}{2} \sin \frac{2\pi t}{k} + (\frac{\pi^2}{3} - \frac{4}{3^3}) \sin \frac{3\pi t}{k}\right.$$

$$\left. - \frac{\pi^2}{4} \sin \frac{4\pi t}{k} + \left(\frac{\pi^2}{5} - \frac{4}{5^3}\right) \sin \frac{5\pi t}{k} + \ldots \right] \qquad (0 < t < k) \qquad (30)$$

$$t^2 = \frac{k^2}{3} - \frac{4k^2}{\pi^2}\left[\cos \frac{\pi t}{k} - \frac{1}{2^2} \cos \frac{2\pi t}{k} + \frac{1}{3^2} \cos \frac{3\pi t}{k} \right.$$

$$\left. - \frac{1}{4^2} \cos \frac{4\pi t}{k} + \ldots \right] \qquad (-k < t < k) \qquad (31)$$

$$1 - \frac{1}{3} + \frac{1}{5} - \frac{1}{7} + \ldots = \frac{\pi}{4} \qquad (32)$$

$$1 + \frac{1}{2^2} + \frac{1}{3^2} + \frac{1}{4^2} + \ldots = \frac{\pi^2}{6} \qquad (33)$$

$$1 - \frac{1}{2^2} + \frac{1}{3^2} - \frac{1}{4^2} + \ldots = \frac{\pi^2}{12} \qquad (34)$$

$$1 + \frac{1}{3^2} + \frac{1}{5^2} + \frac{1}{7^2} + \ldots = \frac{\pi^2}{8} \qquad (35)$$

$$\frac{1}{2^2} + \frac{1}{4^2} + \frac{1}{6^2} + \frac{1}{8^2} + \ldots = \frac{\pi^2}{24} \qquad (36)$$

Laplace Transforms

If $f(t)$ is a piecewise continuous real-valued function of the real variable $t\,(0 \le t < \infty)$, and $|f(t)| < Me^{\sigma t}\,(t > T; M, \sigma, T$ positive constants), then the *Laplace transform* of $f(t)$, given by

$$\mathscr{La}[f(t)] = F(s) = \int_0^\infty e^{-st} f(t)\, dt, \tag{1}$$

exists in the half-plane of the complex variable $s = \sigma + j\omega$, for which the real part of s is greater than some fixed value s_0, i.e., $Re(s)s_0$.

The *inverse transform* is defined as

$$f(t =) \mathscr{La}^{-1}[F(s)] = \frac{1}{2\pi j} \int_{\sigma - j\infty}^{\sigma + j\infty} F(s)\, e^{st}\, ds \tag{2}$$

where $\sigma > s_0$ is chosen to the right of any singularity of $F(s)$.

The property of Laplace transform given by

$$\mathscr{La}\left[f^{(r)}(t)\right] = \int_0^\infty e^{-st} \left(\frac{d^r f}{dt^r}\right) dt = s^r F(s) - \sum_{n=0}^{r-1} s^{r-1-n} f^{(n)}(0^+)$$

or

$$\mathscr{La}\left[\frac{d^r f(t)}{dt^r}\right] = s^r F(s) - s^{r-1}(0^+) - s^{r-2}\frac{df}{dt}(0^+) - \ldots - \frac{d^{r-1} f}{dt^{r-1}}(0^+) \tag{3}$$

makes the Laplace transform very useful for solving linear differential equations with constant coefficients, and many boundary value problems.

A summary of properties of Laplace transformation and a table of Laplace transform pairs are given below.

Summary of Properties of Laplace Transformation

Property	Time Function	Laplace Transform
Linearity	$a_1 f_1(t) \pm a_2 f_2(t)$	$a_1 F_1(s) \pm a_2 F_2(s)$
Differentiation	$f'(t)$	$sF(s) - f(0^+)$
	$f^n(t)$	$s^n F(s) - s^{n-1} f(0^+) - s^{n-2} f'(0^+) - \ldots - f^{n-1}(0^+)$
Integration	$f^{-1}(t) = \int_0^t f(\tau)\, d\tau$	$\dfrac{F(s)}{s} + \dfrac{f^{-1}(0^+)}{s}$
	$f^{-n}(t)$	$\dfrac{F(s)}{s^n} + \dfrac{f^{-1}(0^+)}{s^n} + \dfrac{f^{-2}(0^+)}{s^{n-1}} + \ldots + \dfrac{f^n(0^+)}{s}$

$$\int_0^t f(\tau)g(t-\tau)\,d\tau \qquad\qquad F(s)G(s)$$

Multiplication by t (frequency differentiation)	$tf(t)$	$-\dfrac{d}{ds}[F(s)]$
Division by t (frequency integration)	$\dfrac{1}{t}f(t)$	$\displaystyle\int_s^\infty F(s)\,ds$
Time delay or shift	$f(t-T)\cdot u(t-T)$	$e^{-sT}F(s)$
Periodic function $f(t)=f(t+nT)$	$f(t),\ 0\le t\le T$	$F(s)/[1-e^{Ts}],\ \text{where } F(s)=\displaystyle\int_0^T f(t)e^{-st}\,dt$
Exponential translation (frequency shifting)	$e^{-at}f(t)$	$F(s+a)$
Change of scale (time scaling)	$f(at),\ a>0$	$\dfrac{1}{a}F\!\left(\dfrac{s}{a}\right)$
Initial value	$f(0^+)=\lim\limits_{t\to 0} f(t)$	$\lim\limits_{s\to\infty} sF(s)$
Final value	$f(\infty)=\lim\limits_{t\to\infty} f(t),$	$\lim\limits_{s\to 0} sF(s)$
	where limit exists	$[sF(s)$ has poles only inside the left half of the s-plane.]

Table of Laplace Transform Pairs

$f(t)=\mathscr{La}^{-1}[F(s)]$	$F(s)=\mathscr{La}[f(t)]$
$\delta(t)$ (unit impulse or delta function)	1
$\delta(t-T)$	e^{-sT}
a	$\dfrac{a}{s}$
$u(t)$ or 1 (unit step function)	$\dfrac{1}{s}$
$u(t-T)$	$\dfrac{e^{-sT}}{s}$
t	$\dfrac{1}{s^2}$
$(t-T)\,u(t-T)$	$\dfrac{e^{-sT}}{s^2}$
$tu(t-T)$	$\dfrac{(1+sT)e^{-sT}}{s^2}$
$t^n,\ (n-\text{integer})$	$\dfrac{n!}{s^{n+1}}$
$t^{n-1}/(n-1)!,\ n$ an integer	$\dfrac{1}{s^n}$
e^{-at}	$\dfrac{1}{s+a}$
te^{-at}	$\dfrac{1}{(s+a)^2}$
$e^{-at}t^n$	$\dfrac{n!}{(s+a)^{n+1}}$
$\dfrac{e^{-at}-e^{-bt}}{b-a}$	$\dfrac{1}{(s+a)(s+b)}$
$\sin\omega t$	$\dfrac{\omega}{s^2+\omega^2}$

$\sin(\omega t + \theta)$	$\dfrac{s\,\sin\theta + \omega\,\cos\theta}{(s^2 + \omega^2)}$
$\cos \omega t$	$\dfrac{s}{s^2 + \omega^2}$
$\cos(\omega t + \theta)$	$\dfrac{s\,\cos\theta - \omega\,\sin\theta}{(s^2 + \omega^2)}$
$e^{-at}\,\sin \omega t$	$\dfrac{\omega}{(s+a)^2 + \omega^2}$
$te^{-at}\,\sin \omega t$	$\dfrac{2\omega(s+a)}{[(s+a)^2 + \omega^2]^2}$
$e^{-at}\,\cos \omega t$	$\dfrac{s+a}{(s+a)^2 + \omega^2}$
$te^{-at}\,\cos \omega t$	$\dfrac{[(s+a)^2 - \omega^2]}{[(s+a)^2 + \omega^2]^2}$
$\sinh at$	$\dfrac{a}{s^2 - a^2}$
$\cosh at$	$\dfrac{s}{s^2 - a^2}$
$\left[k_1 e^{-at}\,\cos \omega t + \dfrac{k_2 - k_1 a}{\omega} e^{-at}\,\sin \omega t\right]$	$\dfrac{k_1 s + k_2}{(s+a)^2 + \omega^2}$
$\dfrac{\omega}{\sqrt{1-a^2}}\, e^{-a\omega t}\,\sin \omega\sqrt{1-a^2}\,t$	$\dfrac{\omega^2}{s^2 + 2a\omega + \omega^2}$
$\dfrac{1}{2\omega} t\,\sin \omega t$	$\dfrac{s}{(s^2 + \omega^2)^2}$
$\dfrac{1}{2\omega}(\sin \omega t + \omega t\,\cos \omega t)$	$\dfrac{s^2}{(s^2 + \omega^2)^2}$

Note that all $f(t)$ should be thought of as being multiplied by $u(t)$, i.e., $f(t) = 0$ for $t < 0$.

Index